Blast and Impact Engineering on Structures and Materials

Blast and Impact Engineering on Structures and Materials

Editors

Ricardo Castedo
Lina M. López
Anastasio P. Santos

MDPI

Basel • Beijing • Wuhan • Barcelona • Belgrade • Novi Sad • Cluj • Manchester

Editors

Ricardo Castedo
Department of Geological and Mining Engineering
Universidad Politécnica de Madrid
Madrid
Spain

Lina M. López
Department of Geological and Mining Engineering
Universidad Politécnica de Madrid
Madrid
Spain

Anastasio P. Santos
Department of Geological and Mining Engineering
Universidad Politécnica de Madrid
Madrid
Spain

Editorial Office
MDPI
St. Alban-Anlage 66
4052 Basel, Switzerland

This is a reprint of articles from the Special Issue published online in the open access journal *Applied Sciences* (ISSN 2076-3417) (available at: www.mdpi.com/journal/applsci/special_issues/blast_impact_engineering).

For citation purposes, cite each article independently as indicated on the article page online and as indicated below:

Lastname, A.A.; Lastname, B.B. Article Title. *Journal Name* **Year**, *Volume Number*, Page Range.

ISBN 978-3-0365-8977-0 (Hbk)
ISBN 978-3-0365-8976-3 (PDF)
doi.org/10.3390/books978-3-0365-8976-3

Contents

About the Editors

Ricardo Castedo

Ricardo Castedo is Associate Professor at Geological and Mining Engineering Department (School of Mines and Energy; Universidad Politécnica de Madrid). B.Sc. (2008) in Geological Engineering, M.Sc. (2014) in Teaching and Education, Ph.D (2012) in Engineering (Universidad Politécnica de Madrid). One year research stay at New Mexico Tech (USA), and three research stays at University of Leeds (UK), University of Windsor (Canada) y Politecnico di Torino (Italy). Author of more than 35 publications in scientific journals and more than 40 contributions to national and international conferences. Ricardo has received several awards for his scientific contributions from various institutions. Reviewer and editor of scientific indexed (JCR) journals and International Conferences. Research interests: Numerical Modelling, Dynamic Behavior of Structures; Safety and Security Engineering; testing in structures with explosives, Explosives behavior and characterization, Safety and modeling of Explosives.

Lina M. López

Lina M. López is Professor at Geological and Mining Engineering Department (School of Mines and Energy; Universidad Politécnica de Madrid). B.Sc., Ph.D. in Mining Engineering (Universidad Politécnica de Madrid). Two research stays of at Mines Paris Tech (France) and New Mexico Tech (USA). Research fellowship (1999–2003) funded by Madrid Regional Government-FPI. Author of 4 books, 35 publications at scientific journals and more than 50 contributions to national and international conferences. Lina has received several awards for his scientific contributions from various institutions. Reviewer of Scientific JCR Journals (Elsevier) and International Conferences. Member of the International Society of Explosives Engineers (ISEE) from 2001. Member of the Fragblast International Organizing Committee (FIOC) from 2015. Research interests: Explosives behavior and characterization, Safety and modeling of Explosives, physics of the explosives, thermodynamic calculations and blasting; safety and security Engineering; testing with explosives.

Anastasio P. Santos

Anastasio P. Santos is Associate Professor in Continuum Mechanics at Geological and Mining Engineering Department (School of Mines and Energy; Universidad Politécnica de Madrid). B.Sc. (1987), Ph.D (1996) in Mining Engineering (Universidad Politécnica de Madrid). One year research stay at École Centrale Paris (France). Six years of experience working in industrial companies. Associate Professor in Applied Mathematics on leave of absence (Salamanca University). Author of 25 publications in scientific journals and more than 35 contributions to national and international conferences. Anastasio has received several awards for his scientific contributions from various institutions. Member of the SEMNI (Sociedad Española de Mecánica e Ingeniería Computacionales). Research interests: Numerical Modelling, Structural Mechanics, Dynamic Behavior of Structures; Safety and Security Engineering; testing in structures with explosives.

Preface

This Special Issue aimed to collect and present all breakthrough research on all intentional or unintentional explosions and impact problems. During the last decade, the investigation of these phenomena has been an active area of research in different fields (i.e., civil, defense, mining, aeronautical, naval, etc.), including experimental studies, analytical models, or numerical simulations; and this Special Issue is a faithful reflection of this trend.

A total of twenty-six papers (twenty-five research papers and one review paper) in various fields of blast and impact engineering including blast loading issues over structures, beams, walls; penetration and impact; explosives safety and security; blasting effects on rocks and tunnels; are presented in this Special Issue.

In the topic related with blast loading over structures, seven papers have been published. These articles include papers by Santos et al. [1], Mollaei et al. [2], Shunck and Eckenfels [3], Lim et al. [4], Li et al. [5], Hung et al. [6] and Lukić and Draganić [7] dealing with the effects of explosives (some as improvised explosives devices) on full-scale structures, masonry walls, protective barriers, or the safety of personnel inside tunnel structures.

Eight articles have been published in the topic of penetration and impact. These papers are published by: Yang et al. [8], Imran Latif et al. [9], Pu et al. [10], Yuan et al. [11], Wang et al. [12], Wang et al. [13], Fowler and Teixeira-Dias [14] and Malesa et al. [15]. They deal with different materials such as concrete, ultra-high-performance concrete, recycled aggregate concrete pavements or different phenomena such as penetration of liquid cabin for warships, or the impact of fragments at high speed on wings or on unmanned spacecraft. Most of the papers include tests and numerical simulations.

Six papers published in this issue (Sánchez-Monreal et al. [16], Taylor [17], Marín et al. [18], Chen et al. [19], Filice et al. [20] and Traná et al. [21]) have focused on the development of numerical or empirical models that predict the effects of the shock waves of explosives. The papers are varied, such as the development of computational tools to approximate the effects over structural elements, estimation of attenuation laws, or peak pressure predictions.

Finally, five papers (Alsabhan et al. [22], Ko et al. [23], Dong et al. [24], Choi and Lee [25], Huo et al. [26]) have been published on experiments and numerical models focused on the effects of explosives on rocks. These include topics as varied as the effect of surface impact of projectiles on tunnels, the use of different stemming materials, or models for predicting the peak particle velocity from blasting-induced vibrations.

The editors would like to thank all the authors and reviewers, as well as the MDPI staff (especially Amy An), for their valuable contributions to this Special Issue.

Ricardo Castedo, Lina M. López, and Anastasio P. Santos
Editors

applied sciences

MDPI

Article

Reinforced Concrete Building with IED Detonation: Test and Simulation

Anastasio P. Santos [1], Ricardo Castedo [1,*], Lina M. López [1], María Chiquito [1], José I. Yenes [2], Alejandro Alañón [3], Elisa Costamagna [4] and Santiago Martínez-Almajano [2]

1 E.T.S.I. Minas y Energía—Universidad Politécnica de Madrid, 28003 Madrid, Spain; tasio.santos@upm.es (A.P.S.); lina.lopez@upm.es (L.M.L.); maria.chiquito@upm.es (M.C.)
2 Escuela Politécnica Superior del Ejército—Ministry of Defense, 28071 Madrid, Spain; jyengal@et.mde.es (J.I.Y.); smalmajano@et.mde.es (S.M.-A.)
3 Escuela Politécnica Superior de Ávila—Universidad de Salamanca, 05003 Ávila, Spain; alajua@usal.es
4 Department of Environment, Land and Infrastructure Engineering (DIATI), Politecnico di Torino, 10129 Torino, Italy; elisa.costamagna@polito.it
* Correspondence: ricardo.castedo@upm.es; Tel.: +34-910676518

Abstract: There is growing concern about the possibility of a suicide bomber being immolated when the army forces or the law enforcement agencies discover the place where they prepare their material or simply find themselves inside a building. To study the possible effects that these improvised explosive devices (IEDs) would have on the structures, eight tests were carried out with various configurations of IEDs with vest bombs inside a reinforced concrete (including walls and roof) building constructed ad hoc for these tests. These vests were made with different explosives (black powder, ANFO, AN/AL, PG2). For the characterization of these tests, a high-speed camera and pressure and acceleration sensors were used. The structure behaved surprisingly well, as it withstood all the first seven detonations without apparent structural damage. In the last detonation, located on the ground and with a significant explosive charge, the structural integrity of the roof and some of the walls was compromised. The simulation of the building was carried out with the LS-DYNA software with a Lagrangian formulation for the walls, using the LBE (based on CONWEP) module for the application of the charge. Despite the difficulty of this simulation, the results obtained, in terms of applied pressures and measured accelerations, are acceptable with differences of about 20%.

Keywords: numerical modeling; LS-DYNA; IEDs; field test; reinforced concrete

Citation: Santos, A.P.; Castedo, R.; López, L.M.; Chiquito, M.; Yenes, J.I.; Alañón, A.; Costamagna, E.; Martínez-Almajano, S. Reinforced Concrete Building with IED Detonation: Test and Simulation. *Appl. Sci.* **2022**, *12*, 7803. https://doi.org/10.3390/app12157803

Academic Editor: Jong Wan Hu

Received: 10 July 2022
Accepted: 31 July 2022
Published: 3 August 2022

1. Introduction

The risk of an attack in the operating area or in the neutral zone has increased in the last decades. Many of these attacks are carried out using improvised explosive devices (IEDs) which are unconventional weapons that can be easily fabricated. Access to the products and knowledge necessary for the use and creation of IEDs has risen in recent years. As an example, terrorist attacks such as Flight 9268 which covered the Egypt–Russia route (2015), Paris (2015), Belgium (2016), Germany (2016), England (2017) and Spain (2017), all of which resulted in fatalities, demonstrate the urgent need to better understand the possible effects of these devices on people and/or structures [1–3]. Most of the IED attacks over the past 15 years involved small bombs of less than 5 kg [4] or a person-borne improvised explosive device (PBIED) usually containing less than 10 kg of explosives [5,6]. Moreover, terrorist actions may most often be carried out in crowded areas, in urban environments, near critical infrastructure or even inside buildings. For this reason, there has been considerable research on structural damage and blast effects on buildings, and therefore, much literature has been published on blast mitigation and retrofit methods [7,8]. However, many of these works are not based on experimental tests and use numerical modeling to predict the structural response in different scenarios by comparing it with empirical equations [9,10].

Other times, numerical modeling results are validated with experimental data found in the literature [11,12]. In these scenarios, comparisons can only be made with the available data which in some cases are deficient. Numerical modeling is a good alternative and a very useful tool, but in the case of blast loading, it must be calibrated and validated by corresponding field tests.

Since concrete is a construction material widely used in many building structures, its behavior has been extensively studied through experimental tests and numerical simulation. Experimental data are essential to understand the explosive phenomenon and predict the structural response, but this kind of experiment is very difficult to implement and has a high cost. For these reasons, many of the experiments are based on single structural elements such as beams [13,14] or slabs [15–17] which are easier to handle and monitor. The data obtained in this type of trial can be used to calibrate numerical models as well as to check different laws of materials' behavior. However, these results cannot be used to analyze and predict the structural response of a whole building, as the element failure causes loads to be redistributed to the neighboring supporting elements. The failure of individual structural elements can have a decisive influence on whether or not the structure collapses. Progressive collapse of structures has also been studied by numerous researchers, although not many have conducted experimental tests at full scale [18–20], and there are even fewer cases in which, in addition to the structure, non-structural elements such as masonry walls or the roof are represented [21,22].

However, in the last decade, most casualties of terrorism have been caused by shootings, vehicle impacts or PBIED attacks [23]. In these scenarios, there is no need to protect any kind of structure. In addition, existing infrastructure has proven to be highly resilient and robust against blast loadings. On the other hand, there is a research gap related to primary and secondary blast injuries, even though they are the main source of fatalities. Primary blast injuries are caused by the blast pressure wave and generally affect gas-containing organs, usually the eardrums and lungs. The secondary blast injuries result from the direct impact of airborne debris due to the blast wind [24,25]. Therefore, more research is needed to understand casualty risks from bomb fragmentation and blast overpressure hazards, especially from IEDs and PBIEDs produced inside buildings. In this situation, the overpressure is amplified by the reflection of the blast wave on the enclosure walls, and the explosion yield can be increased up to eight times.

In this research, eight tests were carried out with different IED configurations simulating a PBIED inside a building using vest bombs. The building consisted of a small concrete structure of 6.80 × 5 m with a corridor and an inner room. This work focuses on the analysis of the high-speed video, pressures and accelerations recorded during the tests and the development of a suitable numerical model capable of reproducing the behavior of the blast effects inside the building.

2. Test Description and Instrumentation

In this context, the German Federal Office of Criminal Investigation (namely BKA), which is part of the Federal Ministry of the Interior, started a project in 2017 on the effects of IEDs on state security forces personnel. This project consisted of numerous tests carried out in Germany, with different types of explosive charges, with and without shrapnel, and at different targets. During the project, the BKA had the collaboration of the Centre of Excellence against Improvised Explosive Devices (C-IED COE), a member of the NATO Centre of Excellence community, to advise on the creation of the IEDs.

The last phase of testing consisted of creating a reinforced concrete structure that reproduces a possible location where terrorists prepare their material and detonate them before counter-terrorism police can get in and arrest them. This phase was carried out at the Sierra del Retín maneuvering and firing range, Barbate, Cádiz, from 18 to 20 September 2018. The concrete structure was designed by the *"Subdirección General de Proyectos y Obras—DIGENIN"* (part of the Spanish Ministry of Defense), while the instrumentation, measurements and modeling were carried out by the staff of the E.T.S.I. Minas y

Energía (Universidad Politécnica de Madrid—UPM). Finally, the explosive charges and their detonations were prepared by the Explosive Ordnance Disposal Section (SEDEX) of the Amphibious Mobility Group (GRUMA) of the Third Army (TEAR)—Marines in collaboration with personnel from NATO's Counter Improvised Explosive Devices Centre of Excellence (C-IED COE).

The structure built ad hoc for the tests was made of reinforced concrete and consisted of a perimeter corridor and an interior room in which the IEDs were placed. The design of the structure was based on the project requirements suggested by the BKA. The original idea was to have brick enclosures, but these were going to be destroyed after each trial making the project unfeasible in terms of time and money. For this reason, the structure was redesigned with reinforced concrete walls, with greater thicknesses in the exterior walls than in the interior walls. Doors and windows were also aligned to improve the venting of the shock wave. CYPECAD code was used for the design, and it was developed under the Spanish Technical Building Code (CTE) based on Eurocodes required by the European Union. The ground plan dimensions of the structure are shown in Figure 1, with the height between the floor and ceiling of the structure equal to 3 m. The outer walls were built with a thickness of 40 cm while the inner walls were 30 cm thick and the roof slab 25 cm. The concrete used for both the walls and the roof slab had a nominal compressive strength of 40 MPa, a density equal to 2300 kg/m^3, a tensile strength of 3.5 MPa, an elastic modulus equal to 30.9 GPa and a 20 mm maximum aggregate size. The reinforcement of the structure, made of B-500 C corrugated steel, was equally distributed on both sides of the walls and the roof slab in both directions (vertical and horizontal). The vertical reinforced steel of outer walls was constructed with a 12 mm diameter rebar evenly spaced at 300 mm, while in the inner walls, the diameter of the rebar used was 10 mm spaced at 200 mm. The horizontal steel of the outer walls had a diameter equal to 8 mm spaced at 150 mm; although in the inner walls the diameter used was the same, the spacing was increased to 200 mm. The reinforcement of the roof was made symmetrically on both sides, the inside of the cubicle where the detonation was located and the outside, using 12 mm diameter rebars with a square mesh of 150 mm on each side. In height, both reinforcements were 180 mm apart, with the thickness of the slab equal to 250 mm; therefore, a sufficient concrete layer was ensured on both sides. Finally, there was a perimeter reinforcement in all the joints between the walls and the roof, with 16 mm diameter rebars separated in height by 160 mm. The steel was assumed to have a density of 7850 kg/m^3, Young's modulus equal to 200 GPa, yield strength of 500 MPa, Poisson's ratio of 0.3 and tangent modulus of 20 GPa, following the EN 1992-1-1:2004 [26] and EN 1998-2:2:2005 [27]. Finally, the floor of the structure was covered with a concrete-reinforced layer with a steel mesh of #15 $\times$ 15 $\times$ 6 of 15 cm thick.

Eight tests were carried out (Table 1), and a previous (test) shot was performed to verify the operation of the measurement and recording equipment deployed in the area. The explosives used in the tests were black powder, ANFO, AN/AL and PG-2 (like the US C-4). The black powder used has a composition of potassium nitrate (75%), sulfur (10%) and carbon (15%) and is always granulated and graphitized, with particle sizes ranging from 0.1 to 4 mm. ANFO (ammonium nitrate and fuel oil) is the stoichiometric mixture of ammonium nitrate and fuel oil. AN/AL consists of a mixture of ammonium nitrate and aluminum powder. Finally, PG-2 is a military explosive whose composition is mainly RDX embedded in plastic additives.

The IEDs created for these trials were attached to different types of personal vests and in some cases were confined to steel tubes. The design of the explosive charges used in each test was based on the quantities of each of the explosives that can be included in a typical suicide vest configuration: in the case of tests T1 to T6 (black powder, ANFO and AN/AL), explosives inside steel tubes, and in the case of tests T7 and T8 (plastic explosive—PG2), packages directly attached to the inside of the vest. In all vest and tube tests (tests T1 to T6), 0.7 m of 15 g/m detonating cord was used to initiate the main charge. Instead, 3.7 m of detonating cord was used in tests where the explosive was directly stuck to the vest (without tubes—T7 and T8). Note that the explosive mass in test T8 is higher than the PG2 equivalent as the remaining charges were included.

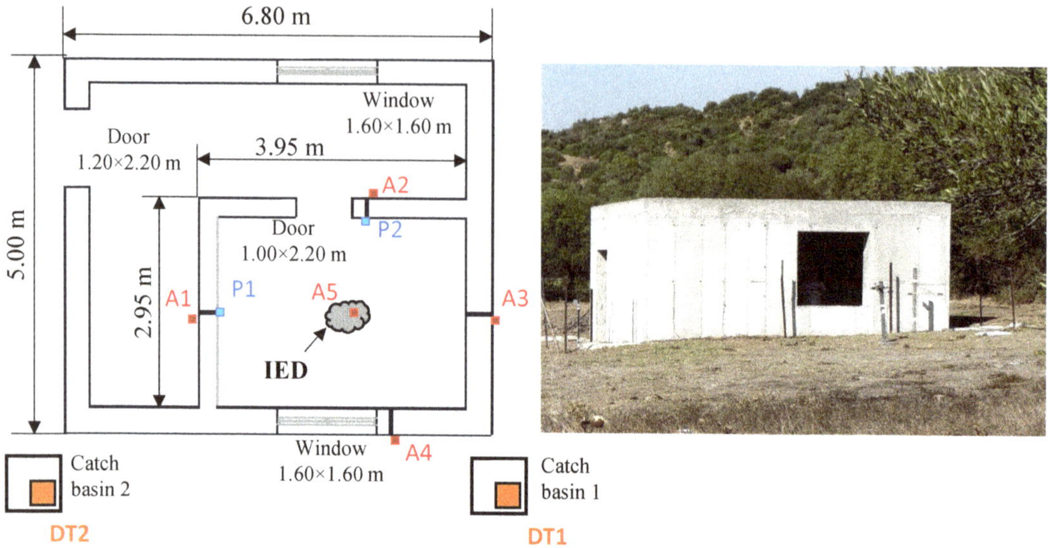

Figure 1. Details of the structure, location of the measuring equipment and photograph of the structure. The letters P refer to pressure sensors, the letters A are accelerometers, and the DTs are Datatrap II recording equipment.

Table 1. Load characteristics during the tests carried out.

Test	Day	Explosive Type	Charge (kg)	PETN (g)	TNT Equivalent Mass (kg)	Confinement
T0	18 September 2018	PG-2	0.10	0	0.14	–
T1	18 September 2018	Black Powder	3.37	10.5	0.79	Steel tubes
T2	18 September 2018	Black Powder	3.27	10.5	0.77	Steel tubes
T3	18 September 2018	ANFO	2.29	10.5	1.48	Steel tubes
T4	19 September 2018	ANFO	2.20	10.5	1.42	Steel tubes
T5	19 September 2018	AN/AL	2.16	10.5	1.88	Steel tubes
T6	19 September 2018	AN/AL	2.25	10.5	1.95	Steel tubes
T7	20 September 2018	PG-2	7.00	55	9.87	Vest
T8	20 September 2018	PG-2	8.20	55	14.21	Vest

The instrumentation of the tests consisted of accelerometers, pressure sensors, recording equipment and a high-speed camera. Figure 1 shows the location of the pressure sensors (P1 and P2), the accelerometers (A1–A5) and the two pieces of recording equipment used (DT1 and DT2).

The two pressure sensors used were 5000 PSI (344.7 MPa) PCB model 102B with ablative protection for the fireball. The sensors were placed with a passing tube on the concrete wall so that the sensing surface was normal to the main direction of the impact. In this way, the first wave registered would be the one reflected by the wall where the sensor is located. These sensors were at a height of 1.51 m and 1.55 m in the case of P1 and P2, respectively. Piezoelectric shock PCB accelerometers located on the opposite side of the wall from the explosive charge of 5000 or 10,000 g measurement limit were used (Table 2). In the accelerometer position called A1, one sensor was used for tests T0 to T4 and a different one for the last three tests (T5 to T7), due to the breakage of the sensor during the T4 test.

No measuring equipment was used in the last test (T8) for fear of complete destruction of the structure or, at least, of compromising its structural stability. Note that the sensor located at position A5 was placed on the roof of the structure on the outside of the structure (Figure 1). Two Datatrap II recorders from MREL were used for data acquisition. This system has up to eight recording channels, with a sampling rate in each channel of 10 MHz with a resolution of 14 bits. It is a portable and very robust piece of equipment prepared to work outdoors, in dust, rain and a wide range of temperatures. Figure 1 shows the location of the data acquisition equipment, inside interconnected chambers (catch basin). Signal conditioners PCB 480E09 were used, necessary to feed and condition both the pressure sensors and the accelerometers. See Figure 2 for more details of the measuring equipment and positions.

Table 2. Characteristics of the accelerometers used and the height at which they were positioned.

#Sensor	Model	Measurement Range (g)	Test	Height (m)
A1	350C23	±10,000	T0–T4	1.375
A1	350C04	±5000	T5–T7	1.375
A2	350C04	±5000	T0–T7	1.370
A3	350C23	±10,000	T0–T7	1.395
A4	350B04	±5000	T0–T7	1.370
A5	350B04	±5000	T0–T7	3.300

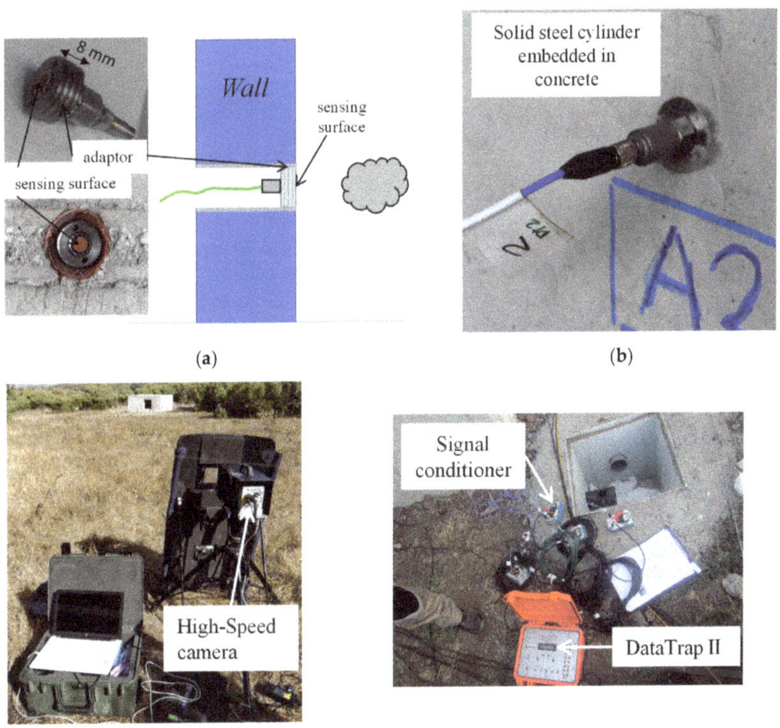

Figure 2. (**a**) Pressure sensors (PCB 102B), details of the P1. (**b**) Accelerometer in position A2. (**c**) Location of the high-speed camera in one of the tests. (**d**) Signal conditioners and DataTrap II recording equipment.

Finally, the high-speed camera (CAV) used was a Photron Fastcam SA3-120k, adapted for explosion testing with a steel case. It reaches a recording speed of 5000 images per second for a resolution of 512 × 512 pixels, reaching up to 120,000 fps for a resolution of 128 × 16 pixels.

3. Numerical Model

The 3D numerical models were made using the LS-DYNA Version 971-R11 software [28], which is based on explicit numerical methods that are suitable for solving problems associated with large deformations subjected to blasting. The destructive effect of these kinds of blast tests, along with the fast structures' reaction and short duration of the explosive event, makes the detailed study of these events very complex.

3.1. Finite Element Model

This model was made of two main critical parts: concrete and steel rebar. In addition, the ground was also introduced into the model but only for visualization purposes (Figure 3). The functionality of "Constrained Lagrange in Solid" was used for the correct operation of both materials as a single assembly. This option can be used as the interaction between parts (steel and concrete) can be presumed to be ideal as the event is almost instantaneous [29,30]. Moreover, the structure was fixed into the ground by using the single point constraint (SPC), canceling displacements and rotations in all directions of space.

Figure 3. Details of the complete 3D model made with LS-DYNA and details of the steel armor.

The concrete was defined with 3D Lagrangian solid elements with reduced integration to decrease the computational time. The element size used for the concrete was 20 mm, based on previous studies of the concrete blasting response under similar conditions and charges [30–32]. The reinforcement was modeled using beam-type elements with a size of 50 mm in length. The number of solids elements was 5,150,187 while the number of beams was 64,266.

To solve the model, LS-DYNA offers two parallel programming methods: symmetric multi-processing (SMP) and massively parallel processing (MPP). SMP runs on a computer with multiple identical cores with the cores and memory connected via a shared data bus, being scalable up to 8 CPUs. MPP uses various separate CPUs running in parallel, each with its own memory to execute a single analysis, performing a domain decomposition of the problem and then distributing the sub-domains to different cores. This solver is scalable over a wide range of CPUs. Although the MPP method allows a reduction in computation time, the size of the model, the lack of symmetries and the complexity of the problem to be solved are considerable. The simulation time was lengthened to 2.5 s, which in SMP resulted in 306 h and 53 min, while in MPP, this time can be reduced to 151 h and 25 min. The computer used for these simulations has two Intel XEON E5-2630 v4 processors at 2.20 GHz (10 cores each, 2 threads per core), with 64 GB of RAM and a Windows 10 operating system.

Moreover, dynamic relaxation (DR) was included in the model. DR is the recommended way to preload a model before the application of dynamic loads in the subsequent transient analysis. This technique makes it possible to achieve a steady-state preload condition free of dynamic oscillations (or nearly free). It is important in cases like this work to apply gravity before the transient analysis (detonation) to avoid an unstressed state at the beginning of the analysis. If the gravitational load is suddenly applied, dynamic oscillations could be enhanced that would invalidate the utility of any calculation. The application of gravity is performed by the Load Body command (see [28] for more details). In this command, a first curve is defined for the quasi-static analysis (dynamic relaxation) of gravity. For this purpose, a curve is created where the acceleration rises linearly from zero to the constant value (gravity) for a short period of time and then remains constant. Then, a second curve of constant value (gravity) is created which will be used for the rest of the simulation time. The Control Dynamic Relaxation card is also used but with the LS-DYNA default values.

Acceleration in LS-DYNA can be measured with the use of sensors at certain coordinates [28]. The processing of these data is sometimes complex and does not usually work well when the blast processing is performed with tabulated values such as the Load_Blast_Enhanced command. Another alternative is using the *Database_History_Node command to explicitly track the history of features of that node (i.e., acceleration, velocity, displacement, etc.). A shorter time interval between acceleration data in those nodes (i.e., 1×10^{-6} s) than the normal one determined between drawings (0.01 s on the D3Plot card) can be used which improves performance, computation time and hard disk space occupied.

3.2. Blast Implementation

Explosive charge implementation can be handled from two different approaches: using the parameters of the explosive material and its equation of state [17]; or by using a TNT equivalent for the load and its implementation with the load blast enhanced (LBE) function. This last option is usually simpler to implement, as well as computationally faster, producing very good results [33–35].

As for the application of the load, the LBE was used here, which is the way LS-DYNA introduces the CONWEP [36]. This can be used assuming that the steel tubes would have a potentially lethal effect on people but are quite harmless to the structure. With the LBE instructions, the necessary input parameters to calculate and apply the generated pressure (incident and reflected) on the concrete elements are the type of shock wave, the equivalent mass of TNT, the coordinates of the load center and the concrete sides where the pressure wave will be applied. The software applies the pressures following Friedlander's equation to calculate the pressure curve, including the negative phase. It should be noted that with this methodology only the pressure peak set by the measured signal is reproduced, and it is not possible to reproduce reflections of the wave or shrapnel produced during the explosive detonation inside the tubes. In addition, the pressures were only recorded in the room where the IEDs were located. This makes that the pressures were only applied with LBE on all faces of this room (including the ceiling) and not outside of it such as the corridor.

3.3. Materials

LS-DYNA offers more than 25 models that can be used to describe the concrete, some require many input parameters while others work with reduced data, but not all of them perform well under blasting events [37–39].

In this research, the continuous surface cap model (CSCM) concrete was used to describe the concrete behavior. The automatic generation of parameters was based on introducing the values of the compressive strength, the aggregate maximum size (maximum) and the density. The model is plasticity based with the implementation of shear failure surfaces corresponding to the elastic limit, residual strength and failure. This model works based on an isotropic elastic behavior before cracking to move to a plastic behavior limited by the failure surfaces. This model implements an internal calculation of the damage that allows the erosion of the elements when they reach 99% of the damage limit [28,40] and the maximum principal strain in the element exceeds a value defined by the user, known as ERODE. The default value of 1.05 was used for the ERODE parameter [40]. The CSCM includes a dynamic increase factor (DIF), governed by specific data from the CEB-FIP design code using the Duvaut–Lions overstress formulation based on time rather than strain rate [41]. The material properties used in this model are shown in Table 3.

Table 3. Concrete and steel properties used in the numerical modeling.

Property	Concrete	Steel
Density (kg/m^3)	2300	7850
Uniaxial compressive strength (MPa)	40	-
Maximum aggregate size (m)	0.02	-
Young modulus (MPa)	-	2×10^5
Poisson's ratio	-	0.3
Yield stress (MPa)	-	5×10^2
Tangent modulus (MPa)	-	2×10^4

The steel used in the reinforcement was the classic B-500 S, introduced in the model as the material model "Piecewise Linear Plasticity". In this work, the option of defining the rupture based on the effective plastic deformation was chosen, as opposed to the rupture based on the time step of the numerical model achieved by the convergence of the method. The value entered was equal to 0.075 [30], i.e., when the plastic strain reaches this value, the element is deleted from the calculation. In addition, to define the stress–strain behavior, a bilinear stress–strain curve was applied by using the tangent modulus. Moreover, the strain rate effects were included in the steel model based on the scale yield stress by using the Cowper–Symonds model (C equal to 25.36 s^{-1} and P equal to 2.52) [28,37,42].

4. Results and Discussion

4.1. High-Speed Camera

The images were captured at a speed ranging from 3000 to 5000 fps. Figure 4 shows a sequence of images of the video obtained in the T2 test with black powder in steel tubes. It shows the extension of the fireball that reaches up about two meters outside the cubicle in the vicinity of the window. The powder generated a significant volume of gases that are expelled practically simultaneously through the two openings to the outside of the cubicle: the window directly to the charge and the window at the rear of the image connected to the main room by a door. If the images of the gunpowder test are compared with those recorded in the ANFO test (T4) shown in Figure 5, important differences are observed in terms of the extension of the fireball and the volume of gases generated. The fireball did not reach the outside in the case of the ANFO test, and the volume of gases was clearly lower. Figure 6 shows a sequence of 12 images obtained in a test with AN/AL (T6). The fireball extended considerably more than in the ANFO test due to the aluminum in its composition.

Figure 7 shows a similar sequence of images for test T7 in which a vest without steel tubes was fired with 7 kg of PG2 plastic explosive. The extension of the fireball reaches the facade completely on both the front and rear faces of the cubicle. The first images just after the start of the detonation show a large white glow indicating very high temperatures. The escape of gases that comes at the same time as the fireball takes place through the openings mentioned above (windows) and gas escape can be seen in the upper part of the cubicle, probably due to the displacement of the upper slab.

4.2. Pressure Signals

In some cases, the registration of the different pressure–time signals presented an important level of noise that can mask the real signal. In these cases, filtering the signal is necessary to obtain the parameters of the shock wave. Details of the procedure followed can be found in the work published by Chiquito et al., 2019 [43]. In the first test with ANFO (T1) there was a problem with the trigger of one of the DT2 recording systems, and therefore, no data were obtained from the P2 sensor. The P1 sensor in that test suffered the impact of fragments, and therefore, it did not record anything either. In the following tests (T2 onwards), no more sensors were placed.

Figure 4. Sequence of images obtained with HSC in trial T2: black powder. Time in milliseconds; the reference time is the first frame of the video where the detonation is observed.

Pressure measurements were recorded with some reliability in tests T0 to T2 (see Table 4). The simulation values are compared with the average (field) values when there is more than one signal. See Figure 8 as an example of the pressure application. The sequence shows how the wave first reaches the inner wall of the door (area closest to the IED) and then the ceiling. It then reaches the rest of the surfaces and expands in a very similar manner. The expansion pattern is the classic one in a shock wave of this type from the center to the sides, ending at the corners. Once the pressure front passes a point, the pressure decays to the initial pressure. As shown in Table 4, the values simulated with the LBE card are quite similar to those obtained in the field, with relatively low errors given the nature of the phenomenon. As mentioned in Section 3.2, LBE only reproduces the first pressure peak, and this is what is compared in Table 4. This is obviously an important limitation of the simulation, but the other available techniques (i.e., SPH, ALE or PBM), which might be able to reproduce the behavior more realistically, become unfeasible due to the resources required (meshing, number of elements, computational time, etc.). The results show differences of about 11%. This shows that the simulation is relatively reliable and that the TNT equivalent used in the description of the explosives was quite accurate.

Figure 5. Sequence of images obtained with HSC in trial T4: ANFO. Time in milliseconds; the reference time is the first frame of the video where the detonation is observed.

Figure 6. Sequence of images obtained with HSC in trial T6: AN/AL. Time in milliseconds; the reference time is the first frame of the video where the detonation is observed.

Figure 7. Sequence of images obtained with HSC in trial T7: PG2. Time in milliseconds; the reference time is the first frame of the video where the detonation is observed.

Table 4. Pressure sensor results. Friedlander adjustment.

Test	Explosive	Sensor	P_r (kPa)	P_r (LS-DYNA) (kPa)	Relative Dif. (%)
T0	PG-2	P1	88.72	99.67	−12.34
T0	PG-2	P2	129.80	136.02	−4.79
T1	Black Powder	P1	193.11	181.21	6.71
T2	Black Powder	P1	195.38		
T1	Black Powder	P2	372.27	309.64	17.93
T2	Black Powder	P2	382.36		

All logs show multiple reflections on the various walls, but no sustained gas pressure is observed due to the large vent provided by the window and access door, as shown in Figure 4. Figure 9 shows how the simulation with LBE only reproduces the first peak of the signal recorded by the sensors. It can also be seen how well the model is able to reproduce the shape (duration and impulse) of the recorded shock wave. Moreover, the pressures recorded in sensor P1 are much lower, almost half, than those of sensor P2. This may be due to the orientation of the explosive device focused more directly toward P2, not having a direct "view" of the sensor located at P1. However, the different reflections to which the sensor is subjected are greater in P1, which makes sense, since it is farther away from the large vents that are the doors and windows.

4.3. Acceleration Signals

Peak acceleration values comparing all the simulations performed and the measurements with the different sensors (A1–A4, Figure 1 and Table 2) can be found in Table 5. A filter was applied to the acceleration signals to eliminate noise and electrical peaks. The filtering applied was the Butterworth low-pass type of order 4; in addition, if the signal presented an offset, it was also corrected. Sensors that were not measured in the field are not reflected in Table 5, which is used to show the differences between measured and simulated values. Measurement failures were sometimes due to poor sensor coupling, failures in the trigger system or measurements that did not make sense because of the extreme (high/low) values obtained.

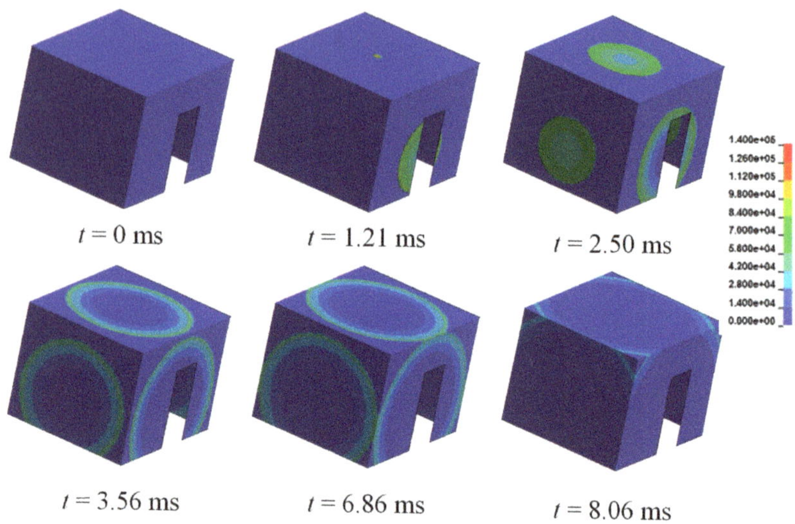

Figure 8. Pressure isovalues (view from the inner door) on the interior faces of the room where the detonations were carried out for the T0 test. The color scale is the pressure values in Pa.

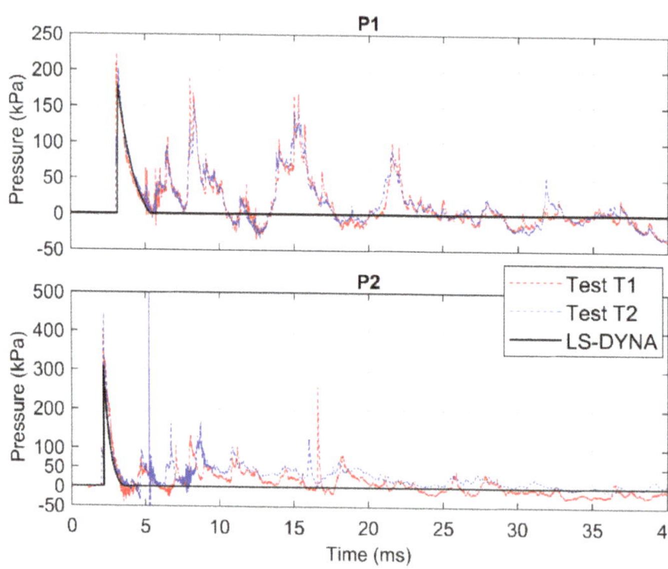

Figure 9. Pressure signals (P1 and P2) for tests T1 and T2: black powder, and the simulated signal with LS-DYNA.

Table 5. Peak acceleration values for different sensors and trials. BP means black powder.

Test	Explosive	#Sensor	Measured Acceleration (g)	Model Acceleration (g)	Difference (%)
T0	PG2	A2	44.4	26.3	40.77
		A3	77.1	57.8	25.03
		A4	29.6	30.2	−1.94
T1	BP	A1	207.2	152.8	26.25
		A3	699.3	416.5	40.44
		A4	69.2	79.9	−15.49
T2	BP	A1	105.2	75.4	28.30
		A3	430.3	403.6	6.21
		A5	298.8	354.8	−18.76
T3	ANFO	A1	1048.9	852.4	18.73
T4	ANFO	A1	1113.6	928.3	16.64
		A3	1048.2	809.9	22.74
		A4	259.8	349.4	−34.50
		A5	1646.4	1056.0	35.86
T5	AN/AL	A1	998.9	1115.2	−11.64
		A3	3786.5	4192.0	−10.71
		A4	4685.5	4896.0	−4.49
		A5	2683.10	1691.5	36.96
T6	AN/AL	A1	903.4	1126.5	−28.68
		A3	5600.6	4305.0	23.13
		A4	4828.6	5094.0	−5.50
		A5	1996.25	1518.4	23.94
T7	PG2	A1	902.7	652.0	27.77
		A3	1507.90	1521.8	3.12
		A4	5171.8	3464.4	33.01
		A5	1342.3	1505.0	−12.12

As can be seen in Table 5 for sensors A1 to A4, the acceleration data obtained are quite large for the ANFO, AN/AL and PG2 tests, with values between 1000 and 5000 g, while in the black powder tests, the acceleration values are around 400 g. This clearly indicates that accelerations increase with the use of more powerful charges, as expected. Given the non-linear nature of the phenomenon, as well as the limitations of the simulation itself, the differences between the model and the real data are quite good with an average absolute value of 20%. Therefore, the model can reproduce with some reliability the acceleration peaks. The highest difference is found in the only useful measurement of sensor A2 and sensor A3 in test T1. On the other hand, the lowest value is also found at sensor A3 in test T7, followed by sensor A4 in tests T5 and T6. The lowest values are generally found in sensor A4, but it is also the sensor that shows the largest deviations from the mean value. The highest mean values are found in sensor A1 but with the smallest deviations. Considering the results obtained, it can be deduced that the differences between measured and simulated values on sensor A1 are the most important. This may be since the behavior of this interior wall is not well reproduced, being more rigid in the model than it should be. The opposite is true for the outer wall where sensor A4 is located, where the model reproduces quite faithfully the behavior of sensor A4.

However, sensor A5, located on the roof of the structure (see Figure 1), shows somewhat different results (see Table 5 and Figure 10). In general, the model can reproduce the results measured in the field with errors averaging (and in absolute value) around 25%. Figure 10 shows how the accelerations look similar in all cases. In the T2 test, the peak accelerations were not recorded with the first arrival of the wave as in the other cases, as always happens in the modeling. Therefore, in this test, the acceleration peaks between LS-DYNA and the tests do not coincide in time, although they do coincide in peak values.

Despite the increase in the explosive load, the maximum accelerations recorded on the roof slightly decreased, contrary to the rest of the sensors. This was especially noticeable in the case of test T7, with 7 kg of PG2. This may be a consequence of the decrease in the stiffness of the structure due to the accumulated damage after performing the tests consecutively without intermediate reinforcement or support actions on the structure.

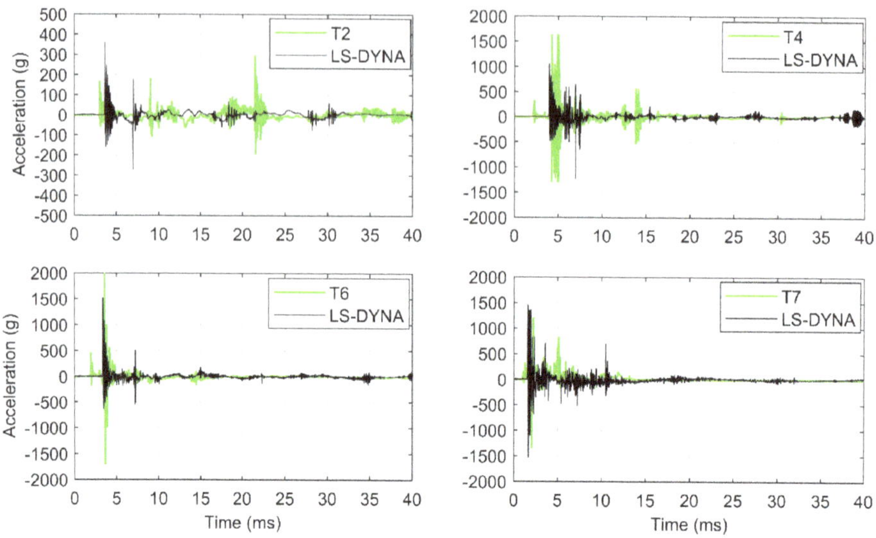

Figure 10. Accelerations from tests T2, T4, T6 and T7, in position A5 (see Figure 1) and their comparison with LS-DYNA.

The extra stiffness provided by the double reinforcement used in the roof slab may have lost its effect after so many tests, but it was probably the one that prevented an earlier collapse of the structure. To reinforce this idea, in the model (Figure 11), it is easy to observe how stress accumulations occur in the window and in the joints of the roof slab with the walls, and therefore, these are the areas that withstood the most stresses throughout the tests. The sequence of images shows how in two seconds the stresses produced in the structure have already stabilized. It can be seen in Figure 8 how the effect of the detonation is only 4 ms when the sequence of images in Figure 11 is every second. It can also be observed that after the T7 test, the stress state of the structure is considerably higher than before.

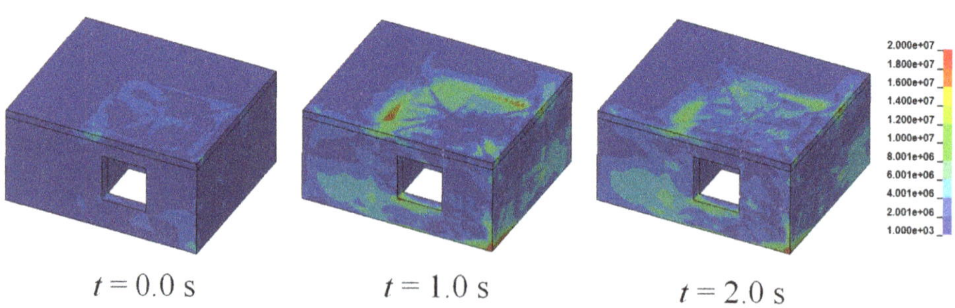

Figure 11. Effective stress (scale in Pa) of trial T7. The first image with time zero corresponds to the stress state when the T7 test load is detonated in the model.

4.4. Final Test E8

In this test, the explosive charge was placed on the ground in the corner between the walls of the access door and where the P1 sensor was located (see Figure 1). Figure 12 shows the result of the structure after the last test (T8). It can be seen how the structure is destroyed on the window wall and on the adjoining one on the DT1 side (see Figure 1). The gases try to exit through that area (as seen in the previous test T7, Figure 7), projecting most of the shock wave energy on this side of the already weakened structure. Consequently, the walls and part of the roof collapse, leaving the reinforcement exposed. In the model, something similar happens: although the roof seems to be somewhat more damaged than the real structure, the side wall shows significant damage as it happened in the test.

Figure 12. Final test (T8) results: numerical model and photographs.

4.5. Effects of the IED Type

In general, the type of IED used can have a major impact on human casualties related to the air blast wave or shrapnel impact. However, it is not very common that they can affect entire structures or parts of them, especially in cases of small charges. This work, although with repeated explosive charges on a structure accumulating some damage, can serve as a small case study.

The results show that the IEDs used in the tests (T1–T6), where explosives that are not too powerful and with low charge are confined in tubes, can produce high accelerations. These pressures are surely produced by the confinement of the explosive in a steel tube since its attempted detonation in air would probably produce milder effects. This fact results in greater damage to the structure given the high accelerations. See Figure 13 for details of the interior parts of the structure affected by successive detonations. It can be seen how the black powder tests hardly affect the structure or the concrete (Figure 13A). In the case of ANFO, whose charge is more powerful, some spalling of the concrete near the interior door can be observed (Figure 13B). The same happens in the case of AN/AL, leaving even the first reinforcements of the structure visible, indicating that the erosion has already been significant (Figure 13C). The case of the plastic explosive (T7–T8) is slightly different because it is a very powerful explosive that does not need confinement to improve its performance. It is known that these explosives are used by security forces and corps for the destruction or demolition of parts of a structure or even the full structure. In this case, it is no different, producing significant damage to the structure as seen in the cracks generated in the interior wall (Figure 13D), which although in principle does not have to compromise the stability of the structure, does leave it very damaged. As discussed above, in the last test (T8), the load was appreciably higher, and this caused the roof to decouple from the walls (Figure 13E), in addition to the obvious collapse occurring in the area of the exterior window.

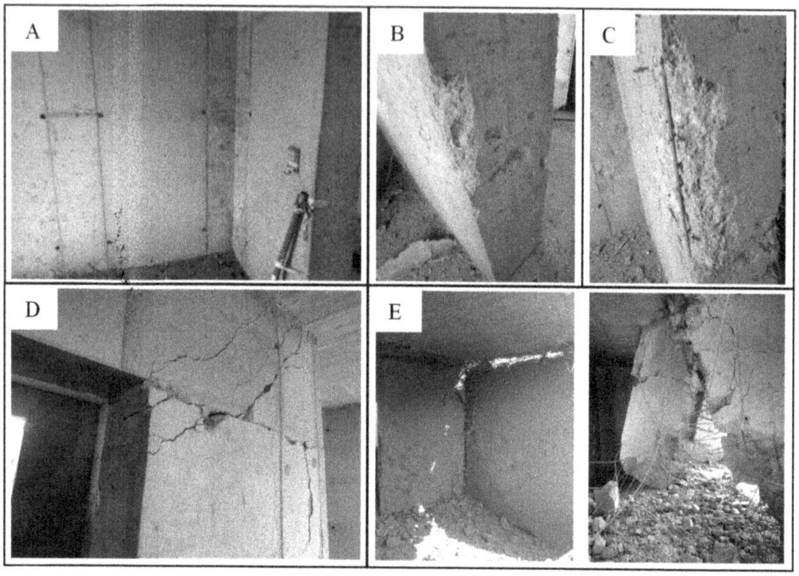

Figure 13. Different effects of IEDs on the interior structure: (**A**) after the two black powder tests (T1–T2); (**B**) after the ANFO tests (T3–T4); (**C**) after the AN/AL tests (T5–T6); (**D**) after the first PG2 test (T7); (**E**) after the second PG2 test (T8).

5. Conclusions

A total of eight tests were carried out with different types of IEDs on the same reinforced concrete structure, simulating a scenario where charges are detonated at the entrance of the State security forces and bodies. Some conclusions can be extracted as follows:

- The high-speed camera images allow us to see the correct detonation of the explosive, while the pressure log allows us to validate the model input data.
- The acceleration recorded at the roof of the structure decreases as more tests are performed due to the loss of stiffness of the structure.
- IEDs of relatively low power (with homemade explosives or low-TNT equivalent), although they cause significant accelerations in the structure, do not compromise its structural stability, while more powerful IEDs (plastic explosives), although with similar accelerations, do put the structural stability of the building at risk.
- A solid element model using LBE offers, even in complex cases such as this one, a reasonable reproduction of the behavior of a structure reducing testing costs by being able to reproduce with some certainty different scenarios.

Author Contributions: Conceptualization, R.C., A.P.S., L.M.L. and J.I.Y.; methodology, R.C., L.M.L. and M.C.; software, A.P.S., A.A. and S.M.-A.; validation, A.P.S., S.M.-A., L.M.L. and M.C., formal analysis, R.C., A.A. and E.C.; investigation, A.P.S., R.C., L.M.L. and M.C.; writing—original draft preparation, R.C., M.C. and E.C.; supervision, R.C., L.M.L. and J.I.Y.; project administration, R.C.; funding acquisition, R.C., A.P.S. and J.I.Y. All authors have read and agreed to the published version of the manuscript.

Funding: This research was funded by the "Subdirección General de Proyectos y Obras—DIGENIN" of the Spanish Ministry of Defense, grant number 340/OD (SEF), File 1004218000499, project code P1806490081.

Institutional Review Board Statement: Not applicable.

Informed Consent Statement: Not applicable.

Data Availability Statement: The data presented in this study are available on request from the corresponding author. The data are not publicly available due to sensible information and the misuse that can be made of them.

Acknowledgments: The authors of this work would like to thank the different institutions involved (BKA, C-IED COE, DIGENIN, GRUMA, TEAR) for their good work in the development of the corresponding tasks and the good atmosphere generated during the trials.

Conflicts of Interest: The authors declare no conflict of interest.

References

1. Kamel, H. Review of design techniques of armored vehicles for protection against blast from improvised explosive devices. In Proceedings of the International Mechanical Engineering Congress and Exposition, Salt Lake City, UT, USA, 11–14 November 2019; Volume 10227, pp. 1–18.
2. Fan, W.; Zhou, N.; Jiao, Q.; Shi, J.; Tang, K. Investigation on the explosive characteristics and damage mode of cylindrical improvised explosive devices. *J. Appl. Mech. Tech. Phys.* **2020**, *61*, 1024–1032. [CrossRef]
3. Liu, H.; Huang, G.; Guo, Z.; Feng, S. Fragments Velocity Distribution and Estimating Method of Thin-Walled Cylindrical Improvised Explosive Devices with Different Length-to-Diameter Ratios. *Thin-Walled Struct.* **2022**, *175*, 109212. [CrossRef]
4. Williams, D.S. Enhancing the Applicability of Blast Modelling and Advice. *Int. J. Prot. Struct.* **2015**, *6*, 701–710. [CrossRef]
5. Price, M.A.; Nguyen, V.; Hassan, O.; Morgan, K. An Approach to Modeling Blast and Fragment Risks from Improvised Explosive Devices. *Appl. Math. Model.* **2017**, *50*, 715–731. [CrossRef]
6. Sielicki, P.W.; Stewart, M.G.; Gajewski, T.; Peksa, P.; Al-rifaie, H.; Studzi, R. Field Test and Probabilistic Analysis of Irregular Steel Debris Casualty Risks from a Person-Borne Improvised Explosive Device. *Def. Technol.* **2021**, *17*, 1852–1863. [CrossRef]
7. Draganić, H.; Gazić, G.; Varevac, D. Experimental Investigation of Design and Retrofit Methods for Blast Load Mitigation—A State-of-the-Art Review. *Eng. Struct.* **2019**, *190*, 189–209. [CrossRef]
8. Goswami, A.; Adhikary, S. Das Retrofitting Materials for Enhanced Blast Performance of Structures: Recent Advancement and Challenges Ahead. *Constr. Build. Mater.* **2019**, *204*, 224–243. [CrossRef]
9. Tai, Y.S.; Chu, T.L.; Hu, H.T.; Wu, J.Y. Dynamic Response of a Reinforced Concrete Slab Subjected to Air Blast Load. *Theor. Appl. Fract. Mech.* **2011**, *56*, 140–147. [CrossRef]
10. Zhao, C.F.; Chen, J.Y.; Wang, Y.; Lu, S.J. Damage Mechanism and Response of Reinforced Concrete Containment Structure under Internal Blast Loading. *Theor. Appl. Fract. Mech.* **2012**, *61*, 12–20. [CrossRef]
11. Lin, X.; Zhang, Y.X.; Hazell, P.J. Modelling the Response of Reinforced Concrete Panels under Blast Loading. *J. Mater.* **2014**, *56*, 620–628. [CrossRef]
12. Jayasooriya, R.; Thambiratnam, D.P.; Perera, N.J.; Kosse, V. Blast and Residual Capacity Analysis of Reinforced Concrete Framed Buildings. *Eng. Struct.* **2011**, *33*, 3483–3495. [CrossRef]
13. Zhang, D.; Yao, S.J.; Lu, F.; Chen, X.G.; Lin, G.; Wang, W.; Lin, Y. Experimental Study on Scaling of RC Beams under Close-in Blast Loading. *Eng. Fail. Anal.* **2013**, *33*, 497–504. [CrossRef]
14. Codina, R.; Ambrosini, D.; Borbón, F. De Alternatives to Prevent the Failure of RC Members under Close-in Blast Loadings. *EFA* **2016**, *60*, 96–106. [CrossRef]
15. Hajek, R.; Fladr, J.; Pachman, J.; Stoller, J.; Foglar, M. An Experimental Evaluation of the Blast Resistance of Heterogeneous Concrete-Based Composite Bridge Decks. *Eng. Struct.* **2019**, *179*, 204–210. [CrossRef]
16. Zhao, C.; Lu, X.; Wang, Q.; Gautam, A.; Wang, J.; Mo, Y.L. Experimental and Numerical Investigation of Steel-Concrete (SC) Slabs under Contact Blast Loading. *Eng. Struct.* **2019**, *196*, 109337. [CrossRef]
17. Reifarth, C.; Castedo, R.; Santos, A.P.; Chiquito, M.; López, L.M.; Pérez-Caldentey, A.; Martínez-Almajano, S.; Alañon, A. Numerical and experimental study of externally reinforced RC slabs using FRPs subjected to close-in blast loads. *Int. J. Impact Eng.* **2021**, *156*, 103939. [CrossRef]
18. Shi, Y.; Li, Z.; Hao, H. A New Method for Progressive Collapse Analysis of RC Frames under Blast Loading. *Eng. Struct.* **2010**, *32*, 1691–1703. [CrossRef]
19. Bermejo, M.; Santos, A.P.; Goicolea, J.M. Development of Practical Finite Element Models for Collapse of Reinforced Concrete Structures and Experimental Validation. *Shock. Vib.* **2017**, *2017*, 4636381. [CrossRef]

20. Alshaikh, I.M.H.; Bakar, B.H.A.; Alwesabi, E.A.H.; Akil, H. Experimental Investigation of the Progressive Collapse of Reinforced Concrete Structures: An Overview. *Structures* **2020**, *25*, 881–900. [CrossRef]
21. Kernicky, T.P.; Whelan, M.J.; Weggel, D.C.; Rice, C.D. Structural Identification and Damage Characterization of a Masonry Infill Wall in a Full-Scale Building Subjected to Internal Blast Load. *J. Struct. Eng.* **2015**, *141*, D4014013. [CrossRef]
22. Xiao, W.; Andrae, M.; Steyerer, M.; Gebbeken, N. Investigations of Blast Loads on a Two-Storeyed Building with a Gable Roof: Full-Scale Experiments and Numerical Study. *J. Build. Eng.* **2021**, *43*, 103111. [CrossRef]
23. Stewart, M.G.; Mueller, J. Terrorism Risks, Chasing Ghosts and Infrastructure Resilience. *Sustain. Resilient Infrastruct.* **2020**, *5*, 78–89. [CrossRef]
24. Dante, D.; William, Y. Primary Blast Injuries—An Updated Concise Review. *World J. Surg.* **2012**, *36*, 966–972. [CrossRef]
25. Mathews, Z.R.; Koyfman, A. Blast Injuries. *J. Emerg. Med.* **2015**, *49*, 573–587. [CrossRef]
26. *EN 1992-1-1*; Eurocode 2: Design of Concrete Structures—Part 1-1: General Rules and Rules for Buildings. European Committee for Standardization: Brussels, Belgium, 2004.
27. *EN 1998-2*; Eurocode 8: Design of Structures for Earthquake Resistance—Part 2: Bridges. European Committee for Standardization: Brussels, Belgium, 2005.
28. Livermore Software Technology Corporation (LSTC). *LS-DYNA Keyword User ́s Manual—R11*; Livermore Software Technology Corporation: Livermore, CA, USA, 2018; p. 3186.
29. Castedo, R.; Segarra, P.; Alañon, A.; Lopez, L.M.; Santos, A.P.; Sanchidrian, J.A. Air Blast Resistance of Full-Scale Slabs with Different Compositions: Numerical Modeling and Field Validation. *Int. J. Impact Eng.* **2015**, *86*, 145–156. [CrossRef]
30. Castedo, R.; Santos, A.P.; Alañón, A.; Reifarth, C.; Chiquito, M.; López, L.M.; Martínez-Almajano, S.; Pérez-Caldentey, A. Numerical Study and Experimental Tests on Full-Scale RC Slabs under Close-in Explosions. *Eng. Struct.* **2021**, *231*, 111774. [CrossRef]
31. Alañón, A.; Cerro-Prada, E.; Vázquez-Gallo, M.J.; Santos, A.P. Mesh Size Effect on Finite-Element Modeling of Blast-Loaded Reinforced Concrete Slab. *Eng. Comput.* **2018**, *34*, 649–658. [CrossRef]
32. Feng, W.; Chen, B.; Yang, F.; Liu, F.; Li, L.; Jing, L.; Li, H. Numerical Study on Blast Responses of Rubberized Concrete Slabs Using the Karagozian and Case Concrete Model. *J. Build. Eng.* **2021**, *33*, 101610. [CrossRef]
33. Abedini, M.; Zhang, C.; Mehrmashhadi, J.; Akhlaghi, E. Comparison of ALE, LBE and Pressure Time History Methods to Evaluate Extreme Loading Effects in RC Column. *Structures* **2020**, *28*, 456–466. [CrossRef]
34. Suhaimi, K.; Sohaimi, R.M.; Knight, V.F.; Sheng, T.K.; Ahmad, M.M.H.M.; Isa, M.F.M.; Sohaimi, A.S.M.; Noordin, M.N.H.; Syaharani, A. Simulation of Hybrid-III Dummy Response Using Three LS-DYNA Blast Methods. *Def. S T Tech. Bull.* **2017**, *10*, 111–120.
35. Hilding, D. Methods for Modelling Air Blast on Structures in LS-DYNA. In Proceedings of the Nordic LS-DYNA Users' Conference, Gothenburg, Sweden, 13–14 October 2016; p. 65.
36. Hyde, D.W. *Microcomputer Programs CONWEP and FUNPRO, Applications of TM 5-855-1, 'Fundamentals of Protective Design for Conventional Weapons'(User's Guide)*; No. WES/IR; Army Engineer Waterways Experiment Station Vicksburg MS Structures Lab.: Vicksburg, MS, USA, 1988.
37. Abedini, M.; Zhang, C. Performance Assessment of Concrete and Steel Material Models in LS-DYNA for Enhanced Numerical Simulation, A State of the Art Review. *Arch. Comput. Methods Eng.* **2021**, *28*, 2921–2942. [CrossRef]
38. Tabatabaei, Z.S.; Volz, J.S.; Baird, J.; Gliha, B.P.; Keener, D.I. Experimental and Numerical Analyses of Long Carbon Fiber Reinforced Concrete Panels Exposed to Blast Loading. *Int. J. Impact Eng.* **2013**, *57*, 70–80. [CrossRef]
39. Brad, D.; Michael, O.; Joseph, M. Modeling Reinforced Concrete Protective Construction for Impact Scenarios. In Proceedings of the International Explosives Safety Symposium and Exhibition, San Diego, CA, USA, 6–9 August 2018; p. 298.
40. Murray, Y.D. *Users Manual for LS-DYNA Concrete Material Model 159*; Federal Highway Administration: Washington, DC, USA, 2007.
41. Brannon, R.M.; Leelavanichkul, S. *Survey of Four Damage Models for Concrete*; Sandia National Laboratories: Albuquerque, NM, USA, 2009.
42. Cadoni, E.; Dotta, M.; Forni, D.; Tesio, N. High Strain Rate Behaviour in Tension of Steel B500A Reinforcing Bar. *Mater. Struct. Constr.* **2015**, *48*, 1803–1813. [CrossRef]
43. Chiquito, M.; Castedo, R.; Lopez, L.M.; Santos, A.P.; Mancilla, J.M.; Yenes, J.I. Blast Wave Characteristics and TNT Equivalent of Improvised Explosive Device at Small Scaled Distances. *Def. Sci. J.* **2019**, *69*, 328–335. [CrossRef]

applied sciences

MDPI

Article

Investigation of Behavior of Masonry Walls Constructed with Autoclaved Aerated Concrete Blocks under Blast Loading

Somayeh Mollaei [1,*], Reza Babaei Ghazijahani [1], Ehsan Noroozinejad Farsangi [2,*] and Davoud Jahani [3]

1 Department of Civil Engineering, University of Bonab, Bonab 55513-95167, Iran
2 Department of Civil Engineering, University of British Columbia (UBC), Vancouver, BC V6T 1Z4, Canada
3 Department of Mechanical Engineering, University of Bonab, Bonab 55517-61167, Iran
* Correspondence: s.mollaei@ubonab.ac.ir (S.M.); ehsan.noroozinejad@ubc.ca (E.N.F.)

Abstract: Autoclaved aerated concrete (AAC) blocks have widespread popularity in the construction industry. In addition to lightness, these materials have other advantages, including fire resistance, low acoustic and thermal conductivity, ease of cutting and grooving, and simple transportation. Since the behavior of AAC under severe dynamic loading conditions such as blast loads has not been adequately studied in the literature, in the current paper, the behavior of masonry walls constructed with AAC blocks was evaluated under blast loading. In this study, after performing experimental testing on materials and obtaining their compressive, tensile, and shear strength values, the finite element (FE) models of AAC-based masonry walls were created in the ABAQUS/Explicit nonlinear platform. Three different wall thicknesses of 15, 20, and 25 cm were simulated, and the models were analyzed under a lateral explosion caused by 5 and 7 kg of TNT at the stand-off distances of 2, 5, and 10 m from the wall face. The stress distributions, displacement responses, adsorbed energy, and crack propagation pattern were investigated in each case. The results showed the inappropriate behavior of these materials against explosion loads, especially at shorter distances and on walls with less thickness. The outcome gives valuable information to prioritize these walls for possible blast strengthening.

Keywords: AAC block; blast loads; masonry wall; finite element; strengthening; ABAQUS

Citation: Mollaei, S.; Babaei Ghazijahani, R.; Noroozinejad Farsangi, E.; Jahani, D. Investigation of Behavior of Masonry Walls Constructed with Autoclaved Aerated Concrete Blocks under Blast Loading. *Appl. Sci.* **2022**, *12*, 8725. https://doi.org/10.3390/app12178725

Academic Editors: Lina M. López, Ricardo Castedo and Anastasio P. Santos

Received: 17 August 2022
Accepted: 28 August 2022
Published: 31 August 2022

1. Introduction

The existing challenges in the construction industry mainly include increasing the speed of the construction process, increasing the useful lifetime of buildings, retrofitting, cost reduction, reducing thermal and acoustic conductivity, reducing the weight of the building, and environmental issues. Efforts to meet these needs led to the invention of autoclaved aerated concrete (AAC) products. AAC is a relatively modern material with a favorable strength-to-density ratio, thermal insulation properties, and other advantages such as lightness, fire resistance, and ease of cutting and application [1–3]. Today, AAC is widely used in the United States, Europe, and many other countries [4]. These materials are considered environmentally friendly construction materials [5]. AAC products are commonly made of cement, water, lime, silica-based materials (silica sand, ash, or silica fume), porosity-generating materials (aluminum powder), and additives [6].

During their service life, buildings may be exposed to several dynamic load conditions, such as earthquakes, explosions, impacts, and wind loads. Explosions caused by terrorist attacks or accidental incidents in urban areas can cause severe human and financial losses. Blast loading experiments and strengthening the structures to reduce the damage caused by explosions are among the most critical topics for researchers and structural engineers [7–10]. In this regard, multiple studies have modeled the effect of blasts on various building materials and structures.

Previous studies on the behavior of structural and nonstructural components made of AAC materials were limited to static loading conditions [11]. A small number of studies have considered the seismic or impulse load conditions. Yankelevsky and Avnon [12] tested AAC exterior walls under impact loading and evaluated the damages patterns [12]. Tanner et al. [13,14] conducted extensive studies on shear walls made of AAC; many of the requirements of ACI regulations on AAC materials [15] are derived from their research. Uddin et al. [16] introduced a new type of sandwich panel using AAC and FRP composite materials. The behavior of the panel was evaluated under low-speed impulse tests [16]. According to the results, the failure patterns and energy absorption of AAC–FRP panels were improved compared to simple AAC units. Moreover, Tomaževič and Gams [17] performed compressive, tensile, and shear strength tests on masonry walls made of AAC blocks. They also tested several reduced-scale structures with AAC walls on the shaking table [17]. In a study conducted by Bayat et al. [18], the behavior of AAC blocks under severe impulse loads was analyzed. They also investigated the ballistic limit velocity of AAC targets under the influence of rigid projectiles. The results showed that the introduced analytical model was in good agreement with the experimental results [18].

TM 5-855-1 [19] can be considered as one of the first instructions provided for nonatomic explosion-resistant structures. In addition, TM 5-1300 [20] instruction was widely used to design explosion-proof structures; TM 5-1300 was more comprehensive than TM 5-855-1 and based on many subsequent theoretical studies. Finally, UFC 3-340-02 [21] guidelines, as an updated version of TM 5-1300, were developed by the US Department of Defense (DOD) and have been widely used as the primary basis for design and research works in this area.

Historically, some studies were conducted on the behavior of masonry building materials under blast loading. In a series of studies, Hao and Wu [22] and Wu and Hao [23] investigated the effect of infilled walls on RC building behavior under explosion loading. Using explicit finite element modeling in LS-DYNA hydrocode, Wei and Stewart [24] reported that increasing the masonry wall thickness reduces the explosion damage to the buildings. In an experimental study, Ahmad et al. [25] tested a cantilever masonry wall consisting of clay bricks under blast loads. Pandey and Bisht [26] and Pereira et al. [27] investigated the dynamic performance of the brick masonry walls against blast loading. Shi et al. [28] studied local and global damage to a reinforced masonry wall under the close-in explosion scenario. According to the results, instead of bending or shear failure of the wall, the close-in explosion caused local damage by punching [28]. Parisi et al. [29] reported the explosive resistance of a stone wall. Keys and Clubley [30] and Badshah et al. [31] investigated failure patterns of masonry walls through real blast loading tests. Zeng et al. [32] applied 3D finite element models to simulate the out-of-plane behavior of un-reinforced masonry walls constructed with bricks under static and dynamic loadings.

According to ASCE 51-11 [33], the fragmentation of building elements and thrown fireballs have the most dangerous impact in an explosion event. Strengthening methods to prevent the destructive effects of explosions have been an area of interest for some researchers. The most common explosives strengthening techniques in masonry walls include the use of fiber-reinforced polymers (FRP), polyurea, and polyurethane coatings, using steel sheets, aluminum foam, and engineered cementitious composites [34–41].

Previous studies on the behavior of AAC materials under blast loads are limited. In particular, studies investigating the effect of blast loads on structural elements made with AAC are scarce. Xu et al. [42] numerically modeled infilled walls constructed with AAC blocks under gas explosion in LS-DYNA. Li et al. [43] investigated the performance of an autoclaved masonry wall under methane explosion. This study was performed using field tests and numerical simulations [43]. In an experimental study, Wang et al. [44] evaluated retrofitted masonry walls consisting of clay bricks and autoclaved aerated concrete blocks under explosion. They used polyurea layers to increase the explosion resistance of the considered walls [44]. In addition, Liu et al. [45] studied the effect of high strain loading conditions on the properties of AAC materials. Sovják et al. [46] determined the ballistic resistance of AAC against projectile penetration.

AAC lightweight concrete blocks are considered among the first alternatives in construction, especially in the reconstruction of urban areas damaged in the Middle East wars. There is a knowledge gap in the assessment methods and priorities of masonry components [47]. This matter is even more notable in the building constructions with AAC units. Since very few studies have been performed in this field so far, the material properties of autoclaved aerated concrete are not comprehensively known, especially under severe loading conditions. Therefore, investigation of the behavior of building elements constructed with AAC units under blast loading seems necessary. Understanding the behavior of these blocks under explosion and providing solutions to increase their explosive capacity can be an interesting topic for researchers in this field.

The present study aimed to identify, investigate, and analyze the behavior of masonry walls made of AAC lightweight concrete units under the effect of blast loading. The crack growth, displacements, stress distribution, and energy absorption of different models of this type of wall were investigated using FE modeling in the ABAQUS/Explicit package. The main goal of this study was to implement an effective FE procedure in the analysis of masonry models under lateral blast pressure considering different wall thicknesses, since the autoclaved aerated concrete units can be produced with various dimensions.

2. Materials and Methods

2.1. Blast Loading

When an explosion occurs in the open air, a shock wave containing very dense air is propagated radially outwards from the source center at supersonic speeds [48]. Figure 1 shows the schematic time variations of blast pressure. The time history of the pressure is mainly divided into positive and negative phases. The positive phase begins from the moment the blast wave reaches the structure (point B in Figure 1). At this point, the pressure suddenly reaches its highest value and then gradually decreases to the atmospheric pressure during the positive phase. Then, as it decreases relative to atmospheric pressure, it creates a negative or suction state (point C in Figure 1). The magnitude of the overpressure in the positive phase is much higher than that in the negative phase, and except for lightweight structures, the reverse pressure effects in the negative phase zone are assumed to be negligible [20]. Points A and D in Figure 1 represent the normal atmospheric pressure.

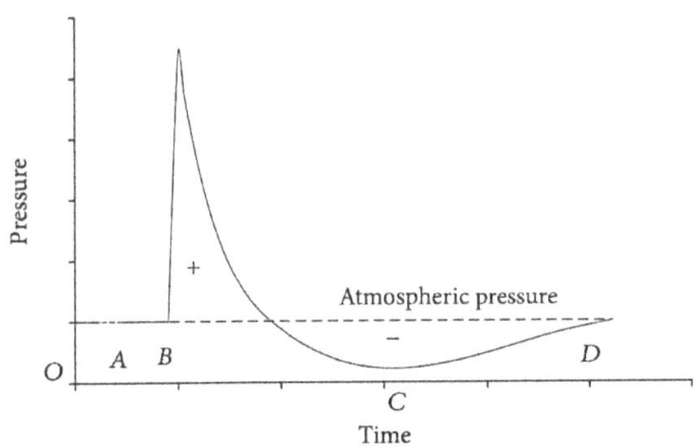

Figure 1. Pressure time history for free explosion [9].

In general, the distance from the source of the explosion (stand-off distance), R, and the explosive charge weight, W, are two crucial factors in determining the specifications of the blast wave. For two different weights of the explosives, if the ratio of the distances from the structure is equal to the ratio of one-third of the power of the charge weight, then the resulting pressure is identical in both cases. This is known as the Hopkinson–Cranz (cube root) scaling law and is expressed according to Equation (1) [49].

$$\frac{R_1}{R_2} = \left(\frac{W_1}{W_2}\right)^{1/3}, \tag{1}$$

where R is the distance from the center of the explosives, and W is the charge weight for two different cases. The scaled distance (Z) is a basis for evaluating the explosion intensity variations (Equation (2)). Scaled distance is one of the most important characteristics that affects all explosion wave parameters.

$$Z = \frac{R}{W^{1/3}} \tag{2}$$

In general, there are three types of explosions, based on the measured distance: contact, close-in, and far-field explosions [20]. In the contact state, blast load usually causes a non-uniform pressure distribution on the face of the structure, and the intensified local pressure causes cracks and ruptures. In the close-in state, blast waves are generated in a high impulse area on the face of the structure. A far-filed blast is a state in which the waves reaching the outside of the building are planar due to the great distance from the structure, and the load distribution can be assumed to be linear or uniform. In this study, the close-in explosion scenario was considered for all the models.

Various experimental relations have been presented in different studies to calculate the explosion wave parameters using the parameter Z [20,21,50–53]. The Conwep module was developed by the US Army Ground Forces Strategic Research Institute following the requirements of TM 5-855-1 Code [54]. The primary purpose of this software is to estimate and apply explosion and impulse loads on the external surface of structures. In this study, the capabilities of this sub-program in ABAQUS were used to calculate the blast load specifications.

2.2. AAC and Grout Materials

Autoclaved masonry walls are defined in MSJC Code [55] as masonry AAC units placed and connected with suitable mortar or adhesives. These walls may be made with or without reinforcement. Equations (3)–(7) estimate the AAC material specifications [55].

$$E = 6500\left(f'_{AAC}\right)^{0.6} \quad (\text{Mpa}), \tag{3}$$

$$f_{t\ AAC} = 0.2\sqrt{f'_{AAC}} \quad (\text{Mpa}), \tag{4}$$

$$f_v = 0.15\sqrt{f'_{AAC}} \quad (\text{Mpa}), \tag{5}$$

$$E_v = 0.4\ E, \tag{6}$$

$$E_g = 500\ f'_g, \tag{7}$$

where f'_{AAC} is the compressive strength, E is the modulus of elasticity, $f_{t\ AAC}$ is the tensile strength, f_v is the direct shear strength, E_v is the shear modulus of AAC materials, f'_g is the compressive strength of adhesive or grout, and E_g is the elastic modulus of adhesive or grout. In this study, the compressive strength of the considered AAC materials and the compressive and tensile strengths of mortar (adhesive) were determined experimentally in the laboratory. Other properties needed for finite element modeling of the materials were estimated using the equations proposed in MSJC. Here, in defining the constitutive behavior of AAC material, the equation proposed by Entezari and Esmaili [56] was used, which is given in Equations (8) and (9).

$$f_c = f'_c \left[\frac{n^{pq\left(\frac{\varepsilon_c}{\varepsilon_0}\right)}}{\left(\frac{\varepsilon_c}{\varepsilon_0}\right)^{npq} + n^{pq-1}} \right], \tag{8}$$

$$q = 1.25 + 0.009 f'_c, \tag{9}$$

For the ascending region, p and q are constants assumed to be 3 and 1, respectively. The quantity of n^{pq} is determined based on the properties of concrete, such as compressive strength, modulus of elasticity, and strain corresponding to the maximum stress. The values of n and p for the descending region are the same as those for the ascending part, and the value of q is determined using the return point of the descending curve. According to the experiments, the stress–strain curve obtained for this study is plotted in Figure 2.

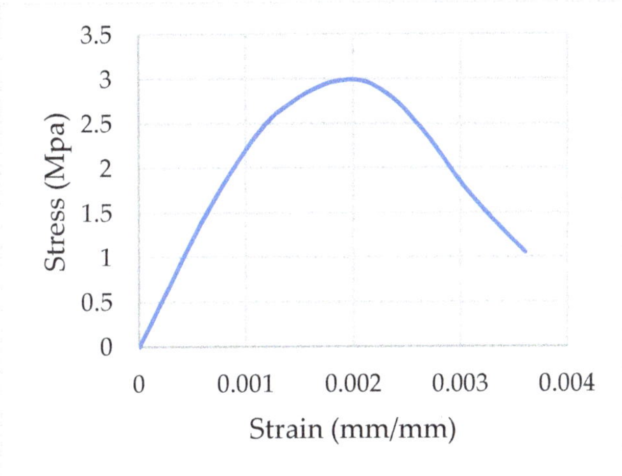

Figure 2. Stress–strain curve of AAC materials with a compressive strength of 3 MPa.

The modulus of elasticity of AAC materials was 1716 MPa, and the Poisson's ratio was 0.2. To define the nonlinear properties of AAC concrete, the concrete damage plastic (CDP) model was used here, the specifications of which are given in Table 1.

Table 1. Parameters of concrete damage plastic (CDP) model used here.

Parameter	Dilation Angle	Eccentricity	Biaxial-to-Uniaxial Compressive Strength Ratio	Shape Factor Function	Viscoelastic Parameter
Quantity	20	0.1	1.16	0.66	0.001

2.3. Numerical Modeling

ABAQUS/Explicit [57] is a finite element package based on an explicit integration approach used to solve extreme nonlinear systems such as high strain rate loadings. In high-velocity dynamic phenomena such as explosive and impulse loads, which apply an intense load in a very short time, it is practically impossible for the FE solution to converge in the implicit approach. Therefore, in this study, the Explicit solver was used to analyze AAC masonry models under explosion loads. ABAQUS includes an extensive library of continuum three-dimensional solid elements which are suitable for modeling solid objects. In this study, C3D20 and C3D8 were used in modeling masonry components made with

AAC units. Based on a mesh sensitivity analysis, a solid elements' meshing size of up to 15 mm was selected.

In ABAQUS, Conwep subroutine can calculate the blast pressure distribution in various structures [57]. In this study, the Conwep feature was used for blast loading. Thus, by entering the explosive charge weight and the stand-off distance, the program automatically calculates the spatial and temporal distribution of the blast pressure on the interaction surface. In this study, the blast event was defined as an air blast in Conwep subroutine, and one side of the wall was considered the blast wave interaction zone.

2.4. FE modeling of Masonry Walls

In addition to the analytical approaches [58] and discrete-element analysis [59], numerical methods can be successfully applied to modeling and analysis of masonry walls. Generally, there are three main methods for developing the FE model of infilled frame walls [60,61], including detailed micro modeling, simplified micro modeling, and macro modeling (Figure 3).

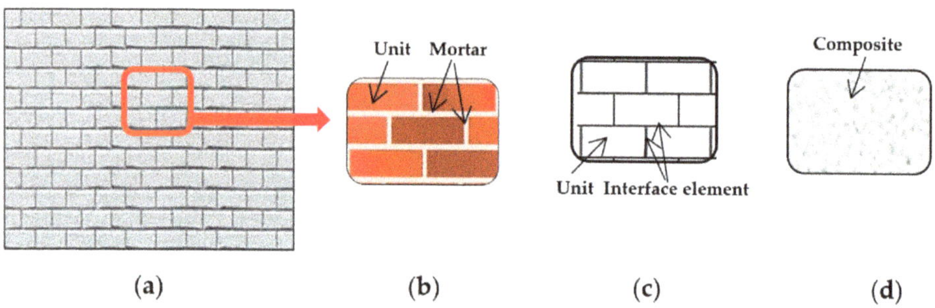

Figure 3. Masonry wall modeling approaches: (**a**) a real masonry wall; (**b**) accurate detailed micro modeling; (**c**) simplified micro modeling; (**d**) macro modeling.

Micro modeling can be performed in accurate or simplified manners. The accurate detailed modeling provides the most realistic state or representation of a masonry wall composite. In the accurate approach, the construction units and joints of the mortar layers are modeled, and the properties of each material are assigned separately. The requirements related to the aspect ratio of the elements in meshing, the low thickness, and the long mortar joints mean the accurate detailed micro model needs a very fine mesh. For this reason, most complex calculations require considerable time [43,62–64].

In simplified micro modeling, bricks (blocks) and mortar are not modeled separately. The mortar is bonded to homogeneous construction units and added to the unit by a zero-thickness interface element. Mortar joints are added to the intermediate elements representing crack and slip surfaces [32,64]. Using this modeling approach, accuracy is expected to decrease to some extent [65]. In this study, a simplified micro modeling approach was used to prepare masonry wall models with AAC units. Therefore, the mortar was not modeled, and its behavior was considered by adding contact elements between the construction units.

In the macro modeling methodology, the whole infilled wall is modeled as a homogeneous material with equivalent properties regardless of its constituent units. The accuracy of this modeling approach is lower, and the analysis speed is much higher than that of the micro models. It should be noted that the mechanical properties of materials have different values for various conditions, i.e., the arrangement of bricks and horizontal/vertical joints of the mortar in the wall cause the varied stiffness values in different directions [66,67].

2.5. Properties of Interface Elements

The elastic properties of mortar joints are determined by normal stiffness (K_{nn}) and shear stiffness values (K_{tt} and K_{ss}). If the interaction between the two pieces is similar to that of adhesive, then adhesive elements can be used. In terms of elastic properties, the relationships between stress and vertical and shear strains can be defined as coupled or uncoupled. The stress–strain relationship for uncoupled and coupled states is in the form of Equations (10) and (11), respectively.

$$\begin{Bmatrix} t_n \\ t_s \\ t_t \end{Bmatrix} = \begin{bmatrix} K_{nn} & 0 & 0 \\ 0 & K_{ss} & 0 \\ 0 & 0 & K_{tt} \end{bmatrix} \cdot \begin{Bmatrix} \varepsilon_n \\ \varepsilon_s \\ \varepsilon_t \end{Bmatrix}, \tag{10}$$

$$\begin{Bmatrix} t_n \\ t_s \\ t_t \end{Bmatrix} = \begin{bmatrix} K_{nn} & K_{ns} & K_{nt} \\ K_{ns} & K_{ss} & K_{st} \\ K_{nt} & K_{st} & K_{tt} \end{bmatrix} \cdot \begin{Bmatrix} \varepsilon_n \\ \varepsilon_s \\ \varepsilon_t \end{Bmatrix}, \tag{11}$$

where t_n, t_s, and t_t are the vertical and shear stresses in two directions, the matrix k is the corresponding stiffness, and ε is the vector of strains for the interface plane. The entries of the main diagonal of the stiffness matrix are the normal and shear stiffness in the main directions of the interface. Here, normal (K_n) and shear ($K_s = K_t$) stiffness were used to define the adhesive behavior according to Equations (12) and (13).

$$K_n = \frac{E_u E_m}{h_m(E_u - E_m)}, \tag{12}$$

$$K_s = K_t = \frac{G_u G_m}{h_m(G_u - G_m)}, \tag{13}$$

where h_m, G_m, and E_m are the thickness, shear modulus, and modulus of elasticity of the mortar, and G_u and E_u are the shear modulus and modulus of elasticity of the block. The coefficient of friction of the layer was also defined as 0.7.

2.6. Mechanical Properties of AAC Materials and Grout

The AAC material mixture considered in this study is summarized in Table 2. With the lack of reliable data related to the mechanical properties of AAC materials, experimental tests were performed on AAC block samples prepared from Aranshahr Aran Polymer Concrete Plant (East Azerbaijan, Iran). The requirements of the ASTM C495 [68] code were used to measure the compressive strength of cubic specimens with the dimensions of 10 cm (Figure 4a). The ASTM C109 [69] code was also used here to determine the compressive strength of adhesive materials (Figure 4b). The dimensions of the cube molds in this experiment were 50 mm, and the samples were treated in water for seven days. The tensile strength of the briquette samples was also determined according to the requirements of the ASTM C 307-3 [70] code (Figure 4c). According to the results, the average compressive strength of the AAC block was about 3 MPa. Moreover, the average compressive strength of the mortar (adhesive) was 10 MPa, and its tensile strength was 1.3 MPa.

Table 2. AAC mix design with a density of 500 kg/m^3.

Materials	Amounts (kg/m^3)
silica sand	350
lime	100
cement	25
aluminum powder	0.5
water	330

(a) (b) (c)

Figure 4. Test setup for (**a**) compressive strength of AAC materials; (**b**) compressive strength of adhesive materials; (**c**) tensile strength of adhesive materials.

In this study, the blast loading was assumed to have occurred as a result of typical suitcase bombs at reasonable distances from the wall face. In addition, the internal pores modeling of the AAC structure was ignored. Experimental data collection and available information from the manufacturer and previous studies were used here to estimate the mechanical properties of AAC material. Moreover, the compressive and tensile strength tests were performed on the standard AAC and the special mortar specimens. The explicit finite element software ABAQUS/Explicit was used for modeling and analysis of masonry walls under explosion loads. Using suitable material models, numerical modeling of the masonry walls made of AAC block units was created and analyzed under various blast loading scenarios perpendicular to the wall face. The cracking, displacement responses, stress distribution, and energy absorption patterns in AAC wall models were investigated and compared.

2.7. Considered Models

In the modeling stage, according to Figure 5, the height and width of the wall were 3 and 2 m, respectively, and the dimensions of AAC blocks were 600 × 250 mm with thicknesses of 15, 20, and 25 cm. The thickness of the mortar layer was also considered to be 10 mm. According to Figure 6, the boundary conditions of the wall were in three different states. The models studied here were subject to the explosions caused by 5 and 7 kg of TNT at the stand-off distances of 2, 5, and 10 m. In all models, the distance of the blast center from the ground (its height from the base of the wall) was considered to be 500 mm. Therefore, in general, 18 models of AAC masonry walls were considered here, with the specifications summarized in Table 3. It should be noted that no axial loading on the walls was taken into account since the considered walls were not assumed to be load-bearing structural components.

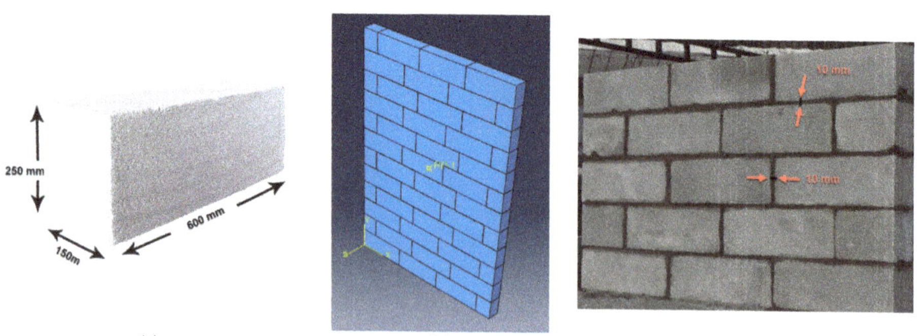

(a) (b) (c)

Figure 5. Specifications of the AAC wall and block model: (**a**) AAC block dimensions; (**b**) FE model of the wall; (**c**) assumed thickness of grout layer.

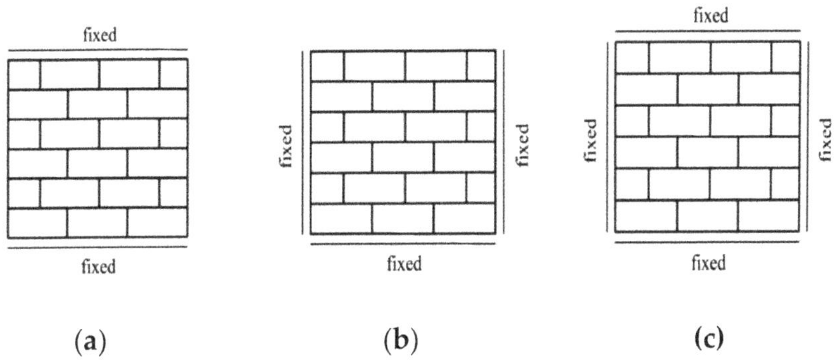

Figure 6. Boundary conditions of the walls: (**a**) fixed top and bottom of the wall; (**b**) fixed sides and bottom of the wall; (**c**) fully fixed BCs.

Table 3. Details of the studied models.

Model	Thickness (cm)	Stand-Off Distance (m)	Charge Weight (kg)	Scaled Distance (m/kg$^{1/3}$)
Model-1	15	2	5	1.170
Model-2	15	5	5	2.924
Model-3	15	10	5	5.848
Model-4	15	2	7	1.046
Model-5	15	5	7	2.614
Model-6	15	10	7	5.228
Model-7	20	2	5	1.170
Model-8	20	5	5	2.924
Model-9	20	10	5	5.848
Model-10	20	2	7	1.046
Model-11	20	5	7	2.614
Model-12	20	10	7	5.228
Model-13	25	2	5	1.170
Model-14	25	5	5	2.924
Model-15	25	10	5	5.848
Model-16	25	2	7	1.046
Model-17	25	5	7	2.614
Model-18	25	10	7	5.228

3. Results and Discussion

3.1. Validation

Kumar et al. [71] conducted a study on the behavior of RC slabs against blast loading. The slab with dimensions of 1000 × 1000 × 100 mm was exposed to explosions with a scaled distance of 0.079–0.527 m/kg$^{1/3}$. Here, to validate the process of blast load calculations in ABAQUS, this concrete slab was modeled, and the blast pressure distribution was compared to the original reference. The *slab-32* model in reference [71] was modeled under the effect of a blast of 2 kg TNT at a distance of 0.5 m (Z = 0.3968 m/kg$^{1/3}$). Here, due to the symmetry of the structure and loading, only one-fourth of the concrete slab was modeled and analyzed. The considered RC plate was a square with 500 mm length of side and 100 mm thickness. The finite element model of the RC slab with its support structure prepared here and the definition of the explosive charge are given in Figure 7.

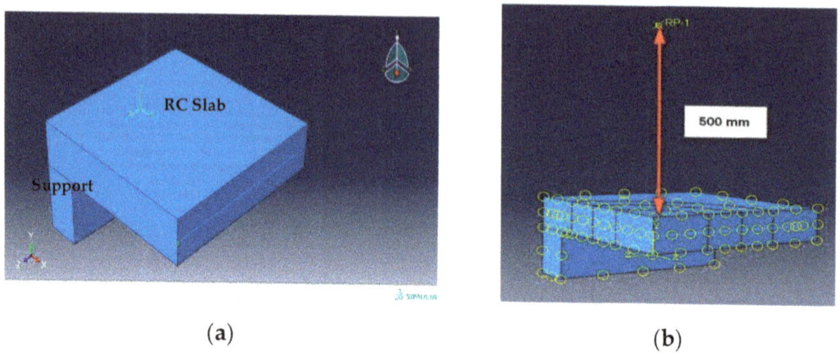

Figure 7. FE modeling of the slab: (**a**) modeling in Abaqus for validation; (**b**) blast loading in Conwep.

Figure 8 shows the time history of the calculated pressure in the middle of the slab in this study compared to the reference [71]. It can be observed that the maximum overpressure difference is equal to $100\% - \frac{42.36}{44.0} \times 100\% = 3.72\%$, which is in an acceptable range.

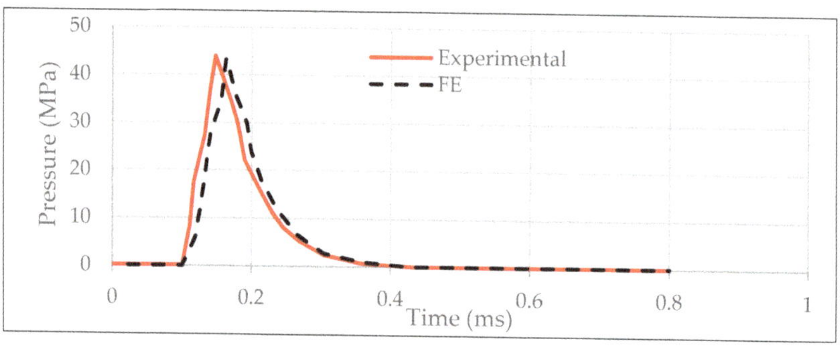

Figure 8. Pressure–time diagram obtained in the present study for validation process.

3.2. Von Mises Stress Distribution

After verifying the modeling procedure, the wall models were simulated, and structural responses were monitored and investigated. Figure 9 shows the maximum von Mises stress distribution in the studied wall models under blast loading. Table 4 also shows the maximum stresses in each model. It can be observed that with a reduction in the stand-off distance, the stress level was increased and distributed over a wider area of the wall. For example, in model-1 with a stand-off distance of 2 m, the maximum stress was 118.56 kN/m^2 higher than that in model-2 with a stand-off distance of 5 m. As the amount of TNT increased, the stress also increased, and more significant damages were observed in the models. For example, in model-3, with an explosive charge weight of 5 kg, the maximum stress was 29.32 kN/m^2 lower than that in model-6, with an explosive charge weight of 7 kg. In addition, as the thickness of the walls increased, the stress generally decreased. For example, in model-13 with a thickness of 25 cm, the maximum stress was 73.55 kN/m^2 lower compared to model-7 with the wall thickness of 20 cm.

Since the considered boundary conditions in these models were fixed at both ends (Figure 6a), the one-way behavior of the wall is obvious in Figure 9. According to Figure 9, shear failure was the dominant failure mode, since the grout shear strength was less than its compressive strength.

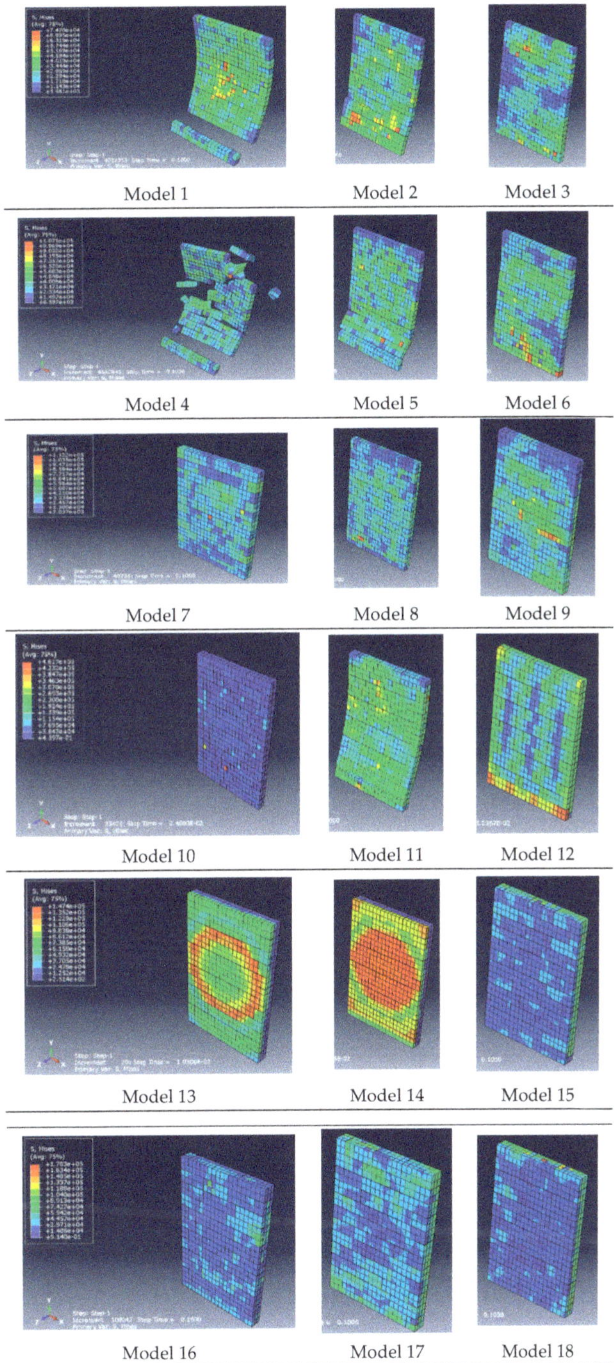

Figure 9. The von Mises stress distribution in some of the simulated wall models.

Table 4. Maximum stress values in the models.

Model	Thickness (cm)	Stand-Off Distance (m)	Charge Weight (kg)	Maximum Stress (kgf/m^2)
Model-1	15	2	5	7.470×10^4
Model-2	15	5	5	6.261×10^4
Model-3	15	10	5	5.494×10^4
Model-4	15	2	7	1.071×10^5
Model-5	15	5	7	9.522×10^4
Model-6	15	10	7	5.793×10^4
Model-7	20	2	5	1.122×10^5
Model-8	20	5	5	6.234×10^4
Model-9	20	10	5	9.829×10^4
Model-10	20	2	7	4.617×10^5
Model-11	20	5	7	6.715×10^4
Model-12	20	10	7	1.367×10^4
Model-13	25	2	5	1.474×10^5
Model-14	25	5	5	3.933×10^4
Model-15	25	10	5	6.075×10^4
Model-16	25	2	7	1.783×10^5
Model-17	25	5	7	5.638×10^4
Model-18	25	10	7	6.591×10^4

3.3. Displacement Responses

According to Figure 10, the displacement time history at the center of the wall was obtained for different thicknesses. It can be seen that the closer the explosives to the wall, the greater the displacement. With an explosive charge weight of 7 kg, for all wall thicknesses, the displacement was more significant than that for the other charge weights. For example, Models 4 and 6 had higher displacements than Models 1 and 3, respectively. According to the results, it can be observed that at a stand-off distance of 2 m (such as in Model-1), the wall models practically failed and had a larger displacement compared to the other models. Therefore, it can be concluded that at short stand-off distances, AAC-based masonry walls do not have enough resistance against blast loading, and a retrofitting scheme is required.

3.4. The Influence of the Boundary Conditions

There were three different cases for the support conditions of the walls. In the first case, the base of the wall was assumed to be restrained. In the second case, the whole perimeter of the wall was restrained, and in the third case, the three sides of the wall (bottom and sides) were restrained. Figure 11 shows the maximum stress distribution for different support conditions in the wall models. Clearly, the greater the restraints of the sides of the wall (case b), i.e., the better the restraining of the wall to the structural components of the building, the lower the stress in the wall. As such, in case "a" with the base of the wall constrained, a large surface of the wall had stress ranging from 2.3×10^4 to 5.2×10^4 kgf/m^2. In case "c" with the base of the wall and the sides restrained, a large surface of the wall had the stress of 1.2×10^4 to 4.8×10^4 kgf/m^2. Finally, in case "b" with the whole sides of the wall restrained, a large surface of the wall had the stress of 3.2×10^3 to 3.3×10^4 kgf/m^2.

According to Figure 11a, shear failure was obvious in the wall with one-way behavior. Two-way performance of the wall, as in Figure 11b, led to a decrease in the deformations compared to the other BCs. However, it increased the induced stress to the elements. As was anticipated, the boundary conditions had significant effects on the response of the walls made with AAC, similar to other masonry walls [22–24].

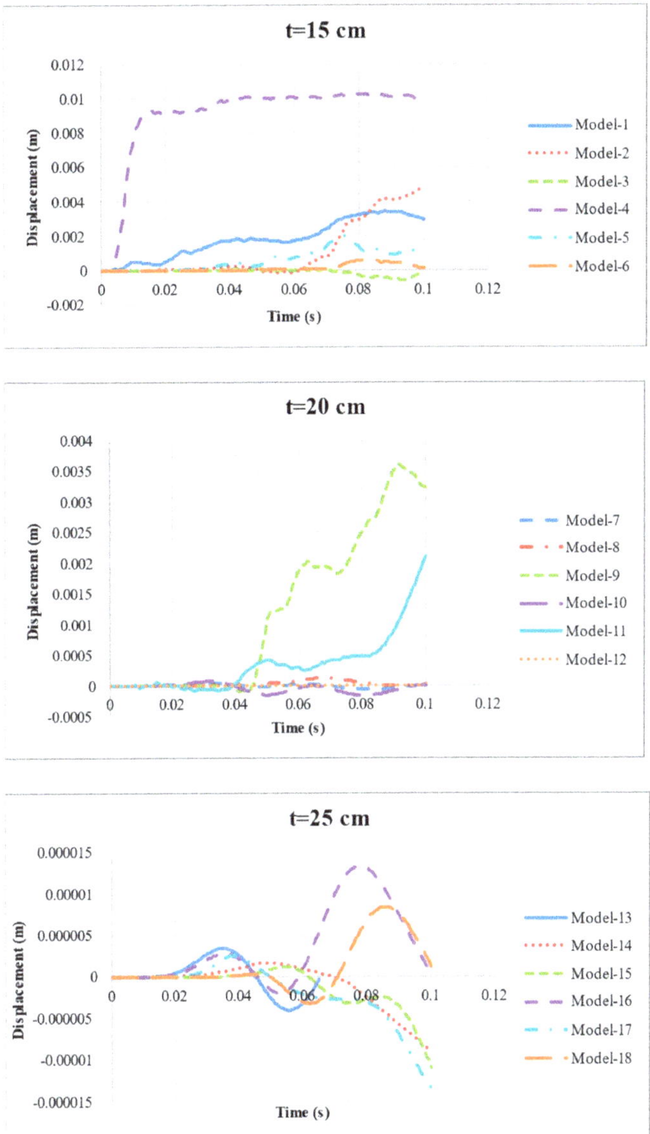

Figure 10. Time–displacement diagrams for wall thicknesses of 15, 20, and 25 cm.

3.5. Base Shear Force

Figure 12 indicates the temporal variations of the base shear in wall models with different thicknesses. It can be seen that with an increase in wall thickness, the force incurred on the base of the wall due to blast loading was significantly reduced. Therefore, the thickness of the walls was very effective in reducing the explosive demand on the AAC-based masonry walls. For example, for a wall with a thickness of 15 cm, the values of the base shear were about ten times higher compared those for a wall with a thickness of 25 cm.

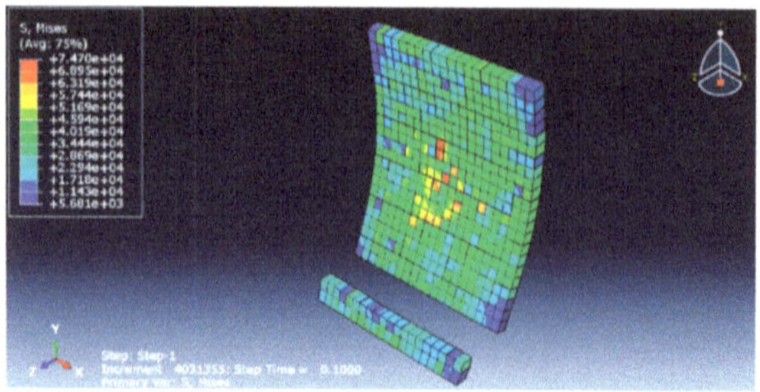

(**a**)

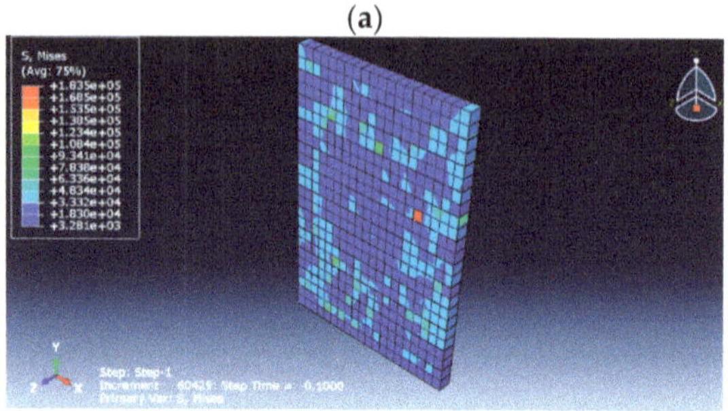

(**b**)

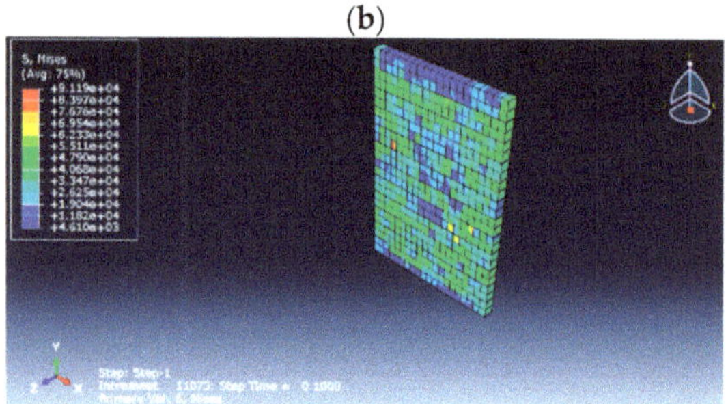

(**c**)

Figure 11. Distribution of stress in different support conditions: (**a**) restrained base; (**b**) restrained whole sides; (**c**) three-sided restrained.

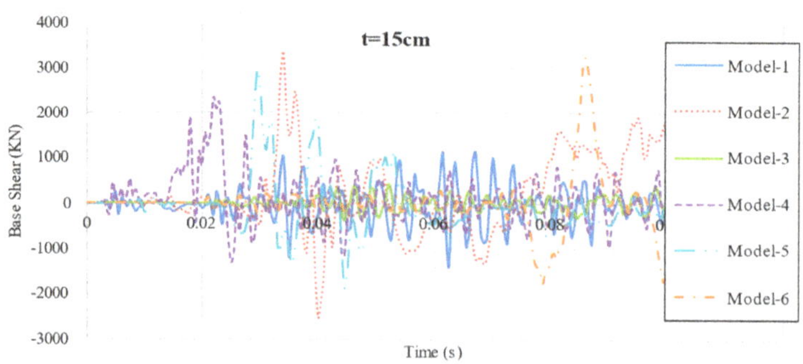

Figure 12. *Cont.*

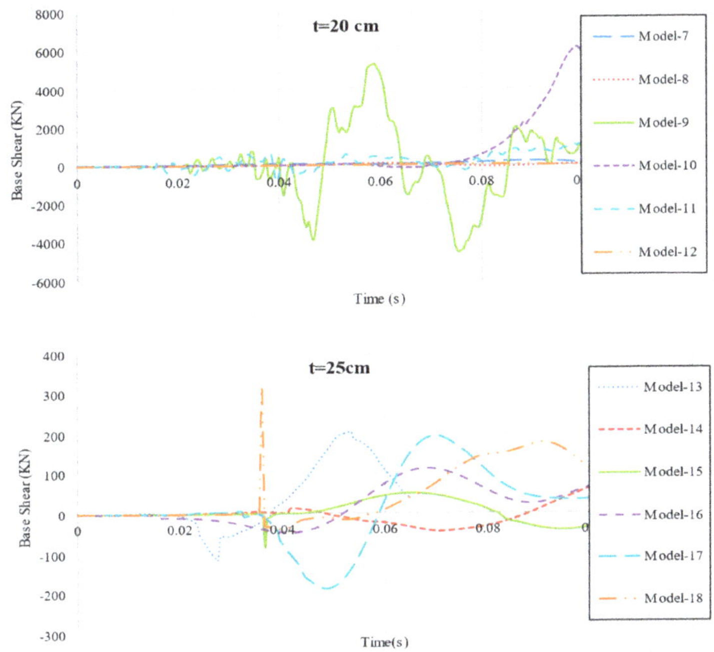

Figure 12. Base shear of the wall with thicknesses of 15, 20, and 25 cm.

4. Conclusions

Autoclaved aerated concrete (AAC) block is used in the construction of load-bearing and masonry walls due to its low thermal expansion coefficient, high fire resistance, and low weight. However, the low strength of materials and the heterogeneity of the material lead to the vulnerability of AAC masonry walls under external loads. To reduce the potential hazards in the structure and enhance the safety level, it is necessary to investigate the dynamic responses and failure of AAC masonry walls under explosive loads. This study aimed to investigate the behavior of AAC walls under blast loading. Therefore, the specifications of the materials required for modeling were first determined through experimental tests. Then, by modeling and analysis of the AAC walls in ABAQUS/Explicit, the behavior of these walls was investigated under blast loads. The main outcomes are as follows:

- Considering the weight of TNT used in a short distance (R = 2 m), it was observed that very large local stresses were created in the wall, which caused the wall to collapse in a very short time. It should be noted that at distances of less than 2 m, the wall models diverged at the very first moments. Therefore, the analysis and presentation of their results were avoided here.
- With the increasing charge weight, wall performance degraded. The stress level in the case of an explosive charge weight of 7 kg TNT increased by about 10% compared to that for 5 kg TNT. It is important to note that the walls modeled in this study under a charge larger than 7 kg TNT had a rapid failure in the initial moments. Therefore, considering the typical values of charge weight in the explosion events of hand grenades and suitcase bombs (about 20 kg-TNT) [72], it can be stated that masonry walls made with AAC do not have a good explosion resistance and would need retrofitting.
- Some retrofitting methods in masonry walls could involve using CFRP coating, steel wire mesh, and laminating. In addition, polyurea and polyurethane coatings, using

steel sheets, aluminum foam, and engineered cementitious composites, are suggested for masonry units that can be applied to walls made with AAC.

- With the increasing charge weight and decreasing stand-off distance, the wall displacement increased significantly, so that at a distance of 2 m, the displacement was several times that for the 5- and 10-m distances. In the walls with thicknesses of 15, 20, and 25 cm, the performance was also observed to be the same. As the amount of TNT increased, the stress values increased, and more damage was observed in the walls.
- The thickness of the walls was very effective in reducing the explosive demand force. For example, for a wall with a thickness of 15 cm, compared to that with a thickness of 25 cm, the base shear values induced by the same explosion were about 10 times higher.

To complete this study and achieve practical findings, complementary studies will be executed on construction units made of AAC blocks, including AAC walls reinforced by various methods such as using horizontal and vertical rebar meshes. The effect of various characteristics of mortar and adhesive on the behavior of AAC walls under blast loads will also be investigated.

It should be noted that the modeling of the walls built with AAC blocks in ABAQUS finite element software requires more extensive data and more detailed experiments. In particular, dynamic properties under high strain rates require further experimental and laboratory studies.

Author Contributions: Conceptualization, S.M. and E.N.F.; methodology, S.M. and D.J.; software, R.B.G.; validation, R.B.G.; formal analysis, S.M. and R.B.G.; writing—original draft preparation, S.M.; writing—review and editing, E.N.F.; project administration, D.J. All authors have read and agreed to the published version of the manuscript.

Funding: This research received no external funding.

Institutional Review Board Statement: Not applicable.

Informed Consent Statement: Not applicable.

Data Availability Statement: All the data used to support the findings of the study are included within the article.

Conflicts of Interest: The authors declare no conflict of interest.

References

1. Korniyenko, S.V.; Vatin, N.I.; Gorshkov, A.S. Heat and power characteristics analysis carried out for the residential building made of autoclaved aerated concrete blocks. *Constr. Unique Build. Struct.* **2016**, *12*, 45–60. [CrossRef]
2. Winkels, B.; Nebel, H.; Raupach, M. Carbonation of autoclaved aerated concrete containing fly ash. *ce/papers* **2018**, *2*, 47–51. [CrossRef]
3. Gyurkó, Z.; Jankus, B.; Fenyvesi, O.; Nemes, R. Sustainable applications for utilization the construction waste of aerated concrete. *J. Clean. Prod.* **2019**, *230*, 430–444. [CrossRef]
4. Saghi, H.; Arefizadeh, N. Quantitative and Qualitatively Evaluation of New Technologies in Perspective of Construction Optimization. *Am. J. Civ. Eng.* **2015**, *3*, 64–68. [CrossRef]
5. Hammond, G.P.; Jones, C.I. Embodied energy and carbon in construction materials. *Proc. Inst. Civ. Eng. Energy* **2008**, *161*, 87–98. [CrossRef]
6. Sherin, K.; Saurabh, J.K. Review of autoclaved aerated concrete: Advantages and disadvantages. In Proceedings of the National Conference of Advanced Structures, Materials and Methodology in Civil Engineering-ASMMCE-2018, Punjab, India, 3–4 November 2018.
7. Momeni, M.; Hadianfard, M.A.; Bedon, C.; Baghlani, A. Damage evaluation of H-section steel columns under impulsive blast loads via gene expression programming. *Eng. Struct.* **2020**, *219*, 110909. [CrossRef]
8. Momeni, M.; Hadianfard, M.A.; Bedon, C.; Baghlani, A. Numerical damage evaluation assessment of blast loaded steel columns with similar section properties. *Structures* **2019**, *20*, 189–203. [CrossRef]
9. Esameelnia Omran, M.; Mollaei, S. Investigation of Axial Strengthened Reinforced Concrete Columns under Lateral Blast Loading. *Shock Vib.* **2017**, *2017*, 94–113. [CrossRef]
10. Mollaei, S.; Babaei, M.; JalilKhani, M. Assessment of Damage and Residual Load Capacity of the Normal and Retrofitted RC Columns against the Impact Loading. *J. Rehab. Civ. Eng.* **2021**, *9*, 29–51. [CrossRef]

11. Deyazada, M.; Vandoren, B.; Dragan, D.; Degée, H. Experimental investigations on the resistance of masonry walls with AAC thermal break layer. *Constr. Build. Mater.* **2019**, *224*, 474–492. [CrossRef]
12. Yankelevsky, D.Z.; Avnon, I. Autoclaved aerated concrete behavior under explosive action. *Constr. Build. Mater.* **1998**, *12*, 359–364. [CrossRef]
13. Tanner, J.E.; Varela, J.L.; Klingner, R.E.; Brightman, M.J.; Cancino, U. Seismic testing of autoclaved aerated concrete shear walls: A comprehensive review. *ACI Struct. J.* **2005**, *102*, 374.
14. Tanner, J.E.; Varela, J.L.; Klingner, R.E. Design and seismic testing of two-story, full-scale autoclaved aerated concrete assemblage specimen. *ACI Struct. J.* **2005**, *102*, 114.
15. ACI Committee 523. *Guide for Design and Construction with Autoclaved Aerated Concrete Panels (ACI 523.4R-09)*; American Concrete Institute: Farmington Hills, MI, USA, 2009.
16. Uddin, N.; Shelar, K.V.; Fouad, F. Impact response of autoclave aerated concrete/FRP sandwich structures. In Proceedings of the Structures Congress 2006: Structural Engineering and Public Safety, ASCE, St. Louis, MO, USA, 18–21 May 2006. [CrossRef]
17. Tomaževič, M.; Gams, M. Shaking table study and modelling of seismic behaviour of confined AAC masonry buildings. *Bull. Earthq. Eng.* **2012**, *10*, 863–893. [CrossRef]
18. Bayat, A.; Liaghat, G.H.; Ghalami-Choobar, M.; Ashkezari, G.D.; Sabouri, H. Analytical modeling of the high-velocity impact of autoclaved aerated concrete (AAC) blocks and some experimental results. *Int. J. Mech. Sci.* **2019**, *159*, 315–324. [CrossRef]
19. U.S. Department of the Army, the Navy and Air Force. *Fundamental of Protective Design for Conventional Weapons-Technical Manual (TM 5-855-1)*; U.S. Department of the Army: Washington, DC, USA, 1986.
20. U.S. Department of the Army, the Navy and Air Force. *The Design of Structures to Resist the Effects of Accidental Explosions-Technical Manual (TM 5-1300)*; U.S. Department of the Army: Washington, DC, USA, 1990.
21. U.S. Department of Defense (DOD). *Structures to Resist the Effects of Accidental Explosions (UFC 3-340-02)*; DOD: Washington, DC, USA, 2008.
22. Hao, H.; Wu, C. Numerical simulation of damage of low-rise RC frame structures with infilled masonry walls to explosive loads. *Aust. J. Struct. Eng.* **2006**, *7*, 13–22. [CrossRef]
23. Wu, C.; Hao, H. Safe scaled distance for masonry infilled RC frame structures subjected to airblast loads. *J. Perform. Constr. Facil.* **2007**, *21*, 422–431. [CrossRef]
24. Wei, X.; Stewart, M.G. Model validation and parametric study on the blast response of unreinforced brick masonry walls. *Int. J. Impact Eng.* **2010**, *37*, 1150–1159. [CrossRef]
25. Ahmad, S.; Elahi, A.; Pervaiz, H.; Rahman, A.G.A.; Barbhuiya, S. Experimental study of masonry wall exposed to blast loading. *Mater. Constr.* **2014**, *64*, e007. [CrossRef]
26. Pandey, A.; Bisht, R. Numerical modelling of infilled clay brick masonry under blast loading. *Adv. Struct. Eng.* **2014**, *17*, 591–606. [CrossRef]
27. Pereira, J.M.; Campos, J.; Lourenço, P.B. Experimental study on masonry infill walls under blast loading. In Proceedings of the 9th International Masonry Conference, Guimarães, Portugal, 7–9 July 2014.
28. Shi, Y.; Xiong, W.; Li, Z.X.; Xu, Q. Experimental studies on the local damage and fragments of unreinforced masonry walls under close-in explosions. *Int. J. Impact Eng.* **2016**, *90*, 122–131. [CrossRef]
29. Parisi, F.; Balestrieri, C.; Asprone, D. Blast resistance of tuff stone masonry walls. *Eng. Struct.* **2016**, *113*, 233–244. [CrossRef]
30. Keys, R.A.; Clubley, S.K. Experimental analysis of debris distribution of masonry panels subjected to long duration blast loading. *Eng. Struct.* **2017**, *130*, 229–241. [CrossRef]
31. Badshah, E.; Naseer, A.; Ashraf, M.; Ahmad, T. Response of masonry systems against blast loading. *Def. Technol.* **2021**, *17*, 1326–1337. [CrossRef]
32. Zeng, B.; Li, Y.; Noguez, C.C. Modeling and parameter importance investigation for simulating in-plane and out-of-plane behaviors of un-reinforced masonry walls. *Eng. Struct.* **2021**, *248*, 113233. [CrossRef]
33. *ASCE/SEI 59-11*; Blast Protection of Buildings. American Society of Civil Engineers: Reston, VA, USA, 2011.
34. Stanley, M.; Metzger, J.; Martinez, R.; Koenig, J. *UL-Like Testing of Commercial Off-the-Shelf Products That Enhance the Blast and Ballistic Resistance of Structures, Quick Look Report 2*; New Mexico Tech, Energetic Materials Research and Testing Center, Karagozian & Case: Burbank, CA, USA, 2005.
35. Tan, K.H.; Patoary, M.K.H. Blast resistance of FRP-strengthened masonry walls. I: Approximate analysis and field explosion tests. *J. Compos. Constr.* **2009**, *13*, 422–430. [CrossRef]
36. Baylot, J.T.; Bullock, B.; Slawson, T.R.; Woodson, S.C. Blast response of lightly attached concrete masonry unit walls. *J. Struct. Eng.* **2005**, *131*, 1186–1193. [CrossRef]
37. Alsayed, S.H.; Elsanadedy, H.M.; Al-Zaheri, Z.M.; Al-Salloum, Y.A.; Abbas, H. Blast response of GFRP-strengthened infill masonry walls. *Constr. Build. Mater.* **2016**, *115*, 438–451. [CrossRef]
38. Bui, T.T.; Limam, A. Out-of-plane behaviour of hollow concrete block masonry walls unstrengthened and strengthened with CFRP composite. *Compos. Part B Eng.* **2014**, *67*, 527–542. [CrossRef]
39. Chen, L.; Fang, Q.; Fan, J.; Zhang, Y.; Hao, H.; Liu, J. Responses of masonry infill walls retrofitted with CFRP, steel wire mesh and laminated bars to blast loadings. *Adv. Struct. Eng.* **2014**, *17*, 817–836. [CrossRef]
40. Ghaderi, M.; Maleki, V.A.; Andalibi, K. Retrofitting of unreinforced masonry walls under blast loading by FRP and spray on polyurea. *Cumhur. Üniversitesi Fen Fakültesi Fen Bilimleri Derg.* **2015**, *36*, 462–477. [CrossRef]

41. Urgessa, G.S.; Maji, A.K. Dynamic response of retrofitted masonry walls for blast loading. *J. Eng. Mech.* **2010**, *136*, 858–864. [CrossRef]
42. Xu, W.X.; Bao, Q.; Li, Z.; Fan, J.Y. Numerical Analysis on the Behavior of Autoclaved Aerated Concrete Block Infilled Wall Subjected to Gas Explosion. *Appl. Mech. Mater.* **2015**, *723*, 259–265. [CrossRef]
43. Li, Z.; Chen, L.; Fang, Q.; Hao, H.; Zhang, Y.; Chen, W.; Xiang, H.; Bao, Q. Study of autoclaved aerated concrete masonry walls under vented gas explosions. *Eng. Struct.* **2017**, *141*, 444–460. [CrossRef]
44. Wang, J.; Ren, H.; Wu, X.; Cai, C. Blast response of polymer-retrofitted masonry unit walls. *Compos. Part B Eng.* **2017**, *128*, 174–181. [CrossRef]
45. Liu, C.; Hou, J.; Hao, Y.; Hao, H.; Meng, X. Effect of high strain rate and confinement on the compressive properties of autoclaved aerated concrete. *Int. J. Impact Eng.* **2021**, *156*, 103943. [CrossRef]
46. Sovják, R.; Koutný, O.; Hála, P. Penetration Resistance of Building Materials against 7.62-mm Armor-Piercing Projectile. *J. Mater. Civ. Eng.* **2021**, *33*, 04021224. [CrossRef]
47. Stepinac, M.; Kisicek, T.; Renić, T.; Hafner, I.; Bedon, C. Methods for the Assessment of Critical Properties in Existing Masonry Structures under Seismic Loads—The ARES Project. *Appl. Sci.* **2020**, *10*, 1576. [CrossRef]
48. Hinman, E. *Primer for Design of Commercial Buildings to Mitigate Terrorist Attacks*; Federal Emergency Management Agency (FEMA): Washington, DC, USA, 2003.
49. Cormie, D.; Mays, G.; Smith, P.D. *Blast Effects on Buildings*, 2nd ed.; Thomas Telford: London, UK, 2009.
50. Brode, H.L. Numerical Solutions of Spherical Blast Waves. *J. Appl. Phys.* **1955**, *26*, 766–775. [CrossRef]
51. Newmark, N.M.; Hansen, R.J. Design of blast resistant structures. *Shock. Vib. Handb.* **1961**, *3*, 04014007.
52. Mills, C.A. The design of concrete structure to resist explosions and weapon effects. In Proceedings of the 1st International Conference on Concrete for Hazard Protections, Edinburgh, UK, 27–30 September 1987.
53. Baker, W.E.; Cox, P.A.; Westine, P.S.; Kulesz, J.J.; Strehlow, R.A. *Explosion Hazards and Evaluation*; Elsevier Scientific: New York, NY, USA, 1983.
54. Hyde, D. *User's Guide for Microcomputer Programs CONWEP and FUNPRO, Applications of TM 5-855-1: Fundamentals of Protective Design for Conventional Weapons*; USA Army Engineers Waterways Experimentation: Vicksburg, MS, USA, 1988.
55. Masonry Standards Joint Committee. *Building Code Requirements and Specification for Masonry Structures: Containing Building Code Requirements for Masonry Structures (TMS 402-11/ACI 530-11/ASCE 5-11), and Specification for Masonry Structures TMS 402-11/ACI 530-11/ASCE 6-11), and Companion Commentaries*; ASCE: Reston, VA, USA, 2016.
56. Entezari, A.; Esmaili, J. Investigation on the Stress Distribution and Flexural Strength of Structural Lightweight Aggregate Concrete. *Concr. Res.* **2010**, *3*, 61–72.
57. *Abaqus v. 6.14 Documentation*; Dassault Systemes Simulia Corporation: Providence, RI, USA, 2011.
58. Sassu, M.; Andreini, M.; Casapulla, C.; De Falco, A. Archaeological consolidation of UNESCO masonry structures in Oman: The Sumhuram Citadel of Khor Rori and the Al Balid Fortress. *Int. J. Archit. Herit.* **2013**, *7*, 339–374. [CrossRef]
59. Pantò, B.; Casapulla, C.; Caliò, I. Discrete rotating links model for the non-linear torsion−shear behaviour of masonry joints. *Proc. Inst. Civ. Eng. Eng. Comput. Mech.* **2021**, *174*, 215–235. [CrossRef]
60. Zucchini, A.; Lourenço, P.B. A micro-mechanical model for the homogenisation of masonry. *Int. J. Solids Struct.* **2002**, *39*, 3233–3255. [CrossRef]
61. Casapulla, C.; Giresini, L.; Argiento, L.U.; Maione, A. Nonlinear static and dynamic analysis of rocking masonry corners using rigid macro-block modeling. *Int. J. Struct. Stab. Dyn.* **2019**, *19*, 1950137. [CrossRef]
62. Maheri, M.R.; Najafgholipour, M.A.; Rajabi, A.R. The influence of mortar head joints on the in-plane and out-of-plane seismic strength of brick masonry walls. *Iran. J. Sci. Technol. Trans. Civ. Eng.* **2011**, *35*, 63–79.
63. Gabor, A.; Bennani, A.; Jacquelin, E.; Lebon, F. Modelling approaches of the in-plane shear behaviour of unreinforced and FRP strengthened masonry panels. *Compos. Struct.* **2006**, *74*, 277–288. [CrossRef]
64. Lourenço, P.B.; Rots, J. *Analysis of Masonry Structures with Interface Elements-Report No. 03-21-22-0*; Delft University of Technology: Delft, The Netherlands, 1994.
65. Lourenço, P.B. Computations on historic masonry structures. *Prog. Struct. Eng. Mater.* **2002**, *4*, 301–319. [CrossRef]
66. Vecchio, F.J.; Collins, M.P. The modified compression-field theory for reinforced concrete elements subjected to shear. *ACI J.* **1986**, *83*, 219–231.
67. Korany, Y.; EL-Haggar, S. Mechanics and modeling of URM structures. In Proceedings of the International Short Course on Architectural and Structural Design of Masonry, Dresden, Germany, 7–18 December 2003.
68. *ASTM C495*; Standard Test Method for Compressive Strength of Lightweight Insulating Concrete. American Society for Testing and Materials (ASTM): Philadelphia, PA, USA, 2002.
69. *ASTM, C109*; Compressive Strength of Hydraulic Cement Mortars (Using 2-in. or 50-mm Cube Specimens). ASTM International: West Conshohocken, PA, USA, 1999; Volume 4.
70. *ASTM C307-03*; Standard Test Method for Tensile Strength of Chemical-Resistant Mortar, Grouts and Monolithic Surfacing. ASTM International: West Conshohocken, PA, USA, 2012.

71. Kumar, V.; Kartik, K.V.; Iqbal, M.A. Experimental and numerical investigation of reinforced concrete slabs under blast loading. *Eng. Struct.* **2020**, *206*, 110125. [CrossRef]
72. National Research Council. *ISC Security Design Criteria for New Federal Office Buildings and Major Modernization Projects: A Review and Commentary*; The National Academies Press: Washington, DC, USA, 2003. [CrossRef]

applied sciences

MDPI

Article

Experimental Study of Explosion Mitigation by Deployed Metal Combined with Water Curtain

Thérèse Schunck * and Dominique Eckenfels

French-German Research Institute of Saint-Louis, ISL, 5 rue du Général Cassagnou, 68301 Saint-Louis, France; dominique.eckenfels@isl.eu

* Correspondence: therese.schunck@isl.eu; Tel.: +33-3-8969-5186

Abstract: In this paper, protective barriers made of perforated plates with or without a water cover were investigated. In urban areas, such barriers could be envisaged for the protection of facades. An explosive-driven shock tube, combined with a retroreflective shadowgraph technique, was used to visualize the interaction of a blast wave profile with one or two plates made of expanded metal. Free-field air blast experiments were performed in order to evaluate the solution under real conditions. Configurations with either one or two grids were investigated. The transmitted pressure was measured on a wall placed behind the plate(s). It was observed that the overpressure and the impulse downstream of the plate(s) were reduced and that the mitigation performance increased with the number of plates. Adding a water layer on one grid contributed to enhance its mitigation capacity. In the setup with two plates, the addition of a water cover on the first grid induced only a modest improvement. This blast mitigation solution seems interesting for protection purposes.

Keywords: blast; mitigation; grid; water curtain

Citation: Schunck, T.; Eckenfels, D. Experimental Study of Explosion Mitigation by Deployed Metal Combined with Water Curtain. *Appl. Sci.* **2021**, *11*, 6539. https://doi.org/10.3390/app11146539

Academic Editor: Ricardo Castedo

Received: 18 June 2021
Accepted: 14 July 2021
Published: 16 July 2021

1. Introduction

The protection of people and structures against the effects of blast waves from terrorist attacks or industrial hazards is of significant interest. It is well known that protective barriers are an effective way to reduce blast loads and to mitigate the adverse effects. Protective solid barriers are usually made of reinforced concrete, concrete masonry unit or steel-concrete-steel composite materials. These barriers are very rigid and have negligible deformation. In urban areas, other protective barriers could be envisaged for facades. The blast wave would be absorbed, deflected or disrupted to be ultimately reduced before it reaches its intended target. Perforated plates disrupt blast waves: it has been shown that grids or perforated plates modify the flow field by introducing new shock waves, regions of vortices and considerable turbulence in which the energy of the incident shock wave can be dissipated [1–3]. Moreover, grids or perforated plates could be modern architectural design elements which can be used outside buildings. The use of water walls for mitigating the damages from blast waves generated by an explosion has also been described [4], although there is only limited literature on this topic. It seems that the mitigation stems from the fact that the blast wave is obstructed, reflected and diffracted by the water wall. Consequently, it could be an advantage to add a water cover to the grids or perforated plates in order to protect building façades or walls.

Recently, the shock wave attenuation performance of protective barriers made of woven wire mesh was investigated [5]. The woven wire mesh was located about 5 m in front of a wall and no obvious mitigation was observed behind the barrier. The woven wire mesh had a very high porosity and this probably explains such results. Two previous studies have been published regarding metal ring meshes, grids or perforated plates combined with a downward-streaming water curtain for blast mitigation [6,7]. Gebbeken et al. [6] tested a stainless steel ring mesh in combination with a flowing water layer. The charges were detonated 5 m in front of the ring mesh with or without a water cover and the side-on overpressure was measured on the shock wave path. The reflected overpressure was also measured on a wall located 5 m behind the ring mesh. For the ring mesh alone, they described an initial side-on overpressure reduction of approximately 17% at 50 cm behind the grid and of around 1–6% at 1.5–2.5 m behind the grid. When the ring mesh was covered by water, the initial side-on overpressure was reduced by 56% close behind the ring mesh and by 20% 5 m behind the mesh. As regards the positive impulse, the ring mesh itself caused no reduction. However, a reduction of about 17–31% was obtained behind the mesh by adding a water curtain. The reflected pressure on the wall was slightly decreased by the ring mesh but the addition of a water cover did not improve the attenuation performance. In the second study [7], a transonic shock tube was used to visualize the interaction of a blast wave profile with a metallic perforated plate or with a metallic perforated plate covered by a layer of water. Free-field air blast experiments were also performed. Three grid types with different porosities were tested. The highest attenuation was obtained with the grid having the lowest porosity. The attenuation was of the order of 17–25% 25 cm downstream from the grid and about 25–30% 3 m downstream from the grid. When the grids were covered by a water layer, the initial overpressure behind the plate was reduced for all grid types. The blast mitigation was improved, especially with the grids having a high porosity. Again, the most pronounced reduction (about 35 to 48%) was obtained with the grid having the lowest porosity. The initial overpressure 3 m downstream was also reduced by about 20–30%. The impulse was also reduced by the grids with or without a water film cover. Consequently, this blast mitigation method appears promising but still needs some further work and improvement. The porosity of the plate is an important factor to take into account in the blast mitigation performance and a water cover on the plate increases even more the blast attenuation. The plate should have a relatively low porosity. The use of a configuration with two plates could also represent an option for improvement. Indeed, it has been shown that shock wave trapping between two perforated plates enhanced the shock wave attenuation downstream from the grids [8].

In this paper, the assessment of this blast mitigation solution made of perforated plates with or without a water cover was further investigated in order to have a better understanding of its mechanism and to improve its performance. Grids made of expanded metal, with or without a water film, served as an obstacle. Expanded metal has an interesting geometry that enhanced the reflection of the shock wave, and a low porosity. First, an explosive-driven shock tube (EDST) was used to visualize the interaction of a blast wave profile with one or two plates made of expanded metal, using a retroreflective shadowgraph technique. Indeed, the propagation of blast waves in complex media is an important topic of shock wave research and there is a need to study wave phenomena in complex environments. EDSTs generates high dynamic loadings compared to conventional shock tubes. Their mitigation capacity could be assessed under a load comparable to that produced by several kilos of high explosives, located several meters from a target. Secondly, free-field air blast experiments were performed in order to evaluate the protection system under real conditions. Configurations with either one or two grids were investigated. The case in which a film of water was added on the grid, or on the first grid in the case of a two-grid configuration, was also studied. The transmitted pressure was measured on a wall placed behind the plate(s) and the shock wave reflection by the plate was assessed using a sensor located on the ground in front of the first grid.

2. Materials and Methods

2.1. Samples

Grids made of expanded metal were investigated (Figure 1). Expanded metal is a metal sheet that has been cut and stretched to form a regular pattern. Expanded metal is stronger than an equivalent weight of wire mesh because the material is deformed, allowing the metal to remain as a single piece. The expanded metal used in the present study had a hexagonal mesh. The characteristics were an open area of 24%, a hole size of 45 mm × 13 mm, a thickness of 3 mm, an apparent thickness of 9 mm and a strand of 5 mm.

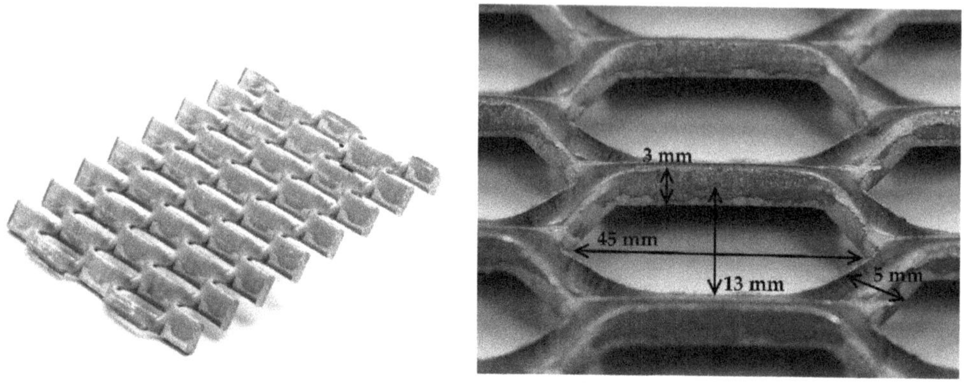

(a) (b)

Figure 1. (**a**) Piece of expanded metal; (**b**) detail: open area of 24%, hole size of 45 mm × 13 mm, thickness of 3 mm, apparent thickness of 9 mm, strand of 5 mm.

2.2. Explosive Driven Shock Tube

The EDST was based on previously published works [9,10]. The shock tube has a square external section of 100 mm × 100 mm, a square internal section of 80 mm × 80mm and a total length of 1750 mm (Figure 2). A pressure sensor (Kulite HKS 375) was used to measure the reflected pressure at a wall behind the plates. A spherical charge of C4 was used (m = 15 g) to produce a planar blast wave. All the charges were cast and detonated, without any container, 50 mm from the shock tube inlet (Figure 2). A part of the initial spherical blast wave enters in the tube. At the tube inlet, the surface of the incident shock has a square curved shape which will be flattened as the shock travels through the tube. The initial spherical blast wave becomes almost completely planar after about 1.5 m of propagation, leading to an initial uniform loading at the outlet of the shock tube. The blast profile obtained in this way is realistic and in line with real threats.

The distance between the outlet of the EDST and the wall was 180 mm. The plate, or the last plate in the case of two plates, was positioned 50 mm in front of the wall. Where two plates were used, the spacing between the plates was 40 mm. The grid holes were always aligned directionally.

2.3. Imaging for EDST

A high speed Photron SA-Z camera was used to record images of the propagation and of the interaction of the shock wave with the plate(s) using a retroreflective shadowgraph technique [11]. An extreme high power light emitting diode (LED) (XHP70.2, CREE), located on an axis with the center of the camera lens, illuminated the outlet of the EDST and a panel covered by a retroreflective material (3M Reflexfolie 4090) which was placed in the background. Figure 3 shows a photograph of the camera with the LED and the EDST and the Figure 2 shows the position of the retroreflective panel relative to the EDST and the camera. A power supply and a purpose-built trigger unit allowed the LED to be pulsed for about 10 ms. The reflection and the transmission patterns of the shock wave through the obstacle could be visualized via their shadow on the panel. Moreover, a piece of retroreflective material was plastered on the wall and a 45-degree inclined mirror was attached to the end of the EDST (Figure 4). The camera filmed the panel placed in the background but also the mirror, allowing for the visualization of the shock wave in the axial direction. The videos were recorded with a frame rate of 100,000 fps at a resolution of 408 × 384 pixels.

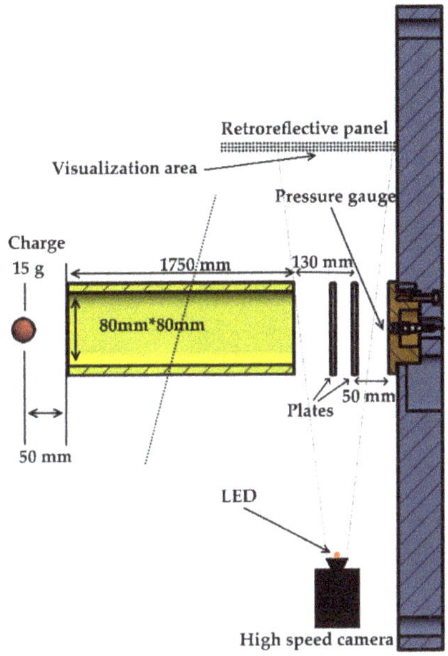

Figure 2. Schematic view of the explosive-driven shock tube and positioning of the pressure sensor and plates. The position of the high speed camera and of the retroreflective panel is also shown.

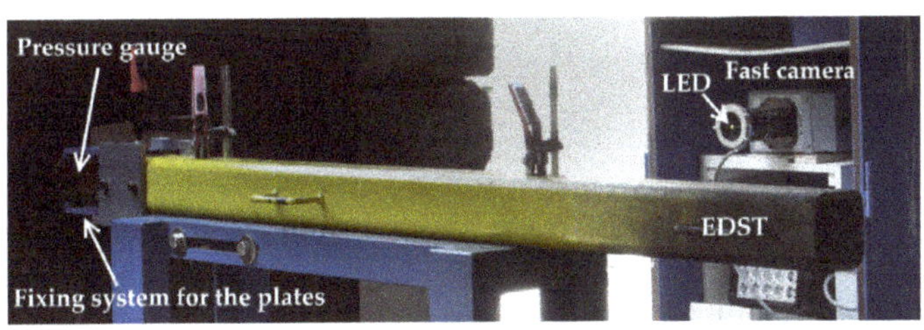

Figure 3. View of the experimental setup, showing the explosive driven shock tube and the high speed camera with the LED.

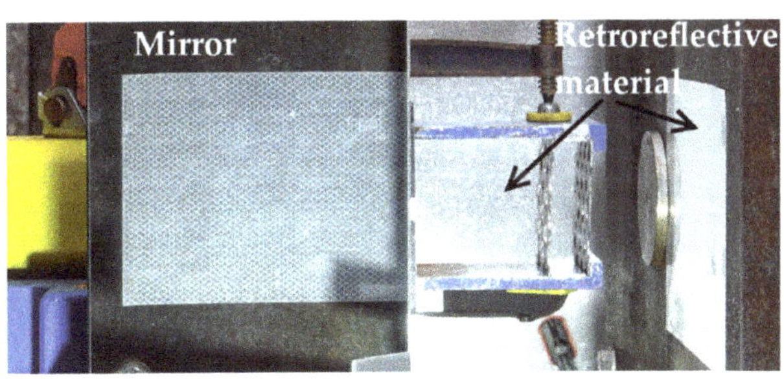

Figure 4. View of the EDST end showing the panel in the background, the piece of retroreflective material plastered on the wall and the 45-degree inclined mirror.

2.4. Free Field

Explosion tests were conducted with spheres of 2 kg C4. The charges were raised by 25 cm and ignited by a high voltage cap (RP 501) (Figure 5). Configurations with either one or two grids were investigated. The addition of a film of water on the grid, or on the first grid in the case of a two-grid configuration, was also tested. For tests with one grid, the charge was positioned 3.8 m from the grid and 4.8 m from the wall (Figure 6). For tests with two grids, the charge was positioned 3 m from the first grid, 3.8 m from the second one and 4.8 m from the wall (Figure 5). When paired, grids were spaced 0.8 m apart. The grid size was 2 m × 3 m and concrete blocks placed at the left and right of the grids were used to mount them (Figures 5 and 6). Each had the following dimensions: length 160 cm, height 40 cm and width 80 cm. One side-on pressure gauge (PCB137A23) (Figure 5), located at a right angle to the shock wave's propagation toward the wall, allowed for verification of reproducibility. One PCB sensor (M102A) was used to evaluate the effect of the grids or, of the grids covered by a film of water, of the reflected pressure on the wall located behind the grid (Figure 7, left). The sensor was positioned at a height of 50 cm. One pressure gauge (M102A) was positioned on the ground in front of the first grid position (Figures 5–7, left). This gauge was located 19.5 cm from the first grid. The height of this gauge was 5 cm. The water layer was generated by a pool fountain (VidalXL) (Figure 8).

(a)

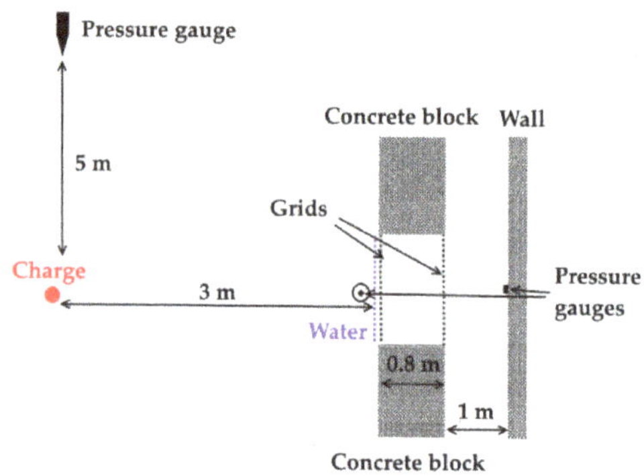

(b)

Figure 5. (**a**) Photography of the experimental setup; (**b**) Schematic diagram showing the charge, grids, concrete blocks, gauge in front of the grid and one side-on pressure gauge, placed at a right angle.

Figure 6. View of the experimental setup showing the one-grid configuration: the grid position, concrete blocks and gauge in front of the grid.

(a) (b)

Figure 7. (**a**) Reflected pressure gauge located on the wall behind the grids; (**b**) pressure gauge located in front of the first grid position.

Figure 8. First grid equipped with a pool fountain.

3. Results

3.1. Explosive Driven Shock Tube

Figure 9 shows the shock wave propagating at the outlet of the EDST when there was no perforated plate. The photographs were taken with high-speed imaging. Thanks to the mirror, set at 45° in front of the camera, imagery could be obtained from the axial direction simultaneously with the direct view. Consequently, events happening axially to the EDST outlet could be observed. The shock wave was not totally planar at the outlet of the tube; indeed, the two metallic plates, fixed parallel to the EDST at the end and which were used to clamp the perforated plates, reflected the shock wave (t = 1440 μs). These two plates had a shoulder at their ends which provided additional thrust for the fixation of the perforated plates. These shoulders can be seen in Figure 4 and the shock wave visibly interacts with them (t = 1520 μs). The shock wave emerging laterally from the EDST was visible in the mirror. It should be noted that the shock wave exited the EDST outlet immediately, but in the beginning the optical setup did not allow the observation of this expansion.

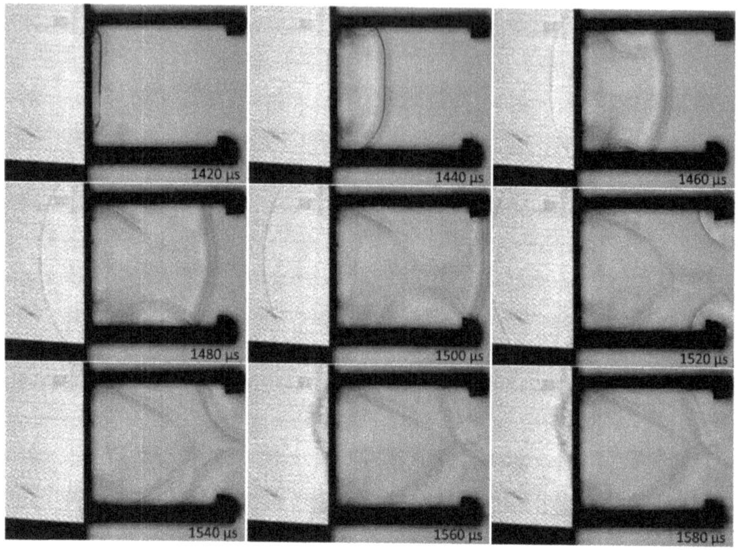

Figure 9. Photographic results of high-speed video recording, showing shock wave propagation at the outlet of the EDST (20 µs between each photograph).

Figure 10 shows shock wave propagation through one deployed metal plate. The plate was positioned at a distance of 130 mm from the tube outlet and at a distance of 50 mm in front of the wall. At t = 1470 µs, the shock wave when passing through the plate apertures split into several shock waves, one for each aperture, and these shock waves recombined further down. At t = 1530 µs, the shock wave behind the plate became almost planar again. A complex structure of turbulence appeared just behind the grid and it persisted for some time; at t = 1510 µs turbulence began to form along the backside the grid, and at t = 1790 µs it was still observable. The reconbinated shock wave behind the perforated plate hit the wall and was reflected at t = 1570 µs. The remaining shock wave, which did not pass through the grid and was reflected by it, became almost planar at t = 1550 µs.

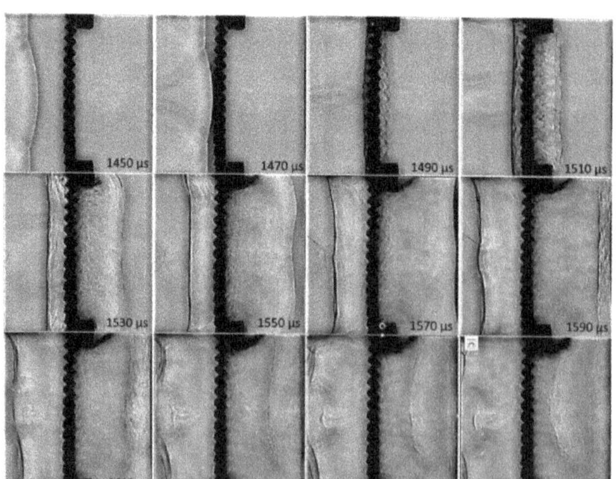

Figure 10. Photographic results of high-speed video recording, showing shock wave propagation through one plate of deployed metal (20 µs between each photograph).

Figure 11 shows shock wave propagation through two deployed metal plates. The first plate was positioned at a distance of 90 mm from the tube outlet and at a distance of 90 mm in front of the wall. Grids were paired 40 mm apart. When passing through the apertures of the first plate (t = 1450 μs), the shock wave split into several shock waves, one for each aperture, and these shock waves recombined further down. The transmitted shock wave was similar to the incident shock wave at t = 1490 μs. At the back of the shock wave, between the two plates, lasting turbulence could be observed, especially immediately behind the first plate. The remaining shock wave, which did not pass through the first plate, was reflected, leading to the pattern captured at the back of this plate. Thereafter, the shock wave transmitted by the first plate impacted the second plate (t = 1510 μs) and, once again, a part of the shock wave was transmitted and part was reflected. The wave reflected by the second plate propagated toward the first plate and, as it arrived near the plate, the shock front and the turbulence became less obvious. We could also see turbulence behind the second plate.

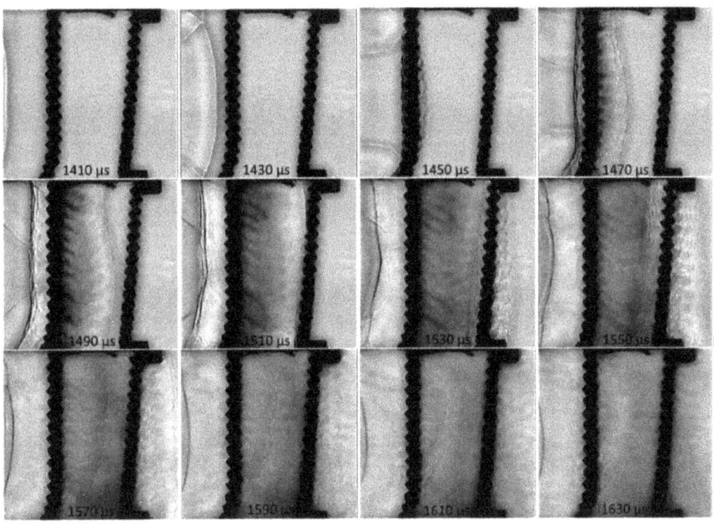

Figure 11. Photographic results of high-speed video recording, showing shock wave propagation through one plate of deployed metal (20 μs between each photograph).

Five reference tests were conducted without any plates and seven experiments were conducted with either one or two grids positioned at the outlet of the EDST. The reflected overpressure versus time was collected by the sensor inserted in the wall behind the plate. The blast wave propagation through the shock tube was computed with Autodyn (ANSYS) and the overpressure at the outlet of the shock tube was obtained. The calculated initial overpressure was 32 bar. The measured value at a distance of 180 mm was ~18 bar, which was consistent with simulation. The impulse, which is the pressure signal integrated over time, as a function of time, was also computed. Figure 12a presents the reflected pressure as a function of time, obtained in a reference test and in one test each using one or two expanded metal plates. When the number of plates increased, the reduction of the initial reflected overpressure increased. The initial overpressure reflected on the wall was reduced by 46% and 72%, respectively, when either one or two plates were positioned in front of the EDST (Table 1). Correspondingly, the maximum impulse was reduced by 68 and 89% (Table 1).

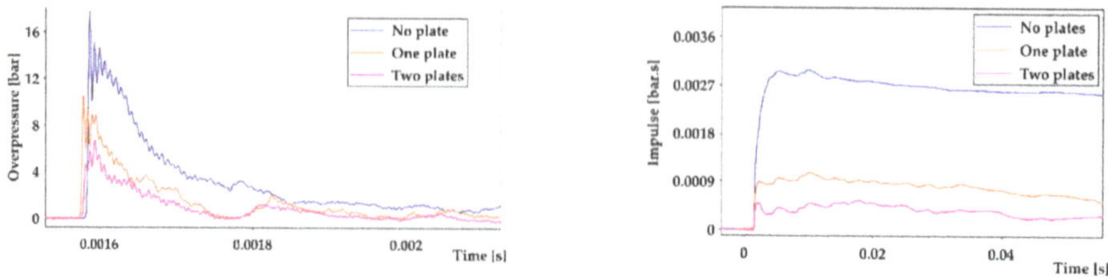

(a) (b)

Figure 12. Reflected pressure (**a**) and impulse (**b**) measured on the wall positioned 50 mm behind either one or two expanded metal plates. One reference test, which was performed without plates, is also shown.

Table 1. Initial reflected overpressure and maximum impulse collected by the sensor inserted in the wall and located 50 mm downstream from the perforated plates for all experiments. The difference from the mean value obtained with the reference tests (no plate) is also given.

Type of Plate	Number of Plates	Overpressure (Bar)	Attenuation (%)	Impulse (Bar·s)	Attenuation (%)
no plate	-	22.82	-	0.0032	-
no plate	-	19.48	-	0.0040	-
no plate	-	15.61	-	0.0035	-
no plate	-	15.65	-	0.0037	-
no plate	-	15.73	-	0.0034	-
expanded metal	1	9.36	48	0.0011	70
expanded metal	1	7.71	57	0.0012	65
expanded metal	1	10.89	39	0.0013	63
expanded metal	1	10.30	42	0.0015	59
expanded metal	2	4.46	75	0.0003	93
expanded metal	2	4.50	75	0.0005	87
expanded metal	2	5.86	67	0.0005	87
expanded metal	2	4.46	75	0.0003	93

3.2. Free Field

The initial overpressure and the maximum impulse obtained through the control gauge for all experiments are presented in Tables 2 and 3, respectively. These measurements showed that there was rather good reproducibility. The side-on overpressure was estimated with Kingery's and Bulmash's formula [12]. The detonation of 2 kg of C4 generates an overpressure of 0.8 bar at a distance of 5 m. The measured values were consistent with this value. Figure 13 presents the overpressure measured on the wall for one test of each configuration. Two or three experiments were conducted for each configuration. Table 2 gives the initial overpressure value obtained in all the tests. According to Kingery's and Bulmash's formula [12], the detonation of 2 kg of C4 generates a reflected overpressure of 2.6 bar at a distance of 4.8 m. The measured values were consistent with this value. The overpressure was reduced when one plate was located in front of the wall, in the order of 32–47%. Adding a second plate on the shock wave path led to a stronger attenuation. The initial overpressure was reduced by about 62–66%. When a water film was used, the reflected overpressure on the wall was reduced even more, especially in the case of the one-plate configuration. The reduction of overpressure was about 69–71% and 66–74% for one-plate configurations and two-plate configurations, respectively. The setup using a water wall alone was also evaluated and the reflected overpressure on the wall was not modified. Figure 14 presents the impulse measured on the wall for one test of each configuration and Table 3 gives the maximum impulse values obtained for all tests. The results were similar to those observed for overpressure.

Table 2. Initial overpressure collected by three sensors (control, inserted into the wall and laid on the ground) for all experiments. The attenuation of the overpressure measured on the wall relative to the mean value obtained by the reference tests (no plate) is also given. Overpressure ground 1 and overpressure ground 2 correspond to the maximum value of the first and the second peak, respectively.

Experiment Type	Overpressure Control (Bar)	Overpressure Wall (Bar)	Attenuation (%)	Overpressure Ground 1 (Bar)	Overpressure Ground 2 (Bar)
no plate	0.90	2.53	-	3.20	-
no plate	0.89	2.61	-	3.22	-
no plate	0.89	2.70	-	-	-
2 expanded metal	0.90	1.05	60	3.08	2.12
2 expanded metal	0.86	0.88	66	2.26	1.96
2 expanded metal	0.89	0.99	62	3.31	2.14
2 expanded metal & water	0.89	0.90	66	1.88	1.89
2 expanded metal & water	0.78	0.68	74	3.17	1.83
1 expanded metal	0.89	1.42	46	3.16	-
1 expanded metal	0.82	1.78	32	3.52	-
1 expanded metal	0.84	1.40	47	3.62	-
1 expanded metal & water	0.80	0.82	69	2.80	-
1 expanded metal & water	0.76	0.75	71	3.08	-

Table 3. Maximum impulse collected by the three sensors (control, inserted into the wall and laid on the ground) for all experiments. The attenuation of impulse measured on the wall relative to the mean value obtained by the reference tests (no plate) is also given. Impulse ground 1 and impulse ground 2 correspond to the maximum value of the first and the second peak, respectively.

Experiment Type	Impulse Control (Bar·s)	Impulse Wall (Bar·s)	Attenuation (%)	Impulse Ground 1 (Bar·s)	Impulse Ground 2 (Bar·s)
no plate	0.00103	0.00247	-	0.001427	-
no plate	0.00105	0.00275	-	0.001388	-
no plate	0.00097	0.00261	-	-	-
2 expanded metal	0.00108	0.00105	60	0.001384	0.002338
2 expanded metal	0.00108	0.00105	60	0.001287	0.002258
2 expanded metal	0.00105	0.00095	64	0.001127	0.001849
2 expanded metal & water	0.00104	0.00077	71	0.001312	0.002546
2 expanded metal & water	0.00098	0.00071	73	0.001143	0.002234
1 expanded metal	0.00107	0.00155	41	0.001411	-
1 expanded metal	0.00100	0.00144	45	0.001395	-
1 expanded metal	0.00096	0.001437	45	0.001289	-
1 expanded metal & water	0.00094	0.00108	58	0.001418	-
1 expanded metal & water	0.00093	0.00110	58	0.001402	-

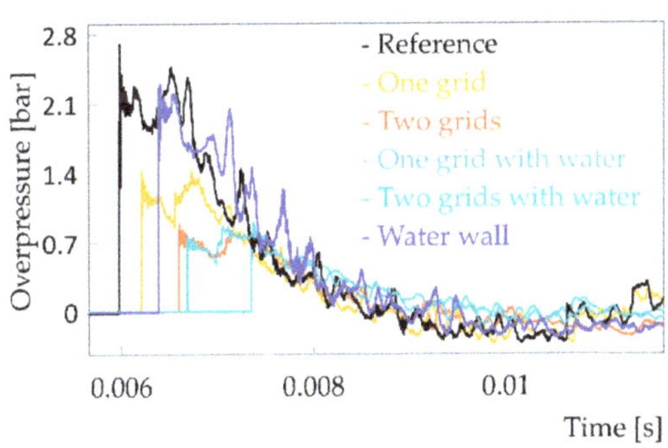

Figure 13. Reflected pressure measured on the wall for one test of each configuration. One reference test is also shown.

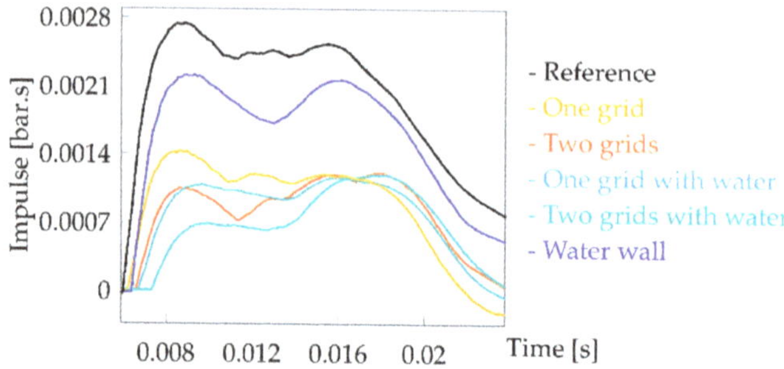

Figure 14. Reflected impulse measured on the wall for one test of each configuration. One reference test is also shown.

Figure 15 presents the overpressure measured by the gauge positioned on the ground at the first plate's position. The reflection of the shock wave, when a plate was placed near this sensor (two-plate configuration), could be observed (see the arrows in the Figure 15). In case of one-plate configurations, the sensor was placed further off the plate and the shock wave reflection was not perceptible. In the case of two-plate configurations, the reflection of the shock wave on the second plate was clearly manifested as a huge overpressure second peak.

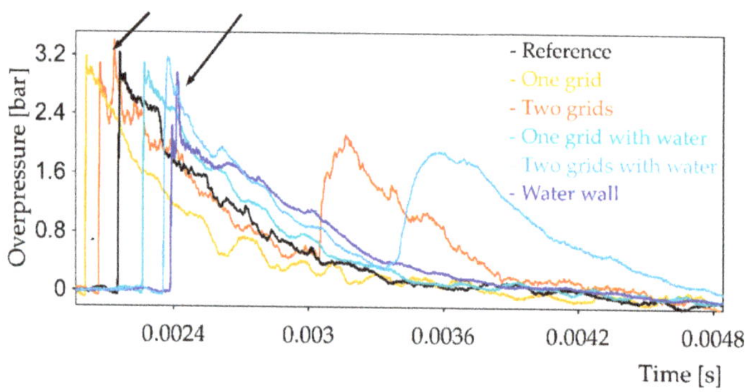

Figure 15. Overpressure obtained by the gauge laid on the ground for one test of each configuration. One reference test is also shown.

4. Discussion

Shock wave propagation through one or two deployed metal plate(s) was observed thanks to the EDST and a retroreflective shadowgraph technique. This experimental set-up was very well suited to this research, since the shock wave interaction with perforated plates could be assessed with high loading and visualized at the same time. This is not possible when using a conventional shock tube, the blast loading being rather moderate. In any event, it is not possible to use high Mach numbers when the air flow in the shock tube is especially blocked, due to risk of damage. Understanding of the complex flow field induced by blast that passes through a complex media is an important aspect in blast mitigation research and could help to design new devices for protection against blast loading. The shock wave, when it passed through the apertures of a deployed metal plate, split into several shock waves, one for each aperture, which recombined further on. A complex structure of turbulence appeared just behind the grid and it persisted for some time. The remaining shock wave, which did not pass through the grid, was reflected by the grid. In the case of two plate configurations, the shock wave had time to reform between the two plates, which impacted the second plate. Once again, a part of the shock wave was transmitted and part was reflected. Turbulence behind the second plate was also visible.

The shock wave reflection and the creation of turbulence led to a blast wave attenuation. Consequently, when two plates served as an obstacle, these phenomena occurred twice and the mitigation was greater. This was confirmed by the reflected overpressure and impulse measured by the sensor inserted into the wall behind the plate(s). When the number of plates increased, these values were reduced. Similar observations were made by O. Ram et al. [13], who assessed the propagation of shock waves through an array of perforated plates in a conventional shock tube. Thus, we can conclude that similar phenomena occurred at low (3 bar) and high loading (20 bar).

The results obtained in free field were consistent with the results obtained by means of the EDST. The reflected overpressure and impulse measured behind one plate on a wall was reduced ion the order of 32–47%. The addition of a second plate on the shock wave path led to a stronger attenuation (62–66%). When a water film was used, the reflected overpressure on the wall was reduced still more, especially in the case of one-plate configurations—the mitigation approached that obtained from two-plate configurations. The water cover had only a small effect on blast mitigation when two plates were used. We can conclude that the water layer's contribution mostly enhanced the reflection of the shock wave by filling the apertures with water. When one perforated plate was covered with a water film, its capacity to reflect the blast wave was enhanced, its performance to mitigate the blast increased and approached that obtained from two-plate configurations. In cases with two plates, the obstruction of the blast wave was rather high, and the addition of a water cover induced only a modest improvement. In [7], thanks to a transonic shock tube, the interaction of a blast wave with a perforated plate with a water cover was imaged and it was observed that the water film disintegrated into droplets significantly after the shock wave front had passed through it. The fragmentation of the water film had little effect on the attenuation, as it broke long after the passage of the shock wave front and there is little extraction of energy from the shock front from water layer fragmentation. In this work, the results have also shown that a water wall alone had almost no impact on the reflected overpressure on the wall. Moreover, the study [4] on blast mitigation using a water wall, in which walls made of plastic bags, filled with water and having a thickness of 5 to 8 cm, it was shown that the mitigation was obtained by obstruction, reflection and diffraction of the blast wave. The mitigation mechanism was comparable to that of a rigid wall, thus the mitigating effect of energy exchange with water was not primarily responsible for the effect. The results obtained here could also be compared to those obtained by Gebbeken et al. [6] and Xiao et al. [5]. In these two studies, blast mitigation observed when using a single grid was very low and this could be explained by the high porosity of the grids used. Indeed, the first study [6] used a stainless steel ring mesh with a porosity of 63% and, in the second, a woven wire mesh having a relative opening fraction of 60.2%. In the setup using ring mesh, adding a water curtain also enhanced the attenuation of both peak overpressure and positive impulse. The authors also claimed that when a blast wave hits ring mesh covered by water, the water layer forms a closed surface that reflects the blast wave to a greater extent.

In the case of a higher-charge explosive, the phenomenon of shock wave transmission/reflection would likely be the same and equivalent mitigation performance would be achieved. However, if the grids are not sufficiently resistant to a high loading, the grids could deform and tear. Some debris could impact the structure behind such grids, and consequently the use of gridded plates could have prejudicial effects. The grids and their attachment system must be sized to guarantee their resistance and their structural integrity with respect to anticipated blast size.

5. Conclusions

In this paper, we have assessed a blast mitigation solution made of perforated plates with or without a water cover. The mitigation of a blast wave after its passage through one or two plate(s) made of deployed metal, covered or not by a water film, was investigated. First, we imaged the interaction of a blast wave with the grids at high loading. Secondly, free-field air blast experiments were performed in order to evaluate the protection system under real conditions. It was observed that the overpressure and the impulse downstream of the grids were reduced and that the mitigation performance increased with the number of plates. Adding a water layer to one grid contributed to its mitigation capacity. However in setups with two plates, the addition of a water cover on the first grid induced only a modest improvement. All in all, this method seems to warrant interest for protection purposes.

Author Contributions: Conceptualization, methodology, experimentations, data analysis, writing, review and editing—T.S. Metrology, methodology, experimentations, data analysis—D.E. Both authors have read and agreed to the published version of the manuscript.

Funding: This research received no external funding.

Institutional Review Board Statement: Not applicable.

Informed Consent Statement: Not applicable.

Data Availability Statement: The data are provided in the paper.

Acknowledgments: We thank our colleagues Yannick Boehrer, Thierry Ottié, Yannick Stehlin and Sylvain Hemmerlin who assisted us in our research by providing technical support.

Conflicts of Interest: The authors declare no conflict of interest.

References

1. Kingery, C.; Pearson, R.; Coulter, G. Shock Wave Attenuation by Perforated Plates with Various Hole Sizes. USA Ballistic Research Laboratory Memorandum Report; 1977; n° 2757, Defense Technical Information Center. Available online: https://apps.dtic.mil/sti/citations/ADA041854 (accessed on 18 June 2021).
2. Britan, A.; Karpov, A.V.; Vasilev, E.I.; Igra, O.; Ben-Dor, G.; Shapiro, E. Experimental and numerical study of shock wave interaction with perforated plate. *J. Fluids Mech.* **2004**, *126*, 399–409. [CrossRef]
3. Britan, A.; Igra, O.; Ben-Dor, G.; Shapiro, E. Shock wave attenuation by grids and orifice plates. *Shock Waves* **2006**, *16*, 1–15. [CrossRef]
4. Chen, L.; Zhang, L.; Fang, Q.; Mao, Y.-M. Performance based investigation on the construction of anti-blast water wall. *Int. J. Impact Eng.* **2015**, *81*, 17–33. [CrossRef]
5. Xiao, W.; Andrae, A.; Gebbeken, N. Experimental investigations of shock wave attenuation performance using protective barriers made of woven wire mesh. *Int. J. Impact Eng.* **2019**, *131*, 209–221. [CrossRef]
6. Gebbeken, N.; Rüdiger, L.; Warnstedt, P. Explosion mitigation by water mist-ring mesh with water curtain. In Proceedings of the 25th MilitaryAspects of Blast and Shock Conference, Hague, The Netherlands, 23–25 September 2018.
7. Schunck, T.; Bastide, M.; Eckenfels, D.; Legendre, J.-F. Explosion mitigation by metal grid with water curtain. *Shock Waves* 2021. [CrossRef]
8. Seeraj, S. Shock Wave Interactions with Porous Plates. Master's Thesis, University of the Witwatersrand, Johannesburg, South Africa, 2007.
9. Louar, M.A.; Belkassem, B.; Ousji, H.; Spranghers, K.; Kakogiannis, D.; Pyl, L.; Vantomme, J. Explosive driven shock tube loading of aluminium plates: Experimental study. *Int. J. Impact Eng.* **2015**, *86*, 111–123. [CrossRef]
10. Ousji, H.; Belkassem, B.; Louar, M.A.; Reymen, B.; Martino, J.; Lecompte, D.; Pyl, L.; Vantomme, J. Air-blast response of sacrificial cladding using low density foams: Experimental and analytical approach. *Int. J. Mech. Sci.* **2017**, *128–129*, 459–474. [CrossRef]
11. Stojko, S.; Freundt, J.; Anderson, J.G.; Delaney, T. Experimental characterization of the interaction of blast waves from multiple high explosive charges. In Proceedings of the 25th MilitaryAspects of Blast and Shock Conference, Hague, The Netherlands, 23–25 September 2018.
12. Kingery, C.N.; Bulmash, G. *Technical Report ARBRL-TR-02555: Air Blast Parameters from TNT Spherical Air Burst and HEMISPHERICAL Burst*; AD-B082 713; U.S. Army Ballistic Research Laboratory: Aberdeen Proving Ground, MD, USA, 1984.
13. Ram, O.; Ben-Dor, G.; Sadot, O. On the pressure buildup behind an array of perforated plates impinged by a normal shock wave. *Exp. Therm. Fluid Sci.* **2018**, *92*, 211–221. [CrossRef]

applied sciences

MDPI

Article

Numerical Simulation on Dynamic Behavior of Slab–Column Connections Subjected to Blast Loads

Kwang Mo Lim [1], Taek Hee Han [2] and Joo Ha Lee [3,*]

1 Korean Peninsula Infrastructure Special Committee, Korea Institute of Civil Engineering and Building Technology, Goyang-si 10223, Korea; limkm@kict.re.kr
2 Coastal Development and Ocean Energy Research Center, Korea Institute of Ocean Science and Technology, 385 Haeyang-ro, Yeongdo-gu, Busan 49111, Korea; taekheehan@kiost.ac.kr
3 Department of Civil and Environmental Engineering, The University of Suwon, Hwaseong-si 18323, Korea
* Correspondence: leejooha@suwon.ac.kr; Tel.: +82-31-220-2159

Abstract: Although many studies on the blast-resistant performance of structures have focused mainly on single members such as beams and columns, there is little research on the behavior of joints that are subjected to blast loads. In this study, the structural behavior of a slab–column connection subjected to blast load was investigated using a numerical analysis method. LS-DYNA was used as a finite element analysis program, and in order to improve the accuracy of numerical analysis, mesh size, material model, and simulation method of blast load were determined through preliminary analysis. The effect of different restraints of the joints, depending on the position of the columns in the slab, on the blast resistance performance was investigated. As a result, the highly confined slab-interior column connection showed better behavior than other edge and corner columns. The drop panel installed between the lower column and the slab was effective in improving the blast-resistance performance of the slab–column connection. For a more accurate evaluation of blast resistance performance, it was suggested that various evaluation factors such as ductility ratio, reinforcing stress, and concrete fracture area can be considered along with the support rotation, which is an important evaluation factor suggested by many standards.

Keywords: blast loads; slab; column; connections; numerical analysis

Citation: Lim, K.M.; Han, T.H.; Lee, J.H. Numerical Simulation on Dynamic Behavior of Slab–Column Connections Subjected to Blast Loads. *Appl. Sci.* **2021**, *11*, 7573. https://doi.org/10.3390/app11167573

Academic Editors: Lina M. López, Ricardo Castedo and Anastasio P. Santos

Received: 23 July 2021
Accepted: 15 August 2021
Published: 18 August 2021

1. Introduction

As explosive terrorism and explosion accidents continue to occur around the world, research on the behavior of structures under such extreme situations is increasing [1–3]. Most of these studies focus on how single members such as beams, columns, and slabs behave under explosive loads. Since the failure or large deformation of the joint can directly lead to the collapse or malfunction of the entire structural system, the study on the joint behavior is no less important than the study on single members. Although many studies have been conducted on the structural behavior of joints under static and dynamic loads, further studies are still needed to improve the understanding of joint behavior under explosive loads [4–6].

In this study, the joints of the columns and slabs were investigated. In particular, the behavior of the joints under blast load was investigated according to the position of the column. In other words, the effect of different restraints of the joints on the blast-resistance performance was observed as the columns were located on the inside, edge, and corner of the slab. In addition, the behavior of slab–column joint with drop panel, which is a square portion provided above the lower column and below the slab, was investigated. Drop panels that provide increased shear strength and moment resistance are expected to be effective in improving the blast-resistance performance of slab–column joints. The experimental approach with explosives is really challenging, expensive, and difficult. Therefore, in this study, numerical analysis, one of the best options to discuss this phenomenon, was performed using LS-DYNA, a general-purpose finite element analysis program whose reliability has been verified through many previous studies [7,8].

2. Literature Review

There are some studies on the structural behavior of the slab–column connections according to the position of columns under static loads. Bianchini et al. (1960) performed tests on a total of 45 specimens of the interior, edge, corner, and isolated columns [9]. As a result, the effective strength of the interior column–slab joint was 75% of the column strength and 1.5 times the slab strength. However, when the column strength exceeds the slab strength by 1.4 times, the effective strength of the column–slab joint at the edge and corner columns is not significantly increased by the restraint of the surrounding slab [9]. McHarg et al. (2000) performed a total of 12 test specimens consisting of column–slab specimens and isolated column specimens [10]. The interior column showed greater axial compressive strength than the isolated column due to the restraint effect of the slab, and also showed ductile behavior [10]. Lee et al. (2007) performed column load transmission experiments depending on the position of the column. As a result, the interior column test specimen has improved ultimate load capacity than the isolated column specimen due to the slab restraint effect [11].

There are also many studies on the structural behavior of slab–column connections under dynamic loads. Some of them experimentally confirmed that the amount of flexural reinforcement affects the seismic behavior of the slab–column joint [12–16]. Some studies have investigated the seismic response of slab–column joints with high-strength concrete (HSC) applied to the slab, and they showed that the specimens with HSC had superior performance in terms of ductility and strength, compared to specimens with conventional normal strength concrete (NSC) [17–20]. Scotta and Giorgi (2016) performed cyclic experiments on four full-scale exterior slab–column connections made of normal concrete and fiber-reinforced lightweight aggregate concrete [21]. They reported that the addition of steel fibers to the concrete mix improved the hysteretic behavior of slab–column connections [21]. Several researchers have studied the progressive collapse and robustness of building structures due to column or joint collapse [22–26]. Setiawan et al. (2019) performed numerical analysis on slab–column connections subjected to cyclic loading and captured the characteristics of cyclic degradation observed in experiments with nonlinear finite analysis and suggested a simplified design method for punching shear [27].

As shown above, studies dealing with the structural behavior of joints under dynamic loads mainly focus on seismic loads. While there are relatively many experimental and numerical investigations of reinforced concrete (RC) slabs subjected to blast loading [28–37], there are few studies of slab–column connections. Shahriari et al. (2021) numerically investigated the blast response and progressive collapse of RC structures equipped with viscoelastic dampers [38]. They found that viscoelastic dampers designed for seismic loads resulted in a suitable performance for reducing the response of structures to blast loads [38]. Krauthammer (1999) reported that plastic hinge control through diagonal reinforcing bars can contribute to the improvement of the blast resistance performance of the connections [39]. Lim et al. (2016) have reported the blast-resistance performance of joints of slab-interior and slab-edge columns [40]. However, there is no study on the blast-resistance behavior for the slab-corner column connection and for the relatively large explosive load. Additionally, the method of transmission of the explosive load has not been verified sufficiently in the numerical analysis. Therefore, it is necessary to investigate the blast-resistance performance of slab–column connections depending on the type of the column and the amount of explosive load.

According to the ASCE/SEI (2011), the connection should be designed to resist shear force, axial force, bending moment, and torsion [41]. The effects of rebound are also considered for all connections. The reinforcements in beam–column connections are supposed to comply with details of earthquake-resistant structures according to ASCE/SEI (2011) [41]. There are no other guidelines for the design of blast-resistant slab–column connections. Especially, the material properties in blast events are different in earthquake conditions because of the difference in strain rate. The strain rate is typically over 10^0 s^{-1} in blast events and 10^{-5} s^{-1} in earthquakes [42]. Therefore, more accurate material properties are needed to design structures subjected to explosive loads comparing to structures subjected to seismic loads.

According to UFC 3-340-02 issued by the US Department of Defense (DoD), support rotation and ductility are selected as criteria for evaluating the structural performance of RC structural members under explosive loads [43]. This criterion assumes that the structural member is effectively resisting the blast load when the support rotation is 2 degrees or less. In addition, ASCE/SEI 59-11 proposes the limit of support rotation for blast-resistant RC structures and also provides a level of protection (LOP) [41]. It is noteworthy that both DoD and ASCE are proposing support rotation as an evaluation factor for the behavior of blast-resistant structures.

3. Numerical Analysis

3.1. Details of Specimens

Details of all slab–column connections are shown in Figure 1. The specimen IC is a statically designed slab-interior column connection. The specimen IC-D has the same reinforcement and shape details as specimen IC, but the drop panel was additionally placed according to ACI 318 (2011) [44] and ACI 352.1R [45]. The specimens EC and CC were designed based on the specimen IC, but the column was located at the edge and corner of the slab, respectively. Accordingly, the specimens EC and CC are confined on three sides and two sides of the connection, respectively, while the specimen IC is confined on all four sides. In general, the effective strength of connections can be improved in static loads when the column in connection is confined by slab [9,46]. To verify the confinement effects, comparative studies were conducted on the behavior of specimens IC, EC, and CC.

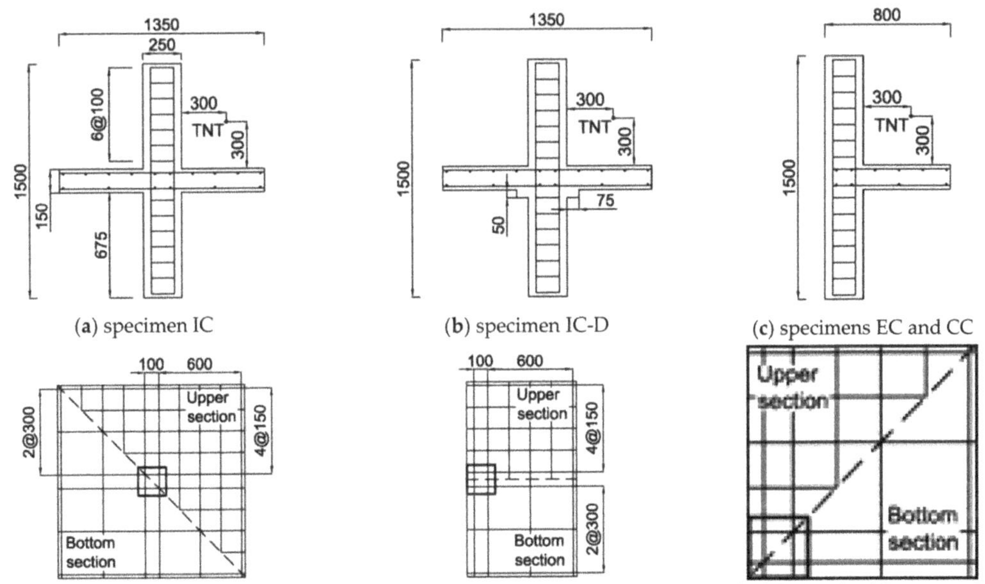

(a) specimen IC (b) specimen IC-D (c) specimens EC and CC
(d) slab details of specimens IC and IC-D (e) slab details of Specimen EC (f) slab details of Specimen CC

Figure 1. Details of specimens: (**a**) specimen IC; (**b**) specimen IC-D; (**c**) specimens EC and CC; (**d**) slab details of specimens IC and IC-D; (**e**) slab details of specimen EC; (**f**) slab details of specimen CC.

3.2. Modeling of Specimens

In order to obtain accurate numerical analysis results, it is necessary to establish an accurate material model. In this study, Mat_072R3 was selected from the concrete material models provided by the analysis program LS-DYNA. This material model reflects the strain-rate effect and has already been found in several studies in the literature to be suitable for analyzing concrete structures under high strain-rate [47–49]. However, Mat_072R3 was unable to exhibit local damage caused by explosions, such as crater spalls, which are associated with structural failure and erosion [50]. Therefore, to simulate these characteristics, LS-DYNA's "Add_Erosion keyword" option was applied to the concrete material model. To model the reinforcing bars, LS-DYNA's Mat_024 was applied, which is defined as an elastic-plastic material with arbitrary stress–strain curve and an arbitrary strain-rate dependency. The fracture of Mat_024 is based on plastic deformation [49].

Numerical analysis results may vary depending on the mesh size of the element [51,52]. According to the previous studies, when simulating a structure subjected to an explosive load, a mesh size of 25 to 30 mm led to the analysis results most similar to the experimental results [33,53]. In this study, before the main analysis was conducted, various mesh sizes were evaluated in terms of accuracy and efficiency of analysis. In the preliminary analysis, the displacement, stress of reinforcing bars, and fracture shape were investigated in the same way as in the main analysis. Considering the analysis results and the time required for analysis, a mesh size of 20 to 25 mm is considered to be the most reasonable. Therefore, in this study, the concrete mesh is composed of 25 mm cubic, 8-node solid elements.

The interaction between concrete and reinforcing bars has a great influence on the behavior of RC structures. In particular, interactions such as bond–slip are very difficult to simulate. A method of tying nodes was recommended to simulate the structure's actual behavior and to provide the simplicity of analysis [48,54]. In this study, the nodes of the reinforcing bar and concrete are connected to each other to provide accurate structural performance.

The one-point integration method used in this analysis is effectively applied to dynamic analysis due to its relatively short analysis time. However, there is a risk of creating a zero-energy state that causes a negative volume or creating an element that behaves differently from the actual behavior. In the numerical analysis, the volume of solid elements is generally reduced when subjected to the compressive pressure, but when zero energy is generated inside the solid elements, the negative volume occurs due to the abnormal operation of the element, resulting in an increase in volume, as shown in Figure 2. In this case, the amount of internal energy loss is called hourglass energy. When the hourglass energy is largely generated, it is difficult to ensure the accuracy of the analysis [55–57]. Therefore, LS-DYNA's "Hourglass" option, which can control the accuracy of analysis due to this phenomenon, was applied to the material model [48,58].

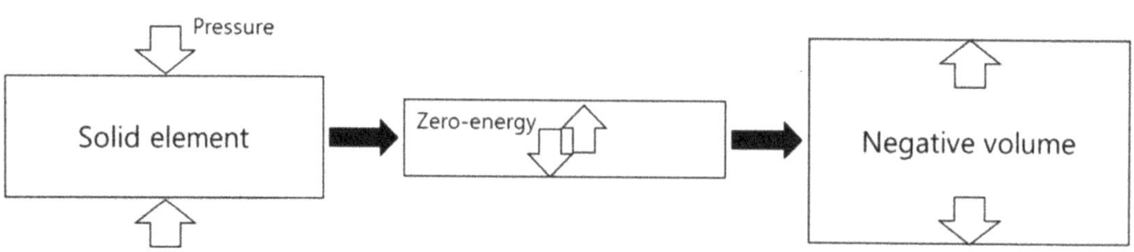

Figure 2. Process of the negative volume.

3.3. Modeling of Blast Loads

In this study, a preliminary analysis was performed to select the analysis method between multi-material arbitrary Lagrangian–Eulerian (MME) and load-blast enhanced (LBE). The concrete walls, which have the same characteristics as the main analysis including materials and element size, were analyzed. As in the main analysis, 4 kg of TNT was placed at a vertical distance of 300 mm from the center of the structure surface. The surfaces for LBE were defined as the front of a wall that was directly affected by the explosion. Table 1 and Figure 3 show the analysis results from both methods of MME and LBE. The maximum pressures of MME and LBE were 3.97×10^{-7} MPa and 3.33×10^{-7} MPa, respectively. The area of the pressure curve of MME was larger than that of LBE, as shown in Figure 3. For the LBE method, the explosive pressure was directly applied to the element surfaces. For the MME, however, since the explosion load at the origin was transmitted through the atmosphere elements, the residual pressure was transmitted through the atmosphere after the maximum explosion pressure. Therefore, as shown in Figure 3, the pressure curve area of the MME was larger than the pressure curve area of the LBE, although the maximum pressure did not show a large difference. Table 1 compared the duration time of both analysis methods. The LBE method could be regarded as a more efficient explosion analysis method since the analysis time of MME was about 90 times longer than that of LBE. As a result, it is considered that the MME method is suitable for understanding the flow and progress of the explosive pressure, and the LBE method is suitable for understanding the effect of the maximum pressure on the structure under the explosive load. Therefore, in this study, the LBE method was chosen considering that the maximum pressure is similar to MME and it is more efficient in terms of analysis time. Moreover, many research studies showed that the LBE method is more efficient than the MME method considering analysis results and time [59–61].

Table 1. Comparison of analysis time of MME and LBE.

Variables	Number of Elements			Analysis Time	Duration of Analysis
	Structures	Air	TNT		
LBE	800	-	-	129 s	100 ms
MME	800	56075	125	11,075 s	

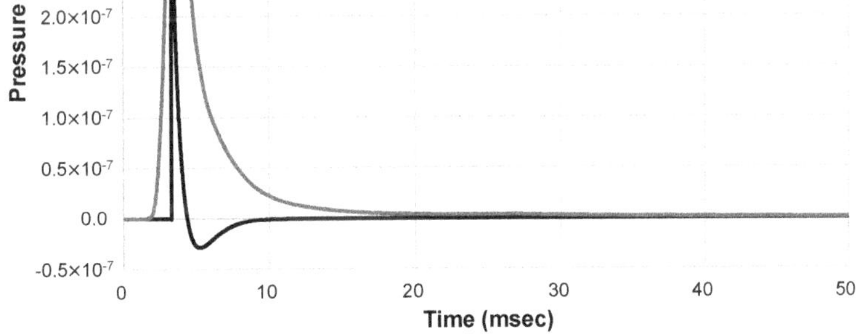

Figure 3. Comparison of pressure histories of MME and LBE.

Based on empirical formulas of blast loads, the Protective Design Center (PDC) of US ARMY releases the Conventional Weapons Effects (ConWep), which could perform a variety of conventional weapons effects from TM 5-855-1 [58,62]. The LS-DYNA applies the ConWep system to LBE. The blast loads of TNTs were located 300 mm from the column and slab. The variables for the amount of TNT were 4 kg and 12 kg. The mass of TNT used in this analysis represents the small and large briefcase bomb, as shown in Table 2 [63]. When 4 kg of TNT was used, which is equal to the amount used in a small briefcase, the blast resistance behavior could be well observed because the slab was not completely destroyed. On the other hand, when 12 kg of TNT was used, which is equal to the amount used in a large briefcase, a significant portion of the slab where the explosive load was placed was destroyed. Table 3 summarized the descriptions of the specimens including blast loads.

Table 2. Typical example of terrorist explosive materials [63].

Explosion Method	Material Type	Loaded Weight
Small briefcase	Military commercial bomb such as TNT	2~4 kg
Large briefcase		4~12 kg
Suitcase		12~22 kg
Bicycle		30 kg

Table 3. Specimen descriptions.

Specimen	Description	Charged Weight of TNT
IC4	Slab-interior—column connection	4 kg
IC12		12 kg
IC-D4	Slab-interior—column connection reinforcing with drop panel	4 kg
IC-D12		12 kg
EC4	Slab-edge—column connection	4 kg
EC12		12 kg
CC4	Slab-corner—column connection	4 kg
CC12		12 kg

4. Analysis Results

From the analysis results, typical forms of pressure distribution were commonly observed in every specimen, as shown in Figure 4. When the explosive load was applied, high compressive forces were generated in the slabs and columns directly affected by the explosion load. Then, the overpressure spread spherically through the slab–column connection. The analysis end time was set to 3000 ms, which is the time at which deformation of all specimens was found stable. As shown in Figure 5a–d, for specimens subjected to 4 kg of TNT, spalling on the rear face of the slab, was severer than that on the front face. These phenomena of pressure development and spalling are quite similar to previous researches [33,43,53]. Looking at the fracture pattern of the specimen under 12 kg of TNT, the part of the slab where the explosion load was placed was completely lost, as shown in Figure 5e–h.

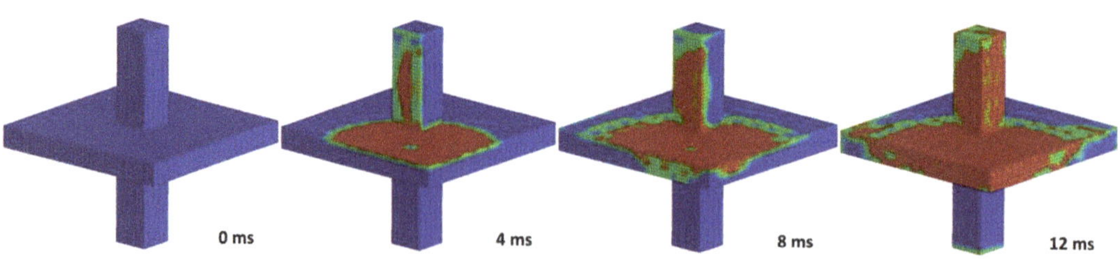

Figure 4. Typical forms of pressure distribution.

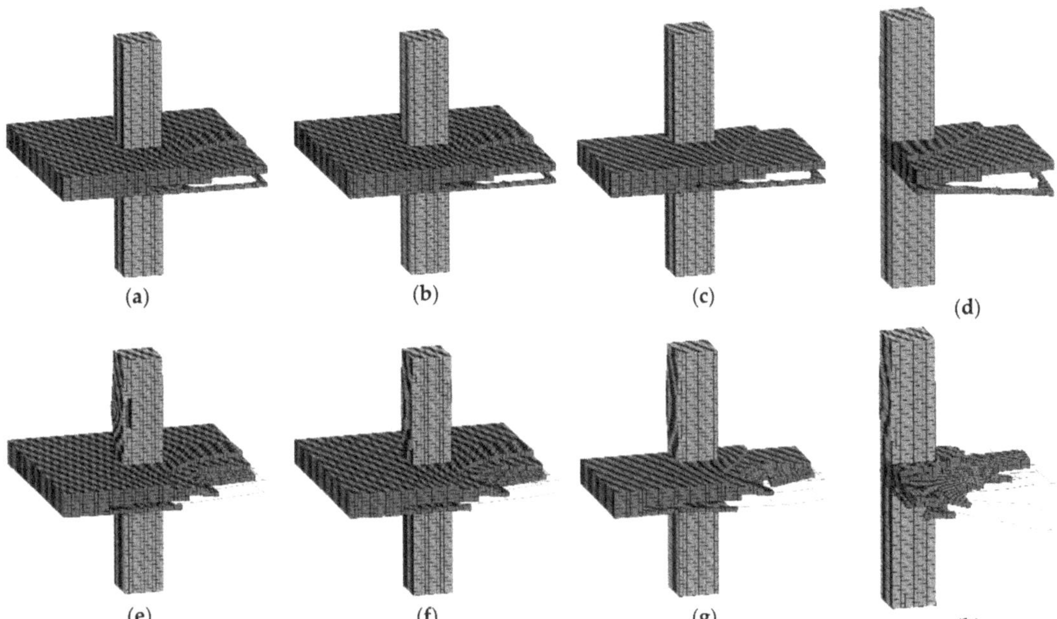

Figure 5. Failure shape of slab–column connections: (**a**) specimen IC4; (**b**) specimen IC-D4; (**c**) specimen EC4; (**d**) specimen CC4; (**e**) specimen IC12; (**f**) specimen IC-D12; (**g**) specimen EC12; (**h**) specimen CC12.

4.1. Slab Deflection

Figure 6 shows the deflection of the slab subjected to TNT 4 kg along the diagonal distance away from the corner of the column. The maximum deflections of the slab occurred similarly in every specimen with TNT 4 kg. The deflection of the slab increased rapidly from the point about 350 mm away from the corner of the column. This point is similar to the point where the fracture of the slab occurred. Figure 7 shows the slab deflections for TNT 12 kg. Large deflections were observed at about 150 mm away from the corner of the column, and beyond that point, fracture of the slab was observed. As the larger explosive load was applied, the fracture area was much larger than that of 4 kg TNT applied specimens.

When comparing the effective deflection of the unbroken part of the slab, it was confirmed that the drop panel slightly reduced deflection. However, for all specimens with TNT of 4 kg and 12 kg, comparing the specimens IC, EC, and CC, the slab deflection according to the position of the column showed no significant difference.

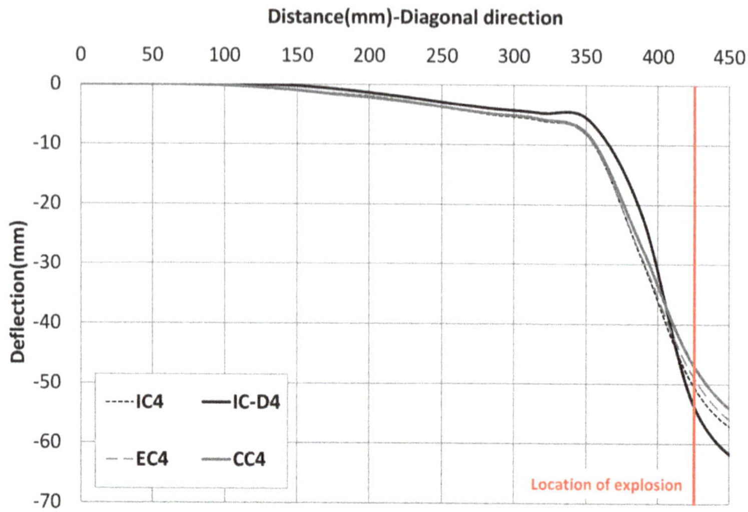

Figure 6. Slab deflection of specimens with TNT 4 kg.

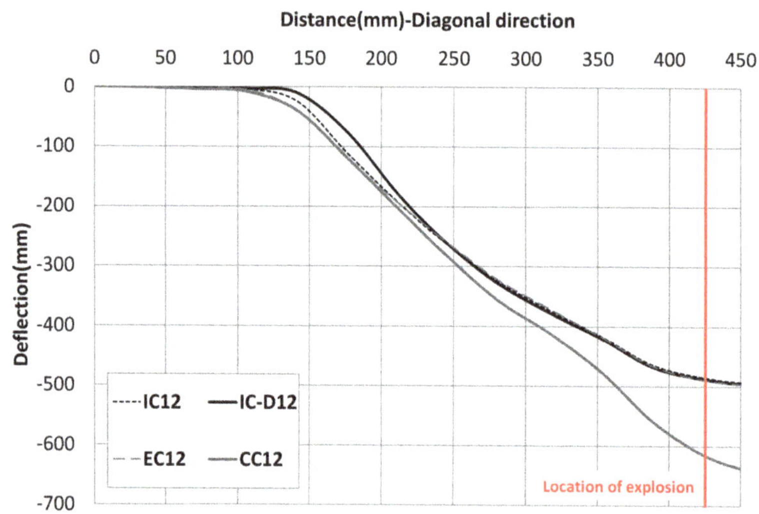

Figure 7. Slab deflection of specimens with TNT 12 kg.

4.2. Critical Section in Slab

In slab–column connections, a section that is 1/2 of the effective depth, d, from the outer surface of the column is regarded as a critical section [45]. Sufficient safety must be ensured for critical sections to prevent the collapse of the entire structure due to large damage of the joints [44,64]. Figure 8 shows failure shapes of slabs for specimens subjected to 12 kg of TNT. For specimens EC and CC, spalling due to the blast load occurred over the critical section, but specimens IC and IC-D showed a more positive structural behavior in which spalling did not spread to the critical section. Table 4 shows the deformation and support rotation at the critical section for all specimens. The deformations in the critical section of the specimens IC-D4 and IC-D12 were 0.040 mm and 3.284 mm, which were the least deformations among the specimens subjected to the same blast load. Figures 9 and 10 show the deflections in the critical section. In Figure 9, the specimens, except for specimen

IC-D4, show similar behavior. The inflection points of deflection curves occurred within the critical sections for all specimens except the specimens reinforced with drop panels. In other words, the safety of the critical section was enhanced by the drop panel. Therefore, the drop panel can be considered as a method to effectively resist blast loads.

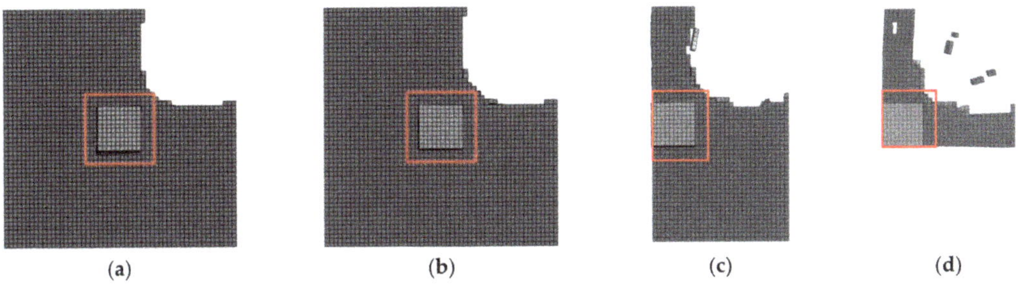

Figure 8. Critical section and slab failure shapes of specimens with TNT 12 kg: (**a**) IC12; (**b**) IC-D12; (**c**) EC12; (**d**) CC12.

Table 4. Maximum deflection and support rotation at critical section.

Specimen	Deflection (mm)	Support Rotation (°)
IC4	0.294	0.159
IC-D4	0.040	0.021
EC4	0.326	0.176
CC4	0.244	0.132
IC12	3.941	1.277
IC-D12	3.284	1.064
EC12	5.190	1.682
CC12	8.297	2.687

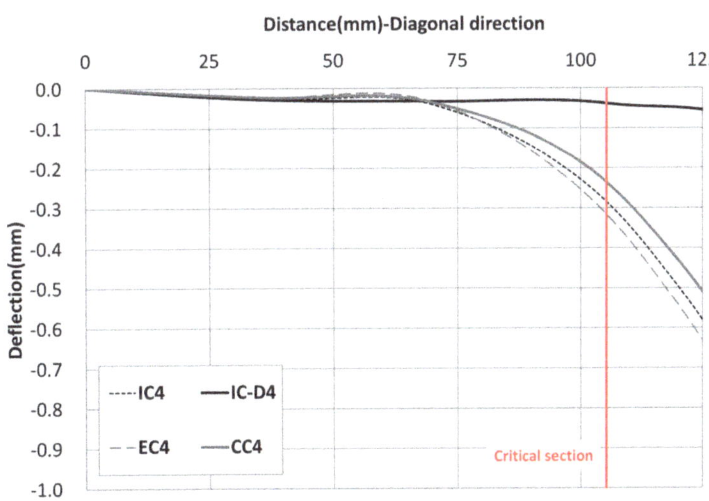

Figure 9. Slab deflection at the critical section with TNT 4 kg.

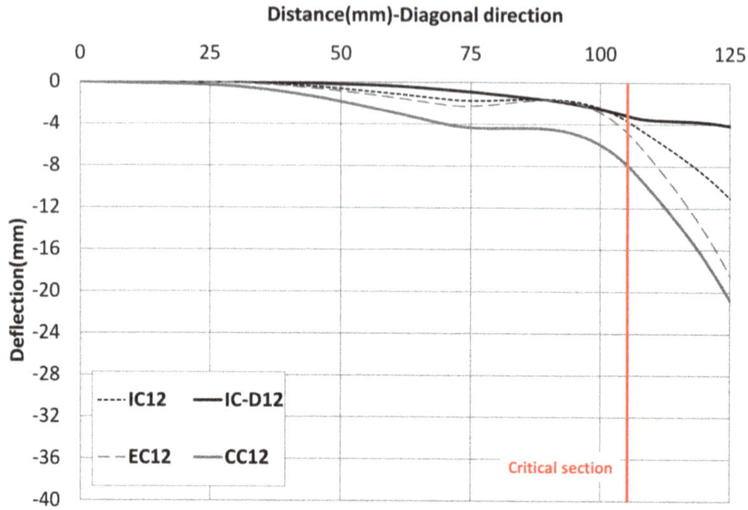

Figure 10. Slab deflection at the critical section with TNT 12 kg.

According to the criteria, the limit of support rotation to effectively resist the moment is two degrees [41,43]. However, according to the analysis results, it was found that this evaluation criterion alone was insufficient to adequately represent the blast-resistant performance of the member. This is because the support rotation in the critical section of all specimens was below the criteria limit of two degrees, but in reality, most specimens showed that the slab was destroyed. Therefore, in order to more accurately and reasonably evaluate blast-resistant performance, it is necessary to consider various evaluation factors in addition to support rotation.

4.3. Steel Stresses in Slab

The peak stresses in the reinforcement of the slab are shown in Table 5. In every specimen with TNT 4 kg and 12 kg, the peak rebar stress occurred near the explosive load. Although a large fracture occurred in the slab, the peak stresses of all reinforcing bars did not reach the maximum strength to which dynamic increased factor was applied. This phenomenon is believed to occur because the concrete is destroyed by the explosive energy at a moment and the energy is not sufficiently transmitted to the reinforcing bar. However, in the previous study, it has been confirmed that the reinforcing bars affect the blast-resistance capacities in the beam–column connections [65].

Table 5. Peak stresses in reinforcements of the slab.

Specimen	Peak Stress (MPa)	
	Top Reinforcing Bar	Bottom Reinforcing Bar
IC4	32.40	157.04
IC-D4	33.93	164.08
EC4	29.52	162.08
CC4	31.17	137.43
IC12	122.62	298.52
IC-D12	128.38	279.72
EC12	176.12	312.99
CC12	440.92	440.48

For most specimens, the stresses of bottom reinforcements of the slab were larger than those of top slab reinforcements. This phenomenon is due to the failure mode in which the rear fracture was more severe than the fracture of the front face to which the explosive load was applied. The effect of the drop panel on the reinforcement stress was not clearly observed, considering that the slab reinforcement stresses of IC and IC-D were similar.

4.4. Column Behavior

Table 6 shows the maximum horizontal deformation of columns. When comparing column behaviors of specimens IC and IC-D, the drop panel was found to be effective in controlling the horizontal displacement of columns under explosive load. Comparing CC, EC, and IC specimens, the larger the constraint of the column by the surrounding slab was, the less horizontal displacement of the column was observed. In the case of CC12, a very large displacement occurred at the end of the analysis, and the column of specimen CC12 seems to have been destroyed as displacement shows a continuous trend of increase. Therefore, it is necessary to pay special attention to the blast-resistance performance for the relatively large explosive load in the case of a corner column having a low constraint by a surrounding slab.

Table 6. Maximum horizontal deformation of the column.

Specimen	Column Deformation (mm)	
	TNT 4 kg	**TNT 12 kg**
IC	0.034	0.104
IC-D	0.033	0.099
EC	0.035	0.171
CC	0.038	12.91

5. Conclusions

The blast resistance of slab–column connection was numerically analyzed. The confinement effect of connection on the blast resistance was investigated through a comparison of the slab-interior column, slab-edge column, and slab-corner column. In addition, the effect of the drop panel on the blast resistance performance was investigated. The conclusions from this numerical study are as follows:

(1) Analysis results showed that the slab-interior column had a better performance than the slab-edge column and slab-corner column in terms of slab failure at critical section and column deformation. The confinement effect seems to be considered in the design of blast-resistant structures. However, the effect of the position of the column on the behavior of the slab such as slab deflection and support rotation under explosive load was not apparent. Further research is needed with the location of the explosive load and the dimensions of columns and slabs as variables.

(2) The drop panel was observed to contribute to the improvement of the blast-resistance performance. For 4 kg and 12 kg of TNT, the drop panel reduced the maximum deflection of the slab at the critical section by approximately 86% and 17%, and the column deformation by approximately 2.9% and 4.8%, respectively.

(3) Although significant concrete fracture occurred in the slab, the maximum stress of the reinforcing bar did not reach the tensile strength. This phenomenon occurs because the concrete is momentarily destroyed by the explosive energy and the energy is not sufficiently transmitted to the reinforcing bars. Further research is needed to ensure that the blast energy can be sufficiently transmitted to the rebar through the concrete.

(4) For most design criteria, the support rotation has been considered as a major criterion for blast-resistant capacities. It is a very simple and good evaluation factor representing the critical behavior of the joint. However, in this study, considerable failure occurred in the slab member even though the support rotations at the critical section were satisfied with the criteria. Therefore, for a more accurate evaluation of blast resistance performance, various evaluation factors such as ductility ratio, reinforcing stress, and concrete fracture area can be considered along with the support rotation.

Author Contributions: Conceptualization, K.M.L. and J.H.L.; data analysis and writing—original draft preparation, K.M.L.; writing—review and editing, K.M.L., T.H.H. and J.H.L.; supervision, T.H.H. and J.H.L. All authors have read and agreed to the published version of the manuscript.

Funding: This research was supported by a grant (21CTAP-C152069-03) from the Technology Advancement Research Program funded by the Ministry of Land, Infrastructure, and Transport of the Korean government.

Acknowledgments: This paper is an extended version of the conference paper presented at the CECAR7 (2016). This research was supported by a grant (21CTAP-C152069-03) from the Technology Advancement Research Program funded by the Ministry of Land, Infrastructure, and Transport of the Korean government.

Conflicts of Interest: The authors declare no conflict of interest.

References

1. Brun, M.; Batti, A.; Limam, A.; Gravouil, A. Explicit/implicit multi-time step co-computations for blast analyses on a reinforced concrete frame structure. *Finite Elem. Anal. Des.* **2012**, *52*, 41–59. [CrossRef]
2. Hajek, R.; Fladr, J.; Pachman, J.; Stoller, J.; Foglar, M. An experimental evaluation of the blast resistance of heterogeneous concrete-based composite bridge decks. *Eng. Struct.* **2018**, *179*, 204–210. [CrossRef]
3. Kumar, V.; Kartik, K.; Iqbal, M. Experimental and numerical investigation of reinforced concrete slabs under blast loading. *Eng. Struct.* **2020**, *206*, 110125. [CrossRef]
4. Abdulsamee, M.H.; Yazan, B.A.T.; Amin, H.A.; George, Z.V. The effect of shape memory alloys on the ductility of exterior reinforced concrete beam-column joints using the damage plasticity model. *Eng. Struct.* **2019**, *200*, 109676.
5. Cumhur, C.; Ahmet, M.T.; Atakan, M.; Turgay, C.; Guven, K. Experimental behavior and failure of beam-column joints with plain bars, low-strength concrete and different anchorage details. *Eng. Fail. Anal.* **2020**, *109*, 104247.
6. Yim, H.C.; Krauthammer, T. Load-impulse characterization for steel connections in monolithic reinforced concrete structures. *Int. J. Impact Eng.* **2009**, *36*, 737–745. [CrossRef]
7. Andrew, R.; Nicola, B.; Giuseppe, C.; Stefano, D.M.; Gianluca, I.; Sonia, M.; Elio, S.; Sara, S.; Gabriel, T. Full scale experimental tests and numerical model validation of reinforced concrete slab subjected to direct contact explosion. *Int. J. Impact Eng.* **2019**, *132*, 103309.
8. Azer, M.; Stijn, M.; Bachir, B.; David, L.; John, V. Blast response of retrofitted reinforced concrete hollow core slabs under a close distance explosion. *Eng. Struct.* **2019**, *191*, 447–459.
9. Bianchini, A.C.; Woods, R.E.; Kesler, C.E. Effect of floor concrete strength on column strength. *ACI J. Proc.* **1960**, *56*, 1149–1170.
10. McHarg, P.J.; Cook, W.D.; Mitchell, D.; Yoon, Y.S. Benefits of concentrated slab reinforcement and steel fibers on performance of slab-column connections. *ACI Struct. J.* **2000**, *97*, 225–234.
11. Lee, J.-H.; Yoon, Y.-S.; Cook, W.D.; Mitchell, D. Benefits of Using Puddled HSC with Fibers in Slabs to Transmit HSC Column Loads. *J. Struct. Eng.* **2007**, *133*, 1843–1847. [CrossRef]
12. Morrison, D.G.; Hirasawa, I.; Sozen, M.A. Lateral-load tests of R/C slab-column connections. *J. Struct. Eng.* **1983**, *109*, 2698–2714. [CrossRef]
13. Marzouk, H.; Osman, M.; Hussein, A. Cyclic Loading of High-Strength Lightweight Concrete Slabs. *ACI Struct. J.* **2001**, *98*, 207–214.
14. Tian, Y.; Jirsa, J.O.; Bayrak, O.; Widianto; Argudo, J.F. Behavior of Slab-Column Connections of Existing Flat-Plate Structures. *ACI Struct. J.* **2008**, *105*, 561. [CrossRef]
15. Drakatos, I.-S.; Muttoni, A.; Beyer, K. Internal slab-column connections under monotonic and cyclic imposed rotations. *Eng. Struct.* **2016**, *123*, 501–516. [CrossRef]
16. Isufi, B.; Rossi, M.; Ramos, A.P. Influence of flexural reinforcement on the seismic performance of flat slab–column connections. *Eng. Struct.* **2021**, *242*, 112583. [CrossRef]
17. Emam, M.; Marzouk, H.; Hilal, M.S. Seismic response of slab-column connections constructed with high-strength concrete. *ACI Struct. J.* **1997**, *94*, 197–205.
18. Marzouk, H.; Emam, M.; Hilal, M.S. Effect of high-strength concrete slab on the behavior of slab-column connections. *ACI Struct. J.* **1998**, *95*, 227–237.
19. Inácio, M.; Isufi, B.; Lapi, M.; Ramos, A.P. Rational use of high-strength concrete in flat slab-column connections under seismic loading. *ACI Struct. J.* **2020**, *117*, 297–310. [CrossRef]
20. Smadi, M.; Yasin, I.B. Behavior of high-strength fibrous concrete slab–column connections under gravity and lateral loads. *Constr. Build. Mater.* **2008**, *22*, 1863–1873. [CrossRef]
21. Scotta, R.; Giorgi, P. Comparative cyclic tests of exterior flat slab–column connections in normal concrete and fiber-reinforced lightweight aggregate concrete. *Mater. Struct.* **2015**, *49*, 4049–4067. [CrossRef]
22. Buitrago, M.; Bertolesi, E.; Sagaseta, J.; Calderón, P.A.; Adam, J.M. Robustness of RC building structures with infill masonry walls: Tests on a purpose-built structure. *Eng. Struct.* **2020**, *226*, 111384. [CrossRef]

23. Adam, J.M.; Buitrago, M.; Bertolesi, E.; Sagaseta, J.; Moragues, J.J. Dynamic performance of a real-scale reinforced concrete building test under a corner-column failure scenario. *Eng. Struct.* **2020**, *210*, 110414. [CrossRef]
24. Mousapoor, E.; Ghiasi, V.; Madandoust, R. Macro modeling of slab-column connections in progressive collapse with post-punching effect. In *Structures*; Elsevier: Amsterdam, The Netherlands, 2020. [CrossRef]
25. Yankelevsky, D.; Karinski, Y.; Feldgun, V. Dynamic punching shear failure of a RC flat slab-column connection under a collapsing slab impact. *Int. J. Impact Eng.* **2019**, *135*, 103401. [CrossRef]
26. Adam, J.M.; Parisi, F.; Sagaseta, J.; Lu, X. Research and practice on progressive collapse and robustness of building structures in the 21st century. *Eng. Struct.* **2018**, *173*, 122–149. [CrossRef]
27. Setiawan, A.; Vollum, R.L.; Macorini, L. Numerical and analytical investigation of internal slab-column connections subject to cyclic loading. *Eng. Struct.* **2019**, *184*, 535–554. [CrossRef]
28. Silva, F.P.; Lu, B. Improving the blast resistance capacity of RC slabs with innovative composite materials. *Compos. Part B Eng.* **2007**, *38*, 523–534. [CrossRef]
29. Schenker, A.; Anteby, I.; Gal, E.; Kivity, Y.; Nizri, E.; Sadot, O.; Michaelis, R.; Levintant, O.; Ben-Dor, G. Full-scale field tests of concrete slabs subjected to blast loads. *Int. J. Impact Eng.* **2008**, *35*, 184–198. [CrossRef]
30. Wu, C.; Oehlers, D.; Rebentrost, M.; Leach, J.; Whittaker, A. Blast testing of ultra-high performance fibre and FRP-retrofitted concrete slabs. *Eng. Struct.* **2009**, *31*, 2060–2069. [CrossRef]
31. Alonso, M.G.; Cendon, A.D.; Galvez, F.; Erice, B.; Galvez, S.V. Blast response analysis of reinforced concrete slabs: Experimental procedure and numerical simulation. *J. Appl. Mech.* **2011**, *78*, 1–12.
32. Wang, W.; Zhang, D.; Lu, F.; Wang, S.-C.; Tang, F. Experimental study and numerical simulation of the damage mode of a square reinforced concrete slab under close-in explosion. *Eng. Fail. Anal.* **2012**, *27*, 41–51. [CrossRef]
33. Thiagarajan, G.; Kadambi, V.A.; Robert, S.; Johnson, F.C. Experimental and finite element analysis of doubly reinforced concrete slabs subjected to blast loads. *Int. J. Impact Eng.* **2015**, *75*, 162–173. [CrossRef]
34. Li, J.; Wu, C.; Hao, H. Investigation of ultra-high performance concrete slab and normal strength concrete slab under contact explosion. *Eng. Struct.* **2015**, *102*, 395–408. [CrossRef]
35. Castedo, R.; Segarra, P.; Alanon, A.; Lopez, M.L.; Santos, P.A.; Sanchidrian, A.J. Air blast resistance of full-scale slabs with different compositions: Numerical modeling and field validation. *Int. J. Impact Eng.* **2015**, *86*, 145–156. [CrossRef]
36. Yao, S.; Zhang, D.; Chen, X.; Lu, F.; Wang, W. Experimental and numerical study on the dynamic response of RC slabs under blast loading. *Eng. Fail. Anal.* **2016**, *66*, 120–129. [CrossRef]
37. Zhou, Q.X.; Kuznetsov, A.V.; Hao, H.; Waschl, J. Numerical prediction of concrete slab response to blast loading. *Int. J. Impact Eng.* **2008**, *35*, 1186–1200. [CrossRef]
38. Shahriari, A.; Birzhandi, M.S.; Zafarani, M.M. Seismic behavior, blast response and progressive collapse of RC structures equipped with viscoelastic dampers. *Soil Dyn. Earthq. Eng.* **2021**, *143*, 106643. [CrossRef]
39. Krauthammer, T. Blast-resistant structural concrete and steel connections. *Int. J. Impact Eng.* **1999**, *22*, 887–910. [CrossRef]
40. Lim, K.M.; Yoon, Y.S.; Lee, J.H. Numerical analysis on dynamic behavior of slab-column connections subjected to blast loads. In Proceedings of the 7th Civil Engineering Conference in the Asian Region (CECAR 7), Waikiki, HI, USA, 30 August–2 September 2016.
41. ASCE/SEI (American Society of Civil Engineers/Structural Engineering Institute). *ASCE/SEI 59-11: Blast Protection of Building*; ASCE: Reston, VA, USA, 2011.
42. Pereira, J.M.; Dias, A.; Lourenco, P.B. Dynamic properties of clay brick at different stain rates. In Proceedings of the 12th Canadian Masonry Symposium, Vancouver, BC, Canada, 2–5 June 2013.
43. Department of Defense. *UFC 3-340-02: Structures to Resist the Effects of Accidental Explosions*; Department of Defense: Arlington, VA, USA, 2008.
44. ACI (American Concrete Institute) Committee 318. *ACI318-11: Building Code Requirements for Structural Concrete & Commentary*; American Concrete Institute: Farmington Hills, MI, USA, 2011.
45. ACI-ASCE (American Concrete Institute-American Society of Civil Engineers) Committee 352. *ACI352.1R-11: Guide for Design of Slab-Column Connections in Monolithic Concrete Structures*; American Concrete Institute: Farmington Hills, MI, USA, 2012.
46. Gamble, W.L.; Klinar, J.D. Tests of High?Strength Concrete Columns with Intervening Floor Slabs. *J. Struct. Eng.* **1991**, *117*, 1462–1476. [CrossRef]
47. Brannon, R.M.; Leelavanichkul, S. *Survey of Four Damage Models for Concrete*; SAND2009-5544; Sandia National Laboratories: Albuquerque, NM, USA, 2009.
48. Crawford, J.E.; Wu, Y.; Choi, H.J.; Magallanes, J.M.; Lan, S. *Use and Validation of the Release III K&C Concrete Material Model in LS-DYNA*; Karagozian & Case Technical Report TR-11-36.5; Karagozian & Case, Inc.: Glendale, CA, USA, 2012.
49. LSTC (Livermore Software Technology Corporation). *LS-DYNA Keyword User's Manual Volume II Material Models*; Livermore Software Technology Corporation: Livermore, CA, USA, 2013.
50. Ling, L. Local Damages and Blast Resistance of RC Slabs Subjected to Contact Detonation. Master's Thesis, Korea University, Seoul, Korea, 2013.
51. Gonzalez, H.A.; Zapatero, J. Influence of minimum element size to determine crack closure stress by the finite element method. *Eng. Fract. Mech.* **2005**, *72*, 337–355. [CrossRef]

52. Krauthammer, T.; Otani, R. Mesh, gravity and load effects on finite element simulations of blast loaded reinforced concrete structures. *Comput. Struct.* **1997**, *63*, 1113–1120. [CrossRef]
53. Foglar, M.; Kovar, M. Conclusions from experimental testing of blast resistance of FRC and RC bridge decks. *Int. J. Impact Eng.* **2013**, *59*, 18–28. [CrossRef]
54. Shi, Y.; Stewart, M.G. Damage and risk assessment for reinforced concrete wall pannels subjected to explosive blast loading. *Int. J. Impact Eng.* **2015**, *85*, 5–19. [CrossRef]
55. Guang, Z.; Hong, N.; Jinbao, C.; Chuanzhi, C.; Heow, P.L. Dynamic analysis of lunar lander during soft landing using explicit finite element method. *Acta Astronaut.* **2018**, *148*, 69–81.
56. Kang, J.J.; Oh, S.I. Hourglass control in rigid-plastic finite element analysis. *Korean Soc. Mech. Eng.* **1996**, *20*, 1290–1300.
57. Kim, K.S.; Cho, J.U.; Choi, D.S. An analysis of plastic stress in square bar impacting plat. *Trans. Korean Soc. Automot. Eng.* **2004**, *12*, 198–204.
58. LSTC (Livermore Software Technology Corporation). *LS-DYNA Keyword User's Manual Volume I*; Livermore Software Technology Corporation: Livermore, CA, USA, 2013.
59. Bae, D.-M.; Zakki, A. Comparisons of Multi Material ALE and Single Material ALE in LS-DYNA for Estimation of Acceleration Response of Free-fall Lifeboat. *J. Soc. Nav. Arch. Korea* **2011**, *48*, 552–559. [CrossRef]
60. Jang, I.H. Sloshing Response Analysis of LNG Carrier Tank Using Fluid Structure Interaction Analysis Technique of LS-DYNA3D. Master's Thesis, Korea Maritime and Ocean University, Busan, Korea, 2007.
61. Zahra, S.T.; Jeffery, S.V. A comparison between three different blast methods in LS-DYNA: LBE, MM-ALE, Coupling of LBE and MM-ALE. In Proceedings of the 12th International LS-DYNA User Conference, Dearborn, MI, USA, 3–5 June 2012.
62. Huh, Y.C.; Chung, T.Y.; Kim, K.C.; Jung, H.J.; Choi, H.H. A study on the modeling techniques of air blast load using LS-DYNA. In Proceedings of the Korean Society for Noise and Vibration Engineering Conference, Daegu, Korea, 26–27 April 2012; pp. 375–376.
63. Yandzio, E.; Gough, M. *Protection of Buildings against Explosions*; The Steel Construction Institute: Ascot, UK, 1999.
64. Krauthammer, T. *Modern Protective Structures*; CRC Press: Boca Raton, FL, USA, 2008.
65. Lim, K.-M.; Shin, H.; Kim, D.-J.; Yoon, Y.-S.; Lee, J.-H. Numerical Assessment of Reinforcing Details in Beam-Column Joints on Blast Resistance. *Int. J. Concr. Struct. Mater.* **2016**, *10*, 87–96. [CrossRef]

applied sciences

MDPI

Article

Corner Convergence Effect of Enclosed Blast Shock Wave and High-Pressure Range

Xudong Li, Haojie Chen, Jianping Yin * and Zhijun Wang

College of Electromechanical Engineering, North University of China, Taiyuan 030051, China
* Correspondence: yjp123@nuc.edu.cn; Tel.: +86-139-9420-8931

Abstract: An explosion inside a cabin will converge at the corners to form high-pressure areas, significantly impacting the destruction of a bulkhead structure. This paper investigates shock wave convergence characteristics at the corners when the explosive detonates at the center of the cabin, based on a combination of the wall reflection law for shock waves and a numerical simulation method. The parameter K represents the aspect ratio of the cabin structure. This study shows that when $1 \leq K \leq 1.19$, the high pressure at the corner is caused by the superposition of Mach waves along both wall surfaces. However, for the initial shock wave, when $1.2 < K \leq 2$, the high pressure is caused by the superposition of Mach waves along the longer wall surface and regular reflected waves on the shorter wall surface; when $2 < K$, the cause are Mach waves along the longer wall surface and the corresponding positive reflection on the shorter wall surface. The influence of K on the range for the high-pressure region at the corner is also analyzed, the functional relationship between the range of the high-pressure area and K is given, and the universality is verified.

Keywords: internal explosion; shock wave; corner; structural dimensions; Mach waves

Citation: Li, X.; Chen, H.; Yin, J.; Wang, Z. Corner Convergence Effect of Enclosed Blast Shock Wave and High-Pressure Range. *Appl. Sci.* **2022**, *12*, 11341. https://doi.org/10.3390/app122211341

Academic Editor: Ricardo Castedo

Received: 30 October 2022
Accepted: 6 November 2022
Published: 8 November 2022

1. Introduction

Internal explosions can cause more significant destruction to structures than air explosions due to the combined effect of the reflection, superposition, and convergence of shock waves [1]. High-pressure shock waves can cause structural damage [2]. The pressure peak resulting from the superposition and convergence effect of a shock wave at the corner of a cabin during implosion is significantly higher than the reflected shock wave at the same distance from the wall in an open environment, which can cause a local tear in the corner of the cabin structure first and then expands to the destruction of the entire bulkhead. Therefore, studying the convergence effect of the shock wave at the corner of the cabin during implosion is critical. It is necessary to understand the high-pressure formation rules and the factors influencing the convergence effect at the corner of the implosion shock wave to guide the design of the protection of the cabin structure against internal explosion, and it is also of importance for shock-wave experiments to determine the Hugoniot and melting curves of metals [3,4].

Explosions inside chambers have been a hot topic of research [5–9]. There are a few specific reports on internal blast wave loading [10]. Shock waves have a significant convergence effect at the corners under internal blast conditions [11]. A combined experimental and numerical simulation study [12,13] of the characteristics and typical destruction modes of cabin structures under implosion loads showed that the intensity of the converging shock waves at the corner of two-wall and three-wall surfaces was, respectively, 5- and 12-times greater than the reflected shock waves on the same region of the wall and that the primary failure mechanism of the bulkhead structure during the implosion of the cabin was tearing failure along the corner. Another numerical simulation study of the load situation under implosion conditions showed that the peak pressure at the corner of the three-wall surface was 9–12 times greater than the pressure at the center of the bulkhead, and the peak pressure at the corner of the two-wall surface was 3–5 times greater than the pressure at the center of the bulkhead [14]. An experimental study measuring the pressure at the corner of a two-wall surface and the peak pressure at the center of the bulkhead showed that the peak pressure at the two walls was smaller than the pressure at the center of the bulkhead [15]. Another study measuring the peak value of the shock wave for three-corner structures (flat plate, concave plate, and convex plate transition connections) at different doses and the peak value of the initial shock wave showed that the corner structure could retard the convergence effect of shock waves at a low dosage [16]; however, when the dosage was higher, the corner structure did not significantly retard the convergence effect. The maximum ratio of the corner converging shock waves to the initial shock waves was 1.24. In addition, another study was reported using an imaging method to explain the convergence effect of corner shock waves [17], whose angle of incidence was the same as the angle of reflection. The actual reflection of the shock wave in the cabin includes the regular oblique reflection and Mach reflection, which was also considered significant. The analysis confirmed the convergence effect of the shock waves at the corner. However, compared to the studies mentioned above, there was a difference between the peak pressures of the convergence of the shock wave at the corner. The study further suggests that the corner convergence occurs at a specific corner area and that the difference in the results is owing to differences in experimental and simulation measurement points. Therefore, it is necessary to investigate the problem of defining the corner convergence area. In addition, the formation of high-pressure areas at the corners and the associated factors have not yet been determined and must be studied in detail.

The characteristics of the explosion load in the enclosed space depend mainly on the spatial dimensions of the structure [18,19]. This paper determines the causes of the convergence phenomenon of shock waves at the corner by 2D cross-sectional analysis with aspect ratio variation. Furthermore, the peak pressure contour map of the corner area at different aspect ratios was plotted through extensive simulation calculations to make a preliminary determination of the range of the high-pressure area at the corner and obtain the functional relationship between the high-pressure area and the size of the structure.

The research in this paper is based on the following three points:

(1) The explosives are in the center of the cabin;

(2) The structure is assumed to be a rigid wall;

(3) The focus is on the peak of the shock wave only.

2. Simulation Model

2.1. Model Design

When the explosive detonates in the center of the cabin, an arbitrary surface is chosen through the location of the explosion point to intersect, giving the 2D rectangular cross-sectional diagram shown in Figure 1. The dual study of the spread of shock waves on the cross-section and the convergence effect at the corners simplifies the calculation and has a general character. Another study [20] adopted the same method in determining the influential factors for implosion loads.

Figure 2 shows the 2D schematic diagram used in the simulation model. Set wall A as the long side and wall B as the short side. The convergence phenomenon at the corners is studied on half of the cross-section, where a and b are halves of the long and short sides, respectively. The red line area in Figure 2 is the corner area, a square with side length b, and the angle between the shock front and the wall surface is Φ. Each side of the corner area is equally divided into 10 parts, and pressure measurement points are set at the intersections, giving a total of 121 side points arranged as shown in Figure 3.

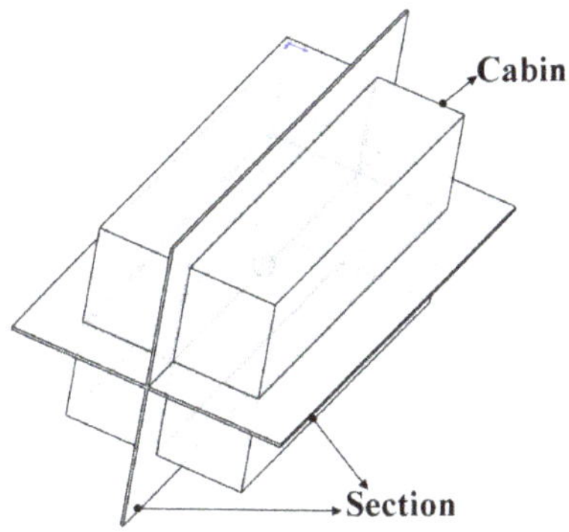

Figure 1. Cross-section through the center of the cabin.

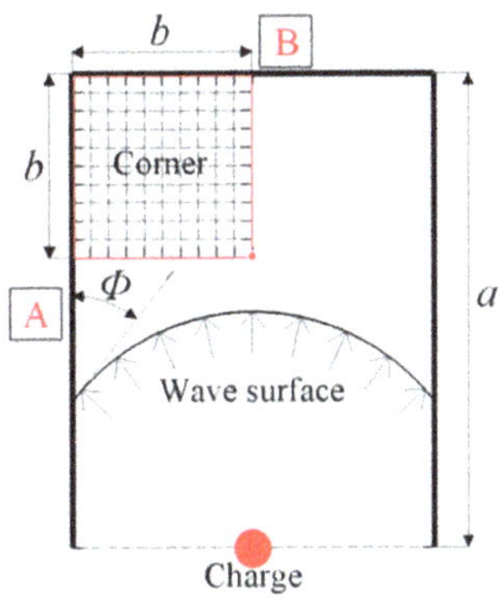

Figure 2. Two-dimensional schematic diagram of the calculation model.

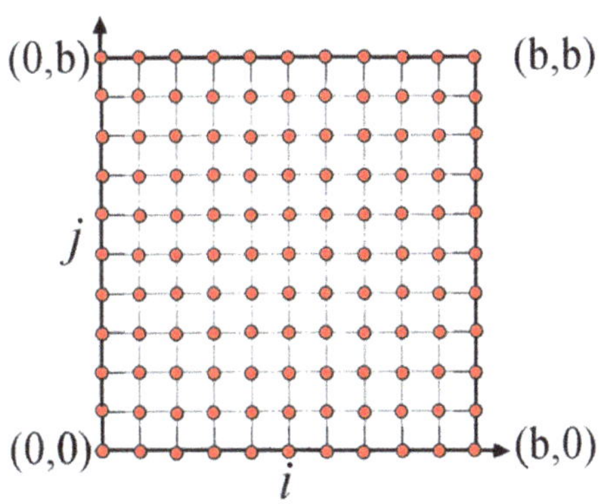

Figure 3. Coordinate distribution of measurement points.

The simulation model was run using AUTODYN-2D. The eulerian unit was used for air and the explosives were packed into the air unit. The initial rigid boundary conditions in AUTODYN are adopted for the air boundary to establish a 2D symmetrical model, as shown in Figure 4. The finite element model uses 0.5 mm × 0.5 mm mesh. Simulations were carried out using 0.5 mm, 1 mm, 2 mm, and 4 mm meshes for the shock wave of a

100 g charge at 1 m, indicating that the simulation results converged when the mesh size was 0.5 mm, as shown in Figure 5. The ideal gas equation of state is used for air:

$$P = (\gamma - 1)\rho e \tag{1}$$

where γ, ρ, and e are the specific heat capacity, density, and internal energy of the air, respectively, and the values used for the simulation are $\gamma = 1.4$, $\rho = 1.225 \times 10^{-3}$ g/cm^3, and $e = 2.068 \times 10^5$ J.

Figure 4. Finite element simulation model diagram.

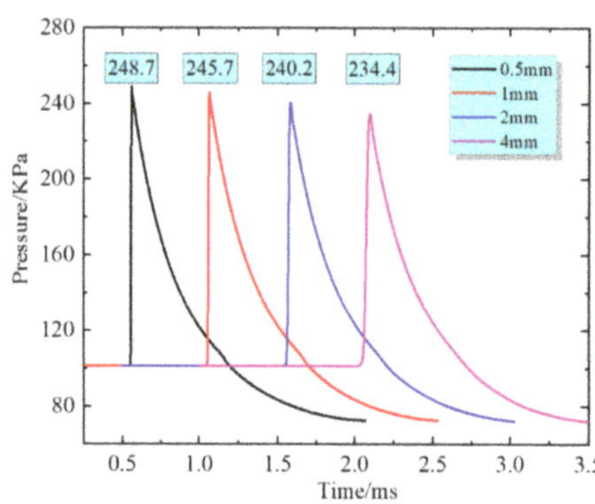

Figure 5. Grid convergence checking.

The Jones–Wilkins–Lee (JWL) equation of state is used for the explosive:

$$P_T = C_1\left(1 - \frac{\omega}{r_1 v}\right)e^{-r_1 v} + C_2\left(1 - \frac{\omega}{r_2 v}\right)e^{-r_2 v} + \frac{\omega e}{v} \tag{2}$$

where C_1, C_2, r_1, r_2, and ω are constants, P_T, v, and e are the pressure, relative volume, and initial energy, respectively. The specific parameters of trinitrotoluene (TNT) are shown in Table 1.

Table 1. Parameters of TNT in the JWL equation of state.

Density, ρ (kg/m^3)	Detonation Velocity, D (m/s)	C-J Pressure (Pa)	C_1 (Pa)
1630	6800	2.10×10^{10}	3.74×10^{11}
C_2	r_1	r_2	ω
3.75×10^9	4.15	0.9	0.35

2.2. Simulation Model Verification

A related study by Isabelle Sochet [21] investigated an explosion in a partially confined space under different boundary conditions using the 0.106 g equivalent of TNT using gas explosives and obtained a time-history curve of shock wave pressure at each measurement point. The experimental arrangement diagram is shown in Figure 6. This paper uses some of these experimental results to verify the simulation model. The simulation determines the time-history curve of the pressure at measurement points A, B, and C when only one, two, and three walls are available. The model parameters and grid size used in the simulation are identical to those used in Section 2.1. A comparison of the simulation results with the experimental results is shown in Figure 7.

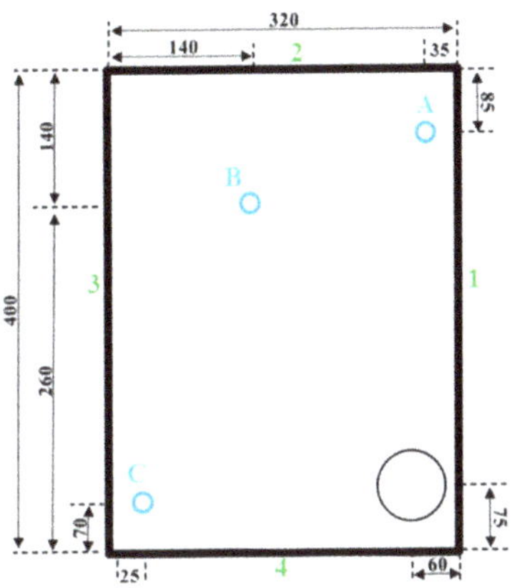

Figure 6. Isabelle Sochet experimental layout diagram.

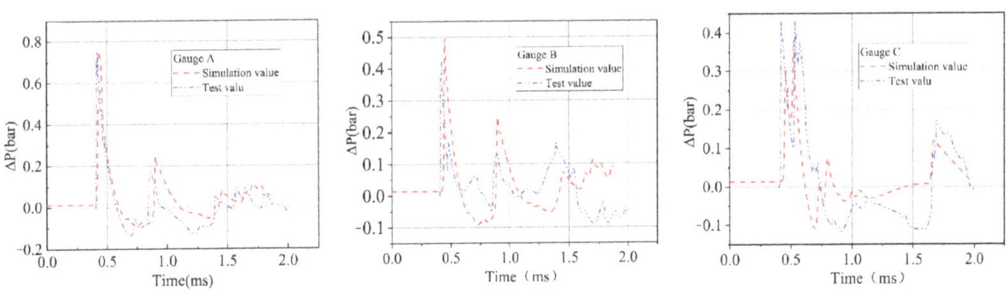

Figure 7. Comparison of numerical simulations with experimental results.

From the comparison of the simulation model and experimental results, the pressure peak and the curve change trend are generally consistent, the pressure peak error at B is larger, and the maximum error is 12% which is within the acceptable range and thus, can verify the reliability of the simulation model.

2.3. Simulation Working Arrangement

The parameter K represents the ratio of the half of the long side, a, to the half of the short side, b, in Figure 2, viz., K = a/b, which is a dimensionless number used to represent the change in size of the structure. In this study, K is in the range of 1 to 5, and the particular values 500 mm and 1000 mm are used for b. Furthermore, the explosive equivalents of 100 g, 200 g, 500 g, and 1000 g TNT are used. Table 2 gives the specific working conditions.

Table 2. Simulation working conditions.

Serial Number	K	a (mm)	b (mm)	W (g)
1	1	500	500	
2	1.2	600	500	
3	1.4	700	500	
4	1.6	800	500	
5	1.8	900	500	
6	2.0	1000	500	
7	2.2	1100	500	
8	2.4	1200	500	100
9	2.6	1300	500	
10	2.8	1400	500	
11	3.0	1500	500	
12	3.4	1700	500	
13	4	2000	500	
14	4.4	2200	500	
15	5	2500	500	
16	1.1	550	500	
17	1.3	650	500	
18	1.5	750	500	
19	1.7	850	500	
20	1.9	950	500	
21	2.1	1050	500	200
22	2.3	1150	500	
23	2.5	1250	500	
24	2.7	1350	500	
25	2.9	1450	500	
26	1.1	1100	1000	
27	1.3	1300	1000	
28	1.5	1500	1000	
29	1.7	1700	1000	
30	1.9	1900	1000	
31	2.1	2100	1000	1000
32	2.3	2300	1000	
33	2.5	2500	1000	
34	2.7	2700	1000	
35	2.9	2900	1000	

3. Mechanism of High-Pressure Formation at Corners

3.1. Theoretical Analysis of Convergence Effects at Corners

The spread of an explosive shock wave inside the cabin is complex, characterized by multiple reflections and superpositions, and follows the wall reflection principle. The shock wave reflection at the wall comprises positive and oblique reflections, with the oblique reflections including both regular and Mach reflections [22]. Figure 8 is a schematic diagram of the wall reflection during an air explosion, where d is the vertical distance from the explosive to the wall, c is the distance from the projection point of the explosive on the wall to the intersection of the shock wave front and the wall, Φ is the angle of incidence of the shock wave on the wall, and θ is the included angle in the vertical direction between the shock wave front and the wall intersection line. The geometric relationship shows that $\theta = \Phi$, thus $\tan\theta = c/d$, $c = d\tan\theta = d\tan\Phi$, and the Mach angle tends to a limiting value of 39.97° [23]. Therefore, when $c/d \geq 0.838$, Mach reflection occurs.

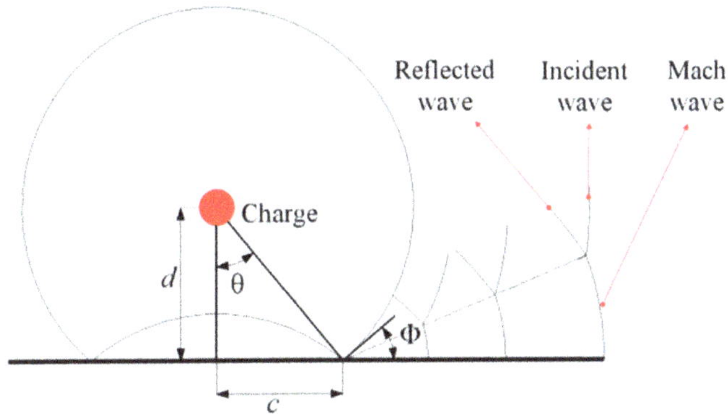

Figure 8. Schematic diagram of shock wave wall reflection.

In analogy to Figure 2, the distance from the explosive to bulkhead A is b, and the distance from the explosive to bulkhead B is a. When $K = a/b = 1 > 0.838$, the reflected shock waves on walls A and B will form Mach reflections before they reach the point (0, b). At point (0, b), the initial shock wave and the Mach reflected waves from walls A and B converge, forming a converging wave. When $K > 1$, the reflected waves from wall A also form Mach waves before they reach the point (0, b). However, for the reflected waves from wall B, the Mach reflection is formed when $b/a \geq 0.838$, i.e., when $a/b \leq 1/0.838 = 1.193$. Thus, when $1 \leq K \leq 1.193$, the initial shock wave at point (0, b) and the Mach reflected waves from walls A and B converge to form a high-pressure region. When $1.193 < K$, there is no Mach reflection on wall B. Therefore, the convergence at the point (0, b) is owing to the initial shock wave, the Mach reflection wave from wall A, and the regular reflection wave from wall B. The Mach wavefront gradually widens during its spread, as shown in Figure 8, thus a value exists for n. When $K \geq n$, the Mach reflection wave along wall A reaches point (0, b) first, while the initial shock wave superimposes with the Mach wave from wall A in its spread towards point (0, b) and spreads along the three-wave line to wall B without converging at the corner. The simulations in the next section were used to verify the above inference and determine the value of n.

3.2. Simulation of Convergence Effects at Corners

Several simulations were conducted with variations in K, as shown in Table 2. As the spread of waves is mainly related to the size of the structure, the convergence clouds of waves at the corner for K = 1, 2, 3, 4, and 5 are exemplified in cases when b = 500 mm and W = 100 g of explosive, as shown in Figure 9.

As observed in the diagram, when K = 1, the converging waves at the corner from the waves reflected at walls A and B and the initial shock waves do not form a noticeable Mach rod phenomenon owing to the short distance between the walls; however, as K increases, the Mach wave on the surface of wall A gradually widens, and a clear Mach rod is observable. As the Mach wave speed is faster than the initial shock wave speed, its wavefront surface gradually flushes with the initial shock wave and surpasses it. Therefore, the high pressure formed near the corner (0, b) comes from the reflection of the Mach wave at the surface of wall B. The above deductions support the theoretical analysis in Section 3.1. For the value of n given in Section 3.1, the simulation shows that when $K \geq 2$, the high pressure at the corner (0, b) mainly comes from the positive reflection of Mach waves from wall A to wall B. An example of the spread of the Mach wave and the initial shock wave, when K = 2.4, is shown in Figure 10, where the black dashed line depicts the three-wave line. The high pressure at the corner (0, b) is formed by the positive reflection of the Mach wave.

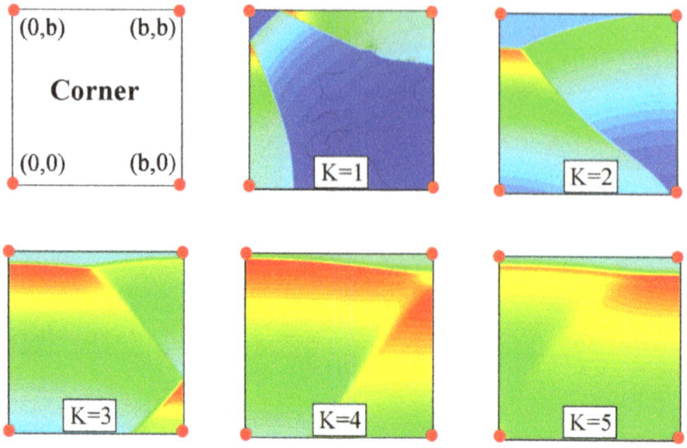

Figure 9. The cloud diagram of the convergence effect of shock waves at the corner at different K values when *b* = 500 mm and *W* = 100 g.

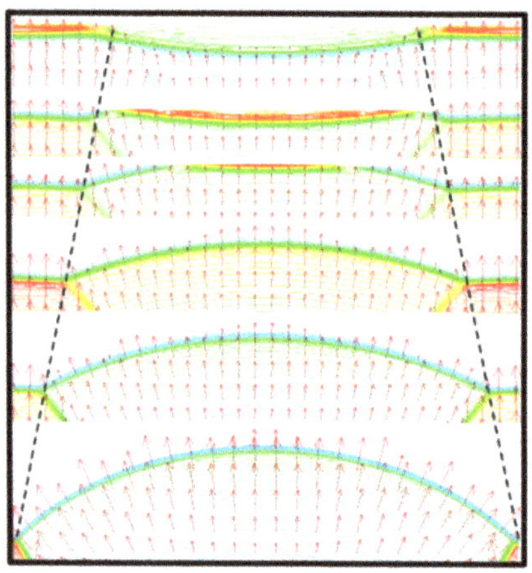

Figure 10. The formation process of the convergence effect of shock waves at the corner when K = 2.4.

In previous experiments [2–5], the peak pressure at points (0, b) and (b, b) was frequently used for comparison. The relationship between the peak pressure at the two measuring points and K when b = 500 mm and W = 100 g and the peak pressure ratio at (0, b) and (b, b) is shown in Figure 11. Figure 11 shows that the value for P (b, b)/P (0, b) tends to be stable when K is in the range of 1 to 2.5, but as K increases, the value for P (b, b)/P (0, b) drops linearly, indicating that the corner convergence effect is weakening.

To better demonstrate the changes in the corner high-pressure area with K, the peak pressure distribution in the corner at b = 500 mm and W = 100 g when K = 2, K = 3, K = 4, and K = 5 is shown in Figure 12. The graph shows that K significantly influences the corner's high-pressure area. As K increases, the high-pressure area at the corner gradually expands and moves towards the vicinity of the short side center (b, b) and finally disappears.

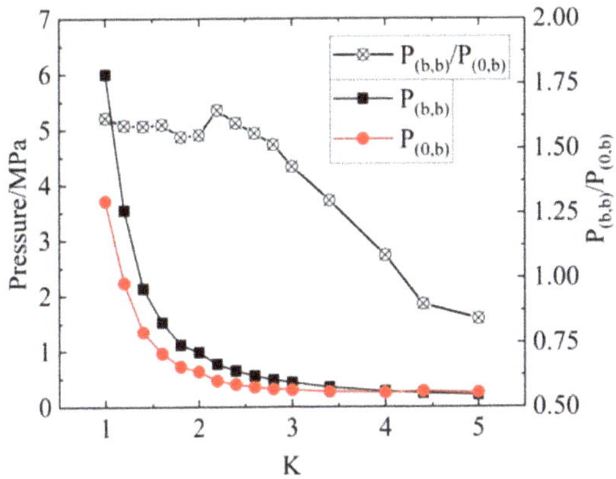

Figure 11. Changes in the pressure and pressure ratio at two lateral points with K when b = 500 mm and W = 100 g.

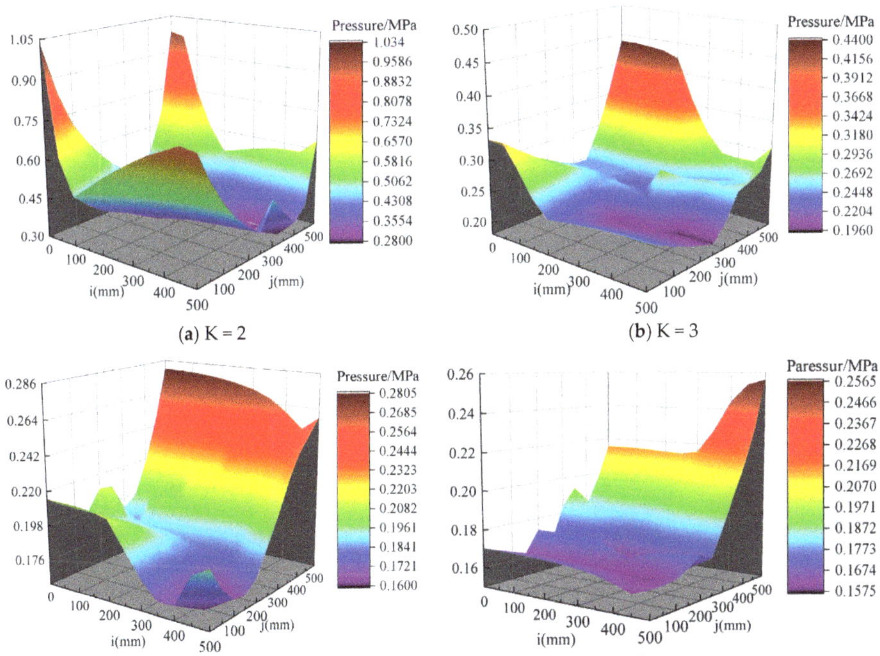

Figure 12. Distribution of peak pressure in the corner area at different values of K.

4. Determination of the Range of the High-Pressure Area at the Corner

The formation of high pressure at the corner has been studied previously but further study into the range of high-pressure areas is also crucial to understanding the convergence effect of shock waves at the corners. Therefore, the pressure peaks at each measurement point in Figure 3 were recorded and pressure contour maps were plotted. Figure 13 shows the pressure contour map when b = 500 mm, W = 100 g, and K = 2, 3, 4, and 5.

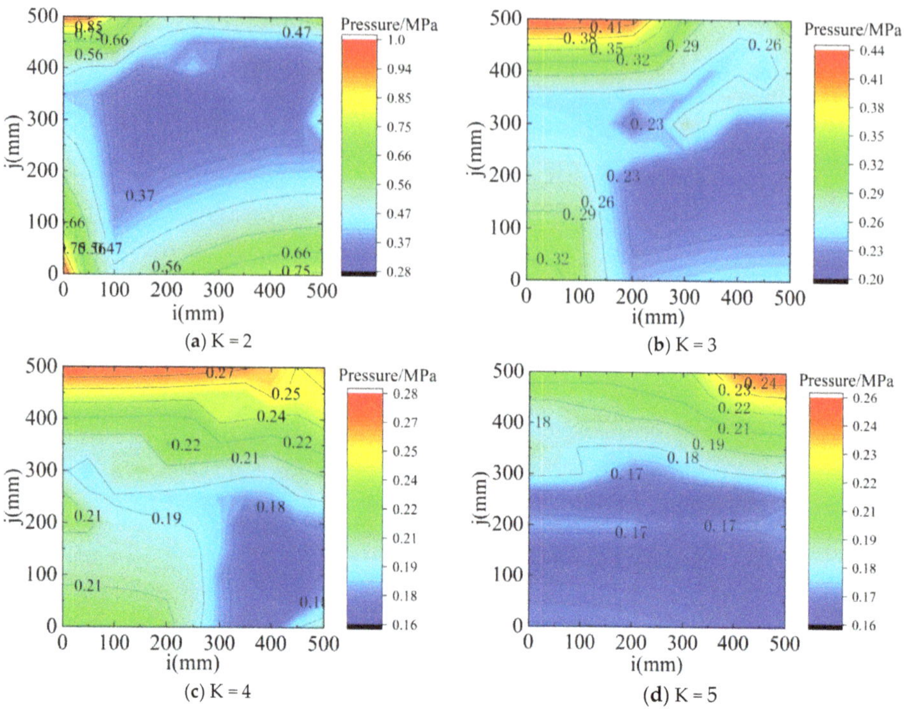

Figure 13. Pressure contour distribution diagram in the corner area.

Figure 13 shows that as K increases, the high-pressure area at the corner gradually widens with little change in height. In addition, the high-pressure area gradually moves towards the center of the surface of wall B and is no longer noticeable at the corner but a high-pressure area is formed at the center of the B wall surface. Figure 13 also shows that when K > 3, the high-pressure area at the corner is not apparent; thus, the K range from 1 to 3 will be discussed next. The high-pressure areas at the corners are individually intercepted along the boundary line, as shown in Figure 14, where the numerical relationship between the size of the area boundary and the value of b is indicated.

Figure 14 clearly shows that the high-pressure area appears triangular when K is small and as isosceles triangles when K = 1. As K increases, the shape of the high-pressure area gradually approximates to a rectangular form; this result is consistent with the high-pressure area formation rule at the corner that has already been discussed. The high-pressure area gradually widens primarily as the Mach-reflected wavefront formed at the surface of wall A widens. The relationship between the range of areas obtained in Figure 14 was represented in a coordinate system with the height and width of the corner high-pressure area a_p and b_p, respectively. Figure 15 shows the data points in the coordinate system with K, with a_p/b and b_p/b as the horizontal and vertical coordinates, respectively.

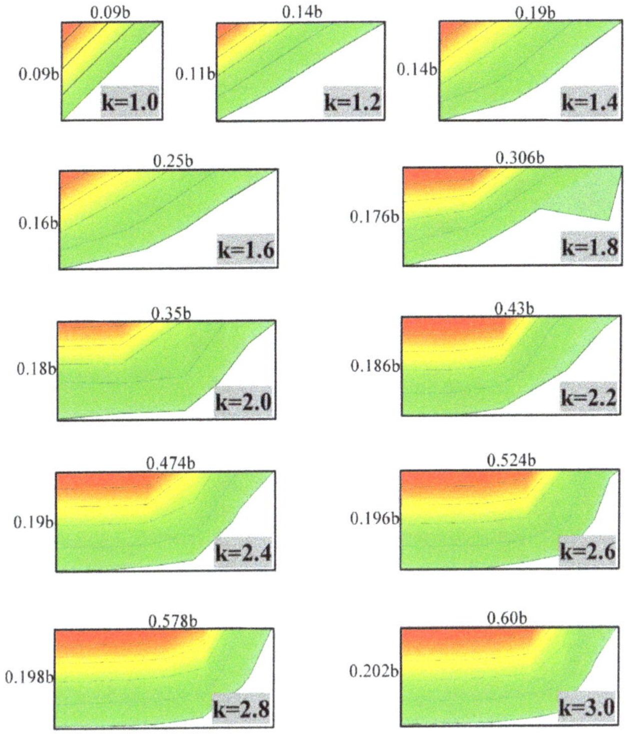

Figure 14. Change of the high-pressure area at the corners with the value of K.

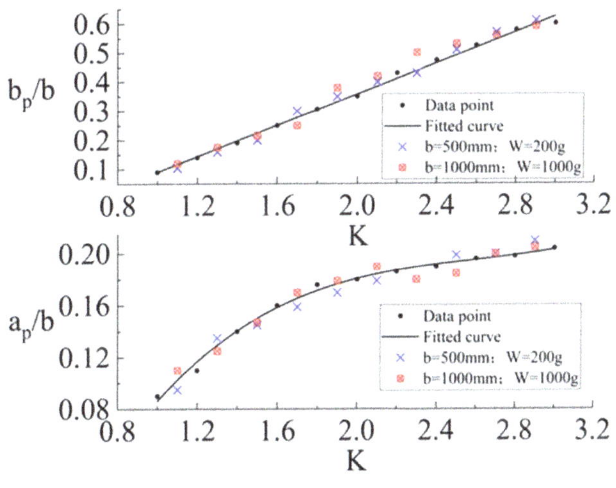

Figure 15. Relationship between the size of the high-pressure area and K.

The polynomial fit performed on the data points in Figure 15 shows that b_p/b increases linearly as K increases. Equation (3) gives the functional relationship between b_p/b and K based on the polynomial fit.

$$b_p/b = -0.17673 + 0.26709K \tag{3}$$

The functional relationship between a_p/b and K is also polynomial, as shown by Equation (4):

$$a_p/b = 0.02015K^3 - 0.15693K^2 + 0.42428K - 0.20127 \quad (4)$$

As the fitted data points were obtained under the conditions b = 500 mm and W = 100 g, simulations were conducted considering the generality of the functional relationship given by Equations (3) and (4). Two sets of data points were obtained at b = 500 mm and W = 200 g, and b = 1000 mm and W = 1000 g, as represented in Figure 15. Both data sets satisfy the functional relationship obtained and show that K is the main factor influencing the range of high-pressure areas.

5. Conclusions

This study investigated the convergence effect of shock waves at the corners of a cabin under implosion conditions using a 2D cross-sectional method. The high-pressure area formation mechanism during implosion shock wave convergence at the corners and the associated change law were determined with the aspect ratio (K). The specific conclusions are as follows:

1. The aspect ratio, K, significantly influences implosion shock wave convergence at the corner and the associated high-pressure area formation mechanism. When $1 \leq K \leq 1.193$, the convergence of the initial shock wave and Mach reflected waves from the surfaces of walls A and B occurs at the corner, creating a high-pressure region. However, when $1.193 < K < 2$, the convergence at the corner comes from the initial shock wave, the Mach reflection wave on wall A, and the regular reflection wave on wall B. When $2 \leq K$, the high pressure at the corner mainly originates from the positive reflection of Mach waves from the surface of wall A to wall B;
2. As K increases, the convergence effect of the shock waves at the corner is no longer noticeable, and the high-pressure region moves towards the center of the short side;
3. The functional relationship between K and the range of the high-pressure region at the corner was obtained when K = 1 to 3 and its universality was verified.

Author Contributions: Conceptualization, X.L. and J.Y.; methodology, X.L.; software, X.L.; validation, X.L.; formal analysis, X.L., J.Y., H.C. and Z.W.; investigation, X.L., J.Y. and Z.W.; data curation, X.L. and J.Y.; writing—original draft preparation, X.L.; writing—review and editing, J.Y.; supervision, J.Y., X.L., Z.W. and H.C.; project administration, X.L. All authors have read and agreed to the published version of the manuscript.

Funding: The authors would like to acknowledge the financial support from the 2021 Basic Research Pro-gram of Shanxi Province (Free Exploration), grant number 20210302123207.

Acknowledgments: The authors would like to thank the editor, associate editor, and the anonymous reviewers for their helpful comments and suggestions that have improved this paper.

Conflicts of Interest: The authors declare no conflict of interest.

References

1. Baker, W.E. *Explosion Hazard sand Evaluation*; Elsevier Pulishing Company: NewYork, NY, USA, 1983.
2. Ma, K.; Ren, F.; Huang, H.; Yang, X.; Zhang, F. Experimental investigation on the dynamic mechanical properties and energy absorption mechanism of foam concrete. *Constr. Build. Mater.* **2022**, *342*, 127927. [CrossRef]
3. Anzellini, S.; Burakovsky, L.; Turnbull, R.; Bandiello, E.; Errandonea, D. P–V–T Equation of State of Iridium Up to 80 GPa and 3100 K. *Crystals* **2021**, *11*, 452. [CrossRef]
4. Baty, S.R.; Burakovsky, L.; Errandonea, D. Ab Initio Phase Diagram of Copper. *Crystals* **2021**, *11*, 537. [CrossRef]
5. Qin, Y.; Yao, X.; Wang, Z.; Wang, Y. Experimental investigation on damage features of stiffened cabin structures subjected to internal blast loading. *Ocean. Eng.* **2022**, *265*, 112639. [CrossRef]
6. Ren, X.; Huang, Z.; Jiang, Y.; Chen, Z.; Cao, X.; Zhao, T.; L, Y. The scaling laws of cabin structures subjected to internal blast loading: Experimental and numerical studies. *Def. Technol.* **2022**, *18*, 811–822. [CrossRef]
7. Qin, Y.; Wang, Y.; Wang, Z.; Yao, X. Investigation on similarity laws of cabin structure for the material distortion correction under internal blast loading. *Thin-Walled Struct.* **2022**, *177*, 109371. [CrossRef]

8. Gan, L.; Zong, Z.; Qian, H.; Lin, J. Experimental and numerical study of damage mechanism of steel box girders under external blast loads. *J. Constr. Steel Res.* **2022**, *198*, 107578. [CrossRef]
9. Yuan, Y.; Zhang, C.; Xu, Y. Influence of standoff distance on the deformation of square steel plates subjected to internal blast loadings. *Thin-Walled Struct.* **2021**, *164*, 107914. [CrossRef]
10. Yuan, Y.; Li, X.; Zhang, C.; Tan, P.J.; Chen, P. Impulse saturation in metal plates under confined blasts. *Int. J. Impact Eng.* **2022**, *168*, 104308. [CrossRef]
11. Li, X.D.; Yin, J.P.; Zhao, P.D.; Zhang, L.; Xu, Y.X.; Wang, Q.; Zhang, P. The effect of stand-off distance on damage to clamped square steel plates under enclosed explosion. *Structures* **2020**, *25*, 965–978. [CrossRef]
12. Hou, H.; Zhu, X.; Mei, Z. Study on the blast load and failure mode of ship structure subject to internal explosion. *Explos. Shock. Waves* **2007**, *27*, 151–158.
13. Hou, H.; Zhu, X.; Li, W.; Mei, Z. Experimental studies on characteristics of blast loading when exploded inside ship cabin. *J. Ship Mech.* **2010**, *14*, 901–907.
14. Yao, D.; Luo, G.; Xie, W.; Li, D. Study on the Law of Explosion Pressure Load in Bare Charge Chamber. *J. Wuhan Univ. Technol.* **2019**, *43*, 168–173.
15. Wang, Q. *Research on Damage Assessment Method of Coupling Effect of Shock Wave and Fragment under Internalbl*; North University of China: Taiyuan, China, 2018.
16. Kong, X.; Wu, W.; Li, J.; Li, X.; Xu, S. Experimental Research of Influence of Corner Structure on Blast Loading under Inner Explosion. *Shipbuild. China* **2012**, *3*, 40–50.
17. Chan, P.C.; Klein, H.H. A study of blast effects inside an enclosure. *J. Fluids Eng.* **1994**, *116*, 450–455. [CrossRef]
18. Hu, Y.; Wu, C.Q.; Lukaszewicz, M.; Dragos, J.; Ren, J.; Haskett, M. Characteristics of confined blast loading in unvented structures. *Int. J. Prot. Struct.* **2011**, *2*, 21–44. [CrossRef]
19. Edri, I.; Savir, Z.; Feldgun, V.; Karinski, Y.; Yankelevsky, D. On blast pressure analysis due to a partially confined explosion: I. experimental studies. *Int. J. Prot. Struct.* **2011**, *2*, 1–20. [CrossRef]
20. Xu, W.Z.; Wu, W.G. Two-dimensional numerical simulation of the factors influencing the explosion overpressure load inside the confined spacz. *Chin. J. Ship Res.* **2019**, *14*, 91–98.
21. Sochet, I.; Sauvan, P.; Trelat, S. Analysis of reflected blast wave pressure profles in a confned room. *Shock. Waves* **2012**, *22*, 253–264.
22. Zhao, X.; Wang, B.; Li, X.; Han, Z. Shockwave Propagation Characteristics of Thermobaric Explosive in an Explosion Chamber. *Chin. J. Energetic Mater.* **2016**, *24*, 231–237.
23. Kinney, G.F.; Graham, K.J. *Explosive Shocks in Air*; Springer: Berlin/Heidelberg, Germany, 1985.

applied sciences

MDPI

Article

Numerical Study of Pressure Attenuation Effect on Tunnel Structures Subjected to Blast Loads

Cheng-Wei Hung [1], Ying-Kuan Tsai [2], Tai-An Chen [3,*], Hsin-Hung Lai [4,5] and Pin-Wen Wu [6]

1 Department of Civil Engineering and Resource Management, Dahan Institute of Technology, No.1, Shuren 5 St., Dahan Village, Sincheng Township, Hualien County 97145, Taiwan; hung.c.w@ms01.dahan.edu.tw
2 Department of Environmental Information and Engineering, Chung Cheng Institute of Technology, 8 National Defense University, 75, Shiyuan Rd., Daxi Dist., Taoyuan 33551, Taiwan; ccitb04007@ndu.edu.tw
3 Department of Harbor and River Engineering, National Taiwan Ocean University, No. 2, Pei-Ning Rd., Zhongzheng Dist., Keelung City 202301, Taiwan
4 Department of Civil Engineering, R.O.C. Military Academy, No.1, Wei-Wu Rd., Fengshan Dist, Kaohsiung 83059, Taiwan; kevin5485xd@hotmail.com
5 Graduate School of Technological and Vocational Education, National Yunlin University of Science and Technology, 123, University Road, Section 3, Douliou, Yunlin 64002, Taiwan
6 Department of Power Vehicle and Systems Engineering, Chung Cheng Institute of Technology, 8 National Defense University, 75, Shiyuan Rd., Daxi Dist., Taoyuan 33551, Taiwan; pinwen1110@gmail.com
* Correspondence: tachen@mail.ntou.edu.tw; Tel.: +886-2-2462-2192

Abstract: This study used experimental and numerical simulation methods to discuss the attenuation mechanism of a blast inside a tunnel for different forms of a tunnel pressure reduction module under the condition of a tunnel near-field explosion. In terms of the experiment, a small-scale model was used for the explosion experiments of a tunnel pressure reduction module (expansion chamber, one-pressure relief orifice plate, double-pressure relief orifice plate). In the numerical simulation, the pressure transfer effect was evaluated using the ALE fluid–solid coupling and mapping technique. The findings showed that the pressure attenuation model changed the tunnel section to diffuse, reduce, or detour the pressure transfer, indicating the blast attenuation effect. In terms of the effect of blast attenuation, the double-pressure relief orifice plate was better than the one-pressure relief orifice plate, and the single-pressure relief orifice plate was better than the expansion chamber. The expansion chamber attenuated the blast by 30%, the one-pressure relief orifice plate attenuated the blast by 51%, and the double-pressure relief orifice plate attenuated the blast by 82%. The blast attenuation trend of the numerical simulation result generally matched that of the experimental result. The results of this study can provide a reference for future protective designs and reinforce the U.S. Force regulations.

Keywords: blast; tunnel; pressure reduction module; LS-DYNA

Citation: Hung, C.-W.; Tsai, Y.-K.; Chen, T.-A.; Lai, H.-H.; Wu, P.-W. Numerical Study of Pressure Attenuation Effect on Tunnel Structures Subjected to Blast Loads. *Appl. Sci.* **2021**, *11*, 5646. https://doi.org/10.3390/app11125646

Academic Editors: Ricardo Castedo, Lina M. López and Anastasio P. Santos

Received: 28 May 2021
Accepted: 15 June 2021
Published: 18 June 2021

Publisher's Note: MDPI stays neutral with regard to jurisdictional claims in published maps and institutional affiliations.

1. Introduction

Tunnels are usually concealed and sheltered by landforms and ground objects to prevent a direct hit from enemy weapons, meaning the transfer of a blast is obstructed and attenuated by orifice plate attenuators, expansion chambers, explosion doors, and tunnel branches.

Studying the dynamic response of structures subjected to air blast loading has received a lot of attention in the last few decades [1–10]. In terms of studies regarding tunnel explosion protection, in 1992, Song et al. [11] used a reduced specimen of a steel ammunition storage magazine, with the internal dimensions of 100 × 50 × 23 cm and loading density of 16.7 kg/m^3; detonated 1.9 kg of C-4 explosives inside the specimen; and then discussed the influence of Straight, Elbow, and Dead-End channels on the blast transfer. In 1993, Scheklinski-Glück [12] used a round-section of a full-scaled tunnel with a diameter of 3.6 m, and 4000 kg, 2000 kg, and 1000 kg cylindrical RDX explosives in a model scale tunnel with a diameter of 9 cm and cylindrical RDX charge weights of 64, 32, and 16 g. The explosives stand outside the entrance in distances from one to five times the tunnel diameters. The direction is from 0° (tunnel axis) to 90° (charge touching the wall) in steps of 30°. The result showed that the blast inside the tunnel attenuated as the distance increased. In 2004, McMahon et al. [13] used a circular tunnel with a diameter of 0.298 m and 54.3 m in length and placed 0.177 kg and 1.77 kg spherical B explosives at the tunnel portal, as well as 60, 30, and 15 cm outside the tunnel portal, in order to perform explosion experiments. The result showed that the blast inside the tunnel attenuated as the distance increased, and the detonation wave impulse inside the tunnel could be regarded as a constant. In the WES (TM 5-855-1, 1998) [5] equation, according to the position of the explosive source, explosions outside a tunnel are divided into end-on and side-on. In the EMI equation (TM 5-855-1, 1998) [14], the proposed empirical equation can be used to estimate the blast inside a tunnel from an explosion outside the tunnel. As proposed by Welch et al., in 2005 [15], the empirical equation can be used for estimating the blast inside a tunnel, as resulted from an explosion outside the tunnel.

In 2006, Cheng et al. [16] used LS-DYNA software to simulate a strip and a bent channel type ammunition storage magazine and analyzed the internal explosion. The simulation result showed that the bent channel was more effective at attenuating the blast than the strip channel. An appropriate channel design could reduce the lethal area of an explosion inside the ammunition storage magazine. In 2007, Ishikawa and Beppu [17] compiled the protective structure explosion experiment results of Johoji et al. from 1965 to 1981. They analyzed the blast transfer attenuation in vertical bar, branch, and mesh tunnels. According to the experimental document review, the aforesaid experiments mainly discussed detonation waves inside the tunnel after an explosion outside the tunnel. This paper discusses the transfer mechanism of a blast resulting from a near-field explosion inside a tunnel. The near-ground and variable tunnel explosion experiments were performed, the numerical simulations and U.S. Force empirical equations were used for analysis and validation, and related empirical equations were established, which are intended to establish a tunnel blast protection evaluation and improvement mechanism to provide a reference for subsequent tunnel building and renovation.

2. Experiment

The aim was to reduce and avoid explosion pressure directly jeopardizing the safety of personnel inside a tunnel structure. This study designed three pressure attenuation models by changing the tunnel's cross-section, namely, an expansion chamber, single-orifice-plate attenuator, and double-orifice-plate attenuator, to investigate the attenuation effect, wave propagation pattern, and pressure distribution. When a blast wave passes through the tunnel, the pressure is expected to be attenuated by diffusion and detour due to the tunnels' cross section change.

In this study, a small-sized rectangular section tunnel specimen was used to demonstrate an underground tunnel structure subjected to external explosions. The tunnel specimen was made of steel plate with a thickness of 0.5 cm, and the size of its cross-section was 30 × 30 cm. The charge used in the explosion test was C-4 explosive. Its appearance is gray to light yellow. The density was between 1.59 and 1.60 g/cm^2, and the detonation speed can reach 8193 m/s.

Two types of pressure transducer produced by PCB company were used in the field test. The first type was pencil type sensor (models: 137A21 and 137A23), and the measuring range was from 345 to 345 MPa. This type of sensor is used to measure the explosion pressure near the ground in the free field; the second type is high-frequency pressure gauge (models: 113B23, 113B27, and 113B28), and the measuring range was from 345 kPa to 69 MPa, which were used to measure the pressure in the rectangular tunnel specimens. The maximum bandwidth of the oscilloscope was 100 MHz, and the maximum sampling rate was 2×10^9 s^{-1}.

2.1. Explosion Experiment on the Pressure Reduction Module Effect

2.1.1. Linear Tunnel with Expansion Chamber

A linear tunnel 140 cm long with a square cross-section of 30 × 30 cm was combined with a 60 × 60 × 60 cm expansion chamber for an explosion experiment. The cross-section dimension of the expansion chamber was four times the section of the linear tunnel. The pressure transducers were mounted on the specimen sidewall at 2, 30, 90, and 170 cm away from the tunnel portal. In order to investigate the pressure reduction effects under different quantities of explosives, we used five quantities of C-4 explosives (100, 150, 200, 250, and 350 g), and the C-4 explosive was hung at 30 cm aboveground and detonated at 60 cm away from the tunnel portal. The experimental configuration is shown in Figure 1. In order to know the blast attenuation characteristic of the expansion chamber, as designed by expanding the cross-section, we analyzed and discussed the transfer of the blast inside the tunnel and the pure linear tunnel explosion experiment.

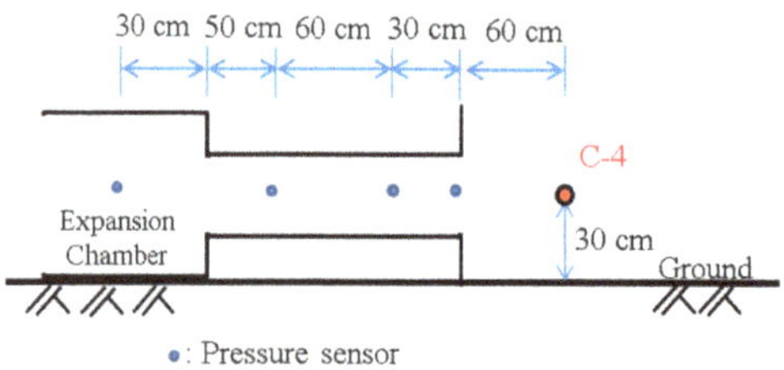

Figure 1. Experimental configuration and specimen of the linear tunnel with expansion chamber.

2.1.2. Linear Tunnel with One-Pressure Relief Orifice Plate

The linear tunnel was a square-section tunnel with a side length of 30 cm—the total length was 200 cm, and the orifice plate (circular orifice in diameter of 12 cm) was mounted at 127 cm away from the tunnel portal. The orifice plate specimen was designed by reducing the scale of U.S. Force regulation UFC 3-340-01 [13] by 2.5 times. The pressure transducers were mounted on the specimen sidewall and at 2, 30, 90, and 170 cm away from the tunnel portal. In order to know the pressure reduction effects under different quantities of explosives, we used five quantities of C-4 explosive (100, 150, 200, 250, and 350 g), and the explosive was hung at 30 cm aboveground and detonated at 60 cm away from the tunnel portal. The experimental configuration and specimen are shown in Figure 2.

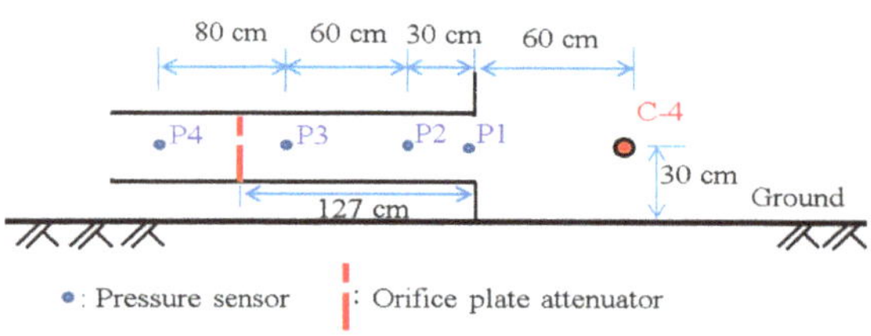

Figure 2. Experimental configuration and specimen of linear tunnel with single-pressure relief orifice plate.

2.1.3. Linear Tunnel with Double-Pressure Relief Orifice Plate

The linear tunnel was a square-section tunnel with a side length of 30 cm—the total length was 300 cm, and the orifice plates (circular orifice in diameter of 12 cm) were diagonally mounted at 127 cm and 134 cm away from the tunnel portal inside the linear tunnel. The pressure transducers were mounted on the specimen sidewall at 2, 30, 90, and 170 cm away from the tunnel portal. In order to know the pressure reduction effects under different quantities of explosives, we used five quantities of C-4 explosive (100, 150, 200, 250, and 350 g), and the explosive was hung at 30 cm aboveground and detonated at 60 cm away from the tunnel portal. The experimental configuration and specimen are shown in Figure 3.

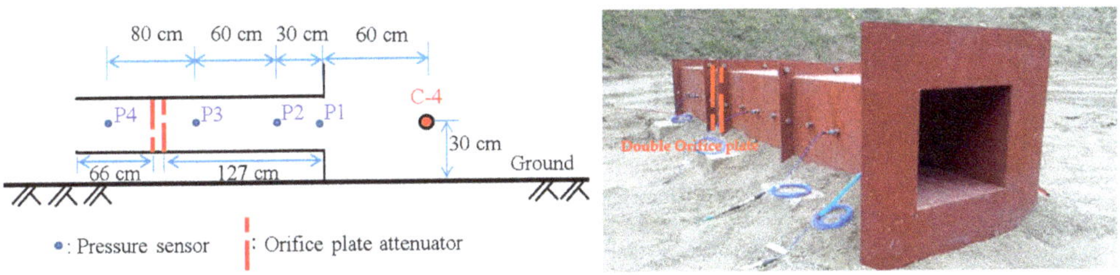

Figure 3. Experimental configuration and specimen of the linear tunnel with double-pressure relief orifice plate.

3. Numerical Simulation

There are three main numerical models in the LS-DYNA program: the Lagrangian numerical model, the Eulerian numerical model, and the ALE (arbitrary Lagrangian–Eulerian) numerical model. As the ALE numerical model has the characteristics of the Lagrangian and Eulerian numerical models, it was used for numerical simulation in this study. It can overcome the problem in that the operation stops as the numerical calculation becomes difficult when the mesh element deformation is too large compared to the Lagrangian system. Eulerian describes the fluid and Lagrangian describes the solid, and it can effectively control and track the motion behavior of the structural boundary. Thus, it is applicable to the dynamic real-time analysis of fluid–solid coupling and it has better computational accuracy than the Eulerian system. However, as the number of grids increases, the analysis model and grid size are limited. In order to solve this problem, we used the LS-DYNA mapping technology to break through the limit.

3.1. Numerical Models

Regarding the building methods of the various pressure reduction modules, as the linear tunnel with expansion chamber was symmetrical, the 1/2 symmetrical simplified numerical model was used for analysis. The linear tunnel with a single-pressure relief orifice plate and the linear tunnel with double-pressure relief orifice plate models were analyzed using full models. The models are shown in Figure 4. Regarding the orifice plate model, as the shell element had coupling directivity in the fluid–solid coupling, this study considered the vortex of the blast through the orifice plate or reflected blast, and in order to avoid analytical errors, the orifice plate was built using entity elements.

3.2. Constitutive Models and Equation of State

Numerical simulation was performed to investigate the pressure attenuation effect on the tunnel models. The constitutive model and material parameters of the air, explosives, and steel plates are described as follows:

3.2.1. Air

Regarding the air part of the numerical model, MAT_NULL material model was provided with the EOS_LINEAR_POLYNOMIAL condition equation, as shown in the following equation:

$$P = C_0 + C_1\mu + C_2\mu^2 + C_3\mu^3 + \left(C_4 + C_5\mu + C_6\mu^2\right)E_0 \tag{1}$$

where P is the pressure composed of initial internal energy, and E_0 is the ratio of current density to initial density, μ, and material parameters, C_0 to C_6. In the present study, because air was assumed to be an ideal gas, C_1, C_2, C_3, and C_6 were set to zero, and C_4 and C_5 were set to 0.4.

3.2.2. Explosive

For explosive material, MAT_HIGH_EXPLOSIVE_BURN material model was applied with the JWL (Jones–Wikens–Lee) equation of state to model TNT explosive with the pressure defined as

$$P = A\left(1 - \frac{\omega}{R_1 V_r}\right)e^{-R_1 V_r} + B\left(1 - \frac{\omega}{R_2 V_r}\right)e^{-R_2 V_r} + \frac{\omega E_0}{V_r} \tag{2}$$

where P is the hydrostatic pressure, and V_r is the relative volume. A, B, R_1, R_2, and ω are material parameters used for the explosives, which can be experimentally determined.

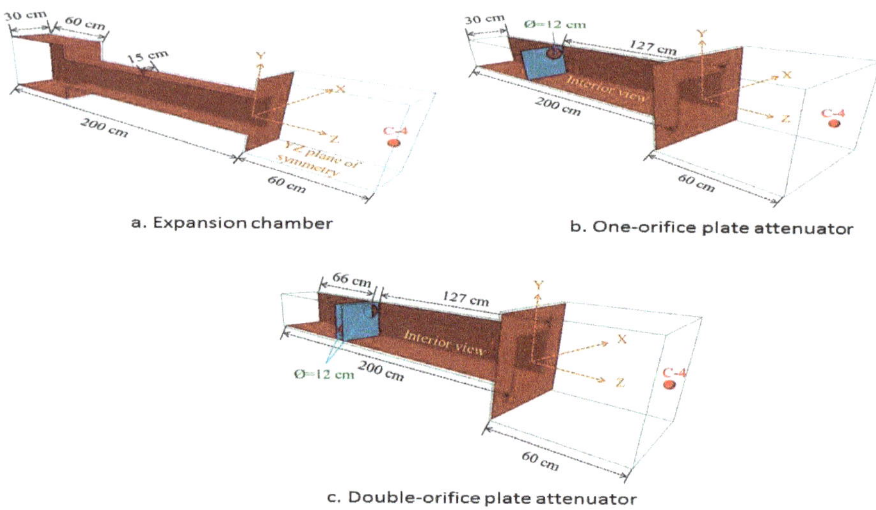

Figure 4. Schematic diagram of numerical models of pressure reduction modules.

3.2.3. Steel Plate

The MAT_PLASTIC_KINEMATIC material model was used to simulate the steel plate structure, which was the tunnel part of the numerical model. For simplified and conservative operations, the idealized stress–strain curve was used, and the strain hardening behavior of the material after plasticization can be controlled by parameter β. If $\beta = 0$, it represents a dynamic plastic hardening material. If $\beta = 1$, it is an isotropic strain hardening material. When unloading occurs, the dynamic plastic hardening curve and isotropic plastic hardening curve unload according to the original slope, the yield stress value of the isotropic plastic hardening curve will increase during reverse loading, and the yield point of the dynamic plastic hardening curve remains. The isotropic plastic hardening curve ($\beta = 1$) is more suitable for large deformation of the material, as resulted from the explosion. The present study assumed $\beta = 0$.

4. Results

4.1. Pressure Reduction Module Effect Analysis

4.1.1. Linear Tunnel with Expansion Chamber Explosion

The variation of blast attenuation inside the tunnel of this experiment is shown in Figure 5. Due to the nature of explosion characteristics, the pressure decayed extremely rapidly with time and space. As a result, for different charge weights, a larger scattering of the pressures could be found at the measurement point P1 compared to P4. According to the experimental results, the transfer of the blast inside the tunnel decreased as the distance increased. In order to know the blast attenuation characteristic of the expansion chamber, as designed by enlarging the section, we analyzed and discussed the blast transfer rate and variation rate of the expansion chamber. In terms of the blast transfer rate, the linear tunnel with expansion chamber was tested, and when the blast was transferred from the smaller tunnel section (pressure transducer P3 position) to the expansion chamber with a larger section (pressure transducer P4 position), the blast transfer attenuation in P3 and P4 was analyzed. In terms of pressure attenuation rate, the linear tunnel with expansion chamber and the pure linear tunnel was tested, and the pressure attenuation in the P4 position was analyzed and discussed.

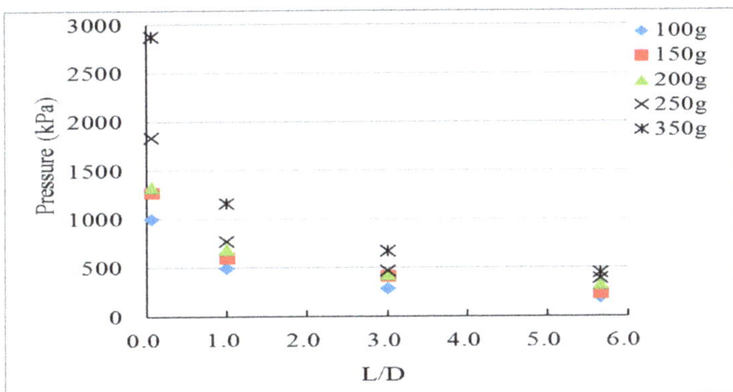

Figure 5. Comparison diagram of explosion experiment results of linear tunnel with expansion chamber.

The effect of the expansion chamber on the pressure transfer rate was described using the pressure transducers in positions P3 and P4. When the quantity of C-4 explosive was 100 g, and the blast had not been transferred to the expansion chamber, the measured blast value (P3) was 282.81 kPa. When the blast was transferred to the expansion chamber (P4), the measured blast value was 189.52 kPa. Therefore, the blast transfer rate in pressure transducer position P4 was 0.67; in other words, when the blast was transferred from position P3 with the smaller tunnel section (side length 30 cm) to position P4 with the larger tunnel section (side length 60 cm), the blast was diffusively attenuated by enlarging the section, and the blast attenuation amplitude was 33%. The blast transfer rate ($R_{transfer}$) was expressed as follows:

$$R_{transfer\_expansion} = \frac{P4_{expansion}}{P3_{expansion}} \tag{3}$$

where $P3_{expansion}$ and $P4_{expansion}$ are the measured blast at the pressure transducer positions, $P3$ and $P4$, respectively.

When the quantity was changed (150~350 g), the blast was transferred from the smaller tunnel section (position P3) to the expansion chamber (position P4), and the range of blast transfer rate was 0.57 to 0.82.

When the quantity of the explosive was ≤350 g, the expansion chamber design mode could reduce the blast transfer rate to 0.7, meaning the blast was transferred from P3 to P4, and the blast could be attenuated by 30% by enlarging the tunnel section.

The effect of the expansion chamber on the pressure attenuation rate was tested using a pure linear tunnel and the linear tunnel with an expansion chamber. The pressure attenuation in position P4 was analyzed and discussed. When the quantity of C-4 explosive was 100 kg, the measured blast value of the expansion chamber (P4) was 189.52 kPa, and the measured blast value of the pure linear tunnel (P4) was 271.77 kPa. Therefore, in pressure transducer position P4, the pressure attenuation rate of the expansion chamber was 0.70, as compared with the pure linear tunnel. The blast attenuation rate ($R_{attenuate}$) is expressed as follows:

$$R_{attenuate\_expansion} = \frac{P4_{expansion}}{P4_{linear}} \tag{4}$$

where $P4_{linear}$ is the measured blast at the pressure transducer positon, $P4$, in the linear tunnel.

When the quantity of the explosive was $\leq$350 g, the pressure attenuation rate in P4 was 0.77, meaning with the expansion chamber, the blast in P4 was attenuated by 23%, as compared with the pure linear tunnel (without an expansion chamber). In addition, according to the comprehensive comparison of pressure transducer positions P1 to P3, before the blast was transferred to the expansion chamber, as the tunnel specimen model was consistent, the blast transfer of the pure linear tunnel was approximate to that of the linear tunnel with an expansion chamber (pressure attenuation rate was 0.95 to 1.06), which matched the estimated result.

Generally speaking, the expansion chamber designed by enlarging the section was very effective on blast attenuation. In terms of the blast transfer rate, the blast was transferred from the smaller section (P3) to the expansion chamber with a larger section (P4), and the transfer rate was 0.70; thus, the blast can be attenuated by 30% by enlarging the section. In terms of the pressure attenuation rate, the pressure attenuation in P4 was discussed according to the experiments on the pure linear tunnel and the linear tunnel with expansion chamber. The findings show that the pressure attenuation rate was 0.77, meaning with the expansion chamber, the blast in P4 could be attenuated by 23%, as compared with the pure linear tunnel (without an expansion chamber).

In terms of numerical simulation, the numerical simulation result of the linear tunnel with expansion chamber and the experimental blast are compared in Table 1. The simulation result shows that the blast inside the tunnel attenuated as the transfer distance increased, and the blast attenuation trend of numerical simulation was similar to that of the experiment; however, the experimental result was a little lower than the numerical simulation. The transfer of the blast inside the tunnel is shown in Figure 6.

Table 1. Comparison of experiment and numerical results of linear tunnel with expansion chamber.

		Weight of C-4 (g)		100	150	200	250	350
Position	P1 (L/D = 0.07)	Experiment	Explosion pressure (kPa)	999.27	1276.14	1327.74	1834.18	2873.28
		Simulation	Explosion pressure (kPa)	889.32	1207.76	1553.81	1725.41	2281.28
	P2 (L/D = 1.00)	Experiment	Explosion pressure (kPa)	492.06	600.50	593.31	766.60	1158.69
		Simulation	Explosion pressure (kPa)	591.04	747.45	872.07	967.59	1133.06
	P3 (L/D = 3.00)	Experiment	Explosion pressure (kPa)	282.81	417.29	427.71	468.46	664.85
		Simulation	Explosion pressure (kPa)	449.52	554.88	639.48	700.86	816.89
	P4 (L/D = 5.67)	Experiment	Explosion pressure (kPa)	189.52	237.99	332.31	384.24	443.45
			Blast transfer rate ($R_{transfer}$)	0.67	0.57	0.78	0.82	0.67
						Average: 0.7		
		Simulation	Explosion pressure (kPa)	244.00	304.74	354.77	391.55	462.05
			Blast transfer rate ($R_{transfer}$)	0.54	0.55	0.55	0.56	0.57
						Average: 0.55		

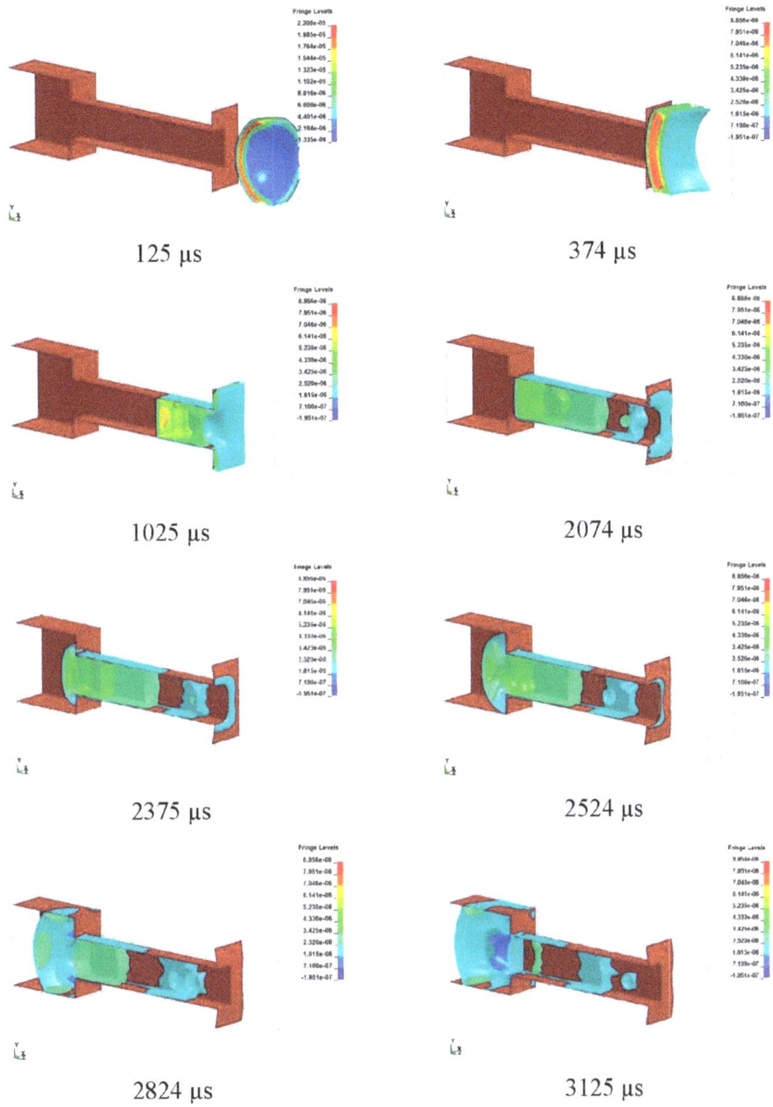

Figure 6. Blast transfer in numerical simulation of linear tunnel with expansion chamber explosion (100 g C-4).

4.1.2. Linear Tunnel with Single-Pressure Relief Orifice Plate Explosion

The variation of blast attenuation inside the tunnel of this experiment is shown in Figure 7. According to the experimental results, the transfer of the blast inside the tunnel will decrease as the distance increases. In order to know the blast attenuation characteristic of the pressure relief orifice plate, as designed by reducing the section, we analyzed and discussed the blast transfer rate and variation rate of the pressure relief orifice plate. In terms of the blast transfer rate, the linear tunnel with a single-pressure relief orifice plate was tested; when the blast was transferred from the larger tunnel section (pressure transducer position P3) to the pressure relief orifice plate with the smaller section (pressure transducer position P4), the blast transfer attenuation in P3 and P4 was analyzed. In terms of the pressure attenuation rate, the linear tunnel with a single-pressure relief orifice plate and the pure linear tunnel were tested, and the pressure attenuation in position P4 was analyzed and discussed. The blast transfer rate ($R_{transfer}$) is expressed as follows:

$$R_{transfer\_single\_orifice} = \frac{P4_{sigle\_orifice}}{P3_{sigle\_orifice}} \quad (5)$$

where $P3_{sigle\_orifice}$ and $P4_{sigle\_orifice}$ are the measured blast at the pressure transducer positon, $P3$ and $P4$, in the linear tunnel with single orifice plate attenuator, respectively.

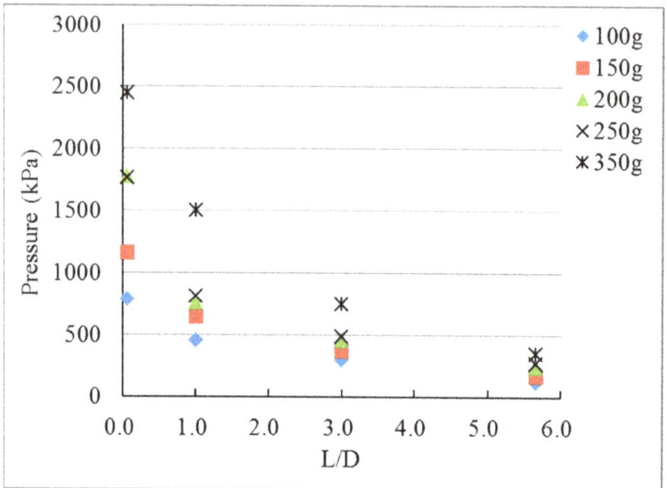

Figure 7. Comparison diagram of explosion experiment results of linear tunnel with single-pressure relief orifice plate.

When the quantity of the explosive was ≤350 g, the one-pressure relief orifice plate design mode can reduce the blast transfer rate to 0.49, meaning the blast was transferred from P3 to P4, and the blast could be attenuated by 51% by reducing the tunnel section.

The effect of the one-pressure relief orifice plate on the pressure attenuation rate was tested using the pure linear tunnel and the linear tunnel with one-pressure relief orifice plate, and the pressure attenuation in position P4 was analyzed and discussed. When the quantity of C-4 explosive was 100 g, the measured blast value of the one-pressure relief orifice plate (P4) was 123.89 kPa, and the measured blast value of the pure linear tunnel (P4) was 271.77 kPa. Therefore, in pressure transducer position P4, the pressure attenuation rate of the one-pressure relief orifice plate was 0.46, as compared with the pure linear tunnel. The blast attenuation rate ($R_{attenuate}$) is expressed as follows:

$$R_{attenuate\_single\_orifice} = \frac{P4_{sigle\_orifice}}{P4_{linear}} \quad (6)$$

According to Table 2, when the quantity of explosive was ≤350 g, the pressure attenuation rate in P4 was 0.56, meaning with the one-pressure relief orifice plate, the blast in P4 was attenuated by 44%, as compared with the pure linear tunnel (without pressure relief orifice plate). In addition, according to a comprehensive comparison of pressure transducer positions P1 to P3, before the blast was transferred to the single-pressure relief orifice plate, as the cross-section was consistent, the blast transfer of the pure linear tunnel was approximate to that of the linear tunnel with one-pressure relief orifice plate (pressure attenuation rate was 0.98 to 1.08), which matched the estimated result.

Table 2. Comparison of experiment and numerical results of steel tunnel with single-pressure relief orifice plate.

		Weight of C-4 (g)		**100**	**150**	**200**	**250**	**350**
Position	P1 (L/D = 0.07)	Experiment	Explosion pressure (kPa)	784.67	1164.00	1777.00	1763.02	2443.99
		Simulation		792.96	1006.73	1623.96	1721.41	2175.51
	P2 (L/D = 1.00)	Experiment	Explosion pressure (kPa)	457.43	642.93	760.66	908.14	1502.18
		Simulation		584.15	739.17	863.49	960.17	1126.15
	P3 (L/D = 3.00)	Experiment	Explosion pressure (kPa)	300.49	367.77	454.65	482.50	747.77
		Simulation		568.95	617.27	655.25	702.23	837.73
	P4 (L/D = 5.67)	Experiment	Explosion pressure (kPa)	123.89	170.62	242.00	272.76	348.09
			Blast transfer rate ($R_{transfer}$)	0.41	0.46	0.53	0.57	0.47
				Average: 0.49				
		Simulation	Explosion pressure (kPa)	117.91	144.36	166.38	182.05	211.96
			Blast transfer rate ($R_{transfer}$)	0.21	0.23	0.25	0.26	0.25
				Average: 0.24				

In terms of numerical simulation, the numerical simulation result of the linear tunnel with a single-pressure relief orifice plate and the experimental blast are compared in Table 2. According to this numerical simulation, when the quantity of explosive was ≤350 g, the blast was transferred from position P3 to position P4, and the blast transfer rate of the one-pressure relief orifice plate was 0.24. Therefore, as predicted by numerical simulation using tunnel section reduction, the blast attenuation amplitude can be 76%. In terms of the blast transfer rates of the experiment and numerical simulation, the blast transfer rate of the one-pressure relief orifice plate (position P4) obtained by experimental analysis was 0.49 (blast was attenuated by 51%). Thus, the blast transfer rate predicted by numerical simulation was 0.24 (blast is attenuated by 76%). In contrast, the blast attenuation predicted by numerical simulation was larger. Both the experimental and numerical simulation results showed that the one-pressure relief orifice plate, as designed by reducing the tunnel section, was surely effective on blast attenuation. The transfer of the blast inside the tunnel is shown in Figure 8.

4.1.3. Linear Tunnel with Double-Pressure Relief Orifice Plate Explosion

The variation of blast attenuation inside the tunnel of this experiment is shown in Figure 9. According to the experimental results, the transfer of the blast inside the tunnel will decrease as the distance increases. In order to know the blast attenuation characteristic of the double-pressure relief orifice plate, as designed by reducing the section, we analyzed and discussed the blast transfer rate and variation rate of the double-pressure relief orifice plate. In terms of the blast transfer rate, the linear tunnel with a double-pressure relief orifice plate was tested, and when the blast was transferred from the larger tunnel section (pressure transducer position P3) to the double-pressure relief orifice plate (pressure transducer position P4), the blast transfer attenuation in P3 and P4 was analyzed. In terms of the pressure attenuation rate, the linear tunnel with a single-pressure relief orifice plate and the pure linear tunnel were tested, and the pressure attenuation in position P4 was analyzed and discussed.

The pressure transducer positions P3 and P4 were taken as examples to describe the effect of a double-pressure relief orifice plate on the blast transfer rate. When the quantity of C-4 explosive was 100 g, before the blast was transferred to the double-pressure relief orifice plate, the measured blast value in pressure transducer position P3 was 272.79 kPa. When the blast was transferred through the double-pressure relief orifice plate (pressure transducer position P4), the measured blast value was 38.18 kPa. Therefore, the blast transfer rate in pressure transducer position P4 was 0.14. In other words, when the blast was transferred from position P3 of the larger tunnel section (side length 30 cm) to position P4 of the smaller tunnel section after the path was changed, the double-pressure relief orifice plate reflected the blast, reducing and detouring the throughput, and the blast attenuation amplitude was 86%. The blast transfer rate ($R_{transfer}$) is expressed as follows:

$$R_{transfer\_double\_orifice} = \frac{P4_{double\_orifice}}{P3_{double\_orifice}} \tag{7}$$

where $P3_{double\_orifice}$ and $P4_{double\_orifice}$ are the measured blast at the pressure transducer position, $P3$ and $P4$, in the linear tunnel with single orifice plate attenuator, respectively.

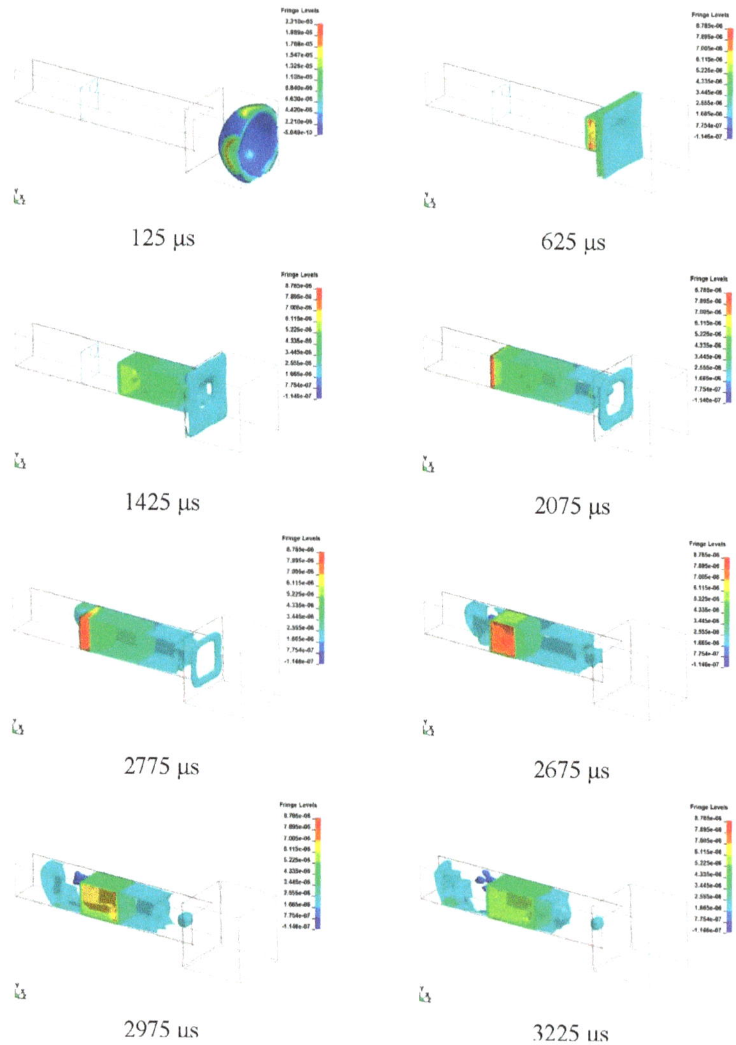

Figure 8. Blast transfer of numerical simulation of linear tunnel with single-pressure relief orifice plate explosion (100 g C-4).

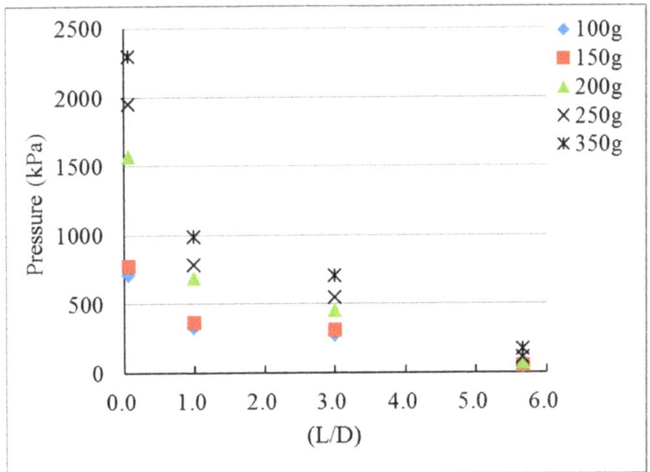

Figure 9. Comparison diagram of double-pressure relief orifice plate explosion experiment results.

When the quantity was changed (150~350 g), the blast was transferred from the larger tunnel section (P3) and through the double-pressure relief orifice plate (P4); thus, the range of the blast transfer rate of the double-pressure relief orifice plate was 0.16 to 0.24.

According to this study, when the quantity of explosive was $\leq$350 g, the double-pressure relief orifice plate design mode could reduce the blast transfer rate to 0.18, meaning the blast was transferred from P3 to P4, and the blast could be attenuated by 82% by reducing the tunnel section and changing the path.

The effect of the double-pressure relief orifice plate on the pressure attenuation rate was tested by using the pure linear tunnel and the linear tunnel with a double-pressure relief orifice plate. The pressure attenuation in position P4 was analyzed and discussed. When the quantity of C-4 explosive was 100 g, the measured blast value of the double-pressure relief orifice plate (P4) was 38.18 kPa, and the measured blast value of the pure linear tunnel was 271.77 kPa. Therefore, in pressure transducer position P4, the pressure attenuation rate of the double-pressure relief orifice plate was 0.14, as compared with the pure linear tunnel. The blast attenuation rate ($R_{attenuate}$) is expressed as follows:

$$R_{attenuate\_double\_orifice} = \frac{P4_{double\_orifice}}{P4_{linear}} \tag{8}$$

According to Table 3, when the quantity of explosive was $\leq$350 g, the pressure attenuation rate in P4 was 0.20; in other words, when the pure linear tunnel was equipped with the double-pressure relief orifice plate, the blast in P4 was attenuated by 80%, as compared with the pure linear tunnel (without the pressure relief orifice plate). In addition, according to a comprehensive comparison of pressure transducer positions P1 to P3, before the blast was transferred to the double-pressure relief orifice plate, as the tunnel specimen model was consistent, the blast transfer of the pure linear tunnel was approximate to that of the linear tunnel with double-pressure relief orifice plate (pressure attenuation rate was 0.86 to 0.94), which matched the estimated result.

In terms of numerical simulation, the numerical simulation result of the linear tunnel with a double-pressure relief orifice plate and the experimental blast are compared in Table 3. According to this numerical simulation, when the quantity of explosive was $\leq$350 g, the blast was transferred from position P3 to position P4, and the blast transfer rate of the double-pressure relief orifice plate was 0.15. Therefore, the blast attenuation amplitude, as predicted by numerical simulation using tunnel section reduction and blast transfer path detour, could be 85%. In terms of the blast transfer rates of the experiment and numerical simulation, the blast transfer rate of the double-pressure relief orifice plate obtained by experimental analysis was 0.18 (blast was attenuated by 82%), and the blast transfer rate predicted by numerical simulation was 0.15 (blast is attenuated by 85%); thus, the blast predicted by numerical simulation was approximate to the experimental result. The transfer of the blast inside the tunnel is shown in Figure 10.

Table 3. Comparison of experiment and numerical results of steel tunnel with double-pressure relief orifice plate.

Weight of C-4 (g)				100	150	200	250	350
Position	P1 (L/D = 0.07)	Experiment	Explosion pressure (kPa)	711.49	779.04	1568.8	1950.37	2294.61
		Simulation		792.96	1006.73	1523.11	1786.34	2204.64
	P2 (L/D = 1.00)	Experiment	Explosion pressure (kPa)	379.33	454.14	686.29	782.28	986.88
		Simulation		584.15	739.17	863.49	960.17	1126.15
	P3 (L/D = 3.00)	Experiment	Explosion pressure (kPa)	272.79	306.57	450.01	544.88	703.04
		Simulation		651.38	780.52	739.47	787.03	843.57
	P4 (L/D = 5.67)	Experiment	Explosion pressure (kPa)	38.18	52.33	71.64	108.96	166.3
			Blast transfer rate ($R_{transfer}$)	0.14	0.17	0.16	0.20	0.24
						Average: 0.18		
		Simulation	Explosion pressure (kPa)	65.40	101.19	116.50	128.89	151.12
			Blast transfer rate ($R_{transfer}$)	0.10	0.13	0.16	0.16	0.18
						Average: 0.15		

Table 4 shows the percentage error between the experimental and numerical results of the three models. Although significant difference was observed in some cases, especially the double-orifice plate at P3, most cases agreed with the test results. It is worth noting that it is challenging to simulate a perfect match with the field test, especially in an explosion test, where highly nonlinear dynamic loading exists. The quality and density variation of explosive charge could also cause inconsistency. In addition, the inherent limitation of the continuum FE model may also cause variation. The results might be improved by upgrading the data acquisition system or further investigating the material parameters used in the model. It can be concluded that the agreement of the trend of the pressure attenuation rate is good, and the model gives reasonable predictions for different tunnel blast attenuation designs.

Table 4. Percentage error of different pressure attenuation models.

	Weight of C-4	100 g	150 g	200 g	250 g	350 g
Expansion chamber	P1	11%	5.4%	17%	5.9%	20.6%
	P2	20%	24%	46%	26.2%	2.21%
	P3	59%	33%	49%	49.6%	22.8%
	P4	29%	28%	6.8%	1.9%	4.2%
Single-orifice plate	P1	1%	13.5%	8.6%	2.3%	10.9%
	P2	27.7%	14.9%	13.5%	5.7%	25%
	P3	89.3%	67.9%	44.1%	45.5%	12%
	P4	4.8%	15.4%	31.2%	33.3%	39.1%
Double-orifice plate	P1	11.4%	29.3%	2.9%	8.4%	3.9%
	P2	53.9%	62.7%	25.8%	22.7%	14.1%
	P3	138%	154%	64.3%	44.4%	19%
	P4	71.2%	93.3%	62.6%	18.3%	9.12%

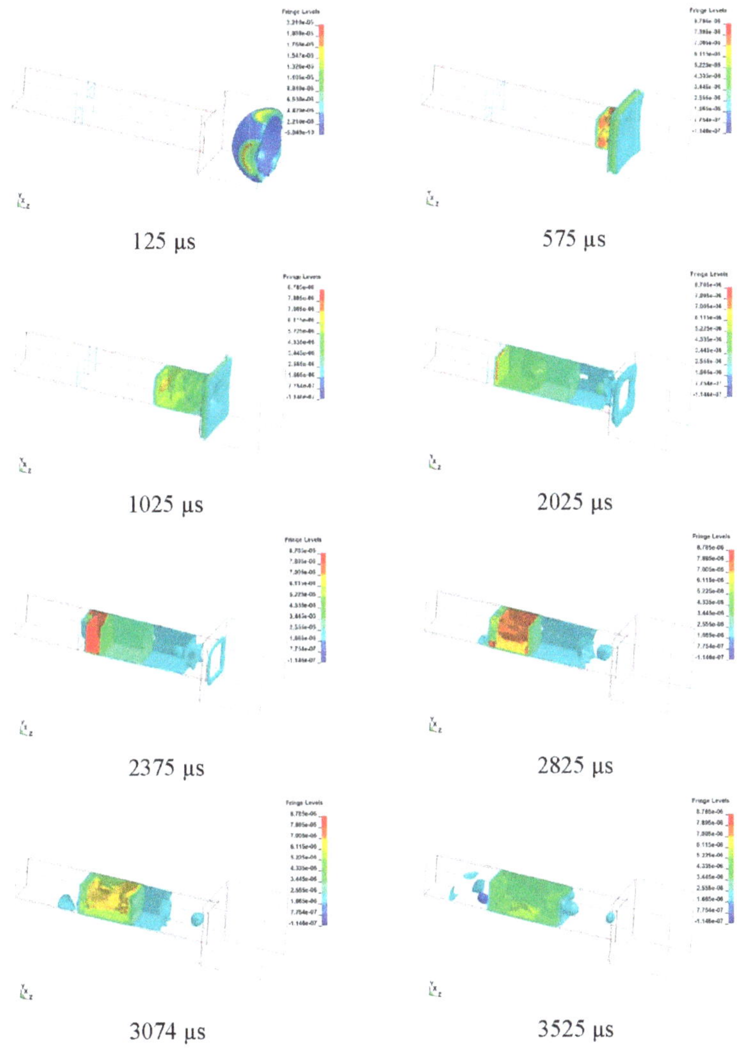

Figure 10. Blast transfer in numerical simulation of linear tunnel with double-pressure relief orifice plate explosion (100 g C-4).

5. Conclusions

(1) The pressure reduction module (expansion chamber, one-pressure relief orifice plate, double-pressure relief orifice plate) changes the tunnel section to diffuse the blast and reduce or detour the transfer; thus, the aforesaid design modes have a blast attenuation effect.

(2) The pressure reduction modules are designed inside the tunnel, and the findings show that the double-pressure relief orifice plate has better blast attenuation effect than the one-pressure relief orifice plate, and the one-pressure relief orifice plate is better than the expansion chamber. The expansion chamber can attenuate the blast by 30%. The one-pressure relief orifice plate can attenuate the blast by 51%. The double-pressure relief orifice plate can attenuate the blast by 82%.

(3) The overall blast attenuation trend in the numerical simulation result of the pressure reduction module matches the experimental result. The results of this study can provide a reference for future protective designs.

Author Contributions: Conceptualization, C.-W.H. and P.-W.W.; data curation, C.-W.H. and Y.-K.T.; formal analysis, C.-W.H. and H.-H.L.; investigation, C.-W.H., Y.-K.T., T.-A.C. and H.-H.L.; methodology, C.-W.H., Y.-K.T. and T.-A.C.; supervision, C.-W.H. and H.-H.L.; writing—original draft, C.-W.H.; writing—review and editing, T.-A.C. All authors have read and agreed to the published version of the manuscript.

Funding: This research received no external funding.

Institutional Review Board Statement: Not applicable.

Informed Consent Statement: Not applicable.

Acknowledgments: The authors would like to thank Sheng-Rong Pi for assistance.

Conflicts of Interest: The authors declare no conflict of interest.

References

1. Smith, P.D.; Vismeg, P.; Teo, L.C.; Tingey, L. Blast wave transmission along rough-walled tunnels. *Int. J. Impact Eng.* **1998**, *24*, 419–432. [CrossRef]
2. Yu, W.F.; Hung, C.W.; Cheng, D.S. E_ect of Subdividing Stacks on Blast Overpressure from Explosion inside Ammunition Storage Magazine. *J. Explos. Propellants* **2008**, *24*, 25–40.
3. Luccioni, B.; Ambrosini, D.; Nurick, G.; Snyman, I. Craters produced by underground explosions. *Comput. Struct.* **2009**, *87*, 1366–1373. [CrossRef]
4. Yu, W.F.; Hung, C.W.; Cheng, D.S. Effect of Blast wall on Safety Distance of Ammunition Storage Magazine Subjected to Internal Explosion. *J. Chung Cheng Inst. Technol.* **2010**, *39*, 131–145.
5. Cheng, D.S.; Hung, C.W. Experiment and Numerical Simulation of Peak Overpressure of C4 Explosives in the Airblast. *J. Explos. Propellants* **2010**, *26*, 75–96.
6. Pi, S.J.; Cheng, D.S.; Cheng, H.L.; Li, W.C.; Hung, C.W. Fluid-Structure- Interaction for a Steel Plate subjected to Non-Contact Explosion. *Theor. Appl. Fract. Mech.* **2012**, *59*, 1–7. [CrossRef]
7. Zakrisson, B.; Wikman, B.; Häggblad, H.Å. Numerical simulations of blast loads and structural deformation from near-field explosions in air. *Int. J. Impact Eng.* **2011**, *38*, 597–612. [CrossRef]
8. Hung, C.W.; Tsai, Y.K.; Chen, T.A.; Wu, P.W. An experimentally validated numerical model for the near-field explosion of an ammunition storage magazine. *Appl. Sci.* **2020**, *19*, 6849. [CrossRef]
9. Hung, C.W.; Lai, H.H.; Shen, B.C.; Wu, P.W.; Chen, T.A. Development and Validation of Overpressure Response Model in Steel Tunnels Subjected to External Explosion. *Appl. Sci.* **2020**, *18*, 6166. [CrossRef]
10. Lai, H.H. Applicability of a Design Assessment and Management for the Current Ammunition Depots in Taiwan. *Appl. Sci.* **2020**, *10*, 1041. [CrossRef]
11. Son, S.Y.; Lee, J.; Ahn, W.; Kim, H.W.; Choi, J.S. Effect of Blast Traps on Air-blast Propagation in Underground Storage. In Proceedings of the 25th Explosives Safety Seminar, 25th DoD Explosives Safety Seminar, Anaheim, CL, USA, 8 August 1992; pp. 85–98.
12. Scheklinski-Gluck, G. Blast in Tunnels and Rooms from Cylindrical HE-Charges Outside the Tunnel Entrance. In Proceedings of the Sixth International Symposium on Interaction of Nonnuclear Munitions with Structures, Panama City, FL, USA, 3–7 May 1993; pp. 68–73.
13. McMahon, G.W.; Britt, J.R.; Patterson, B.C. *Airblast Propagation within Tunnels from Portal Detonations*; MABS 18: Bad Reichenhall, Germany, 2004; Available online: https://www.mabs.ch/spiez-base/mabs-16-to-20/mabs-18/ (accessed on 17 June 2021).
14. Technical Manual TM5-855-1. *Fundamental of Protective Design for Conventional Weapons*; Department of the Army: Washington DC, USA, 1998. Available online: https://www.nrc.gov/docs/ML1019/ML101970069.pdf (accessed on 17 June 2021).
15. Welch, B.; McMahon, G.W.; Davis, K. *Transportation Tunnels and Terrorist Attacks*; ERDC: Vicksburg, MS, USA, 2005.
16. Cheng, D.S.; Chang, Y.L.; Yu, W.F.; Yen, C.P.; Lee, K.H.; Wu, K.H. Simulation of explosions inside ammunition storage magazines of strip and bent channels. In Proceedings of the 16th Seminar on National Defence Science and Technology, Taoyuan, Taiwan, 2008; pp. 1–6.
17. Ishikawa, N.; Beppu, M. Lessons from past explosive tests on protective structures in Japan. *Int. J. Impact Eng.* **2007**, *34*, 1535–1545. [CrossRef]

applied sciences

Review

Blast Loaded Columns—State of the Art Review

Sanja Lukić and Hrvoje Draganić *

Department for Materials and Structures, Faculty of Civil Engineering and Architecture Osijek, Josip Juraj Strossmayer University of Osijek, Vladimira Preloga 3, HR-31000 Osijek, Croatia; slukic@gfos.hr
* Correspondence: draganic@gfos.hr

Abstract: The ever-present threat of terrorist attacks in recent decades gives way to research towards blast-resistant design of structures. Columns, as one of the main load-bearing elements in residential buildings and bridges, are becoming interesting targets in bombing attacks. Research of column blast load behavior leads toward increased safety by identifying shortcomings and problems of those elements and acting accordingly. Field tests and numerical simulations lead to the development of new blast load mitigation technics, either in the design process or as a retrofit and strengthening of existing elements. The article provides a state-of-the-art literature review of filed blast load tests and numerical simulations of a bridge and building columns.

Keywords: blast load; concrete columns; experimental testing; numerical modeling

Citation: Lukić, S.; Draganić, H. Blast Loaded Columns—State of the Art Review. *Appl. Sci.* **2021**, *11*, 7980. https://doi.org/10.3390/app11177980

Academic Editors: Ricardo Castedo, Lina M. López and Anastasio P. Santos

Received: 30 July 2021
Accepted: 26 August 2021
Published: 28 August 2021

1. Introduction

In the last five decades, terrorist attacks have become more frequent. There are different types of terrorist attacks, but according to data provided by the National Consortium for the Study of Terrorism and Response to Terrorism [1], in the last two decades, explosive attacks exceed 50% of the total number of incidents, shown in Figure 1. The attacks on The Twin Towers of the World Trade Center on 11 September 2001 and bridges in California and New York have an impact on the design of structures in the United States and also in the rest of the world [2]. In every country, the transportation system is essential for performing everyday activities, so the Blue Ribbon Panel (BRP) indicates the transportation system as one of the viable targets for a bombing attack. Due to a large number of bridges worldwide, lots of potential casualties, high repair costs, and importance in everyday life, the bridges are increasingly in the focus of terrorist attacks. This is confirmed by the fact that in the last few attacks in Nigeria in 2020, seven bridges were destroyed. It is important to identify which bridges are vulnerable due to their easy accessibility to protect against attacks. Moreover, BRP states that the columns are one of the most critical components on all types of bridges [3]. As there are many types of bridges and many ways to attack the bridge, it is difficult to predict the construction's response to the blast loadings [4]. When detonation of an explosion is under the bridge, then columns are exposed to large lateral forces, depending on standoff distance, which can result in large deformations leading to flexural or shear failures. The contact explosion can breach the column to render it incapable of supporting the dead loads. For small standoff distance, blast waves can cause a serious reduction in a concrete cross-section in terms of spalling and cratering. Since the column failure depends on the position and amount of explosives, all examined attack scenarios were observed.

The main objective of this review article is an extensive literature overview of experimental and numerical research conducted on blast-loaded columns. Both building and bridge columns are considered due to differences in their static and blast behavior. A systematic summary is given of column behavior, possible damage and failure modes, and a review of software used for blast load simulation and analysis.

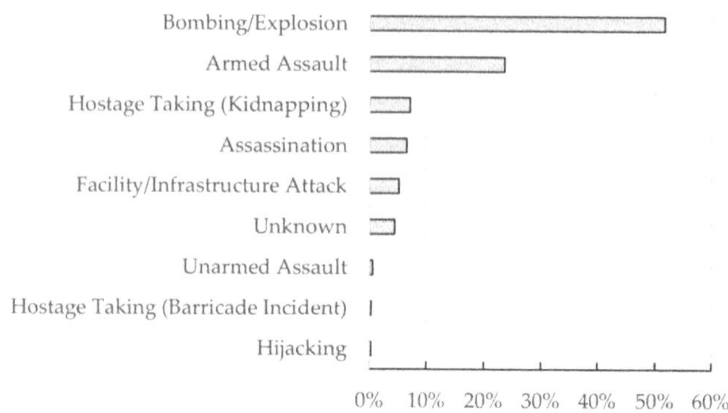

Figure 1. Percentages of terrorist attack types, based on data from [1].

2. Experimental Testing

There are no experiments on real scale specimens of bridge columns due to the experimental setup complexity and high costs, while the building columns are mainly conducted in full scale due to their maximum height of 2.5 to 3 m. Moreover, these types of experiments require special testing ground (usually military field ranges) and trained personnel for handling explosives. Even these special testing grounds have limitations regarding the maximum amount of explosives that can be used in one detonation. This also limits the scale of specimens. Research conducted in recent decades has shown that scale tests provide reliable results and the necessary knowledge to analyze the effects of blast load on full-scale structures [5].

If considering building and bridge columns, except specimen dimensions, there is a difference in their behaviors due to different levels of axial load capacity. Therefore, it is recommended to analyze bridge and building columns separately [4]. A list of conducted experimental research is provided in Table 1.

Table 1. Summary of experimental research.

Author	Year	Structural Element	Experiment Type	Material	Scale
Bruneau et al. [6]	2006	Multicolumn bents	Field	CFCSC	1:4
Fujikura et al. [7]	2008	Multicolumn bents	Field	CFST	1:4
Davis et al. [8]	2009	Bridge column	Field	RC	S. s. + 1:2
Fujikura and Bruneau [9]	2010	Multicolumn bents	Field	RC and RC SJ	1:4
Williamson et al. [4]	2011	Bridge column	Field	RC	1:2
Crawford [10]	2013	Building column	Field	RC + FRP + SJ	1:1
Burrell et al. [11]	2015	Column	Shock tube	SFRC	1:2
Zhang et al. [12]	2015	Building column	Field	CFST	1:1
Aoude et al. [13]	2015	Building column	Shock tube	UHPFRC	1:1
Codina et al. [14]	2016	Building column	Field	RC	1:1
Codina et al. [15,16]	2016	Building column	Field	RC, RC SJ, RC + polyurethane bricks	1:1
Xu et al. [17]	2016	Column	Field	UHPFRC + HSRC	1:1
Echevarria et al. [18]	2016	Bridge column	Field	CFFT + RC	1:5
Fouché et al. [5]	2016	Multicolumn bents	Field	RC MSJ	1:4
Wang et al. [19]	2016	Column	Filed	RPC-FST	1:1
Zhang et al. [20]	2016	Column	Field	CFDST infilled with UHPC	1:1
Zhang et al. [21]	2017	Column	Field	CFDST	1:1
Codina et al. [22]	2017	Building column	Field	RC + reinforced resin panels	1:1
Yuan et al. [23]	2017	Bridge column	Field	RC	1:3
Wang et al. [24]	2017	Building column	Field	CFST	1:1

Table 1. *Cont.*

Author	Year	Structural Element	Experiment Type	Material	Scale
Li et al. [25]	2017	Building column	Field	UHPC + HSRC	1:1
Fouché et al. [26]	2017	Bridge column	Field	CFDST	1:4
Dua et al. [27]	2018	Column	Field	RC	1:1
Dua et al. [28]	2019	Column	Field	RC	1:1
Wang et al. [29]	2020	Bridge column	Field	UHPCC-FST	1:4
Kadhom et al. [30]	2020	Column	Shock tube	RC and RC + CFRP	1:2
Vapper and Lasn [31]	2020	Building column	Filed	RC, RC + GFRP	1:2

Note: CFCSC—Concrete-Filled Circular Steel Columns, CFST—Concrete-Filled Steel Tube, RC—Reinforced Concrete, SJ—Steel Jacket, MSJ—Modified Steel Jacket, FRP—Fiber-Reinforced Plastic, SFRC—Steel Fiber-Reinforced Concrete, UHPFRC—Ultra-High Performance Fiber Reinforced Concrete, HSRC—High Strength Reinforced Concrete, CFDST—Concrete-Filled Double-Skin Tubes, UHPC—Ultra-High Performance Concrete, SFRP—Steel Fiber Reinforced Polymer, UHPFRC—Ultra-High-Performance Fiber-Reinforced Concrete, CFFT—Concrete-Filled Fiber-Reinforced Polymer (FRP) tube, RPC-FST—Reactive Powder Concrete Filled Steel Tubular, HSRC—High Strength Reinforced Concrete, S. s.—small scale, GFRP—Glass Fiber Reinforced Polymer.

2.1. Bridge Columns

The experiments were carried out on standard RC columns, additionally retrofitted columns, and improved composite concrete columns. The columns are exposed to various scenarios of explosive attacks. In addition to the type of column, the scenarios also differ in the position, type, and amount of explosives.

Williamson et al. [32] provided the list of possible terrorist courses of action and indicated that the hand placed explosives on the column and large truck-bomb below the bridge superstructure can destroy columns and cause bridge collapse.

Due to similarities between the effects of the explosions and earthquakes, Bruneau et al. [6] developed a multi-hazard pier concept that they expect to provide a satisfying level of protection against failure under both loadings. All specimens were concrete-filled circular steel columns (CFCSC) with three different diameters (10.16 cm (4″), 12.7 cm (5″), 15.24 cm (6″)) and a minimum steel thickness of 3.2 mm. Specimens were made in 1:4 scale of the prototype bridge columns. Due to security reasons, the actual values of charge weights and standoff distances are not provided. Experimental results showed that even a minimal increase in standoff distance and column diameter significantly reduces column deformation. The CFCS columns showed ductile behavior and high resistance to the effects of the explosion [6]. The same scale and scenario when the explosive was located under the bridge in a car placed near to the column were examined in [7]. They assumed the charge weight similar to the blast weights predicted in FEMA (2003) [33] and FHWA (2003) [34]. Charges are set at heights of 0.25 m and 0.75 m, which correspond to the actual height of 1 m (car bomb) and the half column height, respectively. They concluded that only steel jacketing is not enough to provide adequate resistance to the large shear forces influencing the bottom of the column. Therefore, they found that a better solution is using a fully concrete-filled steel tube (CFST) continuously embedded into the footing. CFST columns provided ductile behavior and sufficient resistance to the lateral forces from earthquakes and explosions. Moreover, the advantage of CFST columns is that they do not have a breach and a spall of concrete, i.e., they do not produce flying debris [7]. Figure 2 shows the connection concept between the foundation beam and the CFST column, which provides the full moment capacity of the column. At a rotation of 3.8° of the bottom of the column, plastic deformation is visible but without cracking of the concrete. The first cracks occur at a rotation of 8.3°, while the fracture of the steel tube occurs at 17°. At the height of the explosive charge, pits and notches appeared on the steel tube, while concrete cracks occurred on the tension side at the bottom and top of the column due to the rigid boundary conditions [7]. They assumed the same blast scenario as Fujikura et al. [7] in their work at the same scale of 1:4, but there were four columns in the test specimen while the bridge prototype has three. Figure 3 shows the test setup for the same blast load scenario but on different types of columns. The RC column exhibited shear failure at the base and cracking of concrete along the column [9], RC SJ shear failure [9], and CFST column flexural failure and buckling [7]. Fujikura and Bruneau [9], in their work, presented experimental and analytical investigations of seismically ductile RC columns and non-ductile RC columns retrofitted with steel jackets. The charge was set to a height of 0.25 m which corresponds to the actual height of 1 m (car bomb), and at this height, the column experienced the maximum deflection. All columns failed in direct shear at the base, but RC columns with steel jackets did not experience any structural damage, and the RC column experienced spalling of concrete at the bottom. Compared to the CFST columns, these columns did not exhibit a ductile behavior.

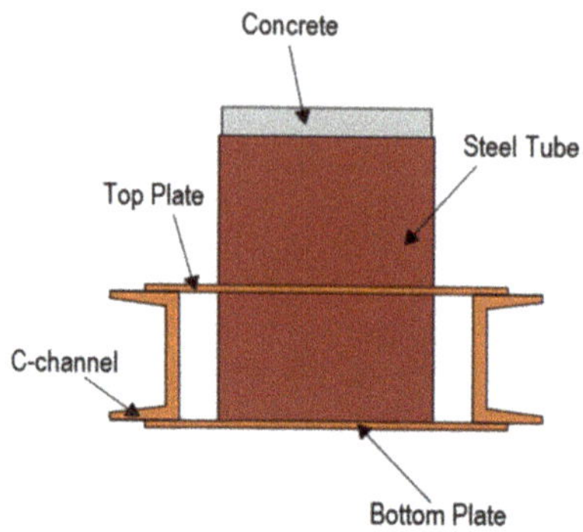

Figure 2. CFST column—details of column-to-foundation beam connection [7].

Davis et al. [8] conducted an experiment in two phases. In the first phase, they tested eight small-scale columns where they first changed the standoff distance and then the amount of explosive charge while the scaled distance was kept fixed. In the second phase, they tested 16 columns in half scale (1:2), where 10 of them were set at a small standoff to observe the mode of failure (flexure or shear) like in [4], and the remaining six were to sustain local damage (spall and breach patterns). In all samples, concrete strength, clear cover, concrete class, and reinforcement grade were unchanged. Boundary conditions for tested columns were assumed to be pinned at the top and fixed at the bottom. The test setup did not include axial load because low levels of axial loads provide greater capacity to the column, and without axial load, tests are on the conservative side. Five experimental observations and guidelines for the design of blast-loaded columns are provided in [4,8,35]:

1. Using protective fences and barriers for vehicles to increase the standoff distance;
2. The circular cross-section can maintain a lower pressure of up to 1/3 concerning a square cross-section of the same dimensions, so the second guideline is to use circular columns, and also, the pressure reducing factors on the circular column were proposed by Winget et al. [2], Marchand et al. [36], and Fujikura et al. [7], respectively, as 0.80, 0.75, and 0.45;
3. Increase in the reinforcement in the column, as this increases the shear capacity, ductility, and confinement of the concrete;
4. Use of continuous reinforcement because discrete hoops can be extracted during a blast load;
5. Placing longitudinal splices away from the charge if they cannot be completely avoided.

They also proposed three design categories (A, B, C) that depend on the scaled distance and require a different approach to designing, i.e., gravity, seismic, blast. In the C category ($Z \leq 0.6$ m/kg$^{1/3}$), columns are exposed to the higher loads than columns in A ($Z > 1.2$ m/kg$^{1/3}$) and B ($0.6 < Z \leq 1.2$ m/kg$^{1/3}$).

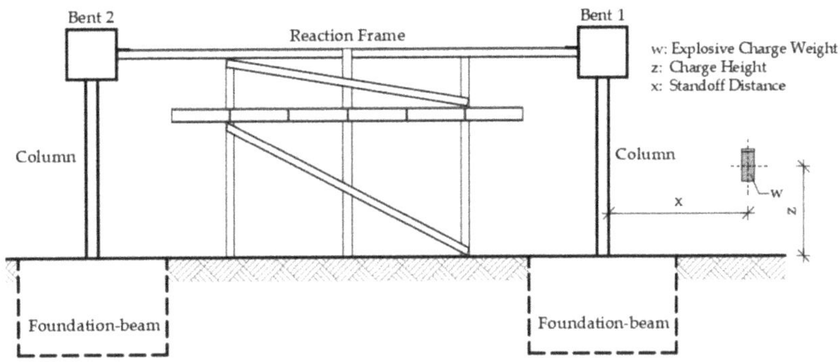

Figure 3. Test setup in [7] and [9].

Williamson et al., in Part II [35], provided a review of experiments represented in Part I [4]. Square columns experienced greater net resultant impulse than circular columns under the same blast loads and also had a larger cross-section area, so less shear occurred at the base. To increase the shear capacity of the column at the base and the ductility, it is necessary to increase the amount of transverse reinforcement. Moreover, columns with continuous spirals had a better performance than columns with discrete ties, which confirms the recommendation given in [8]. Due to changes in column design, the weight of charge, and standoff distance, several levels of damage were obtained [37,38]. The test setup is shown in Figure 4. Superficial damage means that the column performed well and has only surface damage and cracks, while minor damage means spalling of the concrete cover and cracks along the column. Deformations, flexural cracking, and spalling of concrete all together are extensive damage, while the failure of the column means that a shear occurred at the base [4].

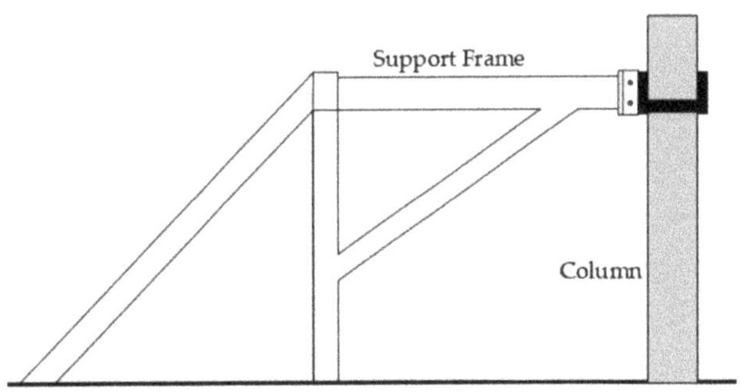

Figure 4. Test setup [4,8].

Echevarria et al. [18] tested RC and Concrete-Filled Fiber-Reinforced Polymer (FRP) tube (CFFT) bridge columns at a scale of 1:5. CFFT columns are reinforced only with longitudinal reinforcement, while RC columns have spiral hoops in addition to longitudinal reinforcement, shown in Figure 5. The quantity and distance of the explosives were not provided for safety reasons. The columns experienced minimal visual damage, but measurements showed that both concrete and steel experienced large loads and strains. In the residual test, CFFT columns showed greater strength and ductility than RC columns.

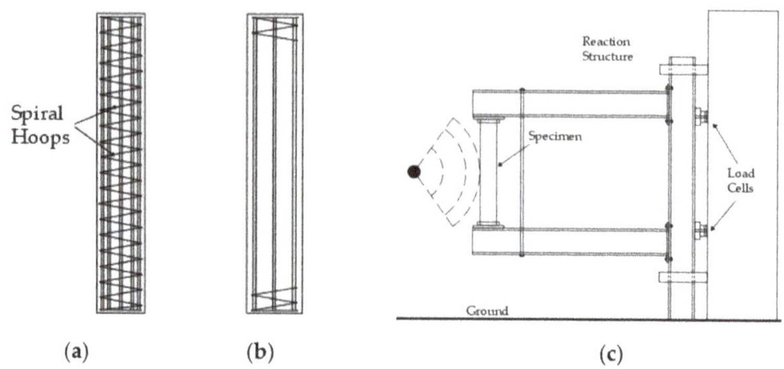

Figure 5. Reinforcement position in specimens: (**a**) RC column and (**b**) CFFT column; specimen position: (**c**) test setup [18].

Fouché et al. [5] made the same prototype of the bridge columns as [9] and exposed them to the same blast load scenario. To improve the behavior of the steel jacked columns to the explosion, they added structural steel collars at the top and bottom of the column. The steel collar at the base of the steel jacked RC column is shown in Figure 2 in **[5]**. They concluded that the modified column was effective in preventing direct shear and had increased ductility. The obtained maximum base rotation ranges from 8.6 to 10.3°, and all the columns showed satisfactory behavior.

Yuan et al. [23] experimentally tested two RC bridge columns in scale 1:3 under 1kg TNT contact explosion placed at the height of 33 cm from the ground. As a retaining structure that prevents the rotation and displacement of the column at the top, they used a wall with the opening that was placed at a distance of 1.4 m from the detonation point. Both columns experienced spallation and crushing of concrete cover, but the stirrup fracture was observed only in the square column. The damages to the front and back sides of the circular and square columns are shown in Figure 13. Therefore, the test results showed that the square column had more severe damage than the circular column.

Wang et al. [29] investigated the impact of contact explosions of 1 kg, 2 kg, and 3 kg of TNT on the mode of failure and original and residual axial capacity of Ultra-High Performance Cementitious Composite Filled Steel Tube (UHPCC-FST) bridge column. The columns were made in 1:4 scale and are tested horizontally, and the views of the test setup are shown in Figure 6. The top of the column is pinned, and the bottom of the column is fixed. The cylindrical explosive is placed at a distance of 25 cm from the lower support, which represents the actual position of the explosive in the vehicle at the height of 1 m. Quantities of explosives of 1 and 2 kg made only a crater in the column, while 3 kg fractured the tube and crushed the core, as shown in Figure 7. In the axial compression test, all columns experienced diagonal shear failure.

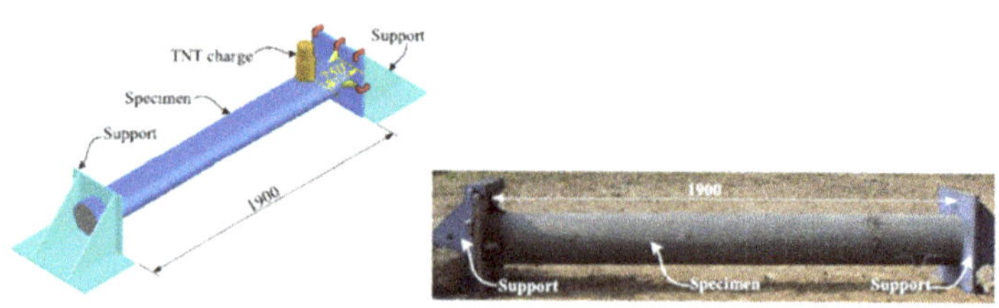

Figure 6. Schematic and field views of test setup [29]. Copyright permission obtained from authors.

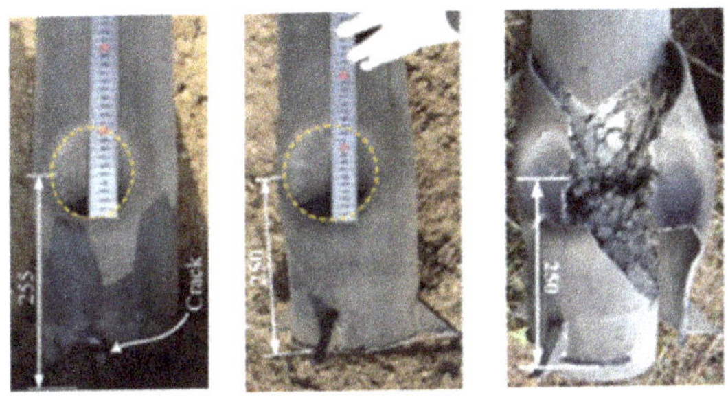

(a) (b) (c)

Figure 7. Damage of UHPCC-FST columns after detonation of: (**a**) 1 kg; (**b**) 2 kg and (**c**) 3 kg of TNT [29]. Copyright permission obtained from authors.

Based on the reviewed literature of bridge columns, maximum support rotations in field blast tests are shown in Table 2. The information can be used as a preliminary, fast damage assessment of blast-loaded columns based on the measured post-blast column rotations.

Table 2. Maximum support rotations (blast at low height).

	Top	Bottom	Crack Patterns of Concrete	Deformation
CFST [7]	1.2° 2.2° 4.9° 18.7°	3.8° 8.3° 17.0° -	No available Tension side Opening of core concrete Blew away	Plastic Plastic On-set of fracture of column Post-fracture of column
RC MSJ [5]	-	8.6–10.3°	Satisfactory ductile behavior	
RC [2,36,39] *	- - -	1.3° 2° 3°	Slight to moderate damage Moderate to heavy damage Lose structural integrity	
RC [40]	- -	2° 4°	Minor damage Collapse	Onset of shear failure at base Shear failure at base
RC (UFC 3-340-02) [41]		2–5° 5–12°	Moderate damage Severe damage	
RC (AISC 341) [42]		2.3°	Highly ductile	

* Based on experimental testing of concrete beam elements in flexure.

2.2. Building Columns

In addition to bridges, interesting targets of terrorist attacks are buildings. By damaging the ground floor columns, the whole building loses stability, so in most of the new buildings, the ground floor columns are designed also considering the blast loads. Building columns differ from bridge columns in the magnitude of the axial loads. Moreover, the dimensions of building columns are significantly smaller, so most of the experiments conducted on building columns were in full or half-scale.

Burrell et al. [11] tested two half-scale reinforced concrete columns and six Steel Fiber-Reinforced Concrete columns (SFRC) with steel fiber content from 0 to 1.5% by volume of concrete (non-seismic and seismic detailing) at shock tube. In the experiment, the axial load equal to 30% load capacity was applied using a hydraulic jack. According to their experiments, columns designed seismically (38 mm distance between transverse reinforcement) have smaller maximum displacements and can withstand larger blast loads. Moreover, SFRC columns with non-seismic detailing (75 mm distance between transverse reinforcement) showed smaller maximum displacements and no secondary blast fragments.

Aoude et al. [13] experimentally tested nine Ultra-High-Performance Fiber Reinforced Concrete (UHPFRC) columns designed with Compact Reinforced Composite (CRC). Tests were performed in a shock tube, and a hydraulic jack was used to input the axial load. The applied blast pressures varied from 69 kPa to 689 kPa. Dimensions of the cross-section were 152 mm × 152 mm, and the free span of the column was 1980 mm. The results showed that the UHPFRC columns reduce secondary blast fragments. Increasing the proportion of fibers from 2% to 4% had a positive effect on the decrease in displacement, but a higher proportion of fibers did not result in improvements in blast behavior.

Zhang et al. [12] experimentally blast tested three square columns and one circular column. The columns were made from steel tubes filled with concrete (CFST). Specimens were placed horizontally with a simple boundary condition at both ends. A pneumatic jack was used to input the axial load, shown in Figure 8. The entire length of the specimen was 2.5 m. In the experiment, an emulsion explosive, which has a TNT-equivalent of 0.7, was used. The maximum and residual column deformations are provided for blast loads utilizing 17.5 to 35 kg TNT equivalence at a standoff distance of 1.5 m. The concrete inside the steel tube reduced local deformations, and the energy was dissipated through the global response of the element. Zhang et al. [20] tested two types of CFDST columns, circular and square, with inner and outer tubes, as shown in Figure 9. They concluded that columns filled with normal strength concrete experienced greater crushing of concrete and higher steel buckling than columns filled with UHPC. Moreover, UHPC proved to be very resistant to spalling or crushing. The tested CFDST and CFST samples have similar oscillation periods and displacements, so it is concluded that they behave similarly. Zhang et al. [21] tested six ultra-high-performance concrete-filled double skin tube columns with square hollow sections. At a standoff distance of 1.5 m, the specimen exposed to 35 kg of TNT did not experience any localized damage or steel buckling. Moreover, the axial load (25% of the maximum load) contributed to the reduction in maximum deflection in the middle of the column. They concluded that the ratio of the cavity and the section influence the overall column deflection and period of oscillation; therefore, it is recommended not to go above 0.5. Wang et al. [19] exposed four circular Reactive Powder Concrete Filled Steel Tubular columns to explosion and fire durations of 0, 60, and 105 min. Fixed supports were simulated, and on one side, an axial load was introduced. Steel tube protects columns against cracks and spalling of concrete. After the detonation of the 17.5 kg explosive, the column experienced bending and after 35 kg bending–shearing deformations. With an increase in the number of explosives and with longer exposure to fire, maximum and residual displacements increased. Wang et al. [24] studied four square and four circular CFST columns under close-range blasts. The column length was 2.5 m, and the thickness of the steel tube was 2.8 mm or 3.8 mm. The standoff distance of the emulsion explosive was 1.5 m, and the charge weight expressed through the TNT equivalence was from 25 to 50 kg. Only a 10% increase in the amount of explosive in square columns increases mid-span deflection by 200%, which is assumed to be caused by a large surface exposed to the blast load. Moreover, increasing the thickness of the steel tube by 1 mm (from 2.8 to 3.8 mm) significantly reduced the displacement (by over 50%). Global failure mode was a flexural failure, and after removing the steel, the square columns sustained spalling and crushing of concrete. Circular columns were broken into several parts. Figure 10 shows the damage to the concrete after removing the steel tube after the blast load.

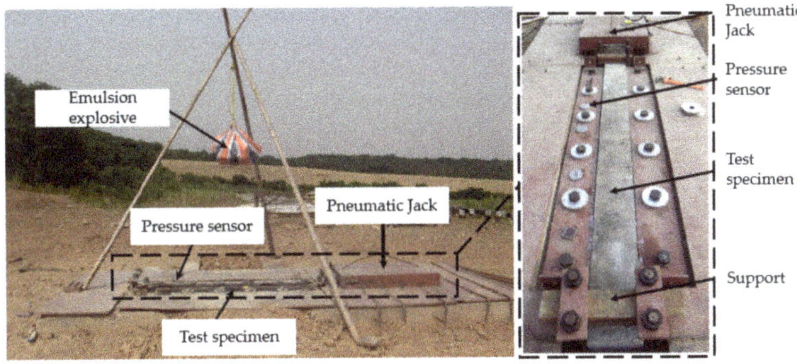

Figure 8. Test setup [24] and test pit [21]. Copyright permission obtained from authors.

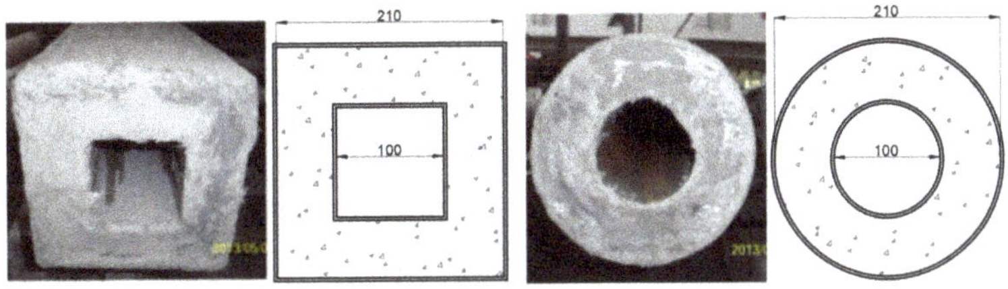

Figure 9. Cross-sections of two CFDST specimens [20]. Copyright permission obtained from authors.

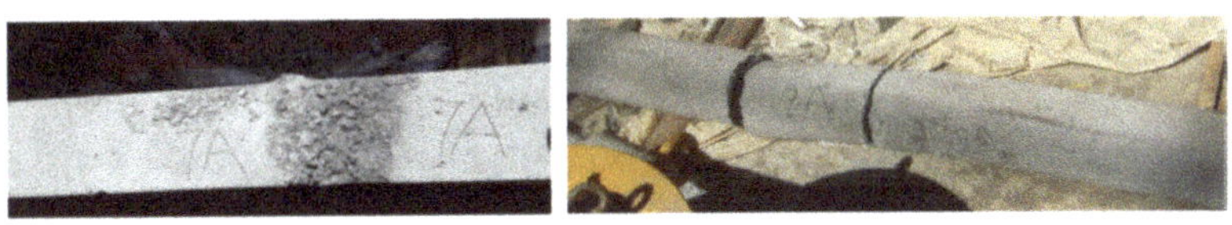

Figure 10. Columns after blast test [24]. Copyright permission obtained from authors.

Codina et al. [14] investigated the effects of close-in blast loading on the full-scale reinforced concrete column. They conducted experimental tests and numerical simulations to calibrate the numerical model. The observed column had a square cross-section of 230 mm × 230 mm and a free height of 2.44 m. The column was tested in a horizontal position, and the standoff distance from the center of the charge (8 kg of equivalent TNT—the used explosive is Gelamon VF65, which is equivalent in a mass to 65% TNT) to column 1 was 100 cm and to column 2 was 60 cm. Both columns experienced flexural damage, spallation of the concrete on the bottom side, and crushing of concrete on the exposed side, shown in Figure 11.

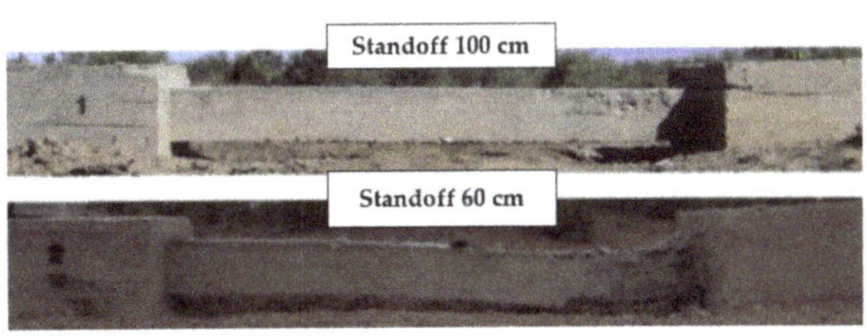

Figure 11. Damage of the RC columns after detonation at 100 cm and 60 cm standoff [14]. Copyright permission obtained from authors.

Codina et al. [15,16] investigated the behavior of RC columns and RC columns with two types of protection (steel jacket and reinforced polyurethane bricks) exposed to a near field explosion. At a standoff distance of 60 cm, 8 kg of TNT equivalent shaped into a cylinder was placed, resulting in a scaled distance Z = 0.30 m/kg$^{1/3}$. Comparing the test results of three types of columns, the steel-jacked column had the best results in residual capacity and in reducing final deflection. Polyurethane bricks are lighter and have cheaper protection for columns but give three times worse results than steel jacketing. It is recommended to set bricks of higher density to improve the effect in the area of blast load. In [16], they also examined reinforced resin panels with insulation layers as a possible improvement of the column. This protection system gave the best results, minimizing column damage and the greatest deflection reduction. The protection system in the field test is shown in Figure 12. Codina et al. [22], after their research on plain RC columns [14], tested RC columns covered with reinforced resin panels with an insulation layer and steel jacketing. From the obtained results, it can be concluded that a significant reduction in damage and displacement was achieved with the cladding system, but the spalling and burst of concrete cannot be prevented. Figure 14 shows a schematic representation of all specimens and their damage covered by the studies of Codina et al.

(a) (b)

Figure 12. RC columns reinforced with: (**a**) polyurethane bricks [15,16] and (**b**) resin panels and insulation layer [22]. Copyright permission obtained from authors.

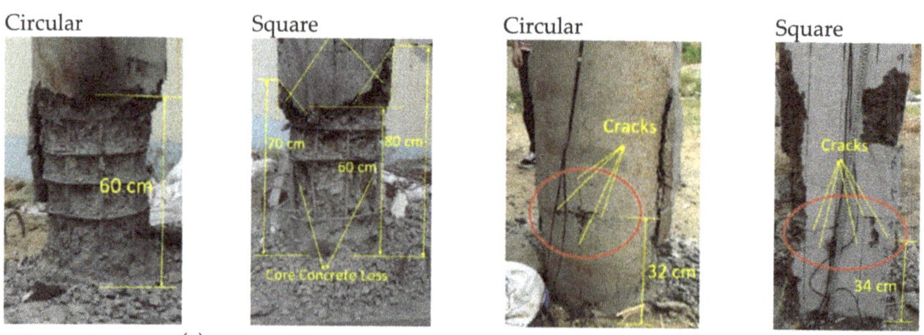

(a) (b)

Figure 13. Comparison of damages of circular and square RC columns after the detonation of 1 kg of TNT: (**a**) front side and (**b**) back side [23]. Copyright permission obtained from authors.

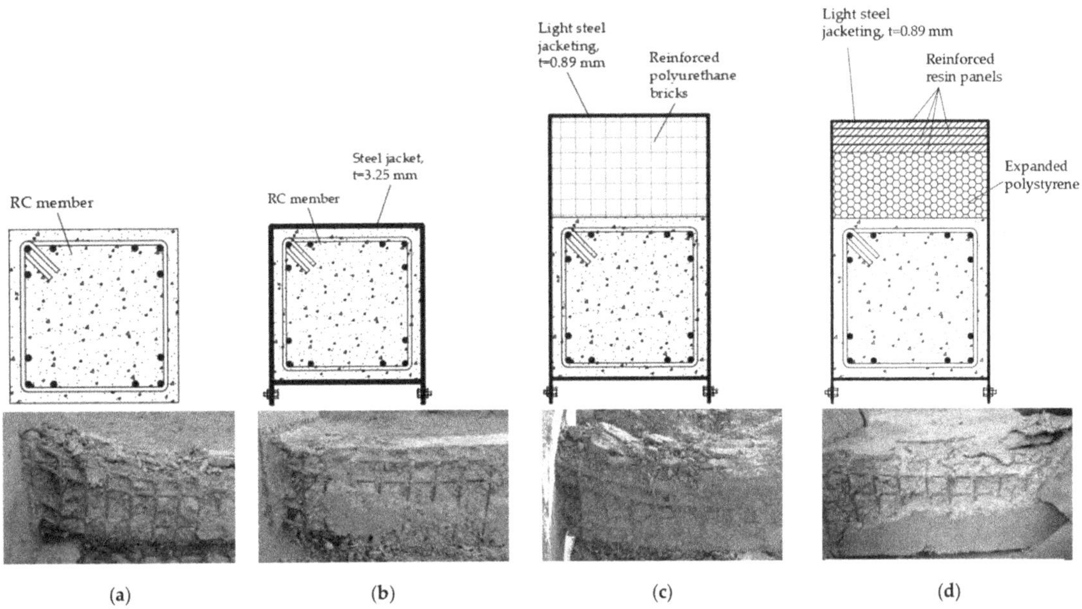

Figure 14. Damage after blast load on RC members: (**a**) without protection; (**b**) with steel jacketing; (**c**) with polyurethane bricks, and (**d**) with reinforced resin panels [15,16,22]. Copyright permission obtained from authors.

Xu et al. [17] subjected four Ultra-High-Performance Fiber Reinforced Concrete (UHPFRC) and four High Strength Reinforced Concrete (HSRC) square columns to the effects of blast loads. Columns were tested under varied charge weights from 1.4 kg to 48 kg of emulsion explosive (TNT equivalence factor is 1.4). The standoff distance was fixed at 1.5 m in all tests. Specimens were placed horizontally, and the axial load was applied using a pneumatic jack. The results showed that UHPFRC columns could better withstand overpressure and shock waves, reducing the maximum displacements.

Li et al. [25] tested 10 Ultra-High-Performance Concrete (UHPC; six reinforced with twisted fiber and four reinforced with micro fiber) and 5 High Strength Reinforced Concrete (HSRC) columns. The length of the specimens was 2.5 m, and the cross-section was square with dimensions 0.2 m × 0.2 m. In experiments, the standoff distances of the explosive from the columns were constantly 1.5 m in all tests, but the explosive weight was changed. For the UHPC columns, weights of 17.5, 25, and 35 kg, and for the HSRC 8, 17.5 and 25 kg were used. Residual load capacity tests showed that UHPC columns after blast loads did not lose much on the axial load capacity. UHCP columns also showed much better load capacity after 35 kg TNT detonation than HSRC columns after 8 kg TNT detonation.

Fouché et al. [26] experimentally tested 12 columns at a scale of 1:4 under the blast load. They varied the void ratio, diameters, and thicknesses of the outer and inner steel tubes. The specimens' cross-section generally experienced denting, and that deformation helped to energy absorption from the overpressure from the near-contact explosions. The inner steel tube played the role of a dowel preventing direct shear failure, and this is the advantage of the CFDST columns over CFST and RC columns. On the tensile side of the column, the concrete was crushed, or horizontal flexure cracks appeared.

Dua et al. [27] tested three RC columns (30 cm × 30 cm × 375 cm) in full scale with the same material and geometrical properties on contact explosion at the bottom of the column. They used 0.1 kg of plastic explosive (PEK—TNT equivalent 1.15) and 0.5 kg and 1 kg of TNT. The TNT charge of 1 kg made a hole in the column; 0.5 kg destroyed the concrete cover, and the remaining core has no residual capacity; 115 g TNT equivalent caused the spalling of the concrete cover. Column damage profiles are shown in Figure 15. A contact explosion causes significant local damage on at least three sides of the column, while a far-field explosion causes the worst damage on the front, exposed side. Dua et al. [28] investigated the same blast load scenario on the column concerning the increase in the cross-sectional width of the column. They experimentally tested columns of dimensions 50 cm × 30 cm, 70 cm × 30 cm, and 90 cm × 30 cm. Rectangular columns showed better behavior under contact explosion than squares. They examined the residual load-bearing capacity of the column and determined the column damage index. When the width dimension of the column subjected to the blast load is greater two and more times from depth, the damage index is lower.

(a) (b) (c)

Figure 15. Column damages after blast loads: (**a**) 1 kg; (**b**) 0.5 kg and (**c**) 115 g TNT [27]. Copyright permission obtained from authors.

Kadhom et al. [30] examined five half-scale RC columns. Three columns are strengthened with unidirectional and woven CFRP laminates while the other two remained unprotected. They were first tested in the shock tube on the induced blast load, and thereafter, their residual axial capacity was examined. The column strengthened with CFRP laminates with ±45° woven fibers showed the best blast behavior and the most ductile response.

Fujikake and Aemlaor [43] investigated how longitudinal and shear reinforcement ratios, concrete strength, and the number of explosives affect the demolition of RC building columns. Their primary research goal is not terrorist attacks but the demolition of dilapidated concrete buildings. They used a Composite 4 (C4) explosive because of its stability and ease of shaping and placed it in the core of the column. They found that shear reinforcement plays a significant role in the residual bearing capacity after blasting; the strength of the concrete also affects the increase in residual compressive and flexural resistance capacities. The quantities of explosives and reinforcement cannot be applied to the external action of the explosion, but certainly, conclusions about the role of reinforcement and the strength of concrete are useful.

Roller et al. [44] observed two scenarios: first is a contact explosion with the amount of PETN that fits in the suitcase, and the second is the close-in scenario when the explosive is in the vehicle. They investigated the impact on RC columns and strengthened columns. Polymer concrete, SIFCON (Slurry Infiltrated Fiber Concrete), DUCON (Ductile Concrete), and Ultra-High-Performance Concrete (UHPC) were used to improve the resistance of the bridge and building columns. The appearances of column damages after contact explosion and the residual load capacities are shown in Table 3. The results showed an increase in residual bearing capacity by up to 70%.

Table 3. Damaged columns after contact detonation and the residual load capacities [44].

	RC	Polymer Concrete	SIFCON	DUCON	UHPC
Type					
Damage				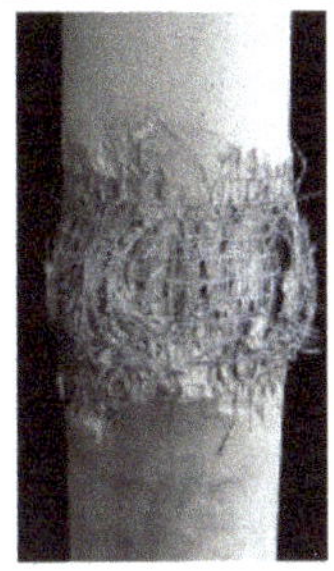	
Residual load capacity	5.5%	68.6%	69.6%	49.3 (coarse)–65.9 (fine) %	-

Xu et al. [45] tested five columns in an explosion containment vessel (ECV). The columns were exposed to an explosion of 40 g charge mass, and the distance of the explosive was changed in each test from a contact explosion to a standoff distance of 50 cm. They installed four smart aggregates (SAs) in each specimen for internal damage detection, shown in Figure 16. The propagation of the stress wave energy decreases with the formation of cracks under the blast load, and hence the amplitude of the time-domain signal recorded by piezometric smart aggregate sensors decreases with the appearance of cracks. This method of detecting internal damage has proven to be useful for completing the picture of the condition of the structure because internal cracks have a greater impact on the damage index of the structure than surface cracks.

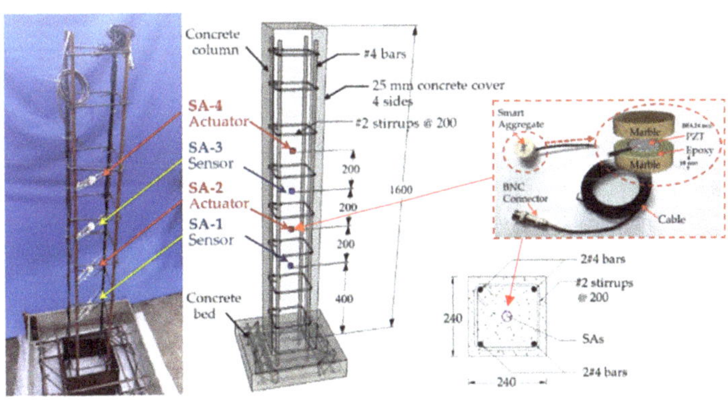

Figure 16. Position and appearance of the SA sensors [45].

Vapper and Lasn [31] examined columns measuring 100 mm × 100 mm × 1000 mm on the action of a different amount of explosion placed at a vertical standoff distance of 300 mm. They tested four types of columns: reinforced concrete columns, columns reinforced with steel fibers, and both types strengthened with Glass Fiber Reinforced Polymer (GFRP).

Plain concrete showed higher compressive strength compared to steel fiber reinforced concrete. GFRP in reinforced concrete columns did not contribute to the increase in the residual strength, while in steel fiber reinforced columns, the contribution was significant. A reduction in surface damage to GFRP-wrapped columns was also observed.

3. Numerical Modeling

Due to the increasing number of terrorist attacks, new challenges were posed to engineers. It is necessary to have a good understanding of computer programs, their capabilities, and their limitations to predict individual attack scenarios. The use of numerical simulations gives a clearer insight into the blast effects on the entire bridge and individual components. It provides the possibility of determining the most critical parts and problems that cannot be numerically simulated and need to be examined experimentally. Thus, one of the most important and difficult parts of this analysis is to properly define air blast loadings. Table 4 summarizes the software used to predict the effects of explosions on different types of columns.

Table 4. Summary of software for prediction and calculation of blast loads.

Author	Year	Structural Element	Software
Ray et al. [46]	2003	Bridge deck and column	ConWep, BlastX, SHAMRC
Marchand et al. [36]	2004	Bridge columns	BlastX, ConWep, SPAn32
Winget et al. [2]	2005	Bridge concrete girders, deck, columns	BlastX, SPAn32, Nonlin
Rutner et al. [47]	2006	Steel and composite bridge columns	MSC.Dytran
Wu et al. [48]	2009	RC and composite building columns	LS-Dyna
Hao et al. [49]	2010	RC building columns	CARLER
Elsanadedy et al. [50]	2011	RC building columns + CFRP	LS-Dyna
Williams et al. [51]	2011	RC bridge columns	LS-Dyna
Williams et al. [52]	2011	RC bridge columns	LS-Dyna
Crawford [10]	2013	RC building columns + FRP + SJ	LS-Dyna
Magali et al. [53]	2013	RC building columns	Abaqus
Eisa [54]	2014	RC building columns	Abaqus
Abladey and Braimah [55]	2014	RC building columns	Autodyn
Li and Hao [56]	2014	RC column	LS-Dyna
Shi and Stewart [57]	2015	RC building column	LS-Dyna
Liu et al. [58]	2015	RC bridge pier-bent model	LS-Dyna, ConWep
Cui et al. [59]	2015	RC column	LS-Dyna
Zhang et al. [12]	2015	CFST building columns	LS-Dyna
Codina et al. [14]	2016	RC building column	Autodyn
Zhang et al. [21]	2016	CFDST columns	LS-Dyna
Arowojolu et al. [60]	2017	RC bridge column	LS-Dyna
Eamon and Aslendi [61]	2017	RC bridge columns + SFRP	LS-Dyna
Kravchenko et al. [62]	2017	RC building columns	LS-Dyna
Kyei and Braimah [63]	2017	RC building columns	LS-Dyna
Yuan et al. [23]	2017	RC bridge columns	LS-Dyna
Abedini et al. [64]	2018	RC building columns	LS-Dyna
Li et al. [65]	2018	CFDST bridge columns	LS-Dyna
Liu et al. [66]	2018	RC bridge piers	LS-Dyna, ConWep
Li et al. [67]	2019	CFDST bridge columns	LS-Dyna
Liu et al. [68]	2019	RC building columns	Autodyn, LS-Dyna
Liu et al. [69]	2019	RC bridge column + CFRP	LS-Dyna
Thai et al. [70]	2019	RC column + SJ	LS-Dyna
Abedini et al. [71]	2019	RC column	LS-Dyna
Dua et al. [72]	2019	RC columns	LS-Dyna
Dua et al. [28]	2020	RC columns	LS-Dyna
Li et al. [73]	2020	CFDST columns	LS-Dyna
Rajkumar et al. [74]	2020	RC columns	LS-Dyna
Vavilala et al. [75]	2020	RC building columns + polymeric foam	Abaqus
Zhang et al. [76]	2020	Segmental CFST column	LS-Dyna
Yuan et al. [77]	2020	RC column	LS-Dyna
Yan et al. [78]	2020	RC columns + CFRP	LS-Dyna
Hu et al. [79]	2021	RC column + CFRP	LS-Dyna

Note: CFRP—Carbon Fiber Reinforced Polymer; FRP—Fiber-Reinforced Plastic; SJ—Steel Jacket; CFST—Concrete-Filled Steel Tube; CFDST—Concrete-Filled Double Steel Tube; SFRP—Steel Fiber Reinforced Polymer.

Conventional Weapon Effects Predictions (ConWep) [80] and BlastX [81] are programs used to calculate the effect of a blast wave from different types of detonation. ConWep is more used for air-blast calculations, including free-field and reflected blast pressure histories from the free-air, surface, and hemispherical burst explosions, and BlastX calculates internal blast pressure histories. BlastX is based on semi-empirical methods, including nonlinear addition laws for blast pressures from multiple reflecting surfaces based on computational fluid dynamics. Second-order Hydrodynamic Automatic Mesh Refinement Code (SHAMRC) [82] is also used to investigate high explosive and blast effects based on finite-difference computational fluid dynamics (CFD) code. Nonlin [83] does not make the empirical adjustments just for blast loads because it is initially designed for earthquake loads. It has similarities with SPAn32 [84] because it performs a nonlinear dynamic response history analysis taking bilinear material properties. Both programs are based on the analysis of a system with a single degree of freedom (SDOF). MSC.Dytran [85] is an explicit finite element analysis (FEA) solution for simulating blast load effects and analyzing the complex nonlinear behavior that structures undergo during detonation. Ansys Autodyn [86], LS-Dyna [87], and Abaqus [88] are programs that provide the ability to simulate detonation, wave propagation from an explosion, interaction with a structure, and nonlinear material behavior, which are known as hydrocode programs specialized for simulations in fluid dynamics.

Ray et al. [46], in their research, compared three methods with three different resolutions for air blast prediction. In the scenario of below-deck detonation, ConWep has a low resolution of air blast prediction, and the charge can be observed as hemispherical or spherical, while BlastX has a medium resolution and considers the shape of the charge and reflections of the blast pressure. The 3D bridge and blast load can be modeled in the SHAMRC because it is an advanced Eulerian-based finite difference code that has high resolution. Research shows that the highest resolution is not always necessary, as it is a mostly low resolution that provides a conservative design. The authors stated that additional analysis is needed to determine the most economical and sufficiently precise tool for a particular problem [46]. The shape of the explosive also drastically affects the resulting pressure and impulse, so it is necessary to use a program that allows the input of charge geometry.

Marchand et al. [36] determined concrete breaching using ConWep and calculated flexural response and support rotation on a reduced diameter column in SPAn32. They concluded that the strength of concrete does not significantly affect the maximum rotation of the support, but it does affect breaching, i.e., the lower strength causes greater breaching.

Winget et al. [2] use SPAn32 to calculate the flexural response of the columns and to define the equivalent SDOF stiffness and mass parameters based on the column properties. For the calculation of the blast load pressure history, the BlastX program was used. Other useful programs are AT Blast [89] and Nonlin. For the calculation of the pressure-impulse history using AT Blast, it is necessary to know the charge weight, angle of incidence, and standoff distance. However, AT Blast does not consider the effects of multiple reflections under the bridge explosions. They list four categories of bridge design concerning their importance, where category 1 represents very important bridges, and category 4, unimportant. Winget et al. did not take the real conditions of the ground and energy absorption by creating craters but the ideal reflecting surface. Footing instability, however, could also result from large ground deformations, and this aspect of behavior must also be addressed.

Wu et al. [48] numerically simulated RC and composite columns in LS-Dyna for the contact-placed TNT charges from 2.5 to 25 kg. In the simulations, they obtained a higher residual bearing capacity of the column when the explosive was placed at the height of 1.5 m from the bottom than when it was placed at the bottom.

Fujikura et al. [7], for calculation of impulse variations per unit length along the height of the column, were using the Bridge Explosive Loading (BEL) [90] program. BEL also considers the reflected pressure of the blast wave on the surface of the superstructure and on the ground.

Rutner et al. [47] studied the behavior of four types of column cross-sections in MSC.Dytran software on blast load: single-cell hollow steel section, multi-cell hollow steel section, single-cell hollow composite column, and multi-cell composite column. Compared to steel columns, the composites showed negligible deflection on blast load. The best stress distribution in the element was achieved with multi-cell composite columns, which is also visible by the displacements shown in Figure 17.

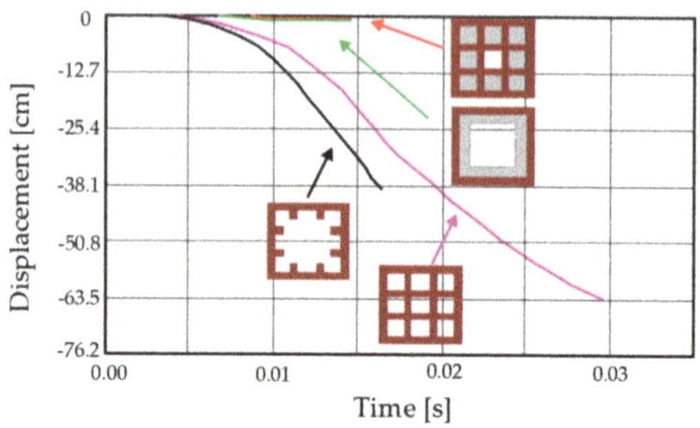

Figure 17. Time history of displacements for 4 column types [47].

Hao et al. [49] analyzed three reinforced concrete columns with the same dimensions, material strengths, and reinforcement ratios but subjected to blast loads of different scaled distances. They wanted to find the failure probability using the computer code CARLER, which is verified with Monte Carlo simulations. They defined four damage levels (D) that depend on the ratio of residual axial load carrying capacity (N_{residual}) when the column is damaged and the axial load of the undamaged column (N_0), shown in Table 5. Through numerical simulations, they concluded that neglecting some of the material properties of the column has minimal impact on the probability of failure, while the random changes in the blast loading have a much greater role.

Table 5. Damage levels of RC column in terms of axial load capacity.

Level of Damage [49]	$D = 1 - \frac{N_{residual}}{N_0}$	Damage Limit States [57]
Low damage	0–0.2	Low damage
Medium damage	0.2–0.5	Repairable damage
High damage	0.5–0.8	Repairable damage
Collapse	0.8–1.0	Collapse

Williams and Williamson [51] emphasized the spalling of side-cover concrete because, in previous works, only the spalling of concrete off the back of reinforced concrete columns was mentioned. The aim of their research was to make and validate a numerical model with respect to the experiment explained in [38] and justified the cross-sectional response mechanisms that cause loss of side-cover concrete. For numerical simulations, they used the LS-Dyna program and the Karagozian and Case concrete (KCC) material model.

Numerical simulations in LS-Dyna showed that the shape of the column cross-section has a large influence on the resulting impulse. The authors developed expressions for calculating column shape factors for circular and square columns. The expressions are used when the R / D ratio is less than 4.5 because they provide sufficiently conservative loads but less than those experimentally determined on the walls [52].

Crawford [10] performed numerical simulations in LS-Dyna of RC columns and columns retrofitted with fiber-reinforced plastic (FRP). FRP increases the resistance of the RC columns to the blast load. For numerical modeling, the choice of concrete material model is very important, and the analysis was performed with four different concrete models (KC, Winfrith, Continuous Smooth Cap, and RHT model). The best results were obtained with the KC model.

Magali et al. [53] performed a parametric numerical analysis in Abaqus to see which of the six varied parameters (section ratio, compressive strength of concrete, column height and thickness, charge radius, and ratio between standoff distance to the charge radius) had the greatest impact on column damage. It was shown that the column thickness, charge radius, and the ratio of standoff distance to charge radius play a significant role in the column response. They give an empirical formula based on the conducted simulations to predict the damage index of the column. Comparing the results obtained by formula and numerical simulations, the deviations are up to a maximum of 15% what is in an acceptable range.

Eisa [54] modeled RC columns in Abaqus. The position of the spherical charge remained unchanged, but the charge weight, column height, longitudinal steel reinforcement, columns aspect ratio, and transverse steel ratio were varied. Fixed boundary conditions are provided at the top and bottom of the column. In addition to column damage, they measured acceleration and displacement concerning varied parameters. Four quantities of explosives were used (45.36, 226.80, 453.59, and 1016.05 kg) and placed at a distance of 4.87 m. The increase in the lateral reinforcement in the column had the effect of reducing the displacement in the middle of the column. It is recommended to examine the influence of the axial force during the blast load and to include additional parameters such as the variation in the standoff distance of the charge.

In numerical simulations using Autodyn, Abladey and Braimah [55] tested three columns designed for different loads in accordance with the Canadian reinforced concrete design code [91]. The first type of column is designed only for gravity loads, and the distance between the transverse reinforcement is 300 mm, and the other two types of columns are designed for seismic loads, but in the second type, the distance between transverse reinforcement is 150 mm, and in the third 75 mm. Column damage is significantly less with denser reinforcement, especially at small-scaled distances. They have proven in research that regardless of the same scaled distance, in a situation where a larger amount of explosive is detonated, the column has a higher deflection.

Li and Hao [56] calibrated the numerical model for RC slab in LS-Dyna according to a previously performed experiment and then used that numerical model for RC column simulation. For good simulation of concrete spallation, the erosion criterion using principle tensile strain of 0.01 was defined. Through simulations, they concluded that denser reinforcement and greater column depth reduced spall damage, i.e., increase the confinement of concrete. The boundary conditions and column height do not play a significant role in the level of spall damage in close-in cases.

Shi and Stewart [57] analyzed a spatial and non-spatial simulation of the blast load on RC columns in LS-Dyna. They used three quantities of ANFO explosives, 50, 100, and 1000 kg, at distances from 0 to 30 m. The analysis is based on axial load-carrying capacity and concluded that the variability of the results in the spatial model is lower, and the probability of damage is significantly higher. They consider the spatial model more reliable and recommended it for future research with an additional assessment of the scale of fluctuation.

Elsanadedy et al. [50] used LS-Dyna for analyzing the behavior of the exterior building RC circular column and strengthened column with Carbon Fiber Reinforced Polymer (CFRP) sheets under blast load. Four different charge weights (100, 200, 500, and 1000 kg) of TNT at three different standoff distances (1, 4, and 15 m) and at 1 m height from the ground were analyzed. They modeled columns with different boundary conditions, first with both fixed ends and a second type with both hinged ends. For calculation of blast load parameters in all assumed scenarios, they used the software ConWep. The use of CFRP is increasing the shear capacity of the column and the strength of the column, which results in less lateral displacement, and more layers of CFRP can undergo more intense blast loads [50]. Moreover, at the scaled distance from 0.50 to 0.68 m/kg$^{1/3}$, columns with CFRP showed better behavior than RC columns.

Ashalekshmi and Subha [92] modeled a bridge column in the Ansys Autodyn software to analyze the impact of concrete grade and spacing of ties under the blast load. They observed total deformation and principal stress for concrete grades M40 and M50 and ties spacings of 10 and 20 cm in the near and far-field. In the near field, the explosive is placed at the same standoff distance of 2 m, and the charge weight varies from 250 to 1500 kg. The influence of the concrete grade on the maximum deformation is visible only at the weight of explosives greater than 750 kg. With concrete M50, the principal stress is higher, but there is no big difference in it when increasing the weight of explosives. In the far-field, the explosive was placed at a standoff distance of 10 m, and the weight also varied from 250 to 1500 kg. There is a slight difference in total deformation concerning the grade of concrete. The principal stress exceeds the strength of concrete, and the difference in stress is visible for the grade of concrete. The effect of tie spacing is visible in the near field only in the increase in the maximum principal stress, while in the far-field, there is no effect on either deformation or principal stress.

Liu et al. [58] modeled the bridge column and bent it in the LS-Dyna for three design categories provided in [35]. They determined six damage mechanisms in the models, four of the column and two of bent. In all three design categories (A, B, C), spalling of concrete and crushing of the bent concrete were observed. Plastic joints in the column and shear of bent occur only in B and C categories. The shear or flexure failure of the column is most probable in category C, i.e., at the highest blast load. They found that increasing the transverse reinforcement reduces damage. In LS-Dyna, they received underestimated blast load, and therefore they used the ConWep program to calculate pressure-time diagrams.

Cui et al. [59] concluded through numerical simulations of columns in LS-Dyna that a larger cross-section and reinforcement ratio, smaller spacing between the stirrups, and a thinner concrete cover for columns exposed to close-in explosions give less damage.

Zhang et al. [12] analyzed circular and square CFST (Concrete Filled-Steel Tube) columns with tube thicknesses of 2.8 and 3.8 mm. The Emulsion explosive was used, with TNT equivalences of (0.7) 17.5, 25, and 35 kg. A numerical simulation of the columns was performed in LS-Dyna, see Figure 18, but to reduce the computation time and model congestion, an air blast model was made in the ConWep program and then imported into LS-Dyna. By comparing the obtained periods of oscillation and maximum displacements, a good match between the numerical model and the experiment was obtained. Differences are found only in residual deflections but are not considered crucial to the accuracy of the model.

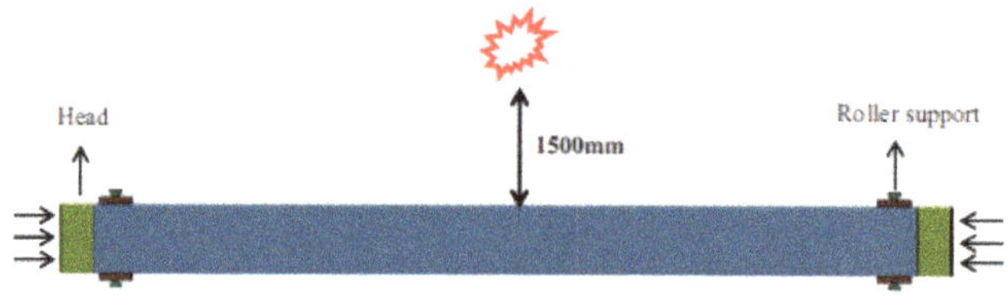

Figure 18. The numerical model of the blast test [12]. Copyright permission obtained from authors.

Zhang et al. [21] numerically modeled CFDST columns with steel-fiber reinforced concrete using LS-Dyna and compared them with experimentally obtained results. For concrete, the KC model was used, and for the steel tube, the Material model 24 was used. The parameters for the concrete model were modified (f_t—tensile strength, B_1—parameter for residual strength, w_{lz}—strain softening, ω—confinement effect, λ, and η—damage parameters) because UHPC was used. Column erosion occurs when the maximum shear strain value reaches 0.045. Numerical research concluded that the axial load up to a certain limit has a favorable effect on the deflection in the middle of the column. It is not recommended to use columns with a hollow section ratio greater than 0.5. Increasing the thickness of the outer steel tube affects the decrease in deflection in the middle of the column. Columns filled with UHPC have less plastic deformation than columns filled with normal strength concrete, shown in Figure 19.

(**a**) (**b**)

Figure 19. Failure mode of CFDST column with: (**a**) normal strength concrete and (**b**) UHPC [21]. Copyright permission obtained from authors.

Codina et al. [14] compared the numerical simulation with an already conducted experiment. The used explosive was Gelamon VF65 (65% TNT equivalence), 8 kg TNT equivalence at distances of 1 m and 0.60 m from the column, which is classified as a near-field range. Overpressure, impulse, deflection, acceleration, and visual damage to the column were measured and compared. For numerical modeling, the Ansys Autodyn program was used in which air and explosives were modeled through an Euler processor and the column through Lagrange. The optimal mesh size of concrete, steel, and the air was 10 mm. The authors made models with default RHT values and with the values proposed in [93] but concluded that the parameters that affect the strength degradation (damage factors D_1 and D_2, and e_{min}^{fail}) and the residual strength, Y^*_{fric} (parameters B and M) should be changed in the model. For good prediction of spallation, the instantaneous geometric strain was used for erosion type with a value of 0.5. The obtained column damage with different RHT parameters is shown in Table 6. The parameters of the RHT material model (shown in Table 3) are validated for scaled distances (Z) from 0.5 to 0.3 m/kg$^{1/3}$.

Table 6. Column damage obtained: (a) experimentally and numerically using (b) default RHT model parameters; (c) parameters provided in [93]; (d) modified parameters by [14].

Parameters (a)	Autodyn (Default) [86] (b)	Tu and Lu [93] (c)	Codina et al. [14] (d)
B	1.6	0.7	0.35
M	0.61	0.8	0.55
	RHT damage model		
D_1	0.04	0.015	0..8
D_2	1	1	1
e_{min}^{fail}	0.01	8.00×10^{-4}	0.03
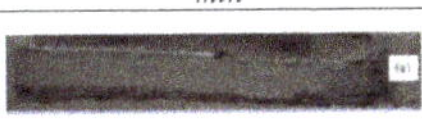	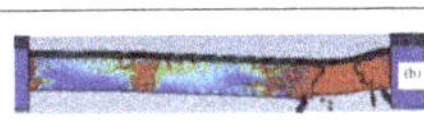	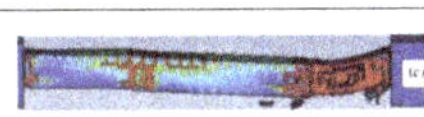	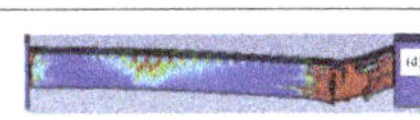

Arowojolu et al. [60] studied, using LS-Dyna numerical models, the influence of axial and blast load on the RC column of the bridge. For the concrete model, they used CSCM (Continuous Surface Cap Model) and for reinforcement MAT 24. Exact quantities of explosives and distances are not given, but scaled distances from 1.77 to 0.45 m/kg$^{1/3}$. They concluded that when an axial load ratio of 0.25 is applied, the displacement in the middle of the column decreases but the damage of the column increases.

Eamon and Aslendi [61] made a numerical model of the column, experimentally tested in [4] using LS-Dyna software. The influence of concrete strength, reinforcement ratio, axial load, and the column wrapping with SFRP (Steel Fiber Reinforced Polymer) was observed. The Johnson–Holmquist–Cook (JHC) model was used for concrete modeling and the elastic–plastic kinematic model for steel reinforcement. The impact of the blast load was determined in the ConWep software, and the detonation point was placed 5 cm above the ground, 40 cm from the column. SFRP proved to be an inexpensive and ductile retrofit. One layer has the largest contribution in blast capacity, while all additional layers have a small effect on increasing the capacity. They obtained a linear relationship between the concrete strength and the increase in the blast load capacity.

Kravchenko et al. [62] performed numerical simulations of the RC column in LS-Dyna. The concrete was modeled using the CSCM concrete model (type 159 material) and reinforced using the plastic–kinematic model (type 3 material). They observed the influence of detonation of 10 kg of TNT at a distance of 1.2 m from the ground and 1 m from the column. They also concluded that the reinforcement ratio has a significant impact on the behavior of columns under the blast load and that the columns in the ground floor and bases need to be better reinforced due to their easy accessibility.

Kyei and Braimah [63] modeled three RC columns in LS-Dyna, which differ in the distance between the transverse reinforcement. They designed the columns according to the instructions for the level of seismicity in the Canadian concrete design code [91]. Concrete was modeled using the Continuous Surface Cap Model (MAT_CSCM_159), and for reinforcement, they used Material Piecewise Linear Plasticity (MAT_024) model, while blast load was calculated in ConWep and then imported with Load Blast Enhanced (LBE) in LS-Dyna. They performed simulations with mesh sizes from 5 to 100 mm, and with 15 mm, they obtained a good ratio of the time spent for the calculation and the accuracy of the results compared with the experiment in [94]. The used explosive was ANFO (100, 250, 500, and 1000 kg) at scaled distances of 0.8, 1.0, and 1.5 $m/kg^{1/3}$. In near-field explosions, the distance between the transverse reinforcement has a significant effect on the reduction in displacement, while in far-field explosions, this effect is negligible. At a high axial load ratio (0.35), the seismically designed columns showed better behavior at scaled distances than the standard ones.

Yuan et al. [23] performed numerical simulations in LS-Dyna of the circular and square columns of the bridge, exposed to the contact explosion of 1 kg of TNT. To reduce the computation time, at the height of 1 m, in the area of the contact explosion, they placed a denser mesh (8 mm), while on the rest of the column, the mesh size was courser (20 mm), shown in Figure 20. The principal strain of 0.5 was used as the erosion criterion. Numerical simulations well described the damage on the front sides of the column, while on the back, there are differences. They concluded that the damage was greater on the square column due to the flat surface and the higher stress concentration than on the circular column. The damage of the columns is shown in Figure 21.

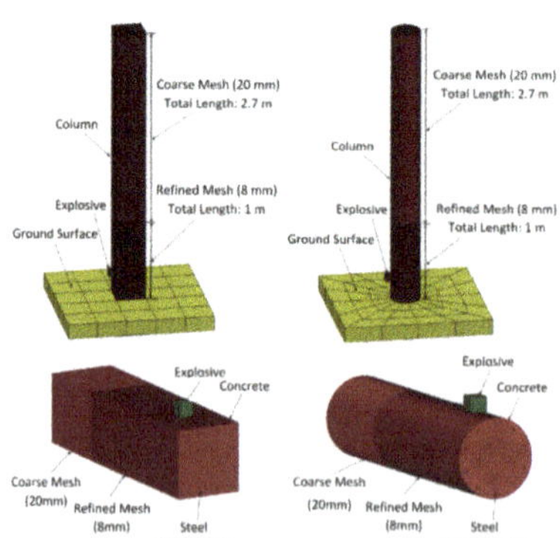

Figure 20. Detailed views of a mesh of 3D column models [23]. Copyright permission obtained from authors.

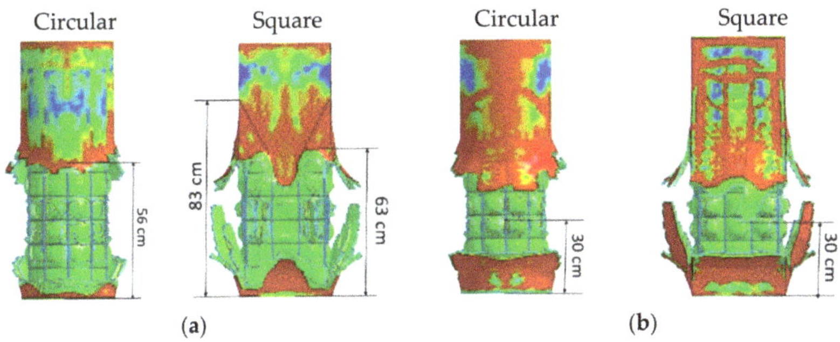

Figure 21. Comparison of damages of circular and square RC columns after the detonation of 1 kg of TNT: (**a**) front side and (**b**) back side [23]. Copyright permission obtained from authors.

Yuan et al. [77] investigated the effect of axial load on RC bridge columns subjected to far-field, close-in, and contact explosion. Columns in the far-field have mainly a flexure response, and the axial load affects the reduction in the maximum displacement in the middle of the column. In the case of a close-in detonation, a shear failure is expected, and the axial load affects the increase in the damage of the column and should not be neglected. In contact detonation, the concrete covers at the front and backside of the column spall off. The axial load reduces the damage of the concrete but increases the stress in the reinforcement and must be considered.

Li et al. [65] conducted numerical investigations on CFDST columns under contact explosion in LS-Dyna. CFDST columns have proven to be good for two reasons: the first is that the confinement of concrete by steel tube allows better energy absorption, and the second is the prevention of the spallation of the concrete cover. In work [73], the researchers concluded that increasing the cross-sectional area and the ratio of reinforcement plays a significant role in the post-blast residual capacity of CFDST columns under contact explosion. In [67], the behavior of CFDST columns subjected to close-in blast loading was studied. They concluded that the influence of the charge shape significantly affects the response and behavior of the column at scaled distances from $0.079\ \mathrm{m/kg^{1/3}}$ to $0.175\ \mathrm{m/kg^{1/3}}$.

Liu et al. [66] performed a dynamic and static analysis of the bridge columns in LS-Dyna. Dynamic analysis is based on the comparison of accelerations and static analysis on the determination of damage through the ratio of the residual to the ultimate axial bearing capacity. They examined the damage to the column concerning the position of the explosives, at the bottom, in the middle, and at the top, shown in Figure 22. In all three cases, bending deformation occurs at the column, but the position of the crack formation differs.

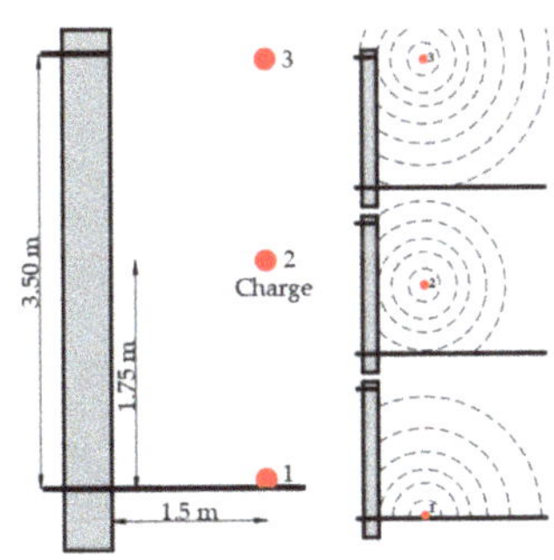

Figure 22. Shock wave propagation depending on the position of the explosive [66].

Liu et al. [68] performed a parametric analysis in LS-Dyna to determine the influence of longitudinal and transverse reinforcement ratios, longitudinal force, and boundary conditions. They concluded that at smaller-scaled distances, the increase in transverse and longitudinal reinforcement reduces the displacement in the middle of the column. It is recommended that the percentage of longitudinal reinforcement does not exceed 6% of the cross-sectional area of the column because too much reinforcement can lead to brittle failure. Analyzing the influence of the axial compressive load, they found that in an amount of up to 40%, it reduces the maximum displacement in the mid-span of the column due to the increase in moment capacity. The conclusions are based on close-in blast loading; a recommendation for future research is to conduct a parametric analysis for near and far filed scenarios.

Liu et al. [69] made numerical models of RC columns strengthened with CFRP in LS-Dyna. The size of the mesh elements for all materials was 10 mm, and for air, 20 mm. The 1 kg and 2 kg TNT charges were placed in contact with the column at the height of 30 cm. The numerical results show that the dragging force of the blast load separates the CFRP from the concrete, but despite this, CFRP protects the column from contact explosion. Column damage is shown in Figure 23.

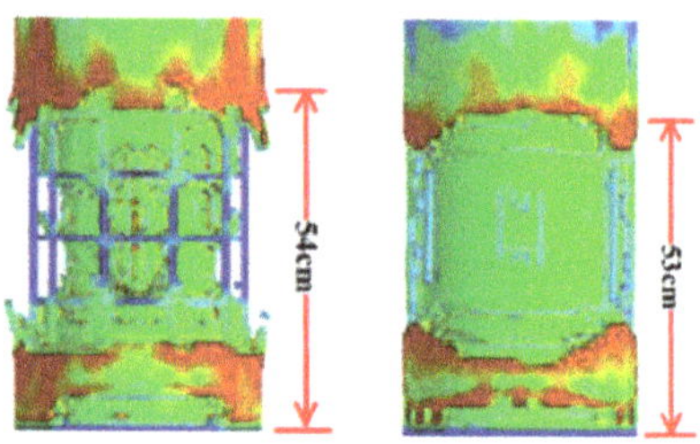

(a) **(b)**

Figure 23. Column damages after detonation of 1kg TNT: (**a**) RC and (**b**) RC + CFRP column [69]. Copyright permission obtained from authors.

Thai et al. [70] modeled in LS-Dyna RC steel jacked columns (25 cm × 25 cm × 360 cm) and observed the influence of steel thickness, scaled distance, and longitudinal compressive force on the behavior of the columns under the blast load. The columns are designed according to Eurocode 2. A 10 mm mesh was used for the column, while 5 mm elements were used for the explosive. Placing 8 kg of TNT in the middle of the column causes less global damage to the column, while placing the same amount at a distance of 32 cm from the ground causes significant local damage. The scaled distance was varied from 0.10 to 0.40 $m/kg^{1/3}$. Increasing the steel thickness from 3 mm to 6 mm did not result in less damage.

Abedini et al. [33] investigated numerically in LS-Dyna the influence of charge and scaled distance on the level of column damage and influence of column width, concrete strength, and reinforcement ratio on the residual axial load capacity.

Dua et al. [72] performed a parametric analysis in LS-Dyna on the RC columns exposed to the contact explosion. They used from 115 g to 1000 g TNT, varied the compressive strength of concrete from 20 to 55 MPa, and reduced the distance of the transverse reinforcement from 200 to 50 mm. Increasing the transverse reinforcement reduces the damage to the concrete core, and higher compressive strength of concrete contributes to the reduction in cracks and peeling. The columns under contact explosion have local damage, while the global damage is negligible. Dua et al. [28] investigated in LS-Dyna the influence of the column cross-sectional width under the contact explosion. They concluded that a larger cross-sectional width (greater by two or more times from the depth) has a favorable effect on the behavior and damage of the column.

Rajkumar et al. [74] examined 45 numerical models of reinforced concrete columns in scale 1:4 (85 mm × 85 mm × 900 mm) in LS-Dyna. In the models, the behavior of four different cross-sections (circular, octagonal, hexagonal and square) on the blast load was examined. The circular column retains the lowest peak pressures and has the smallest deflection in the middle, while the square has the highest pressures and the largest deflection. The edges in cross-sections at small-scaled distances play a significant role in pressure retention, while with increasing scaled distance, the shape influence decreases. An increase in reinforcement in cross-section affects the improvement of the behavior of the columns during the blast load, especially in shapes that retain higher pressures.

Vavilala et al. [75] numerically simulated polymeric foam retrofitted RC columns (35.5 cm × 35.5 cm × 348 cm) in Abaqus. They used a 10 mm mesh for reinforcement and 20 mm for concrete. Columns coated with 5 mm, 8 mm, and 10 mm thick foam were exposed to 10, 25, and 50 kg of TNT. The greatest reduction in displacement in the middle of the column was obtained under 10 kg of TNT when the column was coated with 10 mm thick foam.

Zhang et al. [76] compare in LS-Dyna the behavior of a segmental CFST column with monolithic and prestressed monolithic columns. The columns were exposed to 20 and 50 kg of TNT at a standoff distance of 1.5 m. The segmental column showed a smaller residual displacement, and numerical analysis proved that a larger number of segments had a more favorable effect on the behavior of the column. Moreover, increasing the steel thickness had a beneficial effect on reducing concrete damage.

Yan et al. [78] and Hu et al. [79] used LS-Dyna for numerical simulations of RC columns retrofitted with CFRP subjected to the close-in explosion. CFRP sheets reduce the deformation and spalling of concrete. The CFRP thickness, wrapping, and dimensions ratio of the charge also had a large impact on the damage of the column and peak pressure. Debonding of CFRP is the most common form of failure, but despite this, CFRP has a role in reducing column damage during the blast load, and it may even have a role in changing the failure mode from shear to flexural deformation. The setting of the CFRP on both sides of the column needs to be further investigated because direct shear is possible due to over-reinforcement.

4. Discussion and Conclusions

4.1. *Experimental Testing*

Most of the experiments were conducted for the scenario of an attack by an auto-bomb located near a bridge or building column. Experimental tests on building columns are mostly full-scale, while tests on bridge columns predominate on scales of 1:3 and 1:4, due to the high cost of performing such experiments, the need for trained personnel to handle explosives, large quantities of explosives, and the field where such tests can be carried out.

Experiments showed that even a minimal increase in the cross-sectional dimensions of a column could favorably affect the behavior of the column under the blast load. Moreover, a minimal increase in the standoff distance reduces the impact and intensity of the blast load, and therefore it is necessary whenever possible to fence the column, increase visibility around the column, and reduce its accessibility. In addition to the dimensions of the column, the shape of the column plays a significant role. Circular columns retain less impulse from the blast load than square ones of the same dimensions. Squares have a larger cross-sectional area and, therefore, can better withstand shear. On the action of the contact explosion, the circular column suffered less damage than the square. However, comparing a square and a rectangular column, when the width dimension of the rectangular column is greater two and more times from depth, damage of the column is lower. The shape of the column on the blast load needs to be further investigated. In RC columns, position, quantity, anchoring, and reinforcement overlap have a great impact. Therefore, seismically designed columns have better blast load behavior than standard designed columns.

Comparing steel jacked columns, CFST and CFDST with RC columns, all showed better ductility, less cracking of the concrete, and the absence of flying debris. However,

CFDST showed the best behavior, as the inner steel tube contributes to the prevention of direct shear. With this type of column, it is important that the ratio of the cavity to the column cross-section is not greater than 0.5 and that the thickness of the steel tube is well determined (a thickness of 3.8 mm gave satisfactory results).

Columns with Ultra-High-Performance Concrete and with various Fiber-Reinforced Polymers showed better load-bearing capacity than High-Strength Reinforced Concrete.

The influence of axial load up to 30% of the total load capacity has shown a favorable effect on the reduction in the maximum displacement, but this percentage needs to be further investigated.

Recent research applies smart aggregates to measure internal cracks and internal damage because the column can be damaged and reduced load-bearing capacity without being visible on the outside. Therefore, post-blast tests are performed to determine the residual strength and ductility of the column.

4.2. Numerical Modeling

Numerical simulations make it possible to study the effect of large amounts of explosions (more than 1000 kg) on columns in full-scale. The most widely used software for analyzing the nonlinear behavior of elements on blast load is LS-Dyna. The most accurate simulation of the blast pressure requires as finer mesh as possible, which leads to a long duration of simulations and congestion of the computer processor. Due to the use of a larger mesh size than recommended, the LS-Dyna underestimates the pressures. Most researchers use the ConWep program to calculate pressures and import the resulting pressures into LS-Dyna.

Defining a model of concrete material is the most demanding because many parameters affect its behavior. Karagozian and Case (KC) concrete is mostly used in LS-Dyna. The proper definition of erosion criteria has proven to be very important in modeling column damage.

The position of the explosive plays a significant role in the behavior of the column. Placing the charge in the far-field causes a uniformly distributed load per column and global response, while a charge placed in the near-field, close-in, and contact creates local damage.

Moreover, the columns exposed to the charge placed in the lower half showed greater damage and lowered residual capacity than the columns where the charge was placed in the middle.

In all fields (far, near, close-in), changes in the quantity, shape, and position of the explosives showed a great impact on column behavior. The shape of the column, the ratio of reinforcement, and the concrete grade showed an influence only at small-scaled distances. The concrete grade does not affect the rotation of the column but does affect the reduction in concrete breach and spallation.

Until recently, the axial load on bridge columns was neglected in the calculation because it was considered to be on the safe side. However, numerical simulations showed that axial load has a large impact on increasing damage when the charge is placed near the column or in contact with the column and should not be neglected.

Author Contributions: Conceptualization, S.L. and H.D.; methodology, S.L. and H.D.; resources, S.L. and H.D.; data curation, S.L.; writing—original draft preparation, S.L.; writing—review and editing, H.D.; visualization, S.L.; supervision, H.D.; project administration, H.D.; funding acquisition, H.D. All authors have read and agreed to the published version of the manuscript.

Funding: This research was in part funded by Croatian Science Foundation (HRZZ), grant number UIP-2017-05-7041.

Institutional Review Board Statement: Not applicable.

Informed Consent Statement: Not applicable.

Data Availability Statement: Data available on request due to restrictions, e.g., privacy or ethics. The data presented in this study are available on request from the corresponding author. The data are not publicly available due to limited access to certain literature used in the article.

Acknowledgments: This paper has been supported in part by Croatian Science Foundation (HRZZ) under the project UIP-2017-05-7041 "Blast Load Capacity of Highway Bridge Columns", and support for this research is gratefully acknowledged.

Conflicts of Interest: The authors declare no conflict of interest.

References

1. START. National Consortium for the Study of Terrorism and Responses to Terrorism. *Global Terrorism Database*. 2017. Available online: https://www.start.umd.edu/gtd/ (accessed on 20 May 2020).
2. Winget, D.G.; Marchand, K.A.; Williamson, E.B. Analysis and design of critical bridges subjected to blast loads. *J. Struct. Eng.* **2005**, *131*, 1243–1255. [CrossRef]
3. Cooper, J.D.; Smith, M.C.; Ernst, S.L. *Blue Ribbon Panel Recommendation for Bridge and Tunnel Security*; The American Association of State Highway and Transportation Officials (AASHTO): Washington, DC, USA, 2003.
4. Williamson, E.B.; Bayrak, O.; Davis, C.; Williams, G.D. Performance of bridge columns subjected to blast loads. I: Experimental program. *J. Bridge Eng.* **2011**, *16*, 693–702. [CrossRef]
5. Fouché, P.; Bruneau, M.; Chiarito, V.P. Modified steel-jacketed columns for combined blast and seismic retrofit of existing bridge columns. *J. Bridge Eng.* **2016**, *21*, 4016035. [CrossRef]
6. Bruneau, M.; Lopez-Garcia, D.; Fujikura, S. Multihazard-resistant highway bridge bent. In Proceedings of the Structures Congress 2006: Structural Engineering and Public Safety, St. Louis, MN, USA, 18–21 May 2006; pp. 1–4.
7. Fujikura, S.; Bruneau, M.; Lopez-Garcia, D. Experimental investigation of multihazard resistant bridge piers having concrete-filled steel tube under blast loading. *J. Bridge Eng.* **2008**, *13*, 586–594. [CrossRef]
8. Davis, C.E.; Williams, G.D.; Williamson, E.B.; Marchand, K.A.; McKay, A.E.; Bayrak, O. Design and detailing guidelines for bridge columns subjected to blast and other extreme loads. In Proceedings of the Structures Congress 2009: Don't Mess with Structural Engineers: Expanding Our Role, Austin, TX, USA, 30 April–2 May 2009; pp. 1–10.
9. Fujikura, S.; Bruneau, M. Experimental investigation of seismically resistant bridge piers under blast loading. *J. Bridge Eng.* **2010**, *16*, 63–71. [CrossRef]
10. Crawford, J.E. State of the art for enhancing the blast resistance of reinforced concrete columns with fiber-reinforced plastic. *Can. J. Civil Eng.* **2013**, *40*, 1023–1033. [CrossRef]
11. Burrell, R.P.; Aoude, H.; Saatcioglu, M. Response of SFRC columns under blast loads. *J. Struct. Eng.* **2015**, *141*, 4014209. [CrossRef]
12. Zhang, F.; Wu, C.; Wang, H.; Zhou, Y. Numerical simulation of concrete filled steel tube columns against BLAST loads. *Thin Walled Struct.* **2015**, *92*, 82–92. [CrossRef]
13. Aoude, H.; Dagenais, F.P.; Burrell, R.P.; Saatcioglu, M. Behavior of ultra-high performance fiber reinforced concrete columns under blast loading. *Int. J. Impact Eng.* **2015**, *80*, 185–202. [CrossRef]
14. Codina, R.; Ambrosini, D.; de Borbón, F. Experimental and numerical study of a RC member under a close-in blast loading. *Eng. Struct.* **2016**, *127*, 145–158. [CrossRef]
15. Codina, R.; Ambrosini, D.; de Borbón, F. Alternatives to prevent the failure of RC members under close-in blast loadings. *Eng. Fail. Anal.* **2016**, *60*, 96–106. [CrossRef]
16. Codina, R.; Ambrosini, D.; de Borbon, F. Alternatives to prevent progressive collapse protecting reinforced concrete columns subjected to near field blast loading. *Procedia Eng.* **2017**, *199*, 2445–2450. [CrossRef]
17. Xu, J.; Wu, C.; Xiang, H.; Su, Y.; Li, Z.-X.; Fang, Q.; Hao, H.; Liu, Z.; Zhang, Y.; Li, J. Behaviour of ultra high performance fibre reinforced concrete columns subjected to blast loading. *Eng. Struct.* **2016**, *118*, 97–107. [CrossRef]
18. Echevarria, A.; Zaghi, A.E.; Chiarito, V.; Christenson, R.; Woodson, S. Experimental comparison of the performance and residual capacity of CFFT and RC bridge columns subjected to blasts. *J. Bridge Eng.* **2016**, *21*, 4015026. [CrossRef]
19. Wang, J.; Chen, W.; Guo, Z.; Liang, W. Dynamic responses of RPC-filled steel tubular columns post fire under blast loading. *Open Civil Eng. J.* **2016**, *10*, 236–245. [CrossRef]
20. Zhang, F.; Wu, C.; Zhao, X.-L.; Xiang, H.; Li, Z.-X.; Fang, Q.; Liu, Z.; Zhang, Y.; Heidarpour, A.; Packer, J.A. Experimental study of CFDST columns infilled with UHPC under close-range blast loading. *Int. J. Impact Eng.* **2016**, *93*, 184–195. [CrossRef]
21. Zhang, F.; Wu, C.; Zhao, X.-L.; Heidarpour, A.; Li, Z. Experimental and numerical study of blast resistance of square CFDST columns with steel-fibre reinforced concrete. *Eng. Struct.* **2017**, *149*, 50–63. [CrossRef]
22. Codina, R.; Ambrosini, D.; de Borbón, F. New sacrificial cladding system for the reduction of blast damage in reinforced concrete structures. *Int. J. Prot. Struct.* **2017**, *8*, 221–236. [CrossRef]
23. Yuan, S.; Hao, H.; Zong, Z.; Li, J. A study of RC bridge columns under contact explosion. *Int. J. Impact Eng.* **2017**, *109*, 378–390. [CrossRef]
24. Wang, H.; Wu, C.; Zhang, F.; Fang, Q.; Xiang, H.; Li, P.; Li, Z.; Zhou, Y.; Zhang, Y.; Li, J. Experimental study of large-sized concrete filled steel tube columns under blast load. *Constr. Build. Mater.* **2017**, *134*, 131–141. [CrossRef]

25. Li, J.; Wu, C.; Hao, H.; Liu, Z. Post-blast capacity of ultra-high performance concrete columns. *Eng. Struct.* **2017**, *134*, 289–302. [CrossRef]
26. Fouché, P.; Bruneau, M.; Chiarito, V. Dual-hazard blast and seismic behavior of concrete-filled double-skin steel tubes bridge pier. *J. Struct. Eng.* **2017**, *143*, 4017155. [CrossRef]
27. Dua, A.; Braimah, A.; Kumar, M. Contact explosion response of reinforced concrete columns: Experimental and validation of numerical model. In Proceedings of the Paper presented at the 6th International Disaster Mitigation Specialty Conference, Fredericton, NB, Canada, 13–16 June 2018.
28. Dua, A.; Braimah, A.; Kumar, M. Experimental and numerical investigation of rectangular reinforced concrete columns under contact explosion effects. *Eng. Struct.* **2020**, *205*, 109891. [CrossRef]
29. Wang, Z.; Wu, H.; Fang, Q.; Wu, J. Experimental study on the residual axial capacity of ultra high performance cementitious composite filled steel tube (UHPCC-FST) column under contact explosion. *Thin Walled Struct.* **2020**, *147*, 106515. [CrossRef]
30. Kadhom, B.; Almansour, H.; Saatcioglu, M. Post-blast axial capacity of CFRP strengthened RC columns. In *IOP Conference Series: Materials Science and Engineering, Proceedings of the 4th International Conference on Buildings, Construction and Environmental Engineering, Istanbul, Turkey, 7–9 October 2019*; IOP Publishing Ltd.: Istanbul, Turkey, 2020; p. 012042.
31. Vapper, M.; Lasn, K. Blast protection of concrete columns with thin strips of GFRP overlay. *Structures* **2020**, *25*, 491–499. [CrossRef]
32. Williamson, E.B.; Winget, D.G. Risk management and design of critical bridges for terrorist attacks. *J. Bridge Eng.* **2005**, *10*, 96–106. [CrossRef]
33. Chipley, M. *Reference Manual to Mitigate Potential Terrorist Attacks Against Buildings: Providing Protection to People and Building*; Federal Emergency Management Agency: Washington, DC, USA, 2003.
34. Roberts, J.; Kulicki, J.; Beranek, D.A.; Englot, J.M.; Fisher, J.W.; Hungerbeeler, H.; Seible, F.; Stinson, K.; Tang, M.C.; Witt, K. *Recommendations for Bridge and Tunnel Security*; Federal Highway Administration (US): Washington, DC, USA, 2003.
35. Williamson, E.B.; Bayrak, O.; Davis, C.; Daniel Williams, G. Performance of bridge columns subjected to blast loads. II: Results and recommendations. *J. Bridge Eng.* **2011**, *16*, 703–710. [CrossRef]
36. Marchand, K.; Williamson, E.; Winget, D. Analysis of blast loads on bridge substructures. *WIT Trans. Built Environ.* **2004**, *73*. [CrossRef]
37. National Cooperative Highway Research Program. *Blast-Resistant Highway Bridges: Design and Detailing Guidelines;* Transportation Research Board: Washington, DC, USA, 2010.
38. Williams, G.; Holland, C.; Williamson, E.B.; Bayrak, O.; Marchand, K.A.; Ray, J. *Blast-Resistant Highway Bridges: Design and Detailing Guidelines*; Transportation Research Board; WIT Press: Southampton, UK, 2010; Volume 645.
39. Conrath, E.J. *Structural Design for Physical Security: State of the Practice*; ASCE, cop.: Reston, VA, USA, 1999.
40. Mays, G.; Smith, P.D.; Smith, P.D. *Blast Effects on Buildings: Design of Buildings to Optimize Resistance to Blast Loading*; Thomas Telford: Telford, UK, 1995.
41. US DoD. *UFC 3-340-02: Structures to Resist the Effects of Accidental Explosions*; US DoD.: Washington, DC, USA, 2008.
42. American Institute of Steel Construction. *Seismic Provisions for Structural Steel Buildings*; American Institute of Steel Construction: Chicago, IL, USA, 2010.
43. Fujikake, K.; Aemlaor, P. Damage of reinforced concrete columns under demolition blasting. *Eng. Struct.* **2013**, *55*, 116–125. [CrossRef]
44. Roller, C.; Mayrhofer, C.; Riedel, W.; Thoma, K. Residual load capacity of exposed and hardened concrete columns under explosion loads. *Eng. Struct.* **2013**, *55*, 66–72. [CrossRef]
45. Xu, K.; Deng, Q.; Cai, L.; Ho, S.; Song, G. Damage detection of a concrete column subject to blast loads using embedded piezoceramic transducers. *Sensors* **2018**, *18*, 1377. [CrossRef] [PubMed]
46. Ray, J.; Armstrong, B.; Slawson, T. Airblast environment beneath a bridge overpass. *Transp. Res. Rec. J. Transp. Res. Board* **2003**, *1827*, 63–68. [CrossRef]
47. Rutner, M.P.; Astaneh-Asl, A.; Son, J. Blast resistant performance of steel and composite bridge piers. In Proceedings of the IABSE Symposium Report, Munich, Germany, 29 August–1 September 2005; pp. 47–54.
48. Wu, K.-C.; Li, B.; Tsai, K.-C. The effects of explosive mass ratio on residual compressive capacity of contact blast damaged composite columns. *J. Constr. Steel Res.* **2011**, *67*, 602–612. [CrossRef]
49. Hao, H.; Stewart, M.G.; Li, Z.-X.; Shi, Y. RC column failure probabilities to blast loads. *Int. J. Prot. Struct.* **2010**, *1*, 571–591. [CrossRef]
50. Elsanadedy, H.M.; Almusallam, T.H.; Abbas, H.; Al-Salloum, Y.A.; Alsayed, S.H. Effect of blast loading on CFRP-Retrofitted RC columns-a numerical study. *Lat. Am. J. Solids Struct.* **2011**, *8*, 55–81. [CrossRef]
51. Williams, G.D.; Williamson, E.B. Response of reinforced concrete bridge columns subjected to blast loads. *J. Struct. Eng.* **2011**, *137*, 903–913. [CrossRef]
52. Williams, G.D.; Williamson, E.B. Procedure for predicting blast loads acting on bridge columns. *J. Bridge Eng.* **2011**, *17*, 490–499. [CrossRef]
53. Magali, A.; Alain, R.; Chhim, S. Numerical dynamic simulations for the prediction of damage and loss of capacity of RC column subjected to contact detonations. In *Fracture Mechanics of Concrete and Concrete Structures*; IA-FraMCoS: Toledo, Spain, 2013.
54. Eisa, A.S. Finite element analysis of reinforced concrete columns under different range of blast loads. *Int. J. Civ. Struct. Eng.* **2014**, *5*, 155.

55. Abladey, L.; Braimah, A. Near-field explosion effects on the behaviour of reinforced concrete columns: A numerical investigation. *Int. J. Prot. Struct.* **2014**, *5*, 475–499. [CrossRef]
56. Li, J.; Hao, H. Numerical study of concrete spall damage to blast loads. *Int. J. Impact Eng.* **2014**, *68*, 41–55. [CrossRef]
57. Shi, Y.; Stewart, M.G. Spatial reliability analysis of explosive blast load damage to reinforced concrete columns. *Struct. Saf.* **2015**, *53*, 13–25. [CrossRef]
58. Liu, H.; Torres, D.M.; Agrawal, A.K.; Yi, Z.; Liu, G. Simplified blast-load effects on the column and bent beam of highway bridges. *J. Bridge Eng.* **2015**, *20*, 06015001. [CrossRef]
59. Cui, J.; Shi, Y.; Li, Z.-X.; Chen, L. Failure analysis and damage assessment of RC columns under close-in explosions. *J. Perform. Constr. Facil.* **2015**, *29*, B4015003. [CrossRef]
60. Arowojolu, O.; Rahman, M.K.; Hussain, B.M. Dynamic response of reinforced concrete bridge piers subjected to combined axial and blast loading. In Proceedings of the Structures Congress 2017, Denver, CO, USA, 6–8 April 2017; pp. 98–109.
61. Eamon, C.; Alsendi, A. Resistance of Reinforced Concrete Bridge Columns Subjected to Blast Loads. 2017. Available online: http://www.awarie.zut.edu.pl/files/ab2017/referaty/09/09-02%20-%20Eamon%20C,%20Aslendi%20A%20-%20Resistance%20of%20reinforced%20concrete%20bridge%20columns%20subjected%20to%20blast%20loads.pdf (accessed on 1 June 2021).
62. Kravchenko, G.; Trufanova, E.; Kostenko, D.; Tsurikov, S. Analysis of blast load on a reinforced concrete column in the time domain. *MATEC Web Conf.* **2017**, *106*, 04019. [CrossRef]
63. Kyei, C.; Braimah, A. Effects of transverse reinforcement spacing on the response of reinforced concrete columns subjected to blast loading. *Eng. Struct.* **2017**, *142*, 148–164. [CrossRef]
64. Abedini, M.; Mutalib, A.A.; Raman, S.N.; Akhlaghi, E. Modeling the effects of high strain rate loading on RC columns using Arbitrary Lagrangian Eulerian (ALE) technique. *Rev. Int. Métodos Numéricos Cálculo Diseño Ing.* **2018**, *34*. [CrossRef]
65. Li, M.; Zong, Z.; Liu, L.; Lou, F. Experimental and numerical study on damage mechanism of CFDST bridge columns subjected to contact explosion. *Eng. Struct.* **2018**, *159*, 265–276. [CrossRef]
66. Liu, L.; Zong, Z.; Li, M. Numerical study of damage modes and assessment of circular RC pier under noncontact explosions. *J. Bridge Eng.* **2018**, *23*, 04018061. [CrossRef]
67. Li, M.; Zong, Z.; Hao, H.; Zhang, X.; Lin, J.; Xie, G. Experimental and numerical study on the behaviour of CFDST columns subjected to close-in blast loading. *Eng. Struct.* **2019**, *185*, 203–220. [CrossRef]
68. Liu, Y.; Yan, J.; Li, Z.; Huang, F. Improved SDOF and numerical approach to study the dynamic response of reinforced concrete columns subjected to close-in blast loading. *Structures* **2019**, *22*, 341–365. [CrossRef]
69. Liu, L.; Zong, Z.; Gao, C.; Yuan, S.; Lou, F. Experimental and numerical study of CFRP protective RC piers under contact explosion. *Compos. Struct.* **2020**, *234*, 111658. [CrossRef]
70. Thai, D.K.; Pham, T.H.; Nguyen, D.L. Damage assessment of reinforced concrete columns retrofitted by steel jacket under blast loading. *Struct. Des. Tall Spec. Build.* **2020**, *29*, e1676. [CrossRef]
71. Abedini, M.; Mutalib, A.A.; Mehrmashhadi, J.; Raman, S.N.; Alipour, R.; Momeni, T.; Mussa, M.H. Large deflection behavior effect in reinforced concrete columns exposed to extreme dynamic loads. *Frontiers of Struct. Civ. Eng.* **2020**, *14*, 532–553. [CrossRef]
72. Dua, A.; Braimah, A.; Kumar, M. Contact explosion response of RC columns: Experimental and numerical investigation. *Proc. Inst. Civil Eng. Struct. Build.* **2020**, *173*, 799–820. [CrossRef]
73. Li, M.; Zong, Z.; Hao, H.; Zhang, X.; Lin, J.; Liao, Y. Post-blast performance and residual capacity of CFDST columns subjected to contact explosions. *J. Constr. Steel Res.* **2020**, *167*, 105960. [CrossRef]
74. Rajkumar, D.; Senthil, R.; Bala Murali Kumar, B.; AkshayaGomathi, K.; Mahesh Velan, S. Numerical Study on Parametric Analysis of Reinforced Concrete Column under Blast Loading. *J. Perform. Constr. Facil.* **2020**, *34*, 04019102. [CrossRef]
75. Vavilala, S.; Shirbhate, P.; Mandal, J.; Dass Goel, M. Blast mitigation of RC column using polymeric foam. *Mater. Today Proc.* **2020**, *26*, 1347–1351. [CrossRef]
76. Zhang, X.; Hao, H.; Li, M.; Zong, Z.; Bruechert, J.W. The blast resistant performance of concrete-filled steel-tube segmental columns. *J. Constr. Steel Res.* **2020**, *168*, 105997. [CrossRef]
77. Yuan, S.; Hao, H.; Zong, Z.; Li, J. Numerical analysis of axial load effects on RC bridge columns under blast loading. *Adv. Struct. Eng.* **2021**, *24*, 1399–1414. [CrossRef]
78. Yan, J.; Liu, Y.; Xu, Z.; Li, Z.; Huang, F. Experimental and numerical analysis of CFRP strengthened RC columns subjected to close-in blast loading. *Int. J. Impact Eng.* **2020**, *146*, 103720. [CrossRef]
79. Hu, Y.; Chen, L.; Fang, Q.; Kong, X.; Shi, Y.; Cui, J. Study of CFRP retrofitted RC column under close-in explosion. *Eng. Struct.* **2021**, *227*, 111431. [CrossRef]
80. United States Department of the Army. *Fundamentals of Protective Design for Conventional Weapons*; United States Department of the Army: The Pentagon, VA, USA, 1986.
81. Britt, J.R.; Ranta, E.D.; Ohrt, A.P. *User's Manual for the BlastX Code*; Version 4.0.; U.S. Army Engineer Waterways Experiment Station: Vicksburg, MS, USA, 1998.
82. Crepeau, J. *Second-Order Hydrodynamic Automatic Mesh Refinement Code*; Volume 2: User's Manual; Applied Research Associates, Inc.: Albuquerque, NM, USA, 2001.
83. Advanced Structural Concepts, Incorporated. *NONLIN 7.05*; Advanced Structural Concepts, Inc.: Blacksburg, VA, USA, 2003.
84. US Army Corps of Engineers—Omaha District. *O. Neb. (Distribution Limited to U.S. Government Agencies; Contractors), a.t. SPAn32*; Version 1.2.6.9.; US Army Corps of Engineers—Omaha District: Omaha, NE, USA, 2002.

85. MSC. *Dytran Theory Manual*; MSC. Software Corporation: Newport Beach, CA, USA, 2002.
86. Ansys Inc. *Ansys Autodyn Users's Manual*; Ansys Inc.: Canonsburg, PA, USA, 2010.
87. *LS-DYNA*; Version 971; Livermore Software Technology Corporation (LSTC): Livermore, CA, USA, 2007.
88. Smith, M. *ABAQUS/Standard User's Manual*; Version 6.9; Dassault Systèmes Simulia Corp: Providence, RI, USA, 2009.
89. Applied Research Associates, Inc. *Anti-Terrorist Blast (AT-Blast)*; Version 2.1; Applied Research Associates, Inc.: Albuquerque, NM USA, 2004.
90. U.S. Army Corps of Engineers, Engineer Research and Development Center. *Bridge Explosive Loading (BEL)*; Version 1.1.0.3; Engineer Research and Development Center: Vicksburg, MS, USA, 2004.
91. Canadian Standards Association. *Design and Assessment of Buildings Subjected to Blast Loads*; CSA Group: Mississauga, ON, Canada, 2012.
92. Ashalekshmi, K.G.; Subha, K. Effect of grade of concrete and spacing of ties in the response of RCC bridge pier subjected to ground blast loading. *Int. Res. J. Eng. Technol.* **2018**, *5*, 2023–2029.
93. Tu, Z.; Lu, Y. Evaluation of typical concrete material models used in hydrocodes for high dynamic response simulations. *Int. J. Impact Eng.* **2009**, *36*, 132–146. [CrossRef]
94. Siba, F. Near-Field Explosion Effects on Reinforced Concrete Columns: An Experimental Investigation. Master's Thesis, Carleton University, Ottawa, ON, Canada, 2014.

applied sciences

MDPI

Article

Mesoscale Equivalent Numerical Study of Ultra-High Performance Concrete Subjected to Projectile Impact

Jian Yang [1], Jie Ao [1], Wenzheng Wan [2,]* and Yikang Liu [1]

1 School of Civil Engineering, Central South University, Changsha 410075, China; wyyxaojie@163.com (J.A.); jianyangyy@126.com (J.Y.); liuyikang_csu@csu.edu.cn (Y.L.)
2 Hunan Mingxiang Technology Development Co., Ltd., Changsha 410000, China
* Correspondence: 15717428832@163.com

Abstract: Numerical investigations on the performance of ultra-high performance concrete (UHPC) subjected to projectile impacts have attracted extensive attention, but there are still deficiencies in the accuracy and computational efficiency of related simulation methods. To make up for these deficiencies, a mesoscale equivalent model for UHPC is developed to simulate the response of UHPC under projectile impacts. In this model, an equivalent treatment is conducted on steel fibers to reduce their quantity under the premise that the interfacial shearing force between the fibers and the matrix remains equal. Based on the mesoscale equivalent model, numerical simulations of uniaxial compressive tests and projectile penetration tests on UHPC specimens are performed in LS-DYNA, and the numerical results are compared with the corresponding experimental results to verify the developed model. It is found that the mesoscale equivalent model could accurately reproduce the failure mode and stress-strain curve of UHPC specimens when the amplification factor of steel fibers is lower than 5. When the amplification factor is 5, the computational efficiency of the numerical models for penetration tests is significantly improved, and the maximum relative error between the numerical results of the crater diameter and penetration depth and experimental results is 11.7%. The successful application of the mesoscale equivalent model provides a more precise and in-depth perspective in simulating the response of UHPC with steel fibers subjected to projectile impact. Then, the influence of projectile striking velocities, UHPC compressive strengths, and volume percentages of steel fibers on the depth of penetration (DOP) are further numerically assessed. Based on the simulated data, modifications of the Young equation for predicting the DOP are conducted, and the maximum relative error of the modified equation is 13.9%. This demonstrates that the modified Young equation can accurately predict the DOP of UHPC subjected to projectile impacts.

Keywords: UHPC; mesoscale equivalent model; penetration experiments; parametric analysis; DOP prediction equation

Citation: Yang, J.; Ao, J.; Wan, W.; Liu, Y. Mesoscale Equivalent Numerical Study of Ultra-High Performance Concrete Subjected to Projectile Impact. *Appl. Sci.* **2023**, *13*, 4991. https://doi.org/10.3390/app13084991

Academic Editors: Ricardo Castedo, Lina M. López and Anastasio P. Santos

Received: 19 March 2023
Revised: 6 April 2023
Accepted: 10 April 2023
Published: 16 April 2023

1. Introduction

With the enhancement of the penetration capability of earth-penetrating weapons and the frequent occurrence of local wars, studies on the resistance of new building materials under projectile impacts have aroused wide attention from engineers and researchers. Ultra-high performance concrete (UHPC) is a kind of cement-based composite material with high strength, high toughness, excellent ductility, and good energy absorption capacity, which has a promising application on protective structures that may be subjected to projectile impacts [1,2]. Experimental investigations on UHPC structures against projectile impacts have been carried out over the past few decades [3–7]. However, it is expensive and time-consuming to conduct penetration experiments, which leads to the fact that the current experimental investigations are mainly aimed at small-caliber bullets or reduced-scale projectiles. Meanwhile, it is usually difficult to obtain the expected mechanical data in penetration tests due to the ultra-high strain rate and deceleration, which hinders the in-depth understanding of the penetration mechanism of UHPC.

Recently, efforts have been put into the numerical investigation of the dynamic response of UHPC subjected to projectile impacts, hoping to supplement the deficiencies of experimental investigations and even replace the costly penetration experiments. Prakash et al. [8] carried out numerical investigations on high-velocity projectiles penetrating steel fiber reinforced concrete (SFRC) panels with a volume percentage of steel fibers ranging from 0% to 10% using a modified RHT model. A design chart for determining the optimal panel thickness under different fiber contents and projectile kinetic energies was compiled. Blasone et al. [9] investigated the mechanical behavior of the ultra-high performance fiber-reinforced concrete (UHPFRC) under an armor-piercing projectile using the coupled plasticity-damage model DFHcoh-KST. The damage mechanisms are well simulated by the numerical tool, and it was found that the softening behavior provided by fibers had a significant influence on the damage pattern of UHPFRC targets under projectile impacts. Wan et al. [10] calibrated a set of HJC model parameters for UHPC with steel fiber according to the material test data and then simulated experiments of two different kinds of bullets penetrating UHPC targets using the modified HJC model. The numerical results showed that the modified material model could accurately simulate the depth of penetration (DOP), but it underestimated the damage, especially the tensile damage, inside UHPC targets. Liu et al. [11] presented a numerical study evaluating the performance of UHPC (90–190 MPa) under ogive-nosed projectile impacts with striking velocities from 300 m/s to 1000 m/s, where the effects of the compressive strength, projectile striking velocity, and projectile CRH on the DOP and cratering damage of targets are discussed by applying a calibrated and validated K&C material model. An empirical equation for the DOP prediction was proposed. In this work, the numerical results of projectile impacts lack sufficient experimental comparisons and verifications, and it was found that the empirical equation overestimates the penetration resistance of the UHPC. Zhou et al. [12] established a novel dynamic constitutive model for UHPC based on the KCC model, which was applied to numerically predict the resistance and damage pattern of UHPC subjected to projectile impacts and achieved good accuracy. Liu et al. [13] numerically explored the effects of steel wire mesh on the penetration resistance of reactive power concrete (RPC). An empirical equation was proposed to predict the DOP of steel wire mesh-reinforced RPC targets subjected to high-velocity projectile penetrations.

Generally, UHPC is treated as a homogeneous material when conducting simulations on UHPC targets subjected to projectile impacts, which has shown some shortcomings in reproducing and explaining the dynamic response. In fact, UHPC should be regarded as a two-phase composite material consisting of matrixes and reinforcing fibers. The addition of fibers is an important contributor to the high performance of UHPC [14]. Hence, treating UHPC as a homogeneous material is no longer appropriate, and a three-dimensional model that models the matrix and fibers discretely should be established. Zhang et al. [15], Liang et al. [16], and Peng et al. [17] built a mesoscale model to simulate the mechanical properties and failure characteristics of UHPC under static and dynamic loadings. The numerical results are in excellent agreement with corresponding experimental results, and the mesoscale model provides a mesoscopic perspective to analyze the responses of UHPC. Smith et al. [18] carried out simulations of UHPC slab penetration and perforation experiments using a mesoscale discrete particle model. It was found that the mesoscale model perfectly reproduces cratering damage, spalling damage, and the crack-bridging effect in UHPC slabs. Although the mesoscale model performs better than the homogeneous model in the penetration simulation, the computational cost of the mesoscale model is too high to be used in some large-scale penetration simulations. The main reason of the high cost of the mesoscale model is that the number of established fiber elements is too large. Therefore, some equivalent treatment must be performed on fibers to reduce their quantity in the penetration model to improve computational efficiency.

The mesoscale equivalent treatment on fibers in UHPC has rarely been studied before. In the present study, an efficient mesoscale equivalent model based on the bond-slip constitutive between steel fibers and matrixes is first developed for simulating the dynamic response of UHPC under projectile impacts and is verified by the uniaxial tests and penetration tests. Then, the validated numerical model is applied to numerically investigate the effects of the projectile striking velocity, UHPC compression strength, and volume percentage of steel fibers on the DOP of UHPC targets subjected to projectile impacts. Moreover, a modified Young equation for predicting the DOP of UHPC subjected to the projectile penetration is fitted in terms of the simulated data.

2. Penetration Experiments

2.1. UHPC Targets

In the present tests, two cuboid UHPC targets are cast for the projectile impact under different striking velocities, as shown in Figure 1. The plain sizes of the two targets are both 2.0 m × 2.0 m, while the thicknesses are 0.9 m and 1.5 m, respectively. Each target is wrapped by 1 cm thick steel plates except for the front and back to weaken the lateral boundary effects. Straight copper-coated steel fibers with 0.2 mm diameter and 13 mm long, as shown in Figure 2, are incorporated into the matrix at a volume content (V_f) of 2%. The yield strength of the steel fiber is 2100 MPa, and the density is 7830 kg/m^3.

Figure 1. Cuboid UHPC target.

Figure 2. Straight copper-coated steel fibers.

Quasi-static uniaxial compression tests are carried out on prismatic specimens with dimensions of 100 mm × 100 mm × 300 mm using an electro-hydraulic servo testing machine with a capacity of 300 tons to determine the uniaxial compressive strength of UHPC used for targets and corresponding matrixes without steel fibers, as shown in Figure 3. A total of six specimens are tested, three of which are made of UHPC with steel fibers, and the other three are made of UHPC without steel fibers (which can be named as matrix), complying with the test procedure in Chinese Standard GB/T 50081-2019 [19]. Before the compression tests, all prismatic specimens are steam cured using an automatic control system [20]. The experimental uniaxial compressive strengths of UHPC specimens with steel fibers are 144.1 MPa, 152.3 MPa, and 148.6 MPa, respectively, and the results of matrix specimens are 139.8 MPa, 134.5 MPa, and 145.3 MPa, respectively. In the subsequent discussion, the uniaxial compressive strength of UHPC and matrix are taken as the average values of the corresponding three specimens, i.e., 148.3 MPa and 139.9 MPa, respectively. At the same time, the density of UHPC specimens with steel fibers and matrix specimens is also measured, and the average values are 2430 kg/m^3 and 2237 kg/m^3, respectively.

Figure 3. Quasi-static uniaxial compression test on the prismatic specimen.

2.2. Projectiles

The projectile is a type of tangent ogive-nose projectile with a caliber of 8 cm (d = 8 cm), and detailed dimensions are shown in Figure 4. The projectiles are composed of the outer shell and inner filling, and each weighs 11.9 kg. The outer shell is made of high-strength steel; the yield strength is 1350 MPa. In order to facilitate the launch of the projectiles, the centering ring and sabot are attached to each projectile.

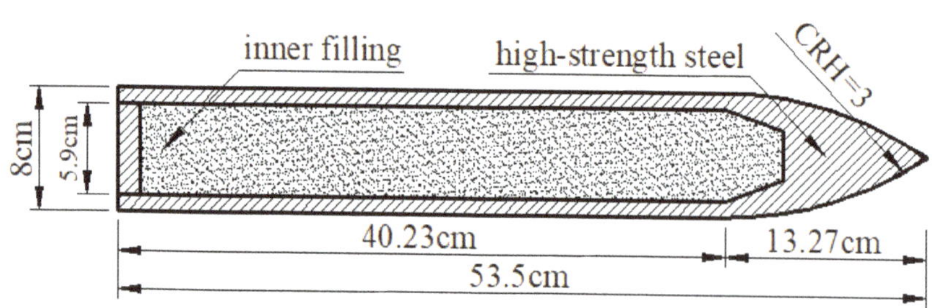

Figure 4. Ogive-nose projectile.

2.3. Experimental Setup

Figure 5 shows the test setup of the penetration tests. A 100 mm smoothbore gun is used to launch the projectiles. The targets are placed vertically on the testing bed. Some thick normal concrete blocks are placed tightly on the back of the targets to restrain the rigid body displacement in the thickness direction of the targets. The smoothbore gun barrel is adjusted to be perpendicular to the front of the targets to ensure that the projectiles can strike in the center of targets vertically. During the flight of projectiles, the centering rings and sabots would automatically separate from the projectiles, so the impact mass (m)

is 11.9 kg. Two sets of high-speed photography equipment are used to measure the actual striking velocity of the projectile (V_0) and to capture the penetration process.

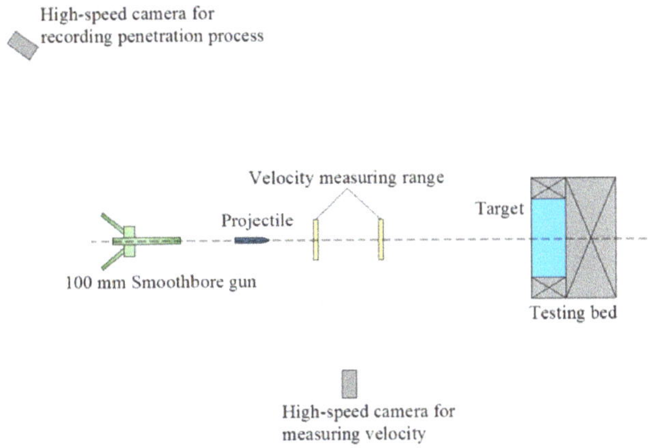

Figure 5. Experimental setup for penetration tests.

2.4. Experimental Results

Figure 6 shows a typical moment that the projectile contacts the target in the penetration process. Detailed experimental data, including the DOP and crater diameter (d_c), are listed in Table 1.

Figure 6. A certain moment in the penetration process.

Table 1. Penetration tests data.

Target	Plain Size (m)	Thickness (m)	f_c (MPa)	V_f	m (kg)	V_0 (m/s)	DOP (cm)	d_c (cm)
U-1	2.0 × 2.0	0.9	148.3	2%	11.9	410	42.1	51.3
U-2	2.0 × 2.0	1.5				664	78.6	78.2

The frontal damages of targets are shown in Figure 7. When the projectile hits the target, concrete around the contact location is ejected out and forms a circular fragment cloud. The projectile with a speed of 410 m/s bounced off the target, while the one at a striking velocity of 664 m/s was stuck in the target. The projectile impact formed an obvious funnel-shaped crater on the impact surface first and then a tunnel deep into the targets. An evident spalling phenomenon could be observed at the edge of the crater. Moreover, the concrete around the tunnel is crushed into powders by ultra-high compressive stress.

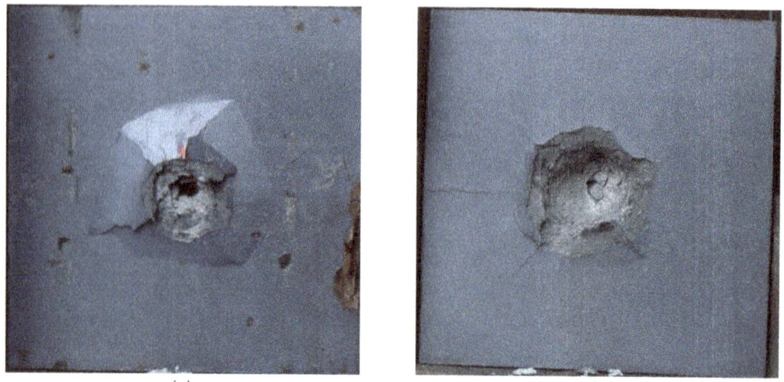

(a) (b)

Figure 7. Frontal damages of UHPC targets: (**a**) U-1 target; (**b**) U-2 target.

As listed in Table 1, the diameter of the crater increases with the growth of the striking velocity of the projectile. In the present penetration tests, The DOP included both the depth of the crater and the depth of the tunnel. The DOP of the projectiles at striking velocities of 410 m/s and 664 m/s is 42.1 cm and 78.6 cm, respectively, which also shows an increasing tendency.

3. Mesoscale Equivalent Model for UHPC

The mesoscale model with explicit modeling of fibers is better at simulating and explaining the response of UHPC under all kinds of loadings. However, building fibers according to actual size will lead to excessive fibers for typical FE models. For example, for the 2.0 m × 2.0 m × 1.5 m cuboid target U-2 used in the above penetration tests, the number of steel fibers would be more than 70 million in a quarter model. Therefore, to reduce the computational cost of simulation, a mesoscale equivalent model is first developed in this section for UHPC with steel fibers.

3.1. Generation of Fibers

In a UHPC specimen, fibers are uniformly and randomly distributed. The generation of random straight round fibers inside a certain specimen volume V can be as follows. First, the number of fibers is calculated by $N = 4V_fV/(\pi d_f^2 L_f)$ according to the fiber diameter d_f, length L_f and volume content V_f. Random points with the number of N are uniformly generated in the specimen space and assigned to be the initial point of each fiber, labeled as (x_{1i}, y_{1i}, z_{1i}). Then, for each fiber, the initial direction is defined by two random spatial angles φ_i and θ_i in the spherical coordinate system with the initial point as the origin. The ending point of a fiber, labeled as (x_{2i}, y_{2i}, z_{2i}), is determined according to the initial point, random angles, and L_f, as shown in Equation (1).

$$\begin{cases} x_{2i} = x_{1i} + L_f \sin\theta_i \cos\varphi_i \\ y_{2i} = y_{1i} + L_f \sin\theta_i \sin\varphi_i \\ z_{2i} = z_{1i} + L_f \cos\theta_i \end{cases} \tag{1}$$

With the initial point and ending point determined, any fiber in the given specimen space can be identified. Equation (1) is cycled N times until all fibers are generated. All generated points and fibers will be written into node keyword file and element keyword file, respectively, which will facilitate subsequent model establishment.

3.2. Bond-Slip Constitutive Model

For building the mesoscale equivalent model for UHPC with steel fibers, the failure mode and interfacial behavior between steel fibers and matrixes must be determined. Deng et al. [21] pointed out that the fiber pullout is the most common failure mode when UHPC is subjected to loading due to the weak fiber-matrix interface and high tensile strength of the steel fiber. Hence, the interfacial behavior between steel fibers and matrixes can be regarded as a kind of bond-slip behavior. Su et al. [22] carried out single steel fiber pullout tests and put forward a bond-slip constitutive model for UHPC with steel fibers. In this model, the interfacial shear stress τ is simplified to be constant at a given relative slip S. For different values of the relative slip, the interfacial shear stress is determined by Equation (2), where τ increases linearly with S in the bonding phase and declines exponentially with S in the debonding phase.

$$\tau = \begin{cases} G_s S & S \le S_{max} \\ G_s S_{max} \times e^{-EXP \times D} & S > S_{max} \end{cases} \tag{2}$$

where G_s is the interfacial shear modulus, EXP is the exponent in the debonding phase, and D is the accumulated plastic displacement in all the integral time steps.

The pivotal constitutive parameters for UHPC (150 MPa) are determined according to the pullout load-slip curve, where G_s is 2393 MPa, S_{max} is 0.00125, and EXP is 0.2. Then, simulations on the static split tension test and SHPB test are performed, where the numerical results agree well with experimental results, and more details can be seen in Reference [22]. This bond-slip constitutive model is proposed, and the determined parameters are validated by static and dynamic tests and are adopted in the present study.

3.3. Equivalent Treatment on Steel Fibers

Since the basic reason for the ultra-high computational cost of the original mesoscale model is the huge number of steel fibers, the key to establishing a mesoscale equivalent model is to reduce the number through the equivalent treatment of steel fibers. Steel fibers play a bridging role in UHPC, which limits the deformation and crack development in the matrix. This kind of bridging effect is reflected by the interaction between steel fibers and matrixes; the essence is the transmission of the interfacial shear force. Therefore, the bridging effect can be equivalent as long as the interfacial shear force can be guaranteed to be unchanged when the equivalent treatment on steel fibers is performed [23,24]. The specific implementation procedure of the equivalent treatment on steel fibers is as follows.

The first step is to amplify steel fibers geometrically under the condition that the aspect ratio of fibers remains unchanged and replace the original fibers with amplified fibers.

$$d_{af} = nd_f, \qquad L_{af} = nL_f \tag{3}$$

where d_{af} and L_{af} are the diameter and length of the amplified fiber, respectively, and n is the geometric amplification factor. According to the premise that the volume percentage of steel fibers V_f remains unchanged, the number of fibers N_{af} after equivalent treatment is determined as:

$$N_{af} = \frac{V_f V}{\pi (d_{af})^2 L_{af}} = \frac{V_f V}{\pi (nd_f)^2 (nL_f)} = \frac{1}{n^3} N \tag{4}$$

At the same time, due to the fibers being uniformly and randomly distributed, the number of fibers in any direction is the same, where the number is labeled as N^κ for original fibers and N_{af}^κ for amplified fibers $N_{af}^\kappa = N^\kappa / n^3$.

For a single original steel fiber embedded in the matrix, as shown in Figure 8, an increment of the axial force in an infinitesimal segment of fiber is:

$$dP = \tau_0 \cdot \pi d_f \cdot dl \tag{5}$$

where τ_0 is the actual interfacial shear stress, the interfacial shear force, i.e., the axial force of the fiber, can be computed as:

$$F = \int_0^{L_e} dP = \pi d_f \int_0^{L_e} \tau_0 dl \quad (6)$$

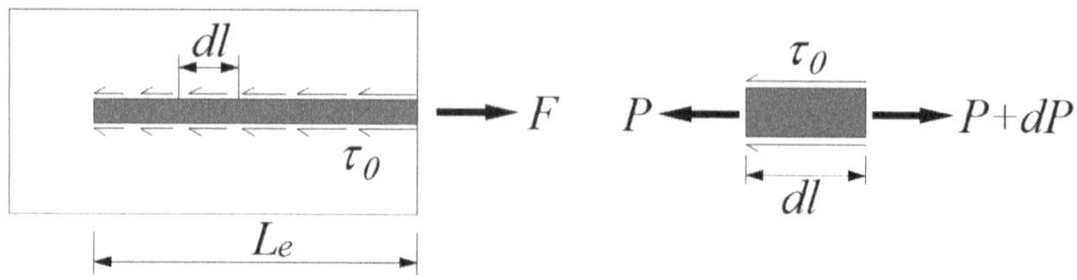

Figure 8. Free-body diagram of the infinitesimal segment of a fiber.

Based on the equivalence of interfacial shear stress distribution [22], Equation (6) is simplified to:

$$F = \tau \cdot \pi d_f \cdot L_e \quad (7)$$

Similarly, the interfacial shearing force of a single amplified fiber is:

$$F_{af} = \tau \cdot \pi(nd_f) \cdot nL_e = n^2 F \quad (8)$$

Then the resultant interfacial shearing force of the original fibers in a direction can be computed as:

$$F^{\kappa} = N^{\kappa} F = N^{\kappa} \tau \cdot \pi d_f \cdot L_e \quad (9)$$

and the resultant force of amplified fibers in the same direction is:

$$F_{af}^{\kappa'} = N_{af}^{\kappa} \tau \cdot \pi(nd_f) \cdot nL_e = \frac{1}{n} F^{\kappa} \quad (10)$$

However, as Equation (10) shows, $F_{af}^{\kappa'}$ is not equal F^{κ}. Therefore, the second step is to modify the interfacial shear modulus G_s to be $G_{saf} = nG_s$ and then the final resultant interfacial shearing force of amplified fibers is:

$$F_{af}^{\kappa} = N_{af}^{\kappa}(n\tau) \cdot \pi(nd_f) \cdot nL_e = F^{\kappa} \quad (11)$$

Equation (11) indicates that the interfacial shearing force of equivalent steel fibers and original steel fibers are ensured to be equal, which means the bridging effect of steel fibers is unchanged and then demonstrates that the equivalent treatment on steel fibers is theoretically reasonable. Ultimately, the mesoscale equivalent model for UHPC with steel fibers is developed based on the equivalent treatment of fibers.

4. Verification on the Mesoscale Equivalent Model

To further verify the developed model, numerical simulations on both uniaxial compression tests and penetration tests presented in Section 1 are conducted in LS-DYNA employing the mesoscale equivalent model. The numerical results are compared with corresponding experimental results, which show that the developed mesoscale equivalent model could well reproduce the behavior of UHPC under static and dynamic loadings.

4.1. Model Verification with Uniaxial Compression Tests

4.1.1. Numerical Model

The typical mesoscale finite element model for prismatic specimens under uniaxial compression is shown in Figure 9. The specimen is placed on a fixed loading plate, and the load is applied to the specimen by imposing displacement on the moving loading plate. An automatic surface-to-surface contact algorithm considering the friction effect is adopted to simulate the contact behavior between the loading plate and the specimen. The solid element SOLID164 is used for modeling the matrix of the specimen and loading plates. The steel fibers are modeled by beam element BEAM161. The keyword *CONSTRAINED_BEAM_IN_SOLID (CBIS) [25] in LS-DYNA is adopted to simulate the bond-slip behavior between steel fibers and matrixes.

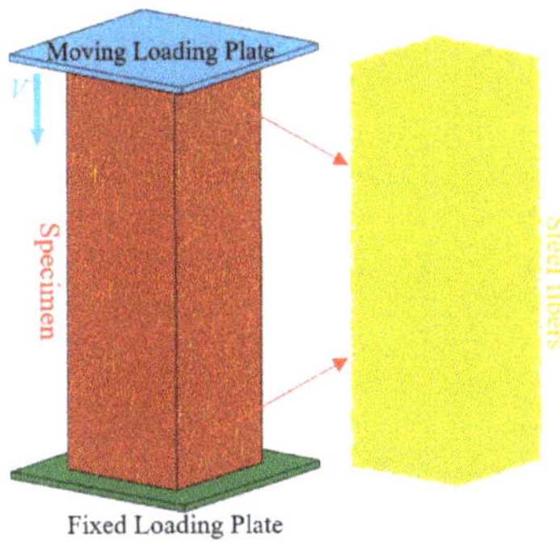

Figure 9. Typical mesoscale FE model for uniaxial compression.

For different amplification factors n, the dimensions and number of steel fibers in the mesoscale equivalent model are different, and the effect of the amplification factor on the numerical needs to be investigated. In this study, numerical simulations on uniaxial compression with different amplification factors n (n = 1, 3, 4, 5, 6, 7, n = 1 means no equivalence treatment on steel fibers) are performed. Figure 10 shows the distribution of steel fibers in a specimen with different n, where the number of steel fiber elements decreases fast with the increasing n. The critical parameters of the CBIS algorithm for the mesoscale equivalent model with different n are listed in Table 2.

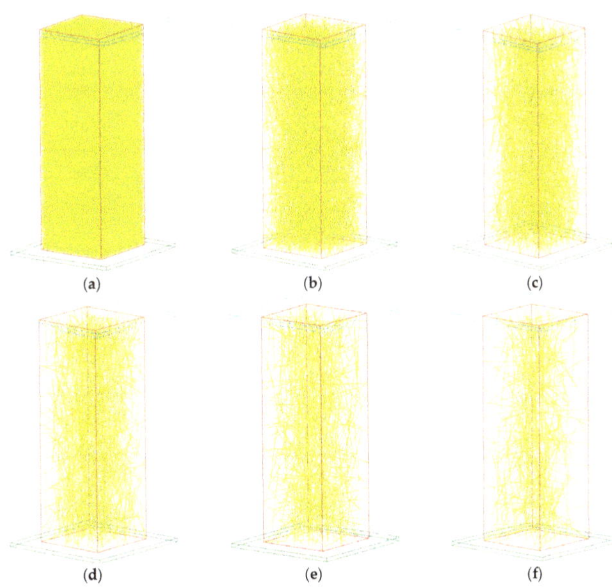

Figure 10. Distribution of steel fibers in matrix with different n: (**a**) n = 1, 146,913 beam elements; (**b**) n = 3, 5442 beam elements; (**c**) n = 4, 2296 beam elements; (**d**) n = 5, 1176 beam elements; (**e**) n = 6, 681 beam elements; (**f**) n = 7, 429 beam elements.

Table 2. Critical CBIS algorithm parameters under different *n*.

d	l	G_S	S_{max}	EXP
0.2·n mm	13·n mm	2393·n MPa	1.25×10^{-3}	0.2

Without adding steel fibers, the UHPC matrix could be regarded as high-strength concrete, modeled by the RHT model [26] in the present study. The critical RHT failure surface parameters A_{fail} and N_{fail} are determined by fitting the triaxial compression test data of 140 MPa HSC in Reference [27], as shown in Figure 11, where A_{fail} and N_{fail} are 1.78 and 0.35, respectively. The other critical parameters are self-defined according to Refs. [10,28,29]. Due to the stiffness of loading plates in practice being much larger than that of the specimen, the loading plates are modeled as a rigid body with the material model *MAT_RIGID. The isotropic and kinematic hardening material model *MAT_PLASTIC_KINEMATIC (*MAT_003) is chosen to build the steel fibers. Parameters of material models used in the uniaxial compression simulations are listed in Table 3.

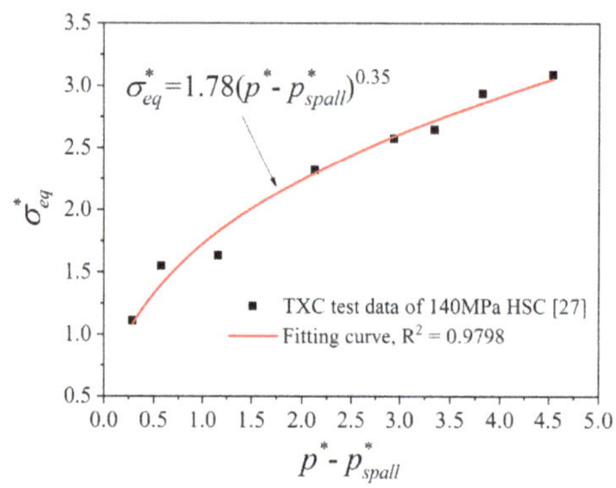

Figure 11. Determination of the parameters A_{fail} and N_{fail}.

Table 3. Parameters of material models for the UHPC matrix, loading plate, and steel fibers (units: cm-g-μs).

Material	Material Model	Input Parameter	Value
UHPC matrix	*MAT_RHT	Shear modulus G	0.19
		Mass density ρ_0	2.273
		Compressive strength f_c	0.0014
		Failure surface constant A_{fail}	1.78
		Failure surface constant N_{fail}	0.35
		Residual Surface constant B	0.8
		Residual Surface constant m	0.3
		Damage constant D_1	0.045
		Damage constant D_2	1.0
		Minimum strain at fracture *EFMIN*	0.011
Loading plate	*MAT_RIGID	Mass density *RO*	7.83
		Young's modulus *E*	2.0
		Poisson's ratio *PR*	0.25
Steel fibers	*MAT_003	Mass density *RO*	7.83
		Young's modulus *E*	2.0
		Poisson's ratio *PR*	0.30
		Yield strength *SIGY*	0.021
		Tangent modulus *ETAN*	0.0021
		Failure strain *FS*	0.2

4.1.2. Results and Discussion

The numerical failure modes of specimens under different amplification factors n are shown in Figure 12. For the cases of n = 1, 3, 4, 5, the specimens exhibit ductile shearing failure modes, where a main oblique crack develops from the end of the specimen to the middle in the direction of about 45 degrees, which is the same as the typical experimental failure mode shown in Figure 12g. However, for the case of n = 6 and n = 7, the failure mode is quite different from the other cases and the experimental result, where the damage occurs firstly in the middle of the specimen and is approximately concentrated in the middle, which displays as brittle splitting failure mode to a certain content. Shearing failure is a type of ductile failure, while splitting failure belongs to brittle failure. Although the volume percentage of steel fibers is the same, the failure mode tends to change from ductile failure to brittle failure with the increase of the amplification factor. The difference between failure modes under different amplification factors indicates that steel fibers greatly influence the ductility of UHPC.

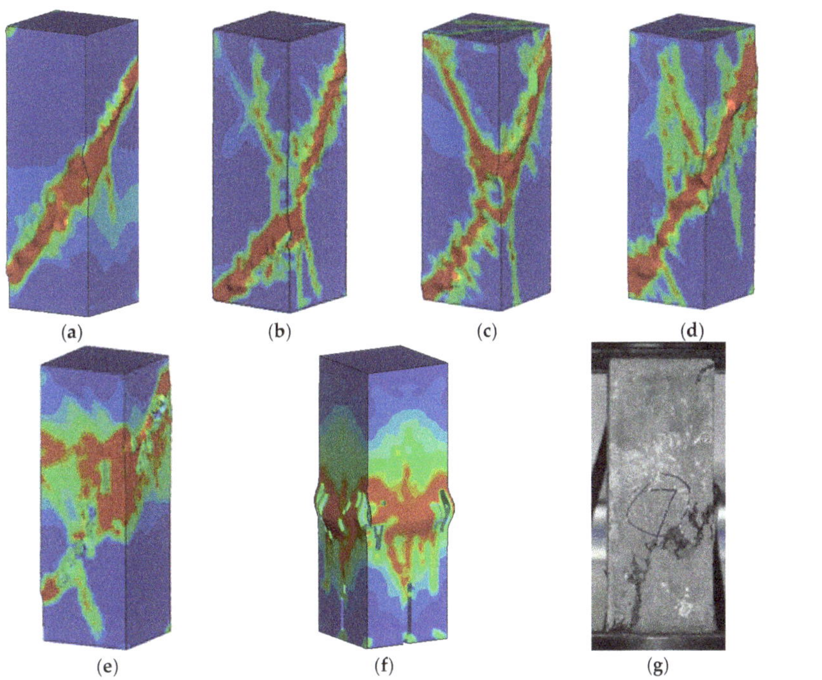

Figure 12. Failure mode of specimens: (**a**) n = 1; (**b**) n = 3; (**c**) n = 4; (**d**) n = 5; (**e**) n = 6; (**f**) n = 7; (**g**) n = 8.

Figure 13 shows the uniaxial compressive curves with different amplification factors n and the average full stress-strain curve obtained from uniaxial compression tests. There is little difference in the ascending branches of stress-strain curves obtained from the numerical simulations under different amplification factors, and all the ascending branches are in good agreement with the test data. As for the descending branches of the stress-strain curves, the numerical results of n = 1, 3, 4, and 5 are in good accord with the test data. Nevertheless, for the cases of n = 6 and n = 7, the numerical descending branches drop more sharply than the other cases and have much lower residual strength, which presents a great difference from the experimental result. If taking the area under the full stress-strain curve of uniaxial compression as an index to characterize the toughness of specimens, Figure 13 shows that the toughness decreases with the increase of the amplification factor.

The variation of the toughness index caused by the change in amplification factor shows that steel fiber influences the toughness of UHPC significantly.

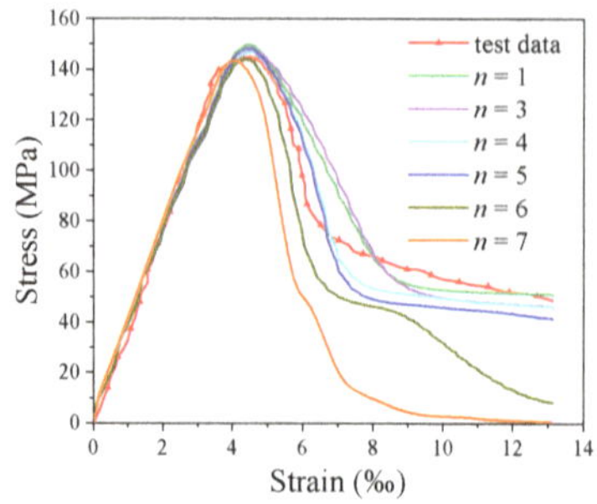

Figure 13. Uniaxial compressive curves of specimens.

Steel fibers improve the ductility and toughness of UHPC by the bridging effect in the matrix. According to the above comparison and discussion, the bridging effect gradually weakens with the increase in the amplification factor. In the present study, the mesoscale equivalent model can accurately simulate the uniaxial compression behavior of UHPC when the amplification factor is lower than 5. However, with further increases in the amplification factor, the mesoscale equivalent model loses its validity. This phenomenon may be because the equivalence of the resultant interfacial shearing force of steel fibers, as shown in Equation (11), is destroyed due to the excessive amplification in the fiber dimension. The equivalent treatment on fibers is based on the assumption that there are enough original fibers in any direction to synthesize a certain number of amplified fibers. When the amplification factor is too large, such as $n = 6$ or $n = 7$, the number of original fibers in some directions may not be enough to synthesize sufficient amplified fibers, leading to the uneven distribution of amplified fibers and causing Equation (11) to fail. Therefore, to ensure the accuracy of the simulation, it is recommended that the amplification factor should not be greater than 5.

4.2. Model Verification with Penetration Tests

Taking the target U-2 as an example, the estimated computation time of penetration simulations under n = 1–5 is listed in Table 4. The computation time falls quickly as the amplification factor increases. Consequently, the amplification factor adopted for penetration simulations is determined to be 5 for balancing the accuracy and computation time of numerical calculation.

Table 4. The estimated computation time of penetration simulations on target U-2 under different amplification factors.

n	1	3	4	5
computation time	1400 h 16 min	51 h 54 min	23 h 18 min	11 h 35 min

4.2.1. Numerical Model

Since no obvious yaw phenomenon is observed in the penetration process, quarter finite element models based on the equivalence method are established, as shown in Figure 14. The keyword *CONSTRAINED_GLOBAL is used to simulate the symmetric boundary condition by defining two global boundary constraint planes. As for the outer surfaces of targets, the fixed boundary condition is imposed according to the experimental setup. Matrix elements (SOLID164) within five times the projectile diameter are refined, and the minimum matrix element size is 0.8 cm. There are 352,512 steel fiber elements (BEAM161) uniformly distributed in the target with 90 cm thickness and 587,571 steel fiber elements in the target with 150 cm thickness. The outer shell and inner filling of the projectile are modeled by the solid element SOLID164 and are connected by joint nodes. The eroding surface-to-surface contact algorithm is used to simulate the contact behavior between the projectile and matrix.

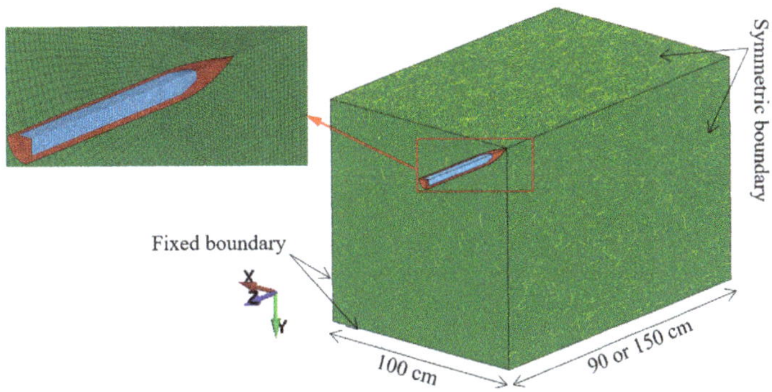

Figure 14. FE model for projectile penetration.

In the process of penetrating, the outer shell of the projectile withstands high temperatures, which will soften the material. Thus, the Johnson–Cook model [30,31] (*MAT_015) applied for metals subjected to large strains, high strain rates, and high temperatures, coupled with *EOS_GRUNEISEN, is adopted to build the outer shell. Material model *MAT_003 is used to build the inner filling of the projectile. The material models and parameters for the projectile are listed in Table 5.

Table 5. Parameters of material models for projectile [3] (units: cm-g-μs).

Material	Material Model	Input Parameter	Value
Outer shell of projectile	*MAT_015	Shear modulus	0.84
		Mass density	7.83
		Poisson's ratio	0.25
		$a/b/n/c/m$	0.01350/0.00477/0.18/0.012/1.0
		Failure stress	−2
		D_1	0.15
		D_2	0.72
		D_3	1.66
		$C/S_1/\gamma/A$	0.4596/1.357/1.71/0.43
Inner filling of projectile	*MAT_003	Mass density *RO*	1.63
		Young's modulus *E*	0.1
		Poisson's ratio *PR*	0.45
		Yield strength *SIGY*	0.0006
		Tangent modulus *ETAN*	0.001
		Failure strain *FS*	3.0

4.2.2. Results and Discussion

Figure 15 shows the numerical results of the penetration process for U-1, where the striking velocity is 410 m/s and the target thickness is 90 cm. The "History Variable #4" refers to the damage parameter D in the RHT model, where $D = 0$ means no damage and $D = 1$ means fracture of the material. It can be seen that severe damage mainly occurs around the ballistic trajectory, and the region of damage distribution expands as penetration depth increases.

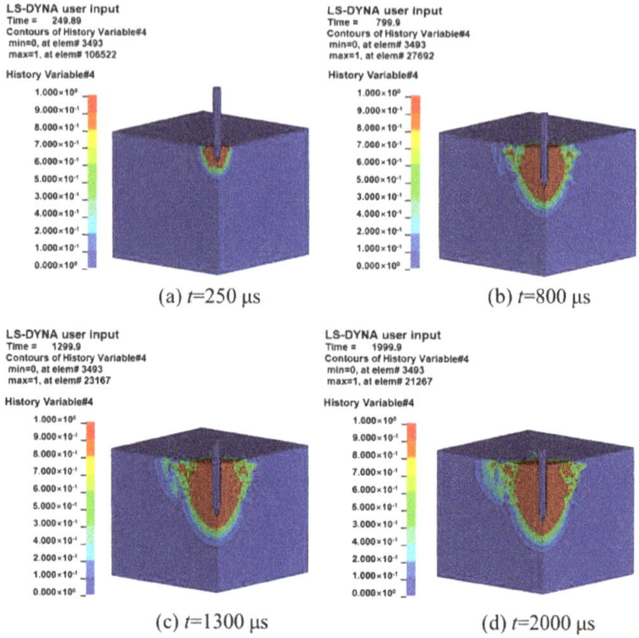

Figure 15. Numerical results of the penetration process for U-1, where the striking velocity is 410 m/s, and the target thickness is 90 cm.

Frontal damage contours of targets U-1 and U-2 are shown in Figure 16. As the striking velocity of the projectile increases, the impact surface of the target tends to be subjected to more severe local damage. The numerical results of the crater diameter on targets U-1 and U-2 are 53.8 cm and 77.3 cm, respectively. Comparing the numerical and experimental results, the maximum relative error in the crater diameter is just 5.9%, showing perfect consistency. In addition to the crater, the projectile impact also forms cracks on the impact surface. The numerical results of cracks show that there are four main cracks on the impact face of target U-1 and eight main cracks on the impact face of target U-2, and these cracks develop radially and orthogonally. The width, number, and distribution range of cracks expand with increased striking velocity. For target U-1, the number and distribution of the simulated cracks are in good agreement with the experimental results. While for target U-2, the difference between the numerical results of the cracks and the experimental results is a little large, which may be caused by the slight yaw of the projectile penetrating into U-2 in the penetration tests.

The time-displacement curves and time-velocity curves of the projectiles are shown in Figure 17. The numerical values of DOPs at striking velocities of 410 m/s and 664 m/s are 40.4 cm and 87.1 cm, respectively. The numerical results and experimental results of DOPs are compared in Figure 18, where the relative errors are −3.8% and 11.7%, respectively. The reason for the slightly larger error in the case of 664 m/s may be that the projectile has a slight yaw, as shown in Figure 7b. Although the yaw is not obvious, it could increase the resistance of the projectile penetration into the target and then reduce the DOP. The maximum relative error of the DOP is lower than 15%, which indicates that the numerical method could reasonably predict the DOP when projectiles penetrate UHPC targets. Figure 17 shows that the velocities of the three projectiles gradually decrease from positive values to negative values with the increase of the DOP. The negative values of the velocities mean that the projectiles bounce in the opposite direction, which is consistent with the experimental phenomenon. The rebound velocities of projectiles at striking velocities of 410 m/s and 664 m/s are 16.7 m/s and 22.6 m/s, respectively.

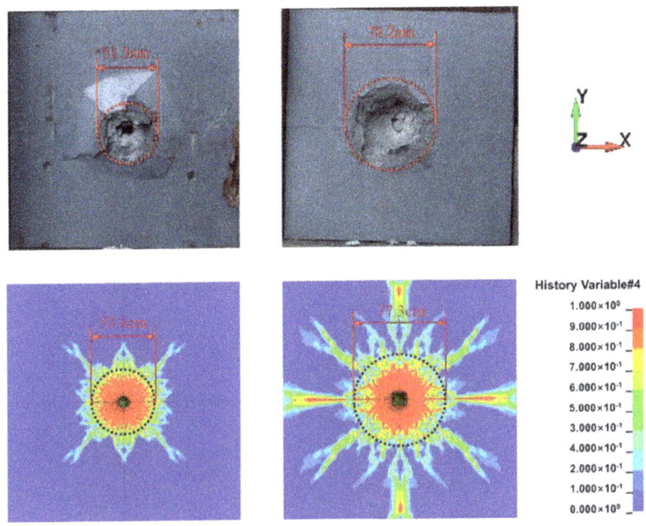

(**a**) (**b**)

Figure 16. Comparison of frontal damage between the experimental and numerical results; (**a**) U-1, 410 m/s; (**b**) U-2, 664 m/s.

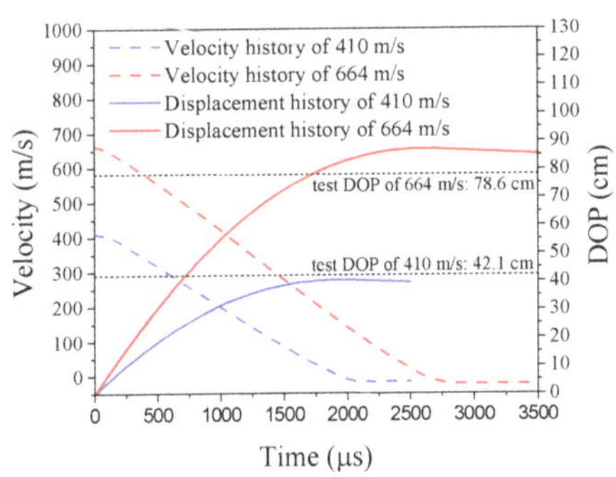

Figure 17. Velocity and displacement histories of projectiles at striking velocities of 410 m/s and 664 m/s.

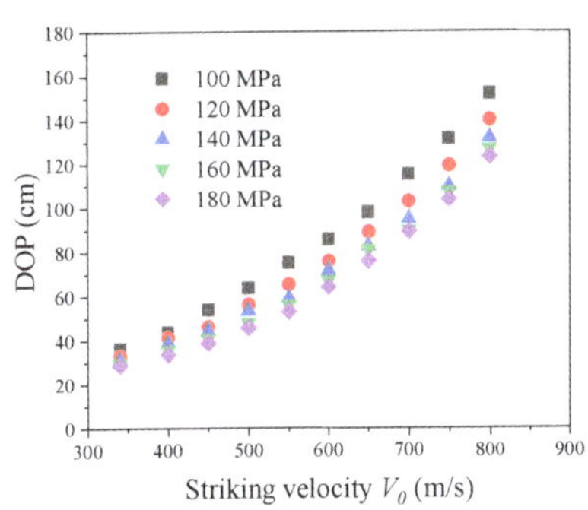

Figure 18. DOP versus striking velocity V_0 for 100 MPa, 120 MPa, 140 MPa, and 180 MPa UHPC targets, where the volume content of the steel fiber is 2%.

Comparing the frontal damage and the DOP between experiments and numerical simulations indicates that the mesoscale equivalent model can reasonably simulate the dynamic performance of UHPC under projectile penetration when the amplification factor is 5.

5. Numerical Investigation on DOP

The DOP is the most important and noteworthy index, which reflects the resistance of concrete defensive structure under the projectile impact, and it is greatly influenced by the projectile's striking velocity and the concrete's compression strength [5]. As for UHPC, steel fibers play an important role in improving tensile strength and toughness, which makes it attractive to study the influence of steel fibers on the DOP. Therefore, based on the validated mesoscale equivalent model, 62 penetration scenarios considering different striking velocities, compression strength, and volume percentage of steel fibers are simulated to investigate the specific effect of these parameters on the DOP. All the simulated results of the DOP are listed in Table 6.

Table 6. Numerical results of the DOP.

Simulation No.	V_0 (m/s)	f_c (MPa)	V_f	Simulated DOP (cm)
1	340	100	2%	36.4
2	400			43.8
3	450			54.0
4	500			63.8
5	550			75.2
6	600			85.7
7	650			98.0
8	700			115.2
9	750			131.3
10	800			151.6
11	340	120	2%	33.2
12	400			41.4
13	450			46.3
14	500			56.3
15	550			65.4
16	600			75.9
17	650			88.9
18	700			103.1
19	750			119.2
20	800			139.9
21	340	140	2%	31.3
22	450			44.5
23	500			53.4
24	550			59.2
25	600			71.7
26	650			82.6
27	700			95.0
28	750			110.0
29	800			131.6
30	400		1%	44.3
31			1.5%	40.9
32			2%	38.8
33			2.5%	37.6
34			3%	36.8
35			3.5%	36.4
36			4%	36.2

Table 6. *Cont.*

Simulation No.	V_0 (m/s)	f_c (MPa)	V_f	Simulated DOP (cm)
37	340	160	2%	30.0
38	400			35.9
39	450			41.0
40	500			48.1
41	550			56.1
42	600			67.1
43	650			81.1
44	700			89.9
45	750			107.4
46	800			126.7
47	340	180	2%	28.7
48	450			39.0
49	500			45.8
50	550			53.2
51	600			64.2
52	650			75.8
53	700			89.1
54	750			103.9
55	800			123.2
56	400		1%	38.7
57			1.5%	36.8
58			2%	35.8
59			2.5%	34.5
60			3%	34.3
61			3.5%	34.0
62			4%	33.8

5.1. *Effect of Striking Velocity*

Figure 18 shows the DOPs of UHPC targets with 2% steel fiber volume content at striking velocities of 340 m/s, 400m/s, 450 m/s, 500m/s, 550 m/s, 600 m/s, 650m/s, 700 m/s, 750 m/s, and 800 m/s. The numerical results manifest that for the UHPC targets of 100 MPa, 120 MPa, 140 MPa, 160 MPa, and 180 MPa, the DOP increases with the increasing striking velocity. Taking the scenario of the 140 MPa UHPC target as an example, the DOPs at striking velocities of 400 m/s, 500 m/s, 600 m/s, 700 m/s, and 800 m/s are 38.8 cm, 53.4 cm, 71.7 cm, 95.0 cm, and 131.6 cm, respectively. The increments between adjacent striking velocities are 14.6 cm, 18.3 cm, 24.3 cm, and 39.0 cm, respectively, and present an exponentially growing trend. For UHPC targets with other compression strengths, the growing trend is similar. The above findings show an exponential relationship between the DOP and striking velocity for UHPC.

5.2. *Effect of Compression Strength*

As shown in Figure 19, the DOP decreases with increasing compression strengths of UHPC targets when the striking velocity is fixed. Comparing the numerical results of UHPC targets under projectile penetration with the striking velocity at 550 m/s, where DOPs of 100 MPa, 120 MPa, 140 MPa,160 MPa, and 180 MPa targets are 75.2 cm, 65.4 cm, 59.2 cm, 56.1 cm, and 53.2 cm, respectively, it can be observed that the increments between the adjacent compression strength are 9.8 cm, 6.2 cm, 3.1 cm, and 2.9 cm, respectively, showing a decreasing trend. Hence, the enhancement of compression strength contributes to improving the resistance of the UHPC target against projectile penetration. However, it should also be noted that the improvement effect diminishes with increasing compression strengths.

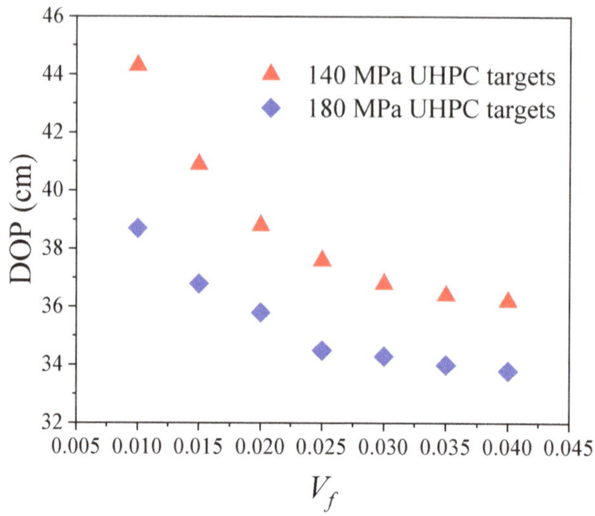

Figure 19. DOP versus volume content of the steel fiber V_f for 140 MPa and 180 MPa UHPC targets, where the striking velocity is 400 m/s.

5.3. Effect of Steel Fiber

Axial force contours of steel fibers for target U-2 in four different moments, along with internal damages and cracks in the matrix, are listed in Table 7. From t = 630 μs to t = 2700 μs, the axial force in steel fibers develops synchronously with the expansion of internal damages and cracks. Since damage is accumulated by the plastic strain, the damage parameter can reflect the deformation of matrix elements. This deformation will cause a relative slip between the matrix and steel fibers embedded in the matrix, resulting in the generation of axial force in the steel fibers. The axial force will increase until the maximum relative slip S_{max} is reached and then decline until total debonding happens. Therefore, taking the regions around the ballistic trajectory at t = 630 μs and regions where the cracks develop at t = 2070 μs as examples, the axial force of steel fibers in the regions with damage is larger than that in the undamaged regions. The above findings show that the steel fibers can exert a bridging effect by limiting the deformation of matrix elements and the development of cracks, improving the resistance of UHPC targets under projectile penetration.

Figure 19 shows the DOPs of 140 MPa and 180 MPa UHPC targets with volume percentages of steel fibers of 1.0%, 1.5%, 2.0%, 2.5%, 3.0%, 3.5%, and 4.0%, where the striking velocity is 400 m/s. No matter the UHPC target of 140 MPa or 180 MPa, the DOP decreases with the increasing volume percentage of steel fibers. DOPs into 140 MPa UHPC targets with 1.0% and 4.0% V_f is 44.3 cm and 36.2 cm, respectively, where the 3% increase in the volume percentage of steel fibers results in an 18.3% decrease in the DOP. For the 180 MPa UHPC targets, the corresponding DOPs are 38.7 cm and 33.8 cm, respectively, and the decrease in the DOP is 12.7%. It can be seen from Figure 19 that when V_f increases from 3% to 4%, the reduction in the DOP is not obvious, which demonstrates that there is a limit to reducing the DOP by adding more steel fibers.

Table 7. Development of internal damages and steel fibers axial force in target U-2.

Time	Internal Damages of Target U-2	The Axial Force of Steel Fibers
$t = 630$ µs	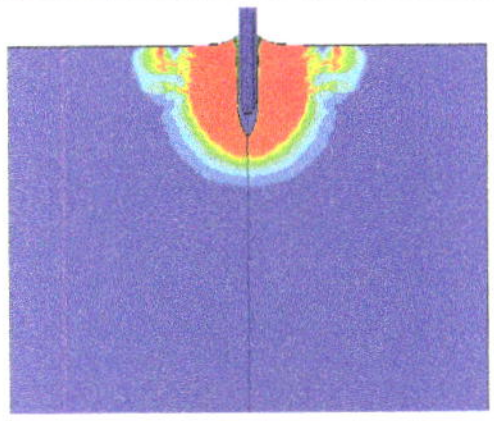	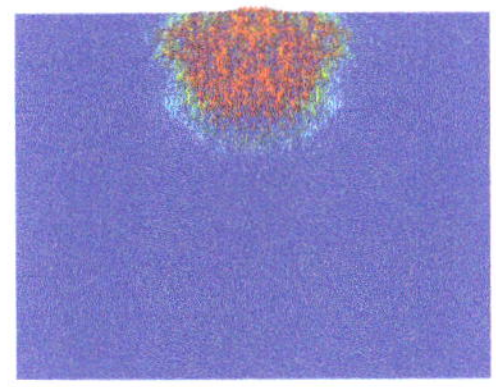
$t = 1350$ µs	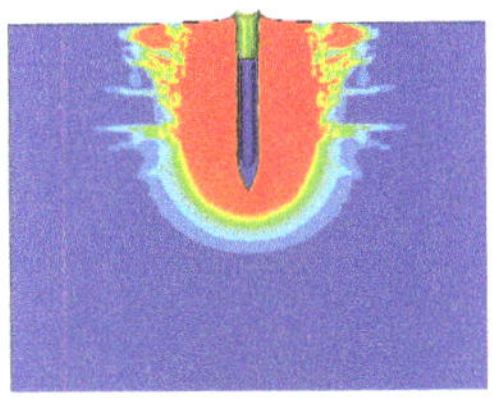	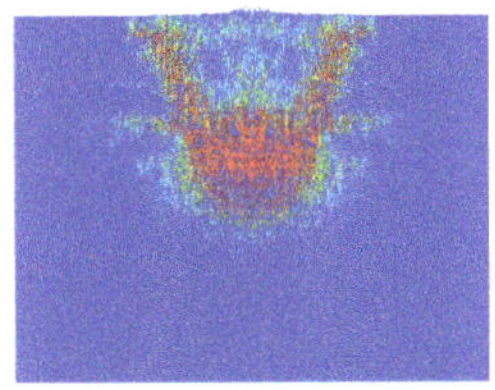
$t = 2070$ µs	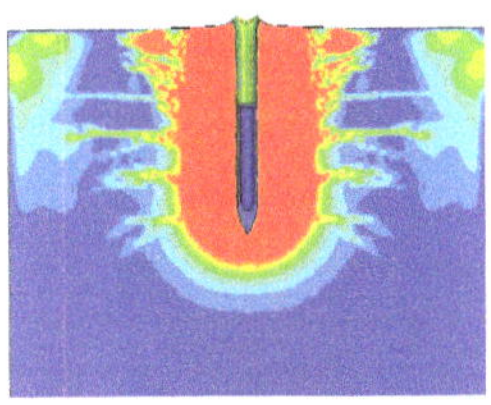	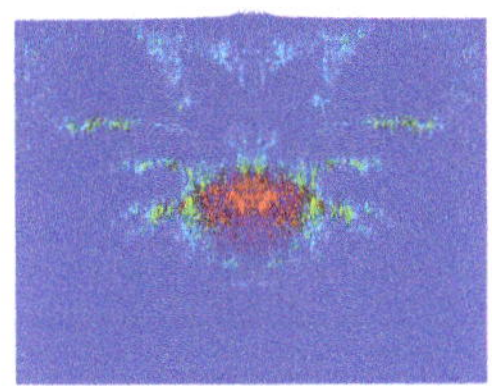
$t = 2700$ µs	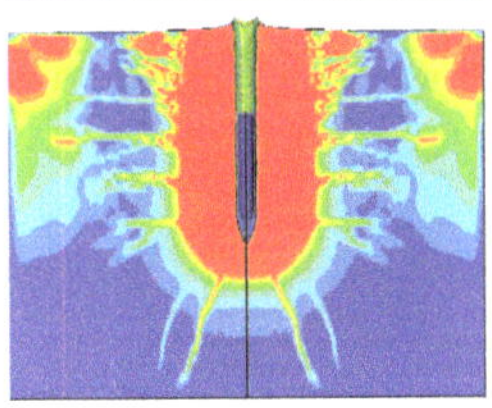	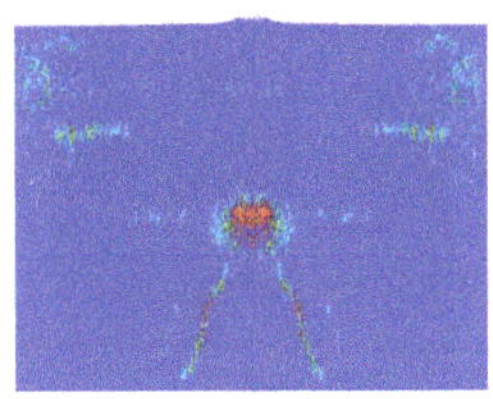

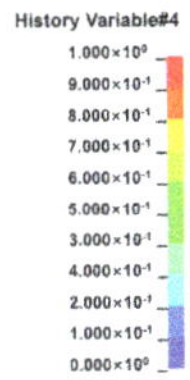

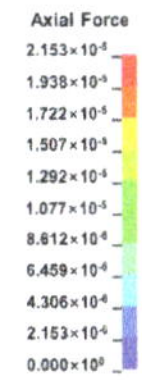

5.4. Modification of Young Equations for DOP Prediction

Based on an extensive experimental database of projectile impact, Sandia National Laboratories proposed the empirical Young penetration equations to predict the DOP into natural earth materials and concrete. With the expansion of the experimental database, Young equations have been constantly updated, and the latest version for concrete is as follows (in SI units):

$$DOP = 0.00000153 \quad N\, K_e K_h (t_c T_c)^{-0.06} (11 - P)(35/f_c)^{0.3} (m\,/A)^{0.7} (V_0 - 30.5), \qquad V_0 \geq 61\ m/s \tag{12}$$

$$N = 0.18(CRH - 0.25)^{0.5} + 0.56, \text{ for tangent ogive nose shapes} \tag{13}$$

$$K_e = (F/\,W_1)^{0.3} \tag{14}$$

$$K_h = 0.46\,(m)^{0.15}, \text{ when } m < 182 \text{ kg; else, } K_h = 1.0 \tag{15}$$

where N is the nose performance coefficient of the projectile, K_e is the correction coefficient for edge effects in concrete targets, and K_h is the correction coefficient for the lightweight projectile. More details about the three coefficients are presented in Reference [32]. t_c is the cure time of concrete, T_c is the thickness of the target, P is the volumetric percentage of rebars in concrete targets, f_c is the compression strength of concrete, m is the mass of the projectile, A is the cross-sectional area of the projectile, V_0 is the striking velocity of the projectile.

The Young equations comprehensively consider the parameters that affect the DOP, which makes them applicable in many penetration scenarios. For normal concrete and high-strength concrete with compression strength under 100 MPa, the accuracy of the Young equations is well validated. However, according to the above parametric analysis, the functional relationship between the DOP and the striking velocity of the projectile for UHPC is different from the linear relationship described in the original Young equations. At the same time, the original Young equations do not consider the influence of steel fibers on the penetration depth. Consequently, the original Young equations are not applicable to predicting the DOP into UHPC with ultra-high strength and steel fibers and should be modified. The parametric analysis shows that the functional relationships between the DOP and striking velocity V_0, compression strength f_c, and volume percentage of steel fibers V_f can be described by an exponential function, power function, and quadratic polynomial, respectively. Hence, the modified Young equation for UHPC without rebar is expressed as:

$$DOP = \alpha \cdot NK_eK_h \cdot (t_cT_c)^{-0.06}(f_c^{\beta})(k_0 + k_1V_f + k_2V_f^2)(m/A)^{0.7}(e^{\gamma V_0}) \quad (16)$$

where N, K_e, and K_h are the same as the original equations, and α, β, k_0, k_1, k_2, and γ are the undetermined coefficients.

According to the numerical results of the DOP listed in Table 6 and the equation form expressed as Equation (13), multivariate nonlinear fitting is performed in MATLAB to determine the pending parameters. The values of α, β, k_0, k_1, k_2, and γ are determined to be 0.00344, −0.43, 292.8, −5048, 77946, and 0.003, respectively. In the end, the modified Young equation is as follows:

$$DOP = 0.00344NK_eK_h(t_cT_c)^{-0.06}(f_c^{-0.43})\,(292.8 - 5048V_f + 77946V_f^2)\,(m/A)^{0.7}\,(e^{0.003V_0}) \quad (17)$$

where f_c is in the unit of "MPa", V_f is a unitless percentage, m is in the unit of "kg", A is in the unit of "m^2", V_0 is in the unit of "m/s", and eventually the DOP is in the unit of "cm". The correlation index R^2 of the fitted function is 0.9977, indicating that the modified equation for the DOP prediction has high goodness of fit.

The DOPs calculated by the modified Young equation and the original Young equation are compared with the experimental data presented in this study and References [3,5,33], as listed in Table 8. It can be seen that the original Young equation greatly underestimates the penetration resistance of UHPC, and the maximum relative error is 26.9%. However, the relative error between the calculated DOPs by the modified Young equation and experimental data is within 14%, which indicates that the modified Young equation proposed in the present study can accurately predict the DOP of UHPC targets subjected to projectile impact.

Table 8. Comparison of the DOPs between results of equation calculation and experimental results for different UHPC targets.

Specimen	f_c (MPa)	V_f (%)	m (kg)	d (cm)	V_0 (m/s)	DOP in Tests (cm)	DOP of Modified Equation (cm)	Relative Error	DOP of Original Equation (cm)	Relative Error
U-1	148	2	11.9	8	410	42.1	39.2	−6.9%	52.7	25.2%
U-2	148	2	11.9	8	664	78.6	84.0	6.8%	92.7	17.9%
UHPC-SF-1 [3]	140	3	0.329	2.53	553	12.9	12.9	−0.2%	16.4	26.9%
UHPC-SF-2 [3]	140	3	0.329	2.53	683	16.6	18.7	12.7%	20.4	23.1%
UHPC-SF-3 [3]	140	3	0.329	2.53	808	20.8	23.7	13.9%	25.8	24.0%
A-5-1 [5]	114	3	0.341	2.53	510	13.4	12.7	−4.9%	16.9	26.1%
1:3 [33]	153	1.5	6.3	7.5	622	51.0	52.2	2.4%	59.7	17.1%

6. Conclusions

In this work, a mesoscale equivalent model is first developed to numerically investigate the dynamic response of UHPC subjected to projectile impacts in a more refined and efficient way. Experiments on UHPC subjected to uniaxial compression and projectile impacts are conducted and are used to validate the developed model. Relying on the mesoscale equivalent model, the influence of projectile striking velocities, UHPC compression strengths, and volume percentages of steel fibers on the DOP is numerically investigated, and a modified Young equation for predicting the DOP is proposed. The main conclusions can be drawn as follows:

(1) The equivalent treatment on steel fibers is to amplify the size of the fibers and the interfacial shearing modulus between fibers and the matrix by n times synchronously. The interfacial shearing force is analytically proven to be equal to that before the equivalent treatment conducted on steel fibers, demonstrating that the equivalent treatment on steel fibers is theoretically reasonable. The mesoscale equivalent model for UHPC with steel fibers is successfully developed based on the equivalent treatment on fibers. When the amplification factor of steel fiber is lower than 5, the proposed model can accurately simulate the uniaxial compression behavior of UHPC specimens. However, when the amplification factor is greater than 5, the model cannot well characterize the ductility and toughness of UHPC.

(2) When the amplification factor of steel fibers is lower than 5, the mesoscale equivalent model can accurately reproduce the failure mode and stress-strain curve of the UHPC specimens under the uniaxial compression. The computational cost of the numerical simulations of penetration experiments decreases rapidly with the increase of the amplification factor. With an amplification factor of 5, the maximum relative error between the numerical results of the cater diameter and penetration depth and experimental results is 11.7%, indicating that the mesoscale equivalent model has high accuracy.

(3) The mesoscale equivalent model provides a more refined and in-depth perspective into numerically investigating the response of UHPC subjected to projectile impacts. The numerical investigation of the DOP shows that the DOP increases exponentially with the increase of the projectile striking velocity. The decreasing relationships between the DOP and the compression strength and volume percentage of steel fibers can be described by the power function and quadratic polynomial, respectively. Steel fibers exert a bridging effect by limiting the deformation of matrix elements to improve the penetration resistance of UHPC, but there is a limit to reducing the DOP by adding more steel fibers.

(4) Based on the simulated data of the DOP, a modified Young equation is proposed for predicting the DOP of UHPC targets subjected to projectile impacts. The maximum relative error between the modified equation and experimental data is 13.9%, showing the proposed equation has high accuracy.

Future studies will focus on establishing a mesoscale equivalent model for UHPC with fibers in different shapes and materials or numerically investigating the dynamic response of UHPC under blast loadings based on the mesoscale equivalent model.

Author Contributions: Conceptualization, J.A., J.Y., W.W. and Y.L.; methodology, J.A. and J.Y.; software, J.A.; validation, J.A., J.Y., W.W. and Y.L.; formal analysis, J.A., W.W. and Y.L.; investigation, J.A.; resources, J.Y. and W.W.; data curation, J.A.; writing—original draft preparation, J.A.; writing—review and editing, J.Y., W.W. and Y.L.; visualization, J.A.; supervision, J.Y. and W.W.; project administration, W.W.; funding acquisition, J.Y. and W.W. All authors have read and agreed to the published version of the manuscript.

Funding: This research received no external funding.

Institutional Review Board Statement: Not applicable.

Informed Consent Statement: Not applicable.

Data Availability Statement: Not applicable.

Conflicts of Interest: The authors declare no conflict of interest.

References

1. Wang, D.; Shi, C.; Wu, Z.; Xiao, J.; Huang, Z.; Fang, Z. A review on ultra high performance concrete: Part II. Hydration, microstructure and properties. *Constr. Build. Mater.* **2015**, *96*, 368–377. [CrossRef]
2. Sharma, R.; Jang, J.; Bansal, P. A comprehensive review on effects of mineral admixtures and fibers on engineering properties of ultra-high-performance concrete. *J. Build. Eng.* **2022**, *45*, 103314. [CrossRef]
3. Liu, J.; Wu, C.; Su, Y.; Li, J.; Shao, R.; Chen, G.; Liu, Z. Experimental and numerical studies of ultra-high performance concrete targets against high-velocity projectile impacts. *Eng. Struct.* **2018**, *173*, 166–179. [CrossRef]
4. Tai, Y.S. Flat ended projectile penetrating ultra-high strength concrete plate target. *Theor. Appl. Fract. Mech.* **2009**, *51*, 117–128. [CrossRef]
5. Wu, H.; Fang, Q.; Chen, X.W.; Gong, Z.M.; Liu, J.Z. Projectile penetration of ultra-high performance cement based composites at 510–1320 m/s. *Constr. Build. Mater.* **2015**, *74*, 188–200. [CrossRef]
6. Feng, J.; Gao, X.; Li, J.; Dong, H.; Yao, W.; Wang, X.; Sun, W. Influence of fiber mixture on impact response of ultra-high-performance hybrid fiber reinforced cementitious composite. *Compos. Part B Eng.* **2019**, *163*, 487–496. [CrossRef]
7. Liu, J.; Wu, C.; Li, J.; Liu, Z.; Xu, S.; Liu, K.; Su, Y.; Fang, J.; Chen, G. Projectile impact resistance of fibre-reinforced geopolymer-based ultra-high performance concrete (G-UHPC). *Constr. Build. Mater.* **2021**, *290*, 123189. [CrossRef]
8. Prakash, A.; Srinivasan, S.; Rama Mohan Rao, A. Numerical investigation on steel fibre reinforced cementitious composite panels subjected to high velocity impact loading. *Mater. Des.* **2015**, *83*, 164–175. [CrossRef]
9. Blasone, M.; Saletti, D.; Baroth, J.; Forquin, P.; Bonnet, E.; Delaplace, A. Ultra-high performance fibre-reinforced concrete under impact of an AP projectile: Parameter identification and numerical modelling using the DFHcoh-KST coupled model. *Int. J. Impact Eng.* **2021**, *152*, 103838. [CrossRef]
10. Wan, W.; Yang, J.; Xu, G.; Liu, Y. Determination and Evaluation of Holmquist-Johnson-Cook Constitutive Model Parameters for Ultra-High-Performance Concrete with Steel Fibers. *Int. J. Impact Eng.* **2021**, *156*, 103966. [CrossRef]
11. Liu, J.; Wu, C.; Chen, X. Numerical study of ultra-high performance concrete under non-deformable projectile penetration. *Constr. Build. Mater.* **2017**, *135*, 447–458. [CrossRef]
12. Zhou, F.; Su, Q.; Cheng, Y.; Wu, H. A novel dynamic constitutive model for UHPC under projectile impact. *Eng. Struct.* **2023**, *280*, 115711. [CrossRef]
13. Liu, J.; Wu, C.; Li, J.; Su, Y.; Chen, X. Numerical investigation of reactive powder concrete reinforced with steel wire mesh against high-velocity projectile penetration. *Constr. Build. Mater.* **2018**, *166*, 855–872. [CrossRef]
14. Yang, J.; Chen, B.; Su, J.; Xu, G.; Zhang, D.; Zhou, J. Effects of fibers on the mechanical properties of UHPC: A review. *J. Traffic Transp. Eng. (Engl. Ed.)* **2022**, *9*, 363–387. [CrossRef]
15. Zhang, X.; Liu, S.; Li, G.; Wang, X. Numerical simulation of failure characteristics of reactive powder concrete with steel fiber. *Front. Phys.* **2021**, *9*, 759531. [CrossRef]
16. Liang, X.; Wu, C. Meso-scale modelling of steel fibre reinforced concrete with high strength. *Constr. Build. Mater.* **2018**, *165*, 187–198. [CrossRef]
17. Feng, T.; Wang, F.; Tan, Y.; Yue, C.; Xu, W.; Liu, Z.; Yang, Z.; Wu, Y.; Jiang, J. Dynamic compression mechanical properties of eco-friendly ultra-high performance concrete produced with aeolian sand: Experimental and three-dimensional mesoscopic investigation. *Int. J. Impact Eng.* **2022**, *164*, 104192. [CrossRef]
18. Smith, J.; Cusatis, G.; Pelessone, D.; Landis, E.; O'Daniel, J.; Baylot, J. Discrete modeling of ultra-high-performance concrete with application to projectile penetration. *Int. J. Impact Eng.* **2014**, *65*, 13–32. [CrossRef]
19. *GBT 50081-2019*; Standard for Test Methods of Concrete Physical and Mechanical Properties. China Architecture & Building Press: Beijing, China, 2019.
20. *DBJ43/T325-2017*; Technical Specification for Reactive Podwer Concrete Structures. China Architecture & Building Press: Beijing, China, 2017.
21. Deng, Y.; Zhang, Z.; Shi, C.; Wu, Z.; Zhang, C. Steel fiber–matrix interfacial bond in ultra-high performance concrete: A review. *Engineering* **2022**, in press. [CrossRef]
22. Su, Y.; Li, J.; Wu, C.; Wu, P.; Tao, M.; Li, X. Mesoscale study of steel fibre-reinforced ultra-high performance concrete under static and dynamic loads. *Mater. Des.* **2017**, *116*, 340–351. [CrossRef]
23. Huo, L.; Bi, J.; Zhao, Y.; Wang, Z. Constitutive model of steel fiber reinforced concrete by coupling the fiber inclining and spacing effect. *Constr. Build. Mater.* **2021**, *280*, 122423. [CrossRef]
24. Xu, Z.; Zhao, Q.; Guo, W.; Zhang, J.; Yu, J.; Wang, D.; Bai, Y. Effect of interfacial transition zone on creep behavior of steel fiber-reinforced concrete. *Constr. Build. Mater.* **2022**, *350*, 128719. [CrossRef]
25. LSTC. *LS-DYNA Keyword User's Manual Revision 11*; LSTC: Livermore, CA, USA, 2018.
26. Riedel, W.; Thoma, K.; Hiermaier, S.; Schmolinske, E. Penetration of Reinforced Concrete by BETA-B-500 Numerical Analysis using a New Macroscopic Concrete Model for Hydrocodes. In Proceedings of the (CD-ROM) 9. Internationales Symposium, Interaction of the Effects of Munitions with Structures, Berlin/Strausberg, Germany, 3–7 May 1999.

27. Hanchak, S.J.; Forrestal, M.J.; Young, E.R.; Ehrgott, J.Q. Perforation of concrete slabs with 48 MPa (7 ksi) and 140 MPa (20 ksi) unconfined compressive strengths. *Int. J. Impact Eng.* **1992**, *12*, 1–7. [CrossRef]
28. Sovjak, R.; Vavřiník, T.; Zatloukal, J.; Maca, P.; Mičunek, T.; Frydrýn, M. Resistance of slim UHPFRC targets to projectile impact using in-service bullets. *Int. J. Impact Eng.* **2015**, *76*, 166–177. [CrossRef]
29. Abdel-Kader, M. Modified settings of concrete parameters in RHT model for predicting the response of concrete panels to impact. *Int. J. Impact Eng.* **2019**, *132*, 103312. [CrossRef]
30. Johnson, G.R.; Cook, W.H. A Constitutive Model and Data for Metals Subjected to Large Strains, High Strain Rates and High Temperature. In Proceedings of the 7th International Symposium on Ballistics, Hague, The Netherlands, 19–21 April 1983.
31. Johnson, G.R.; Cook, W.H. Fracture characteristics of three metals subjected to various strains, strain rates, temperatures and pressures. *Eng. Fract. Mech.* **1985**, *21*, 31–48. [CrossRef]
32. Young, C.W. *Penetration Equations Office of Scientific & Technical Information Technical Reports*; OSTI: Oak Ridge, TN, USA, 1997.
33. Unosson, M.; Nilsson, L. Projectile penetration and perforation of high performance concrete: Experimental results and macroscopic modelling. *Int. J. Impact Eng.* **2006**, *32*, 1068–1085. [CrossRef]

applied sciences

MDPI

Article

A Machine Learning Model for the Prediction of Concrete Penetration by the Ogive Nose Rigid Projectile

Qadir Bux alias Imran Latif [1,*], Zubair Ahmed Memon [2], Zafar Mahmood [3], Mohsin Usman Qureshi [4] and Abdalrhman Milad [1,*]

1 Department of Civil and Environmental Engineering, College of Engineering, University of Nizwa, P.O. Box 33, Nizwa 616, Ad-Dakhliyah, Oman
2 Department of Engineering Management, Prince Sultan University, Riyadh 11586, Saudi Arabia; zamemon@psu.edu.sa
3 Department of Civil and Architectural Engineering, College of Engineering, University of Buraimi, P.O. Box 890, Buraimi 512, Oman; zafar.m@uob.edu.om
4 Faculty of Engineering, Sohar University, P.O. Box 44, Sohar 311, Oman; mqureshi@su.edu.om
* Correspondence: qadir.omran@unizwa.edu.om (Q.B.a.I.L.); a.milad@unizwa.edu.om (A.M.)

Abstract: In recent years, research interest has been revolutionized to predict the rigid projectile penetration depth in concrete. The concrete penetration predictions persist, unsettled, due to the complexity of phenomena and the continuous development of revolutionized statistical techniques, such as machine learning, neural networks, and deep learning. This research aims to develop a new model to predict the penetration depth of the ogive nose rigid projectile into concrete blocks using machine learning. Genetic coding is used in Python programming to discover the underlying mathematical relationship from the experimental data in its non-dimensional form. A populace of erratic formulations signifies the rapport amid dependent parameters, such as the impact factor (I), the geometry function of the projectile (N), the empirical constant for concrete strength (S), the slenderness of the projectile (λ), and their independent objective variable, X/d, where X is the penetration depth of the projectile and d is the diameter of the projectile. Four genetic operations were used, including the crossover, sub-tree transfiguration, hoist transfiguration, and point transfiguration operations on supervised test datasets, which were divided into three categories, namely, narrow penetration ($X/d < 0.5$), intermediate penetration ($0.5 \leq X/d < 5.0$), and deep penetration ($X/d \geq 5.0$). The proposed model shows a significant relationship with all data in the category for medium penetration, where $R^2 = 0.88$, and $R^2 = 0.96$ for deep penetration. Furthermore, the proposed model predictions are also compared with the most commonly used NDRC and Li and Chen models. The outcome of this research shows that the proposed model predicts the penetration depth precisely, compared to the NDRC and Li and Chen models.

Keywords: penetration; machine learning; concrete; rigid projectile; symbolic regression

Citation: Imran Latif, Q.B.a.; Memon, Z.A.; Mahmood, Z.; Qureshi, M.U.; Milad, A. A Machine Learning Model for the Prediction of Concrete Penetration by the Ogive Nose Rigid Projectile. *Appl. Sci.* **2022**, *12*, 2040. https://doi.org/10.3390/app12042040

Academic Editors: Ricardo Castedo, Lina M. López and Anastasio P. Santos

Received: 11 January 2022
Accepted: 11 February 2022
Published: 16 February 2022

1. Introduction

1.1. Research Background

In the 20th century worldwide, various studies were conducted to produce innovative concrete variations [1]. However, since the 19th century, ordinary concrete is still the foremost distinctive practical material and is commonly used to build structures against accidentally-occurring impact loads [1]. These accidentally-occurring impact loads, such as vehicle crashes, plane crashes, tsunami, tornadoes, and flying objects are the main sources of penetration in concrete [2]. Figure 1 shows the illustration of penetration that occurs due to ogive nose rigid projectile impacts on concrete structures [2].

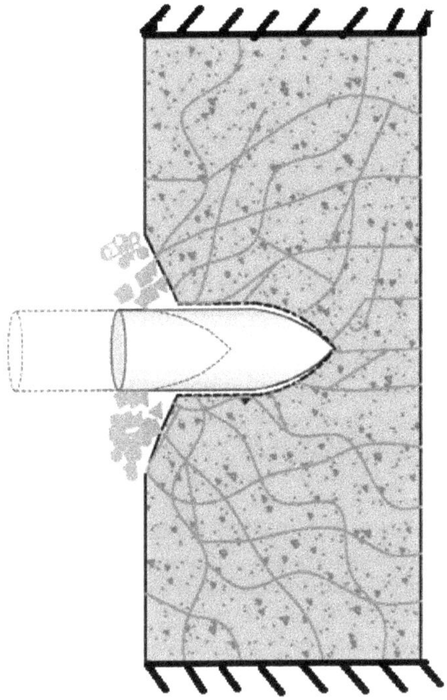

Figure 1. Penetration of concrete caused by ogive nose rigid projectile.

The historical penetration depth prediction model, that was developed based on experimental data, showed that the USA most frequently used the Petry model/the modified Petry model to predict penetration depth into a concrete block [1,3]. The Petry model was initially developed in 1910 and is considered the oldest available empirical model in literature [1,3]. Later, the Petry model was modified by Q.M. Li in the S.I unit [1,3]. In 1941, the Ballistic Research Laboratory (BRL) proposed the penetration prediction model [1,3]. Commencing in 1943, the Army Corp of Engineers also established the ACE model [1,3]. In 1946, the National Defense Research Committee (NDRC) proposed a model founded on the ACE model [1,3,4] The limitation of the NDRC hypothesis was that the NDRC model was used for rear penetration depths [1,3]. The NDRC also suggested a nose shape factor N* for projectiles, as mentioned in Table A1. Later, Ammann and Whitney's model was anticipated to foretell the penetration of concrete traceable to the effect of strenuously precipitated particles at comparatively sophisticated velocities [1,3]. According to Kennedy [5], the Ammann and Whitney model can be used for velocities over 1000 ft/s [1,3]. Whiffen further continued research in the British Road Research Laboratory in the United Kingdom, using experimental data acquired from World War II, and developed a prediction model for penetration depth [1,3]. Kar altered the NDRC model, employing reversion regarding the modulus of elasticity (E), where E is the modulus of elasticity of the projectile, and Es is the modulus of elasticity of steel [1,3]. In the United Kingdom, Barr recommended a UKAEA model by reforming the NDRC model derived from wide-ranging research on nuclear power plant structures [1,3]. The Haldar–Hamieh model [1,3,6] of penetration depth relies on the dimensionless impact factor (Ia), where and N* is the nose shape factor [1,3,6]. Adeli and Amin improved the impact factor (Ia), familiarized by Halder and Hamieh, using regression on Sliter's experimental data [1,3]. Hughes revised the NDRC hypothesis and suggested a model where I_h is a non-dimensional impact factor [1,3] and N_h is a nose shape factor [1,3]. Hughes proposed that the tensile and compressive strength of the concrete (ft/fc) ratio is constant [1,3,7]. Hughes also showed the importance of the strain rate and the dynamic increase factor (DIF), represented by 'S' [1,3,7]. Healy and Weissman introduced a model for penetration depth by revising the NDRC and Kar models [1,3]. The CRIEPI Model, mentioned below in Table A1, assumes the thickness of the concrete barrier Hr = 20 cm (0.2 m) [1,3]. In 1985, the United Kingdom Nuclear Electronics (UKNE) began intrinsic research on the behavior of concrete structures resisting hard projectile penetration by creating the R3 Concrete Impact Working Party [1,3]. The UMIST model for penetration depth (X) improved the form with the reflection of the nose shape [1,3]. Li and Chen [1,3,8] further advanced Forrestal et al.'s [1,3,9] model and proposed a semi-empirical or semi-analytical model for penetration depth (X). The model is in a non-dimensional homogenous form, and these models are valid for an extensive scope of penetration depth, where I is the impact function, and N is the geometry function. $S = 72f_c^{-0.5}$ is an empirical function of fc (MPa). The Li and Chen model is applicable for $X/d \geq 5.0$ [2,4]. Li and Chen [1,3,8,10] recommended $X/d < 5.0$ for small-to-medium penetration depths, where h is the length of the nose of the projectile. In the case of narrow penetrations when $X/d < 0.5$, the penetration depth is given by [1,3,8,10]. Chen and Li [1,3,8,10] recommended a simplified model of $X/d = 0.5(I)$ to predict the penetration depth for deep penetration. The details of the discussed models are given in Appendix A Table A1.

Since 2005, researchers turned their attention to developing models in terms of the required critical impact energy [1,2]. Furthermore, research also turned to the advancement of concrete, creating high-performance concrete (HPC), ultra-high-performance concrete (UHPC), and high-performance fiber-reinforced concrete (HPFRC). However, until today, normal concrete and normal reinforced concrete are still the most commonly used materials for the construction of structures. The revised interest of researchers on concrete penetration depth emerged in 2015, when Husseini and Dalvand implemented an evolution of the statistical model developing technique, the neural network, to develop a model for the prediction of the rigid projectile penetration depth in concrete. Furthermore, in 2021, research was conducted using gradient tree boosting machine learning to predict RC panel failure modes under impact loading. These two kinds of research emerged with the need for the redevelopment of the model, using advanced machine learning tools to develop a prediction model based on an extensive collection of experimental data. Therefore, this research is focused on developing a model based on an extensive collection of 257 experimental test data, using machine learning symbolic regression. Furthermore, the proposed prediction model is validated with the prediction of the NDRC and Li and Chen models, because, in the literature, the most prominent model used to predict penetration depth is the Li and Chen dimensionless model [1].

1.2. Research Motivation

Based on the literature, there is a need to modernize the penetration depth prediction model in recent years. This is mainly due to the complexity of penetration phenomena, where there is a continuous evolution of techniques, such as neural networks and machine learning, especially since concrete is still commonly used in structures [11,12].

In 2015, Husseini and Dalvand used a neural network on 70 experimental datasets taken to estimate the penetration depth in concrete targets using an ogive nose rigid projectile [8,10,13–16]. The neural network models have a very low rate of error and high correlation coefficients compared to the regression-based models [17]. In 2021, Thai et al. used gradient tree boosting machine learning to predict RC panel failure modes under impact loading. The accuracy of the prediction result was not as high as expected, due to the lack of data and the unbalance experimental output features. However, this new approach was recommended for further investigation [18].

1.3. Research Objectives

The main objective of this research is to develop a modernized model for predicting penetration depth using machine learning symbolic regression genetic programming in Python. The model can predict the penetration depth with greater accuracy on a wide range of 257 tests, compared with the NDRC and Li and Chen models. The total of 257 test datasets from [2,8,10,13–16,19] is divided into three categories of narrow penetration depths, where $X/d < 0.5$; intermediate penetration depth, where $0.5 < X/d < 5.0$; and deep penetration, where $X/d > 5.0$. X is the penetration depth and d is the diameter of the projectile. The significance of this research is that a reasonable number of 257 test datasets are trained and tested using machine learning symbolic regression with a genetic programming crossover, sub-tree transfiguration, hoist transfiguration, and point transfiguration to develop a model with better accuracy. The proposed model further compares with the NDRC and Li and Chen equations.

2. Machine Learning Symbolic Regression Genetic Programming Using Python

An interpretable supervised machine learning symbolic regression was used to discover the fundamental scientific equation to describe a Python correlation. Symbolic regression discovers mathematical equations using genetic programming [20]. A populace of erratic formulations is used to signify rapport amid dependent parameters, such as the impact factor (I), the geometry function of the projectile (N), the empirical constant for concrete strength (S), the slenderness of projectile (λ) and their independent objective variable, X/d, where X is the penetration depth of the projectile and d is the diameter of the projectile. In the iterative process, the consecutive propagation of a series of events is progressed from its predecessor's generation by choosing the populace's adequate entities to undertake genetic maneuvers. Genetic encoding yields a sequence of entirely unplanned programs (or models) and further predicts scientific equations to express the relationship between the reliant and sovereign parameters [21]. Four genetic operations used in genetic programming are the crossover, sub-tree transfiguration, hoist transfiguration, and point transfiguration operations, as shown in Figure 2. A sub-tree is randomly selected from the front-runner of an event in the crossover method and is replaced with a randomly selected sub-tree, since it is the front-runner of one more event (Figure 2a). A sub-tree is randomly chosen from the front-runner of an event in the sub-tree mutation method, and it is replaced with a sub-tree that is generated randomly (Figure 2b). The hoist mutation method selects a sub-tree of a randomly selected sub-tree from the winner of a tournament, which replaces the previously chosen sub-tree (Figure 2c). The point mutation method randomly selects some nodes from the tournament winner and replaces them with other building blocks (Figure 2d).

Symbolic regression generates a populace of unplanned scientific formulations with independent data factors as variants [22]. This scientific formulation genus transfigures and emerges as innovative models via genetic encoding [22]. The resulting formulations endeavor to predict trial data by appraising the specified metric (e.g., mean absolute error, root mean-squared error, or mean-squared error) between the predicted and actual values as shown in Figure 3.

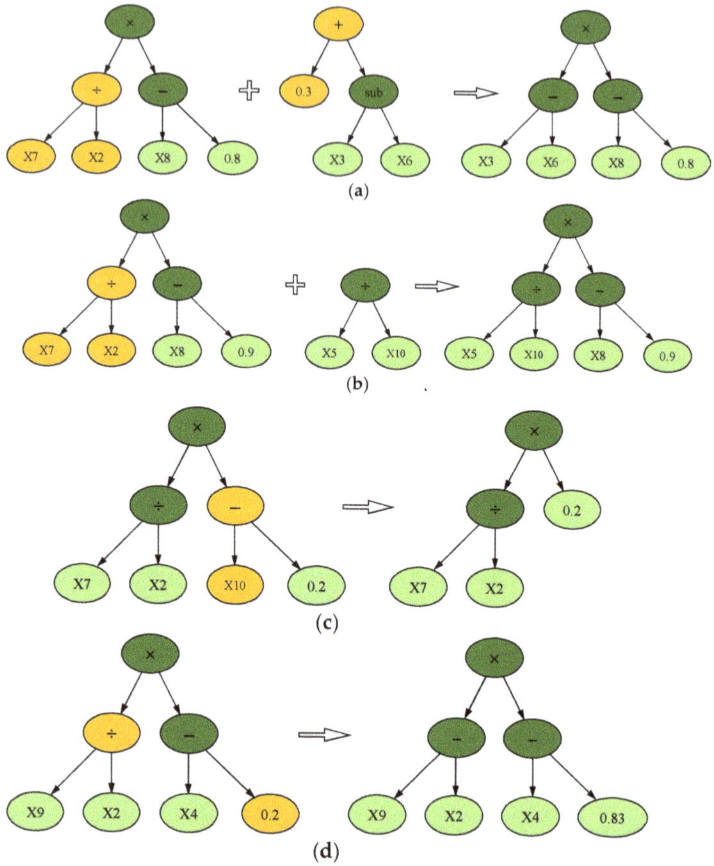

Figure 2. (**a**) The schematic diagrams of crossover method genetic operations. (**b**) The schematic diagrams of sub-tree transfiguration genetic operations. (**c**) The schematic diagrams of hoist transfiguration, genetic operations. (**d**) The schematic diagrams of point transfiguration, genetic operations.

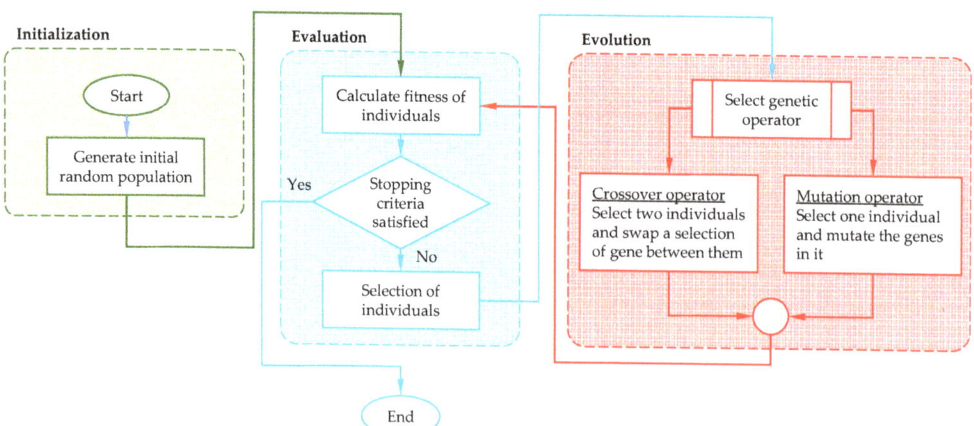

Figure 3. Flowchart of a genetic algorithm.

3. Data Analysis

The penetration of the projectile can be divided into three categories, namely, deep penetration ($X/d \geq 5.0$), intermediate penetration ($0.5 \leq X/d < 5.0$), and narrow penetration ($X/d < 0.5$), where X is the penetration depth and d is the projectile's diameter [10]. Experimental data in dimensionless form, shown in Table 1, is used for the symbolic regression analysis to obtain a mathematical model [2,8,10,13–16,19]. The data's dimensionless parameters include the impact factor $I = \left(\frac{MV_o^2}{f_c d^3 S}\right)$, where M is mass of the rigid projectile, V_o is impacting projectile velocity, fc is the compressive strength of the concrete, d is the diameter of the rigid projectile, and $S = 72 f_c^{-0.5}$ is the empirical constant for concrete strength. The geometry function of the projectile is $N = \left(\frac{M}{l_c d^3 N^*}\right)$, where M is the mass of the rigid projectile, N^* is nose shape factor projectile, l_c is the density of concrete, and d is the diameter of the rigid projectile. The slenderness of the projectile is $\lambda = \left(\frac{M}{l_c d^3}\right)$, where M is the mass of the rigid projectile, l_c is the density of concrete, and d is the diameter of the rigid projectile. The empirical constant for concrete strength is $S = 72 f_c^{-0.5}$, and X/d refers to the dimensionless penetration depth ratio-to-projectile diameter. The dimensionless penetration X/d is taken as a dependent target variable, whereas I, N, λ, and S are independent variables or predictors.

Table 1. Dimensionless data with penetration depths for regression analysis.

λ	S	N	I	$(X/d)_{test}$	λ	S	N	I	$(X/d)_{test}$
15.2	21	143.4	14.45	9.83	5.31	11.81	5.31	0.66	0.86
15.2	21	143.4	36.55	24.15	5.31	11.04	5.31	0.78	0.91
15.2	21	143.4	47.28	27.79	5.31	12.41	5.31	1.33	1.22
15.2	21	143.4	54.73	32.04	5.31	12.07	5.31	1.25	1.29
15.2	21	143.4	93.77	49.54	5.31	11.81	5.31	1.36	1.31
15.2	21	143.4	133.13	65.79	5.31	11.04	5.31	1.16	1.29
15.2	21	200	12.5	8.59	10.75	12.41	10.75	1.47	1.51
15.2	21	200	35.94	24.15	10.75	12.07	10.75	1.43	1.73
15.2	21	200	54.73	33.98	10.75	11.81	10.75	1.45	1.58
15.2	21	200	85.07	51.32	10.75	11.04	10.75	1.37	1.58
15.2	21	200	118.67	66.56	10.75	12.41	10.75	2.77	2.36
19.64	12	125.9	8.45	6.43	10.75	12.07	10.75	2.71	2.89
19.73	12	126.5	17.33	11.52	10.75	11.81	10.75	2.42	2.27
19.66	12	126	18.93	15.28	10.75	11.04	10.75	2.29	2.36
19.77	12	126.7	29.02	17.84	6.12	10.81	6.12	2.41	1.45
19.73	12	126.5	32.62	19.52	6.12	11.39	6.12	1.65	1.47
19.62	12	125.8	36.55	27.1	6.12	11.86	6.12	0.71	1
19.53	12	125.2	33.6	19.07	6.12	10.8	6.12	1.7	1.5
19.57	12	125.5	43.48	22.57	3.14	11.39	3.14	0.35	0.65
19.62	12	125.8	46.02	23.05	3.14	11.17	3.14	0.44	0.77
19.53	12	125.2	64.02	32.19	3.14	11.17	3.14	1.39	1.1
19.6	12	125.6	76.45	35.61	5.85	10.77	5.85	1.91	1.25
19.91	7	127.6	23.15	13.12	6.34	10.77	6.34	1.16	1.2
19.72	7	126.4	24.7	14.28	6.34	11.19	6.34	2.54	1.65
19.93	7	127.8	25.85	16.25					
19.87	7	127.4	25.34	15.69	14.45	15	144.51	1.66	3.15
19.91	7	127.6	39.83	23.42	14.4	15	144.03	2.29	4.07
19.76	7	126.7	38.35	22.49	12.98	11.26	121.91	3.77	3.94
					14.44	15	135.65	3.42	5.51
15.2	15.2	143.4	21.95	13.16	14.31	15	134.43	3.42	5.91
15.2	15.2	143.4	34.62	19.35	12.96	11.26	121.69	5.05	4.99
15.2	15.2	143.4	56.3	34.83	12.93	11.26	237.71	6.47	8.01
15.2	15.2	143.4	75.08	42.57	14.47	15	266.01	4.86	7.61
15.2	15.2	143.4	96	58.05	12.97	11.26	121.79	6.55	5.91

Table 1. *Cont.*

λ	S	N	I	$(X/d)_{test}$	λ	S	N	I	$(X/d)_{test}$
15.2	15.2	143.4	118.24	65.79	14.5	15	136.15	5.36	8.14
15.2	15.2	200	20.28	13.16	14.28	15	134.12	6.62	9.58
15.2	15.2	200	39.47	20.9	14.47	15	135.89	6.66	9.06
15.2	15.2	200	54.45	31.73	19.64	12	125.69	8.46	6.43
15.2	15.2	200	76.9	44.12	12.97	11.26	121.83	9.08	6.96
15.2	15.2	200	99.95	58.82	14.58	15	136.91	6.94	9.97
15.2	15.2	200	128.38	68.11	14.45	15	135.68	9.41	12.6
24.84	8.6	234.3	21.43	14.78	14.49	15	136.09	9.61	12.34
24.84	8.6	234.3	39.63	23.65	14.49	15	136.09	9.71	12.21
24.84	8.6	234.3	71.31	37.44	14.46	15	135.78	9.71	13.39
24.84	8.6	234.3	90.72	46.8	12.97	11.26	238.38	13.37	12.99
24.84	8.6	234.3	103.09	45.32	13.02	11.26	122.23	13.89	12.34
24.84	8.6	234.3	110.97	46.31	19.91	7	127.44	23.15	13.12
24.52	10.5	231.3	17.28	12.13	19.72	7	126.18	24.7	14.28
24.52	10.5	231.3	20.95	13.77	19.87	7	127.16	25.34	16.25
24.52	10.5	231.3	31.29	18.36	19.94	7	127.58	25.93	15.69
24.52	10.5	231.3	44.64	25.57	24.64	10.5	231.41	17.36	12.13
24.52	10.5	231.3	68.09	34.43	14.47	15	266.07	12.26	16.4
24.52	10.5	231.3	70.99	40.33	24.84	8.6	233.31	21.42	14.78
24.52	10.5	231.3	85.31	46.23	14.54	15	136.5	12.32	15.49
24.52	10.5	231.3	107.23	57.38	24.63	8.7	231.3	21.97	14.14
24.52	10.5	231.3	120.29	64.26	24.61	7.9	231.13	24.51	15.08
24.52	10.5	231.3	151.92	66.56	19.73	12	126.25	17.33	11.52
					24.52	10.5	230.26	20.95	13.77
24.63	8.7	232.4	21.97	14.14	19.66	12	125.83	18.93	15.28
24.63	8.7	232.4	41.92	24.19	14.84	20.2	194.91	12.95	8.51
24.63	8.7	232.4	74.7	41.38	15.19	21	199.51	12.5	8.59
24.63	8.7	232.4	114.5	64.04	19.76	7	126.46	38.35	22.49
24.63	8.7	232.4	151.85	78.33	19.91	7	127.44	39.83	23.42
24.63	8.7	232.4	70.37	35.96	14.84	20.2	139.34	14.98	10.06
24.63	8.7	232.4	111.12	57.14	15.19	21	142.63	14.45	9.83
24.63	8.7	232.4	152.64	71.92	14.84	15.2	194.91	20.22	13.16
24.61	7.9	232.2	24.5	15.08	24.52	10.5	230.26	31.28	18.36
24.61	7.9	232.2	42.21	25.9	14.84	15.2	139.34	21.88	13.16
24.61	7.9	232.2	77.91	40.33	24.61	7.9	231.13	42.21	25.9
24.61	7.9	232.2	118.85	63.93	24.84	8.6	233.31	39.63	23.65
24.61	7.9	232.2	121.78	64.26	19.77	12	126.52	29.02	17.84
24.61	7.9	232.2	171.15	87.54	24.63	8.7	231.3	41.85	24.19
					19.73	12	126.25	32.62	19.52
39.69	11.02	39.69	0.18	0.71	19.53	12	125	33.6	19.07
3.48	11.1	3.48	0.15	0.56	19.62	12	125.55	36.55	27.1
3.48	11.1	3.48	0.15	0.66	24.56	10.5	230.69	44.71	25.57
2.45	11.1	2.45	0.15	0.41	19.57	12	125.27	43.48	22.57
3.48	11.1	3.48	0.17	0.61	14.84	15.2	139.34	34.52	19.35
2.45	11.1	2.45	0.15	0.34	19.62	12	125.55	46.02	23.05
2.45	12.22	2.45	0.24	1.31	14.84	15.2	194.91	39.37	20.9
2.93	12.22	2.93	0.15	0.32	24.63	8.7	231.3	70.37	35.96
2.45	12.22	2.45	0.17	0.41	24.84	8.6	233.31	71.31	37.44
2.94	12.22	2.94	0.2	0.78	24.61	7.9	231.13	78.42	40.33
3.48	12.22	3.48	0.27	0.73	24.63	9.04	231.3	68.77	49.75
3.48	12.22	3.48	0.22	0.49	24.63	9.04	231.3	69.81	60.39
2.45	12.22	2.45	0.3	0.57	24.63	8.7	231.3	74.7	41.38
2.45	12.22	2.45	0.08	0.07	24.58	10.5	230.84	68.24	34.43
2.94	12.22	2.94	0.3	0.54	24.52	10.5	230.26	70.98	40.33
3.48	11.34	3.48	0.3	1.47	14.84	20.2	194.91	37.24	23.99

Table 1. *Cont.*

λ	S	N	I	$(X/d)_{test}$	λ	S	N	I	$(X/d)_{test}$
2.45	11.34	2.45	0.33	0.95	15.19	21	199.51	35.93	24.15
2.94	11.34	2.94	0.1	0.1	14.84	20.2	139.34	37.88	23.99
2.94	11.34	2.94	0.22	0.44	15.19	21	142.63	36.55	24.15
2.45	11.76	2.45	0.4	0.75	19.53	12	125	64.02	32.19
13.04	11.76	13.04	0.38	0.5	24.84	8.6	233.31	90.72	46.8
4.6	11.76	4.6	0.25	0.66	14.84	15.2	194.91	54.29	31.73
2.45	11.76	2.45	0.27	0.26	14.84	15.2	139.34	56.12	34.83
13.04	11.76	13.04	0.37	2.15	24.84	8.6	233.31	103.07	45.32
2.45	11.76	2.45	0.1	0.21	24.56	10.5	230.69	85.46	46.23
4.6	11.76	4.6	0.11	0.36	19.6	12	125.41	76.45	35.61
0.83	10.6	0.83	0.09	0.32	24.61	7.9	231.13	116.93	64.26
1.63	11.67	1.63	0.2	0.95	24.61	7.9	231.13	118.86	63.93
0.83	12.13	0.83	0.16	1.04	24.84	8.6	233.31	110.94	46.31
0.69	11.25	0.69	0.08	0.54	24.61	7.9	231.13	121.79	64.26
0.69	11.41	0.69	0.04	0.13	24.63	8.7	231.3	111.12	57.14
1.05	11.1	1.05	0.05	0.34	14.84	20.2	139.34	48.85	27.86
0.53	10.6	0.53	0.06	0.23	15.19	21	142.63	47.14	27.79
0.95	11.41	0.95	0.28	0.34	24.63	8.7	231.3	114.5	64.04
0.95	10.74	0.95	0.35	0.55	24.5	10.5	230.12	107.15	57.38
0.7	11.67	0.7	0.24	0.5	14.84	15.2	139.34	74.85	42.57
0.7	11.25	0.7	0.35	1	14.84	20.2	139.34	56.73	31.73
0.56	11.03	0.56	0.14	0.5	14.84	20.2	194.91	56.73	34.06
96.15	14.28	96.15	5.25	3.6	15.19	21	142.63	54.74	32.04
96.15	14.51	96.15	10.97	5.8	15.19	21	199.51	54.74	33.98
96.63	12.29	96.63	1.14	1.2	14.84	15.2	194.91	76.67	44.12
96.63	10.99	96.63	2.09	1.6	24.66	10.5	231.56	121.02	64.26
96.63	10.96	96.63	2.23	2	24.63	8.7	231.3	151.85	78.33
96.63	11.14	96.63	1.92	1.7	24.63	8.7	231.3	152.64	71.92
5.02	11.13	5.02	0.04	0.09	24.61	7.9	231.13	171.16	87.54
5	12.67	5	0.07	0.25	14.84	15.2	139.34	95.7	58.05
5	12.17	5	0.3	1.13	24.61	7.9	231.13	185.72	92.79
5	12.34	5	0.23	0.38	14.84	15.2	194.91	99.65	58.82
5.05	12.74	5.05	0.39	0.56	24.56	10.5	230.69	152.18	66.56
5.56	15.31	5.56	0.03	0.06	14.84	20.2	194.91	88.14	51.08
5.65	13.93	5.65	0.03	0.05	15.19	21	199.51	85.05	51.32
5.65	13.93	5.65	0.06	0.04	14.84	15.2	139.34	117.88	65.79
5.61	12.9	5.61	0.1	0.12	14.84	15.2	194.91	127.99	68.11
5.65	13.93	5.65	0.14	0.14	14.84	20.2	139.34	97.18	49.54
10.46	13.93	10.46	0.06	0.06	15.19	21	142.63	93.77	49.54
10.46	13.93	10.46	0.1	0.06	14.84	20.2	194.91	122.96	66.56
5.31	12.41	5.31	0.83	0.95	14.84	20.2	139.34	137.97	65.79
5.31	12.07	5.31	0.79	1.02					

The correlation among predictors is shown in Figure 4 as a correlation matrix. The correlations are calculated separately for narrow, medium, and deep penetration. N and λ values are identical in the narrow penetration data. Hence, the data have a perfect correlation of 1.0. S and N are also highly correlated. In the medium penetration data, N is highly correlated with I and λ. In the deep penetration data, the correlation among predictors is weak. A summary of the statistics of predictors for different penetration types is shown in Table 2.

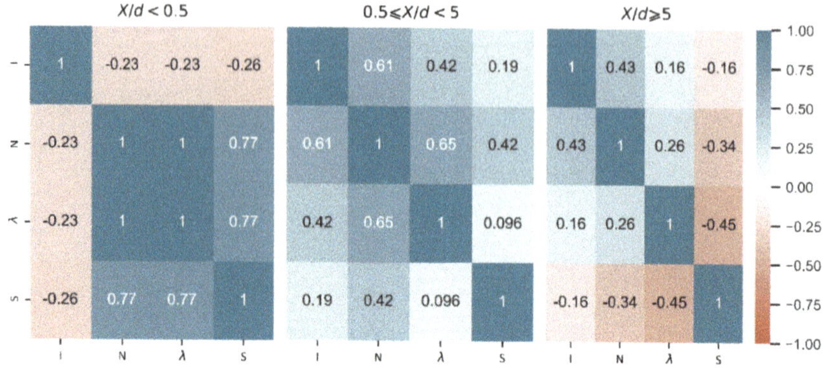

Figure 4. Correlation among N, I, S, λ, and X/d using Python.

Table 2. Descriptive statistics of N, I, S, and λ in narrow, intermediate, and deep penetrations.

Statistics	Independent Variables											
	I			N			λ			S		
	Narrow	Intermediate	Deep	Narrow	Intermediate	Deep	Narrow	Intermediate	Deep	Narrow	Intermediate	Deep
Count	26	59	174	26	59	174	26	59	174	26	59	174
Mean	0.12	1.19	60.05	3.83	22.29	182.81	3.83	14.20	19.96	12.24	11.74	12.75
Std	0.074	1.15	44.29	2.60	39.85	47.24	2.60	25.93	7.30	1.24	0.89	4.55
Cov	0.62	0.97	0.74	0.68	1.79	0.26	0.68	1.83	0.37	0.10	0.08	0.36
Min	0.03	0.08	3.42	0.53	0.56	96.15	0.53	0.56	12.93	10.60	10.74	7.00
25%	0.06	0.30	23.15	2.45	3.14	135.94	2.45	3.14	15.19	11.34	11.10	8.70
50%	0.10	0.79	45.37	2.94	5.31	194.91	2.94	5.31	19.65	11.99	11.67	12.00
75%	0.15	1.68	90.72	5.43	10.75	231.30	5.43	10.75	24.61	12.84	12.15	15.20
Max	0.28	5.25	185.72	10.46	144.51	266.07	10.46	96.63	96.15	15.31	15.00	21.00

Std: standard deviation; Cov: coefficient of variation; 25%: 25th percentile (i.e., 25 percent of data is below this value); 50%: 50th percentile; 75%: 75th percentile.

In order to explore the distribution of predictors, a boxplot (or a box-and-whisker plot) of each predictor for the different penetration ranges is shown in Figure 5. The median is shown by the vertical line inside the box, whereas the left and right sides of the box are the first and third quartiles, respectively. Most of the data lie in the first and third quartiles, and the lines that are referred to as whiskers extend on the right and left sides of the box to indicate the range. There is a clear separation of I, N, and λ in the different penetration ranges. The median of S is the same for different penetration ranges, while there is more deep penetration data. More outliers in all predictors in the medium penetration are observed, as seen by the points plotted beyond the whiskers.

Figures 6–8 show pair plots of the predictors for narrow, medium, and dense penetration ranges, respectively. The diagonals are density plots calculated from the data through a kernel density estimate (KDE) [20,21]. The KDE plots can be considered as smoothed histograms. The plots above the diagonal are scattered plots with bivariate KDE contours overlapping the scatter points. The bivariate KDE plot estimates the probability density of two variables. The shaded contours represent different density levels. The plots below the diagonal are scattered plots with linear regressions among predictors. In the narrow penetration range, the distribution of I, N, and S is skewed, with a high correlation between N and S. In the medium penetration range, the distribution of all predictors is skewed with a high correlation of N with I and λ, respectively. The I and S distributions are skewed in the deep penetration range, whereas N and λ have two modes. Furthermore, all predictors are weakly correlated in the deep penetration range.

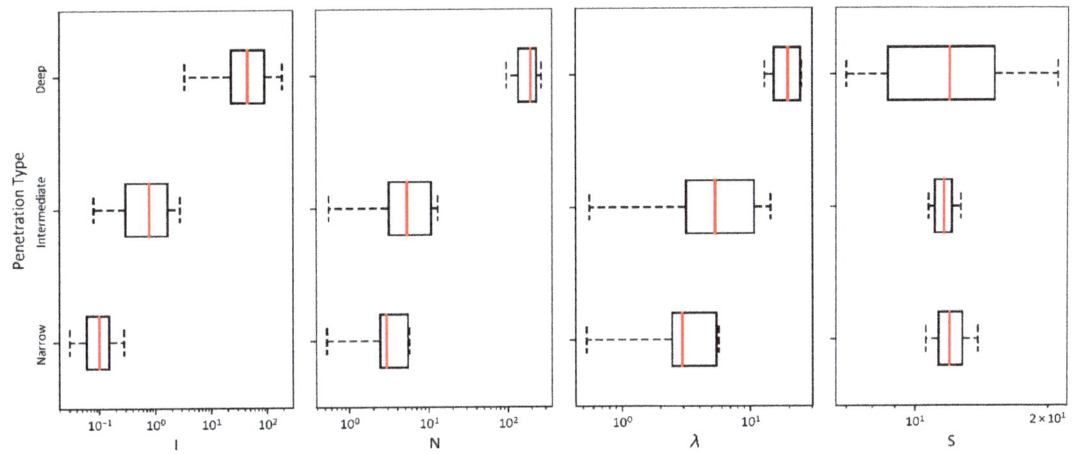

Figure 5. Box-and-whisker plot of N, I, S, and λ, in narrow, intermediate, and deep penetration using Python.

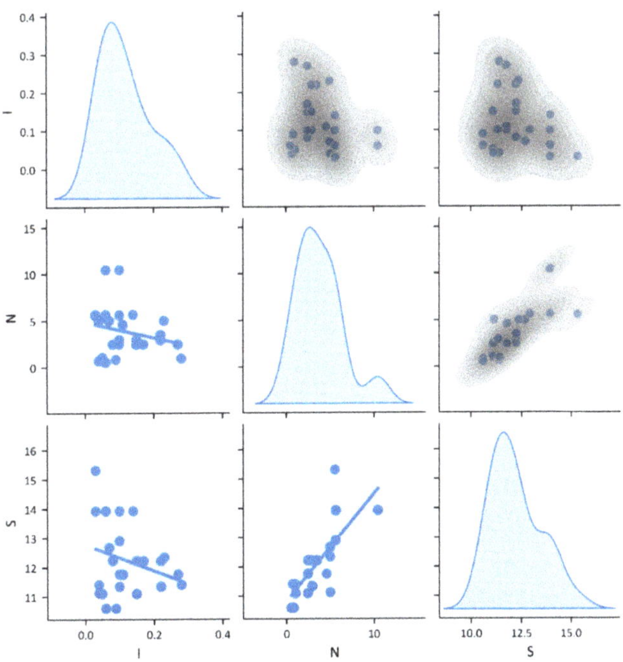

Figure 6. Pair plot of N, I, and S, for narrow penetration data using Python.

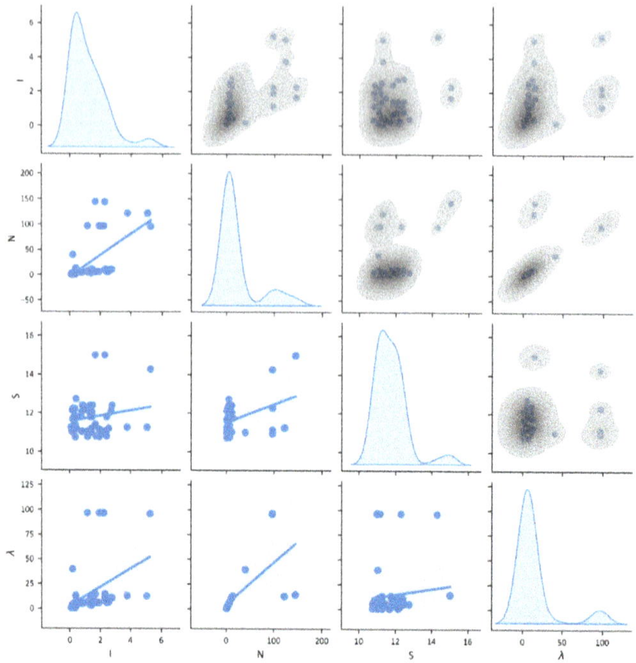

Figure 7. Pair plot of *N*, *I*, and *S*, for intermediate penetration data using Python.

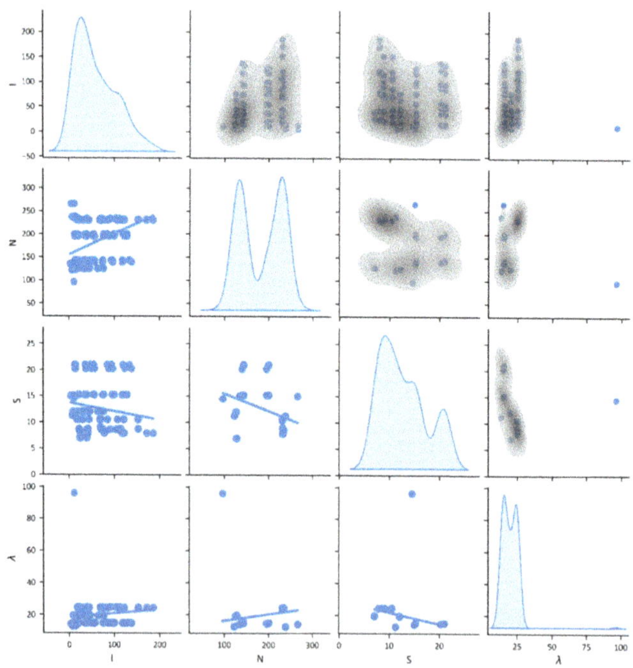

Figure 8. Pair plot of *N*, *I*, and *S*, for deep penetration data using Python.

4. Proposed Model Using Symbolic Regression in Python

Symbolic regression is performed using gplearn [22], which executes genetic encoding in Python through a scikit-learn stimulated and reconcilable application programming interface (API). The hyperparameters used for symbolic regression in gplearn are listed in Table 3. The experimental dataset of 257 observations [2,8,10,13–16,19] as shown in Table 1, is further separated according to the penetration type. Datasets belonging to narrow, intermediate, and deep penetrations consist of 26, 59, and 174 data observations, respectively. Symbolic regression is performed separately for each dataset to obtain the underlying mathematical expressions to describe the best relationship. For intermediate and deep penetration datasets, 70% of data is used to construct the mathematical model (i.e., train the model), and 30% of data is used to test the mathematical model's performance (i.e., test the model). Since data observations in the narrow penetration are small and consist of 26 data observations only, all data is used to construct the mathematical model. The hyperparameters used for symbolic regression in gplearn are shown in Table 3.

Table 3. The hyperparameters for symbolic regression in gplearn.

Parameter	Value		
	EQ 1	EQ 2	EQ 3
population size	5000	5000	5000
generations	60	60	60
stopping_criteria	0.01	0.01	0.01
p_crossover	0.9	0.7	0.7
p_subtree_mutation	0.01	0.01	0.1
p_hoist_mutation	0.01	0.05	0.05
p_point_mutation	0.01	0.1	0.1
function_set	$+, -, \times, \div$	$+, -, \times, \div, \sqrt{}$	$+, -, \times, \div$
tournament size	25	25	25
parsimony_coefficient	0.0003	0.002	0.003
metric	MAE	MAE	MAE
const_range	(−5, 5)	(−5, 5)	(−5, 5)

The following is an explanation of the hyperparameters: population size: number of mathematical formulas in each generation; generations: maximum number of generations; stopping_criteria: MAE value that program stops; p_crossover: crossover probability; p_subtree_mutation: subtree mutation probability; p_hoist_mutation: hoist mutation probability; p_point_mutation: point mutation probability; function_set: building blocks containing mathematical operators; parsimony_coefficient: a constant that penalizes large individuals by adjusting their MAE to make them less favorable for selection; metric: measures how well an individual fits; const_range: the range of constants included in the model.

The mathematical model obtained for narrow, medium, and deep penetration datasets from symbolic regression is shown in the Figures 9–11 as expression tress (ETs).

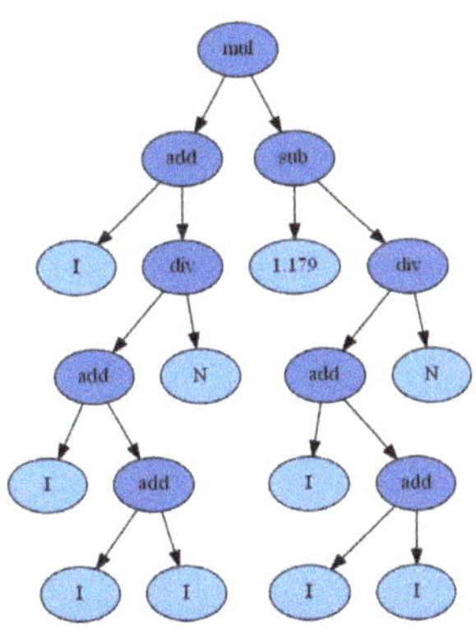

Figure 9. Model equation for narrow penetration, in expression tree form, identified by symbolic regression.

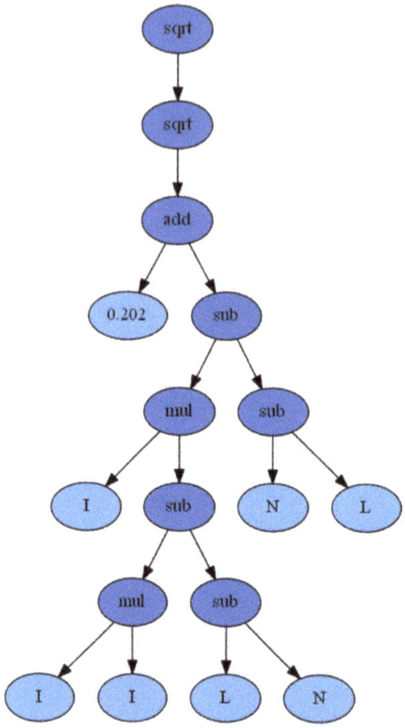

Figure 10. Model equation for intermediate penetration, in expression tree form, identified by symbolic regression.

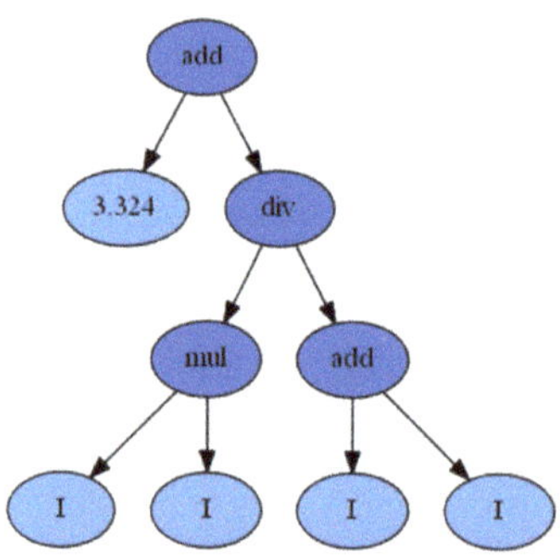

Figure 11. Model equation for deep penetration, in expression tree form, identified by symbolic regression.

The proposed equations, obtained from symbolic regression, can be re-written in mathematical form, as follows.

$$\frac{x}{d} = \left(I - \frac{3I}{N}\right)\left(1.179 - \frac{3I}{N}\right) \text{ for } \frac{x}{d} < 0.5 \tag{1}$$

$$\frac{x}{d} = I^3 + (I - 1.0)(N - \lambda) + 0.202 \text{ for } 0.5 < \frac{x}{d} < 5.0 \tag{2}$$

$$\frac{x}{d} = 0.5I + 3.324 \text{ for } \frac{x}{d} > 0.5 \tag{3}$$

5. Results and Discussion

The mathematical model performances are shown in Figures 12–14 and Table 4 for narrow, intermediate, and deep penetration, respectively. Table 4 shows the R^2, MSE, MAE of proposed model in comparison with NDRC, and Li and Chen model. Figure 12a shows the comparison of the model-predicted values with the actual values of narrow penetration. The coefficient of determination, R^2, is 0.59. Figure 12b illustrates the residual plot of predicted values. The residuals show a constant variance and are evenly spread out. The LOWESS (locally weighted scatterplot smoothing) fit is close to zero but it shows some divergence, as the variance is high at a high predicted value, which is, possibly, an outlier. The frequency distribution of residuals is shown in Figure 12c.

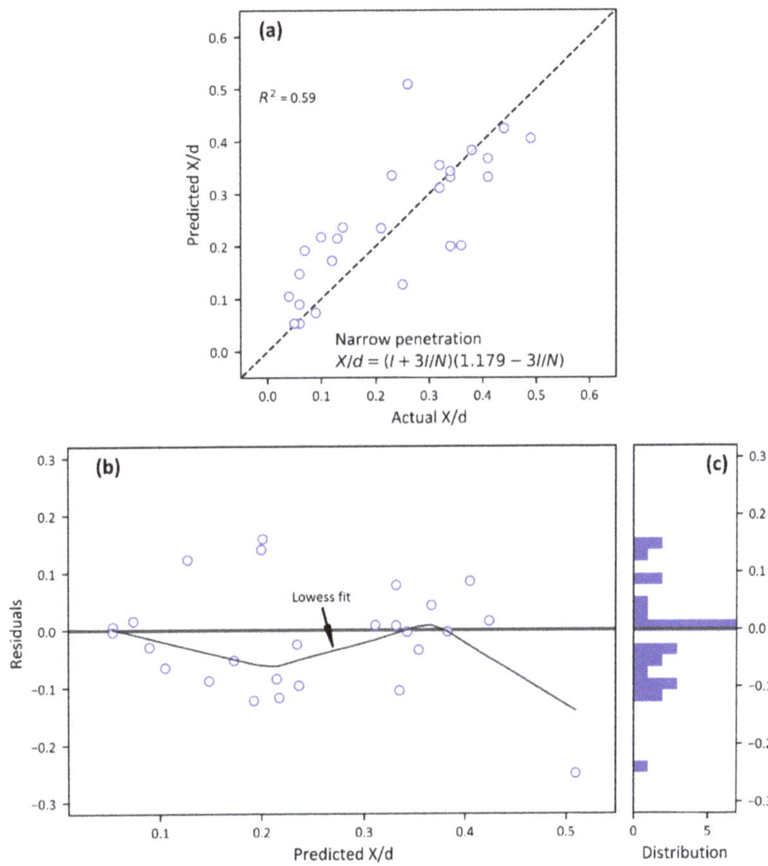

Figure 12. Symbolic regression model for narrow penetration. (**a**) Comparison between actual and predicted values for all data, (**b**) residual plot of predicted values, and (**c**) frequency distribution of residuals.

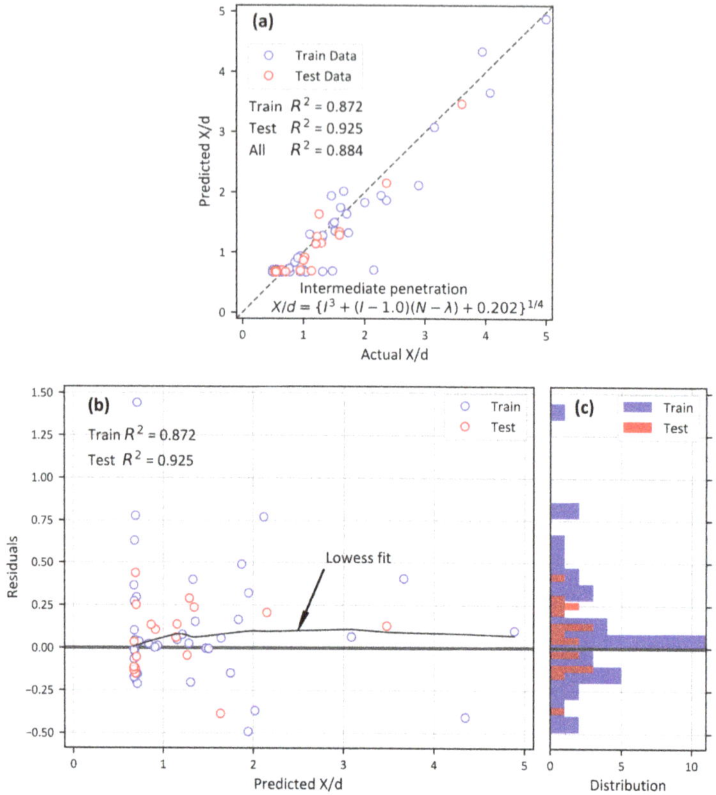

Figure 13. Symbolic regression model for intermediate penetration. (**a**) Comparison between actual and predicted values for train and test datasets, (**b**) residual plot of predicted values, and (**c**) frequency distribution of residuals.

Table 4. Statistics of previous studies compared to the proposed study.

Metric	Comparison of Performance between Different Equations			
	NDRC	**Barr**	**Li and Chen**	**Sym. Reg.[a] (Present Study)**
		Narrow penetration ($X/d < 0.5$)		
R^2	−3.262	0.145	−0.031	0.590
MSE	0.085	0.017	0.021	0.008
MAE	0.270	0.097	0.118	0.068
		Intermediate penetration ($0.5 \leq X/d < 5$)		
R^2	0.746	0.650	0.746	0.884
MSE	0.103	0.142	0.231	0.106
MAE	0.222	0.261	0.330	0.216
		Deep penetration ($X/d \geq 5$)		
R^2	0.565	-	0.963	0.967
MSE	199.369	-	16.877	15.087
MAE	11.295	-	2.799	2.470

[a] Equation (1) for narrow penetration, Equation (2) for intermediate penetration, and Equation (3) for deep penetration.

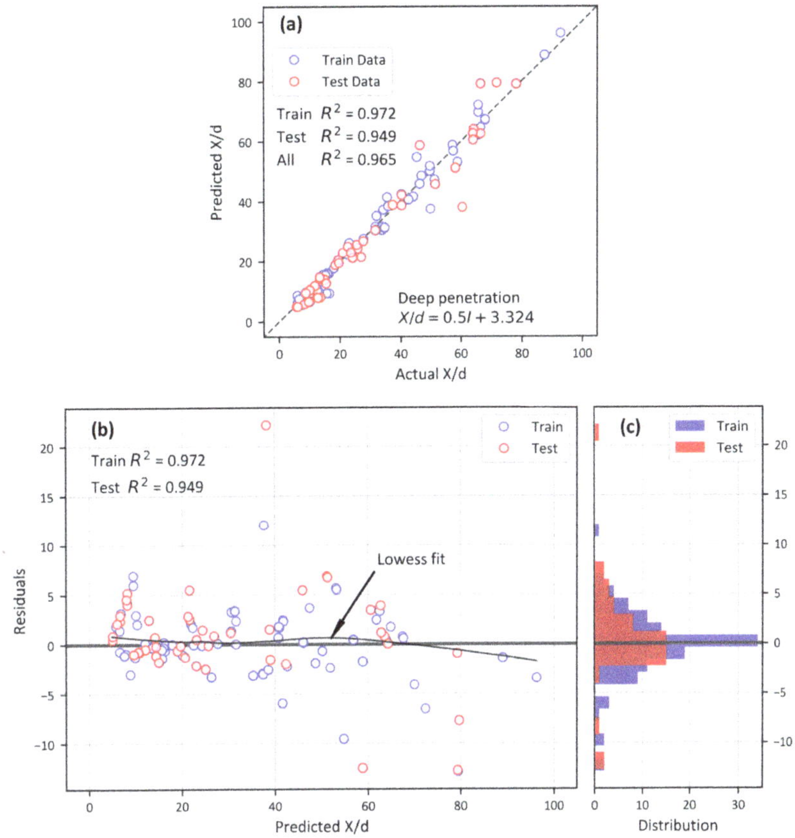

Figure 14. Symbolic regression model for deep penetration. (**a**) Comparison between actual and predicted values for train and test datasets, (**b**) residual plot of predicted values, and (**c**) frequency distribution of residuals.

For the intermediate penetration data, Figure 13a compares model-predicted values with actual values. The prediction of the train and test datasets show that the model performs well on unseen test data. The coefficient of determination, R^2, for all data is 0.88. The residuals of predicted values are evenly distributed, as seen in the Figure 13b,c, which shows constant variance. The LOWESS fit is near zero and does not display diversion at low or high predicted values, as seen in Figure 13b.

For deep penetration data, Figure 14a compares model-predicted values with actual values. The model performs well on unseen test datasets. The coefficient of determination, R^2, for all data is 0.97. The residual of predicted values is evenly distributed at low or high predicted values, exhibiting a constant variance (Figure 14b,c). The LOWESS fit is close to zero, showing an absence of divergence at low or high predicted values.

Furthermore, Figure 15 shows the prediction of the proposed equation for narrow, medium, and deep penetrations compared to the Barr, NDRC, Li and Chen, and L equations. Figure 15a shows that the NDRC and Li and Chen models are almost not applicable in the narrow penetration depths. However, the present model can be useful, compared to other models with less accuracy, due to the complexity of penetration phenomena. Figure 15b,c shows that the proposed model for the prediction of X/d is more accurate, as compared to the NDRC and Li and Chen models within the range of $0.5 \leq X/d < 5.0$ and $X/d \geq 5.0$.

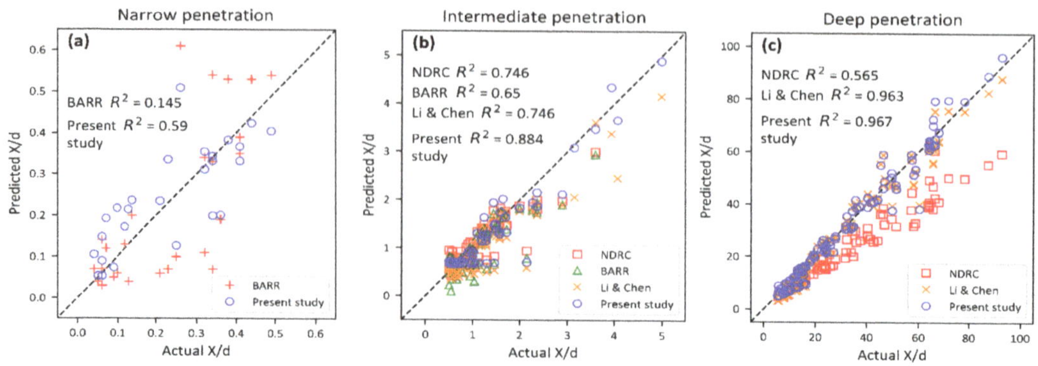

Figure 15. Comparison of performance with different equations for (**a**) narrow penetration, (**b**) medium penetration, and (**c**) deep penetration.

6. Conclusions

The concrete penetration of rigid projectiles is a complex phenomenon that depends on several concrete strength parameters and projectiles. For centuries, continuous research has been conducted to predict penetration with respect to advanced tools and technology. In recent years, machine learning has evolved as an advanced statistical tool that is capable of solving complex phenomena, such as penetration, with acceptable accuracy. This research developed a new model that considers four genetic operations (crossover, sub-tree transfiguration, hoist transfiguration, and point transfiguration operations) using symbolic regression machine learning tools in Python to predict penetration and to compare with the well-established NDRC and Li and Chen models. The three equations are proposed for predicting $X/d < 0.5$, $0.5 \leq X/d < 5.0$, and $X/d \geq 5.0$, respectively. The proposed equations show good relationships between test data and predicted X/d, with $R^2 = 0.88$ for $0.5 \leq X/d < 5.0$, and $R^2 = 0.96$ for $X/d \geq 5.0$. Furthermore, the proposed model is also compared with the predictions of the NDRC and Li and Chen equations. The significance of this research shows that proposed equation predictions are more accurate than the NDRC and Li and Chen models within $0.5 \leq X/d < 5.0$ and $X/d \geq 5.0$. In conclusion, it is recommended to use machine learning tools to achieve great accuracy in complex studies such as penetration, scabbing, and perforation.

Author Contributions: Conceptualization, Q.B.a.I.L. and A.M.; methodology, Z.M. and M.U.Q.; formal analysis, Z.M.; investigation, Z.M. and Q.B.a.I.L.; resources, Q.B.a.I.L., Z.A.M. and A.M.; writing—original draft preparation, Q.B.a.I.L. and Z.M.; writing—review and editing, M.U.Q. and Z.A.M.; visualization, Z.M. and M.U.Q.; supervision, Q.B.a.I.L. and A.M.; project administration, Q.B.a.I.L.; funding acquisition, Z.A.M. All authors have read and agreed to the published version of the manuscript. The data collection was conducted at the University of Nizwa, Oman. The data analysis was performed at Sohar University, Oman. The Model development was done at the University of Buraimi, Oman.

Funding: Prince Sultan University support paying the Article Processing Charges (APC) of this publication.

Institutional Review Board Statement: Not applicable.

Informed Consent Statement: Not applicable.

Data Availability Statement: The article provides all data used in this research.

Acknowledgments: The authors would like to thank Prince Sultan University (PSU) for their financial support.

Conflicts of Interest: The authors declare no conflict of interest.

Appendix A

Table A1. Previous studies equations for prediction of penetration depth with limitations.

References	Equation in S.I. Unit		
Petry model [1,3,5]	$\frac{x}{d} = k\frac{M}{d^3}\log_{10}\left(1+\frac{V_o^2}{19{,}974}\right)$		
Ballistic Research Laboratory (BRL) Model [1,3]	$\frac{x}{d} = \frac{1.33\times10^{-3}}{\sqrt{f_c}}\left(\frac{M}{d^3}\right)d^{0.2}V_o^{1.33}$		
Army Corp of Engineers (ACE model) [1,3,23]	$\frac{x}{d} = \frac{3.5\times10^{-4}}{\sqrt{f_c}}\left(\frac{M}{d^3}\right)d^{0.2}V_o^{1.5}+0.5$		
National Defense Research Committee (NDRC) Model [1,3,4,24]	$G = 3.8\times10^{-5}\frac{N^*M}{d\sqrt{f_c}}\left(\frac{V_o}{d}\right)^{1.8}$	$G = \left(\frac{x}{2d}\right)^2$	$\frac{x}{d} \le 2$
		$G = \frac{x}{d} - 1$	$\frac{x}{d} > 2$
		$\frac{x}{d} = 2G^{0.5}$	$G \ge 1$
		$\frac{x}{d} = G+1$	$G < 1$
Ammann and Whitney model [1,3]	$\frac{x}{d} = \frac{6\times10^{-4}}{\sqrt{f_c}}N^*\left(\frac{M}{d^3}\right)d^{0.2}V_o{}^{1.8}$		
Whiffen model [1,3,25]	$\frac{x}{d} = \left(\frac{2.61}{f_c^{0.5}}\right)\left(\frac{M}{d^3}\right)\left(\frac{d}{a}\right)^{0.1}\left(\frac{V_o}{533.4}\right)^n$	$n = \frac{97.51}{f_c^{0.25}}$	
Kar Model [1,3,26,27]	$G = 3.8\times10^{-5}\frac{N^*M}{d\sqrt{f_c}}\left(\frac{E}{E_s}\right)^{1.25}\left(\frac{V_o}{d}\right)^{1.8}$	$\frac{x}{d} = 2G^{0.5}$	$G \ge 1$
		$\frac{x}{d} = G+1$	$G < 1$
UKAEA model [1,3,28]	$G = 3.8\times10^{-5}\frac{N^*M}{d\sqrt{f_c}}\left(\frac{V_o}{d}\right)^{1.8}$	$\frac{x}{d} = 0.275 - [0.0756 - G]^{0.5}$	$G \le 0.0726$
		$\frac{x}{d} = [4G - 0.242]^{0.5}$	$0.0726 \le G \le 1.06$
		$\frac{x}{d} = G + 0.9395$	$G \ge 1.06$
		$G = 0.55\left(\frac{x}{d}\right) - \left(\frac{x}{d}\right)^2$	$\frac{x}{d} < 0.22$
		$G = \left(\frac{x}{2d}\right)^2 + 0.0605$	$0.22 \le \frac{x}{d} \le 2.0$
		$G = \frac{x}{d} - 0.9395$	$\frac{x}{d} \ge 2.0$
Haldar and Hamieh model [1,3,6]	$\frac{x}{d} = 0.2251I_a + 0.0308$	$I_a = \frac{MN^*V_o^2}{f_c d^3}$	$0.3 \le Ia \le 4.0$
	$\frac{x}{d} = 0.0567I_a + 0.6740$		$4.0 \le Ia \le 21$
	$\frac{x}{d} = 0.0299I_a + 1.1875$		$21 \le Ia \le 455$
Adeli and Amin Model [1,3]	$\frac{x}{d} = 0.0416 + 0.1698I_a - 0.0045I_a^2$	$I_a = \frac{MN^*V_o^2}{f_c d^3}$	for $0.3 \le Ia \le 4$
	$\frac{x}{d} = 0.0123 + 0.196I_a - 0.008I_a^2 + 0.0001I_a^3$		$4 \le Ia \le 21$
Hughes Model [1,3,7]	$\frac{x}{d} = 0.19\frac{N_h I_h}{S}$	$I_h = \frac{MV_o^2}{f_t d^3}$	$I_h < 3500$
		$S = 1.0 + 12.3l_n(1.0 + 0.03I_h)$	
Healy and Weissman Model [1,3]	$G = 4.36\times10^{-5}\left(\frac{E}{E_s}\right)\frac{N^*M}{d\sqrt{f_c}}\left(\frac{V_o}{d}\right)^{1.8}$	$\frac{x}{d} = 2G^{0.5}$	$G \ge 1$
		$\frac{x}{d} = G+1$	$G < 1$
CREIPI Model [1,3,29]	$\frac{x}{d} = \frac{0.0265N^*Md^{0.2}V_o^2\left[114-6.83\times10^{-4}f_c^{\frac{2}{3}}\right]}{f_c^{\frac{2}{3}}}\left[\frac{(d+1.25H_r)H_r}{(d+1.25H_o)H_o}\right]$		
UMIST model [1,3]	$\frac{x}{d} = \left(\frac{2}{\pi}\right)\frac{N^*}{0.72}\frac{MV_o^2}{\sigma_t d^3}$		
	$\sigma_t(Pa) = 4.2f_c(Pa) + 135\times10^6 + \left[0.014f_c(Pa) + 0.45\times10^6\right]V_o$		

Table A1. *Cont.*

References	Equation in S.I. Unit		
Li and Chen Model [1,3,8,10]	$\frac{x}{d} = \sqrt{\frac{\left(1+\left(\frac{k\pi}{4N}\right)\right)}{1+\left(\frac{I}{N}\right)}\frac{4kI}{\pi}}$	$I = \frac{I^*}{S} = \frac{1}{S}\left(\frac{MV_o^2}{f_c d^3}\right)$ $N = \frac{\lambda}{N^*} = \frac{1}{N^*}\left(\frac{M}{l_c d^3}\right)$ $S = 72 f_c^{-0.5}$ $k = \left(0.707 + \frac{h}{d}\right)$	$\frac{x}{d} \leq 5$
	$\frac{x}{d} = \frac{2}{\pi}N \ln\left[\frac{1+\left(\frac{I}{N}\right)}{1+\left(\frac{k\pi}{4N}\right)}\right] + k$		$\frac{x}{d} > 5$
	$\frac{x}{d} = 1.628\left(\frac{\left(1+\left(\frac{k\pi}{4N}\right)\right)}{1+\left(\frac{I}{N}\right)}\frac{4kI}{\pi}\right)^{1.395}$		$x/d < 0.5$
	$\frac{x}{d} = \sqrt{\frac{\frac{4kI}{\pi}}{1+\left(\frac{I}{N}\right)}}$	If $N \gg 1$	for $\frac{x}{d} \leq k$
	$\frac{x}{d} = \frac{2}{\pi}N \ln\left(1 + \frac{I}{N}\right) + \frac{k}{2}$		for $\frac{x}{d} > k$
	$\frac{x}{d} = 1.628\left(\frac{\frac{4kI}{\pi}}{1+\left(\frac{I}{N}\right)}\right)^{1.395}$		$x/d < 0.5$
	$\frac{x}{d} = \sqrt{\frac{4kI}{\pi}}$	When $I/N \ll 1$	$\frac{x}{d} \leq k$
	$\frac{x}{d} = \frac{k}{2} + \frac{2I}{\pi}$		$\frac{x}{d} > k$
	$\frac{x}{d} = 1.628\left(\frac{4k}{\pi}I\right)^{1.395}$		$x/d < 0.5$

References

1. Li, Q.; Reid, S.; Wen, H.; Telford, A. Local impact effects of hard missiles on concrete targets. *Int. J. Impact Eng.* **2005**, *32*, 224–284. [CrossRef]
2. Latif, Q.B.A.I.; Rahman, I.A.; Zaidi, A.M.A.; Latif, K. Critical impact energy for spalling, tunnelling and penetration of concrete slab impacted with hard projectile. *KSCE J. Civ. Eng.* **2014**, *19*, 265–273. [CrossRef]
3. Rahman, I.A.; Ahmad Zaidi, A.M.; Imran Latif, Q.B. Review on Empirical Studies of Local Impact Effects of Hard Missile on Concrete Structures. *Int. J. Sustain. Constr. Eng. Technol. (IJSCET) UTHM* **2010**, *1*, 71–95.
4. NDRC. *Effects of Impact and Explosion*; Summary Technical Report of Division 2; National Defence Research Committee: Washington, DC, USA, 1946; Volume 1.
5. Kennedy, R.P. A review of procedures for the analysis and design of concrete structures to resist missile impact effects. *Nucl. Eng. Des.* **1976**, *37*, 183–203. [CrossRef]
6. Haldar, A.; Hamieh, H.A. Local Effect of Solid Missiles on Concrete Structures. *J. Struct. Eng.* **1984**, *110*, 948–960. [CrossRef]
7. Hughes, G. Hard missile impact on reinforced concrete. *Nucl. Eng. Des.* **1984**, *77*, 23–35. [CrossRef]
8. Li, Q.M.; Chen, X.W. Dimensionless formulae for penetration depth of concrete target impacted by a non-deformable projectile. *Int. J. Impact Eng.* **2003**, *28*, 93–116. [CrossRef]
9. Forrestal, M.; Altman, B.; Cargile, J.; Hanchak, S. An empirical equation for penetration depth of ogive-nose projectiles into concrete targets. *Int. J. Impact Eng.* **1994**, *15*, 395–405. [CrossRef]
10. Chen, X.W.; Li, Q.M. Deep penetration of a non-deformable projectile with different geometrical characteristics. *Int. J. Impact Eng.* **2002**, *27*, 619–637. [CrossRef]
11. Milad, A.; Hussein, S.H.; Khekan, A.R.; Rashid, M.; Al-Msari, H.; Tran, T.H. Development of ensemble machine learning approaches for designing fiber-reinforced polymer composite strain prediction model. *Eng. Comput.* **2021**, 1–13. [CrossRef]
12. Abdolrasol, M.G.M.; Hussain, S.M.S.; Ustun, T.S.; Sarker, M.R.; Hannan, M.A.; Mohamed, R.; Ali, J.A.; Mekhilef, S.; Milad, A. Artificial Neural Networks Based Optimization Techniques: A Review. *Electronics* **2021**, *10*, 2689. [CrossRef]
13. Forrestal, M.J.; Frew, D.J.; Hanchak, S.J.; Brar, N.S. Penetration of grout and concrete targets with ogive-nose steel projectiles. *Int. J. Impact Eng.* **1996**, *18*, 465–476. [CrossRef]
14. Frew, D.; Hanchak, S.; Green, M.; Forrestal, M. Penetration of concrete targets with ogive-nose steel rods. *Int. J. Impact Eng.* **1998**, *21*, 489–497. [CrossRef]
15. Forrestal, M.; Frew, D.; Hickerson, J.; Rohwer, T. Penetration of concrete targets with deceleration-time measurements. *Int. J. Impact Eng.* **2003**, *28*, 479–497. [CrossRef]
16. Frew, D.J.; Forrestal, M.J.; Hanchak, S.J. Penetration Experiments with Limestone Targets and Ogive-Nose Steel Projectiles. *J. Appl. Mech.* **2000**, *67*, 841–845. [CrossRef]
17. Hosseini, M.; Dalvand, A. Neural Network Approach for Estimation of Penetration Depth in Concrete Targets by Ogive-nose Steel Projectiles. *Lat. Am. J. Solids Struct.* **2015**, *12*, 492–506. [CrossRef]
18. Thai, D.-K.; Tu, M.; Bui, Q.; Bui, T.-T. Gradient tree boosting machine learning on predicting the failure modes of the RC panels under impact loads. *Eng. Comput.* **2019**, *1*, 3. [CrossRef]

19. Ahmad Zaidi, A.M.; Imran Latif, Q.B.; Rahman, I.A.; Ismail, M.Y. Development of empirical prediction formula for penetration of ogive nose hard missile into concrete targets with effect of CRH ratio. *Am. J. Eng. Appl. Sci.* **2010**, *7*, 711–716.
20. Duong, T.; Hazelton, M.L. Cross-validation Bandwidth Matrices for Multivariate Kernel Density Estimation. *Scand. J. Stat.* **2005**, *32*, 485–506. [CrossRef]
21. Zambom, A.Z.; Dias, R. A review of Kernel density estimation with applications to econometrics. *Int. Econom. Rev.* **2013**, *5*, 20–42.
22. Afzal, W.; Torkar, R. On the application of genetic programming for software engineering predictive modeling: A systematic review. *Expert Syst. Appl.* **2011**, *38*, 11984–11997. [CrossRef]
23. Chelapati, C.V.; Kennedy, R.P.; Wall, I.B. Probabilistic assessment of hazard for nuclear structures. *Nucl. Eng. Des.* **1972**, *19*, 333–364. [CrossRef]
24. Kennedy, R.P. *Effects of an Aircraft Crash into a Concrete Reactor Containment Building*; Holmes & Narver Inc.: Anaheim, CA, USA, 1966.
25. Bulson, P. *Explosive Loading of Engineering Structures*; E & FN Spon: London, UK, 1997.
26. Bangash, M.Y.H. *Concrete and Concrete Structures: Numerical Modelling and Application*; Elsevier Applied Science: London, UK, 1989.
27. Kar, A.K. Local Effects of Tornado-Generated Missiles. *J. Struct. Div.* **1978**, *104*, 809–816. [CrossRef]
28. Barr, P. *Guidelines for the Design and Assessment of Concrete Structures Subjected to Impact*; Report; UK Atomic Energy Authority, Safety and Reliability Directorate, HMSO: London, UK, 1990.
29. Kojima, I. An experimental study on local behaviour of reinforced concrete slabs to missile impact. *Nucl. Eng. Des.* **1991**, *130*, 121–132. [CrossRef]

applied sciences

MDPI

Article

Analytical Model Formulation of Steel Plate Reinforced Concrete Walls against Hard Projectile Impact

Bo Pu [1], Xiaoming Wang [1], Weibing Li [1] and Jun Feng [2,*]

[1] School of Mechanical Engineering, Nanjing University of Science and Technology, Nanjing 210094, China; 120121023605@njust.edu.cn (B.P.); 202xm@163.com (X.W.); 12011007@njust.edu.cn (W.L.)
[2] National Key Laboratory of Transient Physics, Nanjing University of Science and Technology, Nanjing 210094, China
* Correspondence: jun.feng@njust.edu.cn; Tel.: +86-15850579204

Abstract: Steel plate reinforced concrete (SC) walls can effectively resist projectile impact by preventing the rear concrete fragments flying away, thus attracting much attention in defence technology. This work numerically and analytically investigated the hard projectile perforation of steel plate reinforced concrete walls. Impact resistance theories, including cavity expansion analysis as well as the petaling theory of thin steel plates were used to describe the cratering, tunneling and plugging phases of SC walls perforation. Numerical modeling of SC walls perforation was performed to estimate projectile residual velocity and target destructive form, which were validated against the test results. An analytical model for SC wall perforation was established to describe the penetration resistance featuring five stages, i.e., cratering, tunneling and plugging, petaling with plugging and solely petaling. Analytical model predictions matched numerical results well with respect to projectile deceleration evolution as well as residual velocity. From a structural absorbed energy perspective, the effect of front concrete panel and rear steel plate thickness combinations was also studied and analyzed. Finally, equivalent concrete slab thickness was derived with respect to the ballistic limit of SC walls, which may be helpful in the design of a protective strategy.

Keywords: steel plate reinforced concrete walls; FE simulation; perforation analytical model; cavity expansion analysis; thin plates petaling

Citation: Pu, B.; Wang, X.; Li, W.; Feng, J. Analytical Model Formulation of Steel Plate Reinforced Concrete Walls against Hard Projectile Impact. *Appl. Sci.* **2022**, *12*, 518. https://doi.org/10.3390/app12010518

Academic Editors: Ricardo Castedo, Lina M. López and Anastasio P. Santos

Received: 6 December 2021
Accepted: 1 January 2022
Published: 5 January 2022

1. Introduction

Characterized with easy shaping, efficient fabrication and construction, concrete material structures are widely used for most civilian and military infrastructure, e.g., nuclear power plants, liquefied natural gas storage tanks and civil air defence, which are designed to withstand extreme loading, such as aircraft engine impact as well as internal and external missile impact [1–3]. Under projectile penetration, attaching a relative thin steel plate onto the concrete wall rear (protected) face would better protect the inner inhabitants and vulnerable instruments by preventing rear face ejected fragmentation, which might occur even if the concrete panel is not breached. In practice, steel plate reinforced concrete (SC) walls have been effectively used as primary and secondary shelters in protective structures. It is important to develop a simple but robust analytical model for projectile perforation on SC walls.

SC walls have superior performance in terms of resisting impact loading, since the rear steel plate induces a considerable effect on limiting crater development and preventing the pulverized pieces from flying away. Concerning the impact resistance of the SC walls, experimental studies have compared and analyzed projectile impact tests on concrete slabs with and without the rear steel plate. Remennikov et al. [4,5] investigated the static and impact performance of SC walls in which no shear connectors were utilized to connect the steel faceplates and the concrete core. Owing to the specially designed connection details, the tested panels exhibited tensile membrane resistance at large deformations.

Armour-piercing projectiles impacting high-strength concrete backed by armoured steel were studied by Feng et al. [6], showing spaced composite targets had larger residual penetration depth than segmented ones. Kojima [7] concluded that the rear attached steel plate has little effect on enhancing the impact resistance of SC walls, while it can efficiently restrain scabbing and spalling fragments. Aiming at identifying the influence of the steel plate on local damage of SC walls, Tsubota et al. [8] performed a series of impact tests in which the steel plate was placed on the rear, front and both faces of a concrete panel, respectively. The rear steel plate attached to reinforced concrete (RC) plate could prevent local damage caused by perforation, while the front steel plate attached to the impacted face had a relatively slight effect. Abdel-Kader and Fouda [9] showed that the steel plate placed on the rear face had better impact performance than the front one, which further validated the above conclusion. Hashimoto et al. [10] also showed that a thicker steel plate would result in less local damage under low-velocity impact. Mizuno et al. [11] found that SC panels have better impact resistant performance than RC panels, which enabled reduction of panel thickness by almost 30%.

Bruhl et al. [12] developed an analytical and empirical model of SC walls under rigid projectile impact and proposed a three-step method for designing SC walls against missile impact which can be used to evaluate the ballistic limit of SC walls. For rigid projectile perforation, Grisaro and Dancygier [13] investigated the thickness of the SC composite barrier with respect to energy absorption. Wu et al. [14] performed projectile perforation tests on monolithic and segmented RC panels with a rear steel plate. The ballistic performances of layered RC and SC targets were analyzed quantitatively, which further validated the equivalent approach in Ref. [13]. To assess the core concrete thickness and steel ratios effect on failure modes, Lee and Kim [15] numerically evaluated the impact resistance of SC and RC panels via a LS-DYNA solver, suggesting SC walls can better resist impacting loads than RC panels.

Although extensive studies on SC structures have been conducted both experimentally and numerically, analytical models of hard projectile perforation on SC walls need to be further explored. Cavity expansion analysis and petaling theory were combined to describe the cratering, tunneling and plugging phases of SC wall perforation. Front concrete wall perforation resistance was analyzed by perforation modeling via non-linear transient dynamic solver LS-DYNA. The thin plate petaling theory was utilized to analyze the process of rear face plate damage by projectile impact. With the same ballistic limit, a semiemperical analytical model, converting SC walls to equivalent thickness concrete panels, was developed and validated. This work may shed some light on SC wall ballistic performance related to protective structure design.

2. Impact Resistance Theories

Since the composite structure of SC walls consists of a front concrete plate attached to a rear steel plate, the ballistic performance of SC walls should be roughly separated into three parts: penetration of the front concrete, perforation of the steel plate and interaction with both concrete and steel. When a projectile perforates a concrete panel with a certain thickness it goes through three response stages: front crater scabbing, stable tunneling, and rear crater plugging (spalling) [16]. The impact perforation response of thin steel plates has been successfully analyzed by the petaling theory proposed by Wierzbicki [17].

This work developed a semiemperical analytical model of projectile penetration of steel plate reinforced concrete walls based on five penetration stages, as plotted in Figure 1. Stage 1 is projectile cratering on the front concrete impact face. Stage 2 represents stable tunneling inside the concrete block. Afterwards, rear face plugging occurs in Stage 3 until the projectile starts to hit the rear steel plate. Stage 4 occurs when the projectile interacts with the steel plate and fragments the concrete rear face. Finally, the projectile head has interaction only the steel plate in stage 5. Hence, the penetration resistance consists of concrete resistance and steel plate resistance. Classical resistance equations are introduced below.

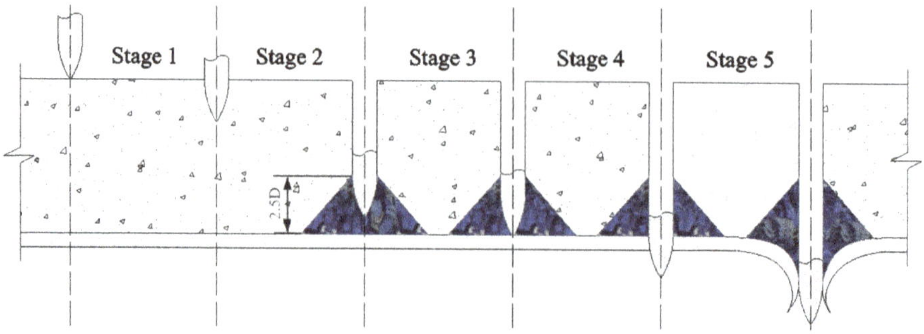

Figure 1. Diagram of a projectile penetrating a steel plate reinforced concrete wall.

2.1. Cavity Expansion Analysis for Concrete

Introduced by Bishop et al. [18], cavity expansion analysis was applied to solve the governing equations of the spherical cavity or cylindrical cavity expansion process in an elasto-plastic incompressible medium. Then Forrestal and Luk et al. [19] extended this to compressible material penetration problems. This classical model has been successfully applied to metal and concrete material penetration analyses.

Projectile penetration inside a solid medium can be regarded as a spherically symmetric cavity expansion process. Embedded in an infinite and isotropic medium, the zero initial radius cavity expands at a constant velocity. Due to the instantaneous rise of dynamic pressure, spherical stress waves are generated to form different response regions corresponding to the concrete constitutive law. According to continuum mechanics, equations of mass conservation and momentum conservation [8] in Euler coordinate for compressible spherical cavity expansion analysis at time t are:

$$\text{div}\,\sigma = \rho a \tag{1}$$

$$\frac{\partial \rho}{\partial t} + \text{div}(\rho v) = 0 \tag{2}$$

where σ is the stress tensor, a is acceleration, and v is velocity. In a spherical coordinate system, the governing equations [9] can be expressed as:

$$\frac{\partial \sigma_{\rm r}}{\partial {\rm r}} + \frac{2(\sigma_{\rm r} - \sigma_{\theta})}{r} = -\rho\left(\frac{\partial v}{\partial t} + v\frac{\partial v}{\partial r}\right) \tag{3}$$

$$\rho\left(\frac{\partial v}{\partial r} + \frac{2v}{r}\right) = -\left(\frac{\partial \rho}{\partial t} + v\frac{\partial \rho}{\partial r}\right) \tag{4}$$

where r is the radial coordinate, and $\sigma_{\rm r}$ and σ_{θ} are radial and circumferential Cauchy stress.

With the corresponding constitutive law for concrete materials, attempts have been made to numerically solve the governing equations with the Runge-Kutta method [20]. The normal stress $\sigma_{\rm n}$, referred as penetration resistance, acting on the projectile nose is usually expressed as:

$$\sigma_{\rm n} = \rho v^2 + R \tag{5}$$

where ρv^2 and R represent the inertial dynamic resistance and static resistance. Forrestal and his coworkers [21–23] deemed the target strength parameter as $R = Sf_c'$ where S is a dimensionless constant and f_c' is the unconfined compressive strength of concrete. After validation with extensive penetration data [21,22,24–26], the semiempericial penetration model gives:

$$S = 82.6\left(f_c'/10^6\right)^{-0.544} \tag{6}$$

Recently, researchers have pointed out that the above penetration resistance equation only applies to medium caliber projectile cases (shank diameter ranging from 12.9 mm to 30.5 mm) [27]. To avoid the concrete size effect in penetration, this work focuses on a 1-inch (25.4 mm) diameter projectile meeting the application range.

2.2. Perforation with Shear Plugging

After cratering and tunneling, the interaction between the projectile and shear plugging fragments can be treated as a collision problem. In Figure 2, the fragments with velocity V_{rf} are pushed away by the projectile, whereby the shear plugging zone is a frustum-of-cone with cone slope angle φ. The projectile residual velocity V_r can be derived from:

$$\frac{1}{2}mV_{r0}^2 = \frac{1}{2}mV_r^2 + \frac{1}{2}\rho_c\Omega V_{rf}^2 \quad (7)$$

where V_{r0} is the initial projectile velocity of the plugging stage, fragments velocity can be estimated by $V_{rf} = \eta V_r$ and $\eta = 0.2$ according to [14], the cone slope angle $\varphi = 62.5°$ is given by Peng et al. [28], and Ω is the volume of the ejected frustum-of-cone fragment, which can be expressed as:

$$\Omega = \frac{\pi}{12}H_p\left(4\tan^2\varphi H_p^2 + 6d_p\tan\varphi H_p + 3d_p\right) \quad (8)$$

where H_p is the shear plugging thickness, and d_p is the projectile diameter.

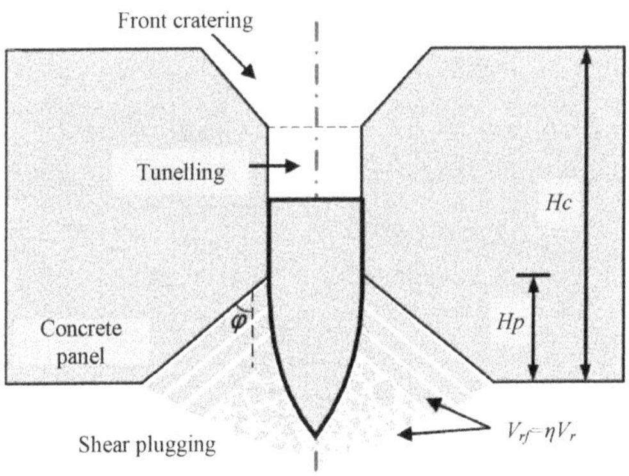

Figure 2. Concrete plate plugging due to perforation [16].

2.3. Thin Plate Petaling

Characterized by multiple symmetric petals forming, petaling damage is a common failure mode of thin metal plates when subjected to localized high intensity loadings, e.g., projectile perforation as shown in Figure 3. The shock wave pressure and the cavity tear the thin shell to generate a much larger radius. The tearing fracture energy is related to the bending energy through the petal local radial curvature and the circumferential curvature. From an energy perspective, the energy consumption of projectile perforation on the thin metal plate results from petaling energy and plastic deformation:

$$E_c = W_L + W_G \quad (9)$$

where W_L denotes the petaling energy, and W_G represents the plastic energy during plate deformation.

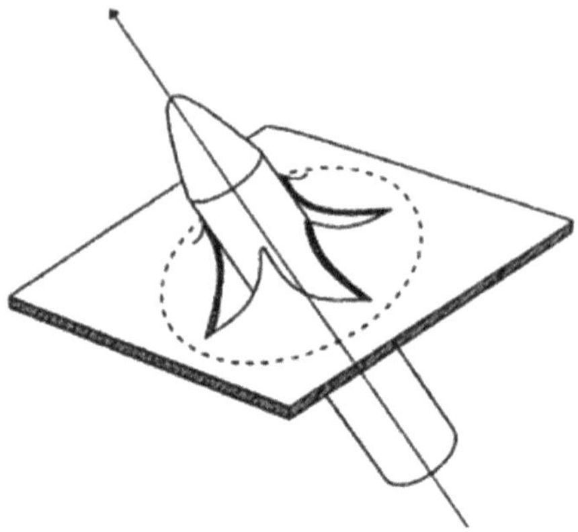

Figure 3. Petaling diagram [17].

According to Wierzbicki [17], the petaling energy due to projectile perforation can be expressed as:

$$W_L = 3.37\sigma_0\, \overline{\delta}^{0.2} h_t^{1.6} d_p^{1.4} \tag{10}$$

where $\overline{\delta} = \delta_t/h_t$, $\delta_t = R_t/\left(\sqrt{3}\sigma_u\right)$, R_t denotes the impact toughness, δ_t is the crack tip opening displacement parameter (CTOD), h_t is the plate thickness, σ_u is the ultimate strength and σ_0 is the nominal strength.

Landkof and Goldsmith [29] measured the plastic deformation energy according to the ballistic tests, hence thin plate perforation energy consumption E_c can be written in the dimensionless form:

$$\frac{E_c}{\sigma_0 d_p^3} = 3.37\overline{\delta}^{0.2}\left(\frac{h_t}{d_p}\right)^{1.6} + 2.8\left(\frac{h_t}{d_p}\right)^{1.7} \tag{11}$$

3. Perforation Model Validation

Recently, Wu et al. [14] conducted penetration experiments to study the ballistic performance of reinforced concrete panels with a rear steel plate which provided valuable data for SC wall perforation analyses. For the sake of analytical model formulation, hypotheses needed to be verified via extensive test data both from experiments and simulations. This section aims to develop the FE numerical model for SC walls perforation which was validated against test data.

3.1. Perforation Test and FE Model

With an ogival-shaped nose, hard projectiles were used for penetration tests in which no apparent erosion occurs. The total projectile mass was 428 g, the shank diameter was 25.3 mm and the caliber-radius-head (CRH) of the ogival nose was 3.0, as shown in Figure 4.

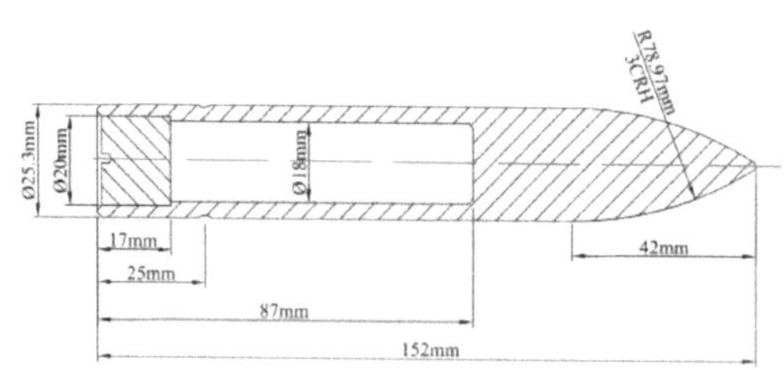

Figure 4. Projectile dimension [14].

Simulations of the SC walls perforation tests were conducted with 200-mm thick concrete panels. Figure 5 depicts the target dimension together with its reinforced mesh. The projectile impact location point is denoted by '×'. For the investigated concrete, the unconfined compressive strength of cylinder sample (f_c') was 41 MPa.

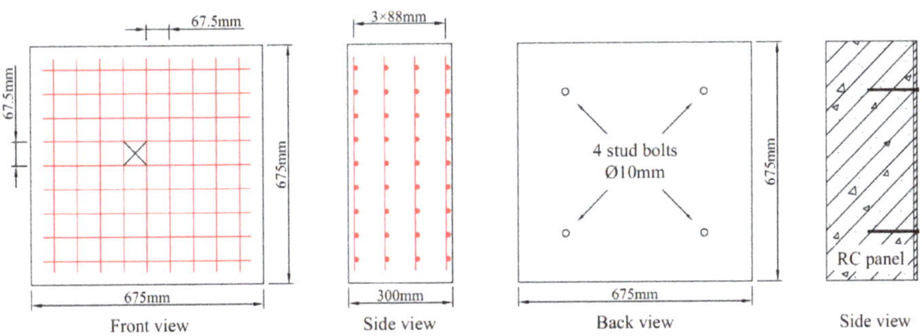

Figure 5. Geometric dimensioning of SC walls target.

For the numerical study, LS-DYNA, the extensively applied explicit solver, was adopted for impact simulations. Since the projectile material was of high strength and great hardness, the projectile was modeled as MAT_RIGID in the simulation. Successfully applied for concrete penetration simulations [30], the Holmquist-Johnson-Cook (HJC) model [31] was chosen to model the concrete material. Originally presented by Holmquist, Johnson and Cook [32], the HJC concrete model was developed for the purpose of impact computations where the material experiences large strains, high strain rates and high pressures. Coupled with isotropic damage, the HJC concrete model is an elastic-viscoplastic model [33] where the deviatoric response is determined by the following constitutive law:

$$\sigma_Y^* = \left[A(1-D) + Bp^{*N}\right][1 + C\ln\dot{\varepsilon}^*] \tag{12}$$

in which $\sigma_Y^* = \sigma_Y / f_c'$ is the normalized equivalent stress, and $p^* = p/f_c'$ is the normalized pressure.

The equation of state in HJC is characterized by three stages: plate elastic (for $p < p_c$) $p = K\mu$; pore collapse (for $p_c \le p \le p_1$) $p = p_c + (p_1 - p_c)(\mu - \mu_1)/(\mu_1 - \mu_c)$; compaction (for $p > p_1$) $p = k_1\overline{\mu} + k_2\overline{\mu}^2 + k_3\overline{\mu}^3$ with $\overline{\mu} = (\mu - \mu_1)/(1 + \mu_1)$; k_1, k_2, and k_3 are constants.

The HJC model includes a scalar damage formulation, where the damage evolution is accumulated from both the equivalent plastic strain increment $\Delta\varepsilon_{eq}^{p}$ and the equivalent plastic volumetric strain increment $\Delta\mu_{eq}^{p}$. The damage evolution is expressed as:

$$\Delta D = \frac{\Delta\varepsilon_{eq}^{p} + \Delta\mu_{eq}^{p}}{\varepsilon_{p}^{f} + \mu_{p}^{f}} \tag{13}$$

where, $\varepsilon_p^f + \mu_p^f = D_1(p^* + T^*)^{D_2}$, ε_p^f and μ_p^f are plastic strain and plastic volumetric strain corresponding to fracture, T^* is the normalized tensile strength, and D_1 and D_2 are damage constants.

The steel plate was modeled by the Johnson-Cook (JC) model [34] for its wide adoption in the metal impact engineering domain. JC is a strain rate and temperature-dependent (adiabatic assumption) visco-plastic material model [35,36]. This model is suitable for problems in which strain rates vary over a large range. The JC model expresses the flow stress with the form:

$$\sigma_Y = \left[A + B\varepsilon_p^N\right][1 + C\ln\dot{\varepsilon}^*] \tag{14}$$

where σ_Y is the effective stress, ε_p is the effective plastic strain, ε^* is the normalized effective plastic strain rate (typically normalized to a strain rate of 1.0 s^{-1}), N is the work hardening exponent, and A, B, C are constants determined by calibration.

With reference to [37], the steel rebar was modeled using MAT_PLASTIC_KINEMATIC. For the concrete HJC model and the steel 1006 JC model, the main parameters are listed in Tables 1 and 2 where those model parameters have been validated against available penetration tests [38,39]. In the test set up, all the top and bottom surfaces of the target were constrained. Element sizes were strictly controlled to guarantee that the steel mesh nodes coincide with concrete element nodes. Figure 6 shows the finite element model developed for SC wall penetration. Table 3 lists the element numbers of projectile, concrete panel, steel mesh as well as steel plate. The refined concrete mesh of the impact area was 3 mm, with 6 mm mesh for the outer region, which have been proven as converged meshes for penetration simulation [40]. The projectile was meshed with 1 to 3 mm size hexahedrons. The meshing sizes of the reinforced rebars were 3 mm and 6 mm. The rear steel plate was modeled with 1 mm × 1 mm × 1 mm for the center area and 1 mm × 2 mm × 2 mm for the outer region.

Table 1. Material parameters of concrete (units: cm-g-μs).

RO	*G*	*A*	*B*	*C*	*N*	*FC*	*T*	*EPSO*	*EFMIN*
2.24	0.1486	0.79	1.6	0.007	0.61	4.1×10^{-4}	4.1×10^{-5}	1×10^{-6}	0.01
SFMAX	*PC*	*UC*	*PL*	*UL*	D_1	D_2	K_1	K_2	K_3
7.0	1.6×10^{-4}	0.001	0.008	0.1	0.04	1.0	0.85	−1.71	2.08

Table 2. Material parameters of steel plate (units: cm-g-μs).

RO	*G*	*A*	*B*	*N*	*C*	*M*	*TM*	*TR*	*EPSO*
7.896	0.818	3.5×10^{-3}	2.75×10^{-3}	0.36	0.022	1	1793	293	1×10^{-6}
CP	*PC*	*SPALL*	*IT*	D_1	D_2	D_3	D_4	D_5	*C2/P*
0.452×10^{-5}	0	2	0	−0.8	2.1	−0.5	0.0002	0.61	1

3.2. Numerical Results, Validation and Discussion

In this work, five penetration simulations with different striking velocities (V_s), were carried out to validate the numerical model. After simulation, numerical predictions of projectile residual velocities ($V_{r,n}$) were compared against test data ($V_{r,e}$) as shown in Figure 7a. As listed in Table 4, the numerical predictions agreed well with the test data. It is also suggested that with a 436 m/s striking velocity, the projectile perforated the SC wall with quite a small residual velocity which was overestimated by the numerical model. Figure 7b shows the projectile velocity history during perforation, implying that it takes less time to perforate the SC wall at higher striking velocity.

Figure 8 compares the post-test target and numerical results from different views. Figure 8a shows the constraints of the target, which was fixed in the steel frame. The actual destructive forms of the rear target are also shown. A three-stage perforation model [16,40] consisting of front impact crater, ballistic tunnel, and a nearly frustum-of-cone shaped rear crater is shown in Figure 8b. In Figure 8c, the steel plate deformation is shown and it was notable that the neighboring area of the plate around the projectile suffers severe deformation. From a cross-section view of the SC walls, Figure 8d illustrates the damage mode of the target in which the foregoing three-stage thick plate model was numerically verified. From the validation results in terms of damage mode and residual velocity, the numerical predictions match well with SC wall perforation tests.

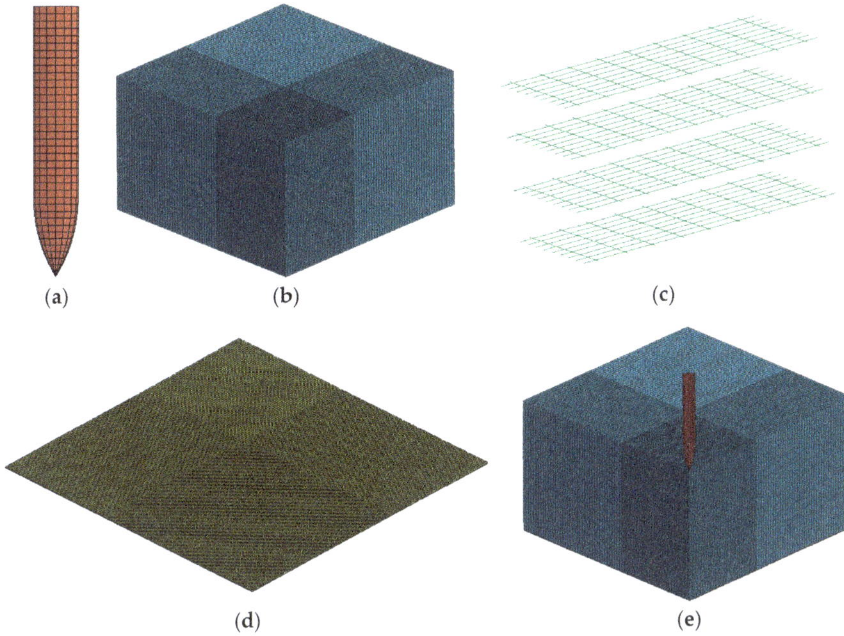

Figure 6. Finite element model. (**a**) Projectile; (**b**) concrete slab; (**c**) reinforcement mesh; (**d**) rear steel plate (**e**) projectile impact SC walls.

Table 3. Element numbers for each component of the SC wall perforation model.

Part	Projectile	Concrete Slab	Steel Mesh	Steel Plate
Number of elements	800	1,822,500	8712	18,496

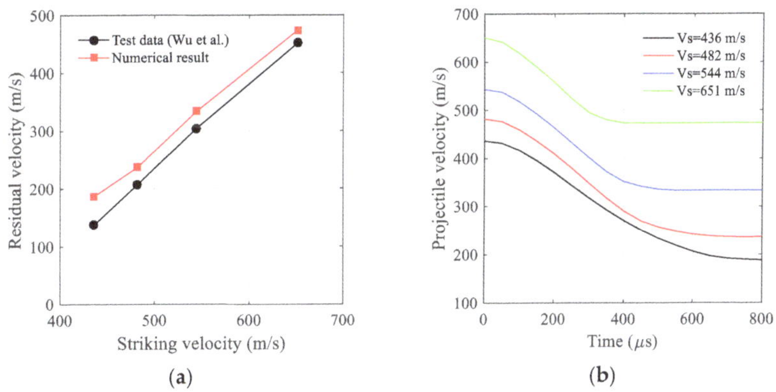

Figure 7. Numerical results of projectile perforation on SC walls. (**a**) Residual velocity comparison; (**b**) projectile velocity evolutions.

Table 4. Comparison between experiment and simulation.

V_s (m/s)	$V_{r,e}$ (m/s)	$V_{r,n}$ (m/s)
436	137	187
482	207	238
544	304	334
651	451	474

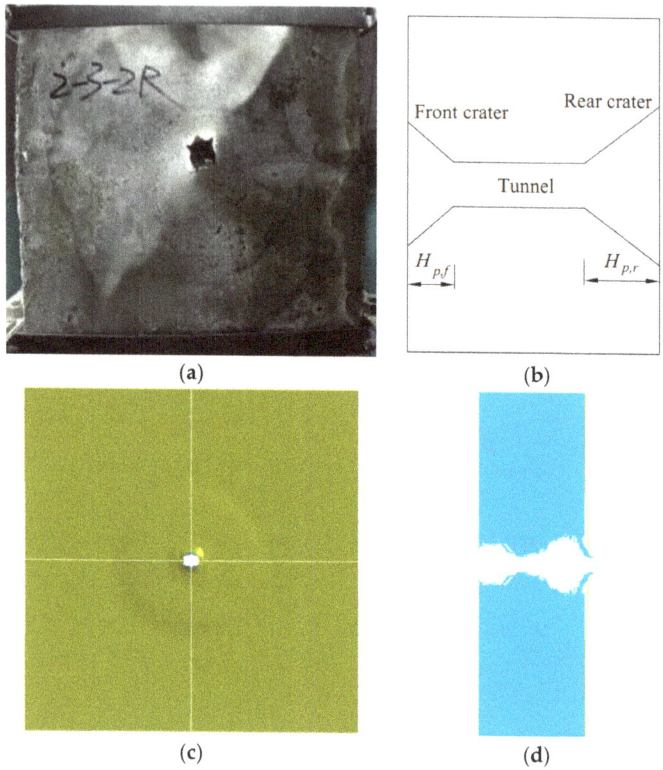

Figure 8. Damaged contour of SC walls. (**a**) Rear view of SC walls; (**b**) section view of concrete panel; (**c**) numerical results of damaged steel plate; (**d**) numerical results of damaged concrete panel.

4. Numerical Study of SC Walls Perforation

This section describes the extensive numerical investigations on SC wall perforations with different concrete panels and steel plates. SC walls with different front concrete panel and rear steel plate thickness combinations are investigated in this section considering protective structure analysis.

4.1. Model Setting

The previously described ogival nose projectile was used as the penetrator. With an 800 mm length and width, the investigated concrete panels were attached with 3 mm to 11 mm thicknesses of the rear steel plate. With the same cross section, the concrete panel thicknesses were selected as 150 mm, 200 mm and 250 mm, respectively. For SC walls with 200-mm thick concrete panel, the striking velocities of the projectile were set at 550, 600, 650, 700 and 800 m/s. Furthermore, a striking velocity of 600 m/s was set for perforation simulations with 150 mm and 250 mm thick concrete panels.

For the sake of computational cost, the SC walls perforation models were developed as quarterly symmetric bodies with symmetric boundaries. Figure 9 shows the FE model in which the grids near the impact region were refined. For the 1/4 model of SC walls with a combination of 11 mm thickness steel plate and 250-mm thick concrete panel, the element numbers for the three parts are given in Table 5. These material models and their parameters are the same as in the validation model. An eroding algorithm was applied to all the interactions between the contact components.

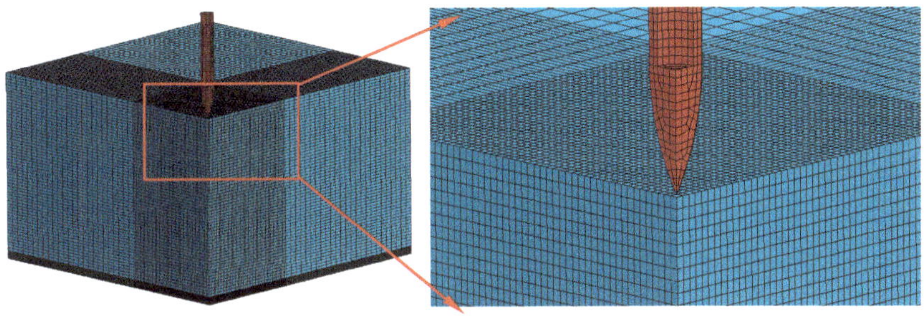

Figure 9. Overview of the meshed FE model.

Table 5. Element numbers for each part.

Part	Projectile	Concrete	Rear Steel Plate
Number of elements	696	376,875	50,000

4.2. Results Discussion of SC Walls Perforation

Seven kinds of thickness combinations of concrete panel and steel plate were used for the SC wall perforation analysis. Figure 10 shows the numerical results for an SC wall with a 250 mm thick concrete panel and an 11 mm thick steel plate. The projectile striking velocity was 600 m/s and the SC wall was perforated as expected. The destructive area of concrete rear surface was larger than its front impact surface due to conical plugging occurring in the back area. In Figure 10b, the von-Mises stress distribution contour exhibits a circular character.

To examine the penetration responses of SC walls under various striking velocities, a 200-mm thick concrete panel supported by rear steel plates with different thickness was numerically studied. Backing steel plates with different thickness were assumed to have various effects on the penetration resistance. Figure 11a illustrates that projectile residual velocity decreased with increasing thickness of the rear steel plate when the striking velocity was close to the ballistic limit [4]. Figure 11b shows the residual velocities for 150, 200 and 250 mm thick concrete panels subjected to a 600 m/s striking velocity impact. With the residual velocity increasing, the curve of velocity over time shows a slightly oscillating character. Under 600 m/s striking velocity impact, the projectile velocity history during SC wall perforation with a 200-mm thick concrete panel is shown in Figure 11c. The early penetration responses were almost the same, due to the fact that the backing steel plate had no influence on impact resistance of the front concrete target. Concerning the later perforation process, results imply that the rear steel plate has a significant effect on penetration resistance. Figure 11d shows the striking velocities and the residual velocities after perforation of SC walls with 200 mm thick concrete panels. With increasing striking velocity, it seems that the rear steel plate had a less pronounced effect on penetration resistance and thus the residual velocities tend to converge.

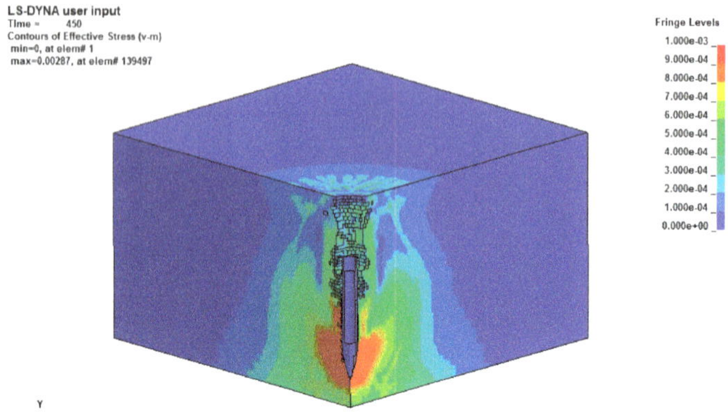

(**a**)

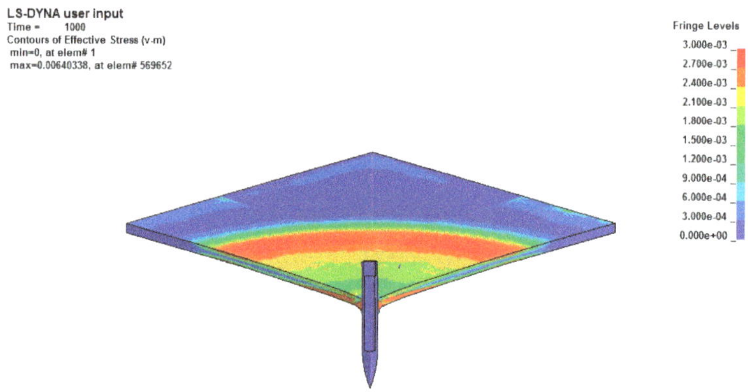

(**b**)

Figure 10. Impact response of an SC wall. (**a**) von-Mises stress distribution in front concrete panel; (**b**) von-Mises stress distribution in rear steel plate.

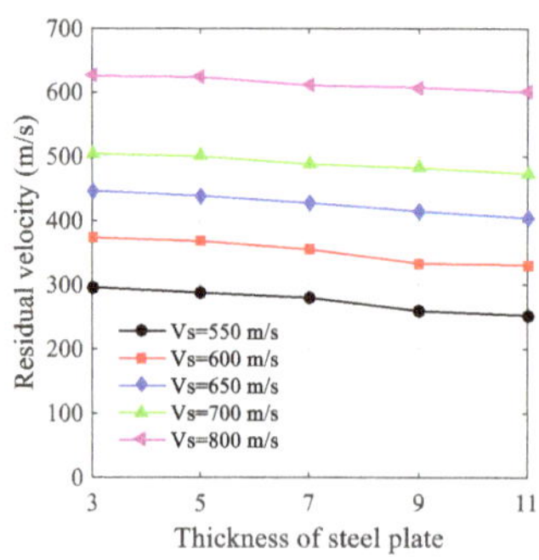

(**a**)

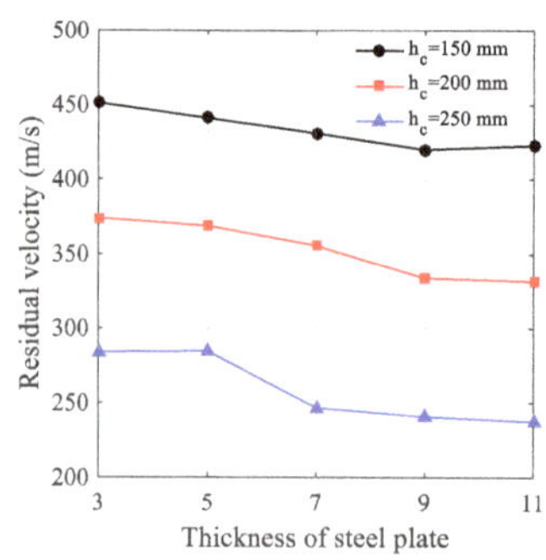

(**b**)

Figure 11. *Cont.*

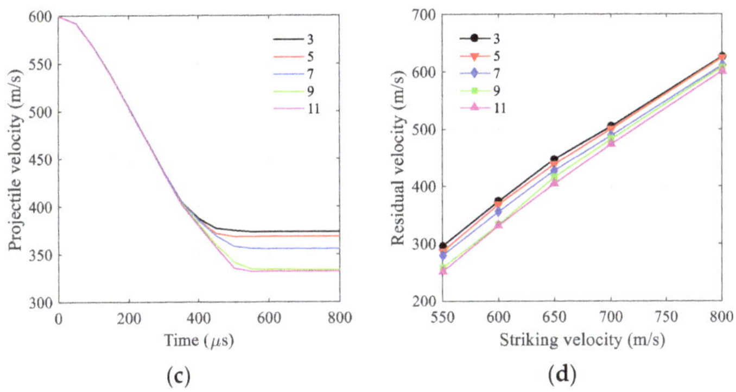

Figure 11. Numerical simulation results. (**a**) Effects of steel plate; (**b**) effects of concrete panel; (**c**) projectile velocity history; (**d**) residual velocity corresponding to striking velocity.

4.3. Free Surface Boundary Effect

Although cavity expansion analysis was successfully applied to projectile deep penetration in concrete, the typical penetration resistance equation proposed by Forrestal et al. [23] was not suitable for a projectile perforation scenario with a concrete panel of limited thickness. The front and back free surfaces might degrade the material strength thus reducing the penetration resistance during cratering and shear plugging, as shown in Figure 12. Therefore, the penetration resistance prior to shear plugging should be revised in the case of a concrete panel with limited thickness.

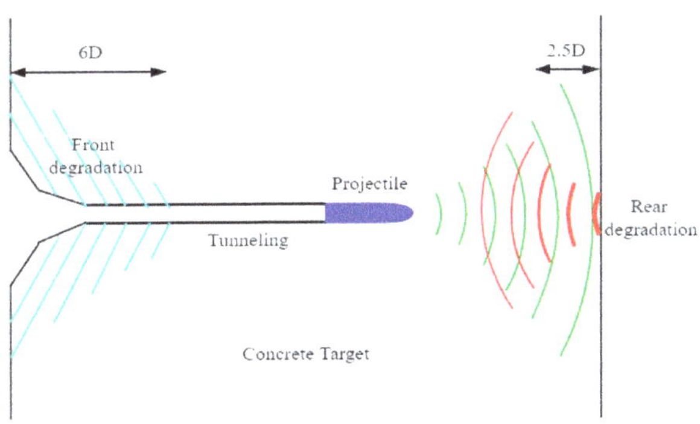

Figure 12. Front and rear free boundary effect on projectile penetration responses.

To develop a penetration resistance equation with respect to penetration velocity, numerical simulations with constant projectile velocity were performed to derive the penetration resistance acting on the projectile nose. Since the front and rear free surface might negatively affect the penetration resistance, concrete panels with thicknesses of 150, 200, 250, 300, 400, 500, 600 and 700 mm were selected for simulation. According to the results plotted in Figure 13, it is interesting that penetration resistant force increased to a plateau. This can be explained by the fact that the front surface degraded the penetration resistance until reaching about $6d_p$. The rear free surface effect was estimated by penetration simulation of panels with 250 mm and 300 mm thickness. Both had a stable plateau during tunneling and started to drop at a position about 68 mm to 70 mm away from the rear surface. The shear plugging height was about $2.5d_p$ which matches well with experimental data in Ref. [34]. For 150 mm and 200 mm thickness concrete panels, both the front and rear free surface affected the penetration resistance, implying no stable tunneling.

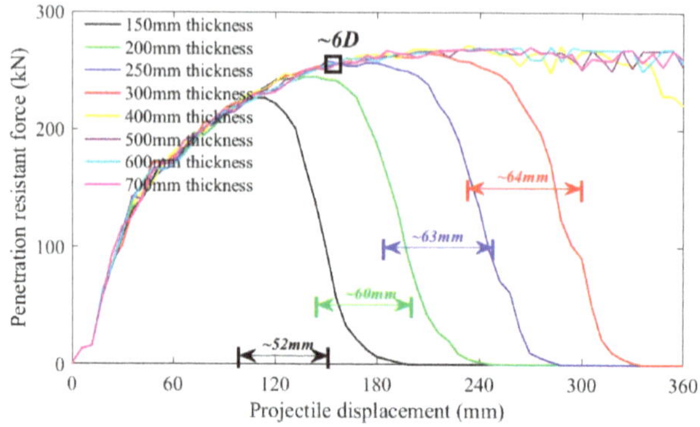

Figure 13. Penetration resistance for panels of different thickness.

5. Theoretical Analyses of Hard Projectile Perforation on SC Walls

According to the foregoing spherical cavity expansion theory, as well as thin steel plate petaling destruction, a semiempirical analytical model was proposed derived from numerical results. The SC wall composite target was composed of concrete and steel; therefore, the penetration resistance acting on the projectile nose was attributed to the concrete and steel plates. The concrete resistance at different penetration stages is related to cavity expansion analysis. The thin steel plate destruction mode due to perforation is generally petaling, hence the penetration resistance may be derived with a petaling model.

5.1. Penetration Resistance Force

The typical ogival-shaped nose projectile is illustrated in Figure 14, where the projectile nose length and shank diameter are denoted as h and d_{p}. Assuming a projectile with a striking velocity V, the normal expansion velocity perpendicular to the nose curve is $v = Vsin\theta$. Take the micro segment length dx to study the stress distribution over the infinite projectile nose surface ds. The resistant force df can be treated as the normal stress σ_n projection along the projectile axial direction:

$$df = \sigma_n sin\theta ds \tag{15}$$

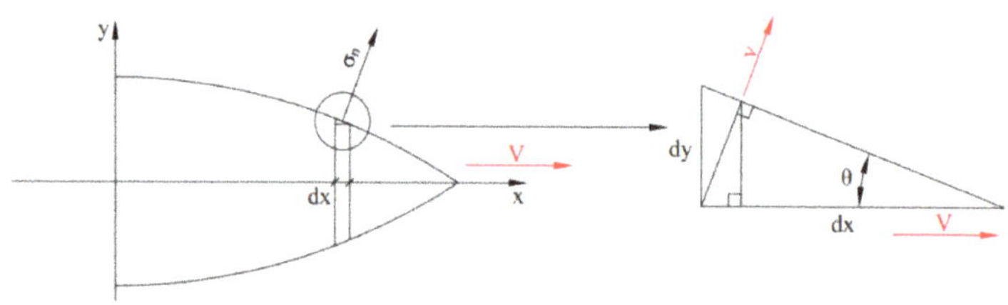

Figure 14. Penetration resistance on the projectile nose.

Integrating the normal stress σ_n over the projectile nose to achieve the axial penetration resistant force:

$$F = \int \sigma_n sin\theta ds \tag{16}$$

$$ds = 2\pi|y|\sqrt{1 + |k|}dx \tag{17}$$

For thick concrete targets, the Forrestal model [41] derived from the empirical formula $S = 82.6\left(f_c'/10^6\right)^{-0.544}$, was applied to describe the static penetration resistance. Hence, the normal resistant stress can be expressed as: $\sigma_n = Sf_c' + \rho v^2$, where f_c' is the unconfined cylinder compressive strength of concrete and ρ is the density of concrete target.

When the ratio of the metal plate thickness h_t and the shank diameter d_p is no larger than 0.5, petaling is generally the damage form of the thin metal target subjected to vertical impact. During the rear steel plate petaling process, the perforation energy E_c in Equation (11) can be regarded as mean resistant force accumulation over the deformation. Therefore, the mean resistant force can be estimated as $F_{mean} = E_c/d$, whereas d is the actual petaling displacement.

5.2. Stages of SC Walls Perforation

Based on the SC wall penetration resistance mechanism, a five-stage semiempirical analytical model was developed, i.e., projectile nose part penetration in concrete (cratering), stable penetration (tunneling), shear plugging, plate petaling with concrete plugging and plate petaling only, as depicted in Figure 1.

For stage 1 and 2, the projectile penetrates the front concrete panel with different contact areas. Hence, a deep penetration model [32,33] is utilized to analyze the first and second perforation stages. For penetration in the concrete panel, the normal stress $\sigma_{n,c}$ acting on the projectile nose can be expressed as $\sigma_{n,c} = Sf_c' + \rho v^2$ where $f_c' = 41$ MPa, $\rho = 2240$ kg/m^3. The static resistance term in the normal direction $Sf_c' = 82.6 \times \left(f_c'/10^6\right)^{-0.544} \times f_c' = 449.17$ MPa and the dynamic term ρv^2 change with actual projectile velocity.

During stage 3, the projectile passes through the concrete fragments until it hits the rear steel plate. For thicker rear plates, the pulverized concrete pieces in the rear crater have more support and provide more penetration resistance to the projectile. Related to its thickness, a certain deformation happens to the rear steel plate. According to previous literature [14,28,42,43], it is assumed that the rear crater depth of the pulverized concrete near rear surface follows the relationship $H_{p,r} = 2.5d_p$. With reference to [44,45], the penetration model with spherical cavity expansion analysis can be modified to describe the penetration response in pulverized concrete. Due to the decreasing static resistance (fragile material), the normal penetration resistant stress $\sigma_{n,c}'$ can be expressed as:

$$\sigma_{n,c}' = (0.1 + 0.3 \times \frac{h_t}{d_p})Sf_c' + \rho v^2 \tag{18}$$

For stages 4 and 5, the projectile penetrates both the pulverized concrete as well as the rear steel plate. Considerable deformation occurs to the rear steel plate with destructive form. The process of a projectile perforating a thin steel plate is complex; therefore, the energy method is used to analyze rear steel plate perforation. The mean penetration resistant force is noted as F_{mean}, and the length is assumed to be $d = h_t + 1.5h$ according to the simulation results. For steel 1006 material, $\sigma_y = 350$ MPa, $\sigma_u = 500$ MPa, $p = 0.36$ and $R = 35$ J/cm^2.

As the projectile perforates the SC wall with a 200-mm thick concrete panel and a 3-mm thick steel plate, the resistant force acting on the projectile nose with a 700 m/s striking velocity is given in Figure 15, where five different stages of SC walls perforation are depicted in different colors for better visualization.

5.3. Analytical Model Validation against FE Simulation

SC walls with a 200-mm thick concrete panel and a 3-mm thick steel plate were studied to validate the analytical model with respect to numerical results in which the projectile striking velocity was 700 m/s.

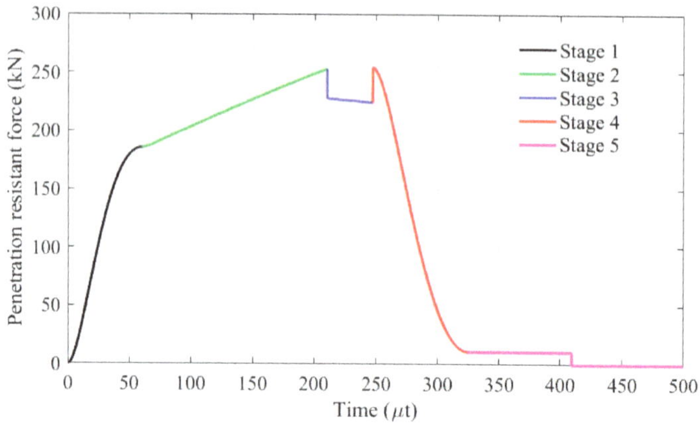

Figure 15. Penetration resistant force for five stages.

The projectile movement process, projectile velocity history, as well as projectile deceleration evolution are plotted in Figure 16. The analytical model has good consistency and regularity with simulation results concerning projectile residual velocity and deceleration history.

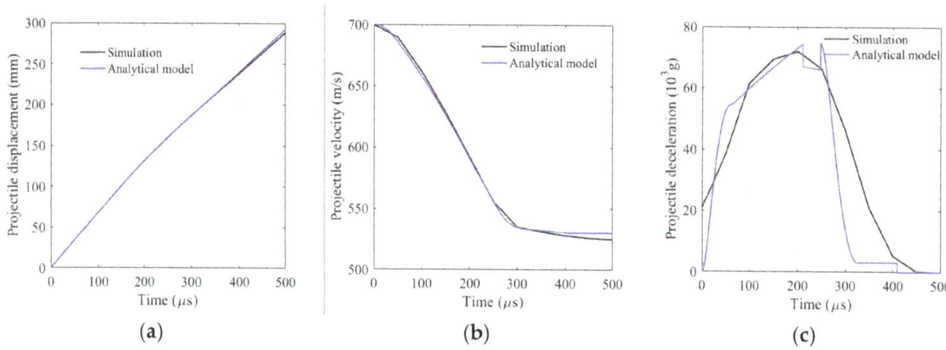

Figure 16. Comparison of analytical model and simulation. (**a**) Projectile displacement history; (**b**) projectile velocities history; (**c**) projectile deceleration history.

It is important to validate the analytical model with different boundary conditions, e.g., various thickness combinations of concrete panel and steel plate, and different projectile striking velocities. SC wall perforation of SC walls of 150, 200 and 300 thickness concrete and 3, 5, 7, 9 and 11 mm thicknesses of steel plate were studied both analytically and numerically. Striking velocities ranging from 650 m/s to 800 m/s were calculated and compared with simulation data. A good match is shown in Figure 17, suggesting the analytical model is validated with FE simulation and thus can be applied to subsequent discussion.

5.4. Rear Steel Plate Effect

For the SC walls with 200 mm thick concrete panels, the penetration responses, i.e., energy consumption and rear steel plate contribution, were explored analytically by taking into account steel plates with different thickness. In Figure 18a, all five curves have the tendency to converge at increasing striking velocity. Under different striking velocities, the absorbed energy of SC walls with a 200-mm thick concrete panel and different thickness steel plates is shown in Figure 18b, where SC wall consumption energy increases with increasing striking velocity. For the SC walls with a 200 mm thick concrete panel, energy consumed by the rear steel plates is depicted in Figure 18c. The absorbed energy increases with increasing rear steel plate thickness. Under penetration impact of different projectile striking velocities, their energy absorption shows a tendency of linear increase.

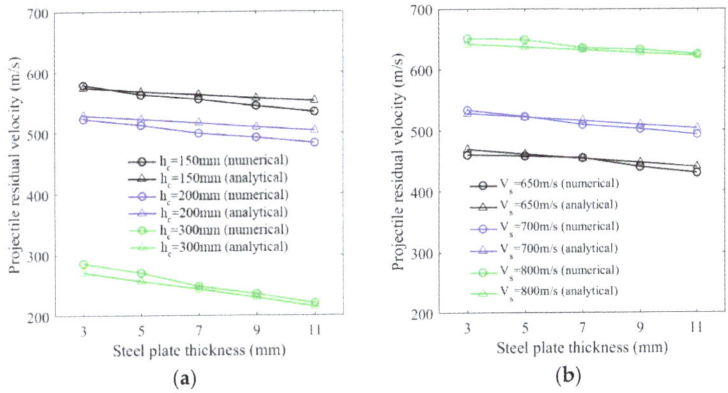

Figure 17. Model validation in terms of residual velocity. (**a**) Different thickness combinations; (**b**) different projectile impact velocity.

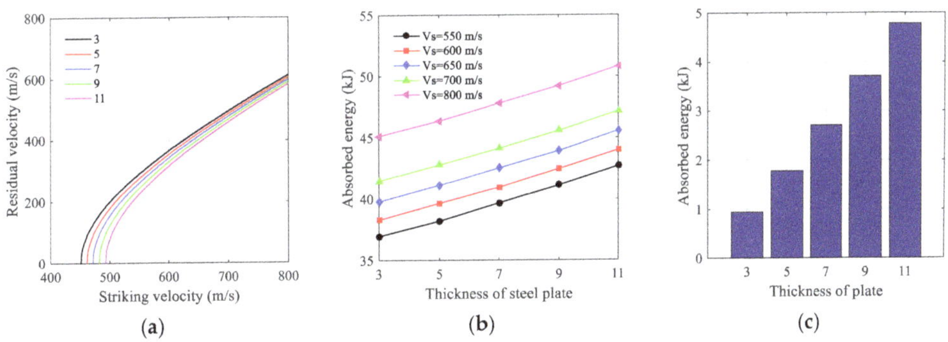

Figure 18. Results of analytical model for perforation on SC walls with h_c = 200 mm. (**a**) Residual velocity; (**b**) absorbed energy by SC walls; (**c**) absorbed energy by steel plate.

The effect of concrete panel thickness on energy consumption during perforation was studied. Figure 19a shows the absorbed energy of SC walls with different concrete panels. Figure 19b shows the equivalent concrete panel thickness determined for various SC wall perforation with the same ballistic limit.

$$h_{eq} = h_c + 3.06h_t \quad (19)$$

To evaluate the penetration resistance of SC walls with front concrete panels and rear steel plates, the concept of equivalent thickness of a concrete panel was considered. It was assumed that the equivalent concrete panel thickness had the same ballistic limit as the corresponding SC walls. After data fitting, Equation (19) was derived for the description of equivalent concrete panel thickness in terms of h_c and h_t of SC walls. This may help researchers to estimate the ballistic limit of SC walls.

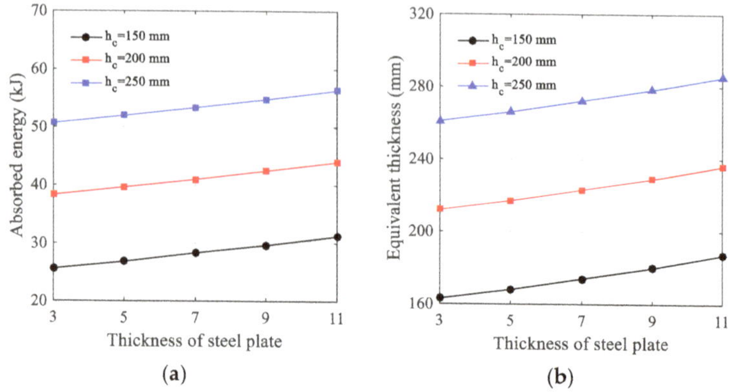

Figure 19. Perforation behavior of SC walls with different thicknesses of concrete panels. (**a**) Absorbed energy for SC walls; (**b**) equivalent concrete thickness.

6. Conclusions

The penetration resistance of SC walls with various thickness combinations of concrete panels and steel plates was explored both numerically and analytically. Through validation against experimental data, the numerical model and semi-empirical analytical model were further investigated to derive the following conclusions. (1) The attached rear steel plate has a positive influence on impact resistance by preventing the pulverized concrete pieces from flying away. (2) With increasing striking velocity, the rear steel plate has a less pronounced effect on the penetration resistance and thus the residual velocities tend to converge (3) Combining spherical cavity expansion analysis and the thin plate petaling theory, an analytical model with five stages of SC wall perforation was proposed and validated against a numerical simulation. (4) With the same ballistic limit, the equivalent concrete panel thickness can be derived with respect to the SC walls as $h_{\rm eq} = h_c + 3.06h_t$.

Author Contributions: Conceptualization, X.W. and W.L.; writing—original draft preparation, B.P. and J.F.; writing—review and editing, J.F.; visualization, B.P.; supervision, X.W.; funding acquisition, W.L. and J.F. All authors have read and agreed to the published version of the manuscript.

Funding: This study was funded by the National Natural Science Foundation of China (No. 11902161). The APC was funded by Open Cooperative Innovation Fund of Xi'an Institute of Modern Chemistry (No. SYJJ200321).

Institutional Review Board Statement: Not applicable.

Informed Consent Statement: Not applicable.

Data Availability Statement: Some or all data, models, or code that support the findings of this study are available from the corresponding author upon reasonable request.

Acknowledgments: Jun Feng thanks the Natural Science Foundation of Jiangsu Province (No. BK20170824) and the Major Science and Technology Projects in Shanxi under Grant 20201102003.

Conflicts of Interest: The authors declare no conflict of interest.

References

1. Thai, D.K.; Kim, S.E.; Bui, T.Q. Modified empirical formulas for predicting the thickness of RC panels under impact loading. *Constr. Build. Mater.* **2018**, *169*, 261–275. [CrossRef]
2. Lee, K.; Shin, J.; Kim, K.; Varma, A. Local responses of steel-plate composite walls subjected to impact loads: Intermediate scale tests. *Eng. Struct.* **2020**, *206*, 110131. [CrossRef]
3. Zou, D.L.; Sun, J.G.; Wu, H.; Hao, Y.F.; Wang, Z.; Cui, L.F. Experimental and numerical studies on the impact resistance of large-scale liquefied natural gas (LNG) storage outer tank against the accidental missile. *Thin-Walled Struct.* **2021**, *158*, 107189. [CrossRef]

4. Remennikov, A.; Ying, K.S.; Uy, B. *Static and Dynamic Behaviour of Non-Composite Steel-Concrete-Steel Protective Panels under Large Deformation*; Research Publishing Services: Singapore, 2010.
5. Remennikov, A.M.; Kong, S.Y. Numerical simulation and validation of impact response of axially restrained steel-concrete-steel sandwich panels. *Compos. Struct.* **2012**, *94*, 3546–3555. [CrossRef]
6. Feng, J.; Sun, W.; Liu, Z.; Chong, C.; Wang, X. An armour-piercing projectile penetration in a double layered target of ultra-high-performance fiber reinforced concrete and armour steel: Experimental and numerical analyses. *Mater. Des.* **2016**, *102*, 131–141. [CrossRef]
7. Kojima, I. An experimental study on local behavior of reinforced concrete slabs to missile impact. *Nucl. Eng. Des.* **1991**, *130*, 121–132. [CrossRef]
8. Tsubota, H.; Yasai, Y.; Noshika, N.; Morikawa, H.; Uchida, T.; Ohno, T.; Kogure, K. Quantitative studies on impact resistance of reinforced concrete panels with steel plates under impact loading part I: Scaled model impact tests. In *Transactions of the 12th SMiRT*; International Association for Structural Mechanics in Reactor Technology, IASMiRT: Stuttgart, Germany, 1993; pp. 169–174.
9. Abdel-Kader, M.; Fouda, A. Effect of reinforcement on the response of concrete panels to impact of hard projectiles. *Int. J. Impact Eng.* **2014**, *63*, 1–17. [CrossRef]
10. Hashimoto, J.; Takiguchi, K.; Nishimura, K.; Matsuzawa, K.; Tsutsui, M.; Ohashi, Y.; Kojima, I.; Torita, H. Experimental study on behavior of RC panels covered with steel plates subjected to missile impact. In *Transactions of the 18th SMiRT*; International Association for Structural Mechanics in Reactor Technology, IASMiRT: Beijing, China, 2005; pp. 2604–2615.
11. Mizuno, J.; Sawamoto, Y.; Yamashita, T.; Koshika, N.; Niwa, N.; Suzuki, A. Investigation on impact resistance of steel plate reinforced concrete barriers against aircraft impact part 1: Test program and results. In *Transactions of the 18th SMiRT*; International Association for Structural Mechanics in Reactor Technology, IASMiRT: Beijing, China, 2005; pp. 2566–2579.
12. Bruhl, J.C.; Varma, A.H.; Johnson, W.H. Design of composite SC walls to prevent perforation from missile impact. *Int. J. Impact Eng.* **2015**, *75*, 75–87. [CrossRef]
13. Grisaro, H.; Dancygier, A.N. Assessment of the perforation limit of a composite RC barrier with a rear steel plate to impact of a non-deforming projectile. *Int. J. Impact Eng.* **2014**, *64*, 122–136. [CrossRef]
14. Wu, H.; Fang, Q.; Peng, Y.; Gong, Z.M.; Kong, X.Z. Hard projectile perforation on the monolithic and segmented RC panels with a rear steel plate. *Int. J. Impact Eng.* **2015**, *76*, 232–250. [CrossRef]
15. Lee, H.K.; Kim, S.E. Comparative assessment of impact resistance of SC and RC panels using finite element analysis. *Prog. Nucl. Energy* **2016**, *90*, 105–121. [CrossRef]
16. Feng, J.; Sun, W.; Wang, L.; Chen, L.; Xue, S.; Li, W. Terminal ballistic and static impactive loading on thick concrete target. *Constr. Build. Mater.* **2020**, *251*, 118899. [CrossRef]
17. Wierzbicki, T. Petalling of plates under explosive and impact loading. *Int. J. Impact Eng.* **1999**, *22*, 935–954. [CrossRef]
18. Bishop, R.F.; Hill, R.; Mott, N.F. The theory of indentation and hardness tests. *Proc. Phys. Soc.* **2002**, *57*, 147–159. [CrossRef]
19. Forrestal, M.J.; Luk, V.K. Dynamic spherical cavity-expansion in a compressible elastic-plastic solid. *J. Appl. Mech.* **1988**, *55*, 275. [CrossRef]
20. Feng, J.; Li, W.; Wang, X.; Song, M.; Ren, H.; Li, W. Dynamic spherical cavity expansion analysis of rate-dependent concrete material with scale effect. *Int. J. Impact Eng.* **2015**, *84*, 24–37. [CrossRef]
21. Forrestal, M.J.; Altman, B.S.; Cargile, J.D.; Hanchak, S.J. An empirical equation for penetration depth of ogive-nose projectiles into concrete targets. *Int. J. Impact Eng.* **1994**, *15*, 395–405. [CrossRef]
22. Forrestal, M.J.; Frew, D.J.; Hanchak, S.J.; Brar, N.S. Penetration of grout and concrete targets with ogive-nose steel projectiles. *Int. J. Impact Eng.* **1996**, *18*, 465–476. [CrossRef]
23. Forrestal, M.J.; Tzou, D.Y. A spherical cavity-expansion penetration model for concrete targets. *Int. J. Solids Struct.* **1997**, *34*, 4127–4146. [CrossRef]
24. Rosenberg, Z.; Dekel, E. The penetration of rigid long rods-revisited. *Int. J. Impact Eng.* **2009**, *36*, 551–564. [CrossRef]
25. Kong, X.Z.; Wu, H.; Fang, Q.; Zhang, W.; Xiao, Y.K. Projectile penetration into mortar targets with a broad range of striking velocities: Test and analyses. *Int. J. Impact Eng.* **2017**, *106*, 18–29. [CrossRef]
26. Feng, J.; Song, M.; Sun, W.; Wang, L.; Li, W.; Li, W. Thick plain concrete targets subjected to high speed penetration of 30CrMnSiNi2A steel projectiles: Tests and analyses. *Int. J. Impact Eng.* **2018**, *122*, 305–317. [CrossRef]
27. Rosenberg, Z.; Vayig, Y.; Malka-Markovitz, A. The scaling issue in the penetration of concrete targets by rigid projectiles-Revisited. *Int. J. Impact Eng.* **2020**, *140*, 103561. [CrossRef]
28. Peng, Y.; Wu, H.; Fang, Q.; Gong, Z.M.; Kong, X.Z. A note on the deep penetration and perforation of hard projectiles into thick targets. *Int. J. Impact Eng.* **2015**, *85*, 37–44. [CrossRef]
29. Landkof, B.; Goldsmith, W. Petalling of thin, metallic plates during penetration by cylindro-conical projectiles. *Int. J. Solids Struct.* **1985**, *21*, 245–266. [CrossRef]
30. Johnson, G.; Beissel, S.; Holmquist, T.; Frew, D. Computed radial stresses in a concrete target penetrated by a steel projectile. In Proceedings of the Structures under Shock and Impact V, Thessaloniki, Greece, 24–26 June 1998; pp. 793–806.
31. Sun, W.; Shi, Z.; Chen, B.; Feng, J. Numerical Study on RC Multilayer Perforation with Application to GA-BP Neural Network Investigation. *Civ. Eng. J.* **2020**, *6*, 806–819. [CrossRef]

32. Holmquist, T.J.; Johnson, G.R.; Cook, W.H. A computational constitutive model for concrete subjected to large strains, high strain rates and high pressures. In Proceedings of the 14th International Symposium on Ballistics, Quebec City, QC, Canada, 26–29 September 1993; pp. 591–600.
33. Polanco-Loria, M.; Hopperstad, O.S.; Børvik, T.; Berstad, T. Numerical predictions of ballistic limits for concrete slabs using a modified version of the HJC concrete model. *Int. J. Impact Eng.* **2008**, *35*, 290–303. [CrossRef]
34. Johnson, G.R.; Cook, W.H. A constitutive model and data for materials subjected to large strains, high strain rates, and high temperatures. In Proceedings of the 7th International Symposium on Ballistics, The Hague, The Netherlands, 19–21 April 1983; pp. 541–547.
35. Kurtaran, H.; Buyuk, M.; Eskandarian, A. Ballistic impact simulation of GT model vehicle door using finite element method. *Theor. Appl. Fract. Mech.* **2003**, *40*, 113–121. [CrossRef]
36. Burley, M.; Campbell, J.E.; Dean, J.; Clyne, T.W. Johnson-Cook parameter evaluation from ballistic impact data via iterative FEM modelling. *Int. J. Impact Eng.* **2018**, *112*, 180–192. [CrossRef]
37. Zhao, F.C.; Chen, Y.J.; Wang, Y.; Lu, J.S. Damage mechanism and response of reinforced concrete containment structure under internal blast loading. *Theor. Appl. Fract. Mech.* **2012**, *61*, 12–20. [CrossRef]
38. Yu, M.; Zha, X.; Ye, J. The influence of joints and composite floor slabs on effective tying of steel structures in preventing progressive collapse. *J. Constr. Steel Res.* **2010**, *66*, 442–451. [CrossRef]
39. Wang, X. Adiabatic shear localization for steels based on Johnson-Cook model and second-and fourth-order gradient plasticity models. *J. Iron Steel Res. Int.* **2007**, *14*, 56–61. [CrossRef]
40. Oucif, C.; Mauludin, L.M. Numerical modeling of high velocity impact applied to reinforced concrete panel. *Undergr. Space* **2019**, *4*, 1–9. [CrossRef]
41. Frew, J.D.; Hanchak, J.S.; Greent, L.M.; Forrestal, J.M. Penetration of concrete targets with ogive-nose steel rods. *Int. J. Impact Eng.* **1998**, *21*, 489–497. [CrossRef]
42. Li, J.; Lu, Z.; Zhang, H.; Huang, F. Perforation experiments of concrete targets with residual velocity measurements. *Int. J. Impact Eng.* **2013**, *57*, 1–6.
43. Hanchak, J.S.; Forrestal, J.M.; Young, R.E.; Ehrgott, Q.J. Perforation of concrete slabs with 48 MPa (7 ksi) and 140 MPa (20 ksi) unconfined compressive strengths. *Int. J. Impact Eng.* **1992**, *12*, 1–7. [CrossRef]
44. Forrestal, M.J. Penetration into dry porous rock. *Int. J. Solids Struct.* **1986**, *22*, 1485–1500. [CrossRef]
45. Shi, C.; Wang, M.; Jie, L.; Li, M. A model of depth calculation for projectile penetration into dry sand and comparison with experiments. *Int. J. Impact Eng.* **2014**, *73*, 112–122. [CrossRef]

applied sciences

MDPI

Article

Dynamic Response of Reinforced Recycled Aggregate Concrete Pavement under Impact Loading

Jifeng Yuan [1,2], Jin Wu [1,*], Tian Su [3,4,5] and Dadi Lin [1]

1 Department of Civil and Airport Engineering, Nanjing University of Aeronautics and Astronautics, 29 Yudao St., Nanjing 210016, China
2 Taizhou Insititute of Science and Technology, Nanjing University of Science and Technology, 8 Meilan East St., Taizhou 225300, China
3 Department of Architectural Engineering, School of Civil and Architectural Engineering, Shandong University of Technology, Zibo 255000, China
4 Department of Architectural Engineering, School of Civil Engineering, Wuhan University, 8 Donghu South Rd., Wuhan 430072, China
5 China Railway 11 Bureau Group Co., Ltd., 277 Zhongshan Rd., Wuhan 430061, China
* Correspondence: wujin@nuaa.edu.cn

Abstract: Airport runway pavements often undergo the direct impact of aircraft landings. For the purposes of designing the structure, it is of great importance to know about the dynamic response of the pavement and its behavior under impact loading. However, the dynamics and failure mechanisms of reinforced recycled aggregate concrete pavements subjected to impact loading are seldom explored in the literature. For this purpose, four reinforced recycled aggregate concrete pavements with different thickness and ratios of reinforcement, and one reinforced normal concrete pavement, were manufactured and tested under impact loading using the drop-weight impact frame system. The impact force characteristics, crack patterns, deformation responses, and strain developments of reinforced concrete pavements subjected to impact loading were evaluated and compared. The above-mentioned study revealed that with an increase in the reinforcement ratio, both the deformation and the steel strain were reduced. Increasing the thickness would reduce the degree of damage and the impact force of reinforced concrete pavement (RCP) but increase the deformation. The results show that under the same compressive strength, the dynamic performance of the reinforced recycled aggregate concrete pavement was worse than that of the reinforced normal concrete pavement because of its lower elastic modulus and weaker interfacial transition zone. The dynamic performance of reinforced recycled aggregate concrete pavement could be improved by increasing the thickness and reinforcement ratio. The use of recycled aggregate concrete (RAC) in RCP is a technically feasible application of the material within the scope of this experimental study.

Keywords: RCP; recycled aggregate concrete; impact loading; impact force; cracking; peak displacement

Citation: Yuan, J.; Wu, J.; Su, T.; Lin, D. Dynamic Response of Reinforced Recycled Aggregate Concrete Pavement under Impact Loading. *Appl. Sci.* **2022**, *12*, 8804. https://doi.org/10.3390/app12178804

Academic Editors: Ricardo Castedo, Lina M. López and Anastasio P. Santos

Received: 16 August 2022
Accepted: 30 August 2022
Published: 1 September 2022

1. Introduction

With the development of national defense construction and civil aviation and the use of a large number of high-speed heavy aircraft, the safety and reliability of airport runway structures are highly sought after. The airport runway system does not only bear the direct impact caused by aircraft landing, but also may encounter large impact loading due to the hard landings of aircraft crashes [1]. Currently, the design specifications of airport pavement structure take the structure under static load as the research object, while impact effects due to hard landings have not been taken into consideration in the design of airport runway pavements [2,3]. It should be emphasized that airport pavements are constructed to provide adequate support for the loads imposed by airplanes, and produce a firm, stable, and smooth surface, and it should be strictly required that there will be no debris or other particles caused by landing, or they could be sucked into the engine and cause serious engineering accidents. In order to satisfactorily meet these requirements, the pavement must be of sufficiently good quality to ensure not failing under the applied load.

Reinforced concrete pavement (RCP) refers to a pavement with embedded steel reinforcing bars in the concrete for crack control. The bars keep the cracks tightly closed, thus allowing longer joint spacing, resulting in an intact and smooth surface that ensures structural integrity and improves the performance of the pavement [4,5]. Reinforced concrete pavement is widely used in airport runway landing areas due to its good performance and low maintenance needs.

The performance of RCP depends on critical stresses and deflections imposed by repeated traffic and environmental loading, and fatigue fracture caused by these stresses is considered to be the limit state in the design of pavement structure. Therefore, in the current design code provisions on pavements, the main task is to determine the thickness of each component of the pavement structure to ensure that it can provide a satisfactory structural life at design fatigue limits. At present, the widely available analysis method of wheel load stress is based on Hertz's elastic thin plate theory [6], and Westergaard [7] has further proposed the solution of Winkler foundation under different load conditions. With the emergence of finite element software, the mechanical stresses are evaluated by a three-dimensional analysis and thermal stresses by two-dimensional analysis [8–11]. However, neither the Federal Aeronautics Administration (FAA) nor the Civil Aviation Administration of China (CAAC) considers impact effects due to hard landings of heavy aircraft in the design of airport runway pavements. The design code recommends that the dynamic effect be taken into account by multiplying the dynamic amplification factor.

A large amount of abandoned concrete is produced during the reconstruction and extension of airport runways, which has a substantial effect on the environment. Sustainability concerns are at the forefront of our society; unfortunately, the abandoned concrete is a non-renewable resource. The use of recycled waste concretes in construction application and pavement construction is one way to promote sustainable development. Consequently, many researchers have investigated the use of recycled concrete aggregate (RCA) in the production of new concrete, which is named recycled aggregate concrete (RAC). Most findings have indicated that the compressive strength, splitting tensile strength, flexural strength, and modulus of elasticity for RAC decrease with an increase in the content of RCA [12–15]. Furthermore, some pieces in the literature have studied RAC structural elements, such as columns, beams, slabs, and pavements; the results show that the incorporation of RAC has negative effects on the performance of these elements [16–21]. In general, the desired reduction caused by using RAC is limited, and further engineering research is encouraged, since satisfactory performance can still be achieved.

Dynamic response of reinforced concrete structural elements under impact loading has been investigated through experiments by many researchers [22–29]. Zineddin et al. [23,24] investigated the effects of different types of slab reinforcements and impact energy on the dynamic response and behavior of reinforced concrete slabs. The addition of steel reinforcement provided substantial strength enhancement to the slab, promoted crack formation on the top surface, and increased the stresses and strains that the concrete and steel materials could safely undergo, especially under higher impact energy. Othman et al. [25] conducted an experiment to investigate the effect of steel reinforcement distribution on the dynamic response of high strength concrete (HSC) plates, taking into account the effects of the main bottom steel reinforcement ratio (1.0, 2.0, and 3.0%) and the steel reinforcement arrangement (single or doubly reinforced plates). The results showed that the change of reinforcement ratio and/or reinforcement arrangement has no significant effect on impulse and absorbed energy values for same impact loading condition, while the impact duration decreased with the increase in reinforcement ratio. The reinforcement arrangement could affect the crack pattern; the HSC plates with single reinforcement typically failed by localized sudden punching, and the HSC plates doubly reinforced typically failed in a ductile punching mode.

Xiao et al. [26] studied the effects of loading rates on the performance of reinforced concrete (RC) slabs. From test results, the damage process, failure mode, strain rate, and energy absorption capacity of RC slabs were similar between the high-loading-rate test and the low-velocity impact test. Therefore, it was suggested that the high load rate test results could be used to analyze the performance of the RC slabs under low-velocity impacts. Both longitudinal and transverse reinforcements were effective in enhancing the maximum strength of specimens. However, the damage to the slab under both high-rate and impact loadings can be more efficiently reduced by adding shear stirrups. In another experimental program [27], five 1200 mm square RC slabs were tested with different nose shapes, diameters of impacted area, drop weights and drop heights in another experimental program; the punching shear failure mode was observed for all the specimens that failed during the test. The damage to the slabs increased with the increase in impact energy, and more impact energy was required to fail RC slabs when the diameter of the impacted area increased.

In order to better understand the effects of supporting conditions on the behavior of RC slabs subjected to impact load, some research has been undertaken. Özgür et al. [28] found that the number of drops until failure was lower for the specimens with four hinge supports than those for the specimens with four fixed supports, but higher than those for the specimens with two opposite hinge supports. The authors also reported that the acceleration, velocity, and displacement decreased due to an increase in the support stiffness. Chiaia et al. [29] studied two-way reinforced concrete slabs over different kinds of yielding supports and concluded that reducing the support rigidity could decrease the displacement and stress of the whole structure. Husem et al. [30] found that the energy-absorbing capacity was decreased by an increase of span size in both fixed and free supported RC slabs, and that the maximum midspan displacement values increased only in free supported RC slabs; however the span size has no considerable effect in fixed supported RC slabs.

Furthermore, some studies were made to improve the impact behavior of RC slabs by blending in other materials, such as steel fibers, carbon fabric, polypropylene fiber, etc. An experimental program by Hrynyk et al. [31] revealed that the increased addition of the steel fibers was effective in increasing slab capacity, reducing crack widths and spacings, and mitigating local damage under impact. The research by Beckmann et al. [32] investigated blending steel fibers; carbon fabric showed substantial advantages in the resistance to the impact load and to the penetration of the impactor but had only a minor influence on concrete strain. AlRousan et al. [33] studied the impact resistance of RC slabs blending with polypropylene fiber; the result showed that the proper quantity of polypropylene fiber could significantly improve the impact resistance of RC slabs, and that a suitable content of polypropylene fiber was 0.90%. Ong et al. [34] studied the impact resistance of concrete slabs blending with four substances (polyolefin, polyvinyl, alcohol, and steel); the result showed that hooked-end steel fiber concrete slabs had the best cracking and energy absorption characteristics compared to other slabs.

However, there are a limited number of studies that comprehensively explore the impact dynamics and failure mechanisms of reinforced recycled aggregate concrete pavement. Wu [35] studied the damage and failure model of rigid concrete pavement under drop weight impact, and obtained the extent of damage, failure mode, deformation, and acceleration. The test results show that the concrete pavement without steel reinforcement was fragmented into three segments and showed brittle failure mode. Cai et al. [36] analyzed the dynamic deflection and velocity response of airport concrete pavement under impact loading and drew the conclusion that the velocity response amplitude decreased with the increase of slab thickness, and that the deflection at the center of the slab decreased with the decrease of pavement slab size.

Most of the studies conducted under impact loading were focused on natural aggregate concrete (NAC) structural elements. Recently, a few studies have been performed on the dynamic response of RAC structural elements under drop weight impact loading [37–40]. Vali et al. [38] studied the behavior of RAC slabs under impact loading with different replacement ratios of RCA and found that the stiffness of RAC slabs decreased with an increasing replacement ratio of RCA, which led to decreases in the punching shear strength at first crack stage and in the ultimate punching shear strength.

In order to design the airport pavement scientifically and to ensure its service life and safe operation, it is necessary to conduct experimental research on the performance of reinforced recycled aggregate concrete pavement under likely impacts. In fact, the excessive pavement damage due to the impact load of hard landings is extremely difficult to measure. Therefore, in order to better understand the behavior of reinforced recycled aggregate concrete pavement subjected to impact loading, an experimental program has been designed and conducted in this paper. The structural dynamic responses are measured during the drop-weight impact tests, and together with the cracking mechanisms, can provide a basis for investigating the impact behavior of reinforced recycled aggregate concrete pavement with different thickness and reinforcement ratios. In the meantime, the drop-weight impact test is also done on the reinforced normal concrete pavement, and the impact force characteristics, crack patterns, deformation responses, and strain developments are compared with those of reinforced recycled aggregate concrete pavement.

This paper presents the details of a well-organized and well-equipped experimental investigation with two main research objectives:

(1) To investigate the effect of steel reinforcement distribution and slab thickness on the impact force characteristics and impact behaviors of RCP;
(2) To evaluate the applicability of reinforced recycled aggregate concrete pavement under impact loads compared to the reinforced natural aggregate concrete pavement.

2. Experimental Methods

2.1. Materials

(1) Cement and Water

The cement used throughout this study was Portland cement (P.O 42.5) conforming to standard GB175-2020 and obtained from Conch Cement Group. The detailed properties of the cement are shown in Table 1. Clean and fresh water was used for casting and curing of the samples and RCP specimens.

Table 1. Properties of cement.

Loss on Ignition (%)	Initial Setting Time (min)	Final Setting Time (min)	Specific Surface Area (m^2/kg)	Compressive Strength (Mpa)		Flexural Strength (Mpa)	
				7 Days	28 Days	7 Days	28 Days
2.35	170	290	337	27.3	45.6	5.6	8.2

(2) Fine Aggregate

The fine aggregate used in the mixes was locally-sourced river sand with a maximum particle size of 4.75 mm and a fineness modulus of 2.60, which met the requirements for a Class II gradation medium sand.

(3) Coarse Aggregate

The coarse aggregate involved two types of natural coarse aggregate (NCA) and recycled coarse aggregate (RCA). The NCA was crushed calcareous limestone from the stone quarries which had continuous gradation with a particle size of 5~20 mm. The RAC was supplied by the local plant from the demolition of twenty- to thirty-year-old concrete pavements, the original strength grade of which was unknown. Prior to the experiment, the aggregates larger than 20 mm were screened, and impurities such as bricks and wood were removed.

Tests were conducted on the NCA and RCA according to the Code of GB/T 25177-2010 and GB/T 14685-2011; the physical properties of the coarse aggregates are shown in Table 2, and Figure 1 shows the sieve analysis of the coarse aggregates. Both NCA and RCA met the specification that ensured the appropriate properties of the fresh and hardened concrete.

Table 2. Physical properties of coarse aggregate.

Type of Coarse Aggregate	Apparent Density (g/m^3)	Clay Content (%)	Water Absorption (%)	Crushing Value Index (%)
NCA	2644	0.6	0.9	8.2
RCA	2567	1.2	3.4	14.5

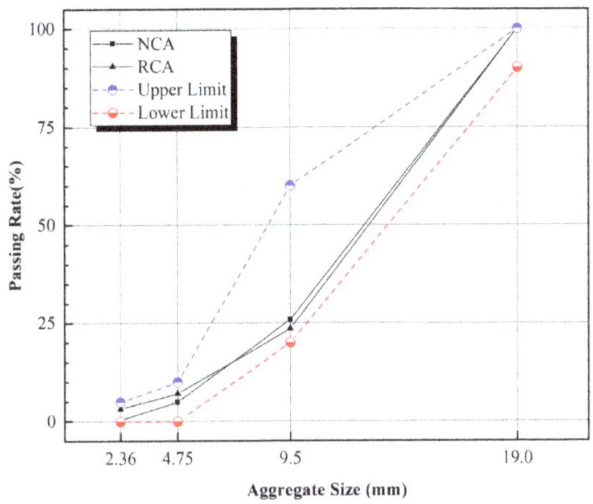

Figure 1. Grading of coarse aggregates.

(4) Steel reinforcement

The diameter of the longitudinal reinforcing bars in the RCP specimens was 8 mm, and the relevant properties were tested by using a tensile test machine. The bar response showed an obvious yield plateau, and the measured yield strength (f_{yk}), ultimate strength (f_{uk}) and elongation after fracture were 426 Mpa, 600 Mpa, and 18.2%, respectively.

2.2. Concrete Mix Design

Two types of concrete were designed in this experiment: Natural Aggregate Concrete (NAC) with natural fine and coarse aggregate, and Recycled Aggregate Concrete (RAC) using natural fine aggregate and RCA of 100% mass replacement. The mix proportion of NAC and RAC were designed to have a similar compressive strength of 45 Mpa, and the concrete mixture proportions for the NAC and RAC are listed in Table 3.

Table 3. Matrix proportions of different mixes.

Type of Concrete	Cement (kg/m^3)	Fine Aggregate (kg/m^3)	Coarse Aggregate(kg/m^3)	Water (kg/m^3)	Water-Cement Ratio	Sand Rate (%)
NAC	18.57	22.10	46.96	7.80	0.42	32
RAC	22.29	21.03	44.69	7.80	0.35	32

Three cube samples of 100 mm and 150 mm size were cast for compressive strength and tensile splitting strength tests, respectively, and three prism samples with 450 mm × 150 mm × 150 mm length were made for the purpose of testing the flexural strength. The mechanical properties of NAC and RAC were measured at 28 days under the standard curing condition according to GB/T 50081-2002, as shown in Table 4 based on average values for three tested samples.

Table 4. Mechanical properties of concrete.

Type of Concrete	Compressive Strength, 28 Days		Split Tensile Strength, 28 Days		Flexural Strength, 28 Days	
	Mean (Mpa)	Standard Deviation	Mean (Mpa)	Standard Deviation	Mean (Mpa)	Standard Deviation
NAC	46.80	1.694	3.55	0.141	5.64	0.303
RAC	48.28	1.236	3.21	0.172	5.44	0.376

2.3. Description of RCP Specimens

Five types of 1000 mm square RCP slab specimens with different longitudinal reinforcement spacing (100 and 150 mm), thickness (60, 70, and 80 mm), and RAC replacement ratio (0 and 100%) were designed for the experimental program. The RCP specimens were named by thickness (cm), type of concrete, the location of the impacting load, and additional information (1 meant the longitudinal reinforcement spacing was 100). For example, two types of concrete were used in the experimental program: NAC was named N, and RAC was named R. M represents that the RCP specimen was subjected to impact load at its mid-point. The details of all RCP specimens are summarized in Table 5, and their reinforcement layouts are shown in Figure 2.

Table 5. Summary of RCP specimen.

RCP Specimen	Thickness (mm)	Type of Concrete	Bar Spacing (mm)	Average Compressive Strength (Mpa)	Maturing Age
6MR	60	RAC	150	48.67	1 year, 11 days
7MR	70	RAC	150	48.73	1 year, 8 days
8MR	80	RAC	150	47.78	1 year, 9 days
7MN	70	NAC	150	47.21	1 year, 11 days
7M1R	70	RAC	100	49.43	1 year, 11 days

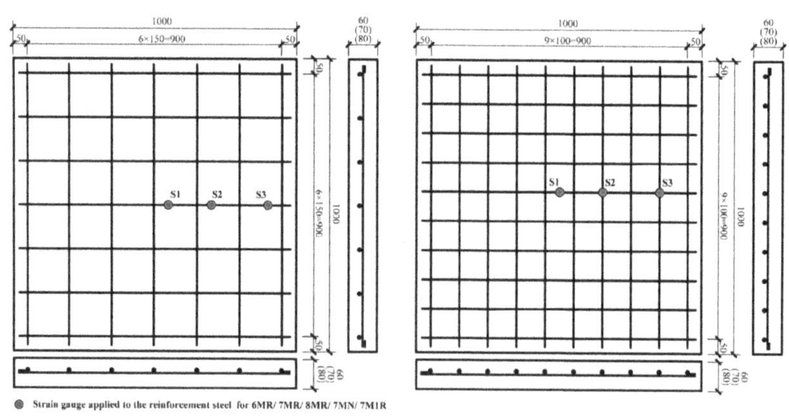

Figure 2. Size and reinforcement layout of RCP specimen (mm).

All RCP specimens adopted the single-layer reinforcement scheme with equal amounts of reinforcement in both planar directions, resulting in two layers of bars. The diameter of reinforcement bars was 8 mm, and the rebar spacing within the RCP specimens ranged from 100 mm to 150 mm. For slab thicknesses of 60 mm, 70 mm, and 80 mm, the thicknesses of concrete protective cover from the bottom surface were 25 mm, 30 mm, and 35 mm to ensure that the steel bars were located in the middle of the slab. According to MT/T 5004-2010, the single-layer steel bars should be located in the lower 1/3~1/2 thickness of the slab [2]. The transverse and longitudinal steel pieces were bundled by steel wires to create a continuous mesh, and those steel meshes were fixed on the cement cushion block during vibration to achieve an accurate positioning in the slab. Short bars of 20 mm length and 8mm diameter were welded at both ends of all reinforcement bars to enhance anchoring capacity and ensure sufficient reinforcement (Figure 3).

Figure 3. Casting of RCP specimen.

To ensure consistency and good quality of each RCP specimen, the concrete was mixed and cured in-house using a 60 L single horizontal-axis forced mixer in the Concrete Materials Laboratory. In addition to the RCP specimens, three 100 mm × 100 mm × 100 mm cube samples were cast from the same batch to characterize the compressive strength of the concrete material. The cube samples and RCP specimens were cured in natural environment conservation, and watered in the first 7 days to ensure the strength of concrete and prevent cracks. The average compressive strength of cube specimens was measured synchronously after the impact experiment, which is listed in Table 5.

2.4. Instrumentation

To fully document the dynamic response of the RCP specimens, various kinds of sensors were installed to monitor specimen displacements, accelerations, concrete strains, and reinforcing bar strains during the test.

Two laser-type displacement sensors were installed to capture the vertical displacement distribution of the RCP specimens; this kind of sensor is generally capable of capturing higher response frequencies than is the linear variable differential transformer, making it more suitable for impact testing applications. Two accelerometers were mounted on the top surface to capture the acceleration distribution of the RCP specimens, and one accelerometer was attached to the drop-weight to estimate the impact force. These accelerometers had the capacity of measuring accelerations within the range of ±5000 g and were used to measure accelerations along the vertical axis of motion. Moreover, three strain gauges with 5 mm gauge lengths were glued to the bottom surface of reinforcement bar prior to concrete pouring, and a total of five strain gauges with 50 mm gauge lengths were arranged on the top and bottom surfaces of the RCP specimens. The range of the magnitude and rate of strain were detected by eight strain gauges applied to the concrete and the reinforcement bar. The arrangement and designation of these sensors are given in Figure 4.

2.5. Test Program

The RCP specimens were placed on the top of a compacted sand-and-gravel layer in a steel strongbox. The net internal size of the steel box was 1020 mm × 1020 mm × 650 mm, which was slightly larger than the size of the RCP specimens; all four edges of the RCP specimens were free edges without constraints. The sand-and-gravel layer was composed of sand and gravel in a mixing ratio of 1:2, and with no clods, roots, or other sundries inside. The maximum size of gravel aggregate was limited to 26 mm, and particle size distributions of the sand-and-gravel layer conformed to the requirements of continuous gradation [41]. The sand-and-gravel layer was compacted to a degree of 0.97, in order to enhance its strength and provide a high-quality subbase.

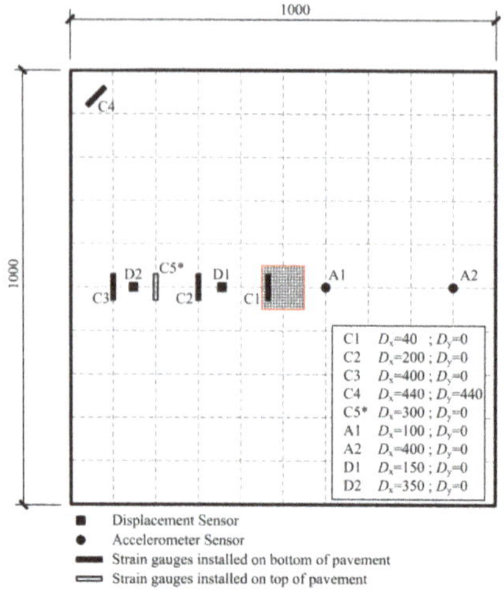

Figure 4. Casting of RCP specimen.

The drop-weight test frame consisted of two columns, the drop-weight, and two vertical guide rails; the rails were used to guide the drop-weight during the fall. The drop-weight in the experimental program was comprised of a cross-beam with a span of approximately 2.5 m and an impactor with a striking surface of a 20 cm diameter hemispherical nose. The total mass of the drop-weight was 200 kg, and the drop-height of the drop-weight was set at 1.0 m above the top surface. A rubber pad with dimensions of 100 mm × 100 mm × 10 mm was placed on the RCP specimen exactly in the contact zone, which simulated the impact cushioning effect of the landing gear. Prior to performing a test, the drop-weight was lifted up along the guide rail to the desired height and secured to an electric clamping style release mechanism. After debugging all related devices and instrumentation, the drop-weight was released in a free-fall condition to generate the impact loads, and all digital data were recorded synchronously. The schematic diagram of the setup and the test configuration is illustrated in Figure 5.

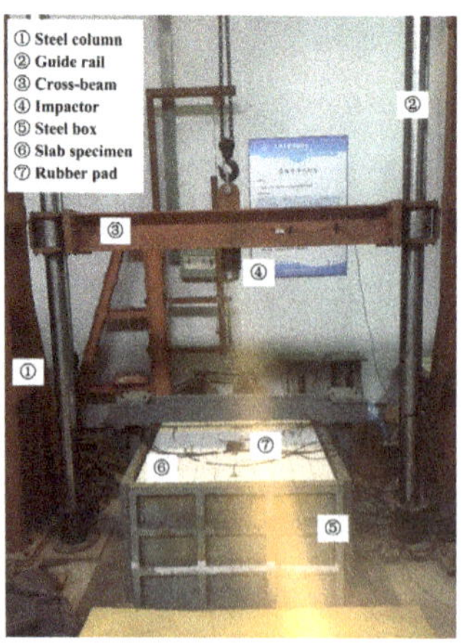

Figure 5. Experimental setup.

3. Experimental Results and Discussion

3.1. Impact Force Characteristics

According to the test program described above, the acceleration-time history of the drop-weight was obtained from the measurement data of the A3 accelerometer, as shown in Figure 6. The acceleration-time history was used to calculate the impact force-time history. The formula is $F(t) = ma(t)$, where m is the mass of the drop-weight that remained at 200 kg consistently in all tests. The impact force-time histories of different tests were similar in shape, showing a high magnitude peak followed by few small magnitude shocks, which were caused by the rebounding of the drop-weight after impacting the RCP specimens. Compared with other subsequent peaks, the magnitude of first impact was very high, and therefore, the impact response under first impact is the most central issue for this research [22,23].

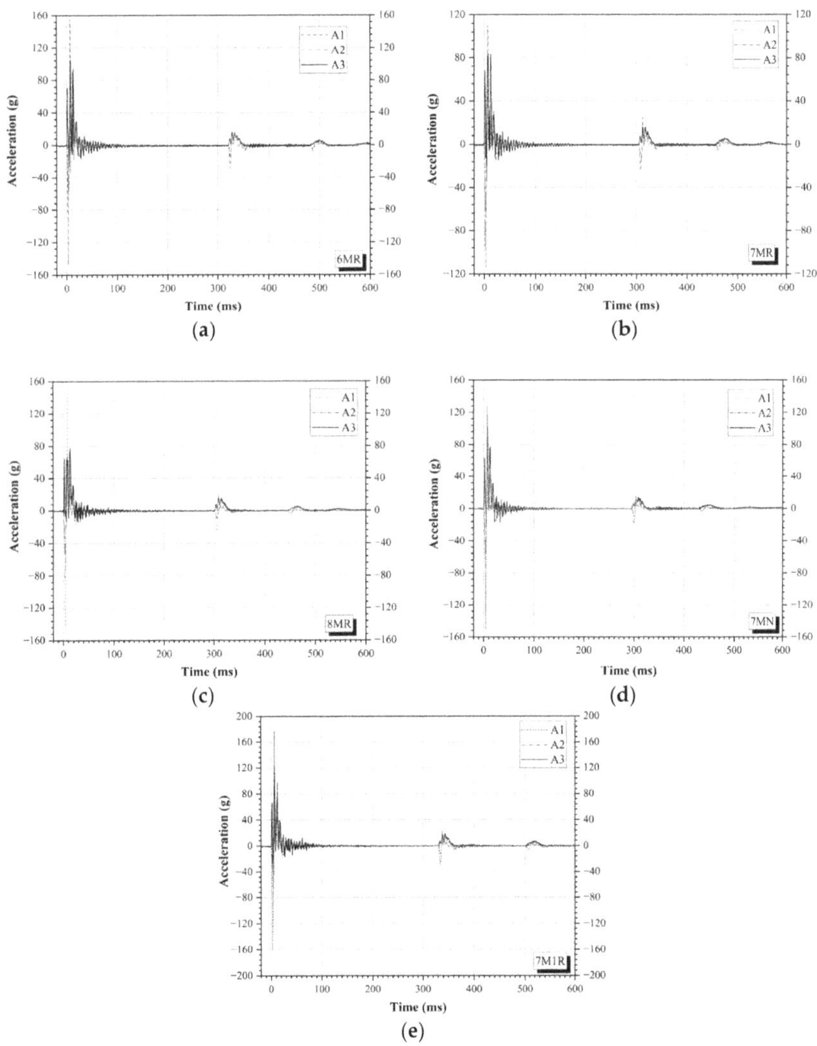

Figure 6. Acceleration-time histories of the (**a**) 6MR specimen, (**b**) 7MR specimen, (**c**) 8MR specimen, (**d**) 7MN specimen, and (**e**) 7M1R specimen.

To assess the inertial force during the impact test, a simple integration approach is proposed; that is, to use the recorded A1 and A2 acceleration data to estimate the inertial force of the RCP specimen [26]. The acceleration-time histories of the A1 and A2 accelerometer are shown in Figure 6. The accelerometer A2 was placed further away from the impact area than was the accelerometer A1, and it was thought that the phase diversity of the acceleration between A1 and A2 reflected the lag in response, owing to the force propagating from the point of impact to the edge [42]. The RCP specimen gained significant downward acceleration immediately after impact and upward inertial force was induced. The RCP specimen was divided into three tributary integration areas according to the positions of accelerometers, as shown in Figure 7. For Area 1, the acceleration was assumed to be uniformly distributed and its value was equal to the value of the A1 sensor. The acceleration was assumed to have a linear distribution for both Area 2 and Area 3. For Area 2, the values of the inner and outer boundaries of acceleration were the values of the A1 and A2 sensors, respectively. For Area 3, since the size of the RCP specimen was much larger than that of the impact area, it could be considered that the acceleration at the far edge of the specimen was sufficiently weak to be ignored. Accordingly, the values of the inner and outer boundaries of the acceleration were the value of the A1 sensor and zero, respectively. The inertial force of the RCP specimen could be calculated by summing up the inertial forces of the three tributary areas.

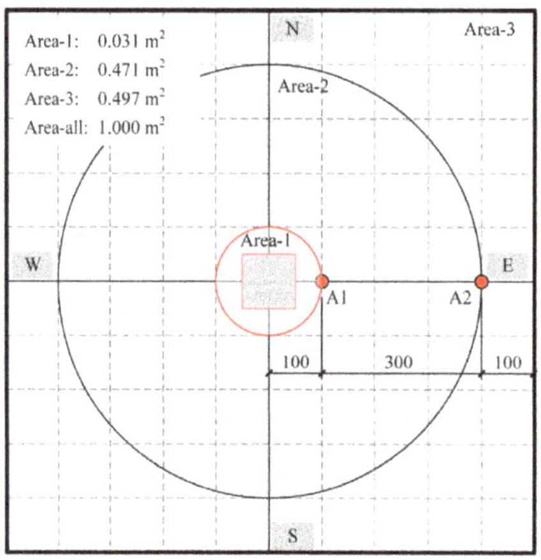

Figure 7. Integration areas of the inertial force.

For all the specimens, impact force and inertial force rose immediately and reached their peak values shortly after the impact began. The impact force suddenly dropped due to the deformation and cracking of the RCP specimen. As the impact force decreased, the inertia force decreased and dissipated. The peak impact force was detected after circa 7–8.5 ms following the first contact between the drop-weight and the RCP specimen, and the result is summarized in Table 6. There was a time lag of circa 1–3 ms between the peak impact force and the peak inertia force, which was due to the stress wave propagation travelling gradually from the impact area to the far edge [25,31,43]. This stress wave traveled at varying speeds within the speed of sound, depending on the mass, the density and the elastic modulus of the concrete type used [44]. Comparing the impact force with the inertia force, it is obvious that the peak amplitude of impact force is greater than that of the inertia force. The reason is that most of the impact force was converted into inertia force, while a portion of impact force was balanced by ground reaction force during impact.

Table 6. Impact force characteristics of RCP specimen.

RCP Specimen	$F_{im,p}$ (kN)	$F_{in,p}$ (kN)	I_p (kN.s)	V_r (m/s)	E_{im} (J)	E_{ab} (J)	E_{ab}/E_{im} (%)
6MR	206.1	85.9	1143.4	1.292	1960.0	1793.0	91.5
7MR	175.9	110.7	1148.5	1.315	1960.0	1787.0	91.2
8MR	150.9	116.1	1153.0	1.338	1960.0	1781.0	90.9
7MN	200.2	105.1	1151.1	1.328	1960.0	1783.5	91.0
7M1R	220.5	111.7	1198.5	1.565	1960.0	1715.0	87.5

$F_{im,p}$: peak impact force; $F_{in,p}$: peak inertial force; I_p: impulse; v_r: rebound velocity; E_{im}: impact energy; E_{ab}: absorbed energy.

All RCP specimens were impacted by a free fall of 200 kg drop-weight from a constant height of 1.0 m, so in the case of ignoring friction and air resistance, when the drop-weight impacted the RCP specimen, the instantaneous impact velocity v_{im} was about 4.4 m/s and the maximum impact energy E_{im} was 1.96 kJ. The reported impulse I_p is the time integration of impact force. The impulse–momentum theorem states that the impulse is equal to the change of momentum [23–25]. Thus, it is expected that the rebound velocity v_r of the drop-weight can be calculated by the formula $I_p = m \times v_{im} - m \times v_r$. Once the rebound velocity v_r is known, the residue kinetic energy of the drop-weight can be calculated. In calculating the energies for the impact test, the current study neglected the energy dissipated in the following mechanisms: the free vibration of the RCP specimen and steel strongbox, and the energy losses due to heat and noise [37,45]. Thus, the energy absorbed by the RCP specimen can be calculated via subtracting the residue kinetic energy of drop-weight from the impact energy [27]. The calculated rebound velocities v_r, absorbed energies E_{ab}, and its ratio over impact energies of different tests are listed in Table 6.

It can be seen from Table 6 that the characteristic value of impact force varies in accordance with the longitudinal reinforcement spacing, the concrete type, and the thickness. Comparing 7MR with 7MN, the peak impact force of 7MR is 13.8% smaller than that of 7MN. According to the contact theory proposed by Hertz [46], the force between two objects in contact is proportional to the relative elastic modulus. The 7MR specimen showed a lower impact force, which can be attributed to a lower modulus of elasticity of the RCA mix compared with the NAC mix. The peak inertial force of 7MN is very similar to that of 7MR, and the difference between the two is less than 5%. This is likely due to the fact that the steel reinforcement contributed more to the stiffness of the RCP specimen in this state, thus reducing the relative influence of concrete on the overall stiffness.

A tendency is observed that the peak impact force increased with the increase in reinforcement ratio. As can be seen from Figure 8, the peak impact force of 7M1R is 25.3% higher than that of 7MR, which is the maximum value among all RCP specimens. Therefore, increasing the reinforcement ratio could improve the stiffness of the RCP specimen and have a significant effect on the impact force [22,23]. In addition, it was observed that the peak inertial force slightly increased as the reinforcement ratio increased.

Compared with 7MR, the peak impact force of 6MR increased by 17.1%, while the peak impact force of 8MR decreased by 14.2%. According to the research results of Xiao [26], the thickness could increase the impact resistance and stiffness of the RCP specimens; therefore, an increase in the peak impact load should be also observed. The reason for this is that, although 6MR, 7MR and 8MR specimens had the same reinforcement layout scheme of D8@150, the reinforcement ratio decreased with the increase in thickness. When considering the peak impact load, the influence of the thicknesses was relatively lower compared with the effect of the reinforcement ratio, which has paramount relevance. In addition, it was observed that the impact force duration slightly decreased as the reinforcement ratio increased in the 8MR, 7MR, 6MR and 7M1R specimens. As can be seen from Figure 8, the peak inertia force variation rules of the RCP specimen were different. The peak inertia force of the 8MR specimen with the largest thickness was maximum, while the peak inertia force of the 6MR specimen with the lowest thickness was minimum.

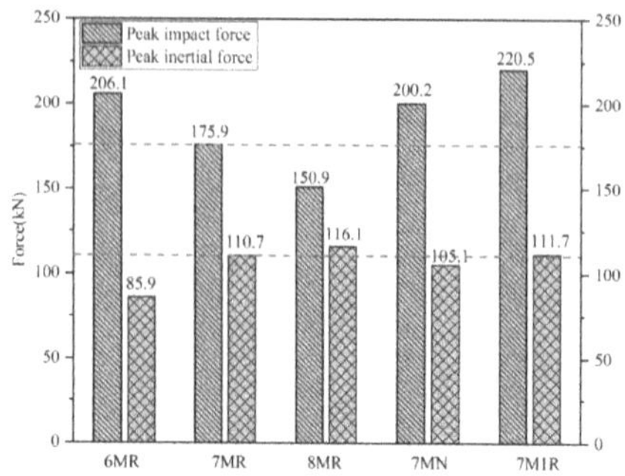

Figure 8. Peak impact load and peak inertial force.

Except for 7M1R, the energy dissipation ratio E_{ab}/E_{im} of all RCP specimens exceeds 90%, implying that the RCP specimens dissipate most impact energies through deformation and cracking. The energy dissipation ratio E_{ab}/E_{im} of 7MR specimen was similar to that of 7MN specimen. For 7M1R specimen, as shown in Figure 9, there was slight damage on the surface after impact, and approximately 87.4% of the impact energy was imparted to the specimen. For the less damaged specimen, more impact energy could be stored through the temporary elastic deformation of the specimen [27]. This stored energy would return to the drop-weight when the elastic deformation recovered, resulting in greater rebound speed v_r. On the contrary, severely damaged specimens had already entered their plastic stage and more impact energy was dissipated in the form of permanent deformation or crack damage. The 6MR specimen was severely damaged and the energy consumption ratio E_{ab}/E_{im} reached 91.5%. As seen in Figure 9, the damage characteristics and crack patterns after the impact also confirmed this phenomenon.

3.2. *Damage Characteristics and Crack Patterns*

Prior to the test, the surfaces of the PCP specimen were painted white and then meshed with spacing of 100 mm grids in order to observe damage characteristics and crack patterns. The "E/S/W/N" symbols were marked at top, bottom and side surfaces of the RCP specimen, and the RCP specimen was divided into four regions according to direction. The cracks that developed after each test were marked, and the crack widths were measured manually by HC-CK101 Concrete Crack Width Meter. For impacting at the mid-point, the sketched cracks profiles of the RCP specimen are shown in Figure 9.

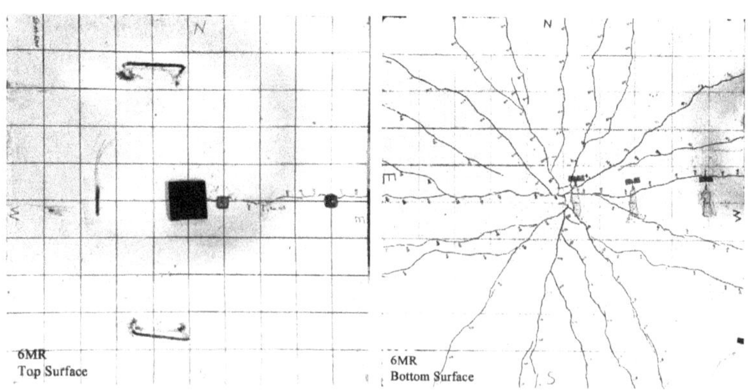

(a)

Figure 9. *Cont.*

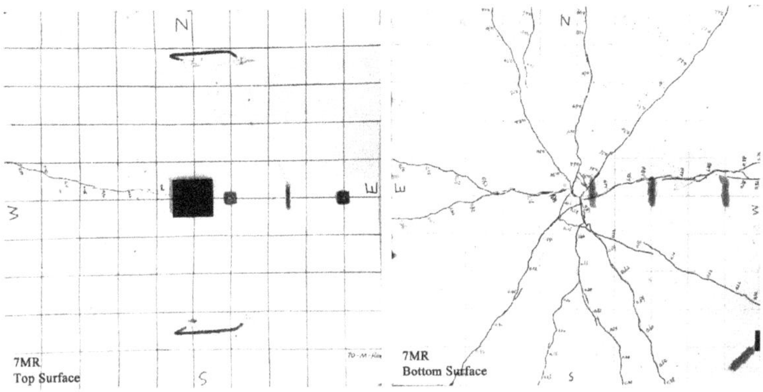

(**b**)

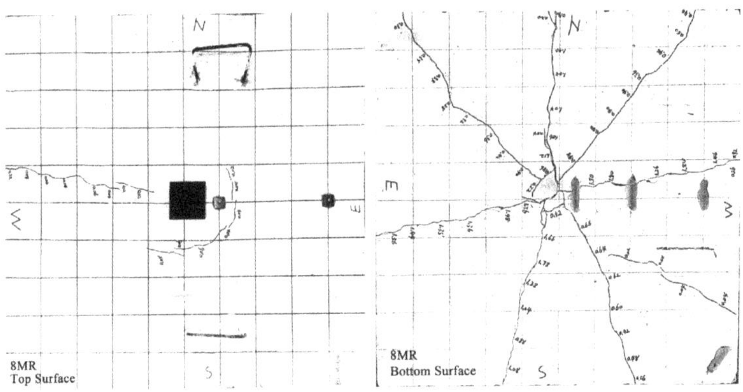

(**c**)

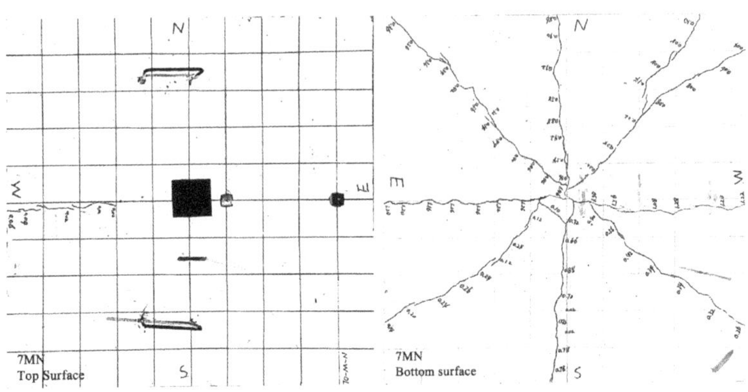

(**d**)

Figure 9. *Cont.*

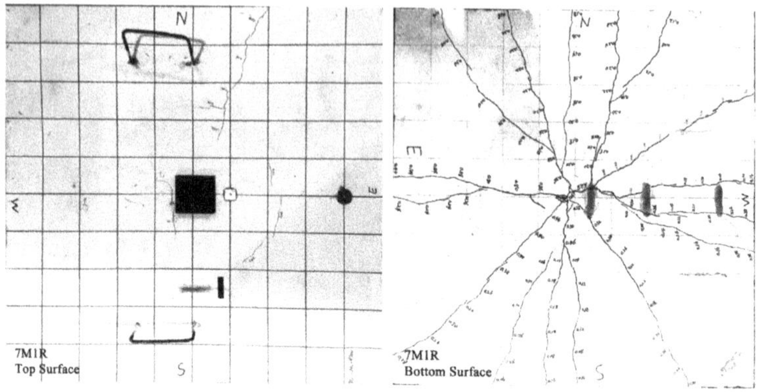

(e)

Figure 9. Sketched cracks profiles of the (**a**) 6MR specimen, (**b**) 7MR specimen, (**c**) 8MR specimen, (**d**) 7MN specimen, and (**e**) 7M1R specimen.

The type of damage and crack development mode on the bottom surface of all RCP specimens are similar. The crack patterns mainly appeared as the radial crack and diagonal crack, indicating that the deformation of the specimen was a global flexural deformation. The major radial crossing cracks and the failure took place simultaneously. The concrete scabbing was quite limited and mainly centralized in the region of 200 mm × 200 mm beneath the impacting point. The maximum residual width of the crack was also found in this region, which reached up to 1.46 mm~1.8 mm. For all RCP specimens, the widths of radial cracks were larger than those of diagonal cracks. This is because the radial crack developed prior to the diagonal crack, and the radial crack could dissipate more impact energy, thus reducing the crack width [27]. There were two different crack patterns on the top surface of the RCP specimen: one was the radial crack propagating from the bottom surface towards the top surface, and the other was the circumferential crack with the impact point as the center. No obvious penetration was observed on the surface of the RCP specimen that was found in Refences [47,48]. The final crack properties of all tested specimens are presented in Table 7.

Table 7. Final crack properties of all RCP specimens.

RCP Specimen	Bottom Surface			Top Surface		
	Crack Pattern	Num of Crack	Maximum Crack Widths	Crack Pattern	Num of Cack	Maximum Crack Widths
6MR	radial crack, diagonal crack	21	1.60	radial crack	1	0.08
7MR	radial crack, diagonal crack	9	1.80	radial crack	1	0.12
8MR	radial crack, diagonal crack	7	1.80	circumferential crack radial crack	2	0.08
7MN	radial crack, diagonal crack	8	1.46	radial crack	2	0.06
7M1R	radial crack, diagonal crack	14	1.50	circumferential crack radial crack	2	0.04

The final damage status of the 7MR specimen is similar to that of the 7MN specimen. In the E-W direction, radial cracks were fully developed, and their widths were in the range of 1.2–1.4 mm for both 7MR and 7MN. While, in the N-S direction, the cracks generated in 7MR are slightly more than those in 7MN, this could be attributed to the character of RCA, whose adhesive mortar and cracks caused by procession have an adverse effect on the behavior of the concrete matrix. Furthermore, due to the high brittleness of RAC, the radial crack propagation was normally unstable [49]. As shown in Figure 10, the radial cracks extending from the bottom to the top run along the W-E direction toward the impact point, simultaneously with the crack widths being gradually reduced. The radial crack widths of 7MR on the top surface were larger than those of 7MN. Furthermore, the radial crack of 7MR extended to the impact point, while the radial crack of 7MN extended only a quarter of the slab span. With the same thickness, reinforcement ratio and concrete grade, RAC has little influence on damage characteristics and crack patterns of RCP specimens; however, the crack resistance of 7MR is slightly lower than that of 7MN.

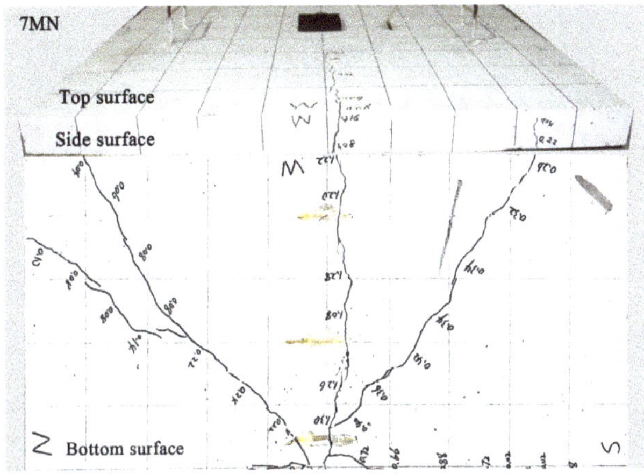

Figure 10. Radial cracks extending from the bottom to the top.

The number of cracks on the bottom surface decreased with the increase in the thickness of the slab. With the reduction of the thickness from 70 mm to 60 mm, multiple tightly spaced hairline cracks formed on the bottom surface, while as the thickness increased from 70 mm to 80 mm, the development of cracks was strongly limited. The crack width widened as the number of cracks decreased. These results suggest that there is an association between crack resistance and slab thickness. The crack patterns on the top surface varied with the change of the thickness of the slab. The radial cracks extending from the bottom to the top were found in the 6MR, 7MR, and 8MR specimens, and the circumferential cracks around the impact area were detected only in the 8MR specimen. These circumferential cracks with a hairline width less than 0.06 mm did not close and developed in the range of a half-circle. The circumferential cracks indicated that localized damage in the form of limited concrete penetration on the impact surface had occurred in the 8MR specimen. The change in crack patterns was due to the stiffness of the specimen increasing as a result of the increase in the slab's thickness, and partial impact energy needing to be dissipated through local damage deformation during the impact [48,50]. Except for the cracks mentioned above, the remaining area on top surface of 8MR specimen was nearly undamaged.

Both circumferential cracks similar to those in 8MR and radial cracks similar to those in 7MR were found on the top surface of 7M1R specimen. In comparison with 8MR, the circumferential cracks of 7M1R had further distributed distance, thinner width, and smaller range. The development of the radial cracks on the top surface of 7M1R was limited when compared to that of 7MR. Based on the observed damage and crack development in tested specimens, it was found that the crack pattern was more affected by the thickness than by the reinforcement ratio. More steel reinforcement would induce a localized failure of concrete [23].

3.3. Displacement Response

The displacement-time histories of D1 and D2 are shown in Figure 11, and it is found that the displacement-time history shapes of all RCP specimens are similar in terms of magnitude, time response, and residual displacement. With each impact event performed, the RCP specimen exhibited progressively increasing peak displacements, and then decreased to a stable residual displacement, followed by few small displacements due to rebounding of the drop-weight after impacting the RCP specimen. It should be recalled that the magnitude of first impact was very high compared with other subsequent peaks; therefore, the peak displacement and residual displacement due to the first impact were recorded in Table 8. The final cumulated residual displacement would affect the performance of the RCP specimen; it was also recorded in Table 8. It can be seen that the displacement at D1 point was larger than that at D2 point for all RCP specimens from Figure 11. This is because when the drop-weight impacted the top surface, due to the limited impact area, sufficient impulse should be provided in this area to prevent the drop-weight from falling until it stopped. Therefore, compared with other areas, the stress around the impact area was greater, the damage was more serious, and the deformation was more obvious.

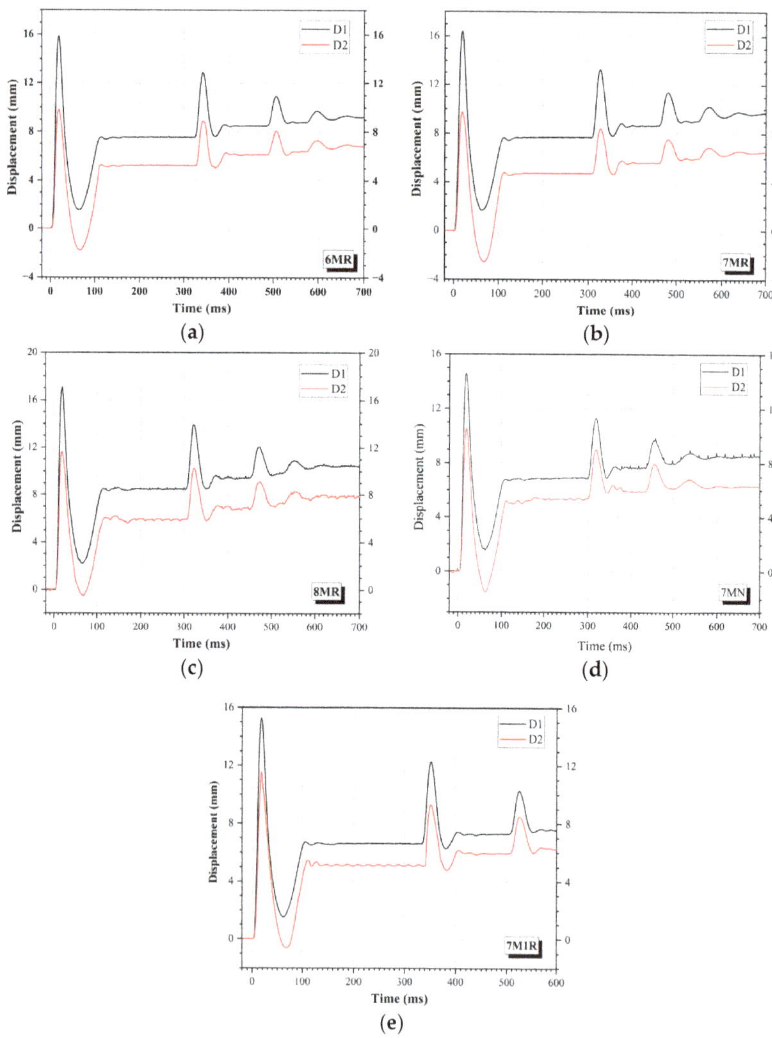

Figure 11. Displacement -time histories of the (**a**) 6MR specimen, (**b**) 7MR specimen, (**c**) 8MR specimen, (**d**) 7MN specimen, and (**e**) 7M1R specimen.

Table 8. Displacement response of all tested specimens.

RCP Specimen	D1			D2		
	ω_{p1}	ω_{r1}	ω_{fr1}	ω_{p2}	ω_{r2}	ω_{fr2}
6MR	15.87	7.46	9.24	9.80	5.04	6.83
7MR	16.40	7.54	9.67	9.75	4.67	5.83
8MR	17.10	8.47	10.42	11.61	5.91	7.89
7MN	14.60	6.84	8.50	10.55	5.27	6.23
7M1R	15.25	6.63	8.16	11.53	5.34	6.66

ω_{p1} and ω_{p2}: the peak displacement at D1 point and D2 point, respectively; ω_{r1} and ω_{r2}: the residual displacement under first impact at D1 point and D2 point, respectively; ω_{fr1} and ω_{fr2}: the final cumulated residual displacement at D1 point and D2 point, respectively.

As can be seen from Table 8, the peak and residual displacements at D1 point of 7MN specimen were lower than that of 7MR specimen, reduced by 11.0% and 5.3%, respectively. According to the literature [51,52], the elasticity modulus of concrete decreased with the increase in RCA replacement ratio. In addition, the micro cracks in adhesive mortar of RCA had a detrimental effect on crack development, which would reduce the stiffness of the specimen. With increased distance from the impact area, both the peak displacement and the residual displacement showed an opposite trend to that of before. The peak and residual displacements at D2 point of 7MN specimen were higher than that of 7MR specimen, increased by 8.2% and 12.9%, respectively. This meant that the difference between D1 and D2 was decreasing, indicating that the deformation on the front surface of 7MN specimen became gentle, and showed more flexural response.

The reinforcement ratio plays an important role in peak deflection and residual displacement [47]. As the reinforcement ratio was increased from 0.48% to 0.72%, the peak and residual displacements at D1 point decreased by 7.0% and 12.2%, respectively. The reason for such behavior may be attributed to the fact that more steel bars could effectively arrest the propagation of cracks inside the concrete, thus improving the stiffness of RCP specimens. Compared with the 7MR specimen, the 7M1R specimen exhibited smaller displacement amplitudes under same impacts and was expected to be able to undergo larger displacement amplitudes before failure. The variation trend of the displacement at D2 point of 7M1R specimen was similar to that of 7MN specimen, and the peak value and residual displacement at D2 point of 7M1R specimen are 18.3% and 12.1% higher than that of 7MR specimen, respectively.

At D1 point, compared with the peak and residual displacements of the 7MR specimen, those of the 6MR specimen had undergone approximately 3.2% and 1.11% decrease, respectively, while those of the 8MR specimen had undergone approximately 4.3% and 12.3% increase, respectively. The peak and residual displacements of 6MR specimen and 7MR specimen at D2 point were not significantly different. In this case, it is thought that before the overall deformation occurred, the impact energy of 6MR specimen would have been dissipated through the development of the dense radial cracks on the bottom surface. The 8MR specimen always maintained a large displacement value at D1 and D2, indicating that when the drop-weight impacted against the 8MR specimen, almost all the impact energy was dissipated through global deformations. As described in Section 3.2, the 8MR specimen had the fewest number of radial cracks on the bottom surface among all RCP specimens. At the same time, as the velocity of drop-weight progressively slowed down with the increase of displacement, the circumferential cracks were formed on the top surface near the impact area.

As can be seen from Figure 11, under first impact, the RCP specimen reached the peak downward displacement and then rebounded upward. The amplitude of D1 and D2 rebound displacement changed differently between specimens. For 7MR, 6MR, and 7MN, the peak upward displacement at D2 point were −2.53 mm, −1.79 mm, and −1.49 mm respectively (downward is positive), while the peak upward displacement at D1 point remained positive. The specimen showed a trend of reverse bending deformation, and the stress wave bounced back from the base to the surface to form tensile stress, which could further explain the radial cracks which appeared on the top surface of these specimens. For 8MR and 7M1R specimens, the upward displacements of D2 point were relatively small, and the radial crack development was limited due to the higher stiffness.

The 8MR specimen with the minimum reinforcement ratio had the maximum final cumulative residual displacement, while the 7M1R specimen with the maximum reinforcement ratio had the minimum final cumulative residual displacement. The final cumulative residual displacement of the RCP specimen was found to correlate with the reinforcement ratio more than with other factors.

Research addressing the displacement shapes of the RCP specimens could provide more information regarding the impact response, which was difficult to directly observe in the displacement analysis at D1 and D2 points. Therefore, the displacement shapes of the RCP specimens were addressed in this paper, which provides a quantitative index for comparing the global impact responses of the RCP specimens. Accelerometers A1 and A2 were arranged along the same axis as displacement sensors D1 and D2, which were 50 mm away from the left and right sides of D1 and D2, respectively, as shown in Figure 4. As described in Section 3.1, the acceleration-time history a(t) at A1 point and A2 point of the RCP specimens were recorded by A1 and A2 accelerometers, respectively. The corresponding velocity v(t) and displacement d(t) responses can be calculated by numerical integration of the acceleration time histories using the Newmark Beta method [22,53]:

$$v_{i+1}(t+\Delta t) = v_i(t) + [(1-\alpha)a_i(t) + \alpha a_{i+1}(t+\Delta t)]\Delta t$$

$$d_{i+1}(t+\Delta t) = d_i(t) + v_i(t)\Delta t + [(1/2-\beta)a_i(t) + \beta a_{i+1}(t+\Delta t)]\Delta t^2$$

The acceleration was assumed to vary linearly between two instants of time in this study; α and β were chosen as 1/2 and 1/6, respectively [22]. Before impact, the initial velocity and initial displacement of the surface were considered to be zero. By assuming symmetric displacement response of the RCP specimen, the deflected shape along the midline of the top surface was plotted by linking the measured displacement data at D1 and D2 points and the calculated displacement data at A1 and A2 points. Uniform displacement with the value calculated by A1 was assumed in the impact area. The deflected shape of all the specimens is plotted at an interval of 2.0 ms and shown in Figure 12.

The value of deformation of the 7MR specimen was small in the initial 4 ms, and then increased rapidly. A global deformation on the top surface could be observed during the impact process, showing elastic flexural behavior. The deflection of the impacted area increased more rapidly than did the deflection of unloaded area, as shown in Figure 12. After reaching its peak displacement at about 14 ms, the impacted area began to rebound, while the unloaded area continued its downward movement for another few millimeters, and then rebounded at 18.5 ms. In the end, the displacement shape of the 7MR specimen flattened out again. In this case, the previously discussion of development of radial cracks observed on the bottom surface are believed to be attributable to the flexural displacements developed in the 7MR specimen. The similar behaviors were also observed in the displacement shapes of other specimens.

The punching shear behavior was observed in the 8MR specimen, indicating that the deflection of the impacted area increased much more rapidly than did the deflection of the unloaded area during 10ms to 16ms. Under the impact events, a slight development of localized displacements was observed to occur on one side of the impact region; however, no significant punching region was observed, and few instances of mass penetration had occurred. By comparing all displacement shapes shown in Figure 12, it can be seen that the displacement shapes of all RCP specimens were uniformly distributed, indicating that the failure of the specimen was mainly caused by the flexural deformation.

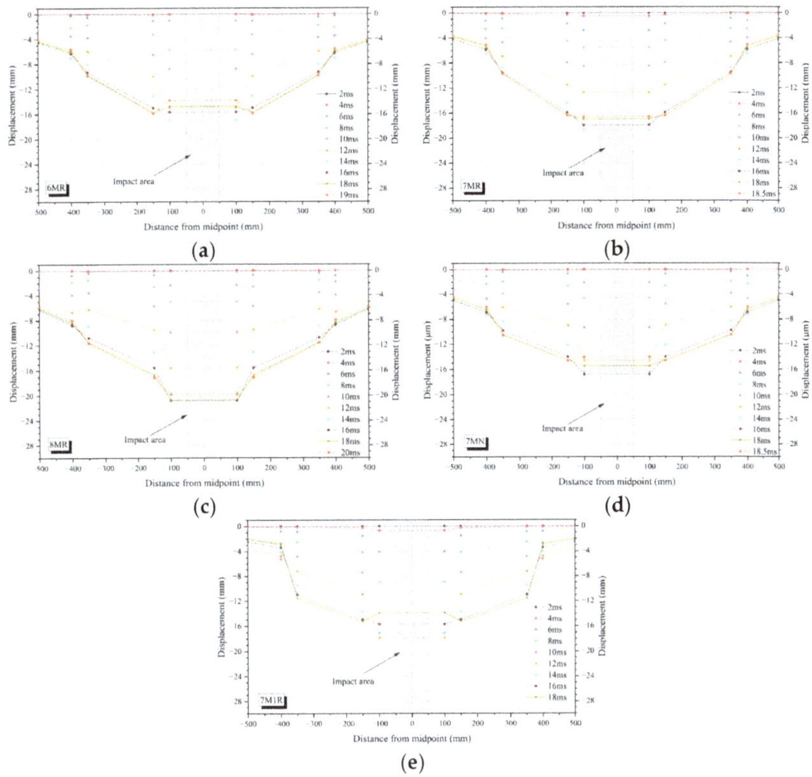

Figure 12. Deflected shapes of the (**a**) 6MR specimen, (**b**) 7MR specimen, (**c**) 8MR specimen, (**d**) 7MN specimen, and (e) 7M1R specimen.

3.4. Strain Due to Impact Load

The material strain was detected by five strain gauges applied to the concrete and by three strain gauges applied directly to the reinforcement steel. Figure 13a shows the strain evolution of the 7MR specimen, used as the reference specimen in comparison with all other specimens. Figure 13b shows a zoomed detail of the graphs under first impact. After the impact, a compressive strain of the concrete on the top can be seen, while a tensile strain of the concrete on the bottom can be observed. The C1 strain gauge was placed in central impact region and was disrupted about 2.0 ms after the first contact of the drop-weight. The failure of the C2 strain gauge could be determined from the horizontal plateau of strain-time history in Figure 13a. The C4 strain on the bottom was very small, indicating that, for the specimen with four free edges, the strain in the corner area of the RCP specimen could be ignored during the impact process, something which could be confirmed by the sketched cracks profiles in Figure 9. The compression strain of C5 on the top lasted for about 30 ms. At 14 ms, the maximum compressive strain of C5 reached about −2248 μ. Corresponding to the compression strain of the concrete mentioned above, the peak value of the tensile strain of S1 was measured 10,292 μ. Regarding the strain of the reinforcement steel, the tensile strain decreased with the increase in distance from the impact area.

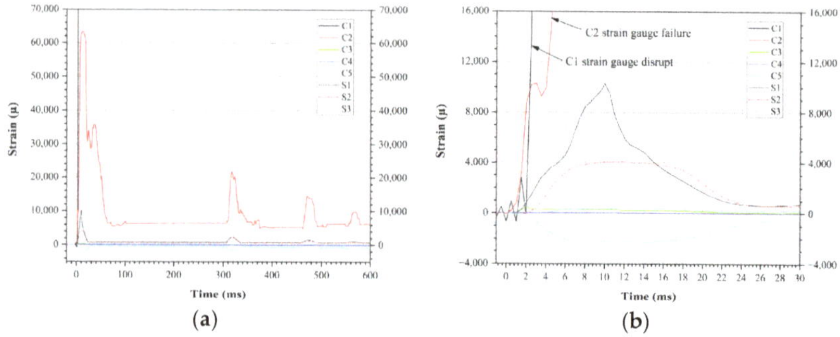

Figure 13. Concrete strain and steel strain of 7MR specimen (**a**) overall, and (**b**) in detail.

In this paper the steel strain of different RCP specimens was compared by means of the strain of the S1, as shown in Figure 14. For 6MR, 7MR and 8MR specimens, the strain values of the S1 strain showed a similar behavior, a rapidly increasing tensile strain followed by a very sharp drop. It can be observed that, with the increase in thickness, the peak S1 strain decreased, and the duration of the tensile strain shortened. For 7MN, the strain-time history showed a slower and smoother strain evolution than did the other specimens. In terms of peak strain values, the difference between the 7MR and the 7MN was 41.5%, while the difference between the 8MR and the 7MN was fairly small. It can be seen from Figure 14 that the steel strain of the 7M1R specimen was a compressive strain in the first 2.5 ms after the impact. This effect is indicative of the local material behaviors due to the impact, which is also described in Refences [32]. After a short duration of 4.0 ms, the S1 strain changes from compressive strain into tensile strain, indicating the transition from local deformation to global flexural behaviors. The more longitudinal reinforcement, the greater the decrease of tensile strain and the smoother the curve shape.

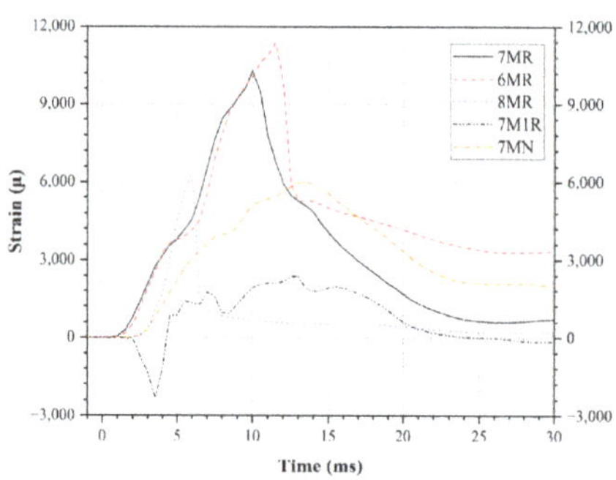

Figure 14. S1 steel strain under first impact.

4. Conclusions

The experimental investigation of five RCP specimens under impact load is presented in this paper. The acceleration, the displacement, and the strain time histories were recorded under constant impact energy in order to determine dynamic response of RCP under impact loading. The following conclusions can be drawn from the experimental study that was conducted:

(1) The peak impact force increased with the increase in reinforcement ratio. The impact force reached its peak value immediately after the impact, but the displacement, concrete strain and steel strain reached their peak value a few microseconds later. Therefore, the peak impact force cannot be directly considered the true impact resistance capacity.

(2) The increase in slab thickness resulted in an increase in the peak inertia force, but it decreased the peak impact force. Moreover, the energy consumption ratio reached 91.5% in 6MR specimen, which been severely damaged.
(3) All RCP specimens had similar crack patterns on the bottom surface, and the number of cracks decreased with the increase in the slab thickness. The reinforcement arrangement could affect the crack pattern; circumferential cracks on the top surface appeared in the 7M1R slab with 100 mm reinforced spacing, and similar cracks were not found in the 7MR slab with 150 mm reinforced spacing.
(4) The reinforcement ratio played an important role in peak deflection and residual displacement. As the reinforcement ratio increased from 0.48% to 0.72%, the peak and residual displacements at D1 point decreased by 7.0% and 12.2%, respectively. The global flexural response could be observed in the RCP specimens. Microscopic punching shear failure modes were observed only in the 8MR and 7M1R specimens.
(5) The 7MN specimen showed lower peak and residual displacement and higher peak impact force compared to the 7MR specimen, but no significant difference was observed between damage characteristics and crack patterns in the 7MR and 7MN specimens.
(6) The influence of using RAC in RCP was relatively small, even at 100% RCA replacement ratio, and the impact of using RCA was diminished for RCP made with 100 mm longitudinal reinforcement spacing.

However, due to the limited investigation conducted here, further research is being recommended to increase the database of test results for the RCP. This study was designed so that the RCP impact occurred at the center of the slab; however, in the real world, the impact can take place at other locations as well, and the response of the slabs under such conditions may be significantly different.

Author Contributions: J.Y. and J.W. collected scientific articles to ascertain the state of the art. J.Y. performed the analyses, analyzed the results, and contributed to the writing of the paper. J.W. organized the work and contributed to the analyses of the results. T.S. and D.L. revised the paper for overall consistency and scope. All authors have read and agreed to the published version of the manuscript.

Funding: This research was funded by the National Key R&D Plan for 13th Five-year Plan of China (2018YFD1101001).

Institutional Review Board Statement: Not applicable.

Informed Consent Statement: Not applicable.

Data Availability Statement: The data presented in this study are available on request from the corresponding author. The data are not publicly available due to privacy reasons.

Acknowledgments: This work was supported by the sub-project of the "13th Five-Year Plan" National Key R & D Plan in China (2018YFD1101001-2), 2018–2022.

Conflicts of Interest: The authors declare no conflict of interest.

References

1. Modarres, A.; Shabani, H. Investigating the effect of aircraft impact loading on the longitudinal top-down crack propagation parameters in asphalt runway pavement using fracture mechanics. *Eng. Fract. Mech.* **2015**, *150*, 28–46. [CrossRef]
2. *MH/T5004*; Specifications for Airport Cement Concrete Pavement Design. CAAC (Civil Aviation Administration of China): Beijing, China, 2010.
3. *AC 150/5320-6E*; Airport Pavement Design and Evaluation. FAA (Federal Aviation Administrator): Washington, DC, USA, 2009.
4. Choi, P.; Kim, D.; Lee, B.; Won, M. Evaluation of structural responses of continuously reinforced concrete pavement (CRCP) using falling weight deflectometer. *Can. J. Civ. Eng.* **2016**, *43*, 28–39. [CrossRef]
5. Salles, L.; Khazanovich, L.; Balbo, J. Structural analysis of transverse cracks in short continuously reinforced concrete pavements. *Int. J. Pavement Eng.* **2020**, *21*, 1853–1863. [CrossRef]
6. Hertz, H. Uber das Gleichgewicht Schwinnender elastischer platten. *Wiedermann's Ann. Cler Phys. Und Chem.* **1884**, *22*, 449–455.
7. Westergaard, H.M. Stresses in concrete pavements computed by theoretical analysis. *Public Roads* **1921**, *8*, 54–60.

8. Khazanovich, L.; Selezneva, O.; Yu, H.T.; Darter, M.I. Development of Rapid Solutions for Prediction of Critical Continuously Reinforced Concrete Pavement Stresses. *Transp. Res. Rec. J. Transp. Res. Board* **2001**, *1778*, 64–72. [CrossRef]
9. William, G.W.; Shoukry, S.N. 3D Finite Element Analysis of Temperature-Induced Stresses in Dowel Jointed Concrete Pavements. *Int. J. Geomech.* **2001**, *3*, 291–308. [CrossRef]
10. Davids, W.; Wang, Z.; Turkiyyah, G.; Mahoney, J.P.; Bush, D. Three-Dimensional finite element analysis of jointed plain concrete pavement with everfe2.2. *Transp. Res. Rec. J. Transp. Res. Board* **2003**, *1853*, 92–99. [CrossRef]
11. Yang, G.; Bradford, M.A. A refined modelling for thermal-induced upheaval buckling of continuously reinforced concrete pavements. *Eng. Struct.* **2017**, *150*, 256–270. [CrossRef]
12. Rahal, K. Mechanical properties of concrete with recycled coarse aggregate. *Build. Environ.* **2007**, *42*, 407–415. [CrossRef]
13. Li, J.; Xiao, H.; Zhou, Y. Influence of coating RA surface with pozzolanic powder on properties of RAC. *Build. Environ.* **2009**, *23*, 1287–1291. [CrossRef]
14. Chen, Y.; Chen, Z.; Xu, J.; Lui, E.M.; Wu, B. Performance evaluation of recycled aggregate concrete under multiaxial compression. *Constr. Build. Mater.* **2019**, *229*, 116935. [CrossRef]
15. Mas, B.; Cladera, A.; Olmo, T.; Pitarch, F. Influence of the amount of mixed RAs on the properties of concrete for non-structural use. *Constr. Build. Mater.* **2017**, *27*, 612–622. [CrossRef]
16. Francesconi, L.; Pani, L.; Stochino, F. Punching shear strength of reinforced recycled concrete slabs. *Constr. Build. Mater.* **2016**, *127*, 248–263. [CrossRef]
17. Rahal, K.; Alrefaei, Y. Shear strength of longitudinally reinforced recycled aggregate concrete beams. *Eng. Struct.* **2017**, *145*, 273–282. [CrossRef]
18. Arezoumandi, M.; Drury, J.; Volz, J.S.; Khayat, K.H. Effect of recycled concrete aggregate replacement level on shear strength of reinforced concrete beams. *Mater. J.* **2015**, *112*, 559–567. [CrossRef]
19. Schubert, S.; Hoffmann, C.; Leemann, A.; Moser, K.; Motavalli, M. Recycled aggregate concrete: Experimental shear resistance of slabs without shear reinforcement. *Eng. Struct.* **2012**, *41*, 490–497. [CrossRef]
20. Brandes, M.R.; Kurama, Y.C. Behavior of shear-critical prestressed concrete beams with recycled concrete aggregates under ultimate loads. *Eng. Struct.* **2018**, *165*, 237–246. [CrossRef]
21. Alnahhal, W.; Aljidda, O. Flexural behavior of basalt fiber reinforced concrete beams with recycled concrete coarse aggregates. *Constr. Build. Mater.* **2018**, *169*, 165–178. [CrossRef]
22. Kantar, E.; Yuen, T.; Kobya, V.; Kuang, J. Impact dynamics and energy dissipation capacity of fibre-reinforced self-compacting concrete plates. *Constr. Build. Mater.* **2017**, *138*, 383–397. [CrossRef]
23. Zineddin, M.; Krauthammer, T. Dynamic response and behavior of reinforced concrete slabs under impact loading. *Int. J. Impact Eng.* **2007**, *34*, 1517–1534. [CrossRef]
24. Zineddin, M. Behavior of Structural Concrete Slabs under Localized Impact. Ph.D. Thesis, The Pennsylvania State University, State College, PA, USA, 2002.
25. Othman, H.; Marzouk, H. An experimental investigation on the effect of steel reinforcement on impact response of reinforced concrete plates. *Int. J. Impact Eng.* **2016**, *88*, 12–21. [CrossRef]
26. Xiao, Y.; Li, B.; Fujikake, K. Experimental Study of Reinforced Concrete Slabs under Different Loading Rates. *ACI Struct. J.* **2016**, *113*, 157–168. [CrossRef]
27. Xiao, Y.; Li, B.; Fujikake, K. Behavior of Reinforced Concrete Slabs under Low-Velocity Impact. *ACI Struct. J.* **2017**, *114*, 643–658. [CrossRef]
28. Özgür Anil Kantar, E.; Yilmaz, M.C. Low velocity impact behavior of RC slabs with different support types. *Constr. Build. Mater.* **2015**, *93*, 1078–1088. [CrossRef]
29. Chiaia, B.; Kumpyak, O.; Placidi, L.; Maksimov, V. Experimental analysis and modeling of two-way reinforced concrete slabs over different kinds of yielding supports under short-term dynamic loading. *Eng. Struct.* **2015**, *96*, 88–99. [CrossRef]
30. Husem, M.; Cosgun, S.I. Behavior of reinforced concrete plates under impact loading: Different support conditions and sizes. *Comput. Concrete.* **2016**, *18*, 389–404. [CrossRef]
31. Hrynyk, T.; Vecchio, F. Behavior of steel fiber-reinforced concrete slabs under impact load. *ACI Struct. J.* **2014**, *111*, 1213–1223. [CrossRef]
32. Beckmann, B.; Hummeltenberg, A.; Weber, T.; Curbach, M. Strain Behaviour of Concrete Slabs under Impact Load. *Struct. Eng. Int.* **2018**, *22*, 562–568. [CrossRef]
33. Al-Rousan, R.Z.; Alhassan, M.A.; Al-Salman, H. Impact resistance of polypropylene fiber reinforced concrete two-way slabs. *Struct. Eng. Mech.* **2017**, *62*, 373–380. [CrossRef]
34. Ong, K.C.G.; Basheerkhan, M.; Paramasivam, P. Resistance of fibre concrete slabs to low velocity projectile impact. *Cem. Concr. Compos.* **1999**, *21*, 391–401. [CrossRef]
35. Wu, J. Development of Advanced Pavement Materials System for Blast Load. Ph.D. Thesis, National University of Singapore, Singapore, 2012.
36. Cai, J.; Wong, L.N.Y.; Hua, W.Y. Dynamic response of airport concrete pavement to impact loading. *Adv. Mater. Res.* **2012**, *594*, 1395–1401. [CrossRef]
37. Guo, J. Research on Dynamic Response of Recycled Aggregate Concrete Beam under Impact Action. Ph.D. Thesis, South China University of Technology, Guangzhou, China, 2016.

38. Vali, N.M.; Babu, B.M. Impact strength assessment of recycle aggregate concrete slabs. *Int. J. Sci. Res. Dev.* **2016**, *9*, 225–228.
39. Ramana NVSubbareddy, Y.V.; Sashidhar, C. Impact resistance of steel fiber reinforced recycle aggregate concrete beams. *Int. J. Res. Appl. Sci. Eng. Technol.* **2017**, *5*, 1919–1927.
40. Zhou, C.; He, W.C.; Kong, X.Q.; Zheng, Z.Y.; Wang, X.Z. Study on impact behavior of steel fiber recycled concrete. *Concrete* **2021**, *379*, 36–41.
41. Weng, X.H. *Design of Airport Pavement*; Communications Press Co. Ltd.: Beijing, China, 2017.
42. Isaac, P.; Darby, A.; Ibell, T.; Evernden, M. Experimental investigation into the force propagation velocity due to hard impacts on reinforced concrete members. *Int. J. Impact Eng.* **2017**, *100*, 131–138. [CrossRef]
43. Soleimani, S.; Banthia, N. A novel drop weight impact setup for testing reinforced concrete beams. *Exp. Tech.* **2014**, *38*, 72–79. [CrossRef]
44. Christoforou, A.; Yigit, A. Effect of flexibility on low velocity impact response. *J. Sound Vibration* **1998**, *217*, 563–578. [CrossRef]
45. Wang, Y.; Qian, X.; Liew, J.; Zhang, M. Impact of cement composite filled steel tubes: An experimental, numerical and theoretical treatise. *Thin-Walled Struct.* **2015**, *87*, 76–88. [CrossRef]
46. Johnson, K.L. *Contact Mechanics;* Cambridge University Press: Cambridge, UK, 1985; pp. 119–124.
47. Othman, H.; Marzouk, H. Impact response of Ultra-High-Performance reinforced concrete plates. *ACI Struct. J.* **2016**, *113*, 1–10. [CrossRef]
48. Micallef, K.; Sagaseta, J.; Ruiz, M.F.; Muttoni, A. Assessing punching shear failure in reinforced concrete flat slabs subjected to localised impact loading. *Int. J. Impact Eng.* **2014**, *71*, 17–33. [CrossRef]
49. Martín-Morales, M.; Zamorano, M.; Ruiz-Moyano, A.; Valverde-Espinosa, A. Characterization of recycled aggregates construction and demolition waste for concrete production following the Spanish Structural Concrete Code EHE-08. *Constr. Build. Mater.* **2011**, *25*, 742–748. [CrossRef]
50. Chen, Y.; May, I. Reinforced concrete members under drop-weight impacts. *Proc. Inst. Civ. Eng. Struct. Build.* **2009**, *162*, 45–56. [CrossRef]
51. Fonseca, N.; de Brito, J.; Evangelista, L. The influence of curing conditions on the mechanical performance of concrete made with recycled concrete waste. *Cem. Concr. Compos.* **2011**, *33*, 637–643. [CrossRef]
52. Kou, S.; Poon, C. Long-term mechanical and durability properties of RAC prepared with the incorporation of fly ash. *Cem. Concr. Compos.* **2012**, *37*, 12–19. [CrossRef]
53. Clough, R.; Penzien, J. *Dynamics of Structures*, 2nd ed.; McGraw-Hill: New York, NY, USA, 1995.

applied sciences

Article

Study on the Penetration Characteristics of Water Entry Rod Projectile into Liquid Cabin at an Attack Angle

Ke Wang, Hailiang Hou, Dian Li * and Yongqing Li *

Department of Naval Architecture Engineering, Naval University of Engineering, Wuhan 430000, China
* Correspondence: lidian916@163.com (D.L.); liyongqing@126.com (Y.L.)

Abstract: The penetration of a projectile into a warship broadside liquid cabin is usually a non-ideal penetration process. To explore the protective effects of the broadside liquid cabin of a large warship against the non-ideal penetration of rod projectiles and to provide reference for the design of new liquid cabin structures, ballistic impact tests of rod projectiles penetrating the liquid cabin at different attack angles were carried out. Combined with numerical calculation, the impact of the attack angle on the water entry and penetration characteristics of the projectile into the liquid cabin as well as their failure modes were studied. The overturning and yawing of the projectile in water were analyzed. The pressure load characteristics in the liquid cabin and the deformation/failure modes of the projectile and the liquid cabin were identified. The results showed that: multiple overturning and yawing occur in the projectile with an initial attack angle during penetration into liquid; the yaw direction is mainly affected by the initial attack angle and projectile attitude; the projectile mainly undergoes four basic failure modes, namely, asymmetric mushrooming at the projectile nose, side erosion, overall plastic bending and fracture; the actual failure of the projectile is a combination of the basic failure modes; the overall plastic bending and fracture are mainly related to the length to diameter ratio, initial attack angle and initial projectile velocity; the front plate of the liquid cabin may undergo tearing along the central plastic hinge line of the plate: at a small attack angle, the tear is "I" shaped, and at a large attack angle, it is "X"-shaped.

Keywords: rod projectile; angle of attack; penetration; liquid cabin; overturn and yaw; trajectory

Citation: Wang, K.; Hou, H.; Li, D.; Li, Y. Study on the Penetration Characteristics of Water Entry Rod Projectile into Liquid Cabin at an Attack Angle. *Appl. Sci.* **2022**, *12*, 10213. https://doi.org/10.3390/app122010213

Academic Editor: Ricardo Castedo

Received: 14 September 2022
Accepted: 5 October 2022
Published: 11 October 2022

1. Introduction

To improve the penetration and destructive capabilities of projectiles, anti-armor weapons are widely equipped with rod projectiles with large length to diameter ratio (L/D ratio), strong anti-interference ability, good flight stability, high energy density and strong armor-piercing ability [1], whereas underwater weapons mostly adopt explosively formed projectiles (EFPs) [2,3]. As the warship broadside can be easily attacked by anti-ship weapons due to its large area, liquid cabin structures are often set on the broadside to protect the inner structure against penetration or armor-piercing damage caused by fragments and debris generated by the warhead shell and the outer plate of the cabin [4]. However, before penetrating into the liquid cabin, the projectile inevitably interacts with the stiffened plate and other obstacles in the empty cabin, changing the ballistic characteristics such as attitude angle of the projectile or causing asymmetric deformation of the projectile, and thus affecting its penetration capability [5]; when attacking underwater targets, torpedoes and projectiles usually enter the water at an attack angle; thus, it is of great significance to study the non-ideal water entry of rod projectiles for warship protection.

Generally, the water entry of projectiles can be divided into four stages: (1) Impact stage. As a projectile impacts the water at a high speed, causing a large impact force on the projectile, the projectile nose is prone to mushrooming deformation, and meanwhile the impact by the projectile leads to the formation of high-speed shock waves in the water, which propagates in a semicircle [6]. (2) Flow formation stage. After being impacted, the water will be separated by the projectile and flow away from its sides as the projectile moves forward, resulting in cavitation. Whip can be easily observed in this stage, where the pressure on the projectile is far less than that in the impact stage. (3) Open cavity stage. The cavity generated after the water entry of the projectile does not disappear immediately. It is still connected with the air at the water surface. As the projectile moves, the cavity gradually expands. When the projectile moves forward in the cavity, the uneven contact between the projectile nose and the water cause projectile swinging, resulting in ballistic instability. At the same time, due to tail swing, the tail contacts the cavity, causing tail flap, which further increases the uneven force on the projectile, leading to greater yaw [7,8]. (4) Closed cavity stage. After formation for some time, the cavity closes at a point on or below the water surface. Cavity closure may generate jet. When the jet strikes the cavity wall and causes its deformation, or hits the projectile and changes its trajectory, the projectile will detach from the cavity, leading to a significant change in the direction of projectile movement [9,10].

At present, the water entry problem of projectiles is generally studied from two perspectives: a projectile penetrating the free water, such as the water entry of aerial torpedo and spacecraft; a projectile penetrating the liquid tank, such as the penetration of a projectile into an aircraft fuel tank, and fragments penetrating the broadside liquid cabin. Researchers have conducted extensive research on the penetration of rod projectiles under ideal conditions.

In terms of the ideal penetration of projectiles into free water, Karman [11] proposed the added mass method to calculate the water entry impact load, and derived the formula of the water entry impact load using the conservation of momentum; Wagner [12] considered the lifting of water surface based on Karmans work, and introduced the wave influence factor to optimize the theory; Cointe [13] established a two-dimensional water entry impact model through progressive matching; Takagi [14] employed the potential method to accurately calculate the added mass, water entry velocity and penetration depth, which were in good agreement with the test results; Mojtaba [15] considered the transient model of cavity shape and established a complete model of the water entry of cylindrical projectile. Alekseevskii [16] and Tate [17] built a theoretical analysis model of the penetration of long-rod projectiles into semi-fluid medium. However, the water entry of projectile is a three-dimensional phenomenon. Both the fluid flow and projectile movement are asymmetric, and the forces involved are quite complex [18]. Chen [19] conducted experiments and found that projectile tumbling and yaw occurred in the cases of both vertical and oblique water entry of the projectile. Projectile tumbling occurs due to the change of the pitch angular velocity when the hydrodynamic force acting on the projectile nose does not pass through the center of gravity of the projectile, resulting in the so-called "whip" phenomenon [20]. Li Tianxiong [21] verified by numerical simulation that symmetrical projectiles also undergo ballistic yawing during vertical water entry.

With respect to the ideal penetration of a projectile into a liquid-filled structure, this process involves not only the interaction between the projectile and water, but also the interaction between water and the liquid tank. As the projectile penetrates the liquid tank, its energy is transferred to the water and tank structure, causing high pressure on them. This phenomenon is called hydrodynamic ram (HRAM) [22]. Researchers have studied the HRAM effect caused by impact on liquid-filled structures from various aspects: (1) Pressure load characteristics. Shi [23] verified that the rise time of initial shock waves under the HRAM effect is in microseconds; Gao [24] experimentally demonstrated that cavitation load is the dominant factor of liquid-filled cell failure; Li [25–27] summarized the load characteristics of rod projectiles penetrating the liquid-filled structure through penetration test and numerical calculation. Then, according to the load characteristics, he divided different areas to establish a simplified calculation model, and compared the structure protection performance; Disimile [28] adopted high-speed photography to show the generation of pressure waves and how the cavitation region expands and collapses. (2) Remaining projectile characteristics. Deletombe [29] proposed that the projectile will undergo overturning, deformation or even fracture when interacting with the water medium; Shen [30–32] analyzed the rule of the mushrooming deformation of the projectile and its influence on penetration resistance by tests and numerical calculations of fragments penetrating the liquid cabin, and proposed the formulas for calculating penetration resistance and velocity considering the influence of projectile deformation. (3) Structural deformation and failure. Artero and Nishida [33,34] found through tests and numerical calculations that water will cause secondary damage to the fluid-filled tubes, thus reducing the structural strength, and that the damage to the rear plate is more serious; David [35] conducted high-velocity impact test at 1000–3000 m/s with fluid-filled aluminum alloy containers, and the wall plate exhibited petal cracking.

In terms of water entry under non-ideal penetration conditions, which includes oblique water entry, water entry with an attack angle and water entry of asymmetric projectiles, currently, a large number of studies are focused on the water entry of asymmetric projectiles and oblique water entry. Takashi Isobe [36] performed oblique water entry tests using projectiles of various shapes, and had the following findings: when the projectile nose is moving in the water, the fluid exerts a lift force on it because of the cavitation caused by projectile motion; due to the pressure difference between the upper and lower surfaces of the projectile, cavitation occurs on the two surfaces is not symmetrical, and then the amount of water displaced by the projectile on the free surface side is less than the other side; thus, the pressure on this side is lower, resulting in asymmetric pressure on the upper and lower surfaces, thus exerting a lift force on the projectile and causing projectile ricochet; the nose shape of the projectile greatly affects the ballistic stability in water: the flat-nosed projectile has better ballistic stability underwater, the sharp-nosed projectile is prone to ballistic instability, and the ogive-nosed projectile tend to overturn in the early stage of water entry [37–40]; compared with symmetric nose, after the asymmetric nose enters the water, a non-axial component exists in the hydrodynamic force on the nose [41], and an overturning moment acts on the projectile, changing its attitude angle. This further affects the hydrodynamic force on the projectile, leading to the nonlinear increase in the overturn and yaw of the projectile [42,43]. The more asymmetric the projectile nose is, the worse the attitude and trajectory stability of the projectile is, and the projectile is prone to instability and inclination, which leads to yaw.

There are much less studies on the water entry by a projectile at an attack angle. Li [44] used numerical calculation to simulate the vertical water entry by a projectile at a small attack angle, and found that under certain attack angles, the projectile tail contacts with the cavity, thus reducing the ballistic stability, and that the spinning of the projectile in water has little impact on ballistic stability. However, Truscott [45,46] found that the spinning motion of the projectile induces a lateral force on the projectile, resulting in a curved trajectory. Yao [47] studied the water entry by the underwater vehicle through numerical simulation and revealed that positive attack angles suppress projectile whipping, whereas negative attack angles aggravate this phenomenon; therefore, ballistic change easily occurs under negative attack angles. Liang Jingqi [48] conducted a study using the LS-DYNA program and found that greater attack angle leads to faster axial velocity attenuation of the projectile, and thus greater overturning angle and projectile velocity. Wang Zhen [49] studied the oblique water entry of projectile at small attack angles with LS-DYNA, and discovered that the attack angle determines the direction and magnitude of the moment on the projectile nose.

To sum up, detailed investigations on the residual characteristics, pressure load characteristics and structural deformation of projectiles have been conducted through the ideal penetration test of the water-entry projectile. During the oblique water entry, entry with an angle of attack and water entry of an asymmetric projectile, ballistic yaw will occur. Among these cases, the penetration at an attack angle has been less investigated, and numerical calculation is the main method adopted. Few experimental studies have been carried out on the water entry by high-velocity projectiles with attack angles because the initial attack angle is difficult to realize in ballistic tests and is not conducive to ballistic control.

Therefore, in the present study, the tests of high-velocity rod-shaped projectiles penetrating the liquid cabin at various attack angles are performed, and the overturning and yawing of the projectile in water, the pressure load characteristics and the deformation/failure modes of the projectile and the liquid cabin are analyzed by combining experimental tests with numerical calculations.

2. Experimental and Numerical Methods

2.1. Ballistic Experiment Design

2.1.1. Water Tank Model

Two types of cylindrical rod projectiles with a diameter of 14.5 mm and lengths of 23.5 mm and 29 mm are employed in the test. The projectile material is 45# steel.

As shown in Figure 1, the water tank is designed with a volume of 300 × 400 × 600 mm, with no cover on the upper side to simulate the free surface of the liquid cabin of an actual warship. The front and rear plates of the water tank have flange structures to facilitate the connection with the target. The front and rear plates and the pressplate are connected with the water tank using bolts with a diameter of 15 mm. A rubber pad is added between the target plate and the water tank to keep water tightness. One side of the water tank is made of 30 mm thick polymethyl methacrylate (organic glass) for the convenience of high-speed photography. On the other side, 50 mm gridlines are drawn for the calculation of the position and velocity of the projectile in water. The front plate is 2 mm thick, and the rear plate 4 mm thick. The material of the water tank and front/rear plates is Q235 steel.

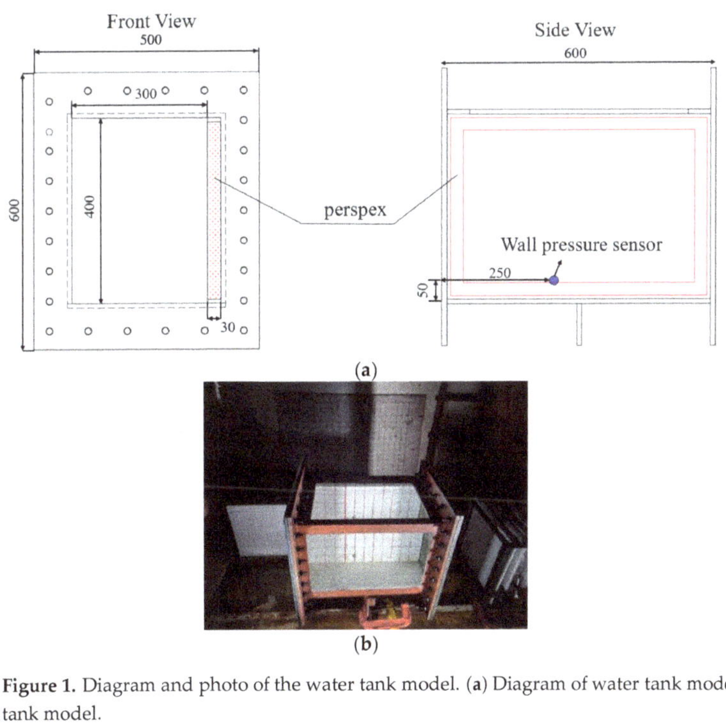

Figure 1. Diagram and photo of the water tank model. (**a**) Diagram of water tank model. (**b**) Water tank model.

2.1.2. Experimental Equipment and Setup

A 14.5 mm ballistic gun is used to fire the projectiles (bullets). The sabot is designed to be not closely fit with the projectile, so that initial disturbances can be generated after the projectile exits the bore, leading to the formation of the initial attack angle as the projectile impacts the target. The initial attack angle is obtained by the combination of high-speed photography and analysis of the breach morphology of the front wall of the water tank. A sabot collector is set to avoid the influence of sabot on the tests.

A laser speed measuring system is installed at the front side of the water tank (liquid cabin) to measure the initial velocity of the projectile (Figure 2a), and a target-net speed measuring system and a projectile recovery box are set at the rear side to measure the residual velocity of the projectile and recover the remaining projectile, respectively. The laser speed measurement system uses Chengdu Test TST6260 transient signal tester, the maximum sampling can reach 20 Msps.

(**a**)

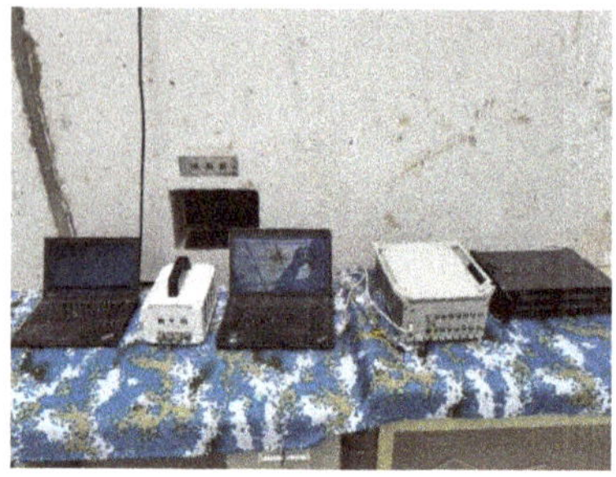

(**b**)

Figure 2. Test equipment and recording devices. (**a**) Laser velocimeter; (**b**) high-speed camera, pressure sensor and target-net velocimeter.

The X213 high-speed camera made by Revealer Company in Hefei, China, is used to observe the projectile attitude in water and the cavitation process (Figure 2b). It shoots at 10,000 frames per second, i.e., one picture is taken every 100 us.

To measure the pressure load in water, a wall pressure sensor with a measuring range of 0~68 MPa and a sampling frequency of 1 MHz is set at a position 5 mm from its center to the bottom of the water tank (Figure 1a).

The experimental setup is shown in Figure 3.

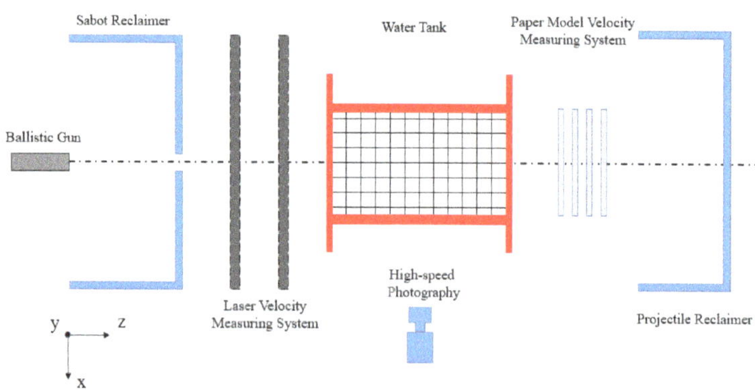

Figure 3. Experiment setup.

2.2. *Brief Experimental Results*

Three penetration tests are conducted with the water tank, and the initial physical state and initial projectile velocity are recorded. Since the rear plate of the water tank have been perforated, the residual velocity of the projectile is not measured. After the tests, the residual projectile mass and its displacement in all directions are obtained through measurements. The test conditions and results are shown in Table 1.

Table 1. Test conditions and results.

Test No.	Y-Axis Attack Angle (°)	Z-Axis Attack Angle (°)	Total Attack Angle (°)	Projectile Length	L/D Ratio	Projectile Weight M (g)	Initial Velocity V_0 (m/s)	Residual Weight of Projectile M (g)	Displacement cm		
									X	Y	Total
1	−30	−40	−50	23.5	1.6	29.9	1016.0	29.08	3.3	−1.9	3.8
2	−15	66	67	23.5	1.6	29.3	991.1	29.56	−0.6	6.8	6.83
3	25	38	45	29	2	37.3	1004.9	36.54	3.5	5.6	6.6

2.3. Numerical Calculation Model and Effective Verification

2.3.1. Numerical Model

The water tank model is simplified as shown in Figure 4. The models are built with SolidWorks, and Hypermesh is used to generate grids. The solid structures such as the projectile and target plate use Lagrangian solid elements; the side walls of the water tank do not involve damage and deformation; therefore, Lagrangian shell elements are used for them. The water domain is constructed in the water tank, the air domain is established outside the water tank; Eulerian elements are adopted for the water and air domains, and the common node approach is applied, so as to realize the flow of water and air.

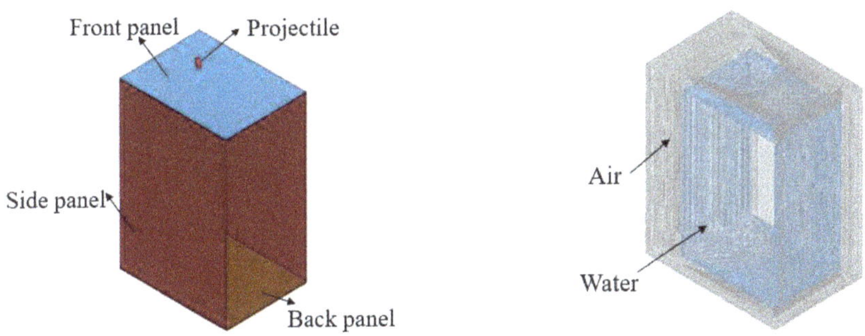

Figure 4. Numerical model of water tank and Eulerian domains.

The projectile is divided into 14 equal parts in the radial direction, 16 equal parts in the circumferential direction, and 1.2 mm grids along the length direction. The grid of the water and air domains is 4 mm in size, and the grids are refined in the central area with a size of 2 mm. The grids of the front and rear plates of the water tank are divided into 1 mm cube elements. Because side walls are thick and can be regarded as rigid walls, their grid size is 5 mm.

2.3.2. Numerical Calculation Method

As the numerical calculation involves the fluid flow and the interaction between fluid and structure, the fluid-solid coupling algorithm in the LS-DYNA R10 software is used. The LAGRANGE_IN_SOLID card is used to realize the structure-fluid coupling. Since gravity cannot be ignored during the projectile's water entry, the LOAD_BODY command is employed to set the gravity field in the Eulerian domain, and the INITIAL_HYDROSTATIC_ALE command is set to define the hydrostatic pressure in the water domain. The contact between the projectile, the steel plate and water tank are set to ERODING_SURFACE_TO_SURFACE.

2.3.3. Material Model

(a) Q235 steel

The water tank and steel plate are made of Q235 steel. The Cowper-Symonds constitutive model is adopted, and its dynamic yield strength σ_d is:

$$\sigma_d = \left(\sigma_0 + \frac{EE_h}{E - E_h}\varepsilon_p\right)\left[1 + \left(\frac{\dot{\varepsilon}}{D}\right)^{1/n}\right] \quad (1)$$

where σ_0 is the static yield strength, E is the elastic modulus, E_h is the strain hardening modulus, $\dot{\varepsilon}$ is the equivalent strain rate, and D and n are the strain rate parameters. Table 2 lists the material parameters.

Table 2. Material parameters of Q235 steel.

Parameter	Value	Parameter	Value	Parameter	Value
Yield strength σ_0 /MPa	235	n	5	Failure strain ε_f	0.28
Strain hardening modulus E_h/MPa	250	$D(s^{-1})$	40.4		

(b) 45# steel

The material of the projectile is 45# steel, and Johnson–Cook constitutive model is selected:

$$\sigma = (A + B\varepsilon_p{}^n)\left(1 + C\ln\frac{\dot{\varepsilon}_p}{\dot{\varepsilon}_{po}}\right)\left[1 - \left(\frac{T - T_0}{T_m - T_0}\right)^m\right] \quad (2)$$

where σ Is the dynamic yield strength of steel, ε_p is the plastic strain, A is the static yield limit, B is the strain hardening modulus, n is the strain hardening index C is the strain rate coefficient, ε_{p0} is the critical strain rate, m is the thermal softening index, T is the temperature, T_m is the melting point of the material, and T_0 is the room temperature.

The J-C failure model is adopted to describe the failure of the materials:

$$\varepsilon_f = \left\{D_1 + D_2\exp\left[D_3\frac{\sigma_h}{\sigma_{eff}}\right]\right\}\left[1 + D_4\ln\dot{\varepsilon}^*\right](1 + D_5T^*) \quad (3)$$

where D_1–D_5 are material parameters, σ_{eff} is the von Mises stress, σ_h is the hydrostatic pressure of the material under the triaxial stress, $T^* = (T - T_r)/(T_m - T_r)$ is the dimensionless temperature, T_r is the room temperature, and T_m is the melting point of the material.

The material parameters of the projectile are presented in Table 3.

Table 3. Mechanical parameters of the projectile.

Parameter	Value	Parameter	Value	Parameter	Value
shear modulus G/GPa	80.8	C	0.0483	T_m/K	1793
A/MPa	335	m	0.804	T_0/K	300
B/MPa	350	D_1	0.8	D_2	0.76
n	0.782	D_3	1.57	D_4	0.005
C_v/J·(kg·K)$^{-1}$	477	D_5	−0.84		

(c) Water

The Grüneisen equation of state (EOS) is adopted for water:

$$P = \frac{\rho_0 C^2 \mu \left[1 + \left(1 - \frac{\gamma_0}{2}\right)\mu - \frac{a}{2}\mu^2\right]}{\left[1 - (S_1 - 1)\mu - S_2 \frac{\mu^2}{\mu+1} - S_3 \frac{\mu^3}{(\mu+1)^2}\right]^2} + (\gamma_0 + a\mu)E \tag{4}$$

where ρ_0 is the density, C is the speed of sound, γ_0 is the Grüneisen parameter, $\mu = \rho/\rho_0 - 1$, a is volume correction, and S_1, S_2 and S_3 are curve fitting parameters.

(d) Air

When air is an ideal gas without viscosity, its EOS is a linear polynomial:

$$P = C_0 + C_1\mu + C_2\mu^2 + C_3\mu^3 + \left(C_4 + C_5\mu + C_6\mu^2\right)E \tag{5}$$

where C_0–C_6 are the parameters and E is the internal energy. Let $C_0 = C_1 = C_2 = C_3 = C_6 = 0$, $C_4 = C_5 = \gamma - 1$, so that it has ideal gas characteristics, where γ is the adiabatic exponent. The material parameters of water and air are shown in Table 4.

Table 4. Material parameters of fluid.

Parameter	ρ_0 (kg/m^3)	v_d	C (m/s)	S_1	S_2	S_3	γ_0	a	C_4	C_5	E_0 (J/m^3)
water	1000	0.89	1448	1.98	0	0	0.11	3			0
air	1.22								0.4	0.4	2.53×10^5

2.3.4. Verification of Calculation Results

Numerical calculation is performed according to the attack angle and initial projectile velocity measured in the tests. The calculation results are compared with the experimental data. With the measured projectile positions in water at different times, the corresponding velocities can be calculated and fitted. The velocity comparison is shown in Figure 5, and the maximum error is within 15%. It can be considered that the velocity attenuation trend of the numerical calculation is in good agreement with the experimental results.

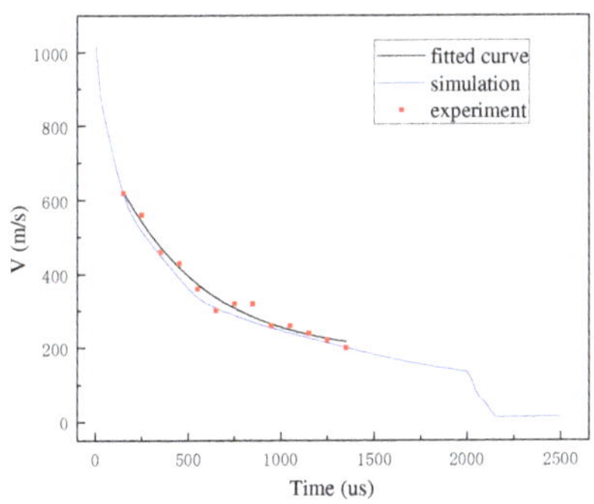

(a)

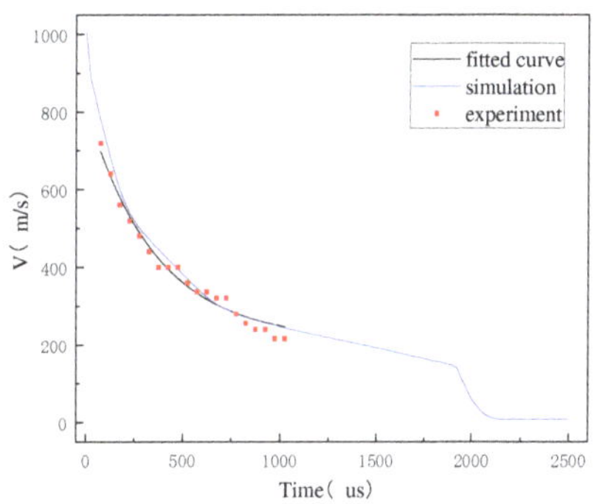

(b)

Figure 5. Comparison of tested and simulated velocities. (**a**) Test 1 velocity curve; (**b**) test 3 velocity curve.

Figure 6 gives a comparison between the measured load curve in the projectile water entry test and numerically calculated loads. The shock wave peak overpressure measured in Test 1 is 19.1 MPa, and the initial shock wave pressure in the simulation is 15.6 MPa, with an error of 18%, whereas the error between simulation and experimental results in Test 3 is 22%. The reason for this difference is: the grids in the water domain are slightly larger, so the shock wave attenuation speed is faster than the actual situation; however, reducing the grid size will increase the computing time at a geometric rate. Therefore, it is considered that the numerical calculation gives reasonable results on the premise of ensuring accuracy, and the error is within the acceptable range.

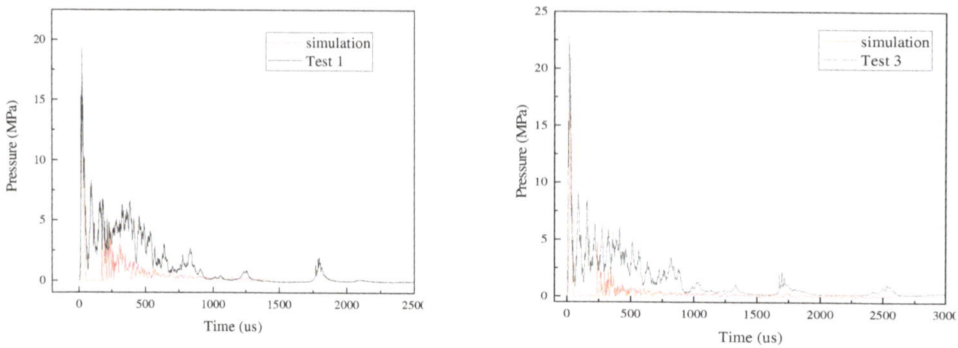

Figure 6. Comparison of experimental and simulated loads.

Figure 7 and Table 5 show a comparison of the remaining projectile after the test and numerical simulation. The side erosion unique to the water entry of the projectile at an attack angle and the asymmetric shape of the nose are successfully simulated, and the error between the calculated and experimental residual mass is very small, demonstrating that the failure modes of the projectile obtained by numerical calculation agrees well with the experimental data.

Figure 7. Comparison of projectile failure morphology between the test and numerical simulation.

Table 5. Comparison between experimental and calculated residual mass of the projectile.

Test No.	Projectile Mass M (g)	Residual Mass of Projectile in Test M (g)	Residual Mass of Projectile in Simulation M (g)	Error Value (%)
1	29.9	29.08	28.24	2.84
2	29.3	29.56	29.03	1.81
3	37.3	36.54	36.52	0.05

To sum up, the proposed numerical calculation methods and numerical models can well simulate the water entry of the projectile, and check with the experimental data, striking a good balance between accuracy and efficiency.

2.4. Test Conditions

To study the influence of the initial attack angle and initial velocity on the deformation morphology of the liquid cabin and the projectile after the penetration of the projectile into the liquid cabin, 30 working conditions for the projectile with a diameter of 14.5 mm and a length of 69.6 mm (L/D ratio = 4.8) penetrating the liquid cabin are investigated using the numerical calculation method. TEST/FEM indicates that ballistic impact test is used in this working condition, supplemented by numerical calculation method, whereas FEM uses the finite element algorithm verified in Section 2.3 for calculation. The specific working conditions are listed in Table 6.

Table 6. Test conditions.

Condition No.	Initial Velocity V_0 (m/s)	Attack Angle (°)	Test Method	Condition No.	Initial Velocity V_0 (m/s)	Attack Angle (°)	Test Method
No. 1	1016	−50	TEST/FEM	No. 18	1200	60	FEM
No. 2	991	67	TEST/FEM	No. 19	1600	0	FEM
No. 3	1004	45	TEST/FEM	No. 20	1600	15	FEM
No. 4	400	0	FEM	No. 21	1600	30	FEM
No. 5	400	15	FEM	No. 22	1600	45	FEM
No. 6	400	30	FEM	No. 23	1600	60	FEM
No. 7	400	45	FEM	No. 24	2000	0	FEM
No. 8	400	60	FEM	No. 25	2000	15	FEM
No. 9	800	0	FEM	No. 26	2000	30	FEM
No. 10	800	15	FEM	No. 27	2000	45	FEM
No. 11	800	30	FEM	No. 28	2000	60	FEM
No. 12	800	45	FEM	No. 29	2400	0	FEM
No. 13	800	60	FEM	No. 30	2400	15	FEM
No. 14	1200	0	FEM	No. 31	2400	30	FEM
No. 15	1200	15	FEM	No. 32	2400	45	FEM
No. 16	1200	30	FEM	No. 33	2400	60	FEM
No. 17	1200	45	FEM	No. 34	1200	8	FEM

3. Results and Analysis

3.1. Overturning and Yawing during Underwater Penetration of Projectile at an Attack Angle

Figure 8 shows the photos of the three groups of tests taken by the high-speed camera. After the projectile enters the water with an attack angle, the cavitation region generated by its high-velocity motion is asymmetric and curved. This is because the penetration attitude and trajectory of the projectile with an attack angle are not stable in water, and the penetration direction of the projectile at the moment it enters the water is related to the initial attack angle of the projectile. In the case of positive initial attack angles, the projectile tends to yaw upward such as in Tests 2 and 3; in the case of negative initial attack angles, the projectile tends to yaw downward, such as in Test 1 (the anticlockwise rotation of the attack angle about the axis is defined as positive attack angle and the clockwise rotation as negative one).

The attitude angle of the projectile in water is an important factor affecting the yaw of the projectile. Taking Test 3 as an example, it can be clearly observed from the photos that the attitude of the projectile keeps changing, overturning anticlockwise first and then clockwise. The attitude angle of the projectile on the XY plane, its Y-axis velocity and displacement are read from numerical calculations (Figures 9–11). When the projectile enters the water with its nose up, the nose and lower side of the projectile are the main positions contacting with water. Because the lower side of the projectile is subjected to the dynamic pressure of the water, which is perpendicular to the contact surface, a lift force is exerted on the projectile, generating a Y-direction velocity, leading to the upward yawing of the projectile. At this time, the force exerted by water on projectile nose is the largest. Due to the existence of the attitude angle of the projectile, the force does not pass through the center of mass of the projectile, and the resultant force is on the upper side of the center of mass. This gives an anticlockwise overturning moment to the projectile, leading to the anticlockwise overturning of the projectile. At t = 130 us, the projectile overturns to the maximum incident flow area (around 90°), and the resultant force on the projectile basically passes through its center of mass. However, as the projectile continues to overturn anticlockwise due to inertia, the original "tail" of the projectile turns over to the "nose" position, and the direction of the resultant force on the projectile becomes downward. The Y-axis velocity of the projectile starts to decrease, and the resultant force is below the projectile's center of mass; thus, the projectile is subjected to a clockwise overturning moment, which generates an angular acceleration in the opposite direction to the overturning direction, and the anticlockwise overturning velocity decreases gradually. At t = 342 us, the angular velocity of the projectile declines to 0 rad/s, the projectile starts to overturn clockwise, and the Y-axis velocity of the projectile is increasing smaller. When the projectile overturns clockwise to the maximum incident flow area, the projectile starts to be subjected to an anticlockwise moment; thus, the angular velocity starts to decrease, the resultant force becomes upward, and the Y-axis velocity increases; finally, as the projectile contacts the back plate, the penetration ceases.

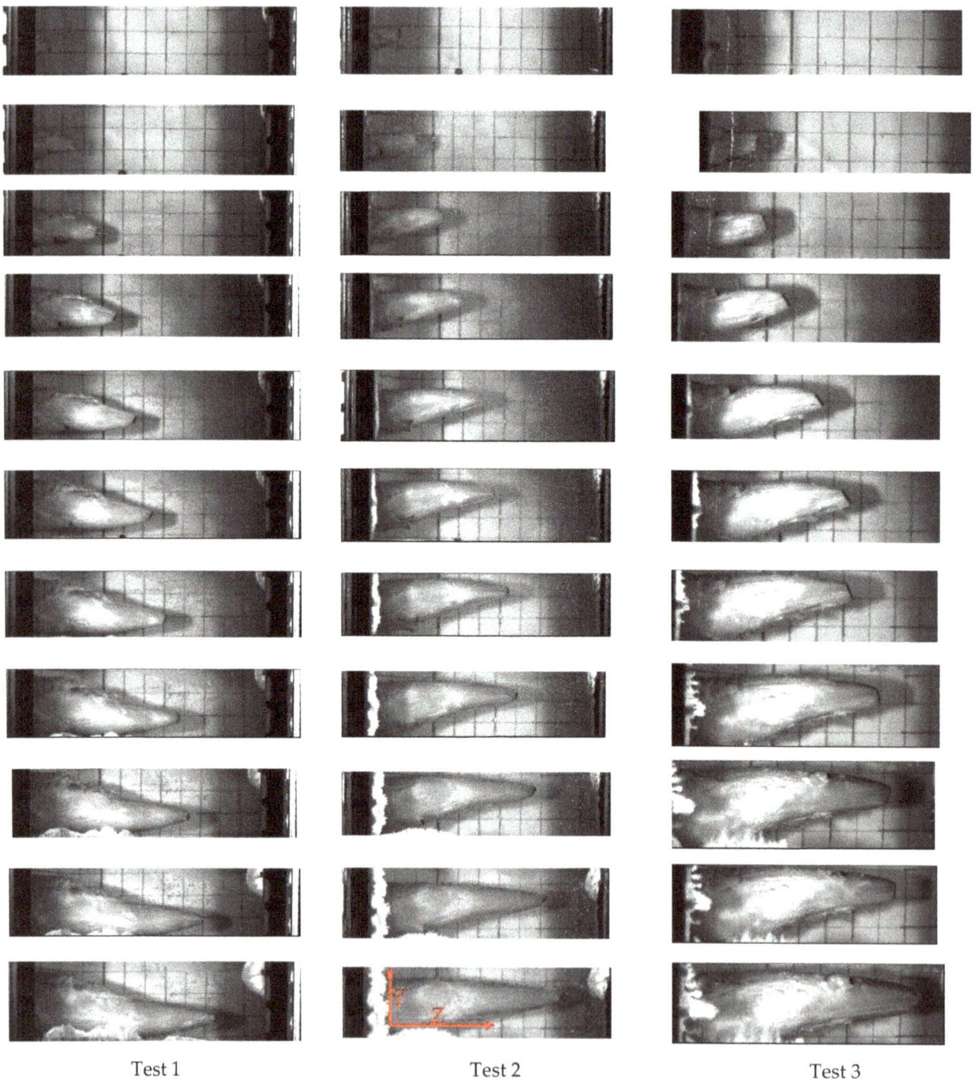

Figure 8. Photographs of underwater penetration process.

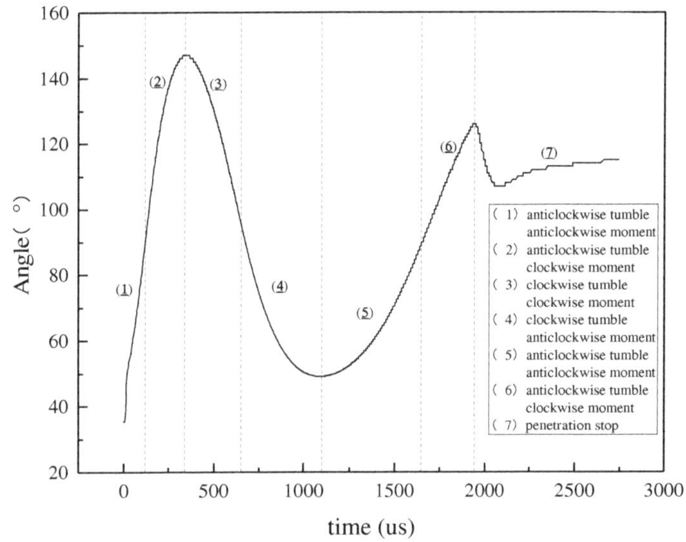

Figure 9. Attitude angle change of projectile in Test 3. (The red dash line is the dividing line of different stages).

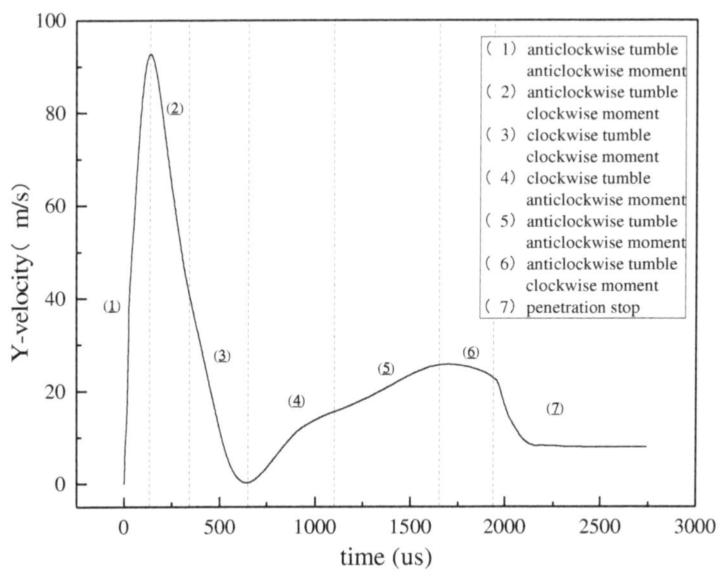

Figure 10. Y-axis velocity curve of projectile. (The red dash line is the dividing line of different stages).

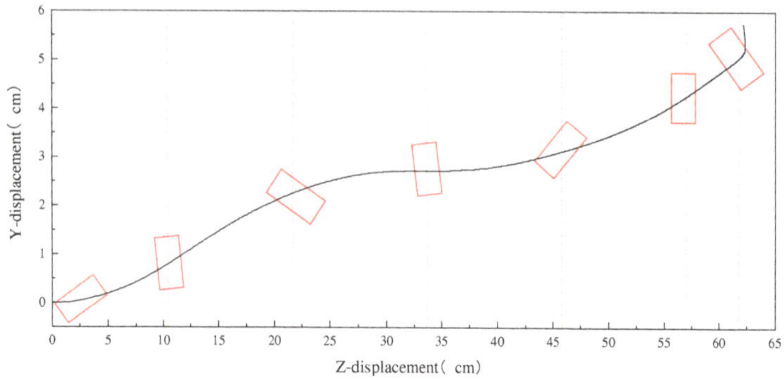

Figure 11. Trajectory of projectile in Test 3. (The red dash line is the dividing line of different stages).

In summary, during projectile penetration into liquid at an attack angle, the projectile is always in the overturning state. Due to the change of the projectile position contacting with water, the projectile will undergo overturning many times. The yaw of the projectile is affected by the attack angle and attitude angle as the two angles determine the direction of the force on the projectile during penetration. As the initial velocity during water entry is large, greater dynamic pressure on the projectile leads to greater Y-direction component. Therefore, the yaw velocity of the projectile is fast at the initial stage of water entry; when the second overturn occurs, the projectile velocity becomes smaller, and the Y-direction component of the projectile is also smaller; thus, the trajectory tends to be stable with a relatively small yaw velocity at the later stage of water entry.

3.2. Analysis of Residual Characteristics of Projectile Penetrating into Liquid Cabin at an Attack Angle

Figure 12 shows the deformation and failure morphology of the projectiles after the penetration tests under various attack angles. Mass abrasion occurs in both radial and axial directions of the projectile (the residual mass increases due to high-temperature fusion of the projectile and front plate fragments in Test 2). Compared with water entry under normal penetration (Figure 4d), the water entry projectile with an attack angle undergoes obvious asymmetric deformation. In the radial direction, one side of the projectile presents overall wavy erosion, whereas the other side shows no deformation; in the axial direction, one side of the projectile undergoes mushrooming deformation, and the other side suffers slight mass loss.

Adiabatic shear failure occurs during the high-velocity projectile impact on the target plate, generating a large amount of heat. As the heat is transferred to the projectile, the yield strength of the projectile is reduced, causing mushrooming and erosion to the projectile nose; the sides of the projectile with an attack angle also impacts the front plate, causing "strip" erosion. The impacted part of the front plate drives the nearby region to move backward, and the velocity of the plate exceeds that of the projectile. Then, a gap is produced between the projectile and the front plate after the initial impact, and as the projectile moves forward, its side impacts the front plate again, causing "strip" erosion to this side again; after multiple impacts, the side shows wavy erosion. After the projectile enters the water, one side and the nose of the projectile are impacted by the water, leading to erosion and mushrooming deformation of the projectile; due to the reduced velocity of the projectile after water entry, its deformation in water is relatively small; cavitation occurs on the other side of the projectile due to its high-velocity motion; the projectile on this side is always in the cavitation region and does not contact with the water; thus, deformation does not occur to this side; when the projectile impacts the rear plate, a small deformation occurs as the velocity at this time has been completely decayed.

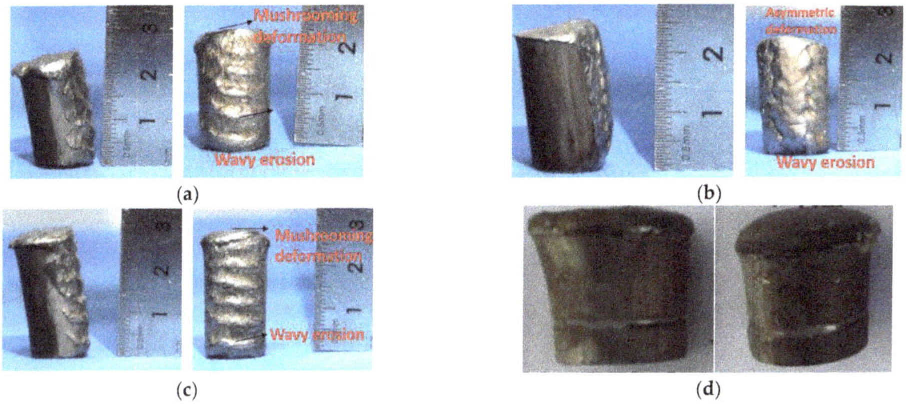

Figure 12. The remaining projectile after the tests. (**a**) Remaining projectile in Test 1; (**b**) remaining projectile in Test 2; (**c**) remaining projectile in Test 3; (**d**) remaining projectile under normal penetration [24].

In conclusion, when the initial velocity of the projectile is 1000 m/s, the penetration into the front plate takes the shortest time but is the main stage when projectile failure occurs, featuring a small deformation and mass abrasion of the projectile.

The failure morphology of the high-velocity rod projectile is studied by numerical calculation. The specific working conditions are shown in Figure 13.

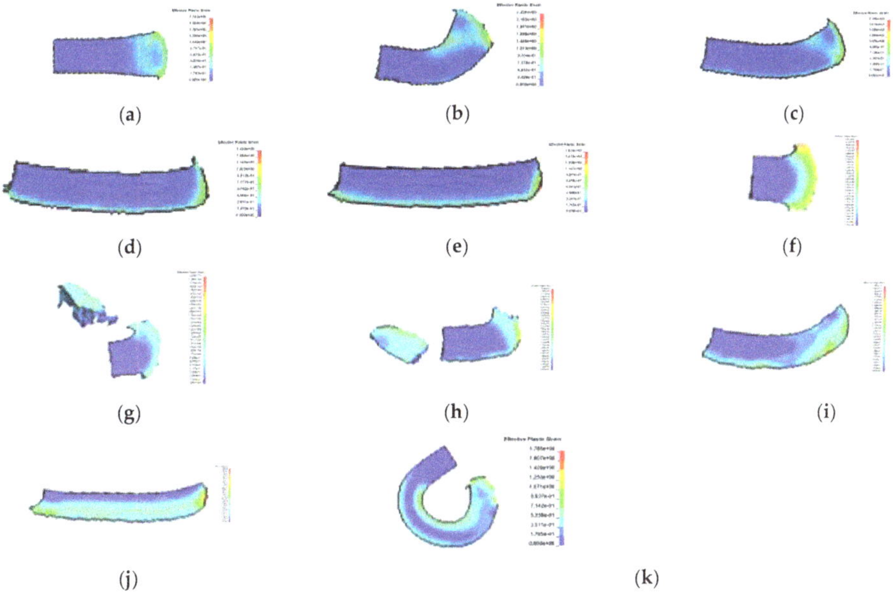

Figure 13. Projectile failure morphology under different working conditions (**a**) No. 19 erosion-mushrooming; (**b**) no. 20 overall plastic bending; (**c**) no. 21 overall plastic bending; (**d**) no. 22 side erosion; (**e**) no. 23 side erosion (**f**) no. 29 erosion—mushrooming; (**g**) no. 30 fracture; (**g**) no. 31 fracture; (**i**) no. 32 overall plastic bending; (**j**) no. 33 side erosion; (**k**) no. 34 bending.

As the front plate is thin, the penetration of the high-velocity projectile into liquid is the main stage of projectile failure. As can be seen from Figure 13, the projectile enters the water at a high velocity, which causes great deformation of the projectile. When projectile water entry is under normal penetration, the projectile undergoes erosion-mushrooming deformation, and the greater the velocity, the more serious the erosion; when projectile water entry is under penetration at an attack angle, the projectile undergoes a large overall deformation at 15~30°, and the overall plastic bending occurs at 1600 m/s. The reason is: an overturning moment is generated on the projectile under asymmetric loads, leading to the bending stress on the cross-section of the projectile; the critical section of the projectile reaches the yield limit under the combined action of axial compressive stress and bending stress; at this time, the plastic hinge line is formed in this section, leading to overall plastic bending deformation of the projectile; when the incident velocity is large enough, such as 2400 m/s, the critical section breaks directly, such as No. 30 and 31; however, when the initial attack angle is greater than 30°, only side erosion occurs during the high-velocity water entry of the projectile, and certain bending deformation also occurs when the incident velocity is high enough (No. 32); when the L/D ratio of the projectile is large enough, a large part of the projectile still undergoes deformation at low velocities and small attack angles; taking No. 33 as an example, the projectile with a L/D ratio of 10 at 1200 m/s undergoes bending when the attack angle is 8°. There are four failure modes of the projectile during its penetration into the liquid cabin: mushrooming, erosion, plastic bending and fracture. The failure phase diagram of the projectile penetrating the liquid cabin at different attack angles and initial velocities can be drawn. Figure 14 is the failure phase diagram of the penetration of the rod projectile with a L/D ratio of 4.8 into the liquid cabin.

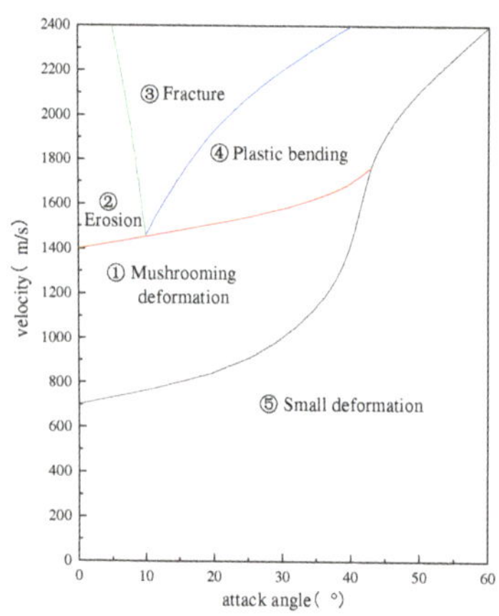

Figure 14. Failure phase diagram of the rod projectile with a L/D ratio of 4.8 penetrating into liquid cabin. ① Mushrooming ② Erosion (Condition 6) ③ Fracture (Condition 7) ④ Plastic bending (Condition 2).

In summary, at low projectile velocities, its penetration into the front plate is the main stage of failure, and at high velocities, the penetration into the liquid is the main stage of failure. The failure modes of the projectile are determined by its initial velocity, attack angle and L/D ratio. The projectile with an attack angle undergoes asymmetric deformation. When the attack angle is in the range of 15~30°, overall deformation is most likely to occur. At low velocities, overall plastic bending deformation occurs and with the increase in velocity, fracture will occur. When the attack angle is greater than 30°, the failure mode of the projectile is side erosion, but with the increasing initial velocity of the projectile, the range of the attack angle that leads to the overall deformation of the projectile also expands. The larger the L/D ratio, the more easily the overall plastic bending occurs to the projectile.

3.3. Analysis of Pressure Load Characteristics of Liquid Cabin

Figure 15 shows the pressure load curve of the bottom of the middle position of the water tank measured in Test 3. As shown in the figure, the pressure loading on the side wall of the projectile during water entry can be divided into three stages: the initial shock wave stage, the cavitation loading stage, and the cavity collapse stage.

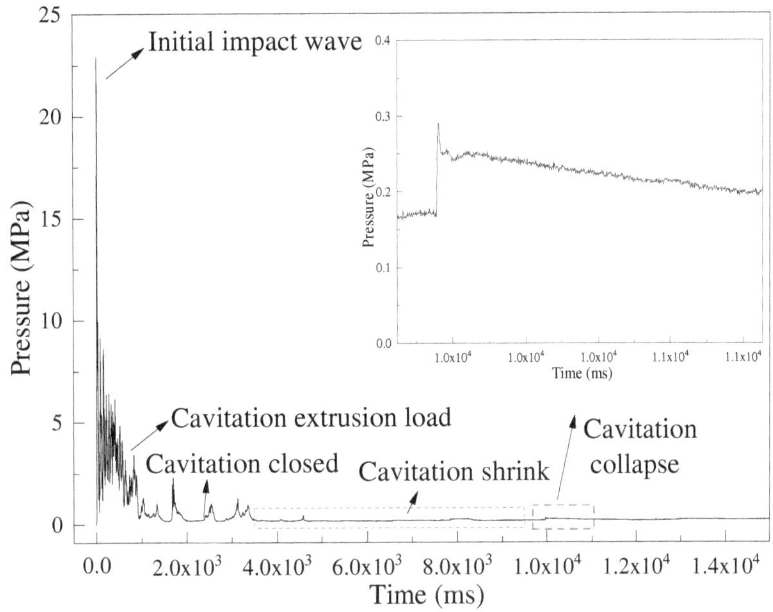

Figure 15. Load curve of measuring points in Test 3.

In the initial shock wave stage, the projectile perforates the front plate and then penetrates into the water, causing a huge acceleration of the previously static liquid relative to the projectile. This acceleration generates a shock wave that propagates in the water in an arc (Figure 16a). Peak value of initial pressure and cavitation load are shown in Table 7. The initial shock wave has the largest peak pressure, which is 19.1 MPa in Test 1 and 22.9 MPa in Test 3 through measurement. Since the L/D ratio and initial velocity of the projectiles in the two tests are different, the ratio of their kinetic energy is 0.821, and the ratio of the two initial shock wave peak pressures is 0.83. This indicates that under the same water tank structure and projectile shape, the initial shock wave peak pressure has a linear relationship with the kinetic energy of the projectile. The larger the kinetic energy, the larger the initial peak pressure.

Then, in the cavitation loading stage, the projectile penetrates into the water at a high velocity and displaces the water, which converts the kinetic energy of the projectile into the kinetic energy of the water. A cavity is formed on the moving path of the projectile, and the water keeps squeezing the water tank due to cavity expansion, causing the cavitation load. In Test 1, the specific impulse of the initial shock wave measured on the side wall is 450 MPa·us, and that of the cavitation load is 2516 MPa·us; in Test 3, the specific impulse of the initial shock wave is 502 MPa·us, and that of the cavitation load is 2617 MPa·us. The peak pressure of the cavitation load is much smaller than the initial shock wave load. However, due to its long duration, the specific impulse of the cavitation load is about five times that of the initial shock wave. Therefore, the cavitation load is the main load causing the side wall failure of the water tank. The ratio of the specific impulse of the initial shock wave to the kinetic energy ratio of the projectile is basically the same. As the kinetic energy increases, the specific impulse of the initial shock wave increases significantly. The reason is: the initial shock wave propagates at a speed in the water close to the speed of sound, and the initial shock wave is affected by the rarefaction waves in all directions; thus, the shock wave pressure decays quickly with a short action time, and the shock wave duration is the same. Therefore, the larger peak value of the shock wave leads to larger specific impulse, but there is little difference between the cavitation load and the specific impulse.

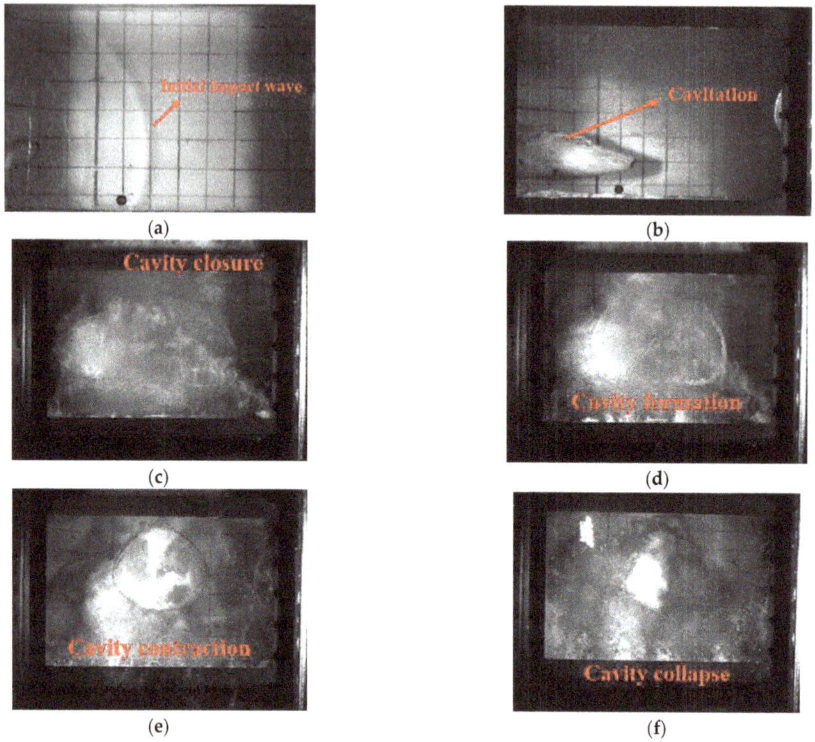

Figure 16. Pressure loading stage. (**a**) 100 us; (**b**) 250 us; (**c**) 2300 us; (**d**) 3300 us; (**e**) 8000 us; (**f**) 11,300 us.

Table 7. Peak value and specific impulse of initial pressure and cavitation load.

Test No.	Kinetic Energy (J)	Initial Peak Pressure (MPa)	Specific Impulse of Initial Pressure (MPa·us)	Peak Cavitation Load (MPa)	Specific Impulse of Cavitation Load (MPa·us)
1	15,432	19.1	450	8.3	2516
3	18,799	22.9	502	9.1	2617

As the kinetic energy of the projectile is continuously converted into the kinetic energy of the water, the water level rises. Under the influence of the backflow of disturbed water during cavity expansion, the air flow between the cavitation bubble and the atmosphere is cut off, and the cavity is closed at the entry hole. At this time, there are lots of gas and water vapor in the cavity, hence the formation of the cavitation bubble, which has the largest potential energy. Due to the pressure difference between the inside and outside of the bubble, its wall contracts, and the bubble gradually shrinks; in this process, the potential energy of the bubble is converted into kinetic energy, and negative pressure is generated in the water tank, which lasts for a long time, but the absolute value of the load is small; thus, it barely causes damage to the structure. With the contraction of the bubble, the gas in the bubble is compressed, leading to the increase in pressure and the formation of a high-pressure region, and then the bubble collapses, followed by bubble expansion again. At this time, shock waves are generated, and most of the energy in the bubble is converted into impact energy. Thereafter, the bubble keeps expanding and collapsing until all energy is dissipated. Due to the existence of the free surface in the water tank and the pressure sensor located at the bottom of the side wall of the water tank, the measured shock wave load for bubble collapse is relatively small (Figure 15). The cavity collapse pressure measured by Disimilea was very large, even exceeding the initial shock pressure. On the one hand, because the measuring points were arranged near the ballistic axis, and the cavity collapse also occurred on the ballistic axis, the measured pressure was large; on the other hand, the liquid-filled tank used was a closed tank with a large volume, resulting in a large cavity and higher pressure value generated during cavity collapse.

To investigate the effect of attack angle on the loads during the projectile penetrating the liquid cabin, No. 29–33 are selected for investigation, and the loads on the side walls of the water tank and at the horizontal distance of 7.5 cm between the front/rear plates and the center are measured. According to Figure 17, as the attack angle of the projectile increases, the initial shock wave pressure load on the side walls of the tank tends to increase first and then decrease; meanwhile, due to improved attack angle, the area of the projectile in contact with water increases, resulting in greater water disturbance, so the load on the front plate keeps increasing; on the contrary, the pressure load on the rear plate decreases with the increasing attack angle for the following reasons: when the projectile enters the water at the velocity of 2400 m/s; meanwhile, the wave velocity in the water is only 1500 us, and the projectile separates from the shock wave after water entry for a period of time; at this time, the projectile yaws, and the measuring point on the rear plate is far away from the projectile axis; the increasing attack angle leads to greater projectile yaw and farther measuring point from the axis, and thus the pressure load on the rear plate decreases with the increasing attack angle. The initial shock wave propagates far away in a hemispherical shape, and the pressure along the wave arc decreases with the angle of the wave moving away from the ballistic axis. The measuring point on the front plate is more than 90° away from the axis, and the measuring point on the rear plate is near the axis, so at small attack angles, the pressure load on the rear plate is greater than that on the front plate; as the attack angle increases, the pressure load on the front plate increases but decreases on the rear plate, with the former exceeding the latter. The specific impulse change in the initial shock wave loads on the front and rear plates is consistent with the variation law of the shock wave pressure peak, but the specific impulse change of the cavitation load on the front plate is not significant. The reason is: as the projectile enters the water, a large cavitation region is generated, resulting in the measuring point on the front plate entering the cavitation region only after being subjected to cavitation loading for a short period of time, not the complete cavitation load.

3.4. Analysis of Failure Modes of Liquid Cabin

Figure 18 presents the deformation morphology and failure diagrams and scanned contours of the front and rear plates before and after the tests. As illustrated, the front plate of the water tank undergoes shear plugging failure and thin film bulging deformation during the projectile penetration into the water tank at a high velocity. When the high-velocity projectile impacts the steel plate at an attack angle with water medium as the "dynamic support", the support of water improves the rigidity of the steel plate. As the front plate is thin, the impact on the target plate by the high-velocity projectile causes adiabatic shearing. The light blue color at the edge of the perforation hole is caused by the release of a large amount of heat during the contact between the projectile and the target. Because projectile penetration is at an attack angle, the shape of the hole is not circular, similar to the shape of the projectile nose, but rectangular, similar to the shape of the side of the projectile. The initial shock wave is generated after the projectile impacts the water and enters it. As the shock wave is close to the front plate, it bulges outward after being impacted. Then, due to the long cavitation, the water in the tank moves around and squeeze the front plate, leading to the wide bulging deformation of the front plate. However, when Wu conducted the penetration test with the water tank, the front plate was depressed, which was caused by the negative pressure induced by bubble contraction on the front plate due to the closure of the tank. In the present paper, as the tank is not covered, there is no plate depression. Comparing the plate deflection, we see that the kinetic energy of the projectile increases, but the deflection of the front plate does not increase apparently. This is explained by the fact that the front plate mainly undergoes thin film bulging deformation caused by the cavitation load, and the specific impulse of the cavitation load in the three groups of tests is basically the same.

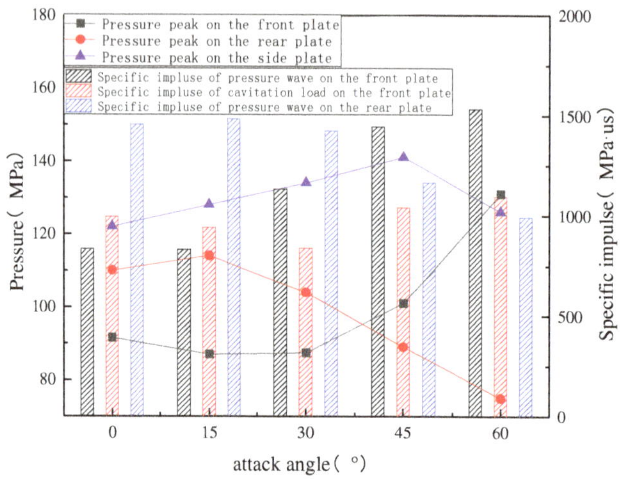

Figure 17. Peak value and specific impulse of pressure wave on each wall of the tank.

Due to the large thickness and stiffness of the rear plate, it has no obvious thin film bulging deformation. Since the projectiles used in Tests 1 and 3 have different masses and L/D ratios, under the same initial velocity, their kinetic energy varies, leading to very different failure modes of the rear plate. In Test 1, the rear plate mainly undergoes bulging and dishing deformation. Due to the long water domain, the incident shock wave generated after the water entry of the projectile is weakened to some extent when reaching the rear plate, having a small influence on the rear plate; as the projectile approaches the rear plate during penetration, a high-pressure region is produced at the position around the projectile nose as the projectile squeezes the water, and this high-pressure region acts on the rear plate, causing dishing deformation; then, the projectile impacts the rear plate, causing bulging deformation. After Test 3, the rear plate has annular breaches and radial cracks. The reason is: as the projectile moves in the water with an attack angle, it keeps overturning; it always has a large attack angle and a certain angular velocity upon reaching the rear plate; as the edge of the projectile nose first touches the target, which is similar to the penetration of a sharp-nosed projectile, the rear plate undergoes severe plastic deformation, and then the material is squeezed toward the cratering position by the sharp nose, causing annular breaches and radial cracks at the edge of the breaches; as the penetration continues, more cracks are generated and developed into petaling failure. In Tests 1 and 3, the kinetic energy ratio of the projectiles is 0.82, and the maximum deflection ratio of the rear plates is 0.60, which indicates that when the rear plate is approaching the ballistic limit, the impact force on the rear plate is increased by the projectile's pushing of the water; thus, the rear plate is subjected to the high pressure of the water and the impact of the projectile, resulting in two failures; the coupling effect of the two failures aggravates the damage to the rear plate.

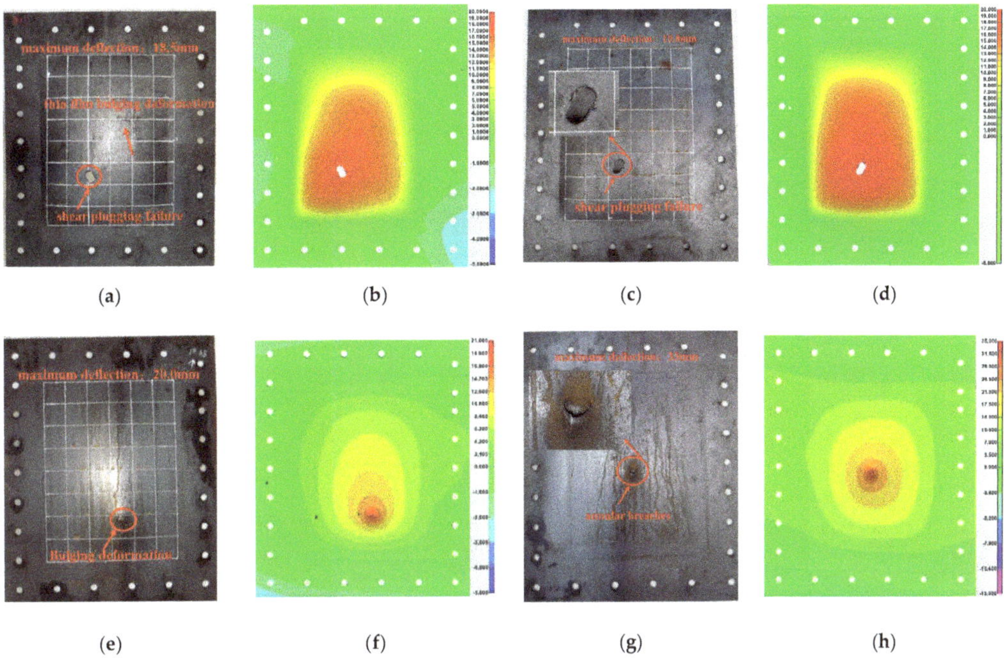

Figure 18. Deformation/failure diagram and deflection contour of the front and back plates. (the deflection unit: mm). (**a**) Deformation and failure diagram of the front plate in Test 1; (**b**) contour of deformation and deflection of the front plate in Test 1; (**c**) deformation and failure diagram of the front plate in Test 3; (**d**) contour of deformation and deflection of the front plate in Test 3; (**e**) deformation and failure diagram of the rear plate in Test 1; (**f**) contour of deformation and deflection of the rear plate in Test 1; (**g**) deformation and failure diagram of the rear plate in Test 3; (**h**) contour of deformation and deflection of the rear plate in No. 3.

Figure 19 shows the failure morphology of the liquid cabin after the penetration of the projectile with a L/D ratio of 4.8 and a velocity of 1600 m/s and at different attack angles (No. 19–23). As the attack angle increases, the breach in the front plate enlarges. During normal penetration, the front plate undergoes shear plugging-thin film bulging deformation, and the breach is circular. When an attack angle exists, the load applied by the projectile changes from a point load to a line load; thus, the area in contact with the target increases, and the shear plugging breach is rectangular. After the water entry of projectile, which is subjected to the shock wave pressure and cavitation load, the front plate undergoes bulging deformation; at the same time, the plate is subjected to the surface load of the water. Based on the classical yield line theory, the plastic hinge line in the rectangular plate is as shown in Figure 20, where the plate tears along the plastic hinge line. Since more water is displaced by the water entry projectile with an attack angle, the initial shock wave pressure and cavitation load on the front plate increase with the increasing attack angle: in the range of 15~30°, the front plate shows "I"-shaped tear along the plastic hinge line; in the range of 45~60°, the front plate is subjected to a greater load, and thus shows the "X"-shaped tear along the plastic hinge line.

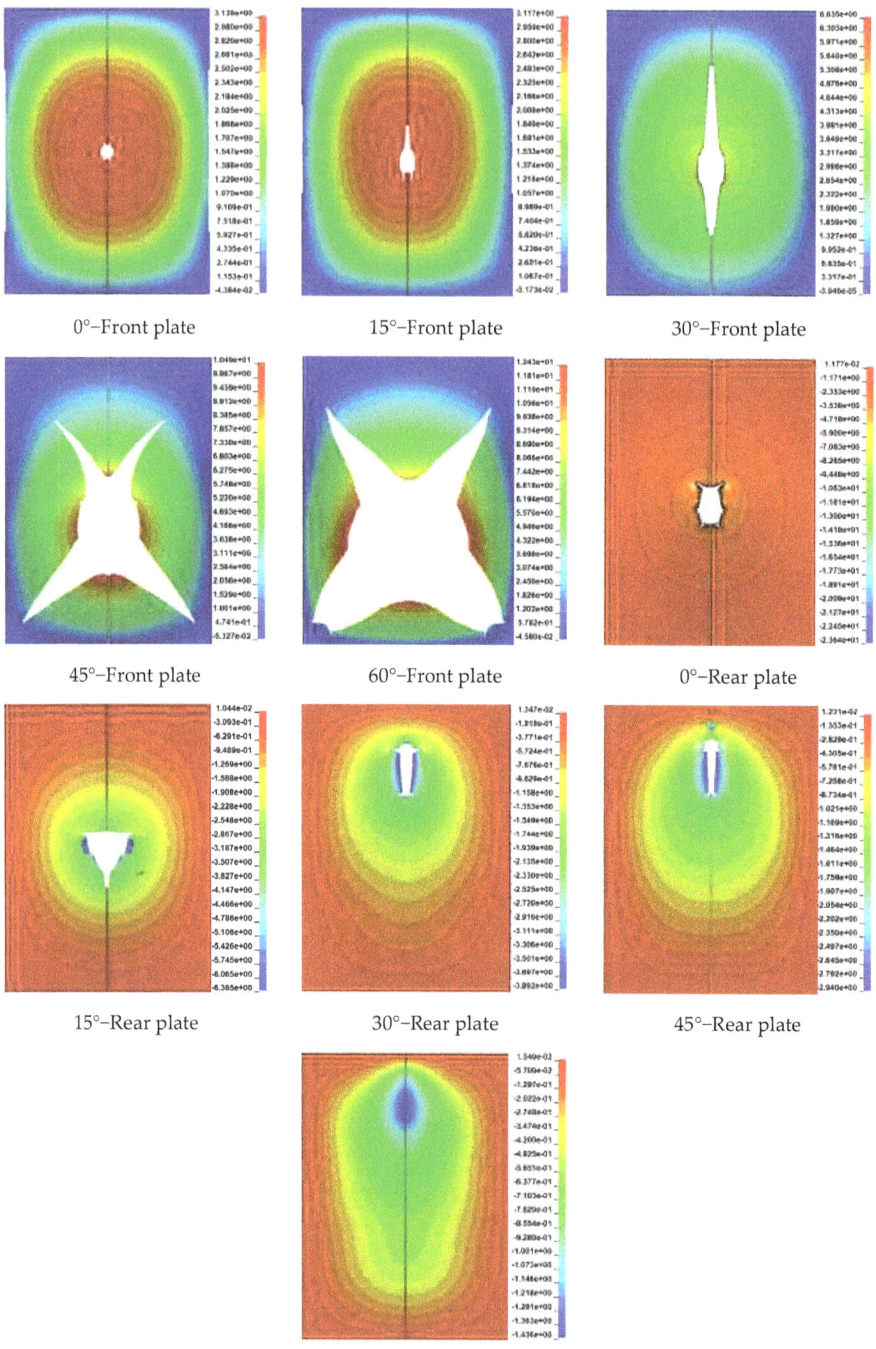

Figure 19. Failure morphology of the front and rear plates at different attack angles (deflection unit: cm).

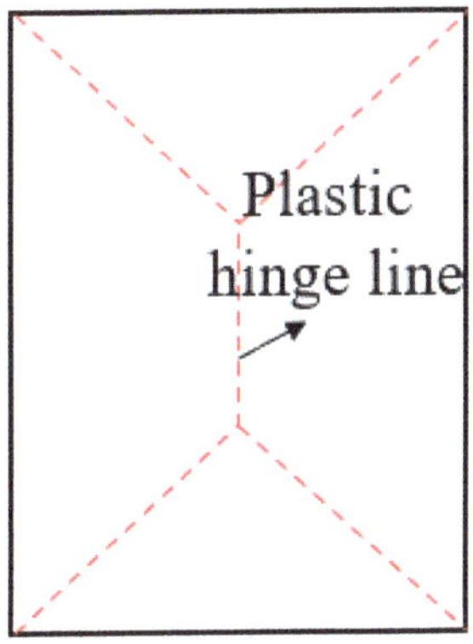

Figure 20. Plastic hinge line in the rectangular plate.

As the attack angle increases, the breach in the rear plate gradually decreases for the following reason: the projectile with an attack angle is subjected to greater drag in water; therefore, the projectile velocity decays faster; moreover, the front plate undergoes overall failure in advance under high attack angles; the liquid pressure is unloaded from the front plate, which reduces the damage to the rear plate; in the case of normal penetration, the rear plate undergoes dishing–petaling deformation; with the increase in the attack angle, the breach in the rear plate changes from the petal shape to the strip shape; in the range of 30~45°, the rear plate undergoes dishing- "I" tearing; at 60°, the projectile cannot perforate the rear plate; therefore, the rear plate exhibits dishing–bulging deformation.

4. Conclusions

In this paper, an effective numerical calculation method was obtained through the model test of the projectile penetrating the liquid tank and the numerical calculation verification with the experimental data. The deformation and failure modes of the projectile after penetration at attack angles, the trajectory and attitude change of the projectile in water were explored, and the load strength on the side walls and the failure modes of the front and rear plates during the penetration of the projectile into the water tank were analyzed. The main conclusions are as follows:

(a) There are four basic failure modes after the projectile penetrating the liquid cabin at attack angles: asymmetric mushrooming at the nose, side erosion, overall plastic bending and fracture. The overall plastic bending and fracture are mainly related to the L/D ratio, initial attack angle and initial projectile velocity; at low velocities, the main failure occurs during the penetration into the front plate; at high velocities, the main failure occurs during the penetration into the water.

(b) In the case of the rod projectile with a L/D ratio of 2 and a velocity of 1600 m/s, the projectile was more prone to overall deformation at the attack angles in the range of 15~30°. At low velocities, the overall plastic bending deformation occurred, whereas with the increasing velocity, fracture failure occurred; when the attack angle was above 30°, the failure mode of the projectile was side erosion, but as the initial velocity increased, the range of the attack angle that led to the overall deformation of the projectile also expanded. The larger the L/D ratio, the more easily the projectile undergoes overall plastic bending failure.

(c) After the projectile entered the water at an attack angle, the overturning moment was generated due to the uneven force on the projectile, and the projectile was at a state of constant overturning. Due to the change of the position contacting with the water, the projectile overturned many times, and projectile yaw occurred; the yaw direction was mainly affected by the initial attack angle and projectile attitude.

(d) After the water entry of the projectile, the side walls of the water tank were mainly affected by three stages of loading: the initial shock wave pressure loading, the cavitation loading and the cavity collapse loading. As the attack angle increased, the peak load of the initial shock wave pressure on the front plate increased gradually, whereas the initial shock wave pressure load on the rear plate decreased gradually.
(e) After the penetration of the projectile into the water tank, the failure modes of the front plate were mainly shear plugging, thin film bulging deformation and tearing failure. At certain velocities, with the increasing attack angle, the front plate underwent tearing along the plastic hinge line; under small attack angles, the tear was "I" shaped, and under large attack angles, it was "X" shaped; the rear plate mainly underwent dishing-bulging deformation; when the rear plate approached the ballistic limit, annular breaches and circumferential cracks were produced.

Author Contributions: Literature review, K.W. and Y.L.; writing—original draft preparation, K.W., H.H. and D.L.; writing—review and editing, K.W., D.L. and H.H.; experiment performance, Y.L. and K.W.; experiment analysis, K.W. All authors have read and agreed to the published version of the manuscript.

Funding: This research was funded by National Natural Science Foundation of China, grant numbers 51979277, 52271338, 52101378.

Institutional Review Board Statement: Not applicable.

Informed Consent Statement: Not applicable.

Data Availability Statement: Not applicable.

Acknowledgments: The authors would like to thank the editor, associate editor, and the anonymous reviewers for their helpful comments and suggestions that have improved this paper.

Conflicts of Interest: The authors declare no conflict of interest.

References

1. Jiao, W.J.; Chen, X.W. Review on long-rod penetration at hypervelocity. *Adv. Mech.* **2019**, *49*, 312–391.
2. Zu, X.D.; Huang, Z.X.; Zhu, C.S.; Xiao, Q.Q. Study of detonation wave contours in EFP warnose. *Def. Technol.* **2016**, *12*, 129–133. [CrossRef]
3. Wu, F.D. *Study on Damage Effect of Shaped Charge-Blasting Warhead on Multiple Hulls of Submarine*; North University of China: Taiyuan, China, 2019.
4. Zhu, X.; Zhang, Z.H.; Liu, R.Q.; Zhu, Y.X. Experimental study on the explosion resistance of cabin near shipboard of surface warship subjected to underwater contact explosion. *Explos. Shock. Waves* **2004**, *24*, 133–139.
5. Wu, Z.Q. *Study on Damage of Ship Target Penetrated by Anti-Ship Missle with Projectile-Target Combination*; Harbin Engineering University: Harbin, China, 2019.
6. Mcmillen, J.H.; Harvey, E.N. A Spark Shadow graphic Study of Body Waves in Water. *J. Appl. Phys.* **1946**, *17*, 541–555. [CrossRef]
7. Hu, M.Y.; Zhang, S.; Meng, Q.C.; Yin, W.B. Hydrodynamic characteristics of projectile oblique entry into water and hydrodynamic impulse load of projectile body. *J. Nav. Univ. Eng.* **2021**, *33*, 6.
8. Hao, B.; Dai, H.; Lyu, C. Research on Water Entry Trajectory and Hydrodynamic Characteristics of High-Speed Projectiles. *J. Ordnance Equip. Eng.* **2020**, *41*, 21–26.
9. Xia, W.X.; Wang, G.; Wei, J.J.; Zhang, X.S. Numerical simulation investigation on water-entry cavity of high-speed spinning sphere. *J. Harbin Inst. Technol.* **2018**, *50*, 138–144.
10. Wang, C.; Xia, W.X.; Chen, C.Q. Numerical Investigation on Multiphase Hydrodynamic Characteristics for Cylinder Water Entry. *Astronaut. Syst. Eng. Technol.* **2021**, *5*, 48–58.
11. Von, K.T. *The Impact of Seaplane Floats during Landing*; National Advisory Committee for Aeronautics NACA Technical Notes: Washington, DC, USA, 1929.
12. Wagner, V.H. Phenomena associated with impacts and sliding on liquid surfaces. *Z. Angew. Math. Mech.* **1932**, *12*, 193–215. [CrossRef]
13. Cointe, R. Two-dimensional water solid impact. *J. Shore Mech. Arct. Eng.* **1989**, *111*, 109–114. [CrossRef]
14. Takagi, K. Numerical evaluation of three-dimensional water impact by the displacement potential formulation. *J. Eng. Math.* **2004**, *48*, 339–352. [CrossRef]
15. Mirzaei, M.; Taghvaei, H.; Alishahi, M.M. Mathematical Modeling of the Oblique Water-entry of Cylindrical Projectiles. *Ocean. Eng.* **2020**, *205*, 107257. [CrossRef]

16. Alekseevskii, V.P. Penetration of a rod into a target at high velocity. *Combust. Explos. Shock. Waves* **1966**, *2*, 63–66. [CrossRef]
17. Tate, A. A theory for the deceleration of long rods after impact. *J. Mech. Phys. Solids* **1967**, *15*, 387–399. [CrossRef]
18. Song, Z.J.; Duan, W.Y.; Xu, G.D.; Zhao, B.B. Experimental and numerical study of the water entry of projectiles at high oblique entry speed. *Ocean. Eng.* **2020**, *211*, 107574. [CrossRef]
19. Chen, C.; Yuan, X.; Liu, X.; Dang, J. Experimental and Numerical Study on the Oblique Water-entry Impact of a Cavitating Vehicle with a Disk Cavitator. *Int. J. Nav. Archit. Ocean. Eng.* **2019**, *11*, 482–494. [CrossRef]
20. Gu, J.N.; Zhang, Z.H.; Zheng, X.L.; Jin, L.B. A Review of the Body's Water Entry Ballistics Research. *J. Nav. Acad. Eng.* **2000**, *12*, 18–23.
21. Li, T.X. *Research on the Characteristics of Entry Trajectory Steering*; Nanjing University of Science and Technology: Nanjing, China, 2020.
22. Ball, R.E. Appendix B: Probability Theory and Its Application to Survivability Assessment. In *The Fundamentals of Aircraft Combat Survivability: Analysis and Design*, 2nd ed.; American Institute of Aeronautics and Astronautics: Reston, VA, USA, 2003.
23. Shi, H.H.; Kume, M. An experimental research on the flow field of water entry by pressure measurements. *Phys. Fluids* **2001**, *13*, 347–349. [CrossRef]
24. Gao, S.Z.; Li, D.; Hou, H.L.; Li, Y.Q. Investigation on dynamic response of liquid-filled concave cell structures subject to the penetration of high-speed projectiles. *Thin-Walled Struct.* **2020**, *157*, 107119. [CrossRef]
25. Li, D.; Zhu, X.; Hou, H.L.; Zhong, Q. Finite element analysis of load characteristic of liquid-filled structure subjected to high velocity long-rod projectile penetration. *Explos. Shock. Waves* **2016**, *36*, 8.
26. Li, D.; Zhu, X.; Hou, H.L. Finite element analysis of energy dissipation mechanism of liquid-filled structure subjected to high velocity long-rod projectile penetration. *J. Nav. Univ. Eng.* **2018**, *30*, 6.
27. Wu, X.G.; Li, D.; Wu, G.M.; Hou, H.L.; Zhu, X.; Dai, W.X. Protection ability of liquid-filled structure subjected to penetration by high-velocity long-rod projectile. *Explos. Shock. Waves* **2018**, *38*, 76–84.
28. Disimile, P.J.; Toy, N. Video analysis of high velocity projectile entering fluid filled tank. *Results Eng.* **2019**, *2*, 100013. [CrossRef]
29. Deletombe, E.; Fabis, J.; Dupas, J.; Mortier, J.M. Experimental analysis of 7.62 mm hydrodynamic ram in containers. *J. Fluids Struct.* **2013**, *37*, 1–21. [CrossRef]
30. She, X.L.; Zhu, X.; Hou, H.L.; Zhou, X.B.; Zhao, H.G. Experimental Study on Penetration Properties of High Velocity Fragment into Safety Liquid Cabin. *Chin. J. Ship Res.* **2011**, *6*, 4.
31. She, X.L.; Zhu, X.; Hou, H.L.; Chen, C.H. Finite element analysis of underwater high velocity fragment mushrooming and penetration properties. *Ship Sci. Technol.* **2012**, *34*, 5.
32. She, X.L.; Zhu, X.; Zhao, H.G. Experimental analysis of underwater explosion fragment penetrating characteristic base on high-speed recording device. *Ship Sci. Technol.* **2015**, *4*, 5.
33. Artero-Guerrero, J.A.; Pernas-Sanchez, J.; Varas, D. Numerical analysis of CFRP fluid-filled tubes subjected to high-velocity impact. *Compos. Struct.* **2013**, *96*, 286–297. [CrossRef]
34. Nishida, M.; Tanaka, K. Experimental study of perforation and cracking of water-filled aluminum tubes impacted by steel spheres. *Int. J. Impact Eng.* **2006**, *32*, 2000–2016. [CrossRef]
35. David, T.; Nick, P.; Peter, M.D. Failure of fluid filled structures due to high velocity fragment impact. *Int. J. Impact Eng.* **2003**, *29*, 723–733.
36. Takashi, I. *Research on Underwater Ballistics*; National Defense Industry Press: Beijing, China, 1983; pp. 56–128.
37. Tian, Y.F. *Dynamic Characteristics of High-Velocity Water Entry of Projectile*; North University of China: Taiyuan, China, 2021.
38. Qi, Y. *Study on Ballistic Stability and Cavity Characteristics of High-Velocity Water Entry of Projectile*; Harbin Institute of Technology: Harbin, China, 2016.
39. Huang, K.; Le, S.W. Experimental research on the behavior of water-entry of different nose shape projectile models. *Phys. Exp.* **2016**, *36*, 13–18.
40. Zhang, W.; Guo, Z.T.; Xiao, X.K.; Wang, C. Experimental investigations on behaviors of projectile high-speed water entry. *Explos. Shock. Waves* **2011**, *31*, 6.
41. Hua, Y.; Shi, Y.; Pan, G.; Huang, Q.G. Experimental study on water-entry cavity and trajectory of vehicle with asymmetric nose shape. *J. Northwestern Polytech. Univ.* **2021**, *39*, 1249–1258. [CrossRef]
42. Wang, Y.; Yuan, X.L.; Lv, C. Experimental Research on Curved Trajectory of High-speed Water-entry projectile. *Acta Armamentarii* **2014**, *35*, 1998–2002.
43. Song, L.; Yu, H.Y.; Jie, B.B.; Huang, Z.G.; Chen, Z.H.; Tang, C.C. Experimental Study of Low Speed Vertical Water Entry with Moving Object of Asymmetric nose Type. *J. Ordnance Equip. Eng.* **2021**, *42*, 35–39.
44. Li, R.; Wang, R.; Yao, Z.; Wang, J. The Numerical Simulation of the Vertical Water Entry Process of High Speed Projectile with Small Angle of Attack. *J. Phys. Conf. Ser.* **2021**, *1865*, 042124. [CrossRef]
45. Truscott, T.T.; Techet, A.H. Water entry of spinning spheres. *J. Fluid Mech.* **2009**, *625*, 135–165. [CrossRef]
46. Techet, A.; Truscott, T. Water Entry of Projectiles. *Annu. Rev. Fluid Mech.* **2014**, *46*, 355–378.
47. Shi, Y.; Pan, G.; Yan, G.X.; Yim, S.C.; Jiang, J. Numerical study on the cavity characteristics and impact loads of AUV water entry-ScienceDirect. *Appl. Ocean. Res.* **2019**, *89*, 44–58. [CrossRef]

48. Liang, J.Q.; Wang, R.; Xu, B.C.; Qi, X.B.; Li, R.J. Research on Influence of Angle of Attack on Process of High-Speed Water-Entry Projectile. *J. Ordnance Equip. Eng.* **2020**, *41*, 23–28.
49. Wang, Z.; Wu, M.L.; Dai, K.L. Study on Load Characteristics of High-speed Water-entry of Large Caliber Projectile. *J. Ballist.* **2020**, *32*, 15–22.

applied sciences

MDPI

Article

Analysis of Damage of Typical Composite/Metal Connecting Structure in Aircraft under the Influences of High-Velocity Fragments

Yitao Wang [1], Teng Zhang [1,*], Yuting He [1], Jiyuan Ye [2], Hanzhe Zhang [1] and Xianghong Fan [1]

1 Aviation Engineering School, Air Force Engineering University, Xi'an 710038, China
2 School of Aeronautics, Northwestern Polytechnical University, Xi'an 710072, China
* Correspondence: mhuyjwqqedw@hotmail.com

Abstract: A two-stage light gas gun was used to conduct a high-velocity impact test on the aircraft's typical composite/metal connecting structure (CFRP/AL). The battle damage simulations used for the CFRP/AL connecting structure were carried out under different intersection conditions. Then, the damage morphology and mechanism of high-velocity prefabricated spherical fragments on typical structures, the dynamic process of hyper-velocity impact, and the formation of debris clouds on the secondary damage morphology of different component structures were investigated. Next, based on the X-ray computerized tomography (CT), the typical mode of different damage areas and evolution trends of CFRP under high-velocity impacts were explored. Finally, a simulation model was established for battle damages of typical structures by combining FEM methods, and structural components' energy dissipation capabilities for fragments under different velocities were analyzed. The study results provide a reference and model support for the rapid repair of battle-damaged aircraft and aircraft survivability design.

Keywords: battle damage; X-ray tomography; two-stage light gas gun; high-velocity fragment

Citation: Wang, Y.; Zhang, T.; He, Y.; Ye, J.; Zhang, H.; Fan, X. Analysis of Damage of Typical Composite/Metal Connecting Structure in Aircraft under the Influences of High-Velocity Fragments. *Appl. Sci.* **2022**, *12*, 9268. https://doi.org/10.3390/app12189268

Academic Editors: Ricardo Castedo, Lina M. López and Anastasio P. Santos

Received: 24 August 2022
Accepted: 12 September 2022
Published: 15 September 2022

1. Introduction

The problem of damage repair of aviation weapon equipment on the battlefield has a long history. According to statistics, the number of battle-damaged aircraft is much greater than the number of battle-destroyed aircraft. Hence, rapid repair of battle-damaged aircraft can have a significant impact on the war situation by increasing operational intensity and ensuring sustained combat capability. Indeed, rapid repair of battle-damaged aircraft has attracted great attention as it is the most effective way to restore aircraft combat effectiveness [1].

The battle damage mode of aircraft is to study all possible conditions and combinations of threat sources causing battle damage. The battle damage simulation based on rapid repair is different from weapon effectiveness analysis, and the combat effectiveness and survivability analysis of aircraft. The purpose is to provide guidance on rapid repair techniques and to provide an aid for the analysis of rapid repair resource requirements, usually focusing on the analysis and research of battle damage of aircraft structures. The combat aircraft is the main target of all types of air defense weapons.

Currently, carbon fiber composites have been widely applied in advanced fighters, especially in wing panels, vertical fins, fuselage skins, and rudders [2]. Carbon fiber laminates are the most important aerospace composite structure. The damage of a composite structure under the impact of combat fragments is different from that of a metallic structure, and its damage mechanism is more complex than that of metallic material. On one hand, the reinforced fiber limits the expansion of damage under the composite structure's own load, and on the other hand, the damage modes (such as the delamination of composite laminate structures) are not found in aerospace metal structures. The impact of the combat fragments on the composite laminate is a complex process of impact load evolution with time, structural deformation and damage extension. The impact damage of composite laminate is closely related to the laminate material's properties, lamination method, processing technology and the fragment's velocity, shape and quality. The damage mode and damage range of the laminate are different in different impact conditions. The possible damage mechanisms include fiber shear fracture, fiber tensile failure, matrix cracking, fiber matrix interface delamination and degumming. The damage mechanism and damage mode of a composite structure in a series of velocity ranges were investigated by numerical simulation and experiments. Thomas et al. conducted a hyper-speed impact test on CFRP/Al honeycomb composites and obtained the dynamic response of the structure with different thickness combinations [3]. Miao et al. conducted a hyper-speed impact test and impact damage test analysis to solve the protection effect problem of a spatial debris soft protection structure composed of a soft protection screen made of multilayer soft composites [4]. Phadnis et al. conducted a CFRP-Al/HC sandwich panel hyper-speed impact behavior analysis based on the finite element method. Coles et al. conducted ballistic tests on braided T300 carbon fiber/epoxy composite flat-plate specimens and 3D X-ray computer tomography (CT) was used to image and visualize the resultant damage inside the samples.

Modern aircraft are mostly multilayered thin-walled structures. Aircraft skin will often encounter penetrating damage under the action of high-velocity fragments, which has considerable penetration capabilities after penetrating the outer skin, and the secondary penetration of fragments will cause damage to the internal structure. The damage mode is closely related to the fragment's incident direction, incident velocity, and strike position [5].

The damage mode of aircraft composite/metal connecting structures under high-velocity impact has been little studied. Aiming at the carbon-fiber-reinforced laminate/Al alloy connection frame structure of a certain type of aircraft wing, the damage mode under the action of high-velocity spherical fragments (1600–2400 m/s) was investigated based on the numerical simulation model of simulated impact tests and the display dynamics of the two-stage light gas gun system.

2. Experimental

The multistage light gas gun is commonly used to achieve the high-velocity loading of small-sized projectiles and is often used in tests related to spacecraft collisions with space debris, and its technical indicators can achieve the requirements for fragment intrusion into aircraft structures [6]. The characteristic structure equal ratio test piece, fragment and the corresponding Sabot were prepared. Based on the two-stage light gas gun system, the simulated penetration of prefabricated spherical fragments into aircraft-typical vulnerable structures under different working conditions was realized.

2.1. Instrument

The two-stage light gas gun was used as the launcher, and a 10# steel ball (diameter = 8 mm) was used to simulate the prefabricated fragment of the missile, as shown in Figure 1. The target structure is fixed in the chamber by the fixture, adjusted to the corresponding strike position, and the intended test strike position is marked by the laser pointer, as shown in Figures 2 and 3.

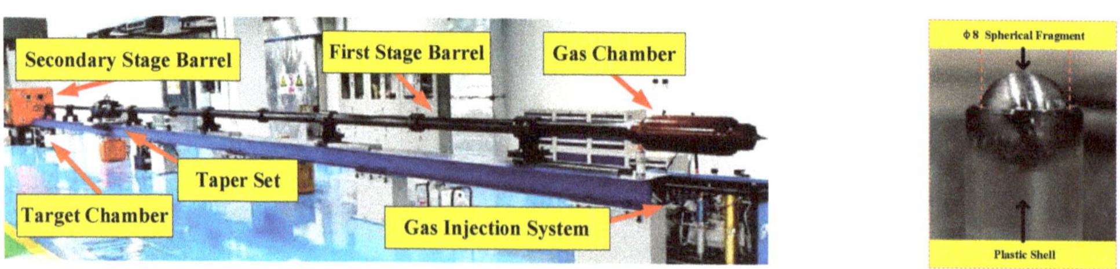

Figure 1. Two-stage light gas gun and simulation of spherical fragments.

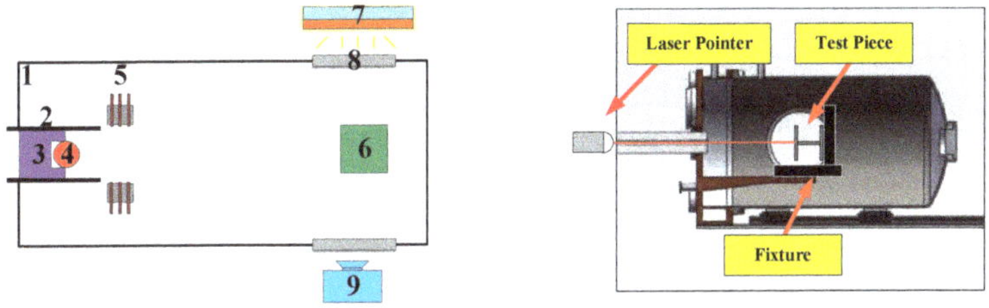

Figure 2. Schematic of the two-stage light gas gun: 1: target chamber; 2: gas gun barrel; 3: sabot; 4: projectile; 5: optical beam blocking (OBB) system; 6: target; 7: flash lamp; 8: windows; 9: high-velocity camera.

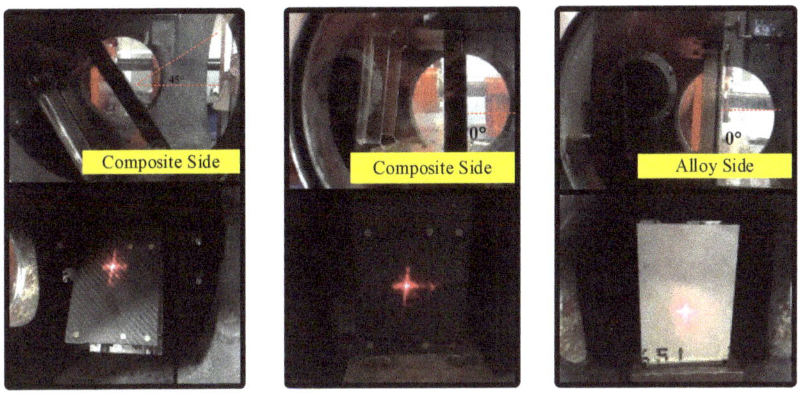

(**a**) (**b**) (**c**)

Figure 3. Tooling settings of the test piece. (**a**) Situation 1; (**b**) Situation 2; (**c**) Situation 3.

The target chamber is connected to the end of the secondary stage barrel, which is equipped with an optical beam blocking (OBB) sensor system to measure the exit velocity of the launched projectile. Three laser beams of similar intensity are placed in the projectile's path and directed to the photoelectric tube connected to the timer. The skimming of the projectile will block the laser beam, and the resulting pulse signal will be recorded on the timer by the photoelectric tube. The projectile velocity is obtained by comparing the time interval of the timer and the distance between the laser beams. The target chamber is equipped with bullet-proof glass observation windows on both sides and a high-velocity camera (Phantom V2512) for capturing the impact process, and a flash for the high-velocity camera to fill in the light. The high-velocity camera uses a 50 mm fixed-focus lens, and records at a pixel of 386 × 216 with a frame rate of 160 kfps, as shown in Figure 4.

Figure 4. Observation section setting of the target chamber.

The YXLON FF85 CT scanning system is equipped with two sets of radiation tubes and a large-sized flat detector for the detection of all types of damage on fiber-enhanced composites. The inspection sample is scanned layer by layer by emitting X-ray, and in combination with the analysis software, the composite can be inspected and global damage 3D modeling and rendering can be achieved. In this study, it is mainly used to implement the internal damage analysis and overall damage assessment of the composite, as shown in Figure 5.

Figure 5. YXLON FF85 CT scanning system.

2.2. Test Piece

The composite/Al alloy joint spacer frame structure test piece, as shown in Figure 6, consists of an Al alloy top skin, rib 1, rib 2, and with a composite bottom skin and stringer. The top skin is connected to rib 1 and rib 2 by 8 mm rivets, and the composite bottom skin is connected to rib 1 and rib 2 by high-locking bolts. The Al alloy is AL7075-T6 produced by Chalco Group, using the GB/T29503-2013 standard, and the composite material is T300/QY8911 epoxy resin-based carbon fiber unidirectional laminate. It contains bottom skin (45/−45/0/−45/0/45/0/45/0/45/−45/0/90/45/90/45/90/0/−45/45/0/−45/45/0/45/0/−45/0/−45/0/−45/45) and stringer (45/−45/0/−45/45/0/−45/0/−45/0/45/0/90/90/0/45/0/−45/0/45/−45/0/−45/45). The laminate was laid orthogonally and symmetrically with plain woven prepreg, as shown in Figure 7.

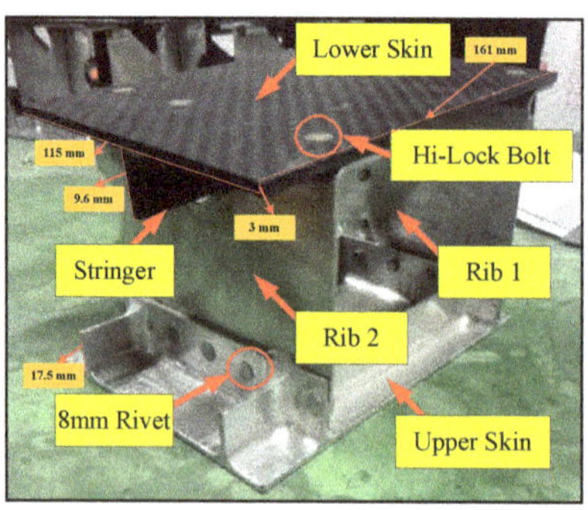

Figure 6. Test piece of composite/Al alloy connecting frame structure.

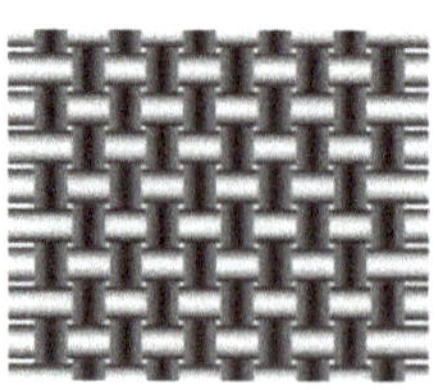

Figure 7. Schematic of composite layup method.

3. Damage Analysis

3.1. Analysis of Impact Process

A high-speed camera (frame rate 160 kfps) is used to capture the dynamic process of the intersection of the spherical fragments with the structure, as shown in Figure 8.

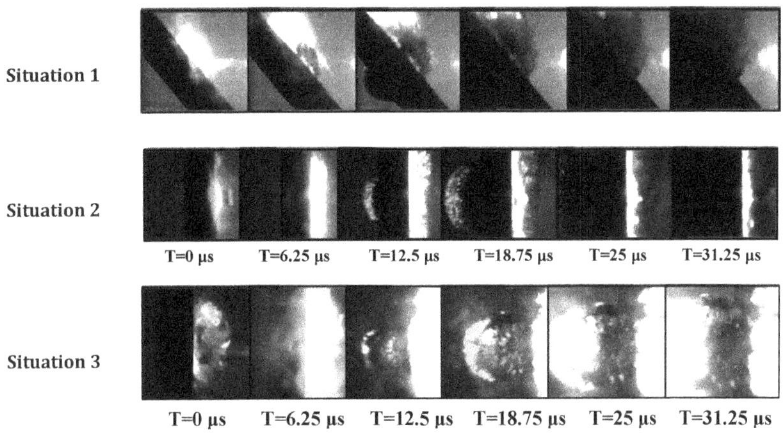

Figure 8. Impact process of test piece of composite/Al alloy lapping frame structure.

Situation 1: The fragment is incident from the bottom skin composite side at an incidence angle of 45°, and the measured fragment exit velocity is 2327 m/s. The moment the fragment intersects with the structure, a strong photo-thermal phenomenon is generated at the impact location, resulting in local overexposure of the high-velocity camera. Subsequently, the fragment penetrates the structure and forms a debris cloud on the outside of the structure, accompanied by a splash of massive fiber. As the invasion progresses, the composite laminate forms a bulge. The rapid rupture of the bulge forms a debris cloud, which is mainly composed of broken small-diameter carbon fiber particles. The outer debris cloud at the incident end expands radially along the vertical direction of fragment incidence, and its main components are bulk fiber debris and fragment metal particles, while the inner debris cloud at the incident end expands radially along the direction of fragment incidence, and its main components are small-diameter carbon fiber debris groups.

Situation 2: The dynamic intersection process of the fragment and structure in Situation 2 is similar to that shown in Situation 1. Due to the change in the direction of incidence, there is a difference in the expansion pattern of the debris cloud between the two. In Situation 2, the difference between the inner and outer debris cloud's highlight phenomena is more obvious, again due to the difference in their main components. The small-diameter carbon fiber debris group has a black main layer with dense spatial distribution and strong light absorption, and the photo-thermal phenomena captured by the high-velocity camera in the inner part of the incident section are weak.

Situation 3: The fragment intruded from the top skin Al alloy side, and the photo-thermal phenomena were stronger both inside and outside. In Situation 1/2, when the projectile intrudes from the composite side, unlike the metal side, the carbon fiber breaks up to generate a large amount of dusty debris cloud, and the debris cloud is dominated by the carbon fiber debris, which will cover the entire high-velocity camera capture area, and the radial velocity is larger than the incident velocity during the expansion of the debris cloud. It is because the composite is made out of anisotropic material that when the impact direction is perpendicular to the fiber layup, the fiber is susceptible to shear fracture, and its normal strength is much lower than that of the metal material, resulting in a difference in the morphology of the bulge formed in the impact process, and therefore, the morphology of debris cloud diffusion formed in the final bulge rupture is also different.

The shock photo-thermal phenomena in Situation 1/2 are not as obvious as in Situation 3, and the main component of the firelight is the high-temperature metal fragments. Due to the violent friction between steel spherical fragments and Al alloys during the impact penetration process, the temperature is extremely high, resulting in the appearance of small metallic debris and the release of part of the heat accumulated by friction in the pattern of luminescence.

3.2. Composite Damage Analysis

In Situation 1/2, the damage morphology of the structural composite top skin and the stringer intrusion by the high-velocity spherical fragments are shown in Figures 9 and 10, respectively. The damage broken hole is mainly ellipsoidal under the 45° oblique impact of the fragment in Situation 1. In Situation 2, under the positive impact of fragment 0°, the damage broken hole is mainly in the pattern of regular spherical rupture. In Situation 1/2, there is fiber spalling on the outer surface of the incidence. The stringer part is thin and shows a band-like broken hole under the shearing effect in both Situation 1/2.

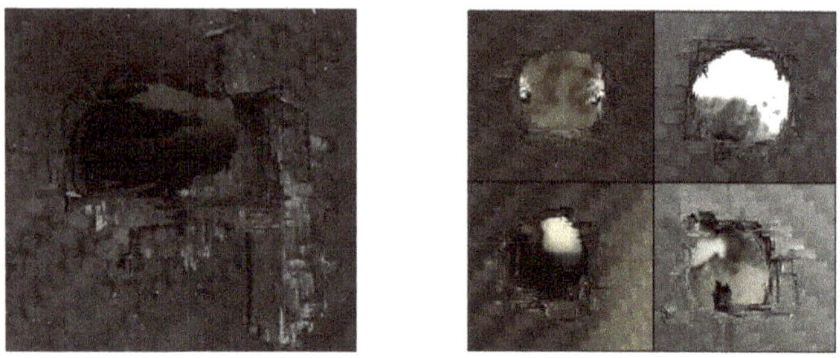

Figure 9. Typical shear damage mode of composite bottom skin.

Figure 10. Typical damage mode of composite stringer.

In Situation 3, the fragment is incident from the Al alloy metal side. According to Figure 11, the fragment formed a debris cloud consisting of a large number of metal particles when it penetrated the top skin. The debris cloud still has a high kinetic energy, and the secondary damage formed at the bottom skin, which is mainly in the pattern of small-diameter broken holes and surface spalling, is widely distributed.

The test piece composite part was scanned by CT, and the damage morphology feature images of the composite test piece were obtained for each directional interface, layer by layer, and then the 3D view of the damage of the test piece composite part under each working condition was obtained by image rendering, as shown in Figure 12.

Figure 11. Damage mode of secondary damage of composite bottom skin.

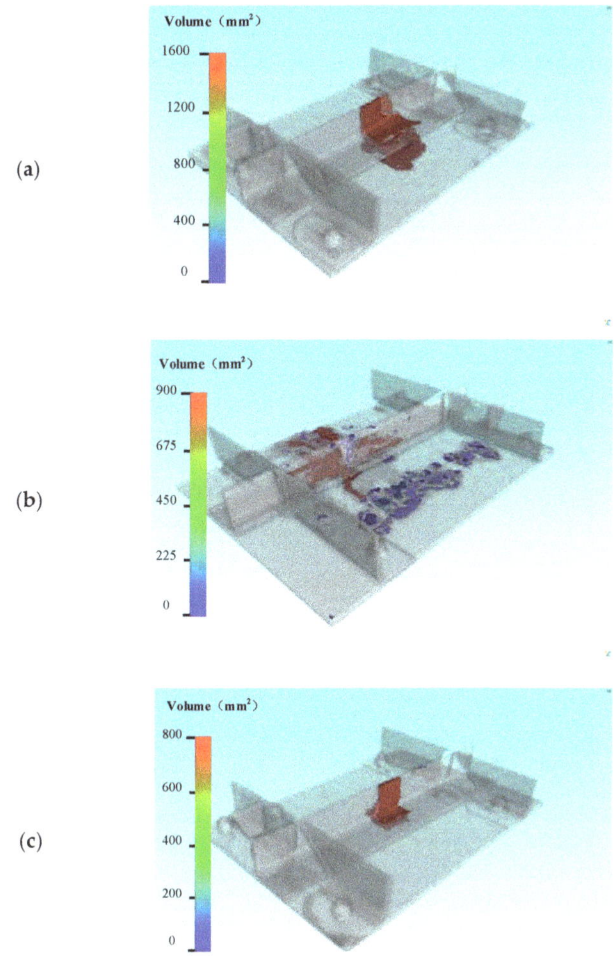

Figure 12. 3D CT damage player diagram. (**a**) Situation 1; (**b**) Situation 2; (**c**) Situation 3.

The section of composite laminate damage area under high-velocity impact of spherical fragments is mainly in the pattern of a combination of cylindrical and circular truncated cones, as shown in Figure 13. In his study on the impact damage of aramid laminate,

Reddy pointed out that the damage aperture of thick plates decreases slightly along the thickness direction and then increases rapidly, while the aperture of thin plates expands in the shape of a circular truncated cone, and Cantwell et al. used the combined area of a cylindrical and circular truncated cone [7–9]. In the early stage of projectile penetration, the front side of the laminate would generate a shear-plugging hole similar to the shape of the projectile contact area, where the matrix material is crushed and loses its support to the fiber, and shear failure occurs between the fiber and the surrounding fiber due to the presence of a large velocity gradient, which is called the shear failure zone (A). With the continuation of the penetration process, the projectile velocity decreases, the target plate bends, and with the continuous expansion of the bending deformation, fiber tensile failure occurs first in the outermost layer of the back side of the impact, and the tensile failure expands from the outer layer to the back side with a certain crack inclination angle and produces delamination, forming a fiber tensile failure zone behind the shear failure zone (B). Meanwhile, there is a small delamination damage zone around the shear failure zone/fiber tensile failure zone (C). In this test, the delamination zone of laminate caused by the high-velocity impact of spherical fragments is limited, and the fiber in the zone is mainly recoverable deformation without significant fiber damage.

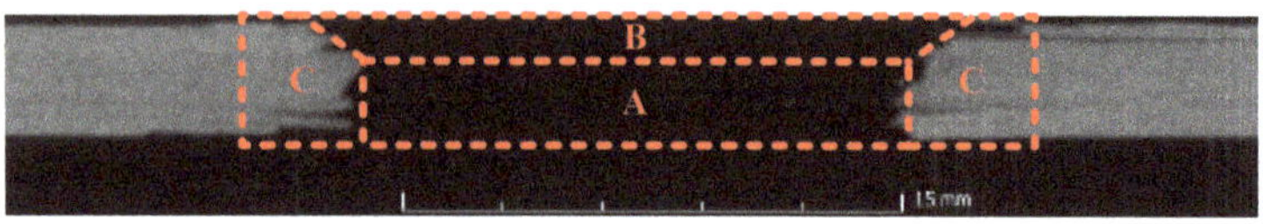

Figure 13. Distribution of damaged area.

The projection of the laminate damage area in the direction of fragment incidence shows different morphological features in different damage areas, as shown in Figure 14. Herein, the position of the blue line is the position of the broken hole section in the vertical plane. On the upper surface of the shear failure zone, the main body of the broken hole section is circular. The laminate-free surface is strongly impacted by spherical fragments, and there are more striated fiber-stripping areas around the circular rupture holes. As the penetration progresses, the breach section inside the shear failure zone is mainly regular circular, and the area of the spalling zone decreases. At the shear failure zone/fiber tensile failure zone intersection, the breach section has common features of both. The shape of the breach section is transformed from a shear failure-oriented circle to a fiber tensile failure-oriented square. As the penetration progresses, the breach section in the fiber tensile failure zone appears as a regular square.

High-speed fragmentation spherical fragment intrusion CFRP, in the process of damage formation, can be divided into four stages as shown in Figure 14. In shear and tensile damage, the compression wave generated by the spherical fragment acting on the target plate propagates faster than the projectile movement, and the strong tensile wave formed by the reflection of the compression wave on the back of the target plate meets the projectile, and the meeting point is the interface between shear damage and tensile damage. When the tensile stress is greater than the bonding strength of the fibers and the substrate or the tensile strength of matrix material, tensile stress is induced at the defective area and is accompanied by partial delamination. Eventually the fibers fracture and splash under the impact, which is in line with the phenomenon captured by the high-speed camera.

As shown in Figure 15, from left to right are the upper surface, shear failure zone, area boundary, tensile failure zone, and lower surface of the broken hole morphology, respectively. The evolutionary pattern of breach section morphology shows a similar pattern under different velocities. Notably, as the fragment impact velocity decreases, the size of the laminate-free surface spalling area decreases significantly.

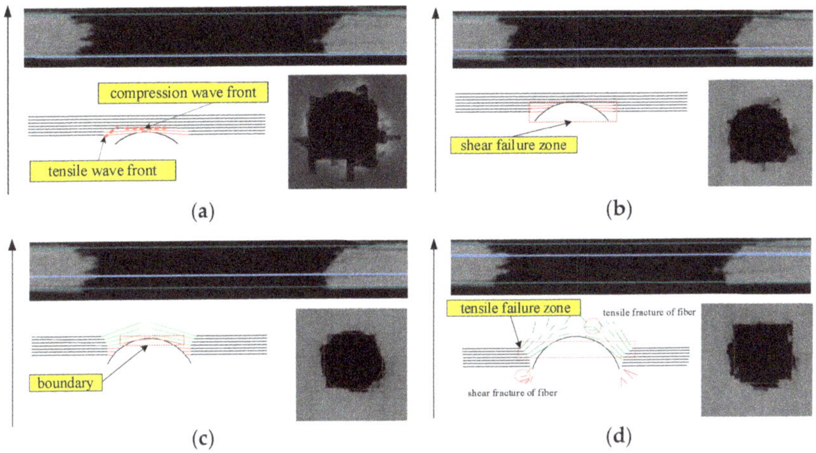

Figure 14. Morphological features of the CFRP damage process. (**a**) Impact crater extrusion, (**b**) Shear intrusion, (**c**) Tensile intrusion, and (**d**) fiber fracture splashing.

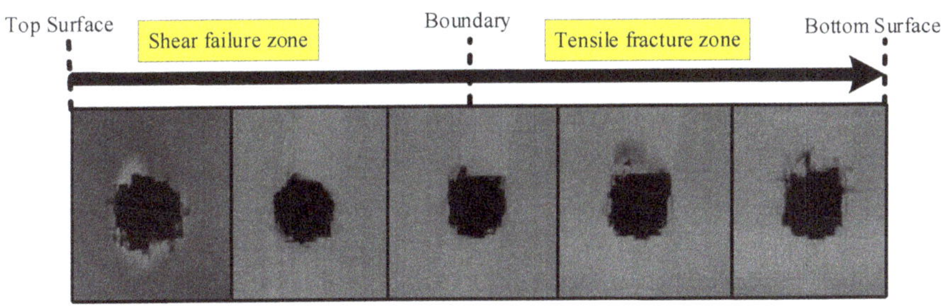

Figure 15. Morphology of damaged area in CFRP.

3.3. Analysis of Al Alloy Damage

Under the Situation 1 incidence condition, the fragment penetrated the bottom skin and a debris cloud hit the rib of the structure. The fragment's main body with small-diameter debris intruded to generate an ellipsoidal rupture hole. The thinner part of the rib is torn and fractured by the impact, as shown in Figure 16.

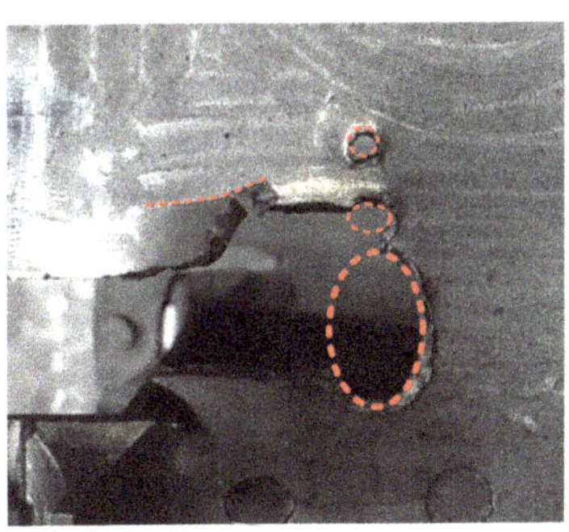

Figure 16. Top skin damage in Situation 1.

For Situation 2's incidence conditions, four sets of velocity gradient variables were set in the range of the actual fragment velocity. The debris cloud composed of small-diameter carbon fiber particles formed by the fragment penetrating the bottom skin composite side does not have the energy to generate a distributed secondary image on the top skin side of the Al alloy. The top skin side damage is mainly caused by residual fragment penetration. The spherical fragments are separated into two parts by the erosion of composite laminate and the shearing effect of stringer, forming shear holes in the top skin. From the contact between the fragment and the top skin, the annular shear stress formed by the impact is much larger than the panel's ultimate strength, the annular shear zone then gradually accumulates and expands to the back of the panel to generate an annular shear surface. The shear punching is completed to generate a bulge, the radial tensile stress at the back of the back panel bulge then rises, with the depth of the intrusion, the tensile stress accumulates and expands, when the accumulation exceeds the ultimate strength of the panel, and the bulge fractures and breaks rapidly, forming a petal-shaped irregular fracture. The greater the initial velocity of the fragment, the greater the reaction force received in the impact process, and the farther apart the two separated parts will be. As the velocity decreases, the two holes are connected and eventually generate a single hole, as shown in Figure 17.

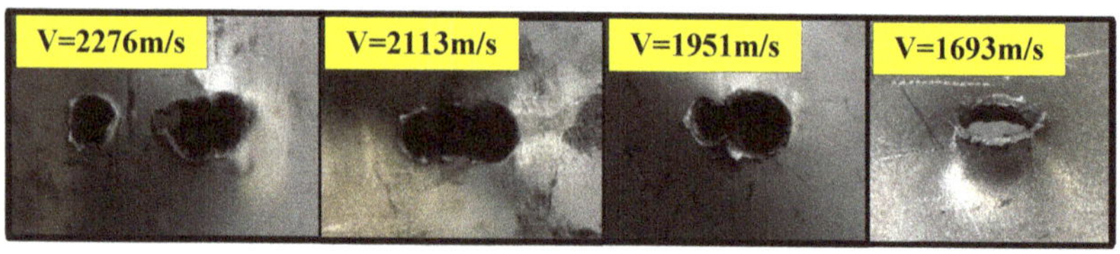

Figure 17. Top skin damage in Situation 2.

Under Situation 3's incidence conditions, the top skin Al alloy exhibits a relatively regular spherical shape in the entire damage area under the high-shearing effect of the fragment's high-velocity impact, and its area is basically the same as the area of the orthogonal projection of the spherical fragments. The reinforced ribs develop a columnar erosion zone under the high-shearing effect of the fragment, and the edges show the typical metal cutting marks under the high-shearing effect and the ablation marks caused by the accumulated heat of impact, as shown in Figure 18.

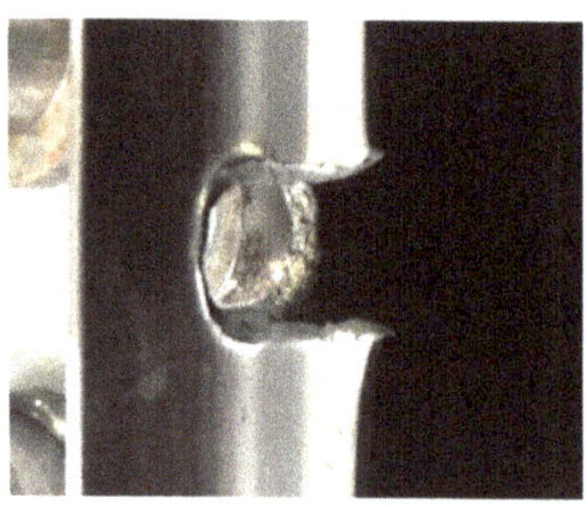

Figure 18. Top skin damage in Situation 3.

3.4. Numerical Modeling

In order to compensate for the limitation of the number of experiments, a fragment impact composite/metal connecting structure model based on LS-DYNA was established. In composite modeling, the * PART_COMPOSITE keyword is used to define the basic physical parameters such as the thickness of each layer of the composite carbon fiber laminate component, and the layup direction. In order to better reflect the loss between different layers independently in the modeling, and to take into account the computational scale and efficiency, a modeling scheme of one 2D shell cell layer is used instead of four to five actual layers, i.e., five 2D shell cell layers (i.e., elements in the meshing) are used in the simulation modeling instead of 25 actual layers of composite material for the bottom skin components, and five 2D shell unit layers are used instead of 22 actual layers of composite material for the stringer components.

The * MAT_COMPOSITE_DAMAGE model is a composite constitutive model commonly used on shell units, containing physical quantities such as density, fiber elastic modulus in three directions, Poisson's ratio, shear modulus and longitudinal tensile strength, and transverse tensile strength, and can be used to define various orthogonal anisotropic materials with brittle fractures, which is applicable to the T300/QY8911 epoxy resin-based carbon fiber unidirectional laminate in this study. The single-layer carbon fiber laminate thickness direction size is much smaller than the other direction size. Hence, it can be analyzed according to the plane stress problem, considering only the in-plane stress state, ignoring the plane normal upward stress. The stress–strain relationship can be expressed as:

$$[\sigma] = [S][\varepsilon] \tag{1}$$

$$\frac{1}{[S]} = \begin{bmatrix} \frac{1}{E_{11}} & \frac{\nu_{12}}{E_{11}} & \frac{-\nu_{31}}{E_{33}} & & 0 & \\ \frac{-\nu_{12}}{E_{11}} & \frac{1}{E_{22}} & \frac{-\nu_{23}}{E_{23}} & & & \\ \frac{-\nu_{31}}{E_{33}} & \frac{-\nu_{23}}{E_{23}} & \frac{1}{E_{33}} & & & \\ & & & \frac{1}{2G_{12}} & 0 & 0 \\ & 0 & & 0 & \frac{1}{2G_{23}} & 0 \\ & & & 0 & 0 & \frac{1}{2G_{31}} \end{bmatrix} \tag{2}$$

where σ is the stress, ε is the strain, E is the elastic modulus, γ is the Poisson's ratio, and G is the shear modulus.

The * MAT_COMPOSITE_DAMAGE model uses the Chang–Chang failure criterion and has the following four failure modes:

(1) If $\sigma_{aa} > 0$, fiber is in the stretched state, when satisfied:

$$\left(\frac{\sigma_{aa}}{X_T}\right)^2 + \beta\left(\frac{\sigma_{ab}}{S_c}\right) - 1 \geq 0 \tag{3}$$

Herein, $E_a = E_b = \nu_{ba} = \nu_{ab} = G_{ab} = 0$ and the fiber undergoes stretching failure;

(2) If $\sigma_{aa} < 0$, the fiber is in compression, when the following conditions are met:

$$\left(\frac{\sigma_{aa}}{X_C}\right)^2 - 1 \geq 0 \tag{4}$$

Herein, $E_a = \nu_{ba} = \nu_{ab} = 0$ and the fiber fails in compression;

(3) If $\sigma_{bb} > 0$, the fiber matrix is in a stretched state, when the following conditions are met:

$$\left(\frac{\sigma_{aa}}{Y_T}\right)^2 + \left(\frac{\sigma_{ab}}{S_c}\right) - 1 \geq 0 \tag{5}$$

Herein, $E_a = \nu_{ba} = G_{ab} = 0$ and the fiber matrix undergoes a stretching failure;

(4) If $\sigma_{bb} < 0$, the fiber matrix is in compression, when the following conditions are met:

$$\left(\frac{\sigma_{bb}}{2S_C}\right)^2 + \left[\left(\frac{Y_C}{2S_C}\right)^2 - 1\right]\frac{\sigma_{bb}}{Y_C} + \left(\frac{\sigma_{ab}}{S_C}\right)^2 - 1 \geq 0 \tag{6}$$

Herein, $E_a = \nu_{ba} = \nu_{ab} = G_{ab} = 0$ and the fiber matrix fails in compression.

The Johnson–Cook constitutive model, Mie–Gruneisen equation of state, and maximum tensile stress damage criterion were used for the Al alloy in the structure. The Johnson–Cook model [10] differs from the common plastic theory in that it characterizes the material response to impact and penetration through parameters such as processing hardening, deformation rate effects and thermal softening. Each parameter is multiplied to characterize the cumulative effect of each effect.

$$\sigma_y = \left[A + B\left(\varepsilon^p_{eff}\right)^n\right](1 + C\ln\varepsilon)\left[1 - (T_H)^m\right] \tag{7}$$

In Equation (7), ε^p_{eff} is the effective plastic strain; $\varepsilon = \frac{\varepsilon^p_{eff}}{\varepsilon_0}$, where ε_0 is the strain rate used to determine A, B, and n; $T_H = \frac{T-T_R}{T_M-T_R}$ is the homologous temperature; T_M is the melting temperature; T_R is the reference temperature; $\Delta T = \frac{1}{\rho C_p}\int \sigma d\varepsilon^p_{eff}$, where ρ is the density, and C_P is the specific heat. The five parameters A, B, n, m and C in the model are basic parameters for characterizing the yield strength, where A is the initial yield strength of the material under the quasi-static strain rate, B and n are the flow stress of the strain-hardening behavior under the quasi-static strain rate, C is the strain rate effect, and m is the thermal softening effect. In addition to the material properties ρ, C_P, and T_M, there are also elastic parameters. Usually, the pressure is defined as a function of the volume strain response, and the shear modulus is integrated along the equation of state [11].

The cumulative damage of the material is used to characterize the failure of the material in the J-C constitutive, as shown in Equation:

$$\varepsilon^F = \left(D_1 + D_2\exp\left[D_3\frac{P}{\sigma_{eff}}\right]\right)(1 + D_4\ln\varepsilon)(1 + D_5 T_H) \tag{8}$$

where $D = \sum\frac{\Delta\varepsilon^P_{eff}}{\varepsilon^F}$ the material failure occurs when D = 1 where ε_{eff} is the effective stress, P is the average stress. The parameters of the Johnson–Cook model for the Al7075-T6Al alloy and the parameters of the Mie–Gruneisen equation of state are shown in Table 1 [12].

Table 1. Al7075-T6 Johnson–Cook model and Mie–Gruneisen EOS parameters.

Parameters	Symbol	7075-T6
	Johnson–Cook model parameters	
Density (kg/m^3)	R0	2.81
Poisson's ratio	PR	0.33
Shear modulus (GPa)	E	0.717
Static yield limit (MPa)	A	0.00546
Strain hardening modulus [13]	B	0.00678
Strain hardening exponent	n	0.71
Strain rate coefficient	C	0.35
Spall type	SPALL	3
Failure parameters D_1	D1	−0.068
Failure parameters D_2	D2	0.451
Failure parameters D_3	D3	−0.952
Failure parameters D_4	D4	0.036
Failure parameters D_5	D5	0.697
	Mie–Gruneisen EOS parameters	
Constants C	C	0.535
Constants S_1	S1	1.34
Constants γ	GAMAO	2.17

T300/QY8911 related material parameters are shown in Table 2.

Table 2. T300/QY8911 material parameters [14].

Parameter	Symbol	T300/QY8911
Density	R0	1.6
Elastic modulus along the a direction	EA	1.32
Elastic modulus along the b direction	EB	0.073
Elastic modulus along the c direction	EC	0.073
ba/ca Poisson's ratio	PRBA/PRCA	0.03
cb Poisson's ratio	PRCB	0.31
Shear strength	SC	0.00079
Tensile strength along the a direction	XT	0.049
Tensile strength along the b direction	YT	4.8
Compressive strength along the b direction	YC	0.002

The simulated spherical fragments are divided by a uniform mesh with a mesh size of about 0.3 mm, using hexahedral eight-node units with a total number of 56,000 units, as shown in Figure 19.

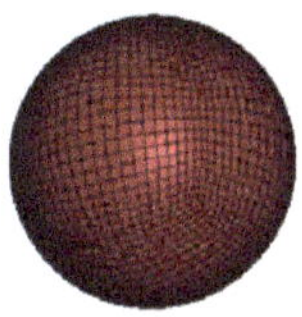

Figure 19. Finite element model of spherical fragments.

The structural metal part of the model mesh uses a hexahedral deca-node unit, and the composite part of the model consists of a 2D shell unit, with a single sub-layer containing three layers of actual layup information. * CONTACT_SURFACE_TO_SURFACE_TIEBREAK [15,16] is used between layers. The total number of model units for the air inlet Al alloy I-beam riveted structure is 241,437, and the total number of model units for the wing composite/Al alloy spacer structure is 1,099,060. In order to improve the overall computational efficiency and ensure the computational accuracy, the local mesh refinement method is used to divide the model into two density meshes, where the impact penetration part is encrypted mesh, and the two are connected by the trapezoidal transition mesh co-node method, as shown in Figure 20.

The fragment is set up with * CONTACT_AUTOMATIC_SURFACE_TO_SURFACE automatic face-to-face contact and * CONTACT_AUTOMATIC_SINGLE_SURFACE automatic single-sided contact between the fragment and the structure.

Typical damage modes of the composite bottom skin and stringer obtained by experiments and simulation are shown in Figures 21 and 22, respectively. In terms of characteristic damage size, the diameter of openings and penetrations obtained from the simulation is close to that of the test. Since the composite simulation model uses 2D shell unit modeling, it cannot simulate the damage morphology of fiber fracture and spalling, and the Mat_Composite_Damage model does not consider the effect of temperature on overall damage. However, there is a small amount of fiber-melting phenomena in the actual test. Therefore, the characteristic damage size of a composite obtained from simulation is relatively small compared with the actual one, but the relative error is not big, and it can meet the requirements of battle damage size prediction to some extent. In terms of the Al alloy side damage morphology, the simulated results are in high agreement with the test, and the difference in feature size is small, as shown in Figure 23.

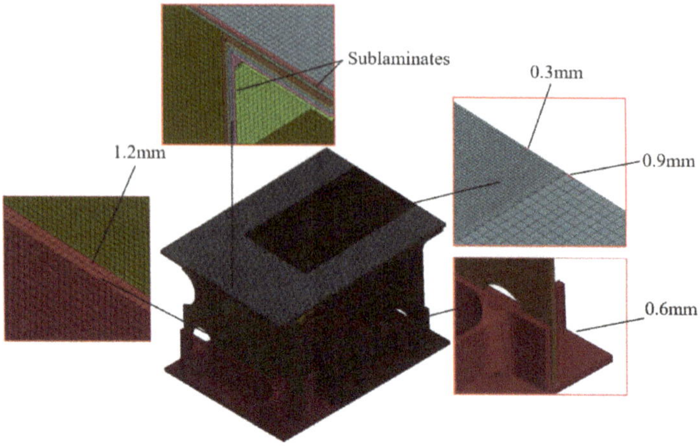

Figure 20. Finite element model of the structure.

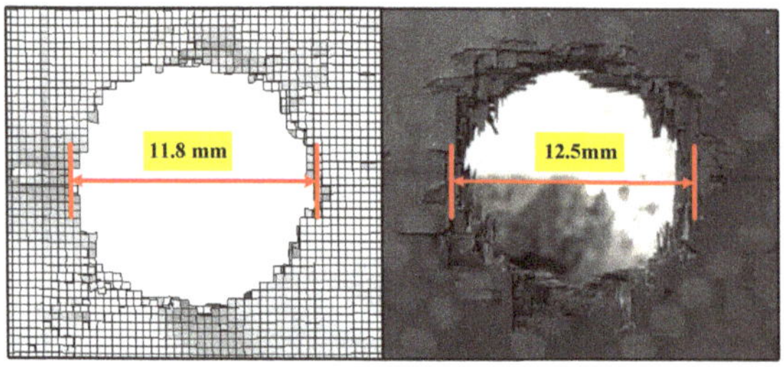

Figure 21. Damages of composite bottom skin damage.

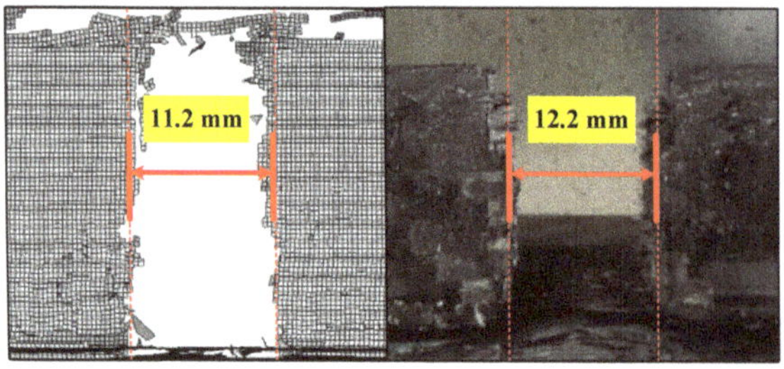

Figure 22. Damages of composite stringer.

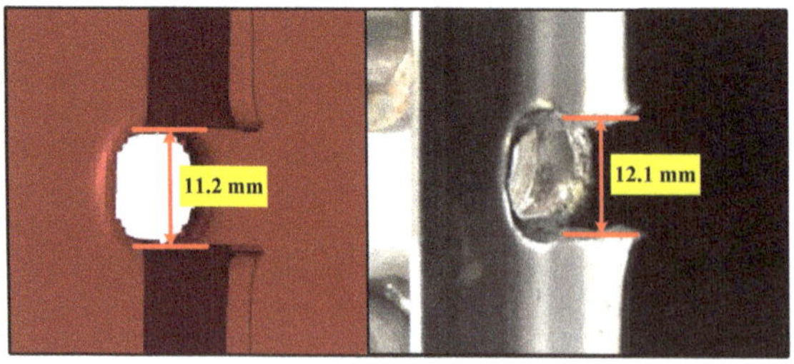

Figure 23. Damages of metallic top skin.

Figure 24 shows the kinetic energy curve of the fragment at a speed of 2400 m/s (kinetic energy = 6 kJ) from the composite side (Situation 2) and the metal side (Situation 3), respectively. The fragment is almost always linearly decaying during the intrusion. At a constant thickness, the kinetic energy dissipation of the fragment is greater for the carbon fiber composite layer, while the Al alloy layer is insensitive to the kinetic energy dissipation of the secondary penetration of the fragment, and the kinetic energy of the fragment decays rapidly to 0 during the secondary penetration of the composite layer.

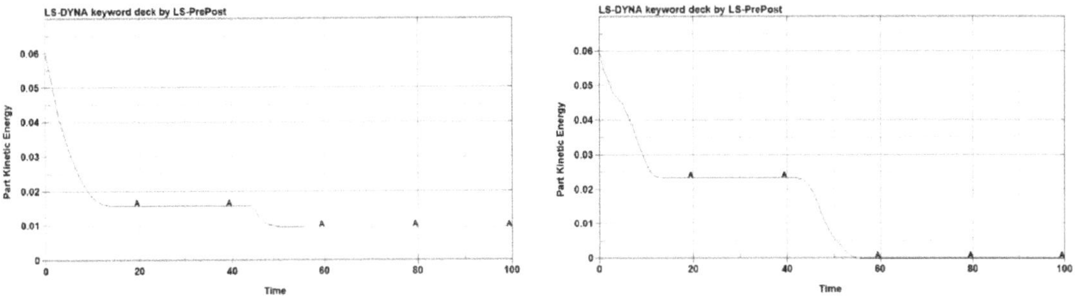

(**a**) (**b**)

Figure 24. Curve of kinetic energy of fragments. (**a**) Curve of kinetic energy of fragments in Situation 2; (**b**) Curve of kinetic energy of fragments in Situation 3.

4. Conclusions

A high-velocity impact test based on a two-stage light gas gun was carried out on an aircraft-typical composite/metal connecting structure (CFRP/AL). The simulated battle damage impact on the typical composite/metal connecting structure of the aircraft under different rendezvous conditions was achieved. This study aims to provide a reference for the rapid repair and assessment of aircraft battle damage and the design of aircraft structural survivability. The following conclusions are drawn from the study:

1. The composite laminate damage is characterized by stages, and its regional profile is mainly in the pattern of a combination of cylindrical (shear failure zone) and circular truncated cones (tensile failure zone), and the upper and lower surfaces will produce different degrees of random spalling phenomena under the action of impact.
2. The established numerical model can well characterize the real damage morphology of both composites and the Al alloy. The damage sizes of predicted results are generally smaller than experimental results, which is within 8% on average.
3. The energy of carbon fiber debris dissipates quickly, while metal debris clouds contain considerable penetration capability, which will cause widely distributed secondary damage to the structure.
4. Different structural components have different energy dissipation capabilities. The kinetic energy of fragments decays by 4.3 kJ and 3.7 kJ, respectively, on the composite part and Al part at the first impact, and decays by 2.3 kJ and 0.4 kJ, respectively, on the composite part and Al part at the second impact. The composite part show stronger energy absorption properties, at the same thickness, than an Al alloy.

Author Contributions: Conceptualization, Y.W. and Y.H.; methodology, Y.W.; software, J.Y.; validation, Y.W.; formal analysis, J.Y.; investigation, T.Z.; resources, T.Z.; data curation, T.Z.; writing—original draft preparation, Y.W.; writing—review and editing, Y.W.; visualization, Y.W. and X.F.; supervision, H.Z.; project administration, Y.H.; funding acquisition, Y.H. All authors have read and agreed to the published version of the manuscript.

Funding: This research received no external funding.

Institutional Review Board Statement: The study was conducted in accordance with the Declaration of Helsinki, and approved by the Institutional Review Board (or Ethics Committee) of Aviation Engineering School, Air Force Engineering University.

Informed Consent Statement: Informed consent was obtained from all subjects involved in the study.

Data Availability Statement: Data will be made freely available on request.

Conflicts of Interest: The authors declare no conflict of interest.

References

1. Horsfall, I.; Austin, S.J.; Bishop, W. Structural ballistic armour for transport aircraft. *Mater. Des.* **2000**, *21*, 19–25. [CrossRef]
2. Sweetman, B. *f-22 Raptor*; Zenith Imprint: Neunkirchen, Germany, 1998.
3. Appleby-Thomas, G.J.; Hazell, P.J. The impact of structural composite materials. Part 2: Hypervelocity impact and shock. *J. Strain Anal. Eng. Des.* **2012**, *47*, 406–418. [CrossRef]
4. Miao, C.; Du, M.; Huang, L.; Zu, Z. Experimental research on hypervelocity impact characteristics of flexible anti-debris multi-shields structure. *Manned Spacefl.* **2017**, *23*, 173–176.
5. Wang, Y.T.; He, Y.T.; Zhang, T.; Fan, X.H.; Zhang, T.Y. Damage analysis of typical structures of aircraft under high-velocity fragments impact. *Alex. Eng. J.* **2022**, *62*, 431–443. [CrossRef]
6. Moritoh, T.; Kawai, N.; Matsuoka, S.; Nakamura, K.G.; Kondo, K.I.; Katayama, M. Hypervelocity impact experiments up to 9 km/s by a compact multi-stage light-gas gun. *Int. J. Impact Eng.* **2003**, *29*, 459–467. [CrossRef]
7. Reddy, P.R.S.; Reddy, T.S.; Srikanth, I.; Madhu, V.; Gogia, A.K.; Rao, K.V. Effect of viscoelastic behaviour of glass laminates on their energy absorption subjected to high velocity impact. *Mater. Des.* **2016**, *98*, 272–279. [CrossRef]
8. Reddy, P.R.S.; Reddy, T.S.; Mogulanna, K.; Srikanth, I.; Madhu, V.; Rao, K.V. Ballistic impact studies on carbon and e-glass fibre based hybrid composite laminates. *Procedia Eng.* **2017**, *173*, 293–298. [CrossRef]
9. Cantwell, W.J.; Morton, J. Comparison of the low and high velocity impact response of CFRP. *Composites* **1989**, *20*, 545–551. [CrossRef]
10. Johnson, G.R.; Cook, W.H. Fracture characteristics of three metals subjected to various strains, strain rates, temperatures and pressures. *Eng. Fract. Mech.* **1985**, *21*, 31–48. [CrossRef]
11. He, Q.G.; Chen, X.W.; Chen, J.F. Finite element-smoothed particle hydrodynamics adaptive method in simulating debris cloud. *Acta Astronaut.* **2020**, *175*, 99–117. [CrossRef]
12. Brar, N.; Joshi, V.; Harris, B. Constitutive model constants for Al7075-t651 and Al7075-t6. In Proceedings of the Aip Conference Proceedings, Nashville, TN, USA, 28 June–3 July 2009; pp. 945–948.
13. Vignjevic, R.; Campbell, J.; Hughes, K.; Orłowski, M.; Garcea, S.; Withers, P.; Reed, J. Soft body impact resistance of composite foam core sandwich panels with unidirectional corrugated and tubular reinforcements. *Int. J. Impact Eng.* **2019**, *132*, 103320. [CrossRef]
14. Zhu, W.; Xu, X. Experiment and finite element simulation in T300/QY8911 laminate under low-velocity impact. *J. Mater. Sci. Eng.* **2013**, *31*, 68–73.
15. Dogan, F.; Hadavinia, H.; Donchev, T.; Bhonge, P.S. Delamination of impacted composite structures by cohesive zone interface elements and tiebreak contact. *Cent. Eur. J. Eng.* **2012**, *2*, 612–626. [CrossRef]
16. Hallquist, J.O. *LS-DYNA Keyword User's Manual*; Livermore Software Technology Corporation: Livermore, CA, USA, 2003; Volume 970, pp. 299–800.

applied sciences

MDPI

Article

Hybrid Shielding for Hypervelocity Impact of Orbital Debris on Unmanned Spacecraft

Kayleigh Fowler and Filipe Teixeira-Dias *

Institute of Infrastructure and Environment (IIE)—School of Engineering, The University of Edinburgh, The King's Buildings, Edinburgh EH9 3FG, UK; k.r.fowler@sms.ed.ac.uk
* Correspondence: f.teixeira-dias@ed.ac.uk

Abstract: The passive shielding of space craft structures is critical due to the increase in demand for lightweight protection, which is required to counter the damaging effects of micro-meteoroid orbital debris (MMOD) on unmanned spacecraft, which have steeply increased in recent years. Research on hypervelocity impact (HVI) led to the development of shield configurations such as the conventional Whipple shield, which consists of two plates separated by a stand-off distance to allow for the fragmentation and dispersion of the debris from the impact. Variations in the Whipple shield have been proposed, where additional layers are included for increased energy dissipation efficiency. In this work, the authors develop, validate and test a numerical model of an orthogonally loaded hybrid Whipple shield, incorporating an aluminium honeycomb core, orientated with hexagonal tubes perpendicular to the direction of proposed debris travel, to mitigate the well-known channelling effect. The debris threat is an A2024-T3 projectile, impacting the structure at a velocity of 6.5 km/s. The proposed model is validated with experimental observations of the debris spread at half-angle and the efficiency of the proposed topology is assessed against a conventional two-plate A2024-T3 shield. The honeycomb core cell density, its position relative to the point of impact, the thickness of the honeycomb shell, and the material of the honeycomb are thoroughly analysed. A hybrid honeycomb structure concept is proposed, which provides a highly efficient alternative to a standard Whipple shield design, without significantly compromising the weight of the structure. The obtained results clearly show that the hybrid Whipple shield exhibits significantly increased the kinetic energy dissipation of the debris from the impactor and shield front plate, with an increase in the dissipated kinetic energy that can reach 86.8% relative to the conventional shield.

Keywords: hypervelocity impact; whipple shield; honeycomb passive shielding; micro-meteoroid and orbital debris (MMOD); Smoothed Particle Hydrodynamics (SPH); finite-element analysis; LSDyna

Citation: Fowler, K.; Teixeira-Dias, F. Hybrid Shielding for Hypervelocity Impact of Orbital Debris on Unmanned Spacecraft. *Appl. Sci.* **2022**, *12*, 7071. https://doi.org/10.3390/app12147071

Academic Editors: Lina M. López, Ricardo Castedo and Anastasio P. Santos

Received: 12 May 2022
Accepted: 12 July 2022
Published: 13 July 2022

1. Introduction

Research into the passive shielding of spacecraft structures has heightened over recent decades owing to an increase in demand for lightweight, cost-effective technology, which is required to counter the damaging effects of micro-meteoroid orbital debris (MMOD) [1]. The amount of MMOD in space has been "steadily rising since the beginning of the space age", as stated in the European Space Agency's Annual Space Environment Report in 2020 [2]. In 2007 alone, a further 32% increase in MMOD was observed as a result of major low-earth-orbit (LEO) collisions [3]. The rise in MMOD significantly increased the risk associated with space exploration and, as a result, the Inter-Agency Space Debris Coordination Committee (IDAC) was established in 1993 to mitigate the damage caused by micro-meteoroid orbital debris.

The design of the International Space Station (ISS) in the 1990s prompted the further development of protective shielding methods. Hypervelocity impact (HVI) research, along with the hydrocode simulations conducted by NASA and other research facilities and groups, led to the development of numerous shield configurations, such as the Whipple, Stuffed Whipple, and metallic foam sandwich Whipple shields [4]. These protective structures are designed with the main aim of mitigating the effects of hypervelocity impacts that, due to their extremely high energy, have the potential to perforate shields in space structures.

The conventional Whipple shield design consists of two plates, usually made of aluminium, separated by a stand-off distance to allow for the fragmentation and dispersion of debris from the impact on the first plate. A schematic illustration of this design principle is shown in Figure 1.

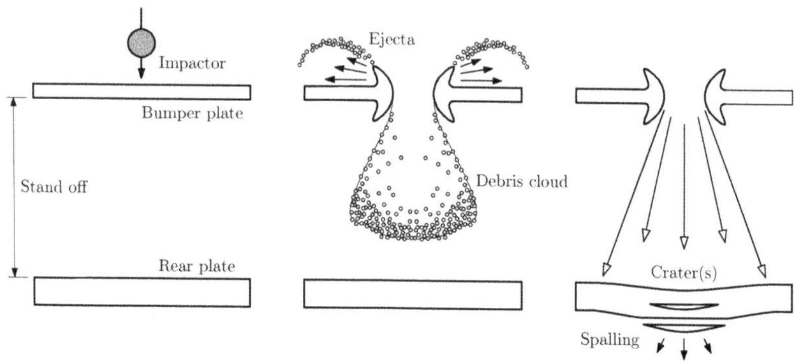

Figure 1. Conventional Whipple shield design: (**left**) Whipple shield configuration, (**centre**) post-impact debris cloud formation and (**right**) effects of impact on rear plate (adapted from [5]).

The Stuffed Whipple, as shown in Figure 2a, incorporates an additional layer, commonly a combination of Nextel or Kevlar/Epoxy, to improve overall shield performance and energy dissipation [6]. The incorporation of metal foams into the Whipple shield design has also been studied in 2017 by Cherniaev and Telichev [6], as shown schematically in Figure 2b. Ryan and Christiansen [5] also demonstrated the potential of such design approaches for space applications, owing to ability to significantly increase the absorption of impact energy compared to more conventional shielding structures. In these studies, it was clearly demonstrated that the two main design characteristics affecting the dissipation of kinetic energy and debris fragmentation are the choice of material(s) (and corresponding material properties) and the geometry of the shield design [1].

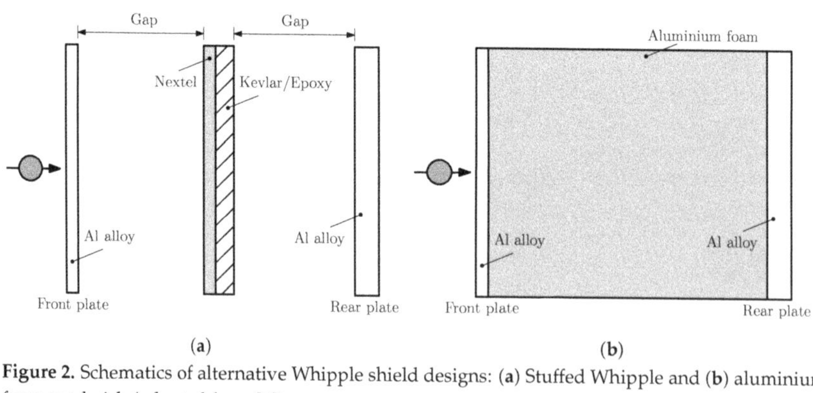

Figure 2. Schematics of alternative Whipple shield designs: (**a**) Stuffed Whipple and (**b**) aluminium foam sandwich (adapted from [6]).

In the late 1990s, Christiansen et al. [7] proposed the use of lightweight, non-metallic materials, such as Nextel ceramic cloth and Kevlar, for the Whipple shield. These materials were configured in a number of different flexible multi-layer designs. Christiansen et al. tested the penetration resistance in terms of hypervelocity impact and successfully demonstrated the potential for this lightweight shielding, going as far as deriving a set of ballistic limit equations for the proposed shields for orthogonal and non-orthogonal low ($v < 2700$ [m/s]), intermediate ($2700 < v < 6500$ [m/s]) and high velocity impacts ($v > 6500$ [m/s]).

Plassard et al. [8] conducted HVI experiments using a two-stage light gas gun. A 3-mm aluminium projectile was fired with a velocity of 4119 m/s at a Whipple shield consisting of an aluminium target plate and a witness plate positioned 30 mm apart. The experimental observations were compared to numerical simulation results in LS-DYNA. The proposed numerical model was found to be a sufficiently accurate representation of the experiment, justifying the use of the hydrocode in further analysis and the shielding design for the hypervelocity impact of orbital debris on unmanned spacecraft.

Research into the material properties of the Whipple shield was more recently conducted by Zhang et al. [9], where the energy absorption efficiency of the combination of a homogeneous aluminium sheet with a Ti-Al-nylon impedance-graded material (IGM) was compared using both laboratory testing and numerical simulations. The experiments were performed using a two-stage light gas gun and focussed on post-impact effects, using 3D scanners to detect physical damage in detail. Zhang et al. proposed a smooth particle hydrodynamics (SPH) numerical model, developed in AUTODYN, and performed a detailed analysis of the relevant kinetic energy dissipation, fragmentation, and pressure distribution. Their results revealed a significant improvement in shield performance using the IGM with regards to both fragmentation and energy dissipation. The increased shock pressure that was experienced, achieved through the interaction between travelling shock waves and reflected rarefractions, allowed for an optimised shield design. The increase in debris spread angle was further confirmation that the material properties of the IGM were suited to HVI shield design.

Recently, the addition of a honeycomb structure to Whipple shield design has been extensively explored by authors such as Carriere and Cherniaev [10,11] and Aslebagh and Cherniaev [12], among others. These researchers adopted an orientation of the honeycomb, where the cells' axis is perpendicular to the front and bumper plates. This allows for a significant reduction in the debris spread angle, but creates a channelling effect, as the honeycomb is orientated parallel to the direction of debris travel, as shown in Figure 3. Although the reduction in the debris spread angle is significant, it creates the adverse effect of concentrating the impact on a smaller area, adversely affecting the energy dissipation of the projectile. Double/multi-honeycomb core configurations were noted as being less prone to channelling effects due to the implementation of a staggered design [10].

More recently, Pai and Shenoy [13] presented a detailed review of recent advances in the Whipple shield design, noting that debris channelling (also referred to as ejecta-tunnelling effect) should be considered in the design process of Whipple shields, and can be detrimental to the energy dissipation and mechanisms.

A solution to the channelling effects, however, would be to change the orientation of the honeycomb cells from parallel to perpendicular to debris travel, with the added advantages that the weight of the structure could remain unchanged. In 2009, Ryan et al. [14] conducted a comparative study between the use of a parallel-orientated honeycomb core and metallic open-cell foam for Whipple shield application. These authors found that the foam had several advantages over the honeycomb due to the elimination of channelling.

Very few studies have been dedicated to determining the effectiveness of a perpendicular honeycomb cell orientation regarding the energy absorption of Whipple shields. Therefore, the proposed research aims to explore the potential benefits of implementing a perpendicularly orientated honeycomb core in relation to the energy dissipation of an impact projectile at hypervelocity, and exploring the effects of the topology of the honeycomb structure on the energy absorption of the shield as a whole.

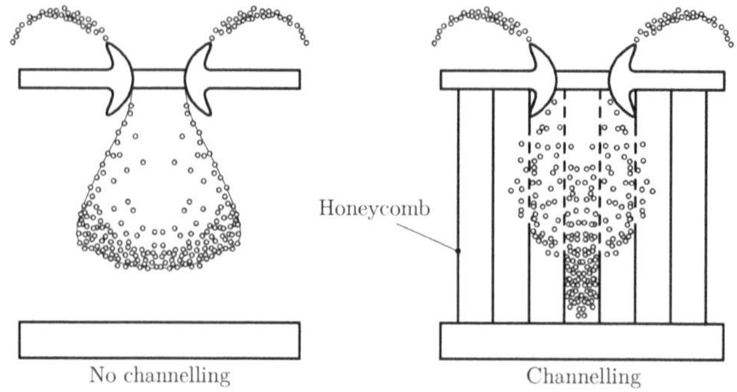

Figure 3. Illustration of the channelling effect on honeycomb core Whipple shields: without honeycomb core (**left**) and with honeycomb core (**right**).

2. Numerical Modelling

The main aim of the proposed research is to explore and optimise the benefits of using lightweight honeycomb structures in the Whipple shield design, with potential application in unmanned spacecraft. A set of numerical models is developed and validated, based on an aluminium honeycomb Whipple shield. These passive shielding models are implemented in LS-DYNA and described in detail in the following paragraphs. The validated models are then used to perform a thorough analysis of the effects of different shield parameters (core density, impact location, etc.) on the energy absorption performance and impact protection of the Whipple shield.

2.1. Model Configuration

The proposed modified Whipple shield has a honeycomb layer—the shield core—between the bumper plate and the rear wall. The honeycomb is orientated with the axis of the cells perpendicular to the impact direction, as shown in Figure 4. The main design principle is that this allows for the sides of the honeycomb cells to maximise the dispersion of fragments at wider angles, also maximising the dissipation of energy further from the back plate and onto a larger area. This orientation also fully eliminates the possibility of channelling effects, which, as previous research suggests, has a highly detrimental effect on reducing the impact kinetic energy of the debris particle [10].

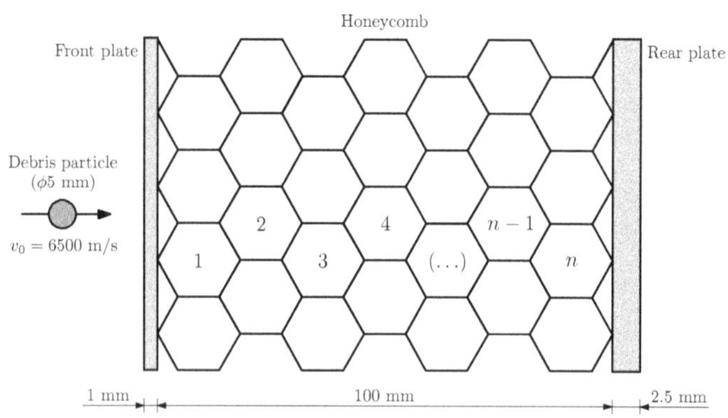

Figure 4. Honeycomb Whipple shield configuration with cell numbering, sequentially from the front towards the back plate.

The numerical models include a spherical projectile, and front and back plates with 1 and 2.5 mm thickness, respectively. The diameter of the projectile is 5 mm, which is consistent with the lower end of the equivalent sphere diameter (ESD) distribution, as reported by a number of different authors [15–17]. The model developed by the authors is validated and optimised based on the effects of a number of design parameters on the energy absorption capacity of the shield. These include the density of the honeycomb (i.e., number of honeycomb cells per unit length), the thickness of the honeycomb shell, the choice of honeycomb material, and the location of impact relative to the honeycomb.

2.2. Finite Elements and SPH Model

The finite-element method is used to model the impact response of the whole system, including the front (bumper) and back plates, and all models were set up in LS-DYNA.

Solid constant stress solid elements are used to model the bumper plate and black wall, and four-node constant thickness shell elements are used in the honeycomb structure. The same mesh size is used in both the bumper and back plates and a thorough mesh convergence analysis is carried out to determine the optimal mesh size. The spherical projectile and the impact area on the front plate are modelled using Smoothed Particle Hydrodynamics (SPH). The convergence analysis was also extended into the SPH domain to determine the optimal particle density, especially as this method is highly computationally heavy and often leads to high CPU times. Specific contacts were implemented to model the interaction between the different model components: (i) tied contacts between the finite element and the SPH particles in the bumper plate, and (ii) automatic contacts between all SPH particles—both from the impactor and the impacted area—and all other finite-element components in the model (the back plate, the front plate and the honeycomb). In LS-DYNA, these contacts were implemented with the keywords `*CONTACT_AUTOMATIC_NODES_TO_SURFACE` and `*CONTACT_TIED_NODES_TO_SURFACE`, respectively.

To optimise computational efficiency the proposed models explored symmetries when possible, that is, when all impact, geometrical and boundary conditions were symmetrical. Oxz and Oyz are the two symmetry planes of this model, as can be seen in Figure 5a. One of these planes—symmetry plane Oxz—is used in this research, as shown in Figure 5b. Symmetry plane Oyz, however, cannot be used, as it is not a symmetry plane for the impact location analyses, where symmetry is broken when the debris particle impacts at different locations.

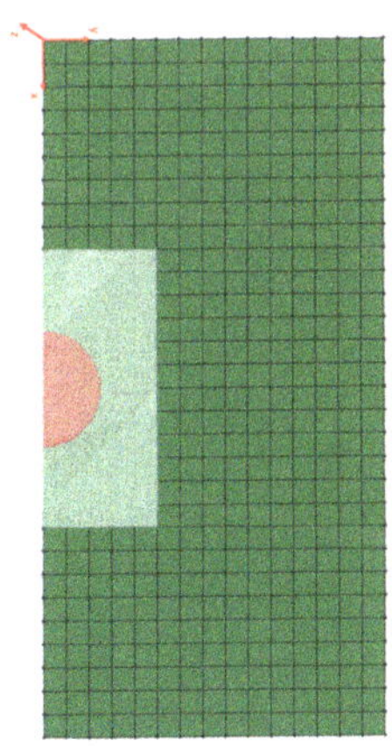

(a)

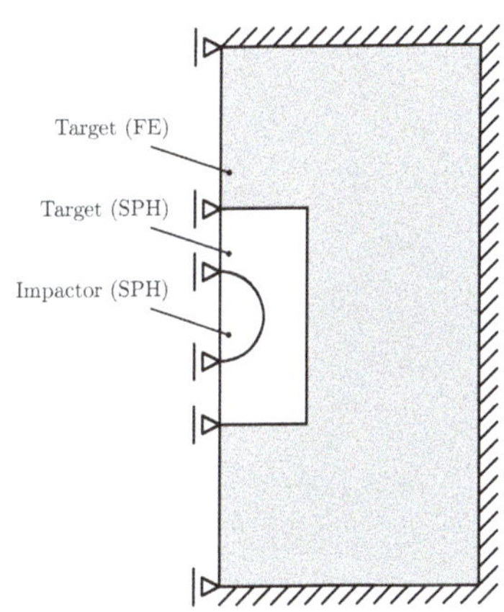

(b)

Figure 5. Simulation set-up showing (**a**) FE mesh, SPH regions and coordinate system; and (**b**) boundary conditions.

The HVI problem being analysed is a kinematics-dominated problem; thus, imposing fixed boundary conditions on the Whipple plates is not strictly necessary. This is supported by the progression of the post-impact radial stress wave on the Whipple front plate. To ensure the consistency of analysis with the experimental results of Zhang et al. [9], these boundary conditions were imposed. The implemented boundary conditions are shown in Figure 5. The external edges of the finite-element meshes on the front and back plate, as well as the honeycomb core, are fixed in all degrees of freedom (linear and rotational). The motion of all SPH particles on the plane of symmetry was also restricted to in-plane motion, that is, fixed in the y-direction, as shown in Figure 5.

An initial constant velocity of 6500 m/s is imposed on the spherical projectile along the direction perpendicular to the bumper plate (z-direction in Figure 5). This velocity is representative of a generic micro-meteoroid orbital debris particle impacting the shield.

2.3. Constitutive Modelling

Due to the high energy involved in the hypervelocity impact, all Whipple shield components, including the honeycomb core and debris particle, were modelled with the Johnson–Cook constitutive law to ensure adequate plasticity, strain-rate and temperature material behaviour is captured. These three terms are explicitly and separately described in Johnson–Cook's constitutive equation

$$\bar{\sigma} = \left[A + B\left(\bar{\varepsilon}^{\mathrm{pl}}\right)^{n}\right]\left[1 + C\ln\left(\frac{\dot{\bar{\varepsilon}}^{\mathrm{pl}}}{\dot{\bar{\varepsilon}}^{\mathrm{pl}}_{0}}\right)\right]\left[1 - (T^{*})^{m}\right] \tag{1}$$

where $\bar{\sigma}$ is the flow stress, $\bar{\varepsilon}^{\mathrm{pl}}$ is the equivalent plastic strain, n is the strain hardening exponent, A, B, C and m are material constants that can be determined experimentally, $\dot{\bar{\varepsilon}}^{\mathrm{pl}}$ is the equivalent plastic strain rate [18], and T^* is the non-dimensional homologous temperature, defined as

$$T^{*} = \frac{T - T_{\mathrm{t}}}{T_{\mathrm{m}} - T_{\mathrm{t}}} \tag{2}$$

where T is the current temperature, T_{m} is the melting temperature and T_{t} is the transition temperature, at or below which there is no temperature dependence for the yield stress.

Damage caused by the impact is modelled using the associated Johnson–Cook damage model, which similarly considers the effects of plasticity, strain-rate and temperature, and can be described by the equation

$$\bar{\varepsilon}^{\mathrm{pl}}_{\mathrm{D}} = \left[D_1 + D_2\exp(-D_3\eta)\right]\left[1 + D_4\ln\left(\frac{\dot{\bar{\varepsilon}}^{\mathrm{pl}}}{\dot{\bar{\varepsilon}}_{0}}\right)\right](1 + D_5T^{*}) \tag{3}$$

where D_i $(i = 1,\ldots,5)$ are the damage parameters measured at or below the transition temperature and η is the stress triaxiality, which represents the ratio of pressure to von Mises stress. The damage parameter is then calculated as

$$D = \sum \frac{\bar{\varepsilon}_{\mathrm{D}}}{\bar{\varepsilon}^{\mathrm{pl}}_{\mathrm{D}}} \tag{4}$$

and damage occurs when the damage parameter D reaches a value of 1.0. After damage initiation, the material stiffness is progressively degraded according to the damage evolution relationship [18].

The developed models of the Whipple shield honeycomb core were tested with two different lightweight materials: an aluminium alloy (AL2024-T3) and a titanium alloy (Ti-6Al-4V). The impact energy absorption efficiency of both materials is analysed and compared. The front plate, the back plate and the debris particle (projectile) are aluminium (AL2024-T3) throughout. The material parameters for all materials in the models are listed in Table 1 [19,20].

Due to the high energy involved in the hypervelocity impact, the compressibility of the materials is modelled using the non-linear Mie–Grüneisen equation of state

$$p = \frac{\rho_0 G^2 \mu \left[1 + (1 - \frac{\gamma_0}{2})\mu - \frac{a}{2}\mu^2\right]}{\left[1 - (S_1 - 1)\mu\right]^2} + (\gamma_0 + a\mu)E \quad (5)$$

where E is the internal energy, $\mu = \rho/\rho_0 - 1$ is the relative density, γ_0 is the non-dimensional Grüneisen constant, a is the volume correction for γ_0 and G, S_1 are material constants that depend on the shock wave and particle velocities. The corresponding parameters are listed in Table 1. This simplified version of the non-linear Mie–Grüneisen equation of state in Equation (5) only considers the first-order dependency of the pressure on the material density ($S_2 = S_3 = 0$).

Table 1. Johnson–Cook constitutive and damage models, and Mie–Grüneisen equation of state parameters for AL2024-T3 (debris particle, shield and honeycomb core) and Ti-6Al-4V (honeycomb core) [21,22].

Constitutive Equation	Material Parameter	AL2024-T3	Ti-6Al-4V
Johnson–Cook	A (MPa)	167	862
	B (MPa)	684	331
	n	0.551	0.34
	C	0.001	0.012
	m	0.859	0.8
Johnson–Cook (damage)	D_1	0.112	−0.09
	D_2	0.123	0.25
	D_3	1.5	−0.5
	D_4	0.007	0.014
	D_5	0	3.87
Mie–Grüneisen (EoS)	G	5240	5130
	γ_0	1.97	1.23
	S_1	1.400	1.028
	a	0.48	0.17

2.4. Convergence and Validation

A standard two-plate Whipple shield was used to validate the methodology and models in this research. To achieve this, the numerical results from the standard Whipple shield (SWS) were compared to experimental observations by Zhang et al. [9], where all model parameters were kept the same to allow for a direct comparison. The SWS simulation setup is shown in Figure 6. A 5-mm aluminium projectile was fired with a velocity of 6500 m/s at a Whipple shield with a front plate of thickness 1 mm and a back plate thickness of 2.5 mm. Figure 7 shows the progression of the simulation in approximately 2-μs intervals.

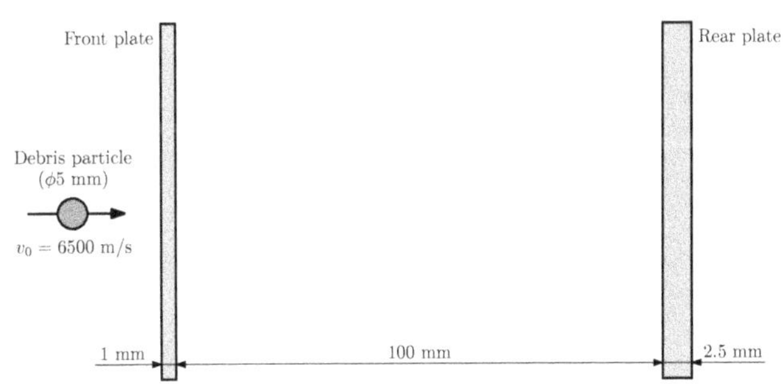

Figure 6. Standard Whipple shield configuration.

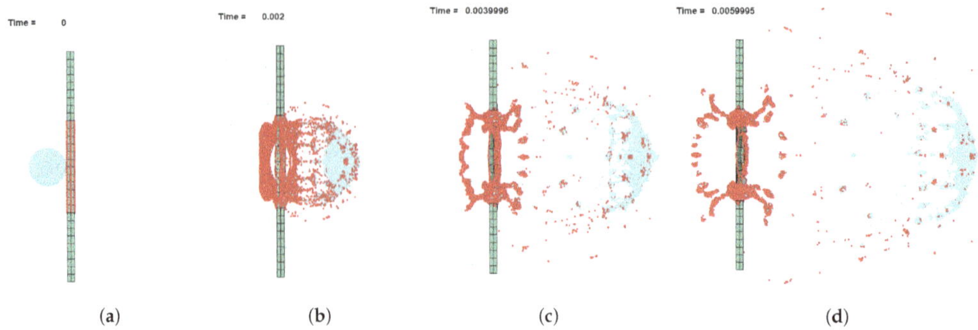

Figure 7. Simulation of standard Whipple shield model shown at: (**a**) $t = 0$ μs, (**b**) $t = 2$ μs, (**c**) $t = 4$ μs, and (**d**) $t = 6$ μs.

2.4.1. SPH Particle Density

Zhang et al. [9] published the results of a pressure analysis conducted on an AL2024 standard Whipple shield. These authors used pressure gauges applied to the front of the projectile to calculate the average impact pressure on the front plate on the first 0.5 μs of the impact. A similar procedure is followed in the numerical model developed in this work. The simulation was run with varying SPH particle densities and the results obtained for $t \in [0, 0.5]$ μs are shown in Figure 8. Datapoints were extracted from the work of Zhang et al. [9] and have also been included in Figure 8 to allow or a comparison and validation to be made, where the average impact pressure p was determined as the average of the pressure distribution for $t \in [0, 0.5]$ μs, for a range of selected SPH particles on the front side (towards the direction of impact) of the projectile, to match the data recorded by Zhang et al. [9].

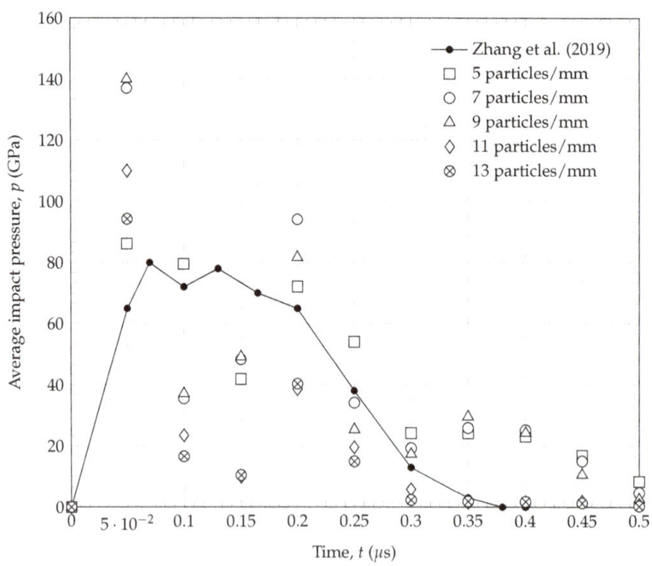

Figure 8. Average impact pressure p for varying SPH density and comparison with observations by Zhang et al. [9].

The general trend of the impact pressure results is similar to the experimental observations of Zhang et al., albeit with a relatively high level of scatter—the largest relative difference in peak average impact pressure between the numerical model and the observations of Zhang et al. [9] is below 43%. The results in Figure 9, showing the average impact pressure p and computational runtime against the SPH particle density, allow for a converging solution to be more easily identified. A clear convergence can be observed from an SPH particle density of 5 particles/mm to 13 particles/mm. The computational runtime is shown to exponentially increase with an increase in the SPH particle density. This clearly suggests that using fewer SPH particles within the model is a more efficient approach. The intersection of best-fit curves in Figure 9 can be used to select the optimum modelling approach and SPH particle density, which, in this case, is below 10 particles/mm. Optimising the computational runtime for the available resources was necessary, leading to an optimal particle density of 5 particles/mm.

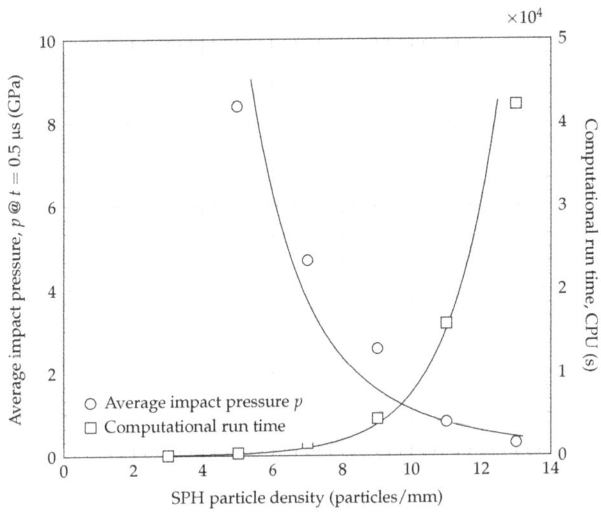

Figure 9. Average impact pressure p at time $t = 0.5$ μs and computational run time (CPU) against the SPH particle density, with corresponding trendlines.

To more robustly validate the proposed models, a thorough comparison of the debris spread angle was also done. Zhang et al. [9] reported that the half-angle obtained using the aforementioned parameters to be 19.8°. The debris spread half-angle for the LS-DNYA model was measured on the Whipple shield model at multiple time intervals and averaged to obtain a half-angle of 21.5°. This corresponds to a relative difference of 7.9%, and is thus considered accurate, further validating the use of the proposed SPH particle density in the model for further analysis.

2.4.2. Honeycomb Mesh Convergence

A detailed mesh convergence analysis was also performed on the honeycomb core structure. The computational run time (CPU) was monitored for finite element meshes of the honeycomb with different element sizes, ranging from 0.4 to 3.2 mm, with the corresponding results shown in Figure 10. The convergence analysis simulation was set up with an SPH density of 5 particles/mm to minimise computational run time, and all remaining models were set up with the optimal 7 particles/mm density. Al2024-T3 was used for all components of the shield, and all parameters other than the mesh element size of the honeycomb shell were kept constant. The results in Figure 10 indicate that the optimum solution in terms of reducing CPU time whilst retaining an accurate solution, is to use a 1 mm element size.

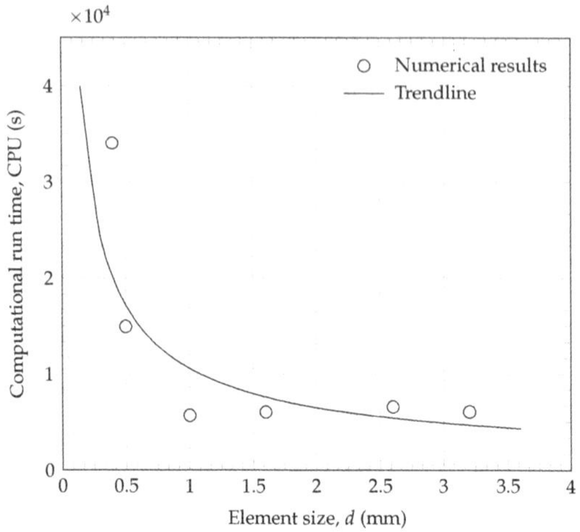

Figure 10. Computational run time (CPU) against finite element mesh element size of the honeycomb core structure.

3. Results and Discussion

This section discusses the results of further analyses done with the models described and validated in the previous sections. These include a number of critical design parameters of the Whipple shield, such as (i) the density of the honeycomb core (i.e., the number of honeycomb cells along the width of the core); (ii) the thickness of the honeycomb shell, both of which will directly impact the weight (areal density) of the final shield; (iii) the honeycomb base material; and (iv) the exact location of the impact relative to the honeycomb. The capacity for the shield to dissipate the kinetic energy of the impact was used as the main design parameter in these studies. The labels and specifications of the developed models are listed in Table 2, along with the key results from each parameter study to analyse the effectiveness of the honeycomb core shield. As an example, Figure 11 shows a 3-dimensional view of the T-50 model, with a core cell density of 6 and a shell thickness $t_h = 0.5$ mm. Table 3 and Figure 12 summarise the main results from all the tests. In order to record the energy dissipated before particles bounce back off the back plate, some results are captured at different times, $t = 17.5$ μs for the honeycomb shell thickness simulations and $t = 20$ μs for all remaining ones.

Table 2. Characteristics and labels of the developed Whipple shield numerical models.

Test	Model Label	Material	Cells	Thickness t_h (mm)	Impact Location
Standard shield	C-0	–	0	–	–
Cell density	C-2	AL2024-T3	2	0.5	Single edge
	C-4		4	0.5	
	C-6		6	0.5	
	C-8		8	0.5	
	C-10		10	0.5	
Shell thickness	T-01	AL2024-T3	6	0.01	Single edge
	T-05		6	0.05	
	T-10		6	0.1	
	T-15		6	0.15	
	T-20		6	0.2	
	T-30		6	0.3	
	T-50		6	0.5	

Table 2. *Cont.*

Test	Model Label	Material	Cells	Thickness t_h (mm)	Impact Location
Material	M-AL	AL2024-T3	6	0.3	Single edge
	M-TI	Ti-6Al-4V	6	0.3	
Impact location	L-SE	AL2024-T3	6	0.3	Single edge
	L-MP		6	0.3	Mid point
	L-DP		6	0.3	Double point

Table 3. Summary of main results and dissipation of kinetic energy for all numerical simulations.

Model Label	Kinetic Energy E_k (kNmm)	Time Stamp t (μs)	Energy Dissipation
C-0	575	20	Reference @ 20 μs
C-2	330	20	42.6%
C-4	202		64.9%
C-6	164		71.5%
C-8	162		71.8%
C-10	146		74.6%
T-0	1286	17.5	Reference @ 17.5 μs
T-01	1285	17.5	0.1%
T-05	1230		4.8%
T-10	1040		19.5%
T-15	773		40.2%
T-20	432		66.6%
T-30	280		78.4%
T-50	172		86.8%
M-AL	244	20	57.6%
M-TI	177		69.2%
L-DP	255	20	55.7%
L-SE	244		57.6%
L-MP	225		60.9%

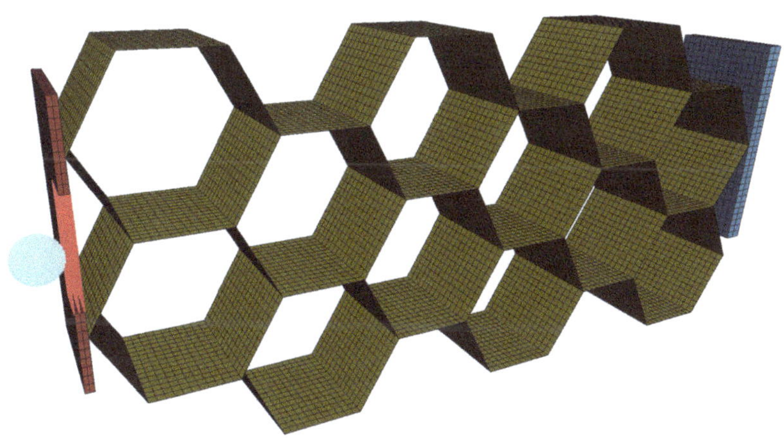

Figure 11. 3-dimensional view of model T-50, with a core cell density of 6 and a shell thickness $t_h = 0.5$ mm (see Table 2).

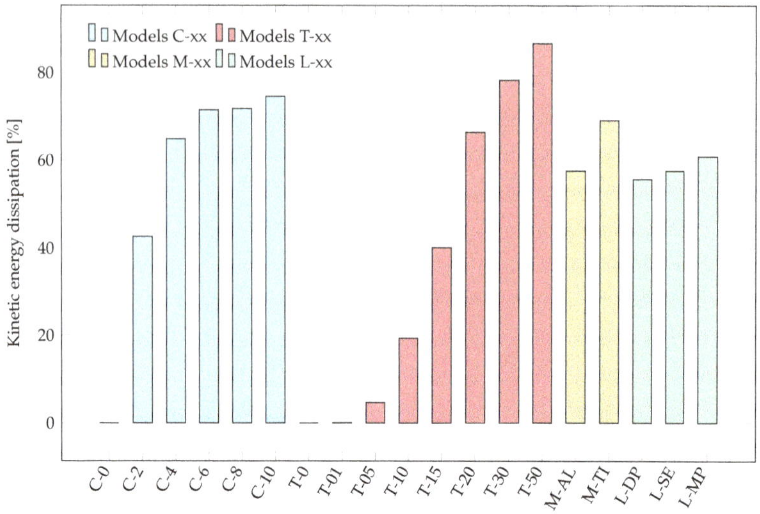

Figure 12. Kinetic energy dissipation for the honeycomb core density, shell thickness, material and impact location models, relative to reference cases (see Table 3).

3.1. Honeycomb Core Cell Density

It was anticipated that the honeycomb core cell density, i.e., the number of cells per unit length across the width of the shield gap, would be a critical parameter influencing the energy absorption efficiency of the shield. In order to test this hypothesis and quantify the influence of the honeycomb core cell density, a number of models were developed using an SPH particle density of 7 particles/mm for the projectile and region of impact, as discussed in Section 2.4.1, and different cell densities. All model parts, including the honeycomb, were set as AL2024-T3 and a finite-element mesh element size of 1 mm was used.

The results in Figure 13 show how the total kinetic energy of the debris particle was dissipated to increase the honeycomb core cell densities (models C-0 to C-10). Model C-0 is included as it represents the standard Whipple shield, i.e., without a honeycomb core. It can be clearly observed that the kinetic energy of the debris particle significantly decreases when a honeycomb component is introduced, corresponding to an increase in the energy dissipation ranging from a minimum of 42.6% to a maximum of 74.6%, as shown in Table 3. Figure 13 also shows that the change in kinetic energy of the debris particle for cores with more than six cells across the width is minimal. The difference in energy dissipation between model C-6 and model C-10 is only 3.1%. Therefore, for the modelled impact conditions, increasing the core cell density above six cells (model C-6) is not beneficial in terms of energy absorption and will clearly be detrimental to the weight of the structure.

The simulation frames in Figure 14, showing the numerical model of each honeycomb configuration at time $t = 16.2$ μs, further support these observations and conclusions, clearly showing that the dispersion of particles becomes more evident when increasing honeycomb core cell density. There is some visible penetration of rogue SPH particles, which is a known issue with SPH modelling. Common strategies to try to mitigate these nonphysical effects include changing the contact algorithm and/or refining the finite-element mesh of the impacted part. In the present case, however, these two strategies were tested and proved to not decrease rogue particle penetration without significantly increasing the cost of the computation. Additionally, the energy of these rogue particles was estimated to be insignificant (less than 1%) compared to the total energy of the problem. A higher level of fragmentation and damage to the honeycomb shell occurred closer to the front plate, which further supports the results shown in Figure 13. A complete absence of channelling effects is also visible.

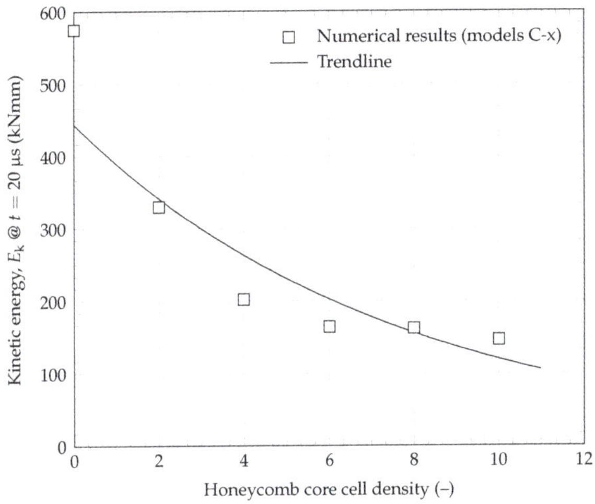

Figure 13. Kinetic energy of the debris particle at $t = 20$ μs against the number of cells across the width of honeycomb core.

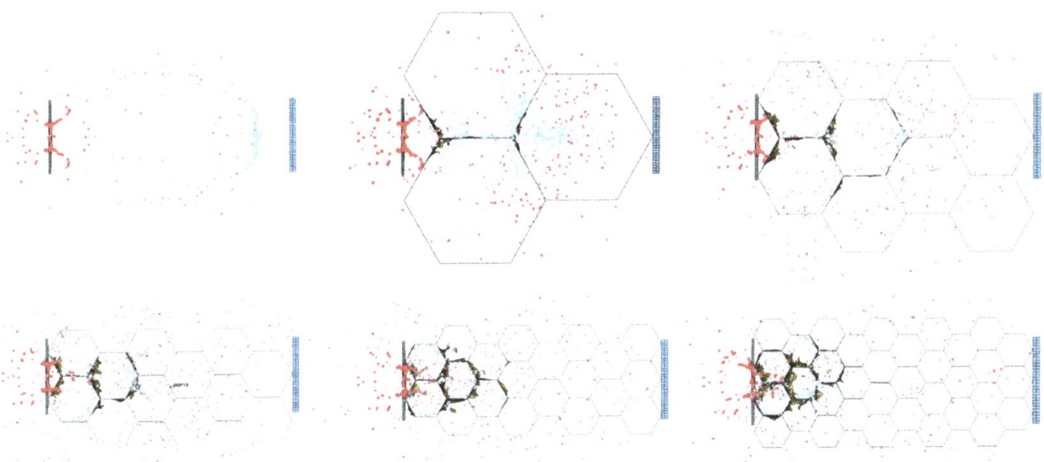

Figure 14. Simulation progress at $t = 16.2$ μs for an increasing number of cells across the width of honeycomb core for model C-0 (**top left**) to model C-10 (**bottom right**).

3.2. Honeycomb Shell Thickness

The impact of the thickness of the honeycomb core shell on the energy dissipation efficiency is analysed by simulating varying values of shell thickness, t_h. Following the conclusions regarding the optimisation of the honeycomb core cell density in Section 3.1 above, model C-6 is used, with an AL2024-T3 honeycomb, and a shell thickness t_h ranging from 0.01 to 0.5 mm, as listed in Table 2. The debris particle and impact zone have an SPH particle density of 7 particles/mm. The results in Figure 15 show a significant decrease in the kinetic energy of the debris particle at 17.5 μs for increasing t_h. This trend shows that, for thicknesses above 0.3 mm, the increase in energy dissipation becomes less evident. The energy dissipation increases only 8.4% for thicknesses between 0.3 and 0.5 mm, compared to a 78.3% increase between 0.01 and 0.3 mm, which corresponds to model T-30 in Table 3.

To further support these conclusions, the simulation images in Figure 16 clearly show that, for shell thicknesses above 0.3 mm, there is no clustering of SPH particles (i.e., debris) reaching the back plate of the Whipple shield. A clear increase in debris fragmentation can also be observed from model T-01 to model T-50 at $t = 16.2$ μs. Models with a smaller shell thickness exhibit a more tightly packed particle spread than models with a larger shell thickness, showing fewer particle clusters. The particle spread also appears much closer to the back plate for a smaller shell thickness. This further supports the results in Figure 15, as the increasing thickness of the honeycomb shell could be expected to positively influence the spread of debris fragments.

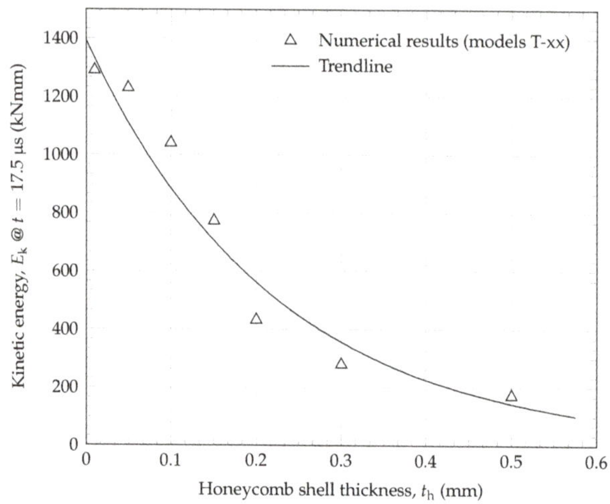

Figure 15. Kinetic energy of the debris particle at $t = 17.5$ μs against thickness of the honeycomb shell.

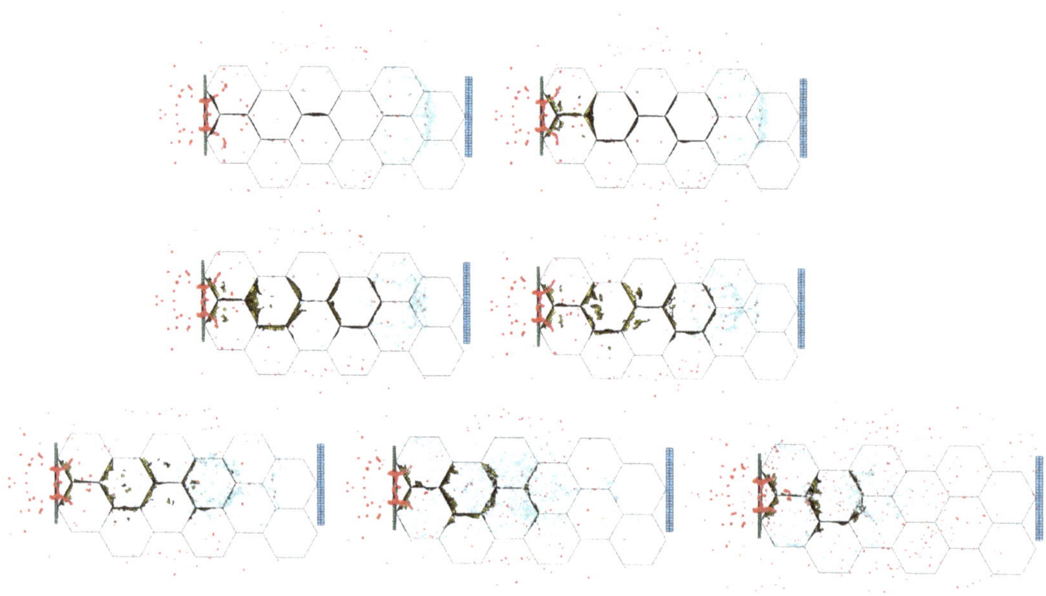

Figure 16. Simulation progress at 16.2 μs for an increasing thickness of honeycomb core from top left (model T-01) to bottom right (model T-50).

3.3. Honeycomb Material

The honeycomb material is a critical design aspect for the optimised Whipple shield as this will not only have a direct impact on the energy absorption, but also on the weight of the structure. Two different numerical models were developed using aluminium alloy (AL2024-T3) and titanium alloy (Ti-6Al-4V) honeycomb cores, corresponding to models M-AL and M-TI in Table 2, respectively. These models are compared to determine the effect that changing the material of the honeycomb core has on the energy dissipation of the debris particle. Figure 17 shows how the kinetic energy is dissipated for both models in the first 20 μs of the impact. These results clearly show that, from early in the impact ($t \approx 1$ μs), model M-TI is more efficient at dissipating the kinetic energy of the debris particle when compared with model M-AL. Although a 27.8% difference can be observed in the kinetic energy of the debris particle between the two different materials at $t = 20$ μs, when comparing to the standard Whipple shield, the energy dissipation efficiency is significantly higher, at 57.6% and 69.2% for the M-AL and M-TI models, respectively.

Figure 18 shows a comparison in debris fragmentation and honeycomb perforation between the two material models, M-AL and M-TI. Although the perforation pattern of the honeycomb is similar between the two models, there is an evident variation, albeit slight, in the distribution of the debris going through the honeycomb structure, with a larger fraction of particles appearing further toward the back plate in model M-AL compared with model M-TI. This agrees with the results in Figure 17, and suggests that the titanium alloy Ti-6Al-4V honeycomb is more effective in dissipating the kinetic energy from the hypervelocity impact.

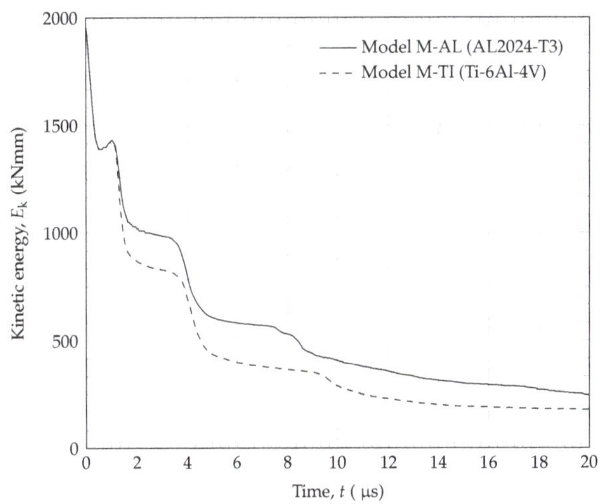

Figure 17. Kinetic energy comparison using Al2024-T3 and Ti-6Al-4V honeycombs.

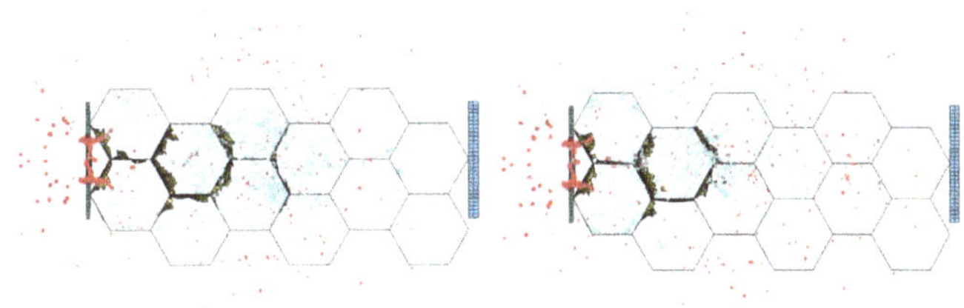

Figure 18. Numerical simulation comparing honeycomb shell materials: Al2024-T3 (model M-AL, **left**) and Ti-6Al-4V (model M-TI, **right**).

3.4. Location of Impact

Given the hexagonal geometry of the honeycomb core when orientated perpendicular to the impact direction, it could be expected that the exact location of the impact might lead to different energy dissipation patterns. To explore this, three distinct models were developed where the location of impact is changed according to the schematics in Figure 19. Model T-30, with an Al2024-T3 alloy honeycomb core, was used to analyse the effect of changing the location of impact on the dissipation of kinetic energy of the debris particle. The debris particle was fired at a velocity of 6500 m/s at a single edge of the honeycomb structure, a double point and at the mid-point. The details of these models are listed in Table 2. The results in Table 3 show that the highest kinetic energy dissipation occurs for model L-MP, where the debris particle impacts the mid-point of the honeycomb. However, the energy absorption history is similar across all three models, as can be seen in Figure 20, which indicates that very little variation is seen between simulations at different debris impact locations. This is additionally supported by the maximum difference in energy dissipation of 5.2% between the three models. Further, Figure 21 shows little visible variation in particle fragmentation, demonstrating that using the honeycomb structure in this orientation is a suitable method for dissipating kinetic energy, irrespective of the exact location of the impact.

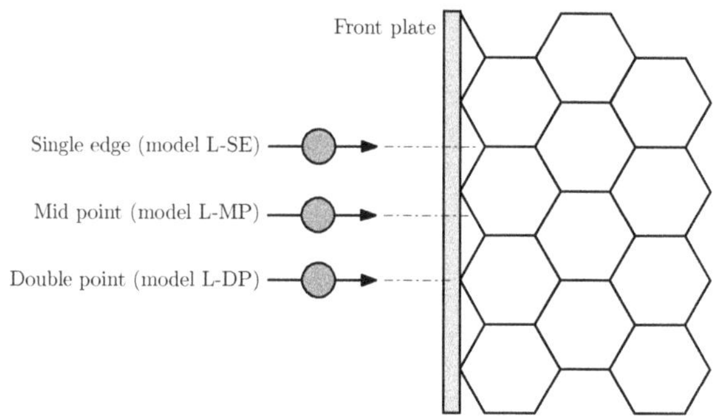

Figure 19. Location of impact on Whipple shield relative to honeycomb.

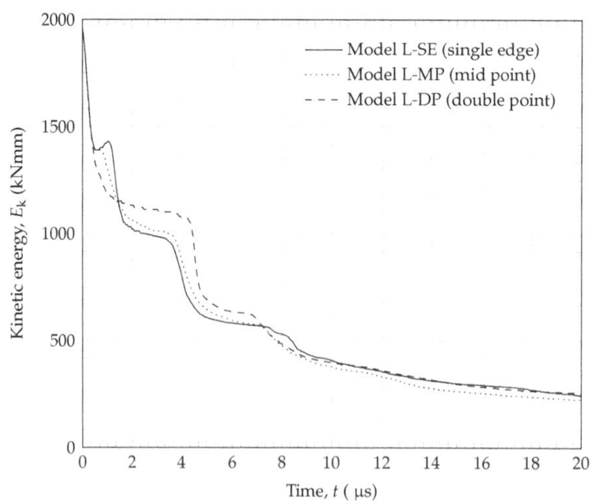

Figure 20. Kinetic energy of debris particle against time for varying debris impact locations.

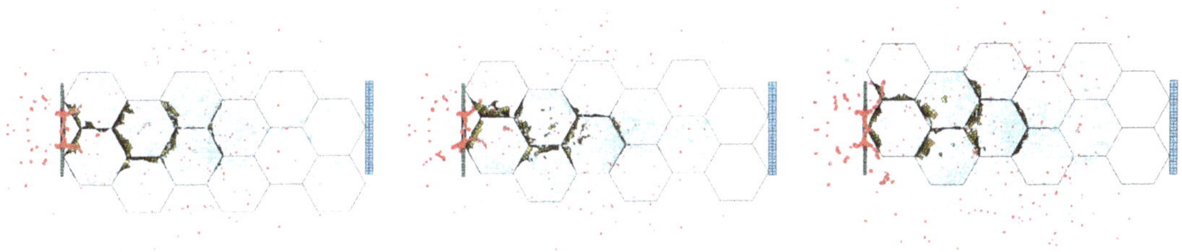

Figure 21. Comparison between varying points of impact: single edge (model L-SE, **left**), mid-point (model L-MP, **centre**) and double point (model L-DP, **right**).

4. Conclusions

This research proposes a set of numerical models to improve the design practice of unmanned hypervelocity impact shields. These are validated with experimental observations, including the debris-spread half-angle. Critical design parameters such as the honeycomb core cell density (i.e., the number of cells per unit length of the standoff gap), the thickness of the honeycomb core shell, the material of the honeycomb and the location of the impact relative to the honeycomb, are thoroughly analysed. The observations and results from this research clearly show that the proposed hybrid Whipple shield exhibits a significantly increased kinetic energy dissipation of the debris from the impactor and shield front plate.

The extent of the effect of altering both the material properties and the location of impact is not comparable to the cases in which the geometry of the honeycomb core was altered (the number of honeycomb cells and the thickness of the honeycomb shell). The dissipation of the kinetic energy of the debris particle is more sensitive to geometrical changes than to changes in the material properties and location of impact. The observed relative increase in energy dissipation can reach 86.8% by including the honeycomb core and increasing the thickness of its shell.

From a shield design perspective, however, there are a number of other factors that cannot be accounted for in the present work. These include, for example, component and assembly level cost, component and assembly level weight and geometrical constraints beyond those considered in this research. The work presented here is not intended to be a multi-parameter optimisation study, but instead a detailed analysis of the main design parameters affecting the shielding performance under hypervelocity impact. With these constraints, the following conclusions can be drawn:

- Adding a honeycomb core can significantly improve the shielding ability.
- Increasing the honeycomb core density also improves the performance of the shield, although these effects are significantly less pronounced for cell densities beyond 4, as can be seen in Table 3 and Figures 12 and 14.
- Increasing the thickness of the honeycomb core shell also increases efficiency, although this parameter has the opposite effect on the cost and weight of the shield.

It should be added that no definite conclusion should be drawn regarding the location of the impact, as this parameter cannot be controlled by design. These observations clearly suggest that the proposed hybrid honeycomb structure concept provides a highly efficient alternative to a standard Whipple shield design without significantly compromising the weight of the structure.

Author Contributions: Conceptualization, K.F. and F.T.-D.; methodology, K.F. and F.T.-D.; validation, K.F. and F.T.-D.; formal analysis, K.F. and F.T.-D.; resources, F.T.-D.; writing—original draft preparation, K.F.; writing—review and editing, K.F. and F.T.-D.; supervision, F.T.-D. All authors have read and agreed to the published version of the manuscript.

Funding: This research received no external funding.

Data Availability Statement: Not applicable.

Acknowledgments: For the purpose of open access, the authors have applied a Creative Commons Attribution (CC BY) licence to any Author Accepted Manuscript version arising from this submission.

Conflicts of Interest: The authors declare no conflict of interest.

Abbreviations

The following abbreviations are used in this manuscript:

CPU	Central Processing Unit
ESD	Equivalent Sphere Diameter
FEM	Finite-Element Method
HVI	Hypervelocity impact
IDAC	Inter-Agency Space Debris Coordination Committee
IGM	Impedance-graded material
ISS	International Space Station
LEO	Low earth orbit
MMOD	Micro-meteoroid orbital debris
NASA	National Aeronautics and Space Administration
SPH	Smoothed Particle Hydrodynamics
SWS	Standard Whipple Shield

References

1. Astromaterials Research and Exploration Science Directorate. *Handbook for Designing MMOD Protection*; Technical Report JSC-64399; National Aeronautics and Space Administration (NASA): Washington, DC, USA, 2009.
2. European Space Agency (ESA) Space Debris Office. *European Space Agency's Annual Space Environment Report*; European Space Agency (ESA) Space Debris Office: Darmstadt, Germany, 2020.
3. National Research Council. *Limiting Future Collision Risk to Spacecraft: An Assessment of NASA's Meteoroid and Orbital Debris Programs*; The National Academies Press: Washington, DC, USA, 2011.
4. Astromaterials Research & Exploration Science, Hypervelocity Impact Technology, Shield Development. National Aeronautics and Space Administration (NASA). Available online: https://hvit.jsc.nasa.gov/shield-development/ (accessed on 2 February 2021).
5. Ryan, S.; Christiansen, E. *Hypervelocity Impact Performance of Open Cell Foam Core Sandwich Panel Structures*; Technical Report TM-2015-218593; National Aeronautics and Space Administration (NASA): Washington, DC, USA, 2015.
6. Cherniaev, A.; Telichev, I. Weight-Efficiency of Conventional Shielding Systems in Protecting Unmanned Spacecraft from Orbital Debris. *J. Spacecr. Rocket.* **2017**, *54*, 1–15. [CrossRef]
7. Christiansen, E.; Kerr, J.; de la Fuente, H.; Schneider, W. Flexible and deployable meteoroid/debris shielding for spacecraft. *Int. J. Impact Eng.* **1999**, *23*, 125–136. [CrossRef]
8. Plassard, F.; Mespoulet, J.; Hereil, P. Hypervelocity impact of aluminium sphere against aluminium plate: Experiment and LS-DYNA correlation. In Proceedings of the 8th European LSDYNA Users Conference, Strasbourg, France, 23–24 May 2011; pp. 142–149.
9. Zhang, P.; Gong, Z.; Tian, D.; Song, G. Study of the shielding performance of Whipple shield enhanced by impedance-graded materials. *Int. J. Impact Eng.* **2019**, *124*, 23–30. [CrossRef]
10. Carriere, R.; Cherniaev, A. Hypervelocity impacts on satellite sandwich structures—A review of experimental findings and predictive models. *Appl. Mech.* **2021**, *2*, 25–45. [CrossRef]
11. Carriere, R.; Cherniaev, A. Honeycomb parameter-sensitive predictive models for ballistic limit of spacecraft sandwich panels subjected to hypervelocity impact at normal incidence. *J. Aerosp. Eng.* **2022**, *35*, 04022039. [CrossRef]
12. Aslebagh, R.; Cherniaev, A. Projectile shape effects in hypervelocity impact of honeycomb-core sandwich structures. *J. Aerosp. Eng.* **2022**, *35*, 04021112. [CrossRef]
13. Pai, A.; Divakaran, R.; Anand, S.; Shenoy, S.B. Advances in the Whipple shield design and development: A brief review. *J. Dyn. Behav. Mater.* **2022**, *8*, 20–38. [CrossRef]
14. Ryan, S.; Hedman, T.; Christiansen, E. Honeycomb vs. foam: Evaluating potential upgrades to ISS module shielding. *Acta Astronaut.* **2010**, *67*, 818–825. [CrossRef]
15. Lambour, R.; Rajan, N.; Morgan, T.; Kupiec, I.; Stansbery, E. Assessment of orbital debris size estimation from radar cross-section measurements. *Adv. Space Res.* **2004**, *34*, 1013–1030. [CrossRef]
16. Rajan, N.; Morgan, T.; Lambour, R.; Kupiec, I. Orbital debris size estimation from radar cross section measurements. In Proceedings of the 3rd European Conference on Space Debris, Darmstadt, Germany, 19–21 March 2001.
17. Cowardin, H.M.; Hostetler, J.M.; Murray, J.I.; Reyes, J.A.; Cruz, C.L. Optical characterization of DebriSat fragments in support of orbital debris environmental models. *J. Astronaut. Sci.* **2021**, *68*, 1186–1205. [CrossRef]
18. Wang, X.; Shi, J. Validation of Johnson-Cook plasticity and damage model using impact experiment. *Int. J. Impact Eng.* **2013**, *60*, 67–75. [CrossRef]

19. Venkatesan, J.; Iqbal, M.; Gupta, N.; Bratov, V. Ballistic characteristics of bi-layered armour with various aluminium backing against ogive nose projectile. *Procedia Struct. Integr.* **2017**, *6*, 40–47. [CrossRef]
20. Zhang, Y.; Outeiro, J.; Mabrouki, T. On the selection of Johnson-Cook constitutive model parameters for Ti-6Al-4V using three types of numerical models of orthognal cutting. In Proceedings of the 15th Conference on Modelling of Machining Operations, Karlsruhe, Germany, 11–12 June 2015.
21. Nor, M.K.M.; Ho, C.S.; Ma'at, N. Modelling shock waves in composite materials using generalised orthotropic pressure. *Contin. Mech. Thermodyn.* **2020**, *32*, 1217–1229.
22. Olleak, A.A.; El-Hofy, H.A. SPH Modelling of cutting forces while turning of Ti6Al4V alloy. In Proceedings of the 10th European LS-DYNA Conference, Wurzburg, Germany, 15–17 June 2015.

applied sciences

MDPI

Article

Numerical Analysis and Experimental Test for the Development of a Small Shaped Charge

Piotr Malesa [1,*], Grzegorz Sławiński [1] and Karolina Pęcherzewska [2]

1 Faculty of Mechanical Engineering, Military University of Technology, Gen. Sylwestra Kaliskiego Street 2, 00-908 Warsaw, Poland; grzegorz.slawinski@wat.edu.pl
2 Ministry of National Defence, 00-911 Warsaw, Poland; kpecherzewska@ron.mil.pl
* Correspondence: piotr.malesa@wat.edu.pl; Tel.: +48-261-83-78-18

Abstract: Currently, shaped charges are widely used in many fields of science and industry. Due to the high efficiency of piercing materials with high strength and hardness, shaped charges are commonly used in mining, military and for structural damage. The main application area of shaped charges is the military industry, where they are used in missiles with warheads (torpedoes, rocket launchers) and for piercing vehicle armor or bunker walls. When analyzing the existing solutions of shaped charges, one can find many typical solutions designed for specific applications. However, there are no universal constructions which, after appropriate regulation, will fulfil their role in a wide range of applications. The subject of this article is a new solution for a shaped charge that is characterized by compact dimensions and a short preparation time. This article presents the results of experimental research and the numerical analyses of such a charge.

Keywords: shaped charge; jet; cumulative charge; numerical simulation; LS-Dyna

Citation: Malesa, P.; Sławiński, G.; Pęcherzewska, K. Numerical Analysis and Experimental Test for the Development of a Small Shaped Charge. *Appl. Sci.* **2021**, *11*, 2578. https://doi.org/10.3390/app11062578

Academic Editor: Ricardo Castedo

Received: 18 February 2021
Accepted: 9 March 2021
Published: 13 March 2021

1. Introduction

Cumulative charges have been widely used for many years, including in military technology [1,2] (mainly in anti-tank weapons) and in the mining industry (drilling holes) [3]. The nature of this phenomenon also allows for its use in the process of developing new design solutions intended for special applications [4].

The analyses carried out in 2017, aimed at identifying the optimal design solution for one such application, showed the need to use a shaped charge which, depending on the need, will enable the pierceability of approximately 80 mm to 200 mm to be obtained. Additionally, such a charge should be as small as possible in weight and dimensions, with a short time to prepare for use and the possibility of detonation with a time fuse.

The analysis of the state of the art in this field has shown that there are known design solutions that enable the adjustment of the distance between the base of the cumulative insert from the surface being destroyed by means of feet (these solutions are protected by patent law) [5]. However, they did not meet the requirements due to the lack of a fuse with a timed electronic system and because of the extended amount of time that it took to prepare the charge for use. Therefore, there was a need to develop a new design solution.

The developed conceptual design assumed the achievement of the required pierceability through the use of a conical, copper shaped liner and a pressed octogen (HMX) explosive in the structure of the charge. The quick adjustment of the height of the load and the distance from the base of the accumulation insert to the destroyed surface was to be ensured by placing the load casing in an additional sleeve in a way that allowed for an abrupt change of the position of both elements in relation to each other. An additional advantage of this solution was the minimization of the dimensions of the load in the transport position. Neodymium magnets, placed in the flange at the base of the sleeve (in the case of mounting the load on steel structures), or the use of a special, universal tape (in the case of the need to mount the load on other types of surfaces), were to ensure the possibility of quick fastening of the load to the destroyed element.

The developed conceptual design also included the construction of a time-type fuse with a self-destruction function. Its block diagram is shown in Figure 1.

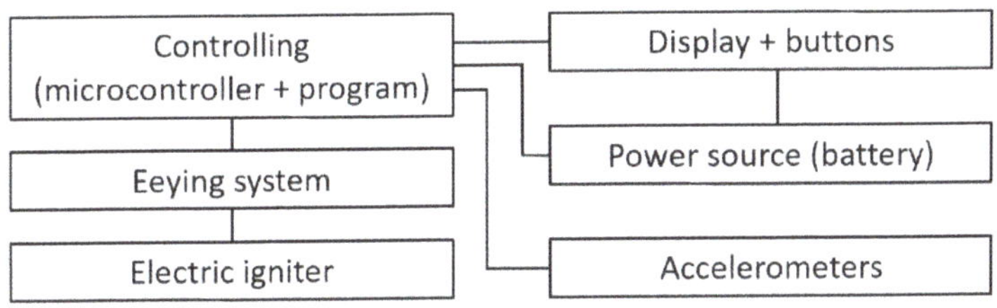

Figure 1. Block diagram of the time-type fuse.

On the basis of the developed conceptual design, a 3D CAD (Computer Aided Design) model of the cargo casing was created, which was then produced using the FDM (Fused Deposition Modeling) 3D printing technique in Figure 2a.

(**a**) (**b**)

Figure 2. View of the structure of shaped charges: (**a**) first design of the shaped charge prototype and (**b**) final design of the shaped charge construction.

In 2018, the first preliminary tests were conducted at the Military University of Technology to verify the developed cargo design concept, which confirmed its correctness.

The next stage of work was the optimization of the structure, aimed at minimizing the weight and dimensions of the load. As a result, modifications to the housing structure were introduced. The number of components was reduced so that the structure consists of a spacer sleeve and a housing. The view of the final cargo structure after the modification is shown in Figure 2b.

The developed solution of the final shaped charge was then subjected to experimental tests and multi-variant numerical analyses, which are presented in the following chapters.

2. Materials and Methods

With the use of modern technologies, it is possible to model many physical processes, the observation and research of which are hampered by various factors, such as process dynamics. All kinds of issues related to research with the use of explosives [6,7] or shaped charges, which are the main topic of this paper [8,9], constitute an example of such processes.

One of the main factors determining the choice of a research method is undoubtedly the economic factor. Computer modeling is usually much cheaper than the corresponding experiment. For this reason, FEM has found wide application in many fields of science, and this proves that it is a method as useful as experimental research.

Modeling of dynamic interactions of bodies [10–12] with the use of numerical analyses has been developed over many years. As a result of the work carried out so far, a set of mathematical and physical models and computer codes has been developed that can be effectively adapted to the research, optimization and evaluation of parameters of various types of loads.

For metal charge elements, i.e., a shaped charge liner, mathematical–physical models are used based on the theory of elastic plasticity supplemented with semi-empirical equations of state and dependencies describing the changes of the plastic flow limit as a function of temperature, pressure, density, plastic deformation and plastic deformation velocity (Johnson–Cook model) [13]. The Johnson–Cook material model is one of the most popular and most frequently used material model for the problems of modeling the cumulation process [14,15].

Reference [16] is one of the most interesting collective works devoted to the modeling of dynamic processes in the LS-Dyna program with the use of axisymmetric models, including the cumulative charge. The model presented in the paper is a 2D axisymmetric model, with an innovative use of a new formulation of elements in this field. The authors of this publication point to the possibility of modeling the problems of cumulation in an easy way, using for this purpose axisymmetric models and the 2D Arbitrary Lagrangian–Eulerian (ALE) element in the Eulerian formulation.

Multi-Material Arbitrary Lagrangian–Eulerian (MM-ALE) formulation is a two-step process [17–19]. The first step of the ALE procedure consists in carrying out the classical Lagrange step, which describes the deformation of the solid state (stiffness matrix with new initial-boundary conditions). The mesh moves with the flowing matter (fluid), thus fulfilling the principle of conservation of mass. The velocity and displacements of the mesh are determined, and the nodes of the deformed elements return to their original position [18,19].

The second stage of the procedure is to carry out the advection step, which includes:

1. Deciding which nodes to move.
2. Displacement of extreme nodes.
3. Displacement of nodes inside.
4. Recalculating all variables related to the elements.
5. Recalculating momentum values and updating speed.

When determining the velocity and fluid displacements, the equations of the conservation of mass, torque and energy are implemented [18,19]:

$$\frac{dM}{dt} = \frac{d}{dt}\int_{V(t)} \rho \, dV = \int_{S(t)} \rho\left(\omega - v\right)\cdot n \, dS, \tag{1}$$

$$\frac{dQ}{dt} = \frac{d}{dt}\int_{V(t)} \rho v \, dV = \int_{S(t)} \rho v \left(\omega - v\right)\cdot n \, dS - \int_{V(t)} \nabla p \, dV + \int_{V(t)} v g \, dV, \tag{2}$$

$$\frac{dE}{dt} = \frac{d}{dt}\int_{V(t)} \rho e \, dV = \int_{S(t)} \rho e \left(\omega - v\right)\cdot n \, dS - \int_{S(t)} p v \cdot n \, dS + \int_{V(t)} \rho g \cdot v \, dV. \tag{3}$$

where ρ is fluid mass density, p is pressure, g is acceleration of gravity and e is the total specific energy. The quantities M, Q and E are total mass, total momentum and total energy, respectively, of control volume $V(t)$, bounded by surface S, which moves in the fluid (gas–air) with arbitrary velocity ω which may be zero in Eulerian coordinates or v in Lagrangian coordinates. The vector n is the outwards normal to the surface S.

The Johnson–Cook (JC) constitutive model was used to describe the proper dynamic behavior of the cumulative insert and the steel block affected by the insert. It is an elastic–plastic model of the material. Plastic deformation in the material is modeled using isotropic material hardening. The yield point of the material is described by the following relationship:

$$\sigma_Y = \left(A + B\varepsilon_p^n\right)\left(1 + Cln\dot{\varepsilon}^*\right)(1 - T^{*m}) \tag{4}$$

where:

A—static yield point,
ε_p—plastic deformation,
$\dot{\varepsilon}^*$—dimensionless strain rate, $\dot{\varepsilon}^* = \dot{\varepsilon}/\dot{\varepsilon}_0$
$\dot{\varepsilon}_0$—reference strain rate,
T^*—the ratio of the absolute temperature of the sample to its melting point, which is determined by the following relationship:

$$T^* = \begin{cases} 0 & dla \quad T < T_{room} \\ \frac{T - T_{room}}{T_{melt} - T_{room}} & dla \quad T_{room} \leq T \leq T_{melt} \\ 1 & dla \quad T > T_{melt} \end{cases} \tag{5}$$

T_{room}—temperature at which the experiment was carried out,
T_{melt}—material melting point,
n—parameter determining the material's susceptibility to hardening by deformation,
m—thermal plasticization exponent,
B—hardening constant,
C—strain rate constant.

Johnson and Cook proposed that fracture strain typically depends on the stress triaxiality ratio, the strain rate and the temperature. The strain at fracture is given by:

$$\varepsilon^f = max\left([D_1 + D_2 expD_3\sigma^*]\left[1 + D_4 ln\dot{\varepsilon}^*\right][1 + D_5T^*], EFMIN\right) \tag{6}$$

where σ^* is the ratio of pressure divided by effective stress:

$$\sigma^* = \frac{p}{\sigma_{eff}} \tag{7}$$

Fracture occurs when the damage parameter:

$$D = \sum \frac{\Delta\bar{\varepsilon}^p}{\varepsilon^f} \tag{8}$$

reaches the value of 1.

The data of the copper insert material [20] with the EOS (Equation of State) model, and the steel material model [21] with the EOS data [22] from which the target was made, are presented in Table 1.

Table 1. Material data for cumulative insert and steel target [20–22].

Parameter	Symbol	Unit	Shaped Charge Liner	Steel Target
Density	ρ	Kg/m^3	8940	7860
Shear modulus	G	GPa	-	81.8
Young modulus	E	GPa	126	209
Poisson's ratio	ν	-	0.335	0.28
Yield stress	A	MPa	99.7	792
Hardening constant	B	MPa	262.8	510
Hardening exponent	N	-	0.23	0.26
Strain rate constant	C	-	0.029	0.014
Thermal softening exponent	M	-	0.98	1.03
Room temperature	Tr	K	293	300
Melting temperature	Tm	K	775	1790
Ref. strain rate	EPSO	s^{-1}	1.0	1.0
Specific heat	Cp	J/kgK	875	477
Johnson Cook failure				
Failure parameter	D_1	-	0.13	0.05
Failure parameter	D_2	-	0.13	3.44
Failure parameter	D_3	-	−1.5	−2.12
Failure parameter	D_4	-	0.011	0.002
Failure parameter	D_5	-	0	0.61
EOS_LINEAR_POLYNOMIAL				
	C_0	-	20.790	-
	C_1	-	1.337 ′ 10^5	-
	C_2	-	1.256 ′ 10^5	-
	C_3	-	1.454 ′ 10^5	-
	C_4	-	1.940	-
	C_5	-	0.585	-
	C_6	-	1.125	-
EOS_GRUNEISEN				
	C	m/s	-	4570
	S_1	-	-	1.49
	γ_0	-	-	1.93
	A	-	-	0.5

For the JC model, it is necessary to define a polynomial equation of state describing the relationship between pressure, volume and the internal energy in a material. The linear polynomial equation of state was used for the model of the copper insert material. This equation is expressed as:

$$P = c_0 + c_1\mu + c_2\mu^2 + c_3\mu^3 + \left(c_4 + c_5\mu + c_6\mu^2\right)E_0 \quad (9)$$

where: $c_0 \div c_6$, state equation parameters; μ, compression factor $\mu = \rho/\rho_0$ expressed as the ratio of the actual density ρ to the original density ρ_0; E_0, internal energy.

The polynomial equation of state in a simplified form is used to describe the gas (air) medium surrounding the explosive charge and the tested object:

$$P = (c_4 + c_5\mu)E \quad (10)$$

where: $\mu = \rho/\rho_0$, C_4 and C_5, equation coefficients; ρ, density; ρ_0, starting density; E, internal energy.

The Grüneisen equation of state was defined for the steel material model. The equation defines the pressure in the shock-compressed material as:

$$p = \frac{\rho_0 C^2\mu\left[1 + \left(1 - \frac{\gamma_0}{2}\right)\mu - \frac{a}{2}\mu^2\right]}{\left[1 - (S_1 - 1)\mu - S_2\frac{\mu^2}{\mu+1} - S_3\frac{\mu^3}{(\mu+1)^2}\right]^2} + (\gamma_0 + a\mu)E \quad (11)$$

whereas for the expanded material as:

$$p = \rho_0 C^2 \mu + (\gamma_0 + a\mu)E \quad (12)$$

C—bulk speed of sound,
γ_0—Grüneisen gamma,
S_1—linear coefficient,
S_2—quadratic coefficient,
S_3—cubic coefficient,
a—first order volume correction to γ_0,
μ—volume parameter, expressed as $\mu = (\rho/\rho_0) - 1$,
ρ—actual density,
ρ_0—initial density,
E—internal energy per unit of mass.

The MAT_HIGH_EXPLOSIVE_BURN material model was selected to describe the octogen (HMX) explosive. The material data for the HMX explosive and presented in Table 2.

Table 2. Pressed octogen (HMX) explosive data with equation of state [23].

Parameter	Symbol	Unit	Value
MAT_HIGH_EXPLOSIVE_BURN			
Density	ρ	Kg/m^3	1890
Detonation velocity	D	m/s	9110
Chapman–Jouget pressure	PCJ	GPa	42
EOS_JWL			
	A	GPa	778.3
	B	GPa	7.071
	R_1	-	4.2
	R_2	-	1
	ω	-	0.3

The Jones–Wilkins–Lee (JWL) equation of state was used to describe the relationship between the parameters of the thermodynamic system for the explosive. The equation of state of gaseous products of detonation of condensed explosives takes the following form:

$$p = A\left(1 - \frac{\omega}{R_1 V}\right)e^{-R_1 V} + B\left(1 - \frac{\omega}{R_2 V}\right)e^{-R_2 V} + \frac{\omega E}{V}, \quad (13)$$

A, B and E have units of pressure. R_1, R_2, ω, and V_0 are dimensionless. E—internal energy per unit volume and ω—the relative volume of the explosive.

The air domain was modeled using the MAT_NULL material model and the EOS_LINEAR_POLYNOMIAL equation of state, for which the material data summarized in Table 3 were used.

Table 3. Air material data with the equation of state [24].

Parameter	Symbol	Unit	Value
MAT_NULL			
Density	ρ	Kg/m^3	1.29
EOS_LINEAR_POLYNOMIAL			
	C_4	GPa	0.4
	C_5	GPa	0.4
	E_0	GPa	2.5×10^{-4}
	V_0	-	1

The explosive casing, which in real conditions was made of Acrylonitrile Butadiene Styrene (ABS) was modeled using the MAT_PIECEWISE_LINEAR_PLASTICITY material model, for which the data are summarized in Table 4.

Table 4. Acrylonitrile Butadiene Styrene (ABS) material data [25].

ρ [Kg/m^3]	**E [GPa]**	**ν [-]**	**SIGY [MPa]**
1040	2.2	0.35	34.17
EPS1 [%]	**EPS2 [%]**	**EPS3 [%]**	**EPS4 [%]**
0	0.8	1.4	2.8
EPS5 [%]	**EPS6 [%]**	**EPS7 [%]**	**EPS8 [%]**
5.7	6.5	7.1	7.7
ES1 [MPa]	**ES2 [MPa]**	**ES3 [MPa]**	**ES4 [MPa]**
34.17	34.52	34.72	35.13
ES5 [MPa]	**ES6 [MPa]**	**ES7 [MPa]**	**ES8 [MPa]**
35.49	35.53	35.54	35.55

All numerical simulations were performed using the LS-Dyna code with an implementation of the Multi-Material Arbitrary Lagrangian–Eulerian (MM-ALE) formulation.

3. Subject and Scope of the Tests

The numerical model of the stand for the penetration test with the use of the shaped charge was developed based on the solid model of the analyzed system. During the experimental tests, a charge consisting of a casing in which the explosive was compressed, a sleeve regulating the distance between the target and the charge and a shaped charge liner were used. The view of the analyzed system for extreme variants of the distance between the load and the target is shown in Figure 3.

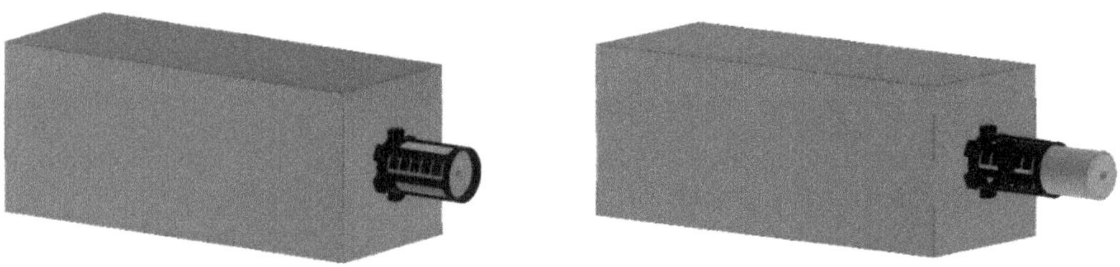

(a) (b)

Figure 3. View of two variants of shaped charges: (**a**) charge closest to the target and (**b**) charge farthest from the target.

The distance between the face of the casing with the explosive and the liner insert was adjusted with the use of a solution based on a system of five serrations enabling insertion and locking of the sleeve within them. The solution allows for a six-step adjustment of the distance of the load from the target, with the minimum distance in step 1 being 7.5 mm and changing every 12.5 mm up to the maximum distance of 70 mm.

Based on the solid model, an axially symmetric shell model was developed, consisting of a steel block, housing, charge and a shaped charge liner placed in the air domain. The axisymmetric model covered half of the whole system due to the applied axial symmetry.

The discretization of the geometry made it possible to freely change the distance between the load and the target. In the initial phase of the analysis, two models were built for the two extreme distances as shown in Figure 4.

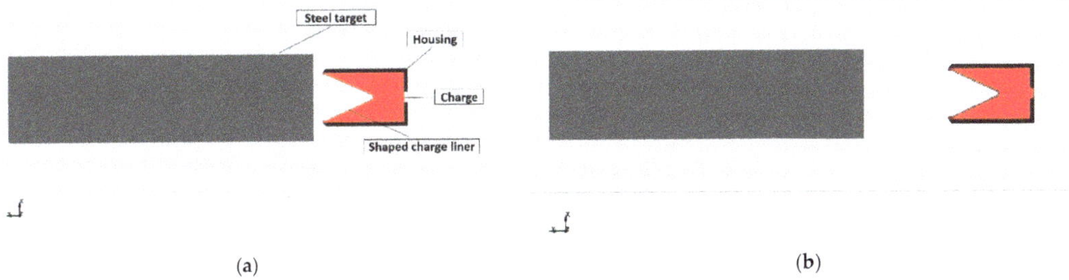

Figure 4. View of two variants of the shaped charge model: (**a**) charge closest to the target and (**b**) charge farthest from the target.

The ALE procedure requires that the finite element mesh be constructed from the smallest possible elements to properly implement the explosion phenomenon. For this reason, the models consist of many elements, which translates into a time-consuming calculation. The analyzed models consisted of finite elements with the dimensions of 0.25 mm × 0.25 mm—158,455 for the model with the load closest to the target and 199,351 for the load farthest from the target.

The calculations were performed at time t = 0.5 ms. For numerical calculations, an explicit type of algorithm was used to solve the structure dynamics equations in the non-linear range.

4. Model Validation, Results and Discussion

At the beginning, numerical analyses were carried out for two extreme variants of the shaping charge displacement against the target. Numerical analyses of the propagation of the shaped charge in the Euler domain and for the penetration of the steel block were carried out. A graphical summary of the subsequent steps of the analysis for two variants was obtained. The results are presented in Table 5.

Table 6 shows a cross-sectional comparison view of a steel block that has been subjected to a shaped charge for experimental and numerical tests. The obtained values of the penetration depth of the shaped charge in the steel block were compared with the results of the experimental study. Table 7 presents the quantitative results of the penetration depth obtained through numerical analyses and experimental tests.

For the charge placed closest to the target, a large-diameter hole was obtained at the entrance of the charge into the material, with a small diameter at its end. In the case of the load placed over the longest standoff from the target, the hole has a more regular shape with a diameter taper toward its end. As can be seen from the obtained cross-sections and numerical analyses, the displacement of the charge from the target causes the shaped charge that hits the target to be formed better, which translates directly into the effectiveness of its penetration.

Table 5. View of the propagation process of detonation products and the penetration process of the shaped charge inside the steel target.

Time	Penetration Process for 7.5 mm Standoff	Penetration Process for 70 mm Standoff
T = 0 ms		
T = 0.01 ms		
T = 0.05 ms	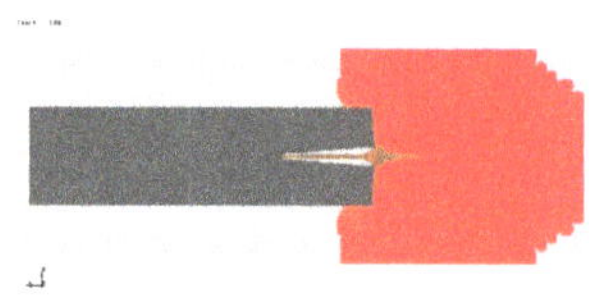	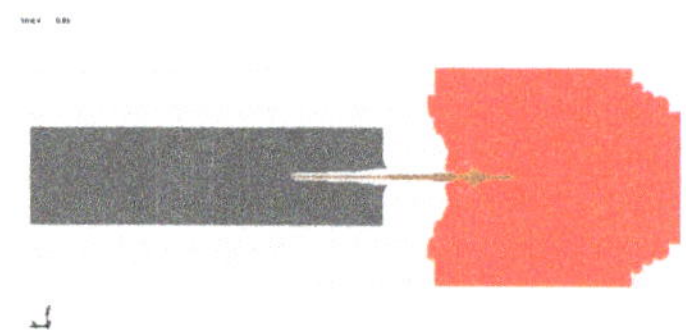
T = 0.1 ms	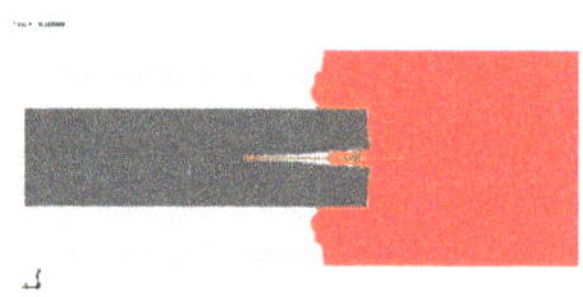	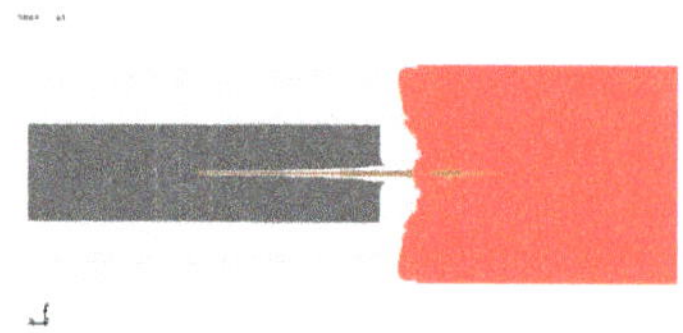
T = 0.5 ms	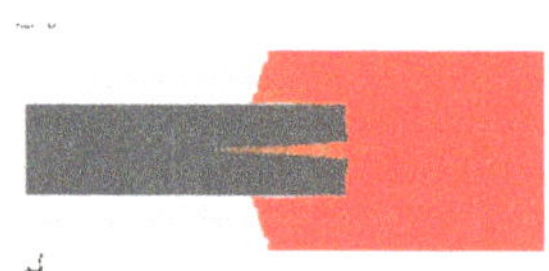	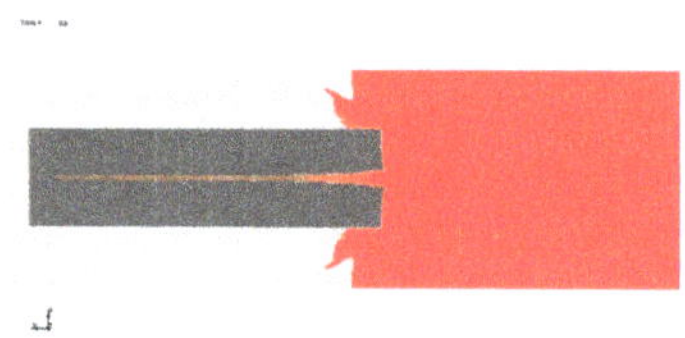

Table 6. View of the section of the steel block after shooting with the shaped charge for experimental and numerical tests.

Cross-Section of Steel Target for 7.5 mm Standoff	Cross-Section of Steel Target for 70 mm Standoff
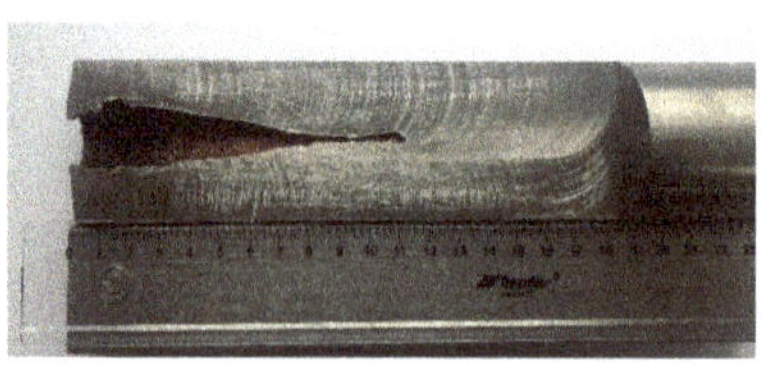	

Table 7. The values of the penetration depth of the steel block with the shaped charge.

	Depth of Penetration for 7.5 mm Standoff	Depth of Penetration for 70 mm Standoff
Experimental test	110 mm	220 mm
Numerical analysis	106 mm	244 mm
Difference	4%	8%

By analyzing the obtained results of the hole depth in the steel block material, it is possible to notice a double penetration value for the charge away from the target. This difference results from the different focusing of the stream of the shaped charge. The distance of the cumulative charge from the target is a very important parameter because its proper selection makes it possible to maximize the energy transferred to the obstacle by the cumulative flux [26] and at the same time reduce energy losses to the environment, thus increasing the effectiveness of the work done by the charge.

To investigate more accurately the effect of moving the charge away from the target on the penetration capabilities of the shaped charge, numerical simulations were carried out for all six variants of the removal of the charge from the target. The variants corresponded to a six-step adjustment by means of a system of teeth and sleeves connecting the load casing with the distance sleeve.

The following variants of the analyses were implemented:

- Variant 1—7.5 mm standoff closest to the target.
- Variant 2—20 mm standoff.
- Variant 3—32.5 mm standoff.
- Variant 4—45 mm standoff.
- Variant 5—57.5 mm standoff.
- Variant 6—70 mm standoff farthest from the target.

Table 8 shows the results of the impact of the shaped charge for all six variants for the analysis times t = 0.1 ms, 0.25 ms and 0.5 ms.

Based on the obtained values of the maximum penetration of the shaped charge in the steel block, the characteristics of the dependence of the puncture efficiency depending on the distance between the charge and the target were determined. The results are shown in Figure 5.

Depending on the distance between the charge and the target, the value of the velocity of the cumulative stream front changes. For the charge in question, the value of the maximum speed at the time of formation of the cumulative stream was 7180 m/s. Figure 6 shows the view of the cumulative flux with a map of the resultant velocity value at the moment of contact between the flux front and the target.

The following values of the cumulative flux velocity at the moment of contact with the target and the diameter of the holes formed in the steel block were obtained for the individual variants of moving the charge from the target.

Table 8. View of the propagation process of detonation products and the penetration process of the shaped charge.

Variant	Penetration Process for Theanalysis Time of 0.1 ms	Penetration Process for Theanalysis Time of 0.25 ms	Penetration Process for Theanalysis Time of 0.5 ms
Standoff = 7.5 mm	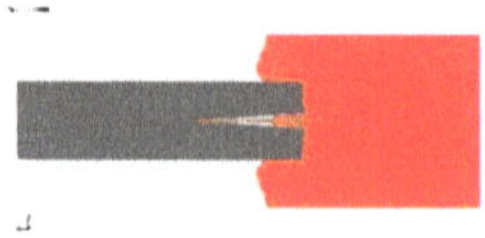		
Standoff = 20 mm	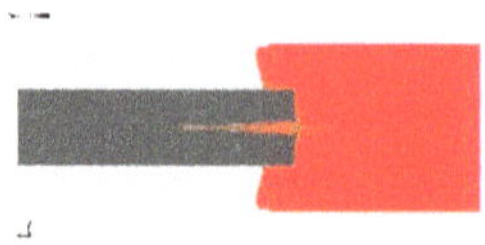		
Standoff = 32.5 mm	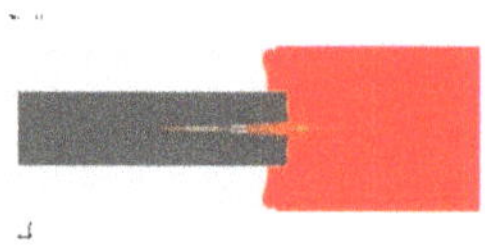		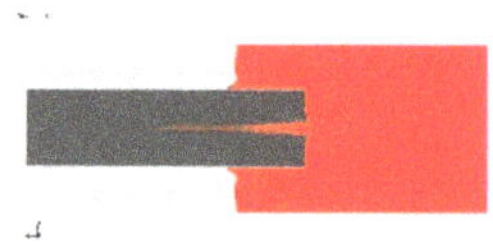
Standoff = 45 mm	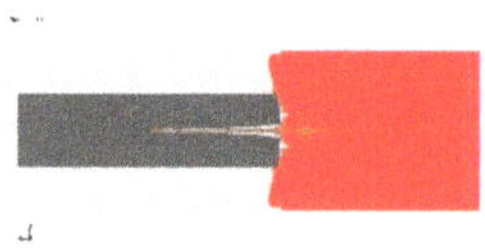		
Standoff = 57.5 mm	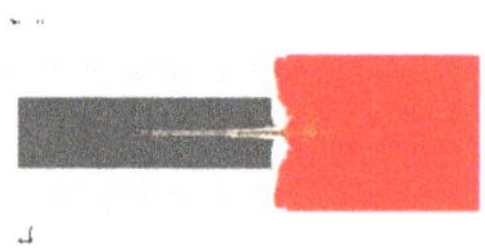		
Standoff = 70 mm	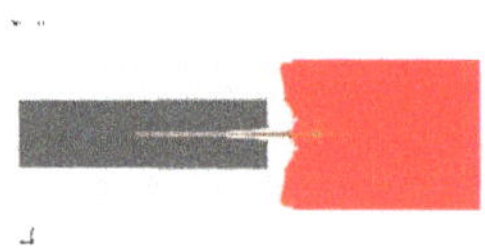	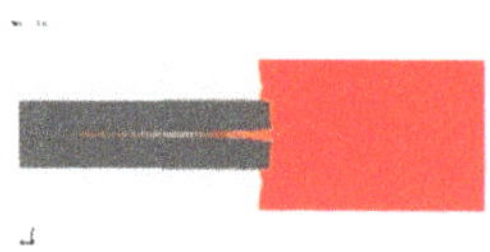	

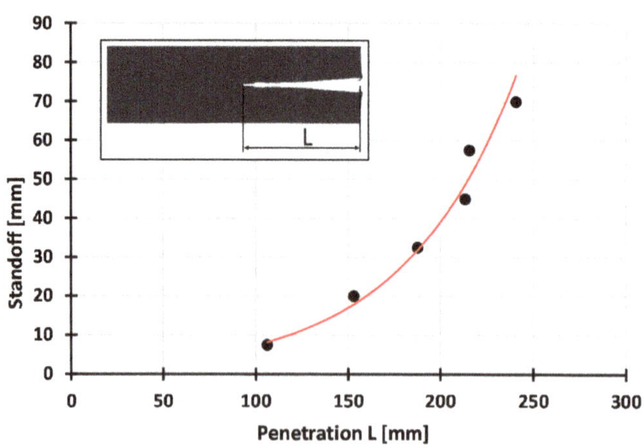

Figure 5. Characteristics of the change of penetration depth of the shaped charge in the steel block depending on the distance from the target.

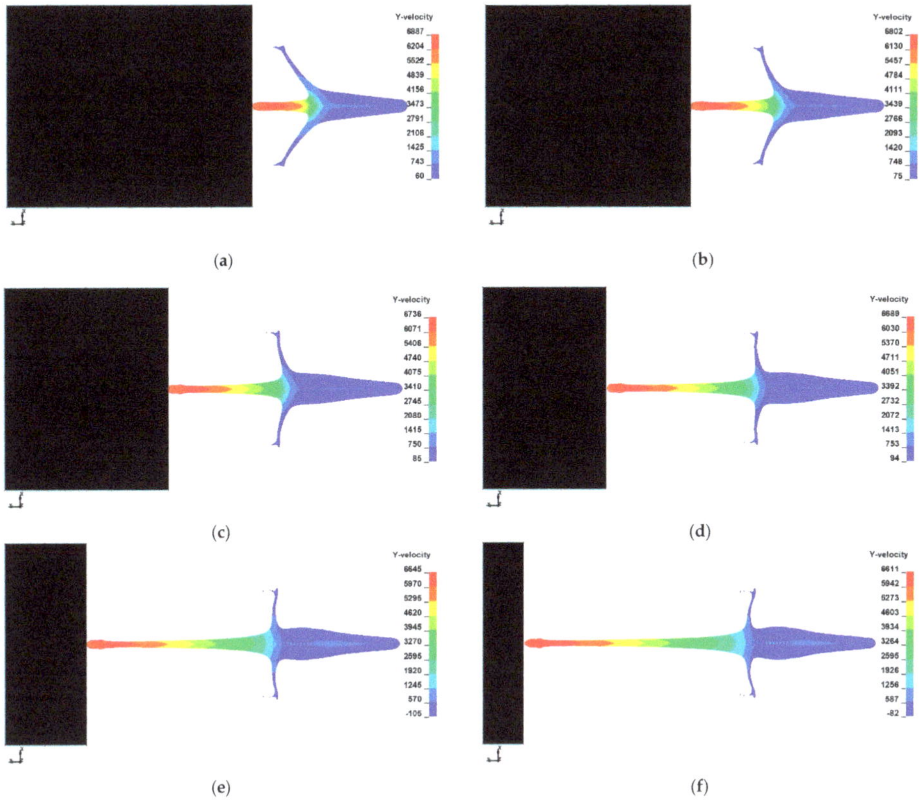

Figure 6. View of the shaped charge jet with the marked map of the resultant velocity at the moment of contact with the target: (**a**) 7.5 mm standoff, (**b**) 20 mm standoff, (**c**) 32.5 mm standoff, (**d**) 45 mm standoff, (**e**) 57.5 mm standoff and (**f**) 70 mm standoff.

5. Conclusions

The main purpose of the paper was to test and analyze the effectiveness of small-shaped charges in terms of their ability to penetrate a steel block. The conducted experimental tests allowed us to determine the real ability to penetrate a steel target with a shaped charge at a variable value of its distance from the target. On the basis of the experimental test, the breakthrough values of 110 and 220 mm were obtained for two variants of the charge standoff from the target, determined by the proprietary solution in the form of a spacer sleeve.

As a result of the numerical analyses carried out, the process of validation of the numerical model in the axisymmetric approach was conducted using the Multi-Material Arbitrary Lagrangian–Eulerian (MM-ALE) formulation method. The performed numerical model made it possible to regulate the distance of the charge from the target in accordance with the actual conditions, while maintaining all dimensions of the components of the load, i.e., casing, cumulative insert and the explosive used there. The obtained results of the penetration depth obtained by means of numerical analyses and experimental tests were characterized by a difference at the level of 4–8%, which made it possible to adopt the method used as effective for modeling the cumulative charge.

After a successful validation process, further numerical analyses were carried out for the remaining variants of the load positioning in relation to the target, using the distance adjustment system of the sleeve. The method of formation of the cumulative flux and the ability to break through another material in terms of the distance between them were checked via numerical analyses. A graph of the change of the penetration capacity of a shaped charge jet in a steel block depending on the distance between it and the target was developed. The exponential trend line shown in Figure 3 illustrates this relationship. Another phenomenon that was verified by numerical analysis was the velocity of the jet reached at the moment when the top of it contacts the target. Based on the conducted analyses, the results for all tested variants of the retraction were collected, as shown in Table 9.

Table 9. Summary of the shaped charge jet velocity at the moment of contact with the target.

Variant	Jet Velocity [m/s]	Hole Diameter [mm]
Standoff = 7.5 mm	6887	12.5
Standoff = 20 mm	6802	12.5
Standoff = 32.5 mm	6736	12.5
Standoff = 45 mm	6689	12
Standoff = 57.5 mm	6645	12
Standoff = 70 mm	6611	11.5

To sum up, the assumption of this paper was to carry out the modeling and validation process of the shaped charge initiation and propagation process in interaction with a steel target. The method of modeling the phenomenon of this problem in the axisymmetric approach was used, which faithfully reflects the conditions of real experimental research. The obtained results constitute the basis for further numerical research based on a correctly functioning and validated numerical model. Subsequent papers will focus on analyzing the effectiveness of a small-shaped charge jet penetration with other types of materials.

Author Contributions: Conceptualization, P.M., G.S. and K.P.; methodology, P.M., G.S. and K.P.; software, P.M. and G.S.; validation, P.M. and G.S.; formal analysis, P.M., G.S. and K.P.; investigation, P.M., G.S. and K.P.; resources, P.M., G.S. and K.P.; data curation, P.M., G.S. and K.P.; writing—original draft preparation, P.M., G.S. and K.P.; writing—review and editing, P.M., G.S. and K.P.; visualization, P.M., G.S. and K.P.; supervision, P.M., G.S. and K.P.; project administration, G.S.; funding acquisition, G.S. All authors have read and agreed to the published version of the manuscript.

Funding: The article was written as part of the implementation of the university research grant supported by the Military University of Technology (No. UGB 22-746/2020).

Institutional Review Board Statement: Not applicable.

Informed Consent Statement: Not applicable.

Data Availability Statement: Not applicable

Acknowledgments: Special thanks for organizing and conducting experimental studies of the impact of the shaped charge on the steel block to Waldemar Trzciński from Military University of Technology.

Conflicts of Interest: The authors declare no conflict of interest.

References

1. Cui, P.; Wang, D.; Shi, D.; Gao, X.; Xu, J.; Zhen, J. Investigation of penetration performance of Zr-based Amorphous Alloy Liner Compared with Copper. *Materials* **2020**, *13*, 912. [CrossRef] [PubMed]
2. Wu, H.; Hu, F.; Fang, Q. A comparative study for the impact performance of shaped charge JET on UHPC targets. *Def. Technol.* **2019**, *15*, 4. [CrossRef]
3. Wilk, Z.; Zygmunt, B. Zastosowanie ładunków kumulacyjnych do perforacji odwiertów geologicznych. *Biul. Wojsk. Akad. Tech.* **2007**, *56*, 245–258.
4. Available online: http://cama.pl/realizacje/wybuchowe-ciecie-stalowej-konstrukcji-zwalowarki-turow (accessed on 20 January 2021).
5. Śliwiński, J.; Ludas, M.; Czubaty, Ł. Opis ochronny wzoru użytkowego nr. 68492. In *Urząd Patentowy Rzeczypospolitej Polskiej*; data zgłoszenia: 12.03.2014; 2014.

6. Sławiński, G.; Malesa, P.; Świerczewski, M. Analysis regarding the risk of injuries of soldiers inside a vehicle during accidents caused by improvised explosive devices. *Appl. Sci.* **2019**, *9*, 4077. [CrossRef]
7. Baranowski, P.; Małachowski, J. Numerical study of selected military vehicle chassis subjected to blast loading in terms of tire strength improving. *Bull. Pol. Acad. Sci. Tech. Sci.* **2015**, *63*. [CrossRef]
8. Pyka, D.; Kurzawa, A.; Bocian, M.; Bajkowski, M.; Magier, M.; Sliwinski, J.; Jamroziak, K. Numerical and Experimental Studies of the ŁK Type Shaped Charge. *Appl. Sci.* **2020**, *10*, 6742. [CrossRef]
9. Panowicz, R.; Nowak, J.; Konarzewski, M.; Niezgoda, T. Parametric studies of directed fragmentation warhead used for combat shaped charges. *Eng. Trans.* **2015**, *63*, 181–190.
10. Malachowski, J.; Baranowski, P.; Gieleta, R.; Mazurkiewicz, Ł. Rubber structure under dynamic loading—Computational Studies. *Eng. Trans.* **2013**, *61*, 33–46.
11. Morka, A.; Kędzierski, P.; Gieleta, R. Selected aspects of numerical analysis of layered flexible structures subjected to impact of soft core projectile. *Arch. Mech. Eng.* **2015**, *62*. [CrossRef]
12. Sybilski, K.; Małachowski, J. Impact of disabled driver's mass center location on biomechanical parameters during crash. *Appl. Sci.* **2021**, *11*, 1427. [CrossRef]
13. Johnson, G.R.; Cook, W.H. A constitutive model and data for metals subjected to large strains and high temperatures. In Proceedings of the 7th International Symposium on Ballistics, Hague, The Netherlands, 19–21 April 1983.
14. Huang, J.; Wang, W.; Zhang, W.; Huang, K. The 3D numerical simulation of the shaped charge jet penetration through the steel target. In Proceedings of the 2nd International Conference on Computer Application and System Modeling, Cochin, India, 20–21 October 2012. [CrossRef]
15. Costa, E.; Ferrante, E.; Trevisi, A.; Bozzolo, A. Shaped charge: A comparative study of jet penetration into a multi-layered target. In Proceedings of the International CAE Conference and Exhibition, Vicenza, Italy, 8–9 October 2018.
16. Van Dorsselaer, N.; Lapoujade, V. A contribution to new ALE 2D Method Validation. In Proceedings of the 11th International LS-Dyna Users Conference, Dearborn, Michigan, 6–8 June 2010.
17. Małachowski, J. *Modelowanie i badania interakcji ciało stałe-gaz przy oddziaływaniu impulsu ciśnienia na elementy konstrukcji rurociągu*; Bel studio: Warszawa, Poland, 2010.
18. Baranowski, P.; Małachowski, J.; Mazurkiewicz, Ł. Numerical and experimental testing of vehicle tyre under impulse loading conditions. *Int. J. Mech. Sci.* **2016**, *106*. [CrossRef]
19. Hallquist, J. *LS-Dyna: Theory Manual*; Livemore Software Technology Corporation (LSTC): Livermore, CA, USA, 2019.
20. Kołodziejczyk, D. Numeryczno-eksperymentalne badanie wpływu uszkodzeń lub deformacji wkładki kumulacyjnej na tworzenie się strumienia kumulacyjnego. Rozprawa Doktorska, Wojskowa Akademia Techniczna, Warszawa, Poland, 2015.
21. Rashed, A.; Yazdani, M.; Babaluo, A.; Hajizadeh, P.P. Investigation on high-velocity impact performance of multi-layered alumina ceramic armors with polymeric interlayers. *J. Compos. Mater.* **2015**, *50*, 25. [CrossRef]
22. Kędzierski, P.; Morka, A.; Sławiński, G.; Niezgoda, T. Optimization of two-component armour. *Bull. Pol. Acad. Sci. Tech. Sci.* **2015**, *63*. [CrossRef]
23. Panowicz, R.; Trypolin, M.; Konarzewski, M. Numerical comparison of blast waves generated by cylindrical explosive charges with varying shapes and materials. *Comput. Assist. Methods Eng. Sci.* **2016**, *23*, 4.
24. Sławiński, G.; Malesa, P.; Niezgoda, T.; Świerczewski, M. A comparison between two different methods of blast modelling. *Lect. Notes Mech. Eng.* **2017**. [CrossRef]
25. Kucewicz, M.; Baranowski, P.; Małachowski, J.; Popławski, A.; Płatek, P. Modelling, and characterization of 3D printed cellular structures. *Mater. Des.* **2018**, *142*. [CrossRef]
26. Wojewódka, A.; Witkowski, T. Metodyka modelowania procesu formowania strumienia kumulacyjnego wydłużonych ładunków materiałów wybuchowych. *Chemik* **2011**, *65*, 28–35.

applied sciences

MDPI

Article

SimEx: A Tool for the Rapid Evaluation of the Effects of Explosions

Juan Sánchez-Monreal †, Alberto Cuadra, César Huete and Marcos Vera *

Departamento de Ingeniería Térmica y de Fluidos, Escuela Politécnica Superior, Universidad Carlos III de Madrid, 28911 Leganés , Spain
* Correspondence: marcos.vera@uc3m.es; Tel.: +34-91-624-99-87
† Current address: Institute of Engineering Thermodynamics, German Aerospace Center (DLR), Pfaffenwaldring 38–40, 70569 Stuttgart, Germany.

Abstract: The dynamic response of structural elements subjected to blast loading is a problem of growing interest in the field of defense and security. In this work, a novel computational tool for the rapid evaluation of the effects of explosions, hereafter referred to as SimEx, is presented and discussed. The classical correlations for the reference chemical (1 kg of TNT) and nuclear (10^6 kg of TNT) explosions, both spherical and hemispherical, are used together with the blast wave scaling laws and the International Standard Atmosphere (ISA) to compute the dynamic response of Single-Degree-of-Freedom (SDOF) systems subject to blast loading. The underlying simplifications in the analysis of the structural response follow the directives established by UFC 3-340-02 and the Protective Design Center Technical Reports of the US Army Corps of Engineers. This offers useful estimates with a low computational cost that enable in particular the computation of damage diagrams in the Charge Weight–Standoff distance (CW–S) space for the rapid screening of component (or building) damage levels. SimEx is a computer application based on Matlab and developed by the Fluid Mechanics Research Group at University Carlos III of Madrid (UC3M). It has been successfully used for both teaching and research purposes in the Degree in Security Engineering, taught to the future Guardia Civil officers at the Spanish University Center of the Civil Guard (CUGC). This dual use has allowed the development of the application well beyond its initial objective, testing on one hand the implemented capacities by undergraduate cadets with the end-user profile, and implementing new functionalities and utilities by Masters and PhD students. With this experience, the application has been continuously growing since its initial inception in 2014 both at a visual and a functional level, including new effects in the propagation of the blast waves, such as clearing and confinement, and incorporating new calculation assistants, such as those for the thermochemical analysis of explosive mixtures; crater formation; fragment mass distributions, ejection speeds and ballistic trajectories; and the statistical evaluation of damage to people due to overpressure, body projection, and fragment injuries.

Keywords: effects of explosions; blast loading; SDOF systems; thermochemistry of explosives; fragments; crater formation; damage to people

Citation: Sánchez-Monreal, J.; Cuadra, A.; Huete, C.; Vera, M. SimEx: A Tool for the Rapid Evaluation of the Effects of Explosions. *Appl. Sci.* **2022**, *12*, 9101. https://doi.org/10.3390/app12189101

Academic Editors: Ricardo Castedo, Lina M. Lopez and Jeong Ik Lee

Received: 22 July 2022
Accepted: 6 September 2022
Published: 10 September 2022

1. Introduction

Unlike the slow energy release exhibited by deflagrations, the instantaneous energy deposition associated with the detonation of a high explosive produces an extremely rapid increase in temperature and pressure due to the sudden release of heat, light, and gases [1]. The gases produced by the explosion, initially at extremely high temperatures and pressures, expand abruptly against the surrounding atmosphere, vigorously pushing away any other object that may be found in their path. This gives rise to the two most notable effects of explosions: the aerial, or blast, wave [2], and the projection of shell fragments or other items (i.e., secondary fragments) located in the surroundings of the charge [3].

If the explosive device is located at a ground level, a fraction of its energy is effectively coupled to the ground, generating seismic waves and a well distinguished surface crater that results from the ejection of the shattered ground materials in direct contact with the charge [4]. Quantifying these phenomena and assessing their effect on the environment, including structural elements, vehicles, objects, or people located around the blast site, is a highly complex task that requires a thorough knowledge of the physical-chemistry of explosions [1,5–7] and their dynamic interactions with nearby structures [8] or the human body [9].

As a result of the growing terrorist threat experienced in the last few decades [10], estimating the effects of explosions has become a critical issue in the design, protection, and restoration of buildings and infrastructures, both civil and military [11]. However, this task is far from trivial, in that it involves transient compressible flows, nonlinear structural response, and highly dynamic fluid–structure interactions. These phenomena can be described with some accuracy using multiphysics computational tools, also known as hydrocodes [12], such as Ansys Autodyn, LS-Dyna, or Abaqus, based on the explicit finite element method [13]. In the simulations, all the critical components are modeled, including the detonation of the explosive charge, the resulting blast wave, the induced dynamic loads, and the nonlinear structural response. However, the enormous computational effort required to complete detailed computational analyses, which includes not only the calculation time itself, but also complex pre- and post-processing stages, remains a critical issue. For instance, simulating the effect of an explosive charge on a full-scale bridge may require more than 10 million finite elements [14]. For this reason, most engineering analyses still make use of simplified models for determining the explosive loads and estimating the resulting dynamic structural response in a timely manner. This enables the fast computation of damage diagrams in the Charge Weight–Standoff distance (CW–S) space, of utility to determine the level of protection provided by an input structural component loaded by blast from an input equivalent TNT charge weight and standoff [15].

In this regard, the American Unified Facilities Criteria UFC 3-340-02 [16], which supersedes the former ARMY TM 5-1300, establishes the requirements imposed by the US Department of Defense in the tasks of planning, design, construction, maintenance, restoration, and modernization of those facilities that must be protected against explosive threats. In the absence of similar regulations in other countries, UFC 3-340-02 [16] is widely used by engineers and contractors outside the US, as it provides a valuable guide for calculating the effects of blast-induced dynamic loads, including step-by-step procedures for the analysis and design of buildings to resist the effects of explosions.

To facilitate the application of the procedures set forth in the UFC 3-340-02 [16], as well as other analyses established in classic references of explosives engineering [3,5,6,9,17–19], fast evaluation software tools have been developed that incorporate the vast amount of data available as tables or graphs in the literature [7]. For instance, the United States Army Corps of Engineers (USACE) has developed and provides support for a series of software packages related to the design of explosion-resistant buildings [20]. Those tools were developed with public funding, and therefore there are regulations that restrict distributing those products outside of the United States. In addition, given the critical nature of this knowledge, access to these packages is severely limited to US government agencies and their contractors, with use only authorized to US citizens.

The inability to access these software packages motivated the authors to develop their own computational toolbox for the rapid evaluation of the effects of explosions. The result was the SimEx platform to be presented in this work. Conceived initially for educational purposes, the main goal was to develop a virtual software platform with an easy and intuitive Graphical User Interface (GUI) to be used in the computer lab sessions of the Explosion Dynamics course of the Degree in Security Engineering, taught at the University Center of the Civil Guard (CUGC) in Aranjuez, Spain. The Civil Guard is the oldest and biggest law enforcement agency in Spain. Of a military nature, its competencies include delinquency prevention, crime investigation, counter-terrorism operations, coastline and border security, dignitary and infrastructure protection, as well as traffic, environment or weapons and explosives control using the latest research techniques. The paradigm of the Civil Guard's capacity is its outstanding role in the defeat of the terrorist group ETA, the longest-running terrorist group in Europe and the best technically prepared. In this context, the main target of the Degree in Security Engineering is the training of Guardia Civil cadets (i.e., the Guardia Civil's future officer leadership) in the development, integration, and management of last generation civil security systems.

The purpose of SimEx was initially limited to the blast damage assessment on simple structural elements [21], such as beams, columns, pillars, or walls, following the Single-Degree-of-Freedom (SDOF) system analysis established by UFC 3-340-02 [16]. The tool has been successfully used since its initial inception in 2014 in both the computer lab sessions of the Explosion Dynamics course, and as a research tool for the development of a number of Bachelor and Master's theses on explosion dynamics and blast effects. This double use as end-users and software developers by the Civil Guard cadets and students from other UC3M degrees has enabled the development of the application well beyond the initially planned objectives [22]. As a result, the current version of SimEx incorporates advanced topics in blast wave propagation, such as the prediction of cleared blast pressure loads due to the generation of rarefaction waves, as well as confined blast loading in vented structures [23]. It also includes several other calculation assistants for the thermochemical analysis of explosive mixtures [5,7,24]; crater formation [4,6,25]; fragment mass distributions, ejection speeds and ballistic trajectories [3,26–28]; and the statistical evaluation of damage to people due to overpressure, body projection and fragment injuries [9,29,30].

2. SimEx Capabilities

This section presents the current capabilities of SimEx, starting with the main interface used for computing the dynamic response of SDOF systems subjected to blast loading, and following with the description of the remaining calculation assistants.

2.1. Single-Degree-of-Freedom System Analysis

In many situations of practical interest, the response of structural elements to blast loading can be reduced, in first approximation, to that of an equivalent spring-mass SDOF system. As sketched in Figure 1, this system is made up of a concentrated mass subject to external forcing and a nonlinear weightless spring representing the resistance of the structure against deformation [8]. The mass of the equivalent system is based on the component mass, the dynamic load is imposed by the blast wave, and the spring stiffness and yield strain on the component structural stiffness and load capacity. Generally, a small viscous damping is also included to account for all energy dissipated during the dynamic response that is not accounted by the spring-mass system, such as slip and friction at joints and supports, material cracking, or concrete reinforcement bond slip [31].

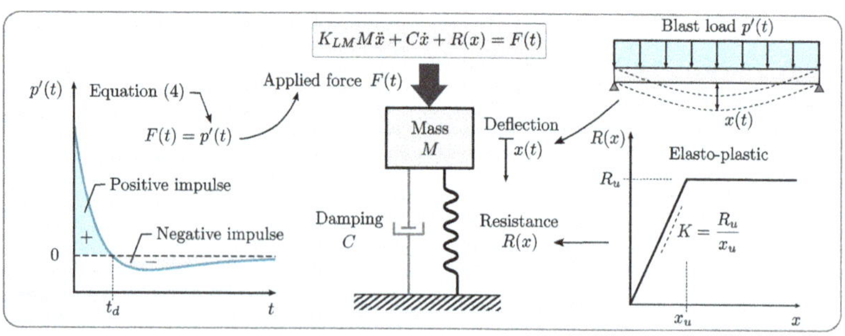

Figure 1. Sketch of the equivalent SDOF system showing the different terms involved in its mathematical description. **Left**: forcing term; **right**: resistance term; **center**: equivalent spring-mass SDOF system and its associated differential equation.

If the system properties are properly defined, the deflection of the spring–mass system, $x(t)$, will reproduce the deflection of a characteristic point on the actual system (e.g., the maximum deflection). The system properties required for the determination of the maximum deflection are the effective mass of the equivalent SDOF system, M_e, the effective viscous damping, C_e, the effective resistance function, $R_e(x)$, and the effective load history acting on the system, $F_e(t)$. To systematize the calculations, the effective properties are obtained using dimensionless transformation factors that multiply the actual properties of the blast-loaded component, respectively, M, C, $R(x)$, and $F(t)$ [32]. These factors are obtained from energy conservation arguments in order to guarantee that the equivalent SDOF system has the same work, kinetic, and strain energies as the real component for the same deflection when it responds in a given, assumed mode shape, typically the fundamental vibrational mode of the system [31].

In the analysis of blast-loaded SDOF systems, it is therefore of prime importance to identify the fundamental vibrational mode of the structural element. This procedure is not trivial, since obtaining the fundamental mode can entail certain difficulties, in which case its shape must be approximated in some way [32]. To determine the equivalent properties of the SDOF system, it is also necessary to determine the type of structure (beam, pillar, frame, etc.) and how the load is applied (typically, a uniform load is assumed). The elastic behavior of the material is often modeled as perfect elasto-plastic, probably the simplest of all nonlinear material models. This assumes that the initial response follows a linear elastic behavior described by an apparent elastic constant K, but once the yield strain is reached, $x \geq x_u$, the material behaves as plastic, flowing at a constant stress with an ultimate resistance $R_u = Kx_u$, i.e.,

$$R(x) = \begin{cases} Kx & \text{for} \quad |x| < x_u \\ R_u & \text{for} \quad |x| \geq x_u \end{cases} \tag{1}$$

Although more complex models could be used, they are not considered here due to the heavy simplifications introduced in the formulation of the problem.

The mass transformation factor, K_M, is defined as the ratio between the equivalent mass M_e and the real mass M of the blast-loaded component; the load transformation factor K_L is defined as the ratio between the equivalent load $F_e(t)$ and the actual load $F(t)$, and usually coincides with the resistance and damping transformation factors; and finally the load-mass factor K_{LM} is defined as the ratio between the mass factor and the load factor

$$K_M = \frac{M_e}{M}; \quad K_L = \frac{F_e(t)}{F(t)} = \frac{R_e(x)}{R(x)} = \frac{C_e}{C}; \quad K_{LM} = \frac{K_M}{K_L} = \frac{M_e}{M} \cdot \frac{F(t)}{F_e(t)} \tag{2}$$

Although all these factors are easy to obtain, even through analytical expressions in some cases, most of them can be found tabulated in the UFC-3-340-02 [16].

The linear momentum equation for the equivalent SDOF system then takes the form [32]

$$K_{LM} M\ddot{x} + C\dot{x} + R(x) = F(t) \tag{3}$$

where, as previously discussed, C represents the viscous damping constant of the blast-loaded component. This constant is often specified as a small percentage, z, of the critical viscous damping, $C = (z/100)C_{cr}$, with a damping coefficient $z = 2$ being a good value when not otherwise known (for further details see [31]). Note, however, that damping has very little effect on the maximum displacement, which typically occurs during the first cycle of oscillation, so the actual value of z is not of major relevance. The inhomogeneous term, $F(t)$, appearing on the right-hand side of Equation (3) represents the dynamic load associated with the blast wave, to be discussed in Section 2.1.1 below.

SimEx provides an easy and intuitive GUI environment for the study of the dynamic response to blast loadings of a variety of structural elements that can be modeled as SDOF systems. Figure 2 shows the main SimEx interface, divided into three calculation assistants for the three basic elements that make up the SDOF system: a module for calculating the properties of the blast wave (forcing term, $F(t)$), a module for calculating the equivalent mechanical properties (resistance term, $R(t)$), and a module for the numerical integration of the problem, which includes the post-processing of the results and their graphic representation in the form of displacements, forces, and deformation diagrams (see the bottom plots of Figure 2) and of CW–S damage charts, to be discussed in Section 3.3.

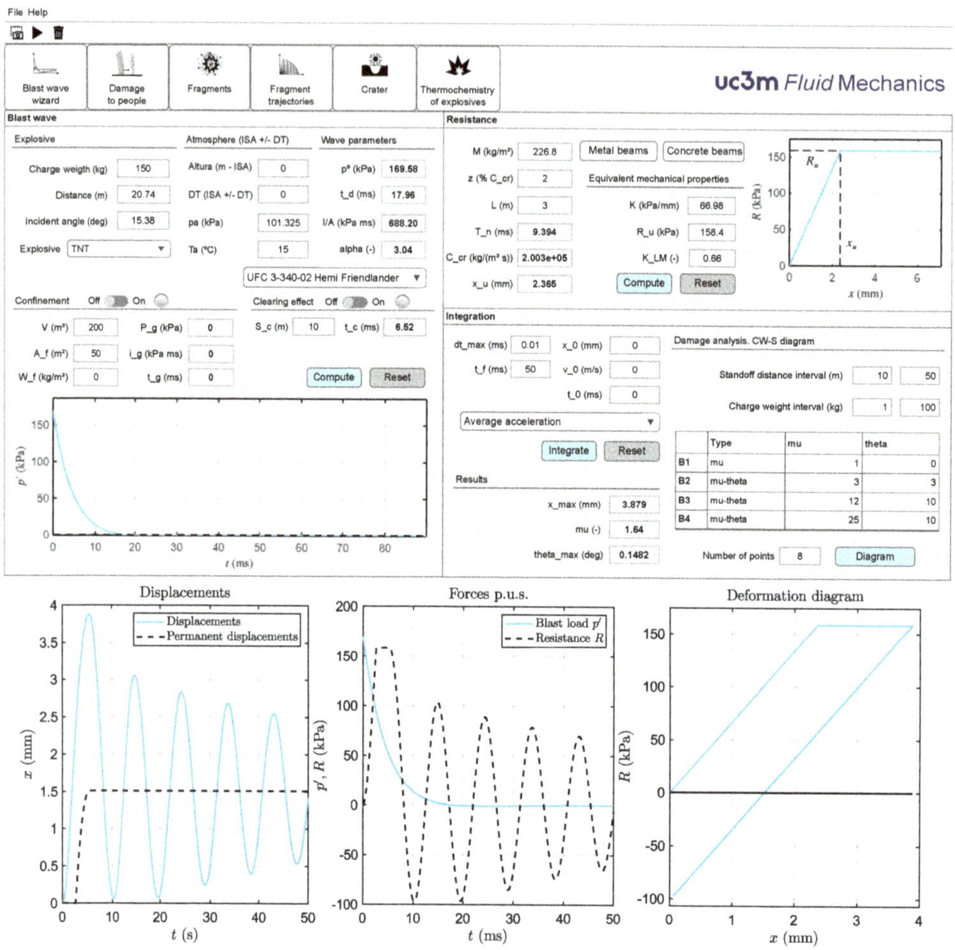

Figure 2. Main interface of SimEx showing the "Blast wave", "Resistance", and "Integration" assistants for the computation of the structural response of perfect elasto-plastic SDOF systems under blast loading. The access buttons to the other calculation assistants are seen under the top toolbar. The bottom plots show the post-processing pop-up window that displays the results of the numerical integration in terms of displacements, forces, and deformation diagrams (for a detailed discussion of these diagrams, see Section 2.1.4).

As a final remark, it is important to note that, following standard practice, the SDOF analysis carried out by SimEx uses the load defined in terms of pressure, $F(t) = p'(t)$ (Pa), so that both the mass M (kg/m^2), the damping coefficient C (kg/(m^2 s)) and the ultimate resistance R_u (Pa) must all be introduced as distributed values per unit surface (p.u.s.) in the different calculation assistants.

2.1.1. Forcing Term

As previously discussed, the blast wave overpressure defined in Equation (4) below can be used directly in Equation (3) as forcing term, $F(t) = p'(t)$, as long as the analysis is formulated per unit surface and uses distributed masses and forces. In order to determine the blast parameters (arrival time, peak overpressure, positive phase duration, impulse per unit area, waveform parameter, etc.), classical correlations [1,2,17–19,33,34] in terms of scaled distance are used together with the scaling laws for spherical or hemispherical blast waves [1,17,35,36], which allow their evaluation for arbitrary CW–S pairs. It is interesting to note that the standoff distance is defined as the minimum distance from the charge to the structural element under study (e.g., a wall). However, the actual distance to a given point of that element, e.g., the centroid (or geometric center), which may be considered the most representative point of the structure, may be slightly different due to the incidence angle being larger than 0 at that point.

The local atmospheric pressure, p_a, and temperature, T_a, are determined using the International Civil Aviation Organization (ICAO) Standard Atmosphere (ISA) [37] with a temperature offset (ISA $\pm \Delta T$). The user must specify the geopotential height, in meters, and the non-standard offset temperature $\pm\Delta T$, although arbitrary ambient temperature and pressure can also be introduced directly [38]. TNT is used as reference explosive, although the results can be extrapolated to other compositions using either the equivalence tables included in SimEx for selected explosives [39], or the thermochemical calculation assistant, to be presented in Section 2.2.1, for less conventional formulations or explosive mixtures.

To estimate the dynamic load exerted by the blast wave, the angle of incidence of the incoming shock wave must be considered, the worst-case conditions being usually those of normal incidence. UFC 3-340-02 [16] contains scaled magnitude data for both spherical and hemispherical blast waves. It also provides methods to calculate the properties of the blast wave with different incidence angles, including both ordinary and Mach reflections for oblique shocks. The time evolution of the blast wave overpressure $p'(t')$ at a fixed distance, d, sufficiently far from the charge (at least, larger than the fireball scaled distance) is approximated using the modified Friedlander's equation, which captures also the negative overpressure phase [1,17,40]

$$p'(t') = p(t') - p_1 = p^\circ \left(1 - \frac{t'}{t_d}\right) \exp\left(-\alpha \frac{t'}{t_d}\right) \tag{4}$$

where $p^\circ = p_2 - p_1$ represents the peak overpressure measured from the undisturbed atmospheric pressure $p_1 = p_a$, with p_2 denoting the peak post-shock pressure, $t' = t - t_a$ is time measured from the blast arrival time, t_d is the positive phase duration, and α is the waveform parameter, closely related to the impulse per unit area of the positive phase $I/A = \int_0^{t_d} p'(t')\,dt'$ (area under the positive phase of the overpressure-time curve) according to $I/A = p^\circ t_d \left[1/\alpha - (1 - e^{-\alpha})/\alpha^2\right]$. SimEx performs by default the complete integration of the Friedlander waveform, but the equivalent triangular pressure pulse can also be used without significant errors [32]. This simplified waveform has the same maximum peak overpressure, p°, but a fictitious positive duration computed in terms of the total positive impulse and the peak over pressure, $t_d = 2(I/A)/p^\circ$.

The "Blast wave" calculation assistant allows the activation of the effects of clearing and confined explosions, which increases the computational capabilities to more realistic situations. The clearing effect takes into account the time required for reflected pressures to clear a solid wall that has received the impact of a blast wave as a result of the propagation of rarefaction waves from the edges of the wall. In the case of confined explosions, SimEx implements the procedure outlined in UFC 3-340-02 [16] to estimate the gas phase peak overpressure and duration of the equivalent triangular pressure pulse in terms of the chamber's total vent area and free volume. These effects can be activated on the lower part of the "Blast wave" calculation assistant.

2.1.2. Resistance Term

The "Resistance" calculation assistant provides a means to define the equivalent mechanical properties (i.e., structural mass, damping coefficient, and structural strength) of the SDOF system under study modeled as a perfectly elasto-plastic system with elastic stiffness K until the yield strain, as given in Equation (1). The characteristic length, L, of the structural element must also be provided, as it is required to determine the maximum rotation angle at its boundaries, often referred to as support rotation, θ. For the equivalent SDOF system, the assistant computes the fundamental natural period, $T_n = 2\pi\sqrt{K_{LM}M/K}$, the critical damping, $C_{cr} = 2\sqrt{K_{LM}KM}$, and the deflection at which plastic deformation initiates in the system, x_u. Direct access to calculation assistants that compute the equivalent properties (M, K, K_{LM}, R_u) required for the calculations is also provided for various types of systems. Currently, standard European wide flange "metal beams" [41] and reinforced "concrete beams" are included (see Section 3.2), although it could be possible to incorporate additional assistants for other elements, such as metal panels/plates, open-web steel joists, reinforced concrete slabs, reinforced/unreinforced masonry, or wood panels/beams. The metal beams assistant also provides the possibility of studying custom (i.e., non-normalized) profiles and materials in order to widen the computation capabilities.

2.1.3. Numerical Integration

Once the characteristics of the equivalent SDOF system have been defined, the resulting ordinary differential equation that models the transient nonlinear response of the equivalent structural system (3) must be integrated numerically. The integration module implements the two numerical methods recommended by UFC-3-340-02 [16], namely the "Acceleration-Impulse-Extrapolation Method" and the "Average Acceleration Method" [16], which can be selected from a drop-down menu. Text boxes are also included to set the initial conditions (displacement and initial speed, which are zero by default) as well as the final integration time. Since both numerical methods use constant time steps, a sufficiently short time increment, typically of the order of a few percentage of the natural period or the positive phase duration (usually, fractions of a millisecond), should be used in order to ensure the numerical convergence of the integration.

2.1.4. Post-Processing

After integration, three plots appear in a pop-up window and a summary table is provided at the bottom left corner of the main window. The left plot shows the instantaneous displacement (solid line) and the permanent displacement, or deformation (dashed line). The central plot shows the temporal variation of the forcing term (i.e., the blast pressure wave, solid line) together with the resistance strength of the SDOF system (dashed line). The right plot shows the displacement–resistance graph, in which it is possible to determine more clearly whether permanent deformations occur or not. Finally, the table of results shows the maximum displacement obtained, x_{max}, along with two damage indicators: the ductility ratio, $\mu = x_{max}/x_u$, defined as the ratio of the peak deflection to the ultimate elastic deflection, and the maximum support rotation, θ, whose calculation depends on the type of structure under study.

By integrating different combinations of charge weights and standoff distances for the same structural element, damage level diagrams can be rapidly obtained in the CW–S distance space. SimEx has a function for it located in the central part of the integrator module. One can select the range of charge weights and standoff distances, the number of intermediate values and the type of damage in terms of the quantitative indicators μ and θ [15]. From the two quantitative indicators, the structural damage level can be classified qualitatively into: superficial, moderate, heavy, hazardous failure, and blowout, with response limit boundaries between these levels denoted respectively by B1 (superficial to moderate), B2 (moderate to heavy), B3 (heavy to hazardous failure), and B4 (hazardous failure to blowout). Convenient limits for the boundaries of component damage levels for common structural components in terms of μ and θ are provided in [15]. An example of a damage level diagram for the façade of a conventional building subject to blast loading computed with SimEx will be presented in Section 3.3 .

2.2. Other Calculation Assistants

The main SimEx interface gives access to several other calculation assistants. These include: a module for the calculation of the theoretical (i.e., thermochemical) properties of explosives and explosive mixtures; a module for estimating the initial velocity, mass distribution and ballistic trajectories of primary fragments; a crater formation calculator; and a module for estimating damage to people, including both primary and tertiary injuries. The fragment assistant also provides estimations of the secondary injuries due to the impact of primary fragments on people. In this section, we shall briefly present and discuss the above-mentioned assistants.

2.2.1. Assistant for the Calculation of the Thermodynamic Properties of Explosives

For the calculation of the theoretical thermodynamic properties of explosives and explosive mixtures, SimEx includes an extensive database of pure CHNO propellants and explosives extracted from Kinney and Graham [1], updated with data from Meyer [7] and Akhavan [5] for more recent explosives. From the properties of pure explosives, the thermochemical assistant estimates the properties of explosive mixtures formed by two or more components by specifying the mass fractions and the density of the mixture.

First, it computes the apparent chemical formula of the explosive mixture along with its molecular weight and maximum density. For the calculation of the decomposition reaction in nominal products, which provides the heat of explosion and the volume of gases generated, one can choose different calculation hypotheses: Kamlet–Jacobs (KJ), Kistiakowsky–Wilson (KW), Modified Kistiakowsky–Wilson (modified KW), Springall–Roberts (SR), or chemical equilibrium [5]. In the latter case, SimEx determines the composition of the product mixture following the chemical equilibrium approach considering a constant–volume explosion transformation that uses the ideal gas Equation of State (EoS) for the products according to the norm UNE 31-002-94 [42], as illustrated in Figure 3.

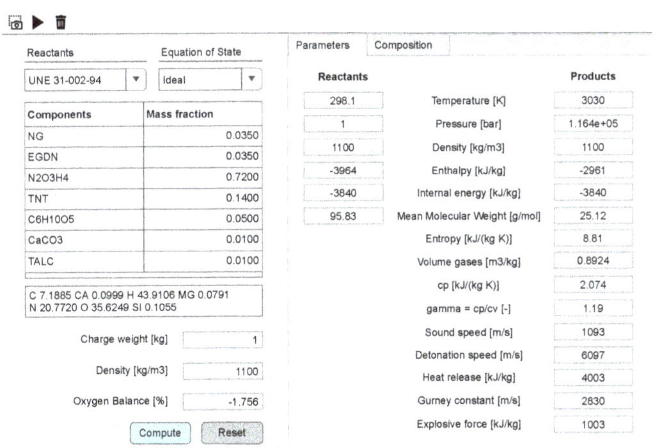

Figure 3. Interface of the assistant for the calculation of the theoretical thermodynamic properties of explosives and explosive mixtures.

More complex computations based on the European Standard EN 13631-15 [43], which use the semi-empirical Becker–Kistiakowsky–Wilson (BKW) EoS [44–46] or the Heuzé (H9) EoS [47] for the products, are also supported in the last version of SimEx. As sample results of these computations, Table 1 shows the detonation properties obtained by SimEx for different explosive mixtures (see Table 2 for its composition) compared with the results reported in the European Standard EN 13631-15 [43], and obtained with the W-DETCOM code [48,49], which computes directly the Chapman–Jouguet state.

Table 1. Comparison of the calculated temperature at constant volume, T, detonation pressure, p_{CJ}, detonation velocity, v_{CJ}, heat release at constant volume, Q_v, and explosive force, F_e, with the results provided by the European Standard EN 13631-15 [43] and by the thermochemical code W-DETCOM [49] for different explosive mixtures using the BKW–S EoS.

Explosive	Source	T [K]	p_{CJ} [GPa]	v_{CJ} [m/s]	Q_v [kJ/kg]	F_e [kJ/kg]
ANFO	CT	2592	7.14	5353	3845	943
	EN 13631-15	2586	-	-	3820	945
	W-DETCOM [1]	2919	6.62	5326	3849	-
ANFO-Al	CT	3026	7.38	5442	4666	1009
	EN 13631-15	3060	-	-	4642	1020
	W-DETCOM [1]	3370	6.55	5215	4655	-
Emulsion	CT	2112	15.3	6549	3263	766
	EN 13631-15	2099	-	-	3236	771
	W-DETCOM [1]	2438	13.9	6758	3214	-
Dinamite I	CT	4173	25.03	7960	6452	1147
	EN 13631-15	4130	-	-	6338	1138
Dinamite II	CT	3165	23.58	7729	5049	987
	EN 13631-15	3151	-	-	4989	984

[1] Calculation performed assuming Chapman-Jouguet detonation.

Table 2. Composition [mass %], density, and oxygen balance of different explosive mixtures tested.

Component	ANFO	ANFO-Al	Emulsion	Dinamite I	Dinamite II
Aluminium	-	5	-	-	-
Ammonium nitrate	94	91	80	-	49
Cellulose	-	-	-	-	3
2,4-Dinitrotoluene	-	-	-	-	4
Nitrocellulose 12%	-	-	10	-	4
Nitroglycerin	-	-	-	45	20
Nitroglycol	-	-	-	45	20
Fuel oil	6	4	7	-	-
Sodium nitrate	-	-	5	-	-
Water	-	-	8	-	-
Density [kg/m^3]	850	850	1300	1500	1500
Oxygen balance [%]	−1.7	0.08	−5.57	−2.26	0.84

The equilibrium calculations are carried out using Combustion Toolbox (CT), an in-house thermochemical equilibrium package developed at UC3M [24,50]. CT determines the equilibrium composition of the product mixture through the Gibbs free energy minimization method by using Lagrange multipliers combined with a multidimensional Newton–Raphson method. The thermodynamic properties (specific heat, enthalpy, and entropy) are computed as a function of temperature derived from NASA's 9-coefficient polynomial fits for combustion of ideal and non-ideal gases and condensed phases.

From the resulting composition of the product mixture at equilibrium, the assistant computes the volume of gases generated, the heat of explosion, the Gurney constant, the detonation pressure, the detonation velocity, and the explosive force (or power index). To estimate the detonation pressure and velocity, the approximate expressions of Kamlet & Jacobs [51,52] are used, whereas the explosive force is estimated using the well-known Berthelot approximation [1]. These data are subsequently used to calculate the TNT equivalent of the explosive composition under study.

2.2.2. Crater

SimEx also has an assistant for the direct and inverse calculation of craters based on the classical correlations for craters reviewed by Cooper [6] (see also Refs. [4,25]), whose interface is shown in Figure 4. With this assistant, one can calculate the radius of the crater generated by the detonation of a certain amount of a given explosive at a certain height above the ground, considering different types of soil. It is also possible to calculate the explosive charge required to produce a crater of a certain size, which may be useful for the forensic analysis of explosions [53]. Buried craters are not yet included in the assistant, but could be incorporated in future versions following the work of Westine [54], as reviewed by Baker et al. [55].

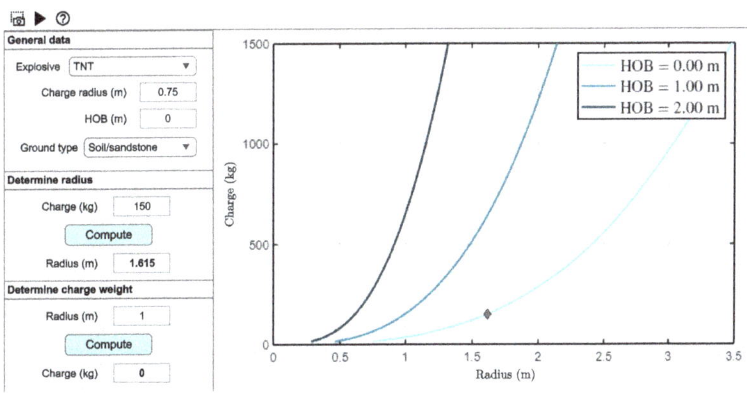

Figure 4. Interface of the assistant for the calculation of craters. HOB denotes the height of burst.

2.2.3. Primary Fragments

SimEx incorporates assistants for calculating the mass distribution, ejection velocity, and ballistic trajectory of primary fragments. The corresponding interfaces are shown in Figures 5–7. The fragment size distribution is estimated using Mott's statistical theory for fragmentation of steel cylindrical shells [3,26–28], as suggested by UFC-3-340-03 [16]. As shown in Figure 5, this model determines the average number of fragments and their average weight. It also provides the size of the largest fragment corresponding to a given Confidence Level (CL). SimEx also includes a ballistic trajectory assistant for primary fragments that, in addition to the flight path, provides the flight time, velocity, and maximum distance, as illustrated in Figure 6. The initial velocity of primary fragments is computed using Gurney's analysis [56] for cylindrical, spherical, and symmetrical/asymmetrical sandwich charges. Although this analysis assumes that all fragments have the same the initial velocity, given the different fragment sizes, both their initial kinetic energy and their subsequent aerodynamic deceleration are different. The assistant thus includes an initial aerodynamic deceleration chart, shown in Figure 7, that provides the fraction of the initial velocity achieved at a certain distance, given the fragment mass and material, and the local air density, specified through the ISA $\pm$ ΔT model. The aerodynamic assistants assume spherical fragments with a variable drag coefficient for all Mach numbers [57], although the model could be extended to account for more realistic (i.e., irregular) fragment shapes in future versions [58]. The results of these models are also used to estimate the lethality risk by the impact of primary fragment in the event of a strike on a person, which is found to depend on the speed and the mass of the fragment, as illustrated by Figure 5.

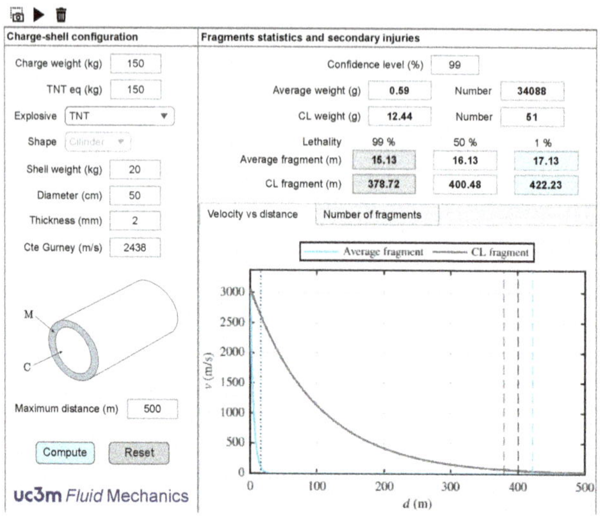

Figure 5. Interface of the primary fragment mass distribution and lethality assistant.

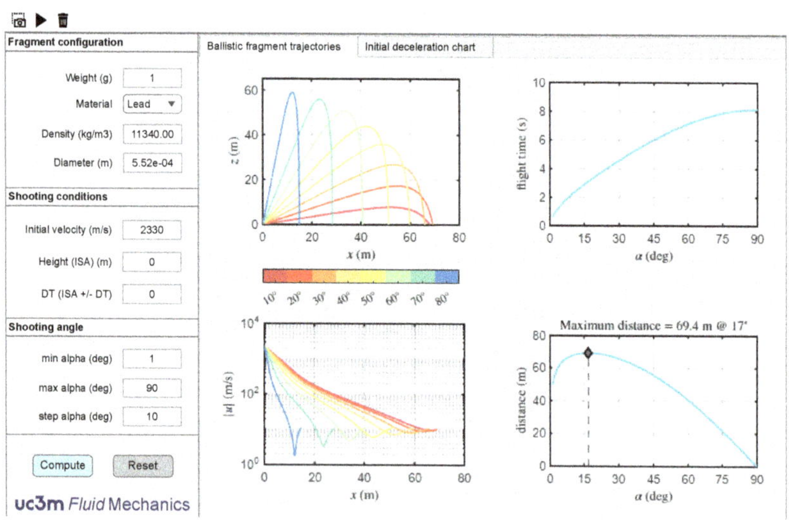

Figure 6. Interface of the primary fragment calculation assistant showing the ballistic fragment trajectory, flight time, velocity, and maximum distance charts. Fragments are assumed spherical.

Non-tabulated explosives or explosive mixtures can also be considered, with the Gurney constant being computed by the thermochemical assistant presented in Section 2.2.1. In this case, the user must select a "custom" explosive, and the thermochemical assistant will open to specify the desired explosive composition. Once the wizard is closed, the Gurney constant is automatically exported to the fragment wizard.

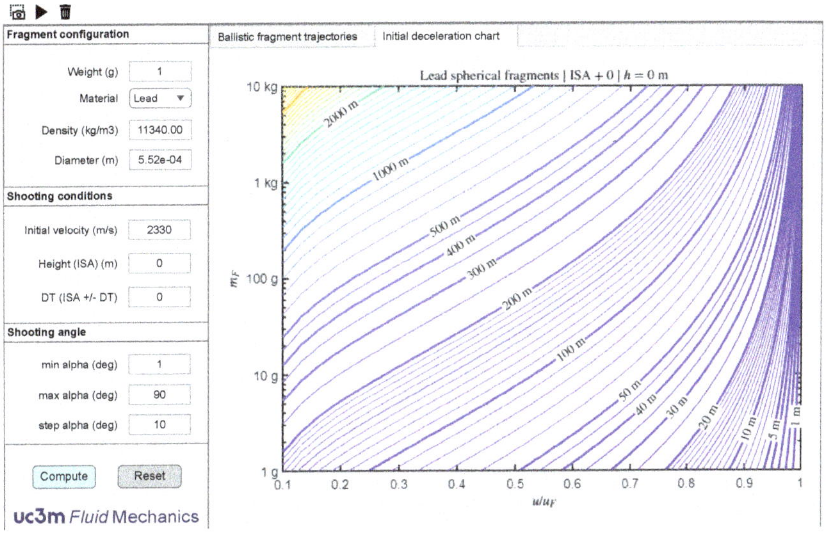

Figure 7. Interface of the primary fragment calculation assistant showing the initial deceleration chart, which provides the fraction of the initial velocity, u/u_f, achieved at a certain distance (contour lines), given the fragment mass, m_f, and material (e.g., lead), and the atmospheric conditions (e.g., ISA mean sea level). Fragments are assumed spherical.

2.2.4. Damage to People

SimEx includes an assistant for estimating damage to people using the widely accepted probit (*probability unit*) functions [58,59] provided by the TNO's Green Book [9] and summarized in Table 3. For each type of injury or cause of death (eardrum rupture, lung injury, etc.), a probit function is defined that depends on the blast parameters: side-on, dynamic or reflected peak overpressure (depending on the body position), impulse per unit area, etc. For primary injuries, lethality due to lung damage is evaluated together with the probability of eardrum rupture. For tertiary injuries, lethality is evaluated for shock-induced body displacement and subsequent direct impact, either with the head or the whole body [29].

Table 3. *Probit* functions used to estimate the probability of different types of primary and tertiary injuries. Pr is the *probit* value, p° [Pa] the peak overpressure, p°_{ef} [Pa] the maximum effective overpressure, depending on the relative orientation of the person with respect to the shock wave, p_1 [Pa] the atmospheric pressure, I/A [Pa · s] the impulse per unit area and m [kg] the weight of the person [9].

Effect	***Probit* Function**
Primary injuries	
Eardrum rupture	$\text{Pr} = -12.6 + 1.52 \ln p^\circ$
Death due to lung damage	$\text{Pr} = 5 - 5.74 \ln\left(\dfrac{4.2}{p^\circ_{\text{ef}}/p_1} + \dfrac{1.3}{i/(p_1^{1/2} m^{1/3})}\right)$
Tertiary injuries	
Death due to displacement and whole-body impact	$\text{Pr} = 5 - 2.44 \ln\left(\dfrac{7380}{p^\circ} + \dfrac{1.3 \times 10^9}{p^\circ i}\right)$
Death due to displacement and skull impact	$\text{Pr} = 5 - 8.49 \ln\left(\dfrac{2430}{p^\circ} + \dfrac{4 \times 10^8}{p^\circ i}\right)$

The appearance of the interface is shown in Figure 8. All necessary parameters can be selected on the left: size, type, and geometry of the explosive charge, as well as the body position relative to the incoming pressure wave, which determines whether side-on, dynamic, or reflected pressure is used to compute the peak overpressure and impulse. The rest of the window presents the results both numerically and graphically, using overpressure–impulse diagrams on the left and CW–S diagrams on the right, with primary injuries shown above and tertiary injuries below. Overpressure–impulse diagrams display the characteristic overpressure–impulse–distance curve for the selected charge weight to facilitate the interpretation of results [60], while CW–S diagrams include a diagonal dashed line indicating the approximated position of the fireball radius, corresponding roughly to an scaled distance $Z = d/W^{1/3} = 1$ m. Above this line, the Freidlander waveform is not valid, and the blast wave parameters are increasingly imprecise [1].

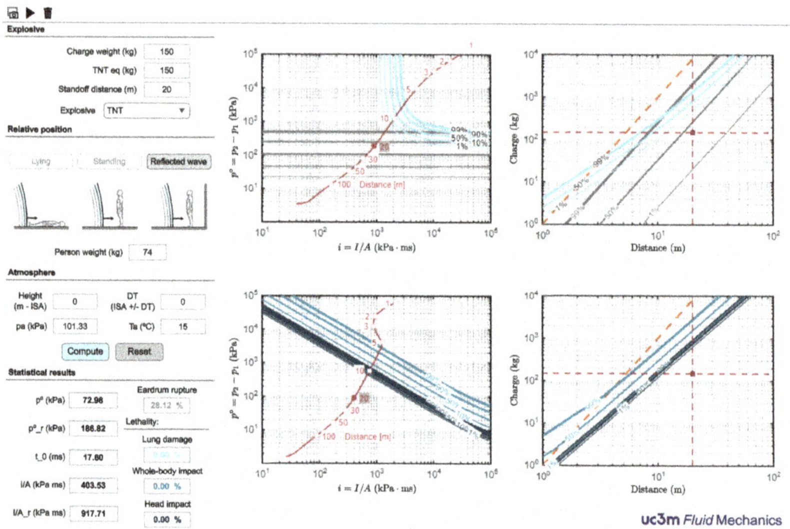

Figure 8. Interface of the assistant for estimating blast-induced damage to people. The CW–S and atmospheric data, along withe the body position relative to the incoming pressure wave, are introduced in the top-left corner, the blast wave parameters and the statistical damage indicators for the chosen CW–S combination appear in the bottom left corner. The right plots represent graphically the statistical damage indicators in the form of overpressure-impulse and CW–S diagrams. Both show the conditions corresponding to the specified CW–S combination with a solid red dot, while the CW–S diagrams include also a diagonal dashed line indicating the approximated position of the fireball radius. Above this line, the Freidlander waveform is not valid, and the blast wave parameters are increasingly imprecise [1].

3. Example of Application: Façade of a Building under Blast Loading

To illustrate the capabilities of SimEx, this section presents a preliminary study to asses the ability of a conventional three-story steel frame building, such as the one shown in Figure 9, to resist three different combinations of charge weight, W, and standoff distance, d, preserving a similar scaled distance, $Z = d/W^{1/3}$. The three CW–S combinations are summarized in Table 4. For simplicity, we assume mean sea level ISA conditions for all the calculations. For illustrative purposes, the figures quoted below show results corresponding to the first floor of the building (hereafter referred to as Level 1) and Case 2 conditions. That is, we shall consider as reference conditions a ground explosion of 150 kg of TNT at a 20 m standoff distance from the front façade of the building, as depicted in Figure 9a.

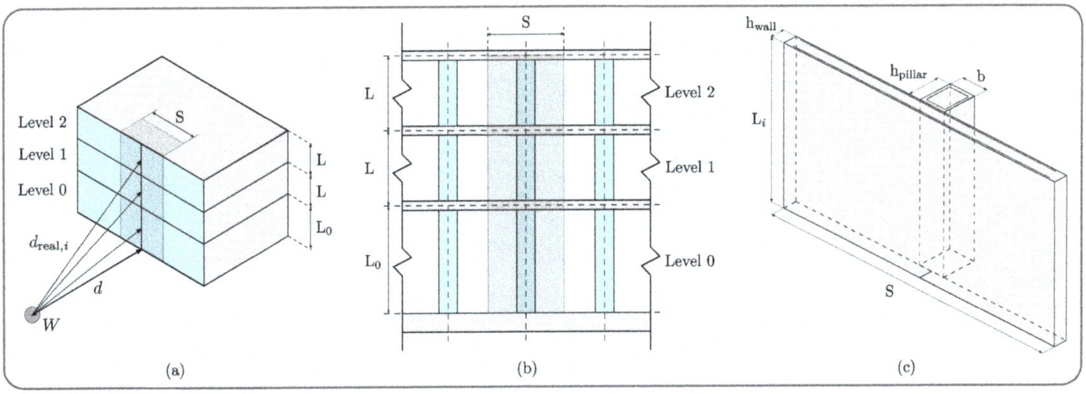

Figure 9. Schematic diagram of the three-story building under study, composed of equally spaced pillars and an outer enclosure wall, including: (**a**) the distances and angles used for the different floor levels ($i = 0$, 1, and 2), including the standoff distance, d, the real distance to the midpoint of the different levels, $d_{\text{real},i}$, and the corresponding angles of incidence, $\delta_{\text{real},i}$; (**b**) schematic of the façade constructive details and dimensions; and (**c**) diagram of the equivalent façade element used in the SDOF analysis. L_i denotes the height of Level i, representing the length of the pillars, and S is the spacing between pillars, representing the tributary loaded width.

Table 4. Standoff distance, d, explosive charge, W, and scaled distance, Z, of the different case studies. The reference case is shown in blue.

Variables	Case 1	Case 2	Case 3
d (m)	12	20	25
W (kg)	30	150	300
Z (m/kg$^{1/3}$)	3.86	3.76	3.73

3.1. Incident Load

As previously discussed, SimEx allows the user to enter directly the desired CW–S combination to define the incident blast load. Figure 2 shows the results corresponding to the reference conditions (Level 1, Case 2). For a more detailed analysis of the load induced by the blast wave, the "Blast Wave" calculation assistant shown in Figure 10 allows a fast evaluation of all blast parameters as a function of the standoff distance. To this end, the user must provide the following input data: the ground distance from the explosion to the point of calculation, d, the elevation of the explosive charge, h_c, and the elevation of the calculation point, h_0, both measured from the ground.

For $h_c = 0$, a hemispherical surface burst computed from Kingery and Bulmash parameters for TNT [18] is considered, although other correlations for hemispherical explosions [1] could also be selected. For $h_c > 0$, hemispherical or spherical blasts are both available, letting the user decide what is the best option based on the height of burst. The code does not include correlations for more complex configurations, such as air bursts producing regions of regular and Mach reflections that eventually modify the incident shock wave. The user must also introduce the angle formed by the normal to the structural element at the point of calculation with the horizontal projection of the line joining the center of the explosion with that point, δ, which is identically zero in our case studies if we assume a symmetric configuration with a pillar in the center of the front façade. These distances and angles are employed for simplicity in obtaining in-field measurements.

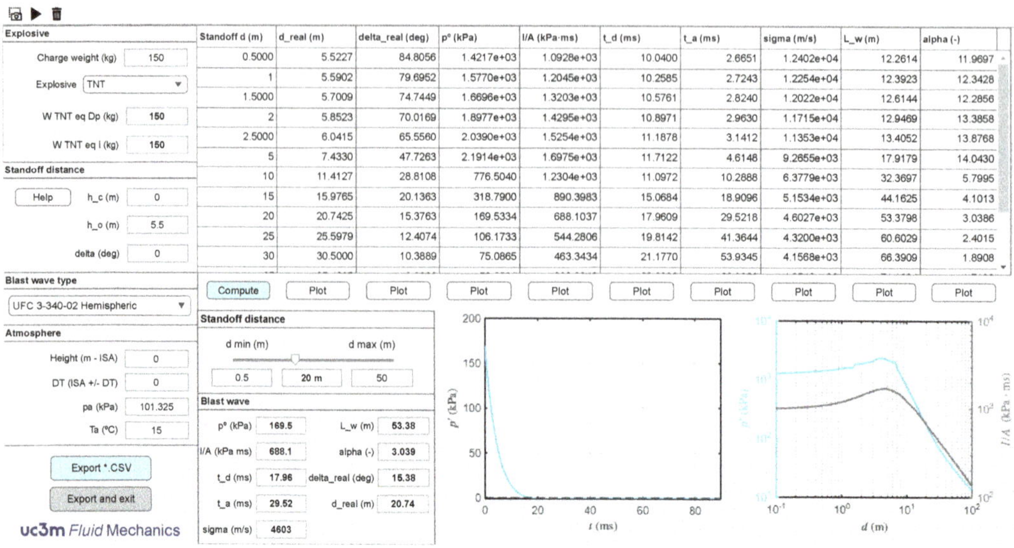

Figure 10. Interface of the Blast Wave calculation assistant for a charge weight of 150 kg of TNT at the ISA mean sea level, showing the variation of the blast parameters with the standoff distance from the front façade (top table). The lower part of the assistant shows the blast parameters calculated at a point located at $d = 20$ m standoff distance and $h_0 = 5.5$ m above the charge.

With these data, the wizard is able to compute the real distance and incidence angle, thereby providing the peak overpressure, p°, the impulse per unit area, I/A, the duration of the positive phase, t_d, the blast arrival time, t_a, the average speed of the pressure front, $\sigma = d_{\mathrm{real}}/t_a$, the positive phase length, L_w, and the waveform parameter, α. The results are presented in a table for several standoff distances, d, which also gives the real distances, d_{real}, and angles of incidence, δ_{real}. The maximum and minimum distances that appear in the table can be easily modified by the user, who can select any intermediate value using a slider bar to compute the blast parameters at a fixed specified distance. A button has also been included to graphically represent the variation of any of the blast parameters as a function of the distance to the center of the explosion. The results are also exportable as a "comma-separated-value" format for further postprocessing.

For more qualitative information, two exportable graphs are presented in the lower part. The graph on the left displays the time evolution of the overpressure at a fixed horizontal distance. The user can change this distance easily with the slider bar. All the characteristics of the blast wave are shown for the particular distance chosen by the user. The graph on the right represents the maximum overpressure and the impulse per unit area as a function of the horizontal distance. As previously indicated, the range of distances is also adjustable by the user. Using the "Export and exit" button, the module is closed and the weight and type of explosive, the distance to the charge, and the real angle of incidence to be used in the integration of the SDOF system are exported to the main SimEx module. Figure 10 shows the calculation of the blast parameters for an explosive charge of 150 kg of TNT on a point at a height of 5.5 m above the horizontal, i.e., the geometric center of the façade of the first floor, corresponding to the reference case (Level 1, Case 2). Other distances are also included in the top table, showing how the angle of incidence tends to become normal as the charge moves away from the target.

3.2. Estimation of the Equivalent SDOF System Response

To study the structural response to an explosive charge, it is necessary to know in detail the type of construction. However, when using a simplified SDOF model, the study can be simplified and generalized for many different cases. In the present example, we will analyze a façade structure like the one in Figure 9b, composed of equally spaced pillars and an outer enclosure wall.

The first element that receives the blast wave is the enclosure of the façade. This, in turn, transmits the load to the rest of the structure. In most constructions, the façade is only an enclosure without structural function (glass façades, brick, etc.). In first approximation, it can be considered that the exterior enclosure transmits the full load received directly to the pillars. The pillars are structural elements whose integrity is considered critical. It will therefore be the first element to be studied since the protection of the supporting structure is pivotal to avoid the potential collapse of the building. The enclosure can be considered as a secondary element in most constructions and therefore a significantly higher level of damage than in primary elements can be allowed.

Figure 9c shows the simplest element in which the façade is to be divided. Each pillar receives loads from a part of the façade corresponding to the distance between pillars and the height between floors. The load generated by the explosion is applied to the pillars crosswise, so they behave in first approximation as bending elements. For the calculation of the equivalent properties, the beam assistants available in SimEx are employed. Either for metal or concrete beams, the length corresponds to the height between floors, while the span is the spacing between pillars. In the case of pillars, the boundary condition between floors is that of embedment on both sides, whereas a free condition is preferred at the roof. As a result, we use fixed-fixed conditions for Levels 0 and 1 and cantilever (or fixed-free) for Level 2. The presence of a roof diaphragm element may require additional considerations regarding the boundary condition at the roof top, but we prefer to use a fixed-free boundary condition for the second floor both for simplicity and for illustrating the effect of considering different boundary conditions on different floors.

In the case of metal beams, it is only necessary to indicate the standard shape of the profile and the size. SimEx uses European cross-section profiles HEB, IPE, and IPN in accordance with Euronorm 53–62 (DIN 1025) [41]. Figure 11 shows the result for a HEB 340 profile with a length of 3 m and a separation between pillars of 5 m. The assistant uses standardized profiles, so if a non-existent measure is introduced, it corrects down to the nearest lower normalized profile. However, it is also possible to select custom profiles and materials. In this case, the area, first moment of area about the bending axis, moment of inertia about the bending axis, density, Young's modulus, and resistance must be provided by the user. Once the structural properties have been introduced, closing the assistant incorporates the computed data into the main SimEx interface. Figure 2 shows the result for the case under study. It should be noted that the additional enclosure mass supported by the pillar when flexed must also be included in the mass of the equivalent SDOF system in the main interface.

If a rectangular reinforced concrete pillar is considered, SimEx requires that the external measurements b and h (perpendicular and parallel to the direction of application of the load, respectively) be introduced. In addition, the properties of the reinforcement should be indicated in a simplified manner, that is, interior spacing, d_c, and reinforcement area, $A_s = n\pi d_{\text{bar}}^2/4$, where n represents the number of steel reinforced bars per side. Figure 12 shows results for a pillar of 45×45 cm^2 with 5 A36 steel reinforcement bars of # 7 (approximately 22.5 mm in diameter) per side, for a length of 3 m and a spacing between pillars of 5 m. The distance d_c must be estimated according to the constructive detail. In this particular case, it is assumed that the reinforcement centers are located at 4 cm from the edge, resulting in an interior reinforcement spacing of $d_c = 37$ cm.

It is worth noting that neglecting axial load can be considered a conservative approach, particularly in the case of columns or pillars. These elements are initially subjected to a significant compression load due to the weight of the supported structure, which reduces the tensile stresses caused by bending. This simplification constitutes a first approximation in the study of the structural response. For a more detailed analysis, the wall should be the next element to be analyzed in order to assure that it is able to fully transmit the blast load to the load-bearing element. If the wall was made of concrete, this could be done using the concrete beam assistant with $b = S$. In this case, the mass of the element under study would be the total mass of the equivalent SDOF system. However, in the case considered here of load-bearing elements (beams or columns/pillars), the total mass can be significantly larger than the mass of the element.

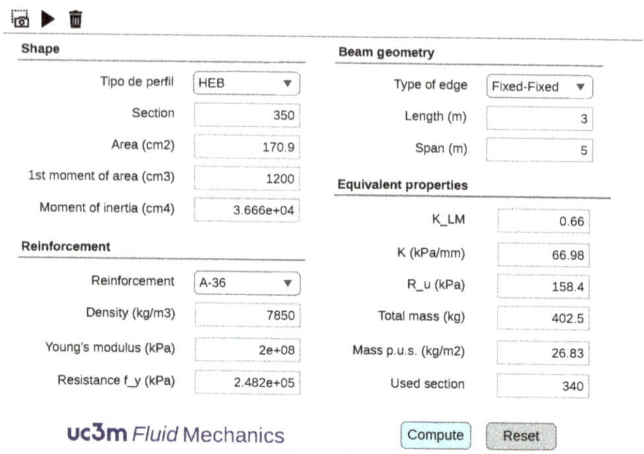

Figure 11. Metal beam calculation assistant showing results for a HEB 340 pillar with a length of 3 m and a spacing between pillars of 5 m. Note that, even though a HEB 350 is requested, which is not included in the norm, the assistant corrects down to the nearest normalized value, HEB 340.

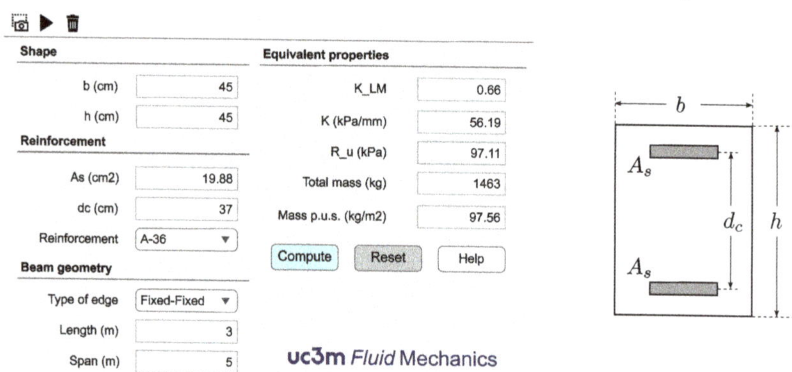

Figure 12. Reinforced concrete beam calculation assistant showing results for a pillar of 45 × 45 cm^2 with a length of 3 m and a spacing between pillars of 5 m. The pillar is reinforced using 5 A36 steel reinforcement bars of 22.5 mm of diameter per side spaced apart 37 cm.

3.3. SDOF System Integration and CW–S Damage Diagrams

Once the user sets the explosive charge and the properties of the equivalent SDOF system, SimEx is ready to integrate the resulting mathematical problem. Figure 2 shows the results for the case of a HEB 340 profile with a 5 m span between pillars. The main results are the maximum deflection, x_{max}, the ductility ratio, μ, and the maximum rotation angle, θ. The two latter parameters are used as indicators to quantify the component damage levels [15]. Assuming that the Level of Protection (LOP) required is very low, in case of a hot rolled compact steel shape for the columns, according to [15], the allowable component damage is heavy (response between B2–B3).

For fixed values of the structural parameters, a parametric sweep can be carried out in CW–S space to obtain damage diagrams such as the ones shown in Figure 13. To this end, it is enough to indicate in the assistant the charge weight and standoff distance ranges to be analyzed and the number of intervals to be used for each parameter. In addition, the desired damage level criteria must be indicated to separate the zones. Figure 2 shows characteristic values of μ and θ for metallic elements, although other values could be selected from [15] for other structural elements and materials. Note that CW–S damage diagrams are presented both in linear and log-log scales.

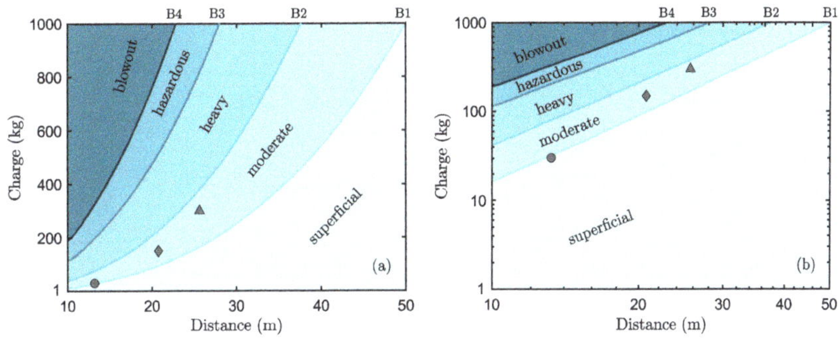

Figure 13. CW–S linear (**a**) and log-log (**b**) damage diagrams for reflected blast load on the façade of the first floor (Level 1): Case 1 (◯), Case 2 (◇), Case 3 (△).

As can be seen, the CW–S damage diagrams shown in Figure 13 include three points corresponding to the three cases considered in Table 4. As the three scaled distances are almost equal, then the damage levels are also very similar, although differences in real distances and incidence angles make them grow from superficial-moderate (B1) to (almost) moderate-heavy (B2) for increasing charge weights and standoff distances. According to the PDC-TR 06-08 [15], a superficial damage level implies "no visible permanent damage", whereas a moderate damage level implies "some permanent deflection" that generally can be repaired. By way of contrast, a heavy damage is associated with "significant permanent deflections" that cause the component to be unrepairable.

To summarize the results obtained in the different case studies, Table 5 reports the incident blast load parameters and the corresponding component damage indicators per floor for Cases 1, 2, and 3. The reference case (Level 1, Case 2) and the worst-case scenario (Level 2, Case 3) are both highlighted for clarity. As can be seen, damage levels are significantly higher in the upper floor (Level 2) as a result of the lowest rigidity imposed by the cantilever boundary condition at the roof top, resulting in heavy damage levels for cases 2 and 3.

3.4. Crater, Fragments, and Damage to People

Figure 4 presents an estimation of the crater generated in the reference case on a sandstone soil, with an approximated radius of 1.6 m. For surface bursts, HOB $= 0$ m, as the one considered here, the equivalent charge radius is irrelevant, as it is only used to determine the dimensionless height of burst, which is identically zero in our example. The figure also shows that, for above-surface bursts, HOB > 0 m, the crater radius is significantly smaller for the same amount of explosive due to the air cushion that exists between the load and the ground, which reduces to a great extent the pressure that reaches the ground surface [6].

Figure 5 shows the interface of the fragment assistant using the input data of the reference case. For the application of Mott's statistical theory for fragmentation of steel cylindrical shells [3,26–28], the explosive charge is approximated to a cylinder of approximately 50 cm diameter surrounded by a steel fragmentation shell with a mass of the order of about 13% of the charge and a thickness of 2 mm.

Table 5. Incident load parameters and component damage indicators per floor. According to the PDC-TR-06-08 [15], the response limits for hot rolled structural steel can be defined in terms of the ductility ratio, μ, and support rotation angle, θ, as follows: B1—superficial $\{\mu,\theta\} = \{1,-\}$; B2—moderate $\{\mu,\theta\} = \{3,3^\circ\}$; B3—heavy $\{\mu,\theta\} = \{12,10^\circ\}$; B4—hazardous $\{\mu,\theta\} = \{25,20^\circ\}$. The reference case and worst-case scenario are indicated in blue and gray, respectively.

Level	Type	Variables	Case 1	Case 2	Case 3
0	Incident load parameters	Δp (kPa)	168.30	182.50	186.80
		I/A (kPa · ms)	406.70	724.40	922.20
		d_{real} (m)	12.17	20.10	25.08
		δ_{real} (deg)	9.46	5.71	4.57
	Damage level indicators	μ (-)	1.60	3.26	4.40
		θ (deg)	0.19	0.39	0.53
1	Incident load parameters	Δp (kPa)	139.40	169.50	178.00
		I/A (kPa · ms)	349.90	688.10	893.00
		d_{real} (m)	13.20	20.74	25.60
		δ_{real} (deg)	24.62	15.38	12.41
	Damage level indicators	μ (-)	0.90	1.64	2.10
		θ (deg)	0.08	0.15	0.19
2	Incident load parameters	Δp (kPa)	110.20	152.00	165.40
		I/A (kPa · ms)	293.00	630.50	845.50
		d_{real} (m)	14.71	21.73	26.41
		δ_{real} (deg)	35.31	23.03	18.78
	Damage level indicators	μ (-)	1.67	5.85	9.26
		θ (deg)	0.87	3.05	4.83

Finally, Figure 8 shows the calculating assistant for estimating damage to people in the reference case. As an illustrative example, the figure presents the results of lethality due to different types of injuries at a distance of 20 m from the origin of the explosion, assuming the worst-case scenario of an average person located close to the façade of the building being attacked. In the pressure-impulse graphs, representative distances are indicated using red dots plotted along the characteristic overpressure–impulse–distance curve [60]. As can be seen, at 20 m standoff distance, lethality due to lung damage or whole-body projection is negligible, but large primary fragments (e.g., CL 99%) may still produce secondary injuries with fatal results, as indicated by Figure 5.

4. Conclusions

SimEx is a computational tool that allows a rapid and easy estimation of the effects of explosions on structural elements and their damage to people. It has been developed in accordance with the specifications of American standard UFC-3-340-02 and other widely accepted directives published in the open literature. It provides assistants for the calculation of the blast-wave load; SDOF dynamic response, including the calculation of the equivalent structural properties of standardized metal and reinforced concrete beams; thermodynamic properties of explosive mixtures; crater formation; projection of primary fragments; and damage to people.

After presenting the main calculating assistants, a preliminary study has been presented to illustrate the full capabilities of SimEx in the assessment of the ability of a building to resist a given explosive charge. The analysis enables the determination of component damage levels for the main structural components, and a further study of the reference case has led to the computation of CW–S damage diagrams for a pillar of the first floor. These diagrams are very useful to provide design guidelines for those facilities that must be protected against explosive threats.

Although still under development, SimEx is being successfully used for research and teaching activities at the Spanish University Center of the Civil Guard. Due to its advanced stage of maturation, it could also be used in other areas within the Army and Law enforcement Agencies involved in the fight against terrorism and the design of blast resistant buildings and structures.

Author Contributions: Conceptualization, J.S.-M. and M.V.; methodology, J.S.-M. and M.V.; software, J.S.-M. and A.C.; validation, all authors; formal analysis, J.S.-M. and M.V.; investigation, all authors; resources, C.H. and M.V.; data curation, J.S.-M. and A.C.; writing—original draft preparation, J.S.-M., A.C. and M.V.; writing—review and editing, C.H. and M.V.; visualization, J.S.-M. and A.C.; supervision, C.H. and M.V.; project administration, M.V.; funding acquisition, C.H. and M.V. All authors have read and agreed to the published version of the manuscript.

Funding: This research was partially funded by UE (H2020-SEC-2016-2017-1) Grant No. SEC-08/11/12-FCT-2016 and by project H2SAFE-CM-UC3M awarded by the Spanish Comunidad de Madrid. The authors would like to thank Henar Miguelez from University Carlos III of Madrid and Col. Fernando Moure from the University Center of the Civil Guard for their continuous support, as well as Com. Miguel Ángel Albeniz from the SEDEX-NRBQ (EOD-CBRN) service of the Civil Guard for many enlightening discussions. Fruitful discussions with Dr. Lina López from the School of Mines at Technical University of Madrid are also gratefully acknowledged.

Institutional Review Board Statement: Not applicable.

Informed Consent Statement: Not applicable.

Acknowledgments: The authors would like to thank Henar Miguelez from University Carlos III of Madrid and Col. Fernando Moure from the University Center of the Civil Guard for their continuous support, as well as Com. Miguel Ángel Albeniz from the SEDEX-NRBQ (EOD-CBRN) service of the Civil Guard for many enlightening discussions. Fruitful discussions with Lina López from the School of Mines at Technical University of Madrid are also gratefully acknowledged.

Conflicts of Interest: The authors declare no conflict of interest. The funders had no role in the design of the study; in the collection, analyses, or interpretation of data; in the writing of the manuscript, or in the decision to publish the results.

SimEx License & Distribution: SimEx is a closed-source proprietary software that may be licensed by the copyright holders, UC3M & Guardia Civil, under specific conditions. Please contact the corresponding author for further information.

Abbreviations

The following abbreviations are used in this manuscript:

BKW	Becker–Kistiakowsky–Wilson EoS
CL	Confidence Level
CT	Combustion Toolbox
CUGC	Centro Universitario de la Guardia Civil
CW–S	Charge Weight–Standoff
EoS	Equation of State
GUI	Graphical User Interface
H9	Heuzé EoS
HOB	Height of Burst
IED	Improvised Explosive Device
ISA	International Standard Atmosphere
LOP	Level of Protection
PDC	Protective Design Center
SDOF	Single Degree of Freedom
SEDEX-NRBQ	Explosive Ordnance Disposal (EOD) and CBRN Defense Service
UC3M	University Carlos III of Madrid
UFC	Unified Facilities Criteria
US	United States
USACE	United States Army Corps of Engineers

References

1. Kinney, G.F.; Graham, K.J. *Explosive Shocks in Air*, 2nd ed.; Springer: New York, NY, USA, 1985.
2. Brode, H.L. Numerical solutions of spherical blast waves. *J. Appl. Phys.* **1955**, *26*, 766–775. [CrossRef]
3. Grady, D. *Fragmentation of Ring and Shells, the Legacy of N.F. Mott*; Springer: Berlin/Heidelberg, Germany, 2006.
4. Ambrosini, R.D.; Luccioni, B.M. Craters produced by explosions on the soil surface. *J. Appl. Mec.* **2006**, *73*, 890–900. [CrossRef]
5. Akhavan, J. *The Chemistry of Explosives*, 3rd ed.; Royal Society of Chemistry: Cambridge, UK, 2011.
6. Cooper, P.W. *Explosives Enginering*; Wiley-VCH, Inc.: New York, NY, USA, 1996.
7. Meyer, R.; Köhler, J.; Homburg, A. *Explosives*, 6th ed.; Wiley-VCH Verlag GmbH: Weinheim, Germany, 2007.
8. Ngo, T.; Mendis, P.; Gupta, A.; Ramsay, J. Blast loading and blast effects on structures—An overview. *Electron. J. Struct. Eng.* **2007**, *1*, 76–91. [CrossRef]
9. *TNO Green Book Methods for the Determination of Possible Damage to People and Objects Resulting from Releases of Hazardous Materials (CPR 16E)*; CIP-data of the Royal Library: The Hage, The Netherlands, 1992.
10. START. National Consortium for the Study of Terrorism and Responses to Terrorism. Global Terror-Ism Database. 2017. Available online: https://www.start.umd.edu/gtd (accessed on 3 September 2022).
11. US Department of Homeland Security (DHS). *Reference Manual to Mitigate Potential Terrorist At-Tacks Against Buildings*, 2nd ed.; Federal Emergency Management Agency: Washington, DC, USA, 2011.
12. Luccioni, B.; Ambrosini, D.; Danesi, R. Blast load assessment using hydrocodes. *Eng. Struct.* **2006**, *28*, 1736–1744. [CrossRef]
13. Anderson, C.E., Jr. An overview of the theory of hydrocodes. *Int. J. Impact Eng.* **1987**, *5*, 33–59. [CrossRef]
14. Han, Y.; Liu, H. Finite element simulation of medium-range blast loading using LS-DYNA. *Shock Vib.* **2015**, *2015*, 631493. [CrossRef]
15. *Protective Design Center Technical Report PDC-TR 06-08*; Single Degree of Freedom Structural Response Limits for Antiterrorism Design. US Army Corps of Engineers: Washington, DC, USA, 2008.
16. *Unified Facilities Criteria (UFC) 3-340-02*; Structures to Resist the Effects of Accidental Explosions. US Department of Defense: Washington, DC, USA, 2008.
17. Baker, W.E. *Explosions in Air*; University of Texas Press: Austin, TX, USA, 1973.
18. Kingery, C.N.; Bulmash, G. *Airblast Parameters from TNT Spherical Air Burst and Hemispherical Surface Burst*;. US Technical Report ARBRL-TR-02555; Aberdeen Proving Ground, Ballistic Research Laboratories (BRL): Aberdeen, MD, USA, 1984.
19. Needham, C.E. *Blast Waves*; Springer: Berlin/Heidelberg, Germany, 2010.
20. Protective Design Center Software, US Army Corps of Engineers. Available online: https://www.nwo.usace.army.mil/About/Centers-of-Expertise/Protective-Design-Center/PDC-Software (accessed on 3 September 2022).
21. Sánchez-Monreal, J.; Vera, M. SimEx: SIMulador de EXplosiones mediante sistemas SDOF. In Proceedings of the III Na-tional Congress of R&D in Defense and Security (DESEi+d 2015), Naval Military School of Marín, Pontevedra, Spain, 19–20 November 2015; pp. 993–1000.
22. Sánchez-Monreal, J.; Vera, M. SIMulador de EXplosiones mediante sistemas SDOF (SimEx): Desarrollo de la aplicación y nuevas funcionalidades. In Proceedings of the IV National Congress of R&D in Defense and Security (DESEi+d 2016), University Defense Center, General Air Academy, San Javier, Spain, 16–18 November 2016; pp. 949–955. Available online: https://www.tecnologiaeinnovacion.defensa.gob.es/.../Actas_DESEi+d2016_1.pdf (ac-cessed on 3 September 2022).
23. Sánchez-Monreal J.; Vera M. SimEx: Una herramienta para la evaluación rápida de los efectos de explosiones. *Bol. Obs. Tecnol. Def.* **2017**, *55*, 23–27. Available online: https://publicaciones.defensa.gob.es/.../bot_55.pdf (accessed on 3 September 2022).
24. Cuadra, A.; Huete, C.; Vera, M. *Combustion Toolbox: A MATLAB-GUI Based Open-Source Tool for Solving Gaseous Combustion Problems (v0.9.7)*; Zenodo: Geneve, Switzerland, 2022.
25. Ambrosini, R.D.; Luccioni, B.M.; Danesi, R.F.; Riera, J.D.; Rocha, M.M. Size of craters produced by explo-sive charges on or above the ground surface. *Shock Waves*, **2002**, *12*, 69–78. [CrossRef]
26. Mott, N.F. *A Theory of Fragmentation of Shells and Bombs*; AC4613; Ministry of Supply: Great Britain, UK, 1943.
27. Mott, N.F. *Fragmentation of H.E. Shells; A Theoretical Formula for the Distribution of Weights of Fragments*; AC3642; Ministry of Supply: Great Britain, UK, 1943.
28. Mott, N.F. Fragmentation of shell cases. *Proc. R. Soc.* **1947**, *A189*, 300–308.
29. Sánchez-Monreal, J.; Llamazares-Mendo, F.; Vera M. Estimación de daños de explosiones sobre personas. In Proceedings of the V National Congress of R&D in Defense and Security (DESEi+d 2017), Toledo, Spain, 22–24 November 2017; pp. 1689–1697. Available online: https://publicaciones.defensa.gob.es/.../actas_v_congreso_id_2017.pdf (accessed on 3 September 2022).
30. González Duperón, E.; Martínez Gil, M.; Giménez Francés, B.; González Ferradás, E. Analysis of effects of detonation of explosive substances on humans from characteristic curves explosive mass-distance from the origin of the explosion. *Process. Saf. Prog.* **2016**, *35*, 233–240. [CrossRef]
31. *Protective Design Center Technical Report PDC TR 06-01*; Rev 1, Methodology manual for the Single-Degree-of-Freedom Blast Effects Design Spreadsheets (SBEDS). US Army Corps of Engineers: Washington, DC, USA, 2008.
32. Biggs, J.M. *Introduction to Structural Dynamics*; McGraw-Hill College: New York, NY, USA, 1964.
33. Brode, H.L. Blast wave from a spherical charge. *Phys. Fluids* **1959**, *2*, 217–229. [CrossRef]
34. Needham, C.E.; Crepeau, J.E. *The DNA Nuclear Blast Standard (1 kT)*; Report No. DNA 5648-T.; Systems, Science and Software, Inc.: Alburquerque, Spain, 1981.

35. Hopkinson, B. *British Ordnance Board Minutes 13565*; The National Archives: Richmond, UK,1915.
36. Sachs, R.G. *The Dependence of Blast on Ambient Pressure and Temperature*; Technical Report 466; Ballistic Research Laboratories: Aberdeen, MD, USA, 1944.
37. *ISO 2533:1975*; Standard Atmosphere. International Organization for Standardization: Geneva, Switzerland, 1975.
38. Gallo, E. Quasi static atmospheric model for aircraft trajectory prediction and flight simulation. *arXiv* **2021**, arXiv:2101.10744.
39. *International Ammunition Technical Gudeline (IATG) 01.80 Formulae for Ammunition Management*; UNODA Office for Disarmament Affairs: New York, NY, USA, 2013.
40. Friedlander, F.G. The diffraction of sound pulses I. Diffraction by a semi-infinite plane. *Proc. R. Soc. London Ser. Math. Phys. Sci.* **1946**, *186*, 322–344.
41. *Euronorm 53–62*; European Wide Flange Beams. CEN: Brussels, Belgium, 1993.
42. *Norma UNE 31-002-94*; Cálculo de las Principales Características Teóricas de los Explosivos. Asociación Española de Normalización UNE: Biddeford, Maine, 1994.
43. *European Standard EN 13631-15*; Explosives for Civil Use-High Explosives—Part 15: Calculation of the Thermodynamic Properties. UNE: Biddeford, Maine, 2005.
44. Arnold, W.; Rottenkolber, E.; Hartmann, T. DRAGON—The German Thermo-Chemical Code Based on the Becker-Kistiakowsky-Wilson Equation of State. *Propellants Explos. Pyrotech.* **2022**, e202100329. [CrossRef]
45. Hobbs, M.L.; Baer, M.R. Nonideal thermoequilibrium calculations using a large product species data base. *Shock Waves* **1992**, *2*, 177–187. [CrossRef]
46. Mader, C.L. *Detonation Properties of Condensed Explosives Computed Using the Becker-Kistiakowsky-Wilson Equation of State*; Report LA-2900; Los Alamos Scientific Laboratory: Los Alamos, NM, USA, 1963.
47. Heuzé, O. Equations of state of detonation products: Influence of the repulsive intermolecular potential. *Phys. Rev. A* **1986**, *34*, 428–432. [CrossRef] [PubMed]
48. López, L.M. Evaluación de la Energía de los Explosivos Mediante Modelos Termodinámicos de Detonación. Ph.D. Thesis, Universidad Politecnica de Madrid, ETSI Minas, Madrid, Spain, 2003.
49. Sanchidrián, J.A.; López, L.M. Calculation of the energy of explosives with a partial reaction model. Comparison with cylinder test data. *Propellants Explos. Pyrotech.* **2006**, *31*, 24–32. [CrossRef]
50. Cuadra A.; Huete, C.; Vera M. Desarrollo de un código termoquímico para la evaluación de las propiedades teóricas de explosivos (CT-EXPLO) y la estimación del rendimiento de motores cohete (CT-ROCKET). In Proceedings of the IX National Congress of R&D in Defense and Se-curity (DESEi+d 2022), Army's "General Morillo" Base , Pontevedra, Spain, 15–17 November 2022, *accepted*.
51. Kamlet, M.J.; Jacobs, S.J. Chemistry of detonations. I. A simple method for calculating detonation proper-ties of CHNO explosives. *J. Chem. Phys.* **1968**, *48*, 23–55. [CrossRef]
52. Politzer, P.; Murray, J.S. Some perspectives on estimating detonation properties of C, H, N, O compounds. *Cent. Eur. J. Energetic Mater.* **2011**, *8*, 209–220.
53. Beveridge, A. *Forensic Investigation of Explosions*, 2nd ed.; CRC Press: Boca Raton, FL, USA; Tyalor & Francis Group: Boca Raton, IL, USA, 2012.
54. Westine, P.S. Explosive cratering. *J. Terramechanics* **1970**, *7*, 9–19. [CrossRef]
55. Baker, W.E.; Westine, P.S.; Dodge, F.T. *Similarity Methods in Engineering Dynamics: Theory and Practice of Scale Modeling*; Fundamental Studies in Engineering No. 12; Elsevier Science Limited: Amsterdam, The Netherlands, 1991.
56. Gurney, R.W. *The Initial Velocities of Fragments from Bombs, Shells and Grenades*; Report No 405; Aberdeen Proving Ground, Ballistic Research Laboratories (BRL): Aberdeen, MD, USA, 1943.
57. Carter, R.T.; Jandir, P.S.; Kress, M.E. Estimating the drag coefficients of meteorites for all Mach number regimes. In Proceedings of the 40th Lunar and Planetary Science Conference, (Lunar and Planetary Science XL), The Wood-Lands, TX, USA, 23–27 March 2009.
58. Häring, I. Damage Analysis I: Probit Functions and Probability Distributions. In *Risk Analysis and Management: Engineering Resilience*; Springer: Singapore, 2015; pp. 187–204.
59. Finney, D.J. *Probit Analysis*, 3rd ed.; Cambridge University Press: Cambridge, UK, 1971.
60. Alonso, F.D.; Ferradás, E.G.; Pérez, J.F.S.; Aznar, A.M.; Gimeno, J.R.; Alonso, J.M. Characteristic overpressure–impulse–distance curves for the detonation of explosives, pyrotechnics or unstable substances. *J. Loss Prev. Process. Ind.* **2006**, *19*, 724–728. [CrossRef]

applied sciences

MDPI

Article

Numerical Analysis of Blast Effects and Mitigation in the Far-Field from Small Explosions

Adam G. Taylor

Lawrence Livermore National Laboratory, Computational Engineering Division, Livermore, CA 94550, USA; taylor265@llnl.gov

Abstract: Requirements for explosive safety are often given in terms of a "*K*-Factor", correlating incident blast effects with the distance and TNT equivalent weight of an explosion. Traditionally, this is conducted using empirical correlations to experimental measurements (e.g., the Kingery–Bulmash equations). In the far-field, empirical verification of incident overpressure and impulse magnitudes can be difficult; extrapolations from data give expected values at reasonable standoff distances that sometimes are too small to measure on available equipment but are larger than some regulations require. The present paper describes the results of numerical hydrocode analysis to verify the expected incident overpressure and impulse from small hemispherical ground charges of TNT at these relatively large distances. Furthermore, the dynamic effect of incident blast waves on lightweight, modular mitigation barriers is studied to gauge their effectiveness at providing safety standard compliance.

Keywords: blast effects; mitigation; hydrocode analysis

Citation: Taylor, A.G. Numerical Analysis of Blast Effects and Mitigation in the Far-Field from Small Explosions. *Appl. Sci.* **2022**, *12*, 8824. https://doi.org/10.3390/app12178824

Academic Editors: Ricardo Castedo, Lina M. López and Anastasio P. Santos

Received: 16 July 2022
Accepted: 29 August 2022
Published: 2 September 2022

1. Introduction

Understanding the structure of blast waves and the dynamics of their interactions with structures is key for mitigation and safety. The formation and propagation of these shock waves is a highly nonlinear dynamic process; thus, prediction of the incident waveforms and their corresponding blast overpressure and impulsive loads for a given scenario can be difficult. It is common for various government, military, and scientific institutions to prescribe criterion for safety from blast effects in terms of "*K* factors":

$$K = R/W^{1/3} \tag{1}$$

Here, *R* is the distance from the explosive source and *W* is the net explosive (TNT equivalent) weight. Allowable exposure for personnel, nearby structures, and withdrawal distances can be given in terms of these *K* factors, which have been empirically correlated to values of incident overpressure and impulse. A figure regularly encountered in explosive safety documentation is the K328 criterion, often referred to as the "Public Withdrawal Distance"; calculated using units of $ft/lb^{1/3}$ this corresponds to a peak incident overpressure of 0.0655 psi, (0.4516 kPa) and is said to be a condition under which there is no probability of harm. Different safety guidelines have different requirements for personnel, but they are very commonly given in terms of these *K* factors.

The empirical nexus of *K* factor correlation appears to be the work of Kingery [1,2]. The original data came from quite large (5, 20, 100, and 500 ton) hemispherical TNT events. Instrumentation at various distances measured arrival time, peak overpressure, and the duration of the positive pressure phase and the positive impulse. This same data set was later reinterpreted and extrapolated to include reflected pressures/impulse and shock velocities by Kingery and Bulmash [3]. It is these fits which became the basis for more widespread application, and thus the empirical equations are often referred to as Kingerly-Bulmash (KB) curves. Swisdak [4] provides a good overview of this history, along with improved equations fitting the same data. More recent fits by Jeon et al. [5] claim to further simplify the curves with the same accuracy.

There is definite uncertainly in the accuracy of the KB curves and other analytic and empirical tools for predicting blast overpressure in a given case. Karlos et al. [6–8] have investigated the structure of blast waves and their parameters for scaling and decay, including variations in explosive type, weight, and configuration on the resulting incident pulses. A recent review of analytical and empirical prediction methods by Ullah et al. [9] shows a very large spread in the predicted blast overpressures and wave structures from various accepted sources. Recent repeated blast measurements from Stewart et al. [10] show large variability in the measured results from what are ostensibly the same experiments. The recent experiment of Filice et al. [11] provides more data and KB comparisons and variances for relatively nearby (2–5 m) and relatively small (100–400 g) explosives. In a review of the experimental literature vs. KB predictions, Rigby et al. [12] state that the variation in experimental predictions is so large in nominally similar experiments that there is a valid question as to whether blast phenomena are inherently deterministic, or whether they should be viewed as fundamentally stochastic processes. Under this lens, KB and others may be viewed as useful only at predicting the order of magnitude of blast effects.

The question arises: can direct physics-based calculation of blast wave parameters provide more detailed and accurate predictions for a given case of interest?

The classical analytical result for the prediction of the evolution of a very strong explosion is the so-called Taylor–von Neumann–Sedov solution [13–16]. This applies only to spherical (1D) blasts and is derived under assumptions (point source, zero ambient pressure) that leave it applicable only for intermediate distances. Some of the earliest published attempts to simulate explosions under real conditions (i.e., into non-zero ambient pressure conditions) were performed by Brode [17,18] and Goldstine and von Neumann [19].

More recently there have been various simulations performed in modern software packages aimed at the prediction of the evolution of blast waves. Ding et al. [20,21] recently presented the results of numerical simulations of very large TNT equivalent blasts and their resulting effects on near and far-field structures. Xue et al. [22] modeled the whole process of explosive shockwave formation and propagation from relatively smaller blasts over larger distances. Sung and Chong have produced a fast-running semi-empirical method for the prediction of blast effects behind shielding barriers; this work includes uncertainty estimations when using KB-type charts [23]. Giodo et al. compared empirical and numerical approaches to investigating the effects of free far-field blasts on masonry wall [24]. Vannucci et al. [25] provide analysis of a blast and shock propagation inside a monumental structure. Draganic and Varevac [26] have provided a useful parametric study on the effects of numerical mesh size on the blast wave parameters.

It is easy to imagine situations (involving explosive training, demolitions, etc.) where relatively small explosions (comparable to 1 kg TNT) send overpressure waves towards personnel relatively far away (30–40 m). These blasts are very small compared to the conditions studied in the published literature or in the data informing KB-type predictions, but nevertheless may induce pressures and impulses in excess of safety guidelines (e.g., the Public Withdrawal Distance). Furthermore, the incident overpressure will be far below the ambient atmospheric pressures and will be difficult or impossible to accurately measure using easily available pressure gauges. Given that there are reasons to question the accuracy of KB-type predictions under these circumstances, research is needed to clarify the situation.

The purpose of this paper is to use numerical tools to investigate cases where very small charges produce relatively small incident overpressure at large distances which still exceed the safety guidelines of public withdrawal distance. The goals here are two-fold:

1. to predict the structure and magnitude of the incident pressure waves in these cases and to compare to the available empirical blast curves;
2. to investigate the efficacy of lightweight, modular barriers at mitigating incident overpressure waves to the desired levels.

Towards the first goal, free-field explosions of small hemispherical ground TNT charges into air are simulated out to a range of 40 m. Wave profiles obtained from the free-field simulations are subsequently employed as boundary conditions for dynamic wave-structure interaction models which investigate the second goal.

It is noted that a few different sets of units were used in the preparation of this work. Much of the original work conducted in blast load estimation was conducted in English units (ft/lb/ms/psi) (see for example the original Kingery report [1]). For that reason, explosive range operators and field experts tend to think in terms of these units, and regulations often give quantities such as K factors in these units. On the other hand, ALE3D hydrocode analyses are traditionally conducted in a special set of units (cm/g/μs/Mbar). The simulations described herein follow in this tradition. For the sake of consistency, all units in this paper will be given in terms of Si units (m/kg/ms/kPa). In some cases, English units will be listed concurrently.

2. Materials and Methods

Simulations presented in this work were performed in ALE3D, a multi-physics software package which utilizes an Arbitrary Lagrangian/Eulerian (ALE) numerical scheme [27]. The numerical simulations performed are of two types: (i) free-field explosions of various weights of TNT in air at atmospheric pressure, and (ii) the dynamic interaction of incident blast waves with simple mitigation barriers. The remainder of this section will describe the material models implemented, and provide further details into the setup of each type of simulation.

2.1. Material Models

Three material models were employed for the three separate material components simulated in this work, namely the TNT explosive, the surrounding air, and the Lexan structural barrier. Only the TNT and air appear in the free-field simulations, and only the air and Lexan appear in the blast mitigation simulations. For the explosive TNT, a simple Jones–Wilkens–Lee (JWL) equation of state [28] is used:

$$P(v,e) = A\left(1 - \frac{\omega}{R_1 v}\right)\exp(-R_1 v) + B\left(1 - \frac{\omega}{R_2 v}\right)\exp(-R_2 v) + \frac{\omega}{v}e \tag{2}$$

Here, P is the pressure, $v = V/V_0 = \rho_0/\rho$ the relative volume, and e is the material energy per reference volume. V, ρ are the volume and density, respectively, while V_0, ρ_0 are the initial (reference) values of these properties. The parameter ω is the Grüneisen coefficient; A, B, R_1, and R_2 are free parameters. ω, R_1 and R_2 are dimensionless, while A and B have units of pressure. The parameter values used in simulations for Equation (2) are given in Table 1.

Table 1. JWL parameters for TNT.

A (kPa)	B (kPa)	R_1	R_2	ρ_0 (g/cm^3)	ω
3.712×10^8	3.231×10^6	4.150	0.950	1.630	0.30

The equation of state of air is given by a simple Gamma-law:

$$P(\rho,e) = (\gamma - 1)\frac{\rho}{\rho_0}e \tag{3}$$

The only free parameter γ is dimensionless and typically has a value of 1.4 for air. The initial (atmospheric) pressure P_0 is obtained through Equation (3) by prescribing and initial energy per unit volume:

$$e_0 = \frac{P_0}{\gamma - 1}$$

The parameter values used in simulations for Equation (3) are given in Table 2.

Table 2. Gamma law parameters for air.

γ	ρ_0 (g/cm^3)	P_0 (kPa)
1.40	1.225×10^{-3}	1.0135×10^{2}

The Lexan mitigation barrier is modeled using a power law constitutive model:

$$\sigma = k(\epsilon_0 + \epsilon)^{y_c}, \quad (4)$$

$$\epsilon_0 = \left(\frac{E}{k}\right)^{\frac{1}{y_c - 1}} \quad (5)$$

Here, σ is the current yield stress and ϵ an equivalent plastic strain. ϵ_0 is an initial yield strain determined by parameters k, E and y_c. E is a standard Young's modulus with dimensions of pressure, k the yield stress coefficient with dimensions of pressure, and y_c is a dimensionless strain-hardening coefficient. An additional equation of state relates pressure P to the bulk modulus K and the relative volume v:

$$P = K\mu \quad (6)$$

Here, $\mu = (1/v) - 1$, and the bulk modulus is derived from the Young's modulus and Poisson ratio ν:

$$K = \frac{E}{3(1 - 2\nu)}$$

The values used for Lexan in the present work are given in Table 3. Given the nature of the low pressure incident waves studied in this paper, only small (elastic) deformations of the barrier are expected. Therefore the values used for the barrier material are not expected to have significant effect on the analysis results.

Table 3. Power law parameters for Lexan.

E (kPa)	k (kPa)	ν	ρ_0 (g/cm^3)	y_c
2.344×10^{6}	1.119×10^{5}	0.4	1.218	2.086×10^{-1}

2.2. Free-Field Detonation of TNT

The free-field detonation of hemispherical TNT was simulated under 2D axisymmetric conditions. Figure 1 depicts a cartoon of the setup. The $x = 0$ axis is the axis of rotational symmetry, while the $y = 0$ has symmetry boundary conditions which are used to crudely approximate the ground; however, this approximation causes the simulation to be equivalent to a spherical charge of the same radius exploding in air. The air domain extends from the origin to 40 m in the x and y directions. The outer boundaries have three different boundary conditions applied; "pressure continuous" provides ghost nodes external to the boundary which keeps the pressure constant on the other side, which keeps the initially pressurized gas from expanding and depressurizing as soon as the simulation starts. "Non-reflecting" boundary conditions dampen out any reflected incident waves to minimize boundary effects. The "outflow" condition allows material given outbound velocity to leave the domain.

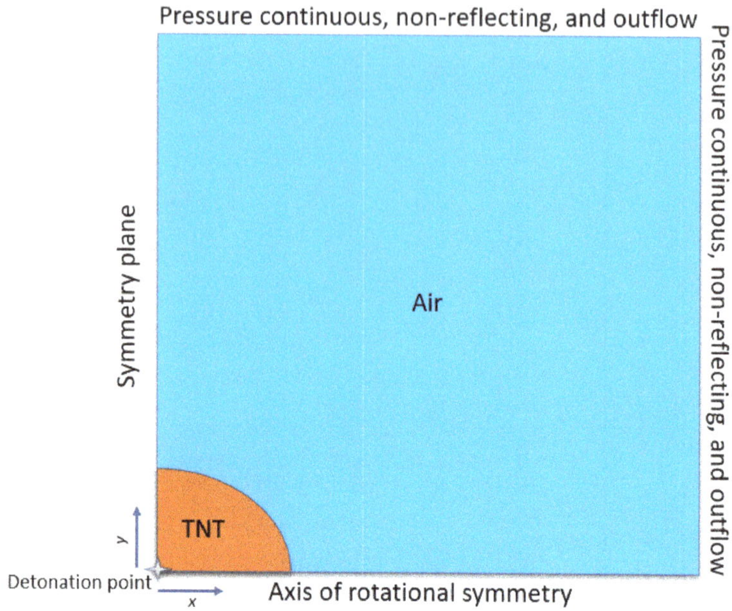

Figure 1. A "cartoon" depiction of the setup of the free-field TNT detonation simulations with materials and boundary conditions labeled (Not to scale).

Four simulations in total were performed with of charges with radius 0.0261 m, 0.051 m, 0.0643 m, and 0.081 m, yielding hemispherical charge weights of approximately 0.06123 kg (0.135 lb), 0.45359 kg (1 lb), 0.90718 kg (2 lb), and 1.81436 kg (4 lb), respectively. Note that the charge radius is around 3 orders of magnitude smaller than the domain length; even in 2D, a uniform Cartesian mesh small enough to adequately resolve the TNT would lead to an intractably-large numbers of zones. Instead, a graded mesh approach was use, coarsening with distance from the origin. Initial zone sizes range from approximately 4.5×10^{-3} m at the center of the charge out to 5.4×10^{-2} m at the outer edge of the domain. The simulations ultimately contained around 7.1 million zones.

Figure 2 shows representative temporal snapshots of pressure in the system as the explosive wave propagates in air. The peak overpressure occurs near the wavefront but rapidly decreases to the ambient pressure and then dips below it for some time before returning. The magnitude of this peak pressure decreases as the wave propagates further from the source. Fixed (Eulerian) pressure tracers were placed every 2 m in the domain just off the y-axis in order to study the structure and evolution of the blast wave. Figure 3 shows the results of these pressure tracer time histories. Each tracer shows a pronounced positive overpressure phase followed by a negative phase where pressure dips below ambient. The effects of these negative pressure phases have been studied and are in general not negligible [29,30].

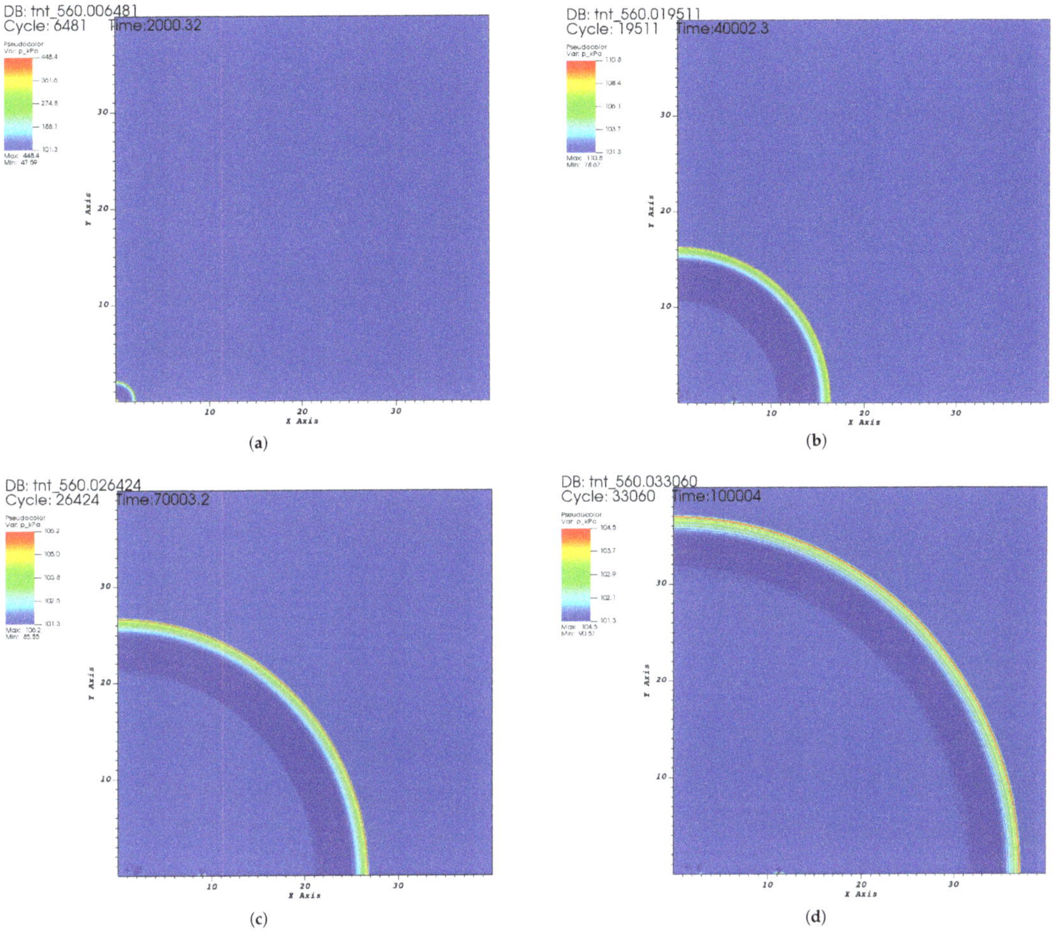

Figure 2. Snapshots of the pressure wave from the 0.90718 kg explosion of hemispherical TNT into air at atmospheric pressure (approximately 101.35 kPa): (**a**) 2 ms, (**b**) 40 ms, (**c**) 70 ms, and (**d**) 100 ms after programmed detonation.

Figure 4 shows the pressure waves recorded at 36.576 m (120 ft) from the source. Note that when compared to some of the larger pressure, early time pressure waves on the left-hand side of the plots in Figure 3, these waves are relatively smooth and have a shallower initial slope as they ramp up to maximum incident overpressure. This corresponds to the fact that at this distance the waves are no longer proper shock waves. In fact, the wave velocity is approximately that of the speed of sound in air. The area under the positive portion of the overpressure wave is the total incident impulse at this point. As will be shown in the next section, pressure time histories of this type are useful in that they can be used as boundary conditions in subsequent simulations to study the dynamic effects of realistic incident waves on structures.

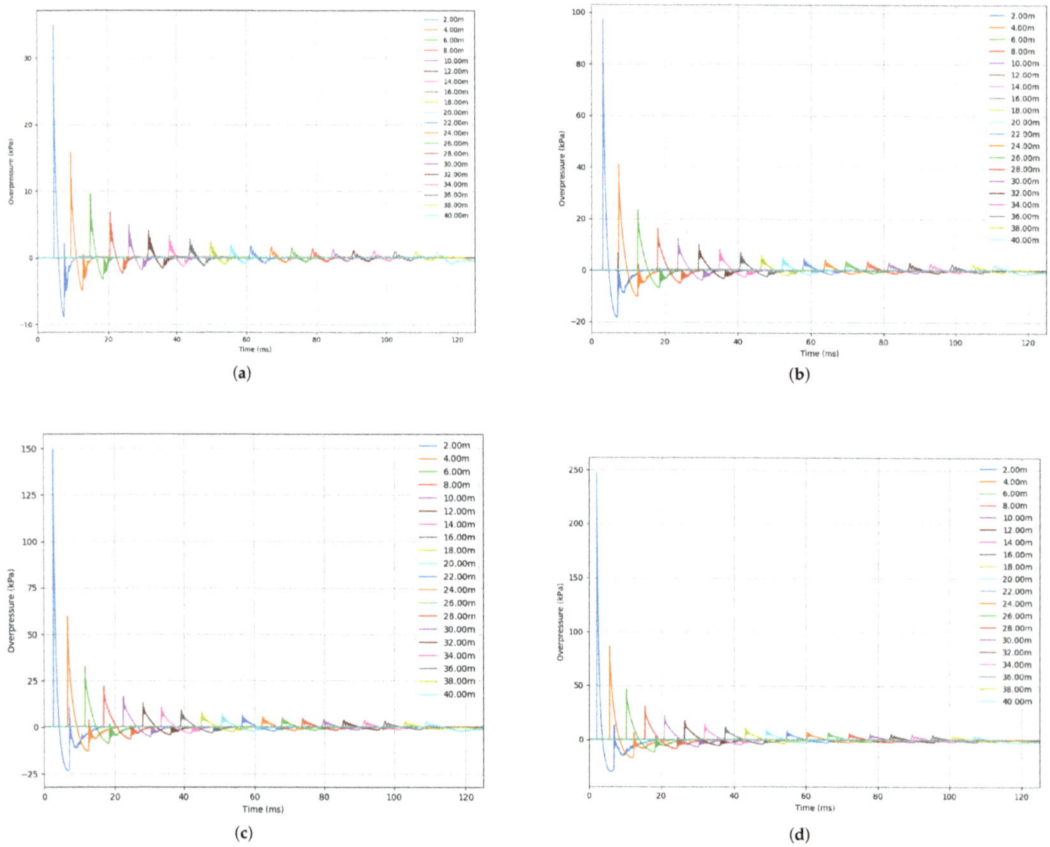

Figure 3. Pressure time histories from four simulations of hemispherical TNT detonation of different weights: (**a**) 0.06123 kg (0.135 lb), (**b**) 0.45359 kg (1 lb), (**c**) 0.90718 kg (2 lb), and (**d**) 1.81436 kg (4 lb).

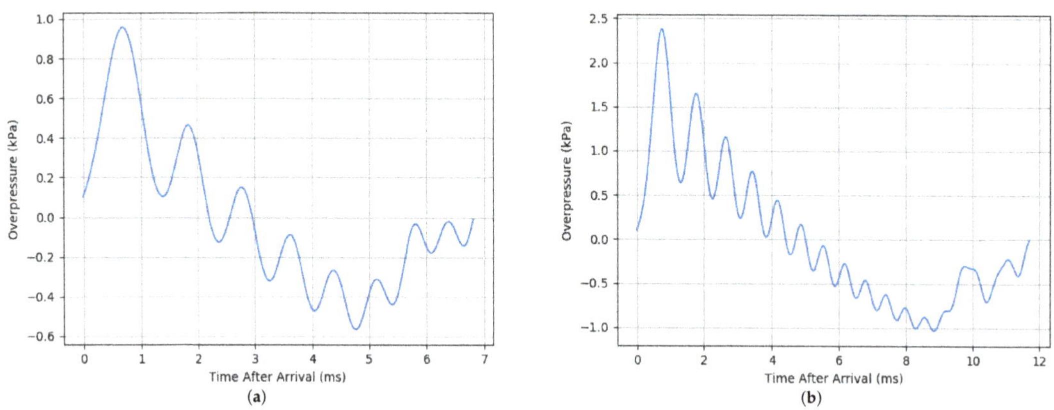

Figure 4. *Cont.*

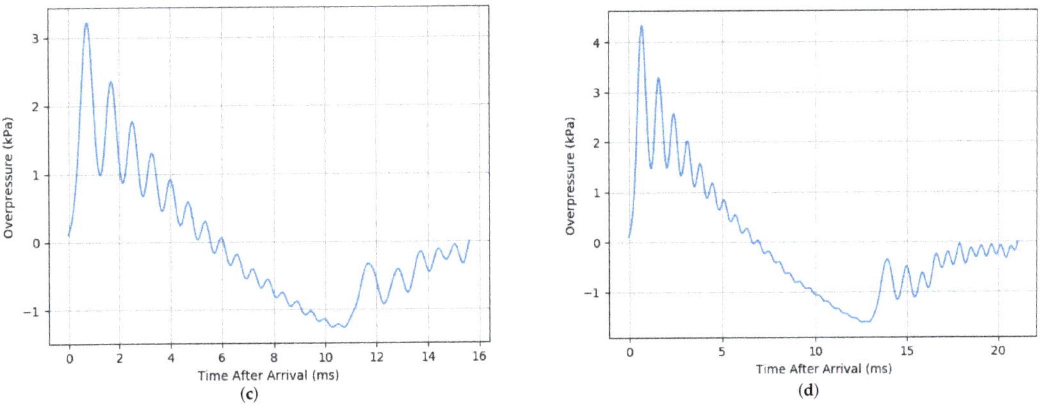

Figure 4. Pressure time histories from four simulations of hemispherical TNT detonation at 36.576 m (120 ft): (**a**) 0.06123 kg (0.135 lb), (**b**) 0.45359 kg (1 lb), (**c**) 0.90718 kg (2 lb), and (**d**) 1.81436 kg (4 lb).

Simulation of Detonation Cord

Further simulations were performed in order to study the variation of effects due to geometry. In particular 1.829 m of detonation cord suspended 1.524 m and parallel to the ground was detonated in a 4.5 m × 4 m × 3 m domain of air under atmospheric pressure (Figure 5). The cord is comprised of a 0.18 cm radius cylinder of TNT, so that ultimately 32.27 g is detonated. Pressure tracers are placed at regular distances from the center of the cord at a height of 1.524 m. Figure 5 shows snapshots of the resulting pressure waves in time.

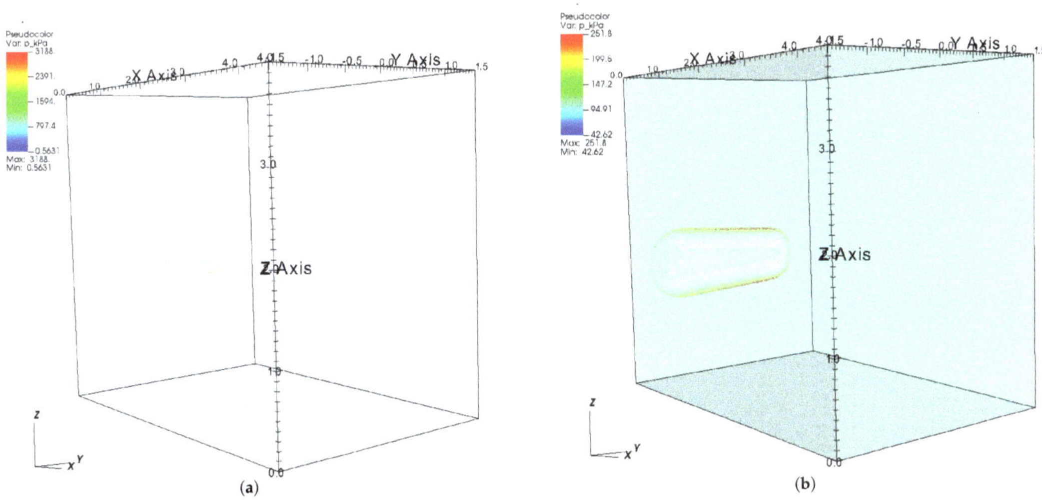

Figure 5. *Cont.*

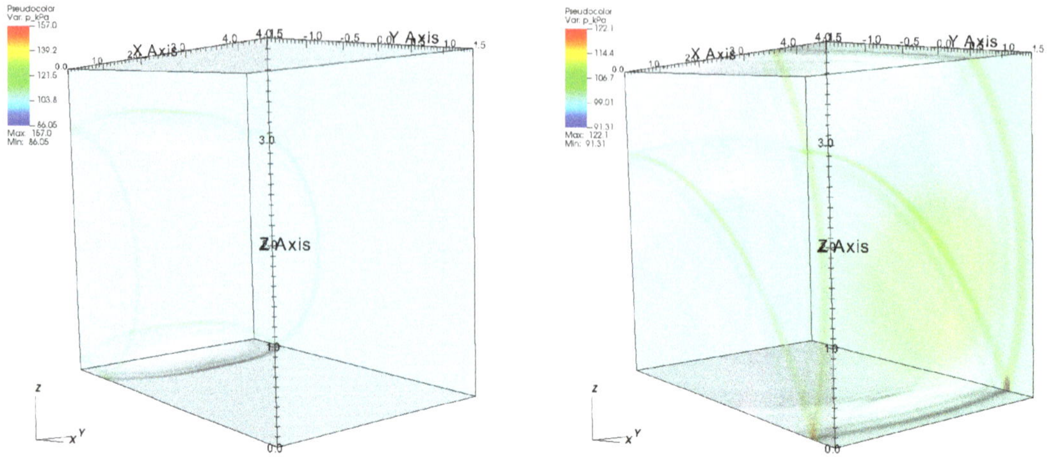

(c) (d)

Figure 5. Snapshots of the pressure wave from explosion of a detonation cord (**a**) 0.27 ms, (**b**) 0.7 ms, (**c**) 4.6 ms, and (**d**) 12 ms after programmed detonation.

2.3. Blast-Barrier Interaction

Simulations of the interaction of incident overpressure waves and lightweight Lexan barriers were performed to gauge the effectiveness of simple modular structures to maintain "Public Withdrawal Distance" conditions where incident overpressures are already quite low. The 2D plane strain simulations were performed, given the assumption that multiple barriers could be placed alongside each other to minimize any edge effects. Further larger 3D cases of interest were explored to visualize and quantify the effects of lateral wraparound for standalone barriers.

Figure 6 presents a "cartoon" depiction of these simulations with labeled boundary conditions. Again the lower boundary is taken as a symmetry plane to estimate ground interactions as perfect reflections. The upper and outer boundaries have pressure continuous non-reflecting conditions. The $x = 0$ plane is given a pressure load curve corresponding to the pressure tracer time histories derived from the free-field blast simulations (Figure 4). It is assumed that in the far field the incident waves are planar. A problem arose in earlier simulations where reflections off of the barrier reached the $x = 0$ plane a re-reflected back into the problem domain before the relevant dynamic events could conclude, causing undesirable boundary effects. It was found that the non-reflecting boundary conditions did not coexist well with the pressure load curves and thus caused numerical issues with the incident pressure waves. To avoid these issues, the barrier was placed at a distance $d = \frac{1}{2}ct_{wave}$, where c is the speed of sound in air (approximately 343 m/s) and t_{wave} is the wavelength (in time) of the incident pressure wave, including positive and negative overpressure phases. Because the far-field waves are traveling at approximately the speed of sound, under these conditions the entire incident wave enters the domain before reflections can return to the boundary. Then, at time t_{wave}, the pressure load curve boundary conditions are replaced with pressure continuous, non-reflecting conditions which eliminate the problem of reflection. The 3D simulations were performed in half-symmetry, so that the $y = 0$ plane was a symmetry plane and the y_{max} also had pressure continuous and non-reflecting conditions.

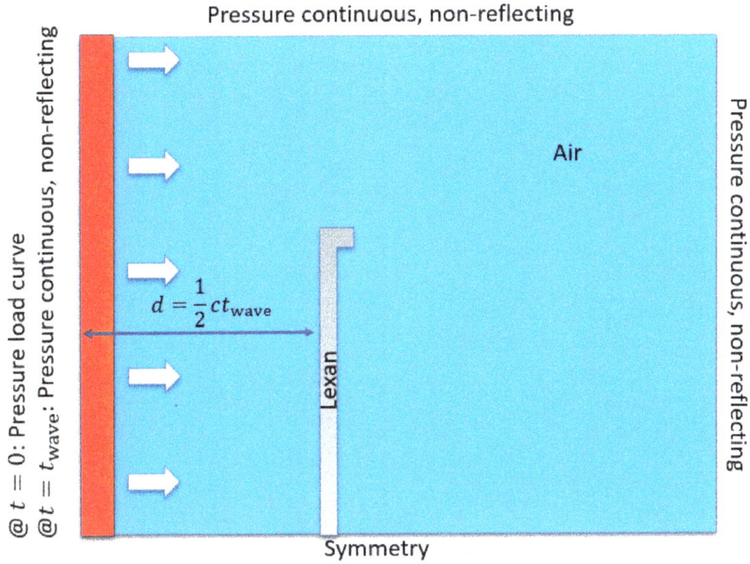

Figure 6. A "cartoon" depiction of the setup of the blast-structure interactions simulations with materials and boundary conditions labeled. (Not to scale)

Various simple designs of mitigation barriers were studied. These included three major types: single fairing, compound fairing, and deep-roof (Figure 7). The barriers are all 1.2 m wide, and 3.8 cm thick. The total height varies with the length and angle of the fairing, but the bases are approximately 2.2 m high.

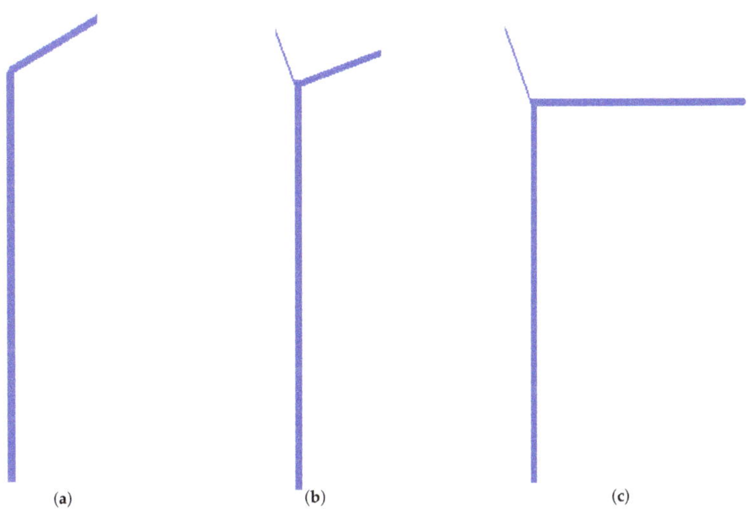

Figure 7. Representative cross-sections of the types of mitigation barriers studied: (**a**) Single-fairing barrier. (**b**) Compound-fairing barrier. (**c**) Barrier with deep roof.

All calculations utilized a graded mesh which was most refined in the area around the mitigation barrier. The 2D plane strain simulations ultimately contained around 1.2 million zones. The 3D simulations in general utilized a coarser mesh that was graded more aggressively, but still contained on the order of 10 million zones per simulation. Figures 8 and 9 show snapshops of the pressure fields in representative 2D and 3D simulations, respectively.

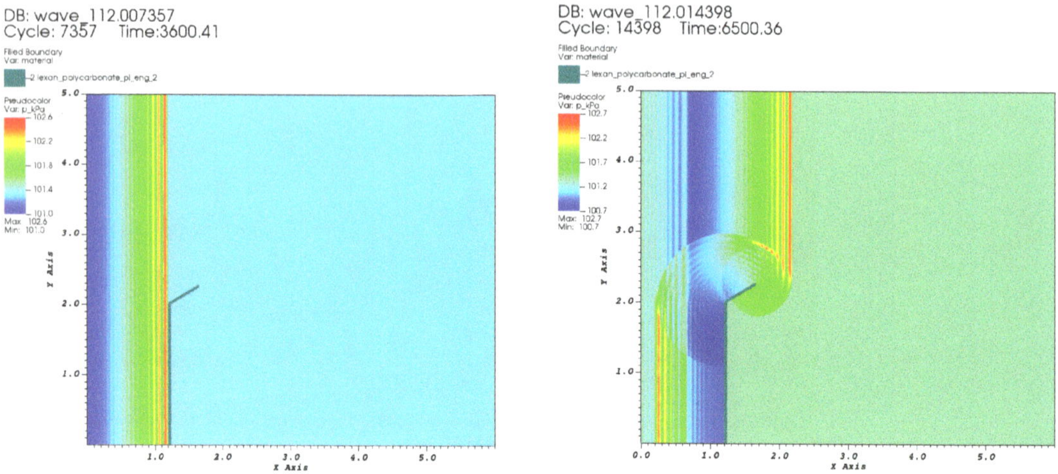

(a) (b)

Figure 8. Images from a representative 2D plane strain blast-barrier mitigation simulation: (**a**) at arrival time of wave at barrier. (**b**) During dynamic interaction event.

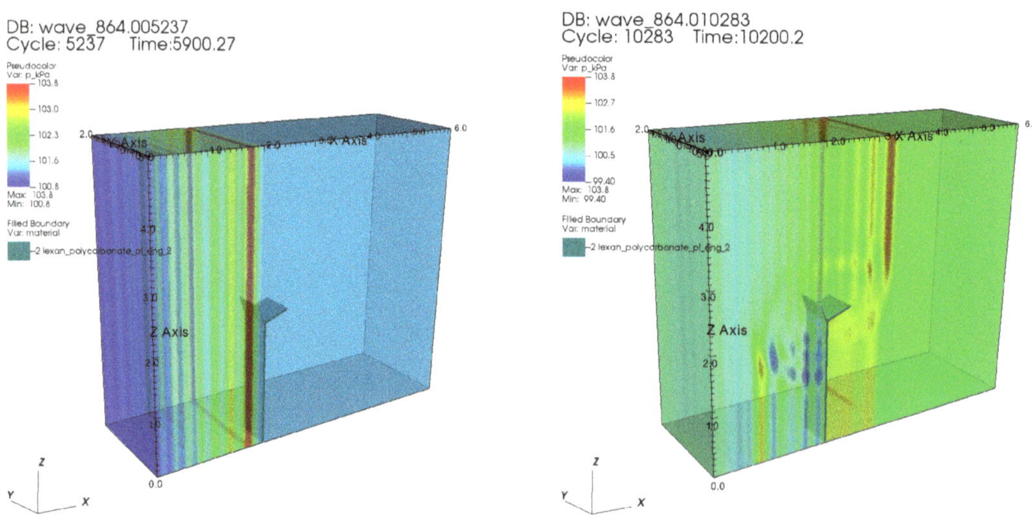

(a) (b)

Figure 9. Images from a representative 3D blast-barrier mitigation simulation: (**a**) at arrival time of wave at barrier. (**b**) During dynamic interaction event.

3. Results

3.1. Comparison of Predicted and Simulated Blast Effects

Of particular interest is the comparison of numerical results with the classical Kingery-Blumash type empirical curves. The fits to these data are most conveniently given by Swisdak [4] in the following form:

$$\exp\left(A + B\ln K + C(\ln K)^2 + D(\ln K)^3 + E(\ln K)^4 + F(\ln K)^5 + G(\ln K)^6\right) \tag{7}$$

Here, K is the K factor given by Equation (1). The curve parameters $A - G$ for Equation (7) fitting peak incident overpressure and positive impulse are given in Tables 4 and 5, respectively.

Table 4. Parameters for Equation (7) for peak incident overpressure (from Swisdak).

K Values	*A*	*B*	*C*	*D*	*E*	*F*	*G*
0.2–2.9	7.2106	−2.1069	−0.3229	0.1117	0.0685	0.0	0.0
2.9–23.8	7.5938	−3.0523	0.40977	0.0261	−0.01267	0.0	0.0
23.8–198.5	6.0536	−1.4066	0.0	0.0	0.0	0.0	0.0

Table 5. Parameters for Equation (7) for incident impulse (from Swisdak).

K Values	*A*	*B*	*C*	*D*	*E*	*F*	*G*
0.2–0.96	5.522	1.117	0.6	−0.292	−0.087	0.0	0.0
0.96–2.38	5.465	−0.308	−1.464	1.362	−0.432	0.0	0.0
2.38–33.7	5.2749	−0.4677	−0.2499	0.0588	−0.00554	0.0	0.0
33.7–158.7	5.9825	−1.062	0.0	0.0	0.0	0.0	0.0

The results for blast overpressure are also compared with predictions from the Taylor–von Neumann–Sedov result. It is shown in [31] that from this solution, the blast radius and corresponding peak pressure are given as a function of time as:

$$R(t) = \beta\left(\frac{Et^2}{\rho_0}\right)^{1/5} \tag{8}$$

$$p(t) = \frac{2}{\gamma+1}\rho_0\left(\frac{2}{5}\frac{R}{t}\right)^2 \tag{9}$$

Here, E is the energy of the explosion, ρ_0 the initial density of the air. γ is the same parameter appearing in Equation (3), and β is a corresponding parameter which has a value of 1.033 for air. Solving (8) for t and substituting into (9) yields an equation for pressure as a function of blast radius:

$$p(R) = \frac{8}{25(\gamma+1)}ER^{-3}\beta^5 \tag{10}$$

This result is valid for a point source explosion in a zero-pressure medium expanding spherically from the origin. In order to compare with our hemispherical results, we compare to a blast having twice the energy of 1 kg TNT; this corresponds with the fact that the symmetry conditions on the floor of our free-field simulations make them numerically equivalent to spherical blasts of the same radius, i.e., twice the weight.

Figure 10 shows the comparisons for peak incident blast overpressure of the free field hemispherical and detonation cord simulations with Equations (7) and (10). Figure 11 shows the corresponding positive impulses calculated from the pressure tracers by numerically integrating the positive portions of the pressure tracers from the hemispherical simulations compared to Equation (7).

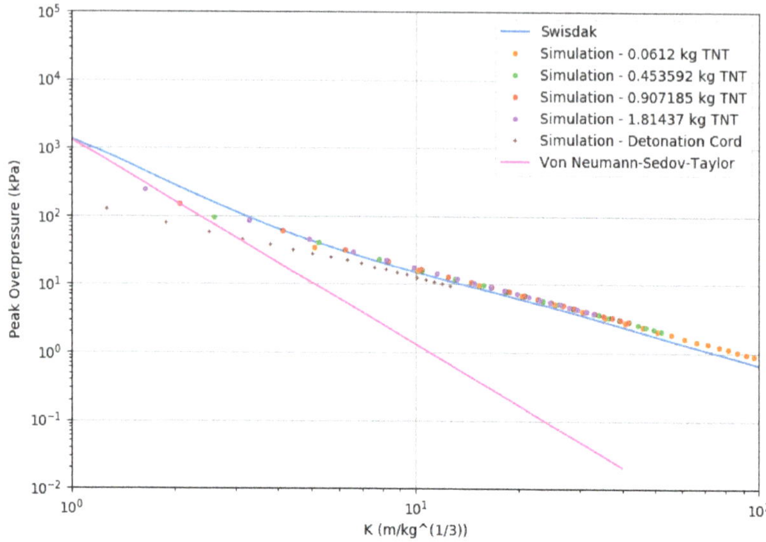

Figure 10. Peak incident blast overpressure versus K factor from the four TNT hemispherical simulations and the detonation cord simulation in comparison with the KB curve from Swisdak and the Taylor–von Neumann–Sedov prediction.

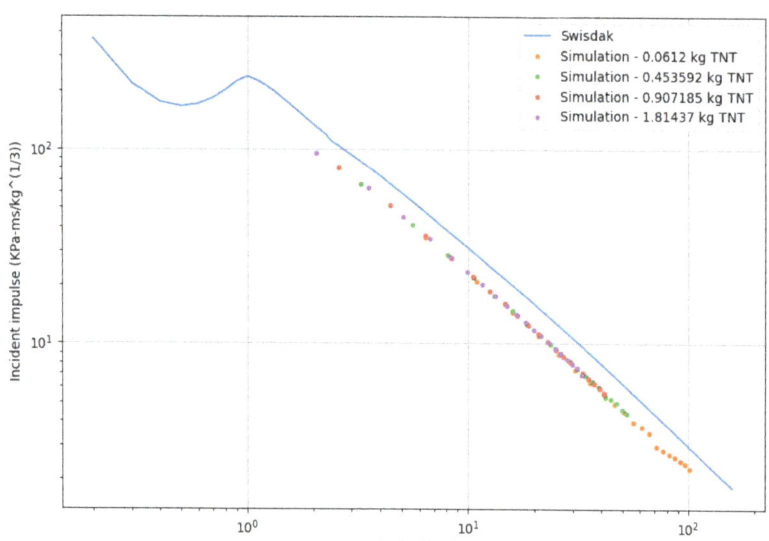

Figure 11. Positive impulse versus K factor from the four TNT hemispherical simulations in comparison with the KB curve from Swisdak.

3.2. *Mitigation Effectiveness*

The effectiveness of the simple Lexan barriers at mitigating incident pressure fields is investigated with particular emphasis on the so called "Public Withdrawal Distance". In $m/kg^{1/3}$ units this corresponds to a K value of 130.12 and a blast overpressure of approximately 0.4516 kPa (0.0655 psi). To gauge mitigation effectiveness, pressure tracers were placed in a uniform grid behind the barriers in the present simulations; the pressure time histories are then queried based on the aforementioned peak pressure criterion, and a "bubble" of space satisfying the maximum desired conditions can be plotted.

Figures 12 and 13 show the analysis of a single fairing mitigation barrier interacting with a wave from 0.06123 kg (0.135 lb) of TNT at approximately 36.576 meters (120 ft). The peak incident overpressure in this case is approximately 0.95 kPa (0.137 psi). Figure 12b shows that in the plane strain case, the pressure is effectively mitigated behind the barrier below 0.4516 kPa for a region over 2 m high and extent of almost 5 m. Figure 13b shows than in the 3D case with a barrier of finite width, there are small localized regions near the edges, center, and ground where edge wraparound and reflections exceed this pressure threshold.

Figures 14 and 15 show the results of plane strain analysis from a 0.45359 kg (1 lb) TNT charge at approximately 36.576 m (120 ft) interacting with a compound fairing and 'deep roof' type barrier. The peak incident overpressures in this case is approximately 2.38 kPa (0.345 psi). In both cases, the incident pressure wave is partially mitigated, so there are still large regions behind the barrier seeing pressures larger than 0.4516 kPa. The 'deep roof' style barrier provides a large 'bubble' for pressures under 0.4516 kPa. It is worth noting that in all cases, the largest pressures behind the barrier occur when the wave which passes over the top reflects back off the ground and the back of the barrier. The incident wave over the barrier has been mitigated below the target pressure, but the reflections exceed it.

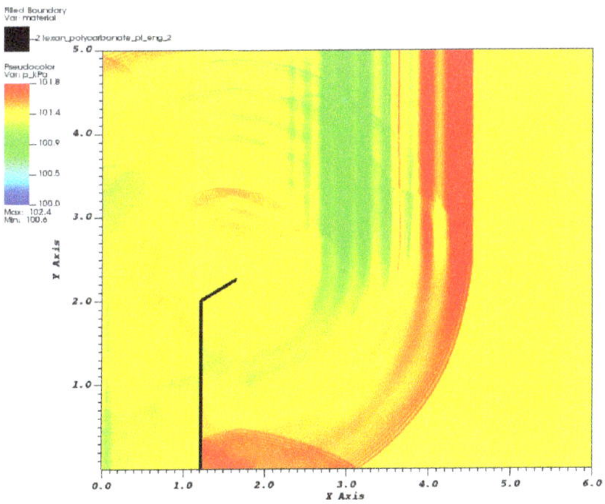

(a)

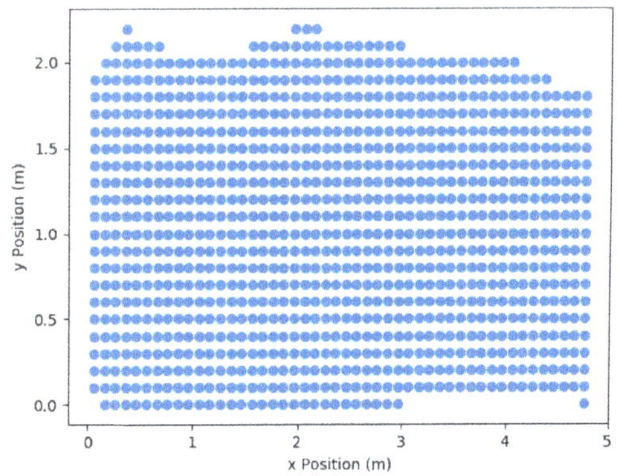

(b)

Figure 12. Results from a plane strain analysis of a single-fairing mitigation barrier loaded by a wave generated from 0.06123 kg of TNT at approximately 36.576 m. (**a**) A snapshot of the wave reflecting over the barrier. The color gradient is set so that max (red) values are above the 0.4516 kPa overpressure threshold. (**b**) The "bubble" behind the barrier for which max overpressure was beneath 0.4516 kPa.

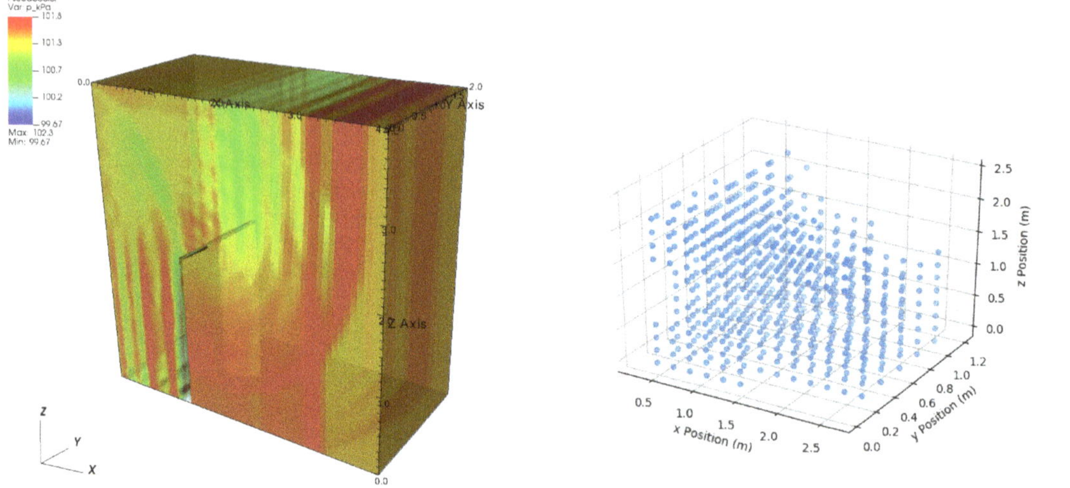

(a) (b)

Figure 13. Results from a 3D analysis of a single-fairing mitigation barrier loaded by a wave generated from 0.06123 kg of TNT at approximately 36.576 m. (**a**) A snapshot of the wave reflecting over the barrier. The color gradient is set so that max (red) values are above the 0.4516 kPa overpressure threshold. (**b**) The "bubble" behind the barrier for which max overpressure was beneath 0.4516 kPa.

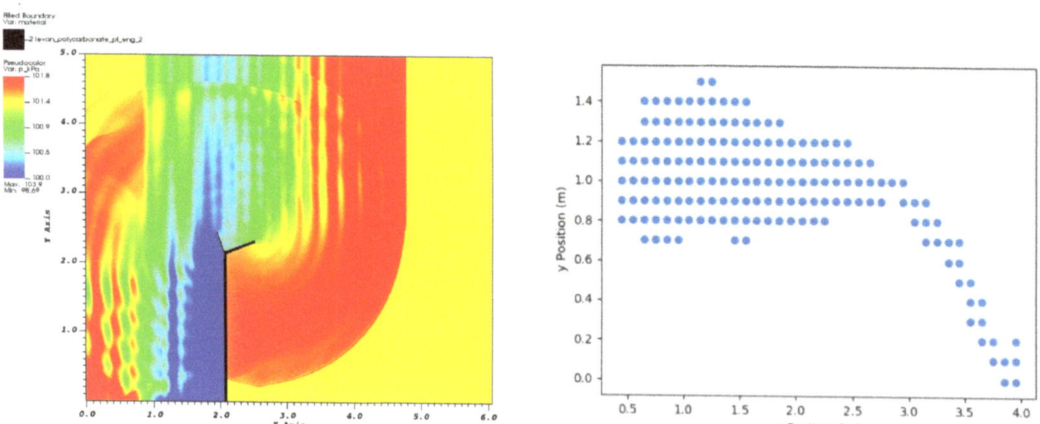

(a) (b)

Figure 14. Results from a plane strain analysis of a compound-fairing mitigation barrier loaded by a wave generated from 0.45359 kg of TNT at approximately 36.576 m. (**a**) A snapshot of the wave reflecting over the barrier. The color gradient is set so that max (red) values are above the 0.4516 kPa overpressure threshold. (**b**) The "bubble" behind the barrier for which max overpressure was beneath 0.4516 kPa.

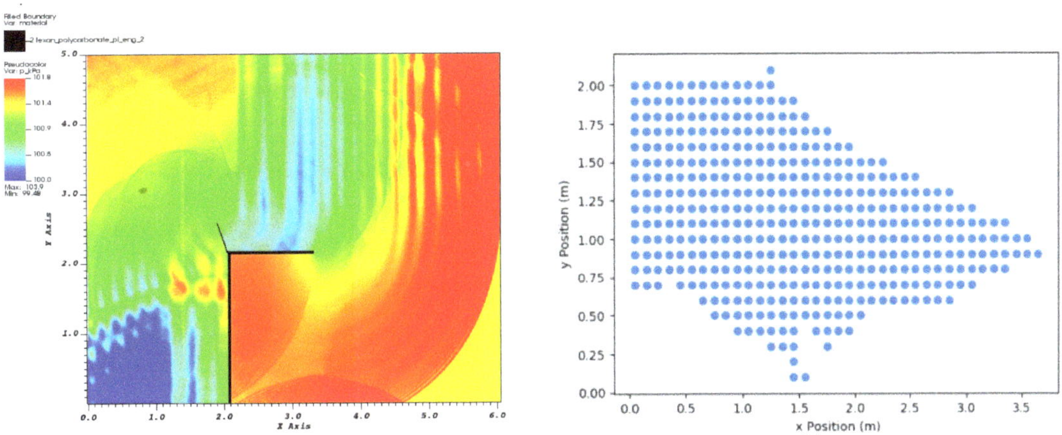

(**a**) (**b**)

Figure 15. Results from a plane strain analysis of a 'deep-roof' type mitigation barrier loaded by a wave generated from 0.45359 kg of TNT at approximately 36.576 m. (**a**) A snapshot of the wave reflecting over the barrier. The color gradient is set so that max (red) values are above the 0.4516 kPa overpressure threshold. (**b**) The "bubble" behind the barrier for which max overpressure was beneath 0.4516 kPa.

4. Discussion

The calculated incident peak overpressures from the free-field hemispherical blast simulations show good agreement with the KB predictions in Figure 10. Interestingly, the largest deviation from the KB curve appears to at the points closest to the explosions yielding the largest overpressure; these first three points lie closer to the von-Neumann-Taylor-Sedov prediction, which rapidly deviates from the Swisdak (KB) curve. The analytical prediction is only valid at an intermediate distance from large explosions; it breaks down near the explosion, as the point source assumption washes out details of the actual detonation event, but also in the very far-field, where the assumption that $p_0 = 0$ in the ambient gas begins to corrupt the results as the incident overpressure approaches the ambient atmospheric pressure. Since the deviation of the data points from the empirical curve is likely within the experimental errors of the original fits, the fact that the data seem to jump from the analytical to the empirical curves may be coincidental. In the very far-field, the KB predictions seem to be doing a reasonable job at predicting the calculated overpressures, despite the fact that it is fit to data from explosions that were orders of magnitude larger.

As expected, the KB predictions do not do well at predicting the overpressures near the detonation cord. The asymmetrical blast wave from a long, thin cylindrical cord lit at one end reaches a nearby point at different times, making the peak pressure smaller than that predicted from a localized (hemispherical) source. However, with greater distance this time delay becomes smaller and the data appears to converge onto the KB curve.

There is a larger discrepancy between the KB-predicted and calculated incident impulses in Figure 11. There is very good agreement in the slope of the data versus the curve, but the free-field simulations appear to uniformly under-predict the impulse relative to the KB curve by a relatively small amount. Given the better agreement in the peak overpressures, there may be some discrepancy in the shape or duration of the whole incident pressure wave. The source of this error could be numerical or physical. There may be low pressure effects to the waves which we not captured in the very large Kingery tests. Note the oscillations that appear in the smaller pressure time histories in Figure 3. These appear after the sharp shock-like pressure spikes decay into more smooth waves traveling at sound speed. While these oscillations could be numerical effects, subsequent calculations were conducted to investigate this by changing mesh size and the position of the tracer nodes, which appeared to have no effects on the oscillations at distance. Thus it is possible that the oscillations in the far-field small pressure waves is in fact a physical phenomenon. This could in part explain the discrepancy in impulse when there is good agreement in the peak pressure magnitude. It is noted that similar but less pronounced oscillations also seem to appear in the farthest-field pressure histories in the works of Xue et al., Ding et al. [20,22]. There do not appear to be other curves available from similarly small charges at distance to compare with Figure 4. Taking experimental measurements of incident pressures and impulses much smaller that atmospheric pressure very far away from small explosives is quite difficult.

It is further noted that the overall predictive accuracy in the subsequent blast-mitigation dynamic simulations is in part dependent on the accuracy of the predicted incident waves. At 36.576 m (120 ft), only the wave from the smallest (0.06123 kg) charge was mitigated down below the "Public Withdrawal Distance" value of 0.4516 kPa consistency behind the barrier under plane strain conditions; a large bubble was confirmed under a larger 3D simulation of a single barrier, though there were small regions near the edge and center where pressure rose higher in this case. The practical suggestion gleaned from this is that when implementing this type of barrier it may be wise to include more than one side by side to approximate the plane strain condition.

It is noted that the relevant hydrodynamics effects are likely more accurately captured in the 3D simulations. For example, mixing and turbulence are fundamentally 3D phenomena. Furthermore, resolution of any smaller-scale effects is inherently limited by the resolution of the simulation at those scales. However, due to the relatively low velocities and pressures these factors are not thought to have much influence in the cases studies here. Recent work has shown that purposefully exploiting wave interference can be useful in blast mitigation for incident strong shocks [32].

Finally, the effectiveness of using TNT equivalence values to compare expected blast effects from different explosives depends on the situation [33]. The present work has employed only a simple model of TNT with the simplest numerical detonation/burn assumptions. This seemed appropriate when gauging effects in the far-field, when the incident waves are sufficiently decoupled from the nuances of the blast and the blast products. Further work should be conducted to verify the accuracy of the KB charts and the predictions made here with other types of explosive, as well as to simulations with more sophisticated burn models (e.g., ignition and growth [34]).

5. Conclusions

The present manuscript lays out two open problems (namely, what incident overpressure and impulses are felt at given distances from relatively small hemispherical ground charges, and how well can certain types of boundaries mitigate the incident overpressure below a certain threshold). It then describes the results of numerical investigations to attempt to answer these questions. A major motivating factor in this research is the uncertainty in the available empirical curve fits (e.g., Kingery–Bulmash). The source of this uncertainty is twofold: there is relatively large error between some of the original data and the available fitting curves, and the original data were taken for explosions that were many orders of magnitude larger than the charges investigated here. A major unknown remains the extent to which the assumed scaling described by Equation 1 (distance by the cubed root of charge weight) holds as weights become small. The free-field blast simulations presented here indicate that the strong shock of the initial blast smoothed out within the distance simulated and continued to propagate near the sound speed. The slowing of the wave speed is in fact predicted by the empirical Swisdak (KB) equations, but there remains uncertainty into how this change in the physics regime and the shape of the waveform effects the ultimate impulse at different scales. A benefit of the direct numerical calculations is the availability of the full waveforms in time at all distances in the simulation domain; this was further leveraged in the subsequent mitigation simulations. The ultimate shape of the incident pressure wave may be another degree of freedom which is not fully captured by the K factor scaling. This may explain why the simulations agree well with the peak overpressure and the slope of the impulse curves from Swisdak, but seem to consistently predict slightly smaller impulse magnitudes.

It was never assumed that the empirical curves would or should be "exact" predictors of incident overpressure and impulse for a given case. While this was a primary motivator for the present attempts for a direct physics-based prediction, it is also not assumed that these predictions will correspond exactly to any field case. Ultimately, the analyst, engineer, or responsible person must weigh uncertainty and risk to assess a given scenario. It is hoped that the present simulations (or others like them) could be used in uncertainty quantification efforts for blasts effects in wider-varying scenarios.

The blast-barrier mitigation simulations presented here were also motivated by this desire to mitigate risk and uphold safety standards. The "Public Withdrawal Distance" or "K328" threshold was taken as a more-or-less arbitrary datum against which to gauge effectiveness. The findings of this work should not be used to indicate whether a given scenario is "safe", but rather to elucidate some of the physical mechanisms of mitigation in a dynamic blast event. Safety standards and acceptable risk vary from scenario to scenario; this work provides a methodology of analyzing the effectiveness of hypothetical tools to decrease risk.

The specific barrier designs presented here were somewhat ad hoc and experimental. The fairings were designed to reflect incident waves and further mitigate overpressure from wraparound over the top. The double fairing was intended to facilitate mitigation further by partially reflecting the incident wave from the backward-facing fairing. The "deep roof" concept was designed to provide even further mitigation. Each subsequent design was found to enhance mitigation. The results indicate that these types of simple barriers are in fact effective at mitigating incident pressure and impulse. They do not, however, eliminate these risks. Ultimately, distance from the source is the surest form of mitigation.

All conclusions herein would be much strengthened by specific field test data taken from experiments with the same charge weights and at the same distances, both in the free-field and behind the proposed barriers. As far as the author knows, no data exists that is a direct match for the scenarios described here. Currently available state-of-the-art instrumentation may be able to reliably measure the small dynamic pressures considered in this study. The experimental verification of these scenarios is outside of the scope of the present work.

Funding: This research received no external funding.

Data Availability Statement: Not applicable.

Acknowledgments: This work was performed under the auspices of the U.S. Department of Energy by Lawrence Livermore National Laboratory under Contract DE-AC52-07NA27344. Special thanks to Mark M. Hart who posed the present problem and provided crucial expertise, intuition, and feedback. Further Thanks to Larry D. McMichael for guidance and expertise in setting up the numerical simulations.

Conflicts of Interest: The authors declare no conflict of interest.

Abbreviations

The following abbreviations are used in this manuscript:

1D	One-dimensional
2D	Two-dimensional
3D	Three-dimensional
ALE	Arbitrary Lagrangian/Eulerian
KB	Kingery–Bulmash

References

1. Kingery, C. *Air Blast Parameters Versus Scaled Distance for Hemispherical TNT Surface Burst*; Technical Report 1334; U.S. Army BRL: Harford County, MD, USA, 1966.
2. Kingery, C.; Pannill, B. *Peak Overpressures vs. Scaled Disatnce for TNT Surface Bursts (Hemispherical Charges)*; Technical Report BRL Memorandum Report 1518; U.S. Army BRL: Harford County, MD, USA, 1964.
3. Kingery, C.G.B. *Airblast Parameters from TNT Spherical Air Burst and Hemispherical Surface Burst*; Technical Report ARBRL-TR-02555; U.S. Army BRL: Harford County, MD, USA, 1984.
4. Swisdak, M. *Simplified Kingerly Airblast Calcualtions*; Technical Report; Naval Surface Warfare Center Indian Head Div: Silver Spring, MD, USA, 1994.
5. Jeon, D.; Kim, K.; Han, S. Modified Equation of Shock Wave Parameters. *Computation* **2017**, *5*, 41. [CrossRef]
6. Karlos, V.; Solomos, G.; Larcher, M. Analysis of the blast wave decay coefficient using the Kingery–Bulmash data. *Int. J. Prot. Struct.* **2016**, *7*, 409–429. [CrossRef]
7. Karlos, V.; Solomos, G. *Calculation of Blast Loads for Application to Structural Components*; Technical Report EUR 26456EN, JRC Technical Report; Publications Office of the European Union: Luxembourg, 2013.
8. Karlos, V.; Solomos, G. *Analysis of Blast Parameters in the Near-Field for Spherical Free-Air Explosions*; Technical Report EUR 27823EN, JRC Technical Report; Publications Office of the European Union: Luxembourg, 2016.
9. Ullah, A.; Ahmad, F.; Jan, H.; Kim, S.; Hong, J. Review of analytical and empirical estimations for incident blast pressure. *KSCE J. Civ. Eng.* **2017**, *21*, 2211–2225. [CrossRef]
10. Stewart, M.G.; Netherton, M.D.; Baldacchino, H. Observed airblast variability and model error from repeatable explosive field trials. *Int. J. Prot. Struct.* **2020**, *11*, 235–257. [CrossRef]
11. Filice, A.; Mynarz, M.; Zinno, R. Experimental and Empirical Study for Prediction of Blast Loads. *Appl. Sci.* **2022**, *12*, 2691. [CrossRef]
12. Rigby, S.; Tyas, A.; Fay, S.; Clarke, S.; Warren, J. Validation of semi-empirical blast pressure predictions for far field explosions—Is there inherent variability in blast wave parameters? In Proceedings of the 6th International Conference on Protection of Structures against Hazards, Tianjin, China, 16–17 October 2014.
13. Taylor, G.I. The formation of a blast wave by a very intense explosion I. Theoretical discussion. *Proc. R. Soc. Lond. A* **1950**, *201*, 159–174. [CrossRef]
14. Taylor, G.I. The formation of a blast wave by a very intense explosion—II. The atomic explosion of 1945. *Proc. R. Soc. Lond. A* **1950**, *201*, 175–186. [CrossRef]
15. von Neumann, J. The point source solution. In *Collected Works*; Pergamon: New York, NY, USA, 1963; Volume 6; pp. 219–237.
16. Sedov, L. Propagation of strong shock waves. *J. Appl. Math. Mech.* **1946**, *10*, 241–250.
17. Brode, H. Numerical Solutions of Spherical Blast Waves. *J. Appl. Phys.* **1954**, *26*, 766–775. [CrossRef]
18. Brode, H. Blast Wave from a Spherical Charge. *Physcis Fluids* **1959**, *2*, 217–229. [CrossRef]
19. Goldstine, H.H.; von Neumann, J. Blast Wave Calculation. *Commun. Pure Appl. Math.* **1955**, *8*, 327–354. 1.1705911. [CrossRef]
20. Ding, Y.; Zhang, X.; Shi, Y.; Zhang, H. Prediction of far-field blast loads from large TNT-equivalent explosives on gabled frames. *J. Constr. Steel Res.* **2022**, *190*, 107120. [CrossRef]
21. Zhang, X.; Ding, Y.; Shi, Y. Numerical simulation of far-field blast loads arising from large TNT equivalent explosives. *J. Loss Prev. Process. Ind.* **2021**, *70*, 104432. [CrossRef]
22. Xue, Z.; Li, S.; Xin, C.; Shi, L.; Wu, H. Modeling of the whole process of shock wave overpressure of free-field air explosion. *Def. Technol.* **2019**, *15*, 815–820. [CrossRef]
23. Sung, S.H.; Chong, J.W. A fast-running method for blast load prediction shielding by a protective barrier. *Def. Technol.* **2020**, *16*, 308–315. [CrossRef]
24. Godio, M.; Williams Portal, N.; Flansbjer, M.; Magnusson, J.; Byggnevi, M. Experimental and numerical approaches to investigate the out-of-plane response of unreinforced masonry walls subjected to free far-field blasts. *Eng. Struct.* **2021**, *239*, 112328. [CrossRef]
25. Vannucci, P.; Masi, F.; Stefanou, I. A Comparative Study on the Effects of Blast Actions on a Monumental Structure. Doctoral Dissertation, UVSQ, Versailles, France, 2017.
26. Draganić, H.; Varevac, D. Analysis of Blast Wave Parameters Depending on Air Mesh Size. *Shock Vib.* **2018**, *2018*, 3157457. [CrossRef]
27. Nichols, A.; Dawson, D. *ALE3D User's Manual*; Technical Report LLNL-SM-800962; Lawrence Livermore National Laboratory: Livermore, CA, USA, 2020.
28. Menikoff, R. *JWL Equation of State*; Technical Report LA-UR-15-29536; Los Alamos National Laboratory: Livermore, CA, USA, 2017.
29. Krauthammer, T.; Altenberg, A. Negative phase blast effects on glass panels. *Int. J. Impact Eng.* **2000**, *24*, 1–17. [CrossRef]
30. Rigby, S.E.; Tyas, A.; Bennett, T.; Clarke, S.D.; Fay, S.D. The Negative Phase of the Blast Load. *Int. J. Prot. Struct.* **2014**, *5*, 1–19. [CrossRef]
31. Landau, L.; Lifshitz, E. *Fluid Mechanics*; Course in Theoretical Physics 6. Translated from the Russian by JB Sykes and WH Reid; Pergamon: New York, NY, USA, 1987; pp. 403–406.

32. Zong, R.; Hao, H.; Shi, Y. Development of a New Fence Type Blast Wall for Blast Protection: Numerical Analysis. *Int. J. Struct. Stab. Dyn.* **2017**, *17*, 1750066. [CrossRef]
33. Cooper, P.W. Comments on TNT Equivalence. In Proceedings of the 20th International Pyrotechnics Seminar, Colorado Springs, CO, USA, 25–29 July 1994.
34. Lee, E.L.; Tarver, C.M. Phenomenological model of shock initiation in heterogeneous explosives. *Phys. Fluids* **1980**, *23*, 2362–2372. [CrossRef]

applied sciences

MDPI

Article

Empirical Attenuation Law for Air Blast Waves Due to the Detonation of Explosives Outdoors

Juan Andrés Marín, Rafael Rodríguez *, María B. Díaz and Saray Antón

Department of Mining Exploitation and Prospecting, School of Mining, Energy and Materials Engineering, University of Oviedo, Independencia 13, 33004 Oviedo, Spain
* Correspondence: rrodrifer@uniovi.es

Abstract: The detonation of explosives in the open air was studied, analyzing different amounts of explosives detonated at different distances, monitoring the overpressure or air blast wave generated with the aim of determining a model, which allows to establish safety zones. A series of tests measuring the air wave with different loads and sensors placed at various distances from the origin of the explosion were carried out. The work was focused on designing full-scale trials that allowed to develop a predictive empirical method based on the calculation model of the equivalent mass of TNT. A total of 18 different gelatinous dynamite charges, placing the sensor at six different distances from the origin of the explosion, produced a total of 90 tests measuring the air wave produced by the detonation of gelatinous dynamite. Later, the outdoor detonation of 10 TNT explosive charges was analyzed to extend the model and improve its scope. With all this, it has been possible to develop a predictive model that allows assessing the overpressure generated by the detonation of a TNT-equivalent explosive charge. The results are useful to predict the air blast wave in common open-air blasts, such as those carried out with shaped charges to demolish metallic structures. On the other hand, the results are also useful to determine the air blast wave overpressure in the case of large explosive charges detonated in the open air, such as accidental explosive detonation or terrorist bombs. It is important to point out the relevance of the results achieved after the detonation of large explosive charges (more than 80 kg) simulating a type of bomb frequently used by terrorists. Reproducing the explosion on a real scale, the results are fully representative of the overpressure produced by an explosion of these characteristics without the need of extrapolating the results of tests with small loads. In addition, the detonation was carried out with TNT, which can serve as a standard to compare with any other type of explosive.

Keywords: detonation; TNT; dynamite; air blast wave; overpressure

Citation: Marín, J.A.; Rodríguez, R.; Díaz, M.B.; Antón, S. Empirical Attenuation Law for Air Blast Waves Due to the Detonation of Explosives Outdoors. *Appl. Sci.* **2022**, *12*, 9139. https://doi.org/10.3390/app12189139

Academic Editors: Lina M. López, Ricardo Castedo and Anastasio P. Santos

Received: 20 July 2022
Accepted: 9 September 2022
Published: 12 September 2022

1. Introduction

1.1. Air Blast Wave

An explosion is a physical phenomenon in which there is a sudden, very rapid release of energy. The phenomenon lasts only some milliseconds, and it results in the production of gas with very high temperature and pressure. During detonation, the hot gases that are produced expand in order to occupy the available space, leading to wave-type propagation through space that is transmitted spherically through an unbounded surrounding medium. Along with the produced gases, the air around the blast (for air blasts) also expands, and its molecules pile up, resulting in what is known as a blast wave and shock front. The blast wave contains a large part of the energy that was released during detonation and moves faster than the speed of sound [1].

This shock wave is characterized by an abrupt pressure rise followed by a relatively slow decrease to a value below atmospheric pressure and with a subsequent return to the positive value [1,2]. This phenomenon, which initially takes a few milliseconds, depends on the explosive mass and the distance to the initiation of the explosion. Subsequently, this waveform derived in a series of damped oscillations.

The study of the air wave produced by the detonation of explosives in the open air inevitably requires analyzing different controlled detonations and measuring the different parameters that characterize the air wave. This experimental level is not at all easy in the civil sphere, since the detonation of explosive substances involves having the availability of both the explosive and the initiator and the appropriate place to carry out the different detonations without affecting the surrounding environment—people, buildings, and communication ways.

The most characteristic effect of an explosion is the sudden increase in pressure that happens in the surrounding air, which propagates in the form of a spherical wave in all directions. The shape, characteristics, and magnitude of the wave depend on the type of explosion, the environment, and the distance from the origin where it was generated.

If the explosion takes place at a point far from the ground, the blast wave expands spherically, and its characteristics (maximum overpressure, duration, impulse, arrival time, etc.) are known as open-air explosion parameters. If the explosion occurs in the vicinity of the ground or on it, the parameters are known as surface explosion. In the first, any point will be affected by two shock waves: first, the incident one from the explosion and then the one reflected from the ground. In the second, the reflection on the ground is linked to the incident wave from the point of explosion, forming a single practically hemispherical wave, whose amplitude, for the same mass of explosive, is considerably greater than in the first case, since the energy must be distributed only in one hemisphere.

1.2. Negative Effects of Air Blast Wave

The air blast wave is an undesirable side effect that occurs in any explosive detonation and consequently has to be studied. The study of the air blast wave due to explosive detonation has been carried out in the last decades from two points of view.

One is the safety point of view, and the other is the environmental impact. The air blast wave is studied from the safety point of view because it has a great destructive effect within a radius that depends on the amount of explosive detonated.

During the second half of the 20th century, a considerable number of experimental and theoretical studies were conducted to understand the effects of blast on buildings and structures [3–7]. The aim was first to study the behavior of air blast waves including the determination of their characteristics and then to investigate the dominant factors influencing the incident waves. Another objective was to investigate the response of the building structure to blast loads [8–13].

The damage caused by the air waves on the structures depends on the overpressure, the impulse, and the formation of projectiles. The level of severity is also influenced by the orientation with respect to the direction of advance of the wave, the geometry of the structure (height/length ratio), and the construction materials. For emergency planning, it is interesting to consider inhabited buildings, due to the greater severity of the consequences.

When a shock wave reaches a structure, it is reflected, with an overpressure at least double that of the incident wave. The wave continues its propagation, reaching a moment in which the entire structure is encompassed by the wave. The explosions produced on the surface cause practically horizontal loads on the structures that they find in their path (except on the roof).

If the structure is small, with few openings, the load results in a homogeneous compression of it; if the structure is large, the load will be markedly different at the front and at the rear, with a greater potential for damage. The existence of openings or the breakage of some part of the structure will result in the homogenization of the pressure between the interior and the exterior of the structure. The calculation of the loads on a structure is carried out by combining the incident pressure and the dynamic pressure and their duration.

Actually, the response of a structure depends not only on the incident overpressure but also on the impulse (which takes into account the duration of the pressure pulse).

In the case of blasting in which the explosive is confined, it generates an air wave with a large proportion of low frequencies that can induce vibrations in buildings, although they are not heard because they are infrasonic. In any case, the effects of the air wave produced by a confined explosive are rarely harmful except in remote cases of glass breakage.

On the other hand, the air blast wave has been extensively studied from the environmental protection point of view. The air blast wave, even of a small intensity, can produce negative effects near the blasting areas. It is very typical of blasting related to mining (quarries or open-pit mines) or civil works (excavation or demolition). For example, the air blast wave can negatively influence the wildlife, which is critical in the case of protected animal species. In the same way, the air blast wave can produce different negative effects on population, from complaints of the neighbors of a village, to small damages to buildings, such as glass breakage or displacement of some tiles on the roof.

1.3. Empirical Prediction Models

Because of the importance of assessing the magnitude of the air blast wave, a lot of prediction models to determine explosion parameters, mainly overpressure, have been developed. These can be based on empirical (or analytical), semiempirical or numerical methods. Empirical methods are essentially correlations with experimental data. Most of these approaches are limited by the range of experiments carried out. The accuracy of all empirical correlations decreases with distance to the source of the explosion.

The use of empirical laws has been extensively studied and has been applied in various recommendations, mostly proposed by military authorities. After the first attempt due to Cranz [14], several methods were proposed [3–7], and due to the relevance of the topic recently, works about this topic have been published [15–17].

In the field of mining and civil engineering, several empirical models have also been proposed to estimate the magnitude of the air blast overpressure as for example [18–20].

In many cases, the air blast wave is given as a function of the scaled distance Z (in m/kg$^{1/3}$):

$$Z = \frac{R}{W^{1/3}} \tag{1}$$

R (m) is the distance from the explosion to the measurement point, and W (kg) is the amount of explosive detonated.

In order to be able to characterize the wave generated by any explosive substance and to be able to compare them with each other to assess their harmful effects after a detonation in the open air, it was important to establish a base explosive. The selected explosive was the Trinitrotoluene (TNT), which has well-known explosive properties. The TNT-equivalent mass is the mass of Trinitrotoluene (TNT) that would release an amount of energy equal to the explosive charge in question. If there is a mass W of a given explosive with an explosion heat Q, the equivalent TNT mass W_{eq} is:

$$W_{eq} = W\,\frac{Q}{Q_{eq}} \tag{2}$$

where Q_{eq} is the explosion heat of TNT Q_{eq} = 4520 kJ/kg.

The relationship (2) is widely accepted for blast-resistant design. It is proposed in documents taken as a reference or guides, such as UFC 3-340-02 [21] or EUR 2645EN [22], which allow to determine the incident and reflected overpressures and impulses of a spherical or hemispherical TNT explosion.

1.4. Research and Objectives

The detonation of explosives in the open air has been studied, analyzing amounts of explosive material and distances at which it detonates, with the aim of establishing safety zones, which implies previously determining the primary characteristic variables, as the air blast wave level.

Experimentation in this field presents great technical and economic difficulties, which is why most evaluations are carried out by extrapolation from small-scale experiences or from computer model results.

In the present study, two sets of full-scale tests were carried out. The first with small/medium explosive charges from 0.2 to 7 kg and the second trial with a large amount of explosive, from 25 to 84 kg (simulating terrorist bombs).

Two factors were taken into account that will fundamentally influence it: the explosive charge and the distance to the focus of the explosion.

To test the influence of these two factors, a campaign of air wave measurement tests was carried out with different charges and with sensors placed at different distances from the point of the explosion. With these tests, the intention was to obtain a model to predict the overpressure or magnitude of the air blast wave that is one of the factors influencing negatively on the environment and, in extreme cases, the main factor that affects the structures in outdoor detonations.

The works were focused on the design of a full-scale test procedure that would allow the development of a predictive empirical method based on the model for calculating the equivalent mass of TNT.

A total of 18 different Riodin explosive charges were formed, placing the sensor at six different distances from the focus of the explosion, with which a series of campaigns were carried out with a total of 90 air wave measurement tests produced by the detonation of gelatinous dynamite. With the results obtained, the pertinent adjustment of the TNT-equivalent mass calculation model was carried out, which was used to predict the effects generated by the air blast wave in the simulation processes of predefined scenarios.

Subsequently, the outdoor detonation of 10 TNT charges was analyzed in order to adjust the model and determine its range. Therefore, the results obtained in this work from the measurement of the air wave pressure peak in 100 full-scale tests are presented and analyzed, in which industrial and military explosives were detonated in the open air, without confinement, in different amounts, the highest that the environment allows without affecting people, communication routes, or buildings, which will conclude with the proposal of a calculation methodology based on the experience.

With all this, it was possible to develop a predictive model that allows assessing the overpressure generated by the detonation of a TNT-equivalent explosive charge. The results are useful to predict air blast waves in common open-air blasts, such as those carried out with shaped charges to demolish metallic structures. On the other hand, the results are also useful to determine the air blast wave overpressure in the case of large explosive charges detonated in the open air, such as accidental explosive detonation or terrorist bombs.

It is important to point out the relevance of the results achieved after the detonation of large explosive charges (more than 80 kg) simulating a type of bomb frequently used by terrorists. Reproducing the explosion on a real scale, the results are fully representative of the overpressure produced by an explosion of these characteristics without the need to extrapolate the results of tests with small loads. In addition, the detonation was carried out with TNT, which can serve as a standard to compare with any other type of explosive.

2. Materials and Methods

2.1. Equipment

For this research, the equipment used for data collection was an Instantel seismograph, Minimate Plus model, which has a channel for a microphone. It is a piece of equipment for monitoring vibrations and overpressure widely used in mining and civil works. Due to the wide range of acoustic pressure values measured, two different microphones were used for data collection. One is the microphone for air overpressure monitoring, which is supplied by default with the Minimate Plus seismograph; it is of the linear or A-weight type (see

Table 1). The other is a high-pressure microphone, which allows to measure pressure waves higher and can reach up to 69 kPa (Table 2).

Table 1. Instantel linear microphone characteristics used to measure air overpressure.

Scale type	Linear or A
Linear range	88 to 148 dB (500 Pa)
Linear resolution	0.25 Pa
Linear accuracy	+/−10% or +/−1 dB, whichever the higher, between 4 and 125 Hz
Linear frequency response	2 a 250 Hz between −3 dB points of roll off
A range	50–110 dBA
A resolution	0.1 dBA

Table 2. Instantel high-pressure microphone characteristics used to measure air overpressure.

Sensitivity	0.0233 V/kPa
Pressure range	0.0345 kPa to 69 kPa
Frequency response	5 to 1000 Hz

2.2. *First Tests: Air Detonation of Dynamite Charges*

The tests consisted of measuring the pressure wave or shock wave produced in a total of 90 explosions of different charges of a commercial explosive. These tests were carried out in the facilities of the Santa Bárbara Foundation, a public nonprofit foundation that works on training and R&D, always acting within the field of applied technology, safety, and technological progress. The foundation has several schools; one of them is located in the municipalities of Folgoso de la Ribera and Torre del Bierzo (León) where the trial was carried out.

For these first tests, gelatinous dynamite was used, specifically Riodin from the Maxam explosives manufacturer. The gum dynamite has a gelatinous consistency due to the greater amount of nitrogelatin in its composition (nitroglycerin/nitroglycol and nitrocellulose; >22%), and a predominant element is the ammonium nitrate. This mixture is even more energetic than nitroglycerin itself. This consistency of the explosive gives it, in general, an excellent resistance to water, as well as a high density. These characteristics, together with their high power and detonation speed, make them suitable for blasting rocks of a medium/high hardness, as well as for bottom loading holes and being essential for underwater blasting. Table 3 shows the main characteristics of Riodin. In order to obtain the amount of dynamite desired, cartridges of 26 mm and 32 mm in diameter (both 200 mm in length) were used in the tests.

Table 3. RIODIN main characteristics.

Packing density	1.45 g/cm^2
Detonation speed	6000 m/s
Heat of explosion at constant volume	4.09 MJ/kg
Gas volume produced	895 L/kg
Residual fume quality	Less than 2.27 L/100 g

To analyze the influence of the two more influencing factors, explosive dynamite charge and distance, a total of 90 airwave measurement tests were carried out. The distances and charges of Riodin-type gelatinous dynamite for each individual test are shown in Table 4.

Table 4. Riodin charge and distance for each test.

Num.	Distance (m)	Charge (kg)	Num.	Distance (m)	Charge (kg)	Num.	Distance (m)	Charge (kg)
1	25	0.238	**31**	25	3.571	**61**	15	3.571
2	25	0.714	**32**	25	4.286	**62**	25	3.571
3	25	1.190	**33**	25	4.762	**63**	40	3.571
4	25	1.190	**34**	25	5.476	**64**	50	3.571
5	25	1.190	**35**	25	5.952	**65**	75	3.571
6	25	2.381	**36**	25	6.667	**66**	15	4.762
7	25	3.571	**37**	25	7.143	**67**	15	5.952
8	25	4.762	**38**	75	2.381	**68**	15	7.143
9	25	5.952	**39**	75	1.190	**69**	15	5.952
10	25	7.121	**40**	75	0.714	**70**	15	4.762
11	25	4.762	**41**	50	2.381	**71**	10	3.571
12	10	2.381	**42**	50	1.190	**72**	10	2.381
13	10	3.571	**43**	50	0.714	**73**	10	1.190
14	15	3.571	**44**	40	2.381	**74**	10	4.762
15	15	4.762	**45**	40	1.190	**75**	25	2.381
16	15	5.952	**46**	40	0.714	**76**	25	3.571
17	25	2.381	**47**	25	2.381	**77**	10	4.762
18	25	2.381	**48**	25	1.190	**78**	10	1.190
19	25	3.571	**49**	25	0.714	**79**	15	1.667
20	25	3.571	**50**	15	2.381	**80**	15	2.381
21	25	4.762	**51**	15	1.190	**81**	25	1.905
22	25	4.762	**52**	15	0.714	**82**	25	3.095
23	25	3.550	**53**	25	0.714	**83**	25	3.571
24	25	0.238	**54**	25	1.190	**84**	25	3.571
25	25	0.476	**55**	25	2.381	**85**	25	4.762
26	25	0.714	**56**	15	1.667	**86**	25	5.714
27	25	1.190	**57**	25	1.905	**87**	25	5.714
28	25	1.905	**58**	40	1.905	**88**	25	5.714
29	25	2.381	**59**	50	1.905	**89**	25	5.714
30	25	3.095	**60**	75	1.905	**90**	25	4.286

2.3. Second Trial: Air Detonation of TNT Charges

The second tests consisted in measuring the pressure wave or shock wave produced in a total of 10 explosions with large charges of TNT.

The test was carried out at the "San Gregorio" Training Center, belonging to the Spanish Army (the General Military Academy, Zaragoza, Spain), which is located in the province of Zaragoza. It is the third largest training site in Europe.

The explosive chosen to be detonated in the open air was TNT. It is a light yellow, solid with a bitter taste, and it is less poisonous than other explosive substances. It has great chemical stability and very little sensitivity to shock. It is not affected by humidity, but by light, under whose action it acquires a dark color. Exposure to sunlight can cause sensitive alterations, and it burns without exploding, producing dense black smoke, unless stored in large quantities. It is the best of military explosives. It is used as a basic constituent of a multitude of explosive mixtures in the loading of projectiles, firecrackers, and multipliers. Its detonation speed is around 7000 m/s.

The mass and configuration of the explosive charge were typical of bombs used by terrorists. The handcrafted geometry of the TNT explosive is very characteristic (Table 5, Figure 1), which provides higher explosive characteristics than a normal configuration, since it deals with directed charges.

Table 5. TNT charge and distance for each test.

Num.	Distance (m)	TNT Charge (kg)
91	25	84
92	50	84
93	50	84
94	30	84
95	25	84
96	25	84
97	25	42
98	25	25
99	25	42
100	25	84

Figure 1. Directed charges of 42 kg of TNT.

Different resistant element designs were subjected to the action of the explosive detonated in the open air. These loads were raised from the ground using wooden supports, the distances at which the loads were separated from the structures between 1.5 and 3 m apart (see Figure 2).

Figure 2. Charge locations in front of the different structures.

Each of the structures was designed to withstand the effects of overpressure of a shock wave generated by the detonation of a TNT charge, directed at a given distance and different charges and separation distances depending on the structural element. The analysis of the behavior of these resistant elements is confidential, and it is out of the scope of the present work.

Nevertheless, we can say that all the results were not satisfactory or as expected. The main problem attributed by most of the calculators was the lack of full-scale tests in sufficient quantity to validate the air wave characterization models used to carry out the different designs. The importance of this air blast wave study can be then understood.

3. Results and Discussion

3.1. Results of the First Tests and Attenuation Law for the Air Overpressure Due to Common Blasts

The detonation of the 90 charges of Riodin-type gelatinous dynamite located at different distances, detailed in Table 4, was carried out on different days. For each detonation, the value of the air overpressure of the detonation was measured in a straight line and was recorded without obstacles using the high-pressure microphone.

In order to analyze the air blast wave values measured in the full-scale tests, the variable scaled distance Z ($m/kg^{1/3}$) defined by Equation (1) was used. This variable includes the influence of the two independent variables that clearly affect the value of the detonation overpressure. The calculated scaled distance and the value of the air blast wave or air overpressure for each detonation are shown in Table 6.

Table 6. Values of scaled distances and air overpressure for each detonated charge.

N	Distance (m)	Charge (kg)	Scaled Distance (m/kg$^{1/3}$)	Overpressure (kPa)	N	Distance (m)	Charge (kg)	Scaled Distance (m/kg$^{1/3}$)	Overpressure (kPa)	N	Distance (m)	Charge (kg)	Scaled Distance (m/kg$^{1/3}$)	Overpressure (kPa)
1	25	0.238	40.35	2.84	**31**	25	2.571	16.36	9.85	**61**	15	3.571	9.81	16.80
2	25	0.714	27.95	4.70	**32**	25	4.286	15.39	10.50	**62**	25	3.571	16.36	12.10
3	25	1.190	23.59	5.95	**33**	25	4.762	14.86	11.40	**63**	40	3.571	26.17	5.91
4	25	1.190	23.59	5.95	**34**	25	5.476	14.18	11.50	**64**	50	3.571	32.17	5.12
5	25	1.190	23.59	6.57	**35**	25	5.952	13.79	13.20	**65**	75	3.571	49.07	2.73
6	25	2.381	18.72	9.79	**36**	25	6.667	13.28	10.50	**66**	15	4.762	8.92	21.90
7	25	3.571	16.36	11.40	**37**	25	7.143	12.98	14.30	**67**	15	5.952	8.28	16.40
8	25	4.762	14.86	11.60	**38**	75	2.381	56.17	2.63	**68**	15	7.143	7.79	20.20
9	25	5.952	13.79	14.90	**39**	75	1.190	70.77	1.80	**69**	15	5.952	8.28	26.10
10	25	7.121	12.99	11.50	**40**	75	0.714	83.9	1.42	**70**	15	4.762	8.92	16.30
11	25	4.762	14.86	12.30	**41**	50	2.381	37.44	3.63	**71**	10	3.571	6.54	23.90
12	10	2.381	7.49	24.96	**42**	50	1.190	47.18	2.73	**72**	10	2.381	7.49	27.10
13	10	6.571	6.54	24.10	**43**	50	0.714	55.93	2.07	**73**	10	1.190	9.44	23.40
14	15	3.571	9.81	21.12	**44**	40	2.381	29.96	5.15	**74**	10	4.762	5.94	32.20
15	15	4.762	8.92	22.86	**45**	40	1.190	37.74	3.53	**75**	25	2.381	18.72	10.20
16	15	5.952	8.28	26.54	**46**	40	0.714	44.75	2.73	**76**	25	3.571	16.36	12.90
17	25	2.381	18.72	7.85	**47**	25	2.381	18.72	10.10	**77**	10	4.762	5.94	32.04
18	25	2.381	18.72	9.58	**48**	25	1.190	23.59	6.98	**78**	10	1.190	9.44	21.14
19	25	3.571	16.36	9.30	**49**	25	0.714	27.97	5.32	**79**	15	1.667	16.65	14.17
20	25	3.571	16.36	10.10	**50**	15	2.381	11.23	17.80	**80**	15	2.381	11.23	16.01
21	25	4.762	14.86	12.10	**51**	15	1.190	14.15	13.70	**81**	25	1.905	20.17	9.44
22	25	4.762	14.86	8.47	**52**	15	0.714	16.78	9.65	**82**	25	3.095	17.15	11.20
23	25	3.550	16.39	10.70	**53**	25	0.714	27.97	5.32	**83**	25	3.571	16.36	12.50
24	25	0.238	40.34	2.46	**54**	25	1.190	23.59	7.09	**84**	25	3.571	16.36	12.40
25	25	0.476	32.01	3.67	**55**	25	2.381	18.72	9.06	**85**	25	4.762	14.86	15.60
26	25	0.714	27.97	4.50	**56**	15	1.667	12.65	14.50	**86**	25	5.714	13.98	15.00
27	25	1.190	23.59	6.46	**57**	25	1.905	20.17	8.30	**87**	25	5.714	13.98	16.00
28	25	1.905	20.17	7.40	**58**	40	1.905	32.27	5.05	**88**	25	5.714	13.98	16.60
29	25	2.381	18.72	8.71	**59**	50	1.905	40.34	3.60	**89**	25	5.714	13.98	15.90
30	25	3.095	17.15	9.16	**60**	75	1.905	60.50	1.76	**90**	25	4.286	15.39	13.90

All the cases are characterized by short overpressure pulses. To illustrate it, the overpressure records obtained in tests no. 17 (S_b = 7.85 kPa) and no. 37 (S_b = 14.3 kPa) are shown in Figure 3 (left and right, respectively). The duration of the positive phase is only a few milliseconds, 5–10 ms. They are in accordance with the results of recently published research [16], keeping in mind that in our case, the explosive charge is on the floor, and consequently the overpressure is approximately twice the overpressure measured by them.

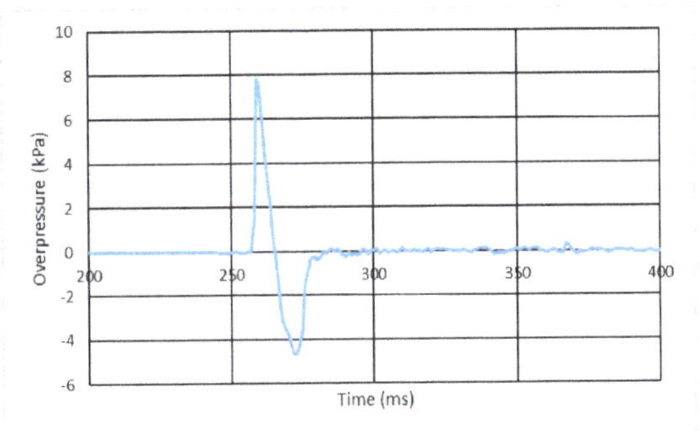

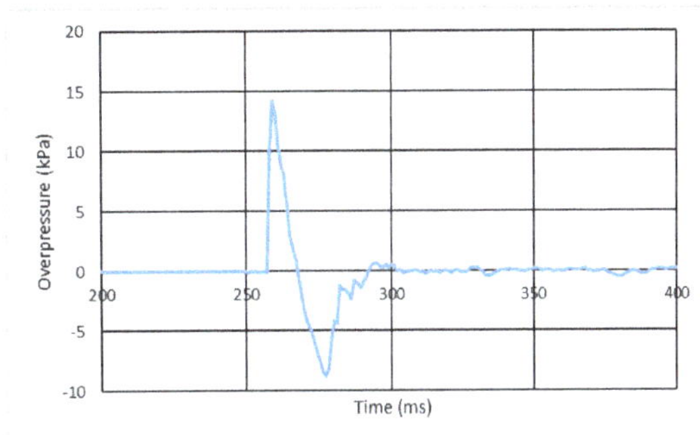

Figure 3. Air overpressure measured in tests no. 17 (**left**) and no. 37 (**right**).

The graph in Figure 4 was obtained by representing the overpressure measured at each detonation against the scaled distance in logarithmic scales. It is clear that there is a linear relationship between the log(S_b) and the log(Z), which means that there is a potential relationship between the variables S_b (kPa) and Z (m/kg$^{1/3}$). By applying logarithms and a least squares adjustment, the following relationship was found:

$$S_b = 309.33 \cdot Z^{-1.216} \tag{3}$$

with a high correlation coefficient r^2 = 0.96. This is in accordance with the first experiences in this field [10].

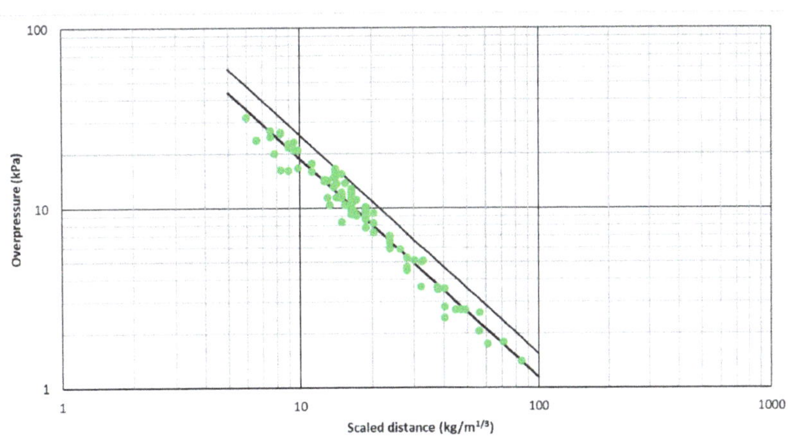

Figure 4. Air overpressure measured as a function of scaled distance with Riodin charges (dots are actual values while lower and upper lines correspond to Equation (3) and Equation (5) respectively).

On the other hand, the formula is quite similar to the prediction model proposed by the manufacturer of the explosive:

$$S_b = 322 \cdot W^{0.56} \cdot R^{-1.3} \quad (4)$$

although the latter gives results lower than the ones obtained from the experiences described here and it is useful only for Z >100.

In the last years, different relationship between air peak overpressure Sb and scaled distance Z, mainly polynomial, have been proposed by several authors [3–5]. We propose the exponential function for coherence with the analysis of air blast wave due to blasting in civil engineering with which this study is most related. On the other hand, it is a simple formula that only needs two empirical parameters. The relationship between the logarithm of the air overpressure $\log(S_b)$ and the logarithm of the scaled distance $\log(Z)$ is linear, and these two parameters can be deduced easily from field data by means of a linear regression. In the present study, the correlation coefficient found is high, r^2 = 96%, demonstrating that it is a sufficiently accurate approach for different analysis.

The point cloud and the regression line are represented in Figure 4. As can be deduced from the same figure, some actual values are higher than the predicted ones. Due to the fact that the aim of the research is safety, a coefficient can be used to assure that any predicted value is higher than the actual one with a given confidence level, i.e., 90% (the predicted value is higher than the actual one in more than 90% of the cases). By using the coefficient of 1.35, the predicted air overpressure fulfils this requirement. The expression deduced in this way is known as the attenuation law:

$$S_b = 417.59 \cdot Z^{-1.216} \quad (5)$$

Equation (4) corresponds to the lower line of the graph, while Equation (5) corresponds to the upper one.

With the values given by Formula (5), we have a predictive model that allows us to characterize the aerial wave generated by the detonation of Riodin-type gelatinous dynamite charges as a function of the distance to the detonation focus. It allows us to assess the overpressure generated by the detonation of a charge of this specific explosive and the possible effects on people or buildings that it will produce. Thus, protection and attenuation mechanisms are established and designed to greatly reduce the consequences of this detonation.

However, the reality is that explosive substances can be of a different nature and composition, not just gelatinous dynamites. For example, a typical blasting work, which produces high air overpressure, is the demolition of metallic structures with shaped charges (Figure 5). It is due to the fact that the explosive is not confined in a blast hole, but it detonates in the open air. In this case, the explosive is pentolite (Riocut), different from dynamite (Riodin), and then the deduced Formula (5) cannot be used directly.

So, in order to be able to characterize the wave generated by any explosive substance and to be able to compare them with each other to assess their harmful effects after a detonation in the open air, the equivalent TNT mass is used.

To apply this calculation method, it is necessary to know the heat of explosion, both of the TNT and of the explosive to be compared. The heat of explosion for TNT is 4520 kJ/kg, and from Table 3, there is a heat of explosion for this Riodin dynamite of 4090 kJ/kg. So, 1 kg of Riodin is equivalent to $1 \times 4090/4520 = 0.905$ kg of TNT. With these explosion heat values, the TNT equivalent of each charge used in the 90 detonations is determined, as well as the reduced distance for each of them with this resulting TNT-equivalent charge (Table 7).

The resulting values from Table 7 are shown in Figure 6 in which the measured overpressure is plotted against the TNT-equivalent scaled distance.

Figure 5. Demolition of metallic structures with shaped charges ((left) metallic silo; (right) large mining stacker).

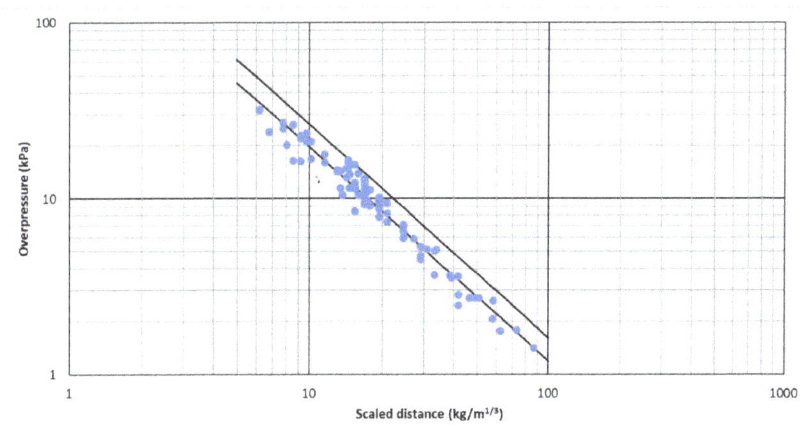

Figure 6. Overpressure measured as a function of scaled distance using the equivalent mass of TNT (dots are actual values while lower and upper lines correspond to Equations (6) and (7)).

Due to the proportionality between the Riodin and TNT explosion heats used, the expression deduced in this case by linear regression is similar to the previous one:

$$S_b = 322.13 \cdot Z^{-1.216} \quad (6)$$

where S_b is the overpressure generated by the wave in kPa, and Z is the reduced distance in $m/kg^{1/3}$. The correlation coefficient for this prediction model is also 96.06%.

By using the coefficient of 1.35, the predicted air overpressure will be higher than the actual one in more than 90% of the cases, and the formula represents the attenuation law of the air wave in the case of TNT explosive:

$$S_b = 434.87 \cdot Z^{-1.216} \quad (7)$$

Equation (6) corresponds to the lower line of the graph, while Equation (7) corresponds to the upper one.

Table 7. TNT-equivalent charge and scaled distance for each Riodin-detonated charge.

N	Distance (m)	Eq. TNT Charge (kg)	Scaled Distance ($m/kg^{1/3}$)	Overpressure (kPa)	N	Distance (m)	Eq. TNT Charge (kg)	Scaled Distance ($m/kg^{1/3}$)	Overpressure (kPa)	N	Distance (m)	Eq. TNT Charge (kg)	Scaled Distance ($m/kg^{1/3}$)	Overpressure (kPa)
1	25	0.215	41.70	2.84	**31**	25	3.232	16.91	9.85	**61**	15	3.232	10.15	16.80
2	25	0.646	28.91	4.70	**32**	25	3.878	15.91	10.50	**62**	25	3.232	16.91	12.10
3	25	1.077	24.39	5.95	**33**	25	4.309	15.36	11.40	**63**	40	3.232	27.06	5.91
4	25	1.077	24.39	5.95	**34**	25	4.955	14.66	11.50	**64**	50	3.232	33.82	5.12
5	25	1.077	24.39	6.57	**35**	25	5.386	14.26	13.20	**65**	75	3.232	50.73	2.73
6	25	2.154	19.36	9.79	**36**	25	6.032	13.73	10.50	**66**	15	4.309	9.22	21.90
7	25	3.232	16.91	11.40	**37**	25	6.463	13.42	14.30	**67**	15	5.386	8.56	16.40
8	25	4.309	15.36	11.60	**38**	75	2.154	58.07	2.63	**68**	15	6.463	8.05	20.20
9	25	5.386	14.26	14.90	**39**	75	1.077	73.16	1.80	**69**	15	5.386	8.56	26.10
10	25	6.444	13.43	11.50	**40**	75	0.646	86.74	1.42	**70**	15	4.309	9.22	16.30
11	25	4.309	15.36	12.30	**41**	50	2.154	38.71	3.63	**71**	10	3.232	6.76	23.90
12	10	2.154	7.74	24.96	**42**	50	1.077	48.78	2.73	**72**	10	2.154	7.74	27.10
13	10	2.331	6.76	24.10	**43**	50	0.646	57.83	2.07	**73**	10	1.077	9.76	23.40
14	15	3.231	10.15	21.12	**44**	40	2.154	30.37	5.15	**74**	10	4.309	6.15	32.20
15	15	4.309	9.22	22.86	**45**	40	1.077	39.02	3.53	**75**	25	2.154	19.36	10.20
16	15	5.386	8.56	26.54	**46**	40	0.646	46.26	2.73	**76**	25	3.232	16.91	12.90
17	25	2.154	19.36	7.85	**47**	25	2.154	19.36	10.10	**77**	10	4.309	6.15	32.04
18	25	2.154	19.36	9.58	**48**	25	1.077	24.39	6.98	**78**	10	1.077	9.76	21.14
19	25	3.232	16.91	9.30	**49**	25	0.646	28.91	5.32	**79**	15	1.508	13.08	14.17
20	25	3.232	16.91	10.10	**50**	15	2.154	11.61	17.80	**80**	15	2.154	11.61	16.01
21	25	4.309	15.36	12.10	**51**	15	1.077	14.63	13.70	**81**	25	1.724	20.85	9.44
22	25	4.309	15.36	8.47	**52**	15	0.646	17.35	9.65	**82**	25	2.801	17.74	11.20
23	25	3.212	16.94	10.70	**53**	25	0.646	28.91	5.32	**83**	25	3.232	16.91	12.50
24	25	0.215	41.7	2.46	**54**	25	1.077	24.39	7.09	**84**	25	3.232	16.91	12.40
25	25	0.431	33.1	3.67	**55**	25	2.154	19.36	9.06	**85**	25	4.309	15.36	15.60
26	25	0.646	28.91	4.50	**56**	15	1.508	13.08	14.50	**86**	25	5.171	14.46	15.00
27	25	1.077	24.39	6.46	**57**	25	1.724	20.85	8.30	**87**	25	5.171	14.46	16.00
28	25	1.724	20.85	7.40	**58**	40	1.724	33.36	5.05	**88**	25	5.171	14.46	16.60
29	25	2.154	19.36	8.71	**59**	50	1.724	41.7	3.60	**89**	25	5.171	14.46	15.90
30	25	2.801	17.74	9.16	**60**	75	1.724	62.55	1.76	**90**	25	3.878	15.91	13.90

3.2. Results of the Second Tests and Analysis of the Air Blast Wave Due to Bombs

Table 8 shows the parameters and results related to the ten explosions with a large amount of TNT explosive. Detonation number 91 was canceled because the microphone did not work properly.

Table 8. Values of scaled distances and air overpressure for each detonated TNT charge.

N	Distance (m)	TNT Charge (kg)	Scaled Distance (m/kg$^{1/3}$)	Overpressure (kPa)
91	25	84	5.71	-
92	50	84	11.42	16.00
93	50	84	11.42	21.90
94	30	84	6.85	45.30
95	25	84	5.71	63.80
96	25	84	5.71	57.60
97	25	42	7.19	36.30
98	25	25	8.55	33.00
99	25	42	7.19	57.00
100	25	84	5.71	54.10

In the case of detonation of TNT charges, two different behaviors can be seen. There is one test in which the air blast wave is moderate, and the shape of the overpressure pulse is similar to that described above. It is rather symmetrical, and the positive and negative parts are approximately of the same magnitude as can be seen in the overpressure record measured in test no. 92 (S_b = 16.0 kPa), Figure 7 (left).

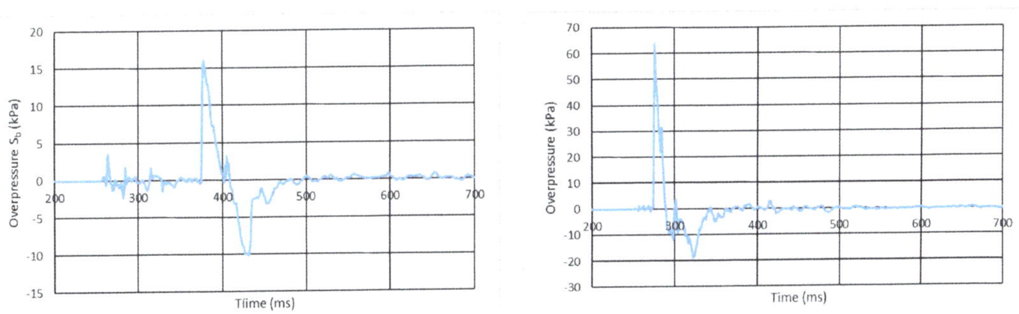

Figure 7. Air overpressure measured in tests no. 92 (**left**) and no. 95 (**right**).

Nevertheless, when the air blast wave is high, the shape of the pulse is equal to the ideal blast wave pressure with the positive part much higher than the negative one. On the other hand, the duration of the positive phase in these tests is significantly higher than in the others. For example, the overpressure measured in test no. 95 (S_b = 63.8 kPa) is shown in Figure 7 (right).

These overpressure results can be drawn together with the results obtained with the TNT explosive equivalent to Riodin dynamite. Then the graph of Figure 8 was obtained.

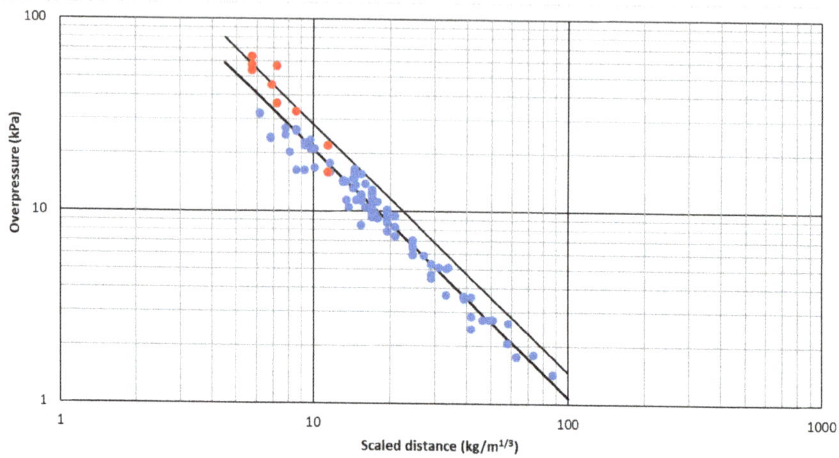

Figure 8. Overpressure measured as a function of scaled distance for the detonation of pure TNT and equivalent TNT (dots are actual values while lower and upper lines correspond to Equation (8) and Equation (9) respectively).

The expression derived from the data set is:

$$S_b = 396.27 \cdot Z^{-1.280} \quad (8)$$

With a correlation coefficient $r^2 = 95.9\%$.

By using the safety coefficient 1.35, the predicted air overpressure will be higher than the actual one in more than 90% of the cases, and the attenuation law of the air wave in the case of TNT explosive is:

$$S_b = 534.96 \cdot Z^{-1.280} \quad (9)$$

Formula (9), or alternatively the graphic of Figure 6, is useful to predict air blast wave overpressure near the explosion even in the case of detonation of a large amount of explosive.

4. Conclusions

The peak pressure value of the air blast wave from a total of 100 records corresponding to the detonation of different explosive charges in the open air was analyzed. These records can be separated into two basic groups: records from open-air detonations of a gelatinous dynamite-type explosive and records from open-air detonations of a TNT-type explosive.

The most important result achieved was the definition of an air wave attenuation law, overpressure S_b as a function of the scaled distance Z, for the determination of the overpressure peak due to the detonation of explosive charges in the outdoors. The law is simpler than others since it only requires the determination of two empirical parameters that can be determined with a smaller number of samples.

The model predicts the peak value of the air blast wave S_b (kPa) from the detonation of a given or equivalent TNT explosive charge in the open air that relates to the value of such variable, S_b, with the scaled distance Z (m/kg$^{1/3}$):

$$S_b = 396.27 \cdot Z^{-1.280}$$

where $Z = R/W_{eq}{}^{1/3}$, that is, the distance R (m) divided by the cubic root of the equivalent TNT mass W_{eq} (kg).

By using a safety coefficient of 1.35, the predicted S_b is higher than actual S_b in more than 90% of the cases:

$$S_b = 534.96 \cdot Z^{-1.280}$$

It has been demonstrated that this law is valid in a wide range of the reduced distance, with Z varying between 5.71 and 86.74 m/kg$^{1/3}$, and in a wide range of the air wave, with S_b between 1.42 and 63.8 kPa. In this way, the attenuation law is useful both for the prediction of the air blast wave due to the detonation of charges of a few kgs of explosives (such as the shaped charges used in civil works for the demolition of metallic structures) and for the prediction of the air wave in the case of the detonation of several tens of kgs of explosives (such as explosive detonations by accident or terrorist bombs).

The model proposed aims to serve as a basis for the design of protection and containment elements, but it is considered necessary to continue testing with full-scale explosives, in order to further limit other parameters involved in the propagation of the resulting wave of a detonation, tests that are difficult to carry out because they are of a destructive nature and because they are controlled materials for which there is authorization for consumption, qualification, and training.

Author Contributions: Conceptualization, R.R. and J.A.M.; methodology, J.A.M. and R.R.; investigation, J.A.M., R.R., M.B.D. and S.A.; data curation, J.A.M.; writing—original draft preparation, R.R., J.A.M. and S.A.; writing—review and editing, R.R. and J.A.M. All authors have read and agreed to the published version of the manuscript.

Funding: This research was carried out in the framework of the project *"Especificaciones de diseño y estructurales para aparcamiento de terminales de transporte con riesgo de ataque terrorista"* funded by the Spanish Ministry for Science and Innovation (No. Exp. 16-08).

Institutional Review Board Statement: Not applicable.

Informed Consent Statement: Not applicable.

Data Availability Statement: The article provides all data used in this research.

Conflicts of Interest: The authors declare no conflict of interest.

References

1. Baker, W.E. *Explosions in Air*; University of Texas Press: Austin, TX, USA, 1973.
2. Baker, W.E.; Cox, P.A.; Westine, P.S.; Kulesz, J.J.; Strehlow, R.A. *Explosion Hazards and Evaluation*; Elsevier Scientific Publishing Company: New York, NY, USA, 1983.
3. Henrych, J. *The Dynamics of Explosion and Its Use*; Elsevier: Amsterdam, The Netherlands, 1979.
4. Kingery, C.N.; Bulmash, G. *Airblast Parameters from TNT Spherical Air Burst and Hemispherical Surface Burst*; Technical Report ARBRL-TR-02555; U.S. Army BRL: Aberdeen, MD, USA, 1984.
5. Kinney, G.F.; Graham, K.J. *Explosive Shocks in Air*; Springer: Berlin/Heidelberg, Germany, 1985.
6. Mills, C.A. The design of concrete structures to resist explosions and weapon effects. In Proceedings of the 1st InternationalConference on Concrete for Hazard Protections, Edinburgh, UK, 27–30 September 1987.
7. Sadovskiy, M.A. Mechanical effects of air shockwaves from explosions according to experiments. In *Geophysics and Physics of Explosion*; Sadovskiy, M.A., Ed.; Selected Works; Nauka Press: Moscow, Russia, 2004.
8. Beshara, F.B. Modelling of blast loading on aboveground structures—I. General phenomenology and external blast. *Comput. Struct.* **1994**, *51*, 585–596. [CrossRef]
9. Brode, H.L. *Numerical Solution of Spherical Blast Waves*; American Institute of Physics: New York, NY, USA, 1955.
10. Remennikov, A.M. A review of methods for predicting bomb blast effects on buildings. *J. Battlef. Technol.* **2003**, *6*, 5–10.
11. Koccaz, Z.; Sutcu, F.; Torunbalci, N. Architectural and structural design for blast resistant structures. In Proceedings of the 14th World Conference on Earthquake Engineering, Beijing, China, 12–17 October 2008.
12. Draganic, H.; Sigmund, V. Blast loading on structures. *Teh. Vjesn.* **2012**, *19*, 643–652.
13. Wu, C.; Hao, H.; Lu, Y.; Zhou, Y. Characteristics of wave recorded in small scale field blast tests in a layered rocksoil medium. *Geotechnique* **2003**, *53*, 587–599. [CrossRef]
14. Cranz, C. *Lehrbuch der Ballistic*; Springer: Berlin/Heidelberg, Germany, 1926.
15. Shirbhate, P.A.; Goel, M.D. A Critical Review of Blast Wave Parameters and Approaches for Blast Load Mitigation. *Arch. Comput. Methods Eng.* **2021**, *28*, 1713–1730. [CrossRef]

16. Filice, A.; Mynarz, M.; Zinno, R. Experimental and Empirical Study for Prediction of Blast Loads. *Appl. Sci.* **2022**, *12*, 2691. [CrossRef]
17. Ding, Y.; Zhang, X.; Shi, Y.; Zhang, H. Prediction of far-field blast loads from large TNT-equivalent explosives on gabled frames. *J. Constr. Steel Res.* **2022**, *190*, 107120. [CrossRef]
18. Siskind, D.E.; Stachura, V.J.; Stagg, M.S.; Kopp, J.W. *Structure Response and Damage Produced by Airblast from Surface Minin*; Report of Investigation 8485; U.S. Bureau of Mines: Washington, DC, USA, 1980.
19. Richards, A.B.; Moore, A.J. Airblast design concepts in open pit mines. In Proceedings of the 7th International Symposium on Rock Fragmentation by Blasting–Fragblast 7, Beijing, China, 11–15 August 2002; pp. 553–561.
20. Kuzu, C.; Fisne, A.; Ercelebi, S.G. Operational and geological parameters in the assessing blast induced airblast-overpressure in quarries. *Appl. Acoust.* **2020**, *70*, 404–411. [CrossRef]
21. *UFC 3-340-02. Structures to Resist the Effects of Accidental Explosions, Unified Facilities Criteria (UFC)*; Manual Number 3-340-02 (This Manual Supersedes US ARMY TM5-1300); US Department of Defense: Washington, DC, USA, 2008.
22. Karlos, V.; Solomos, G. *Calculation of Blast Loads for Application to Structural Components*; JRC Technical Report EUR 26456EN; European Union: Luxembourg, 2013.

applied sciences

MDPI

Article

Forecast of Airblast Vibrations Induced by Blasting Using Support Vector Regression Optimized by the Grasshopper Optimization (SVR-GO) Technique

Lihua Chen [1], Panagiotis G. Asteris [2,*], Markos Z. Tsoukalas [2], Danial Jahed Armaghani [3], Dmitrii Vladimirovich Ulrikh [3] and Mojtaba Yari [4]

[1] Department of Civil Engineering, Chongqing Vocational Institute of Engineering, Jiangjin District, Chongqing 402260, China
[2] Computational Mechanics Laboratory, School of Pedagogical and Technological Equation, 15122 Maroussi, Greece
[3] Department of Urban Planning, Engineering Networks and Systems, Institute of Architecture and Construction, South Ural State University, 76, Lenin Prospect, 454080 Chelyabinsk, Russia
[4] Department of Mining Engineering, Faculty of Engineering, Malayer University, Malayer 65719-95863, Iran
* Correspondence: asteris@aspete.gr

Abstract: Air overpressure (AOp) is an undesirable environmental effect of blasting. To date, a variety of empirical equations have been developed to forecast this phenomenon and prevent its negative impacts with accuracy. However, the accuracy of these methods is not sufficient. In addition, they are resource-consuming. This study employed support vector regression (SVR) optimized with the grasshopper optimizer (GO) algorithm to forecast AOp resulting from blasting. Additionally, a novel input selection technique, the Boruta algorithm (BFS), was applied. A new algorithm, the SVR-GA-BFS_7, was developed by combining the models mentioned above. The findings showed that the SVR-GO-BFS_7 model was the best technique ($R^2 = 0.983$, RMSE = 1.332). The superiority of this model means that using the seven most important inputs was enough to forecast the AOp in the present investigation. Furthermore, the performance of SVR-GO-BFS_7 was compared with various machine learning techniques, and the model outperformed the base models. The GO was compared with some other optimization techniques, and the superiority of this algorithm over the others was confirmed. Therefore, the suggested method presents a framework for accurate AOp prediction that supports the resource-saving forecasting methods.

Keywords: blasting; airblast; input selection; hybrid SVR model; prediction

Citation: Chen, L.; Asteris, P.G.; Tsoukalas, M.Z.; Armaghani, D.J.; Ulrikh, D.V.; Yari, M. Forecast of Airblast Vibrations Induced by Blasting Using Support Vector Regression Optimized by the Grasshopper Optimization (SVR-GO) Technique. *Appl. Sci.* **2022**, *12*, 9805. https://doi.org/10.3390/app12199805

Academic Editors: Ricardo Castedo, Lina M. López and Anastasio P. Santos

Received: 23 August 2022
Accepted: 13 September 2022
Published: 29 September 2022

1. Introduction

Air-overpressure (AOp) or airblast is an unwelcome outcome of blasting in mining operations. The blasting creates temporary air blast pressure waves that persist for some time [1,2]. More than 20% of the explosive energy is used to fracture and replace the rock fragments. More than 70% of this energy is dissipated, which causes AOp and other unwanted phenomena [1]. Various parameters, including terrain circumstances, blast design, and climate, are influential on AOp [2,3]. The enormous shock waves coming out of the blast spot toward the free facade create AOp. Hence, the AOp can be defined as a shock wave deflected laterally by density changes in the air. These AOp waves are released with some audible high- or low-frequency sounds. AOp can cause structural damage and harm to people in the vicinity of quarry sites [4].

Several studies attempted to establish associations for the AOp forecast using its influential factors. Kuzu et al. [5] ascertained an experimental association between AOp and interval among blast planes and monitoring spot and mass of explosive substances (32%). To reproduce ground shock and air explosion tensions deriving from facade explosions,

Wu and Hao [6] connected Autodyn2D to mathematical models, wherein properties of rock materials and free air were involved. A partial-empirical model for forecasting the airwave tension caused by blasting operations outside a tunnel was developed by Rodríguez et al. [7]. They claim that their model works in different situations. Rodríguez et al. [8] proposed a photometric curve and iso-attenuation curves to describe the phenomenon, as well as a proposed charge-distance curve to resolve the query.

Recently, various investigations have applied machine learning (ML) techniques to resolve science and engineering problems [9–19]. Such techniques were used to predict the AOp values and identify the most influential predictors [20–22]. These techniques are both time and cost-saving and can help both researchers and practitioners allocate resources to other necessary operations. Among the ML algorithms, artificial neural networks (ANNs) and an adaptive neuro-fuzzy inference system (ANFIS) were employed more frequently than other techniques in the investigations of AOp [23]. Some studies used tree-based techniques such as XGBoost, random forest (RF), and M5 [24]. Rare studies employed support vector regression to predict the AOp [25]. Additionally, some of these models were optimized with some techniques to improve accuracy and efficiency. Some of these optimization techniques include genetic algorithms (GA) and particle swarm optimization (PSO) [2]. Lastly, different methods, such as fuzzy Delphi methods, were used to choose the inputs before the model was built [26].

Despite the vast application of ML algorithms for AOp prediction, various efficient models for prediction, optimization, and input selection are neglected. In this study, the authors developed a novel prediction model that encompasses SVR as a prediction model, the Grasshopper algorithm for optimizing the SVR's hyperparameters, and the Boruta algorithm (BFS) for input selection. The Grasshopper algorithm is used in this study because it is easy to use, has a structure without gradients, avoids local optimums well, and treats problems as black boxes. SVR is rarely employed for AOp prediction, and to the best of the authors' knowledge, the Grasshopper and Boruta algorithms have not been applied in this domain. The rest of this paper is structured as follows: the next section discusses algorithms used in this study, including SVR, GO, and BFS. Additionally, the case study of this research will be explained in this section. The results and discussions' section describes the data preparations and performance criteria. The results of input selection and model optimization are reported in this section. The paper ends with a conclusion that sums up what was learned from this study and makes some suggestions for future research.

2. Material and Methods

This section thoroughly describes the methods used in this study. These methods include SVR as a well-known prediction technique, GO as a metaheuristic optimization technique and BFS as an input selection approach. The SVR-GO-BFS_n model, which is the result of combining the algorithms mentioned above, is adequately explained. Furthermore, four ML techniques and three optimization techniques were applied to verify the performance of the SVR-GO-BFS_n model.

2.1. Data Collection

Data for this study was collected from a published work by Hajihassani et al. [27]. According to their study, four granite mines were selected and considered for data collection in the Johor area, Malaysia. In total, 62 blasting operations were performed. The main substance of the explosion was ANFO, a widely used bulk industrial explosive, the stemming substance was granular gravel, and the diameters of the blast holes were 75, 89, and 115 mm. The specifications of rock quality designation (RQD) as well as the height of the bench are shown in Table 1. Various blasting parameters, including RQD, burden, hole depth, spacing, powder factor, and stemming length, were evaluated during the data collection.

Table 1. Some additional measurements in blasting sites.

Site	RQD (%)	Height of Bench (m)
Masai	60–84	15–20
Pasir Gudang	67–89	13–25
Pengerang	70–91	10–23
Ulu Tiram	65–88	10–15

The research team observed the AOp employing microphones (L type), which were linked to the AOp grooves of record-keeping elements. The AOp values ranged from 88 dB to 148 dB. To ensure an accurate measurement of overpressures, the operating frequency response of microphones was selected between 2 and 250 Hz. This frequency is suitable for measuring the overpressure for both human hearing and construction.

The minimum values of 10 m, 0.34 kg/m^3, 60 kg, 1.7 m, 1.5 m, 2.65 m, 60%, 12, 300 m, and 89.1 dB were recorded for hole depth, powder factor, maximum charge per delay, stemming, burden, spacing, RQD, no. of hole, distance from the blast face, and AOp, respectively, while the values of 25 m, 0.76 kg/m^3, 171 kg, 3.2 m, 4 m, 91%, 63, 600 m, and 126.3 dB were recorded as maximum amounts of the same variables. More information regarding the data used in this study can be found in the original study [27].

2.2. Preparation of Data

The min/max transformation technique was used to normalize the collected data. The objective of this transformation was to restrict the inputs' possible advantages to those with noticeable numerical values, over those with small values. Handling large-value inputs can be challenging and complicated due to the fact that the kernel quantity relies on vectors' internal multiplication of inputs. Therefore, conquering mathematical complications throughout calculation procedures is another crucial aspect of input normalization. The data were transformed using Equation (1) and the normalized data ranged from zero to one.

$$a_i^n = \frac{a_i - a_{min}}{a_{max} - a_{min}} \tag{1}$$

where the input vectors with the calculated observation points are denoted by a_i. a_{min} and a_{max} refer to the lowest and highest values that relate to the calculated data set. a_i^n is the transformed variant of a_i.

2.3. Support Vector Regression (SVR)

One of the most effective approaches for handling regression complications is SVM, which is a supervised technique [28]. The formation and optimization approach of SVM varies according to the nature of inputs, and ε-SVR is the regression form of SVM. The principal objective of SVR is to acquire a hypothesis whose entire errors of regression forecast are situated within a predetermined threshold, ε. The next aim of the learned function is that this function possesses an excellent achievable generalization capability. This aim is purposely attempted in order that a flat model can be established. The following equations enact the aforementioned aims, forming a typical convex quadratic optimization problem with linear constraints set. The above goals are met by the following equations, which, along with the set of linear constraints, make a typical curved quadratic optimization problem.

$$minimze\ \frac{1}{2}\|\omega\|^2 + c\sum_{i=1}^{k}(\vartheta_i + \vartheta_i^*) \tag{2}$$

$$subject\ to \begin{cases} b_i - \langle \omega, a_i + c\rangle \le \vartheta_i + \varepsilon,\ \forall n \\ \langle \omega, a_i + c\rangle - b_i \le\ \vartheta_i^* + \varepsilon, \forall n \\ \vartheta_i,\ \vartheta_i^* \ge 0, \forall n \end{cases} \tag{3}$$

where, for training points $(a_i, b_i), \ldots, (an, bn)$, k denotes the number of data samples, the vectors of a_i denote values of input, and b_i implies the corresponding output value for a_i. The upper and lower errors of training are represented by $\boldsymbol{\vartheta}_i$ and $\boldsymbol{\vartheta}_i^*$, respectively. The errors are indifferent to a particular margin defined by ε; afterwards, the cost function will be added by penalties. The normal vector is denoted by ω. The regularization parameter $(c > 0)$ regulates the balance of the pair of goals enacted in the above equations. The authors employed Lagrange multipliers to ascertain the SVR's optimization problem expressed by the above equations. Some alterations were performed following that the Lagrangian is calculated until the next equation is obtained:

$$f(a,\cdot\beta_i, \beta_i^*) = \sum_{i=1}^{n} (\beta_i - \beta_i^*) l(a, a_i) - c \tag{4}$$

The equation obtained above is based on theories of optimality constraints, the kernel method, and Lagrange multipliers. While four renowned kernels, including sigmoid, polynomial, linear, and RBF, are available, this study employed RBF. This kernel was intentionally picked due to its computational capability. Typically, this kernel outperforms others [28]. RBF is extremely nonlinear, including possessing some inputs and an unlimited-dimensional space of mapping [29]. The RBF kernel is displayed in the following equation:

$$L(a_i, a_j) = e^{-\gamma(\|a_i - a_j\|^2)} \tag{5}$$

where $\gamma \in R$, $\gamma > 0$ describes the expanse of the radial basis kernel function.

Figure 1 displays the structure of SVR based on Equation (3). This structure admits the requirements of Karush–Kuhn–Tucker for resolving a quadratic optimization query. The values of $(\beta_i - \beta_i^*)$ were used to obtain the decision function. It is worth mentioning that these values were non-null support vectors. One of the most vital steps to develop a profoundly accurate and stable prediction model is to optimize the pair of SVR's hyper-parameters, including C and γ. Adopting optimization methods for ascertaining these parameters' optimal conditions is considered in recent studies.

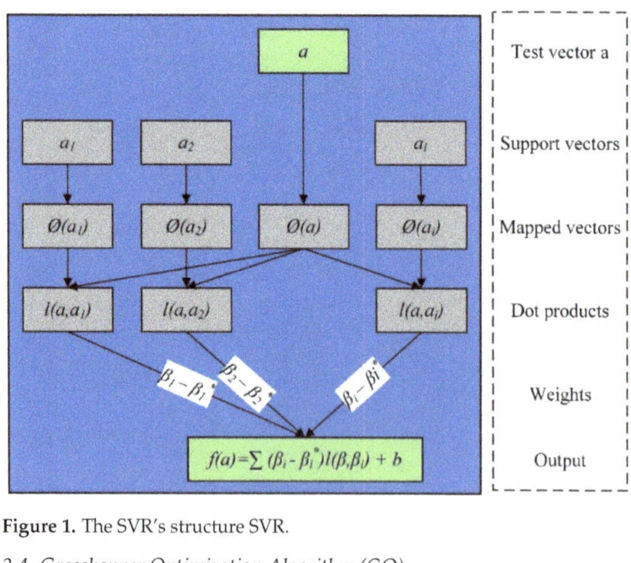

Figure 1. The SVR's structure SVR.

2.4. *Grasshopper Optimization Algorithm (GO)*

A recent swarm intelligence method, which is acknowledged as the grasshopper optimization (GO) algorithm, was employed in this study. Roles of nature are behind this method. This optimization technique was initially developed to deal with complicated optimization queries [29]. GO mimics the behavior of grasshopper colonies for resolving the queries of optimization. The GO involves two search processes: exploitation and exploration. Nymph and adult grasshoppers execute the search procedure. Typically, long intervals are covered by adult grasshoppers. Hence, they can search a whole space (global) to discover more suitable areas, where more foods are provided. In fact, the exploration procedure is performed by them. On the other hand, nymph grasshoppers perform the exploitation task, which means that they aim for a specific or local region. The GO secures an equilibrium between exploitation and exploration. This balance may lead to a slightly more complex algorithm. Mathematically, Saremi et al. [30] achieved a method to represent the colony function of grasshoppers. The following formula shows the mathematical model of the grasshoppers' swarming function.

$$Y_i = A_i + B_i + C_i \tag{6}$$

where Y_i denotes the ith grasshopper's place. A_i indicates the idea of social interplay. So long as B_i expresses the strength of gravity forced on the ith grasshopper, the wind advection is demonstrated by C_i. Remarking that the formula was re-written $Y_i = m_1 A_i + m_2 B_i + m_3 C_i$ to produce arbitrary behaviour, wherein m_1, m_2, and m_3 are accidentally picked figures between zero and one. The three steps of the implementation of GO algorithm are shown in Table 2.

Table 2. Implementation steps of GO algorithm.

1st STEP	$A_i = \sum_{j=1, j \neq i}^{N} a\left(f_{ij}\right) \cdot f_{ij}$	• f_{ij} is a space, which divides the ith and jth grasshoppers, $f_{ij} = \lvert a_j - a_i \rvert$ • $f_{ij} = (a_i - a_j)/f_{ij}$ means the unit vector between the ith and jth grasshoppers	(7)
	$a(m) = z \cdot e^{-m_d} - e^{-m}$	• z is the attraction intensity • d signifies the attractive length scale • $m = \lvert d_{ij} \rvert$	(8)
2nd STEP	$B_i = -b \cdot e_b$	• b is gravitational constant • e_b stands for the unit vector heading to the globe center	(9)
3rd STEP	$C_i = x \cdot e_\omega$	• x is a constant drift • e_ω refers to a unity vector in the wind direction • N signifies the grasshoppers' number	(10)
	$Y_i = \sum_{j=1, j \neq i}^{N} b\left(\left\lvert a_j - a_i \right\rvert\right) \cdot \frac{a_j - a_i}{f_{ij}} - t \cdot e_t + x \cdot e_\omega$		(11)
	$Y_i^f = c\left(\sum_{j=1, j \neq i}^{N} c \cdot \frac{ub_d - db_f}{2} \cdot a\left(\left\lvert a_j^d - a_i^d \right\rvert\right) \cdot \frac{a_j - a_i}{f_{ij}} \right) + O_f$	• db_f and ub_d are lower and upper boundaries in the Fth dimension • O_f is the location of the optimum solution it has found yet	(12)
	$c = c_{max} - d \cdot \frac{c_{max} - c_{min}}{D}$	• c_{min} and c_{max} refers to the minimum and the maximum values of the coefficient c • d implies the existing iteration • D represents the greatest iterations	(13)

In the first step, the concepts of social interaction (A_i) and social forces (a(m)) were determined. The function "a" is able to divide the space amongst two grasshoppers into 3 areas: attraction, repulsion, and comfort. In the second step, the force of gravity imposed on the ith grasshopper was determined (B_i).

The wind advection (C_i) was established in the third step. It is worth mentioning that because the nymph grasshoppers do not have wings, the wind direction strongly impacts their movement. The elements of A_i, B_i, and C_i were replaced in Equations (6) and (11) was formed. Typically, the grasshoppers reach their pleasure areas rapidly, and the group does not gather in particular spots. Thus, Equation (11) is unable to deal with the optimization queries immediately. This equation was amended to solve the optimization issue mentioned above, and Equation (12) was created. As shown in Equation (12), "c" is a lessening coefficient which is utilized on the way to decrease the comfort, attraction, and repulsion areas. In this Equation, the "c" enters double because of the following reasons:

- By expanding the abundance of iterations, the motion of marked grasshoppers is decreased by the initial "c". This parameter equilibrates the whole exploration and exploitation of the target.
- The following "c" decreases the repulsion, attraction, and comfort areas amongst grasshoppers. The aforementioned decline is proportionate to the iterations' abundance.

GO needs to be avoided by becoming stuck in the local optimum. Alternatively, it attempts to acquire a precise calculation of the global optima. Grasshoppers achieve progressive equilibrium between exploration and exploitation because of the diverse pleasure area parameter "c". In each iteration, the "c" can be calculated by Equation (13). In this study, the authors employed great rates of repulsion since this is a crucial method in the GO to circumvent local solutions. The outcomes reveal that great repulsion rates limit grasshoppers to staying at a local optimum. To summarize, Algorithm 1 depicts the processes involved in executing the GO.

Algorithm 1 GO optimization

Initialize the swarm population (grasshoppers) Yi, where $i = (1, 2, \ldots, N)$
Initialize the parameters: c_{min}, c_{max}, D
Calculate the fitness value of each search agent
Assign O to the best search agent (the individual with highest fitness value)
while $d < D$ do
    Use Equation (13) to update c
    for each search agent
        Normalize the distance between grasshoppers within [1,4]
        Update the position of the current search agent by Equation (12)
        Bring the current search engine back when it exceeds the boundaries
    end for
    if there is a better solution, update O
    $d = d + 1$
end while
Return O

2.5. Boruta Feature Selection (BFS) Algorithm

The BFS is an ensemble-based input selection technique that follows the function system of RF with some extra tools to obtain better outcomes [31]. The BFS attempts to identify all the important inputs in both regression and classification queries. The principal concept of this technique is employing analytical measurements and executing various RFs to examine the significance of the original inputs and inputs with an expanded randomness degree. The additional randomness allows a greater understanding of what inputs are significant. Figure 2 shows the running steps of BFS.

The BFS determines all important inputs in the knowledge system and renders the inputs' importance degree. This system also designates significant inputs with numerical rates indicating their significance. Therefore, this may assist scholars in building various input mixtures based on their relevance ranking to determine the optimal input collection. Detailed information about the Boruta input selection technique can be obtained from Kursa and Rudnicki [31].

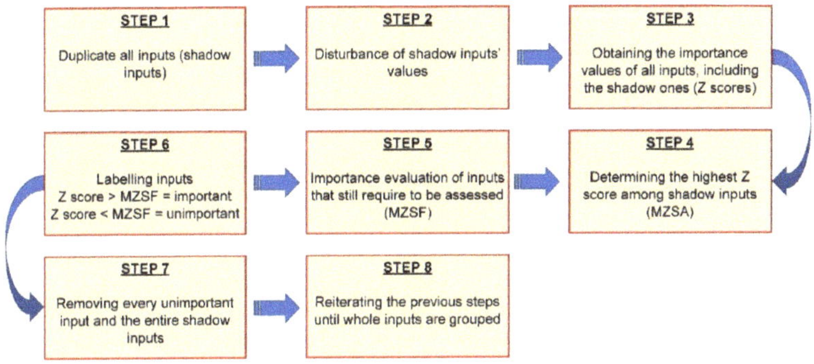

Figure 2. The execution steps of BFS.

2.6. *SVR-GO-BFS$_n$ Model Development*

A hybrid model expressed by SVR-GO-BFS$_n$ was developed to improve the performance of the AOp forecast. The effective algorithms are combined into SVR-GO-BFS$_n$: SVR, GO for optimizing parameters, and BFS for input choice. The parameterization of the SVR-GO-BFS$_n$ model was based on "n", which showed the abundance of inputs engaging in developing the model in line with the "n" greatest importance values assigned by BFS. The most important inputs were selected using BFS. Next, GO was utilized to train SVR and optimize a pair of its hyperparameters (γ and C). Finally, the developed model was used to predict the AOp values. Figure 3 shows a fundamental flowchart of the model developed, which includes the main four steps.

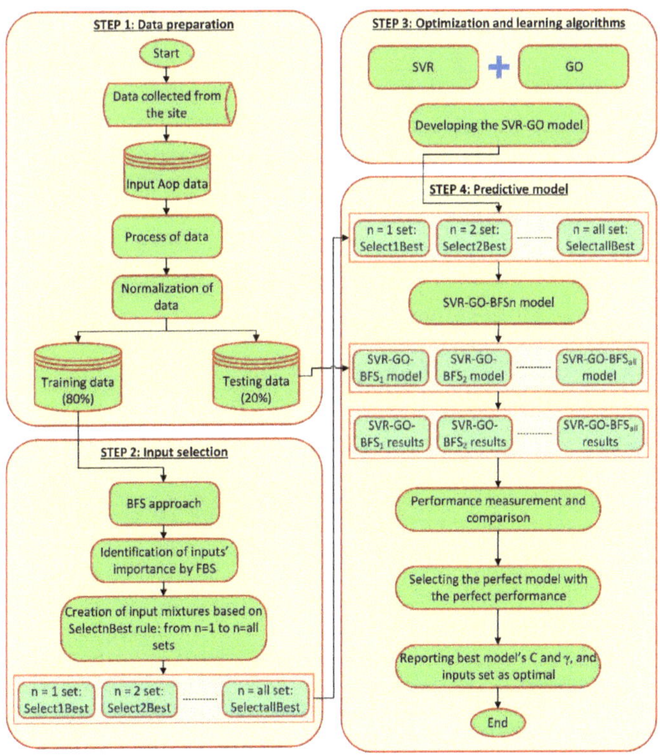

Figure 3. Flowchart of this study.

Step one involved the preparation of data. The dataset included nine candidate inputs. In this step, the data were also normalized as previously described. The second step involved the selection of inputs using the BFS technique. This method contributed to a fair and logical determination and ranking of significant and insignificant inputs from the AOp dataset. In line with the significant input ranking, various input mixtures were built and expressed as n = 1, n = 2, ... , n = all. If the "n" is one, it means that the input collection simply includes one input (the most significant input). If the "n" is two, it refers to an input set that involves the two most significant ones. If "n" equals "all", it means that an input assortment involves all significant ones. Obtaining the minimum optimal input's collection was the main objective of this method.

Step three engaged in the optimization of SVR utilizing GO. A pair of SVR's parameters, including C and γ, were optimized by applying GO. Ultimately, various regression models (SVR-GO-BFS_n) were developed based on various input mixtures.

Step four involved performing predictions using SVR-GO-BFS_n. Utilizing the testing set, the corresponding built predictive models were assessed, and the forecast outcomes were reported. The most suitable model was picked following its precision and error performance. Therefore, the most suitable input set was regarded as the optimal collection of inputs and its C and γ were the optimal SVR values.

2.7. Validation Scheme

The training data employ a different pre-process once the AOp dataset has been randomly partitioned into the training (80%) and test (20%) sets. We call this method the k-fold cross-validation approach. This technique improves the models' flexibility and, consequently, their precision. Therefore, the statistical examination would generalize properly to a particular dataset. In comparison with holdout validation, cross-validation is more suitable for datasets with a small sample size. This method randomly divides the initial data into k equivalent sub-data sets. Following that, the k-1 sub-data are used for training, and one sub-data set is used for testing the model. This procedure is repeated k times. Finally, a single approximation is achieved by averaging the k results from the folds. In this study, the k value is 10 (Figure 4).

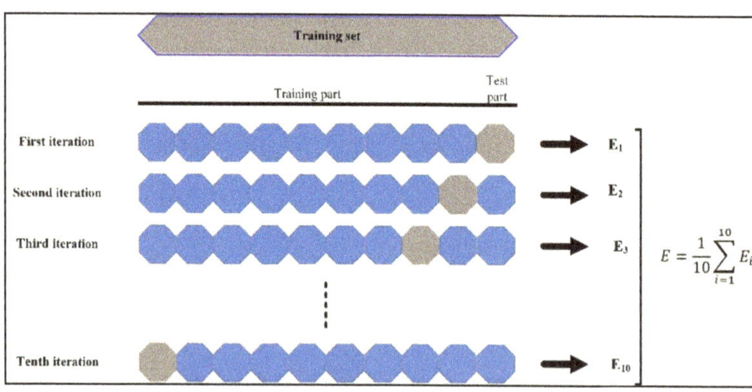

Figure 4. 10-fold cross validation schematic view.

3. Results and Discussion

3.1. Performance Criteria

Various performance criteria were utilized to gauge the performance of the models established in the present study. These metrics included squared correlation (R^2), root mean square error (*RMSE*), mean absolute error (*MAE*) and mean absolute percentage error (*MAPE*). The formulations of these criteria are presented by Equations (14)–(17). These performance indices have been used by some other researchers as well in predictive and classification studies [32–37].

$$R^2 = 1 - \frac{\sum_{h=1}^{N}\left(z_i - \bar{b}\right)^2}{\sum_{h=1}^{N}\left(b_i - \bar{b}\right)^2} \tag{14}$$

$$RMSE = \sqrt{\frac{1}{N}\sum_{h=1}^{N}(z_h - b_h)^2} \tag{15}$$

$$MAE = \frac{1}{N}\sum_{h=1}^{N}|z_h - b_h| \tag{16}$$

$$MAPE = \frac{1}{N}\sum_{h=1}^{N}\left|\frac{z_h - b_h}{b_h}\right| \times 100\% \tag{17}$$

where N is the total quantity of samples; z_h and b_h signify predicted and real values; $\bar{b}$ is the mean value of b.

3.2. Input Selection

The BFS was applied to evaluate the significance of inputs for predicting the AOp. In the beginning, the suggested approach examined nine inputs for the final selection of the inputs, and 100 iterations were used to execute the BFS. The authors did not notice any variations in the research results exceeding 100 runs. The findings of the BFS-based technique are presented in Figure 5.

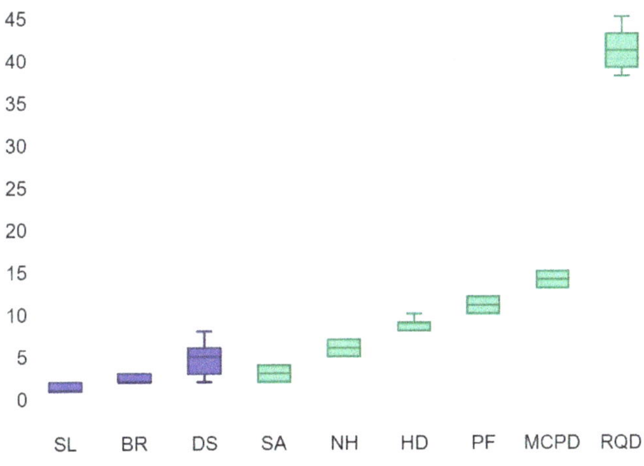

Figure 5. The results of the input selection study on the data.

Box plots in Figure 5 explain the significance of the inputs assessed through BFS. The green plots show those inputs that have more prominent predictability than those indicated by the blue colors. All inputs were classified as significant. Therefore, in developing various input mixtures for the AOp forecast, all nine inputs will be employed. Based on the suggested structure, nine predictive models will be proposed. This approach aims to determine the minimal optimal variable's collection for overcoming the issues of underfitting and overfitting. Furthermore, the inputs indicated in red in the BFS results possess smaller informational potential compared with shadow traits. Hence, these inputs are eliminated from the final collection. Moreover, the yellow inputs show tentative ones. As a result, no inputs appeared tentative or unimportant.

The RQD, MCPD, and PF inputs were confirmed to be the most significant inputs, graded in the same descending form of value for the data obtained from the site. Following these inputs, the HD placed fourth in significance. These findings confirm that using the three characteristics of blasting improves the effectiveness of AOp predictability. Therefore, this study suggests that prospective scholars use these variables as inputs in their models. This algorithm is strong and can produce an unbiased and firm choice of significant and insignificant inputs from a dataset. Since combining more inputs can induce overfitting issues, the novel BFS's capacity to prioritize inputs in decreasing sequence of values can assist scholars in deciding which inputs apply to the AOp forecast. Hence, dropping unnecessary or less correlated inputs may reduce calculation complications and time linked with enhancing the suggested hyperparameters of the scheme.

3.3. *SVR-GO-BFS$_n$ Model Performance*

Following recognizing the importance level of the inputs by BFS, an SVR model kernelized with RBF is employed to carry out the predictive analysis. During running the model, γ and C that are pair hyperparameters of SVR are optimized by the GO algorithm. Nine SVR-GO-BFS$_n$ models (SVR-GO-BFS$_1$ to SVR-GO-BFS$_9$) are developed based on nine various inputs sets (n = 1 set to n = 9). The n = 1 set comprises just the first most significant input, while the n = 9 collection encompasses all nine vital inputs estimated by BFS. To choose among the SVR-GO-BFS$_n$ model structures, this study uses MAPE as the primary criterion. In addition, the RMSE is used as GO's objective function. Prediction of the AOp values is the target of these models.

The AOp database is split into training (80%) and test (20%) sets at random throughout the experiment's run. The training set of data is utilized to develop the forecasting model, while the testing data are utilized to evaluate the predictability. Importantly, all generated models receive the same training and test sets on a regular basis. Following building numerous models, it has been evidenced that as the number of iterations rises, the model computation time grows. Small population sizes, on the other hand, generate inconsistent fitness values. Therefore, multiple groups of 50, 100, 150, 200, and 250 population numbers in the optimization model were chosen for the purposes of the current study, and their iteration curves were made based on the right fitness values.

Concerning the GO, the number of search agents was set as 40, as well as the largest iteration number of developed models was set as 100. Regarding SVR, the lower and upper bounds of γ and C were set to (0.01–50) and (0.01, 100). All inputs were normalized between zero and one for the estimation of performance criteria, as well as to decrease the calculation complications during searching for hyperparameters of models. Table 3 presents the best predictive fulfilment of the models developed in the current investigation. This table corresponds to SVR hyperparameter values optimized using GO and the smallest optimal collection of inputs. It can be seen that the model with seven inputs (SVR-GO-BFS$_7$) achieved the highest accuracy and lowest errors. Figure 6 depicts the outcomes of fitness values for SVR-GO-BFS$_7$ models in forecasting AOp, along with their iteration counts. Furthermore, to minimize the GO's cost function, the RMSE was chosen. This figure shows that the best population size for SVR-GO-BFS$_7$ is 200. Sizable errors in prediction are improbable to have occurred. Only average alternations were adopted up to iteration number 65; following this, no significant difference in the RMSE values was indicated. It should be noted that all models achieved the minimum RMSE in less than 70 iterations, which shows the power of GO in optimizing the SVR hyperparameters.

It was not required to have the full collection of significant inputs (n = nine) to obtain the most reliable predictive performance. Therefore, the authors can draw the conclusion that the effectiveness of SVR-GO-BFS$_n$ in forecasting the AOp is excellent.

The performance of the SVR-GO-BFS$_n$ models based on various mixtures of significant inputs (n = 1 set to n = 9 sets) is presented in Figure 7 through the stacked area. In Figure 7, it is obvious that the MAPE, RMSE, and MAE estimates obtained from all the SVR-GO-BFS$_n$ models were essentially lower than 2.6953, 3.6637, and 3.3083, sequentially, even if only one input was added to the model. For instance, the achieved values of MAPE, RMSE, and MAE were 2.6953, 3.2241, and 2.8476, sequentially, if just one input is employed for the model creation. Furthermore, the SVR-GO-BFS$_2$ model is associated with the poorest performance. This model achieved 0.9209 for R^2, 36637 for RMSE, 3.1152 for MAPE, and 3.3083 for MAE. Instead, the developed models become more precise through employing the three most significant inputs and beyond. For instance, the acquired RMSE varied from 2.1659 to 1.6092 for the models from SVR-GO-BFS$_3$ to SVR-GO-BFS$_9$. Therefore, the authors can assume that employing just the three most significant inputs from the dataset picked and rated by BFS would produce strong prediction outcomes. Moreover, comparable issues were found with R^2, MAPE, and MAE. The scatter plots of the real and predicted AOp values made by the developed models show this trend in Figure 8.

Table 3. Best model performance.

Performance Criterion	
Best model	SVR-GO-BFS$_7$
Inputs No.	7
R^2	0.9826
RMSE	1.3315
MAE	1.2108
MAPE	1.1633
C	9.3119
γ	0.6363

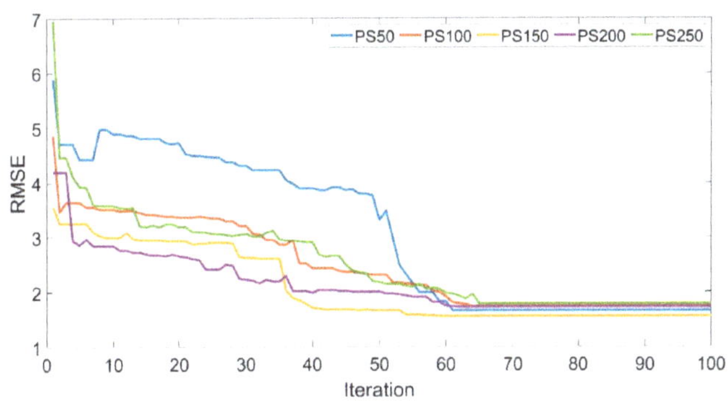

Figure 6. SVR-GO-BFS$_7$ optimization model for different population sizes (PSs).

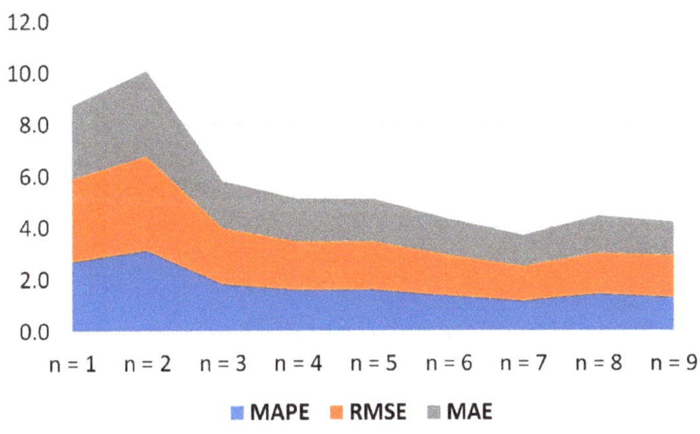

Figure 7. *Cont.*

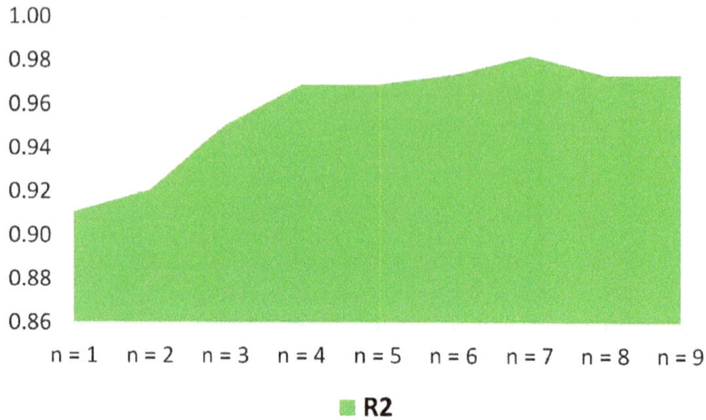

Figure 7. Performance of the SVR-GO-BFS$_n$ models with various inputs.

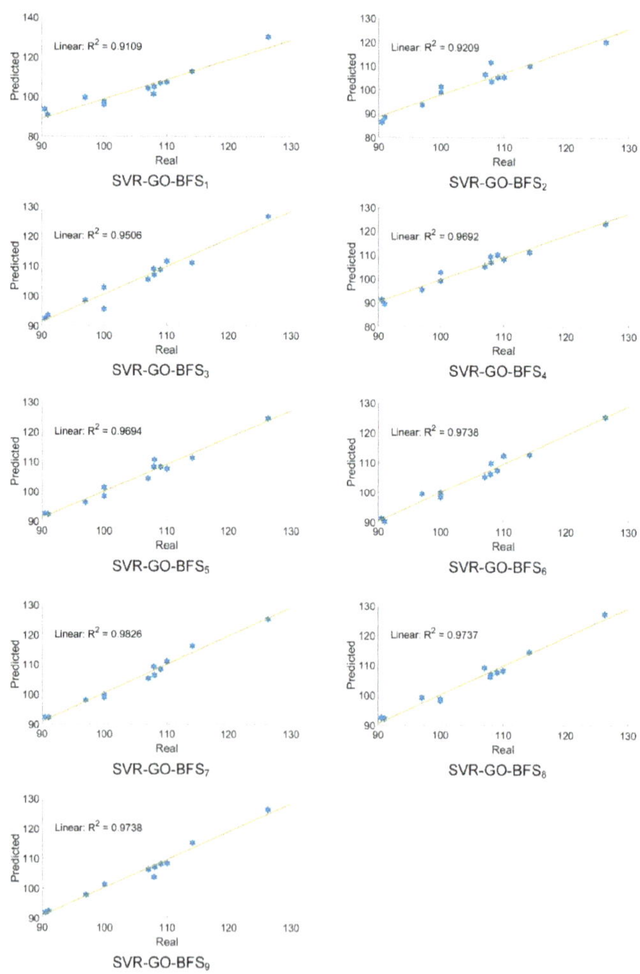

Figure 8. Real and predicted AOp values by various developed models, SVR-GO-BFS$_7$ is the best model.

3.4. Performance Comparison

The authors compared the performance of the developed SVR-GO-BFS_7 model with a single SVR model. All nine variables were used to train the single SVR model. The outcomes of this comparison are presented in Table 4. The SVR-GO-BFS_7 model achieved a notably lower MAPE value compared with the single SVR model. The value of MAPE improved by about 62% when the newly developed model was applied to the data. Furthermore, R^2 was enhanced by approximately 19%. RMSE and MAE were improved by 68.09% and 62.26%, respectively. Hence, for predictive precision, it can be assumed that SVR-GO-BFS_7 particularly beats the single SVR model for AOp forecasting in the selected granite quarry sites in Malaysia. The principal responsible for enhancing the prediction performance of the SVR-GO-BFS_7 model was SVR's parameter optimization by GO and employing BFS for input choice.

Table 4. Performance comparison between single SVR and SVR-GO-BFS_7.

Performance Criterion	SVR-GO-BFS_7	Single SVR
R^2	0.9826	0.8245
RMSE	1.3315	4.1728
MAE	1.2108	3.2083
MAPE	1.1633	3.0527

This study also compared the achievement of the developed model with some well-known ML models, including Random Forest (RF), Artificial Neural Networks (ANN), Extreme Gradient Boosting Tree (XGBT), and Classification and Regression Trees (CART). Nevertheless, BFS and GO were not hybridized with these models. All models were trained using the full set of inputs (nine inputs). For XGBT, Eta and Lambada were set as 0.3, 1.0, and its objective function was reg:linear. For CART, the maximum tree depth was 7. Concerning ANN, a backpropagation procedure by the Levenberg–Marquardt training algorithm was employed for its optimization. Additionally, the ANN structure included a single hidden layer and 11 hidden nodes. Furthermore, the authors used a sigmoid activation function while the value of the learning rate was 0.2. Table 5 shows how these models compare to SVR-GO-BFS_7 in terms of how well they work.

Table 5. Comparison between the SVR-GO-BFS_7 and other models.

Performance Criterion	SVR-GO-BFS_7	ANN	CART	RF	XGBT
R^2	0.9826	0.9767	0.5192	0.8874	0.9342
RMSE	1.3315	1.7206	7.0700	3.3300	2.5632
MAE	1.2108	1.4767	5.3168	2.9808	2.3600
MAPE	1.1633	1.3927	5.2427	2.8692	2.3240

The RMSE, MAPE, and MAE values of the developed SVR-GO-BFS_7 model were less than all benchmark models. Among benchmark models, ANN showed a better performance in terms of both accuracy and errors. Instead, the worst model was CART, which achieved the lowest accuracy and highest errors. While the XGBT obtained better accuracy than the RF, the RF outperformed the XGBT in terms of errors. The results of this comparison confirmed that the developed SVR-GO-BFS_7 was statistically better than the models developed for comparison. For a better explanation, the predictive effectiveness of the developed BA-GO-BFS_7 is demonstrated in Figure 9. The figure showed that the predicted data effectively track the real data with insignificant differences. The results of the performance criteria in Table 4 showed that the values of the error metrics were comparably low. The results of AOp predictions by SVR-GO-BFS_7 and other ML models are presented in Figure 10. The advantage of the developed SVR-GO-BFS_7 model was justified through the outcomes of the comparative evaluation. So, the importance of combining methods (SVR, GO, and BFS) is confirmed in the right way.

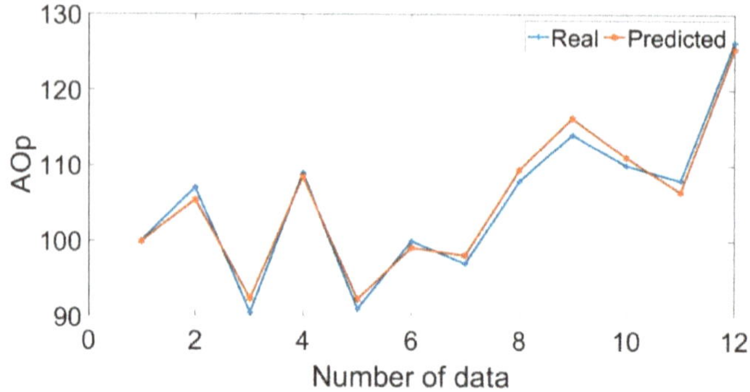

Figure 9. Real and predicted AOp values.

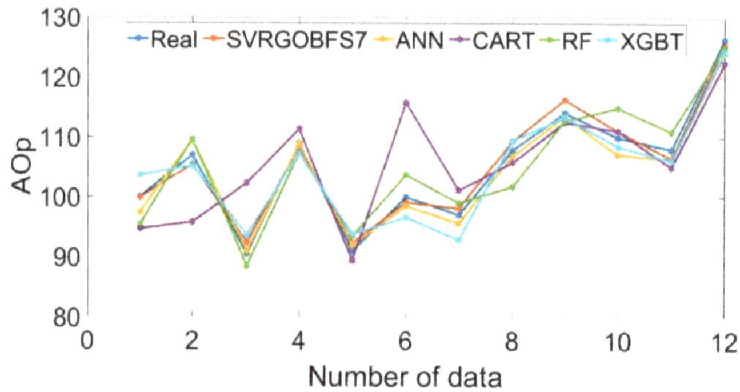

Figure 10. Results of predictions by various models.

3.5. Comparison with Other Optimization Models

In the current section, various examinations were carried out to confirm that the synthesis of BFS, SVR, and GO produces the most reliable returns. This experiment engaged three optimizers to obtain the hyperparameters of SVR. One of these techniques was PSO, which explains the regular optimization performance for adjusting the SVR's hyperparameters. Another method was the Cuckoo Optimization Algorithm (COA), which was broadly employed for fine-tuning the parameters of ML algorithms [38]. The last optimizer was the Neural Network Algorithm (NNA), one of the most advanced optimization techniques [39]. The SVR model was optimized with PSO, COA, and NNA. Seven of the inputs that SVR-GO-BFS_n used were used again when the new optimized models were made.

The precision of the optimized models is presented in Figure 11. The performance results of the tuned SVR models by the optimized techniques are displayed in Table 6. For the granite quarry dataset, the precision of the SVR-GO-BFS_7 model was higher than that of the SVR-PSO-BFS_7, SVR-COA-BFS_7, and SVR-NNA-BFS_7 models. As a result, the capacity of the GO technique to search the SVR's hyperparameters was more effective than NNA, PSO, and COA. Simply put, the SVR-GO-BFS_7 method achieves high accuracy for AOp forecasting and has the best efficiency and consistency among all basic techniques. Overall, the SVR-GO-BFS_7 technique obtained great precision for AOp prediction and had the greatest performance and cohesion amongst all basic methods. Consequently, in this research, the SVR-GO-BFS_n model is used for prediction, and further studies are suggested to utilize this method in other investigations based on the authors' concerns.

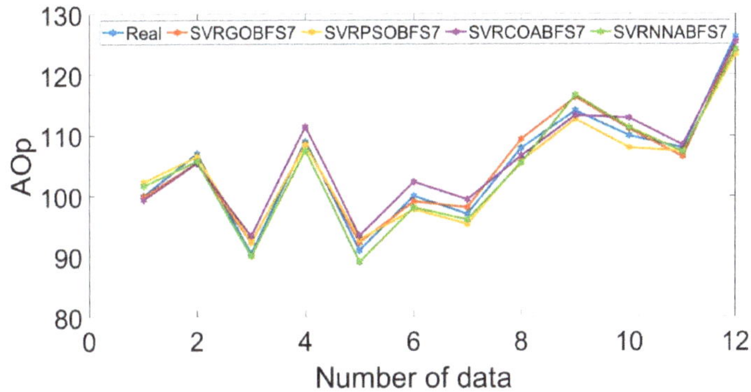

Figure 11. The outcomes of the developed model with various optimization methods.

Table 6. Comparison of various optimizers for AOp prediction.

Performance Criterion	SVR-GO-BFS$_7$	SVR-PSO-BFS$_7$	SVR-COA-BFS$_7$	SVR-NNA-BFS$_7$
R^2	0.9826	0.9796	0.9729	0.9747
RMSE	1.3315	1.8188	1.9547	1.6929
MAE	1.2108	1.6675	1.7475	1.5625
MAPE	1.1633	1.5997	1.7170	1.4740

In comparison with the previous work, the model developed in this study showed better performance. For instance, Hajihassani et al. [27] applied an ANN-PSO to the same inputs, and they did not utilize any input selection technique and only used their model for AOp estimation. The best R^2 that they achieved was 0.8836. The SVR-GA-BFS7 model achieved a better R^2 while using a fewer number of inputs, which decreased the model complexity. The authors of this study believe that the current study and its process and results are able to add value to the available literature.

4. Conclusions and Future Works

Prediction of AOp values is vital because of their negative impacts on people and construction near the blasting zone. In this paper, a hybrid learning model, the SVR-GO-BFS$_n$, was developed to forecast the AOp values. A summary of the significant findings of this study is provided below:

- In incorporation with SVR, the BFS algorithm produced excellent forecasts on the dataset of this research concern.
- The GO algorithm was confirmed to efficiently function in exploring the optimal conditions of SVR's hyperparameters and achieving an accurate AOp prediction.
- The reason for this efficacy was that this algorithm possesses a compelling ability for finding and controlling the optimal answers to multi-scale issues.
- The developed model outperformed other established models in this study, including single SVR, CART, RF, ANN, and XGBT. Hence, the SVR-GO-BFS$_n$ can be a useful procedure for forecasting AOp values. Among the various optimizers used in this study, the GO optimizer outperformed the alternative ones, i.e., PSO, NNA, and COA.

As shown by the findings of the present study, the performance of the model developed was better than others. This approach is usable by other studies in different domains. Nevertheless, the regression issue is to be resolved by the developed model and is confined to the AOp forecast. Moreover, it is vital to note that the outcomes of this model are limited to quarries with similar characteristics. Although GO is a powerful optimizer, this study only used it to tune the SVR hyperparameters. We used only one dataset, which may be a limitation of this study. Future studies should use more datasets to test our proposed model.

Following this discussion, the authors suggest that further investigations could be carried out. Further studies should implement this developed method to deal with other issues in the blasting and mining analysis. Supplementary analysis of AOp predictions in various areas is required to assure the generalization of the outcomes of the developed system. The employment of GO as an input selection technique should be examined, and, subsequently, its efficiency should be analyzed against the BFS.

Author Contributions: L.C. Methodology, Software, Formal analysis, Writing—original draft. P.G.A. Methodology, Writing—review and editing, Supervision. M.Z.T. Writing—review and editing, Supervision. D.J.A., Conceptualization, Software, Formal analysis, Writing—review and editing, Supervision. D.V.U. Writing—review and editing. M.Y., Conceptualization, Writing—review and editing. All authors have read and agreed to the published version of the manuscript.

Funding: Supported by the Science and Technology Research Program of Chongqing Municipal Education Commission (Grant No. KJQN202103415).

Institutional Review Board Statement: Not applicable.

Informed Consent Statement: Not applicable.

Data Availability Statement: The data is available from the corresponding author upon reasonable request.

Conflicts of Interest: The authors declare no conflict of interest.

References

1. Hagan, T.N. Rock breakage by explosives. In *Gasdynamics of Explosions and Reactive Systems*; Elsevier: Amsterdam, The Netherlands, 1980; pp. 329–340.
2. Harandizadeh, H.; Armaghani, D.J. Prediction of air-overpressure induced by blasting using an ANFIS-PNN model optimized by GA. *Appl. Soft Comput.* **2020**, *99*, 106904. [CrossRef]
3. Khandelwal, M.; Kankar, P. Prediction of blast-induced air overpressure using support vector machine. *Arab. J. Geosci.* **2011**, *4*, 427–433. [CrossRef]
4. Konya CJ, W.E. *Surface Blast Design*; Prentice Hall: Englewood Cliffs, NJ, USA, 1990.
5. Kuzu, C.; Fisne, A.; Ercelebi, S.G. Operational and geological parameters in the assessing blast induced airblast-overpressure in quarries. *Appl. Acoust.* **2009**, *70*, 404–411. [CrossRef]
6. Wu, C.; Hao, H. Modeling of simultaneous ground shock and airblast pressure on nearby structures from surface explosions. *Int. J. Impact Eng.* **2005**, *31*, 699–717. [CrossRef]
7. Rodríguez, R.; Toraño, J.; Menéndez, M. Prediction of the airblast wave effects near a tunnel advanced by drilling and blasting. *Tunn. Undergr. Space. Technol.* **2007**, *22*, 241–251. [CrossRef]
8. Rodríguez, R.; Lombardía, C.; Torno, S. Prediction of the air wave due to blasting inside tunnels: Approximation to a 'phonometric curve'. *Tunn. Undergr. Space. Technol.* **2010**, *25*, 483–489. [CrossRef]
9. Hosseini, S.; Monjezi, M.; Bakhtavar, E.; Mousavi, A. Prediction of Dust Emission Due to Open Pit Mine Blasting Using a Hybrid Artificial Neural Network. *Nat. Resour. Res.* **2021**, *30*, 4773–4788. [CrossRef]
10. Khandelwal, M.; Singh, T. Evaluation of blast-induced ground vibration predictors. *Soil Dyn. Earthq. Eng.* **2007**, *27*, 116–125. [CrossRef]
11. Yang, H.; Song, K.; Zhou, J. Automated Recognition Model of Geomechanical Information Based on Operational Data of Tunneling Boring Machines. *Rock Mech. Rock Eng.* **2022**, *55*, 1499–1516. [CrossRef]
12. Yang, H.; Wang, Z.; Song, K. A new hybrid grey wolf optimizer-feature weighted-multiple kernel-support vector regression technique to predict TBM performance. *Eng. Comput.* **2020**, *38*, 2469–2485. [CrossRef]
13. Kardani, N.; Bardhan, A.; Samui, P.; Nazem, M.; Asteris, P.G.; Zhou, A. Predicting the thermal conductivity of soils using integrated approach of ANN and PSO with adaptive and time-varying acceleration coefficients. *Int. J. Therm. Sci.* **2022**, *173*, 107427. [CrossRef]
14. Asteris, P.G.; Rizal, F.I.M.; Koopialipoor, M.; Roussis, P.C.; Ferentinou, M.; Armaghani, D.J.; Gordan, B. Slope Stability Classification under Seismic Conditions Using Several Tree-Based Intelligent Techniques. *Appl. Sci.* **2022**, *12*, 1753. [CrossRef]
15. Barkhordari, M.; Armaghani, D.; Asteris, P. Structural Damage Identification Using Ensemble Deep Convolutional Neural Network Models. *Comput. Model. Eng. Sci.* **2022**. [CrossRef]
16. Asteris, P.G.; Lourenço, P.B.; Roussis, P.C.; Adami, C.E.; Armaghani, D.J.; Cavaleri, L.; Chalioris, C.E.; Hajihassani, M.; Lemonis, M.E.; Mohammed, A.S. Revealing the nature of metakaolin-based concrete materials using artificial intelligence techniques. *Constr. Build. Mater.* **2022**, *322*, 126500. [CrossRef]
17. Zhou, J.; Qiu, Y.; Khandelwal, M.; Zhu, S.; Zhang, X. Developing a hybrid model of Jaya algorithm-based extreme gradient boosting machine to estimate blast-induced ground vibrations. *Int. J. Rock Mech. Min. Sci.* **2021**, *145*, 104856. [CrossRef]

18. Zhou, J.; Dai, Y.; Khandelwal, M.; Monjezi, M.; Yu, Z.; Qiu, Y. Performance of Hybrid SCA-RF and HHO-RF Models for Predicting Backbreak in Open-Pit Mine Blasting Operations. *Nat. Resour. Res.* **2021**, *30*, 4753–4771. [CrossRef]
19. Zhou, J.; Chen, C.; Wang, M.; Khandelwal, M. Proposing a novel comprehensive evaluation model for the coal burst liability in underground coal mines considering uncertainty factors. *Int. J. Min. Sci. Technol.* **2021**, *31*, 799–812. [CrossRef]
20. Zhang, R.; Li, Y.; Gui, Y.; Zhou, J. Prediction of blasting induced air-overpressure using a radial basis function network with an additional hidden layer. *Appl. Soft Comput.* **2022**, *127*, 109343. [CrossRef]
21. Zeng, J.; Jamei, M.; Nait Amar, M.; Hasanipanah, M.; Bayat, P. A novel solution for simulating air overpressure resulting from blasting using an efficient cascaded forward neural network. *Eng. Comput.* **2022**, *38*, 2069–2081. [CrossRef]
22. Hosseini, S.; Poormirzaee, R.; Hajihassani, M. Application of reliability-based back-propagation causality-weighted neural networks to estimate air-overpressure due to mine blasting. *Eng. Appl. Artif. Intell.* **2022**, *115*, 105281. [CrossRef]
23. Hasanipanah, M.; Jahed Armaghani, D.; Khamesi, H.; Bakhshandeh Amnieh, H.; Ghoraba, S. Several non-linear models in estimating air-overpressure resulting from mine blasting. *Eng. Comput.* **2016**, *32*, 441–455. [CrossRef]
24. Murlidhar, B.; Bejarbaneh, B.; Armaghani, D.; Mohammed, A.; Mohamad, E. Application of Tree-Based Predictive Models to Forecast Air Overpressure Induced by Mine Blasting. *Nat. Resour. Res.* **2020**, *30*, 1865–1887. [CrossRef]
25. Hasanipanah, M.; Shahnazar, A.; Bakhshandeh Amnieh, H.; Jahed Armaghani, D. Prediction of air-overpressure caused by mine blasting using a new hybrid PSO–SVR model. *Eng. Comput.* **2017**, *33*, 23–31. [CrossRef]
26. He, Z.; Armaghani, D.J.; Masoumnezhad, M.; Khandelwal, M.; Zhou, J.; Murlidhar, B.R. A Combination of Expert-Based System and Advanced Decision-Tree Algorithms to Predict Air—Overpressure Resulting from Quarry Blasting. *Nat. Resour. Res.* **2021**, *30*, 1889–1903. [CrossRef]
27. Hajihassani, M.; Jahed Armaghani, D.; Sohaei, H.; Tonnizam Mohamad, E.; Marto, A. Prediction of airblast-overpressure induced by blasting using a hybrid artificial neural network and particle swarm optimization. *Appl. Acoust.* **2014**, *80*, 57–67. [CrossRef]
28. Vapnik, V. *The Nature of Statistical Learning Theory*; Springer Science & Business Media: Berlin, Germany, 2013; ISBN 1475732643.
29. Dreyfus, G. *Neural Networks: Methodology and Applications*; Springer: Berlin/Heidelberg, Germany, 2005.
30. Saremi, S.; Mirjalili, S.; Lewis, A. Grasshopper optimisation algorithm: Theory and application. *Adv. Eng. Softw.* **2017**, *105*, 30–47. [CrossRef]
31. Kursa, M.B.; Rudnicki, W.R. Feature selection with the Boruta package. *J. Stat. Softw.* **2010**, *36*, 1–13. [CrossRef]
32. Lu, S.; Koopialipoor, M.; Asteris, P.G.; Bahri, M.; Armaghani, D.J. A Novel Feature Selection Approach Based on Tree Models for Evaluating the Punching Shear Capacity of Steel Fiber-Reinforced Concrete Flat Slabs. *Materials* **2020**, *13*, 3902. [CrossRef]
33. Asteris, P.G.; Argyropoulos, I.; Cavaleri, L.; Rodrigues, H.; Varum, H.; Thomas, J.; Lourenço, P.B. Masonry compressive strength prediction using artificial neural networks. In Proceedings of the International Conference on Transdisciplinary Multispectral Modeling and Cooperation for the Preservation of Cultural Heritage, Athens, Greece, 10–13 October 2018; pp. 200–224.
34. Huang, J.; Asteris, P.; Pasha, S.; Mohammed, A.; Hasanipanah, M. A new auto-tuning model for predicting the rock fragmentation: A cat swarm optimization algorithm. *Eng. Comput.* **2020**, *38*, 2209–2220. [CrossRef]
35. Asteris, P.G.; Gavriilaki, E.; Touloumenidou, T.; Koravou, E.; Koutra, M.; Papayanni, P.G.; Pouleres, A.; Karali, V.; Lemonis, M.E.; Mamou, A. Genetic prediction of ICU hospitalization and mortality in COVID-19 patients using artificial neural networks. *J. Cell. Mol. Med.* **2022**, *26*, 1445–1455. [CrossRef]
36. Asteris, P.G.; Douvika, M.G.; Karamani, C.A.; Skentou, A.D.; Chlichlia, K.; Cavaleri, L.; Daras, T.; Armaghani, D.J.; Zaoutis, T.E. A novel heuristic algorithm for the modeling and risk assessment of the COVID-19 pandemic phenomenon. *Comput. Model. Eng. Sci.* **2020**, *125*, 815–828. [CrossRef]
37. Psyllaki, P.; Stamatiou, K.; Iliadis, I.; Mourlas, A.; Asteris, P.; Vaxevanidis, N. Surface treatment of tool steels against galling failure. In *Proceedings of the MATEC Web of Conferences*; EDP Sciences: Les Ulis, France, 2018; Volume 188, p. 4024.
38. Rajabioun, R. Cuckoo optimization algorithm. *Appl. Soft Comput.* **2011**, *11*, 5508–5518. [CrossRef]
39. Sadollah, A.; Sayyaadi, H.; Yadav, A. A dynamic metaheuristic optimization model inspired by biological nervous systems: Neural network algorithm. *Appl. Soft Comput.* **2018**, *71*, 747–782. [CrossRef]

applied sciences

MDPI

Article

Experimental and Empirical Study for Prediction of Blast Loads

Anselmo Filice [1,2,*], Miroslav Mynarz [3] and Raffaele Zinno [2]

[1] Department of Computer Science, Modeling, Electronics and Systems Engineering, DIMES, University of Calabria, 87036 Arcavacata, Italy

[2] Department of Environmental Engineering, DIAM, University of Calabria, 87036 Arcavacata, Italy; raffaele.zinno@unical.it

[3] Faculty of Safety Engineering, VSB, Technical University of Ostrava, 70833 Ostrava, Czech Republic; miroslav.mynarz@vsb.cz

* Correspondence: anselmo.filice@unical.it; Tel.: +39-333-1371258

Abstract: This paper presents the issue of determining the blast load on an engineering structure. In cases of industrial accidents or terrorist attacks, in many cases it is necessary to determine the necessary explosion parameters to determine the response of the structure, preferably in a simple and time-saving manner. In such a way, the empirical relationships can be used to estimate the selected parameters of the explosion load. Many empirical relationships have been derived in the past, but not all are suitable for different types of explosions. This article compares and validates experimentally determined selected explosion parameters for the chosen explosive with empirical relationships. For comparison, three already verified and frequently used calculation procedures (Kingery, Kinney, Henrych) and one newly derived procedure (PECH) were used. As part of the experimental measurements, blast wave explosion parameters for small charges were determined for near-field explosions. The general-purpose plastic explosive Semtex 10-SE was used for the experiments. The results of the comparative study presented in this article demonstrate the importance of taking these procedures into account for a reliable determination of the effects of blast actions on buildings.

Keywords: blast loads; engineering structures; prediction; validation; Semtex

Citation: Filice, A.; Mynarz, M.; Zinno, R. Experimental and Empirical Study for Prediction of Blast Loads. *Appl. Sci.* **2022**, *12*, 2691. https://doi.org/10.3390/app12052691

Academic Editors: Lina M. López, Ricardo Castedo and Anastasio P. Santos

Received: 30 January 2022
Accepted: 2 March 2022
Published: 4 March 2022

1. Introduction

The effects of explosions were studied more extensively after World War II, mainly because many explosives were developed during this period and are still in use today. For many explosives, demands are made for high stability, safety and effectiveness. Most aromatic nitrolates and a much smaller number of nitric acid esters and nitroamines meet these criteria. The most commonly used explosive is the well-known Trinitrotoluene (tritol or TNT).

To estimate or calculate the response of an object or structure to the effects of an explosion, it is necessary to know the explosion parameters. Of these parameters needed to determine the response, the most important are usually the peak overpressure, the positive phase duration and the positive phase impulse. The positive phase impulse can usually be calculated from the blast curve and is dependent on the overpressure and positive duration. Knowledge of the blast parameters then defines the basic load of buildings or structures. A variety of methods can be used to determine explosion parameters. These can be based on empirical (or analytical), semi-empirical or numerical methods. Empirical methods are essentially correlations with experimental data. Most of these approaches are limited by the range of experiments carried out. The accuracy of all empirical correlations decreases with distance to the source of the explosion. Semi-empirical methods are based on simplified models of physical phenomena. They try to use the essential physical processes in a simplified way. These methods rely on extensive experimental data. Their predictive accuracy is generally better than that of empirical methods.

Numerical methods are based on mathematical equations that describe the basic laws of the phenomena being solved. These methods consider the conservation of mass, momentum and energy, or the physical behaviour of materials by means of constitutive relations. It is the empirical methods in the initial assessment of objects in the context of prevention or severe consequence assessment that can be a very useful tool. When rapid estimates of explosion parameters are needed, often for a large number of objects, these methods are essential. For a more in-depth analysis of the load of a building or to refine the initial load values of a structure, numerical methods are much more suitable.

During the second half of the 20th century, a considerable number of experimental and theoretical studies were conducted to understand the effects of blast on buildings and structures [1–5]. The aim was first to study the behaviour of air blast waves including the determination of their characteristics and then to investigate the dominant factors influencing the incident waves. Another objective was to investigate the response of the building structure to blast load [6,7], based on the analysis of several experimental data, which presented the formulae to compute peak positive overpressure, positive phase duration and positive phase impulse [2,4], and utilised both experimental and theoretical means to obtain the parameters of the blast wave such as overpressure, positive phase duration, blast wave arrival time and positive phase impulse [1]. In 1984, Kingery and Bulmash presented the parameters for air burst in terms of high order polynomials [4,8] and presented the same results as were produced by Kingery, in terms of simplified polynomials functions.

The use of empirical laws has been extensively studied and has been applied in various recommendations, mostly proposed by military authorities. In particular, the two most commonly used empirical models are based upon different but related studies of the U.S. Army Corps of Engineers (USACE): the document [9], containing the model CONWEP, and the Technical Manual TM5-1300 [10], completed by successive documents [11]. In 2013, the Joint Research Centre of the European Union produced a Technical Report [12], substantially referring to these two last documents and to another Technical Report of the U.S. Army [4]. In [12], all the empirical laws of [10] are reproduced using the International System of Units. Most available publications concerning the effects of an explosion on a civil structure regard reinforced concrete structures, and usually the geometries considered are really simple, normally a squared building [13–18], or in some cases bridges, e.g., [19]. Analyses of complex and structurally advanced objects are not published to any great extent. These analyses require the use of sophisticated computational software, usually based on CFD or FEM methods. These methods require a high level of computational expertise and are also very time consuming, both in terms of model building and computational time.

2. Materials and Methods

2.1. Ideal Blast Wave Characteristics

An explosion is a physical phenomenon in which there is a sudden, very rapid release of energy. The phenomenon lasts only some milliseconds, and it results in the production of very high temperatures and pressures. During detonation the hot gases that are produced expand in order to occupy the available space, leading to wave-type propagation through space that is transmitted spherically through an unbounded surrounding medium. Along with the produced gases, the air around the blast (for air blasts) also expands and its molecules pile-up, resulting in what is known as a blast wave and shock front. The blast wave contains a large part of the energy that was released during detonation and moves faster than the speed of sound [20].

Figure 1 illustrates the idealised profile of the pressure in relation to time for the case of a free air blast wave, which reaches a point at a certain distance from the detonation. The pressure surrounding the element is initially equal to the ambient pressure P_0, and it undergoes an instantaneous increase to a peak pressure P_{S0} at the arrival time t_A, when the shock front reaches that point. The time needed for the pressure to reach its peak value is very small and for design purposes it is assumed to be equal to zero. The peak pressure

P_{S0} is also known as side-on overpressure. The peak overpressure $P_{S0} - P_0$ is marked as P_{ti} in the following sections. The value of the peak overpressure as well as the velocity of propagation of the shock wave decrease with increasing distance from the detonation centre. After its peak value, the pressure decreases with an exponential rate until it reaches the ambient pressure at $t_A + t_0$, where t_0 is called the positive phase duration. After the positive phase of the pressure–time diagram, the pressure becomes smaller (referred to as negative) than the ambient value, and finally returns to it. The negative phase is longer than the positive one, its minimum pressure value is denoted as $P_{S0}{}^{-}$ and its duration as $t_0{}^{-}$. During this phase the structures are subjected to suction forces, which is the reason why sometimes during blast loading glass fragments from failures of facades are found outside a building instead of in its interior. This specific type of failure is usually due to a primary failure of the glass filling by the positive phase, or a situation may occur where the negative phase has a significantly higher negative impulse value $i_s{}^{-}$ [21].

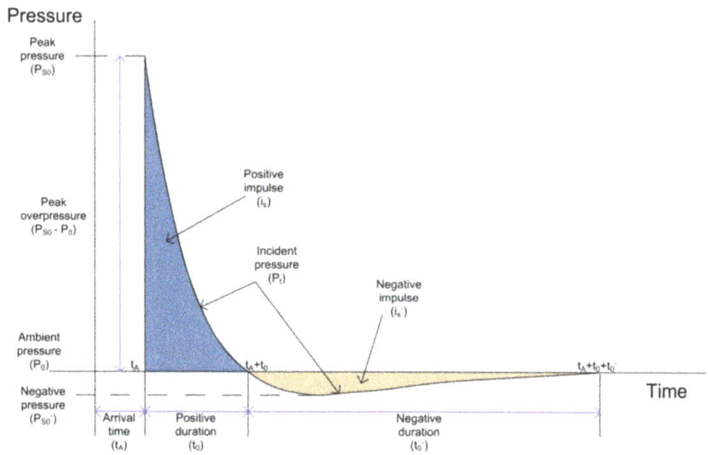

Figure 1. Ideal blast wave's pressure time course.

2.2. *Positive Phase/Shock Wave*

The positive (expanding) phase of a shock wave is created when the very hot expanding gases produced by the detonation compress the surrounding air. These compressed layers of air are sometimes visible as white, rapidly expanding rings called a shock front. The width of the shock front is only very small and represents that part of the atmosphere which is compressed just before it sets itself in motion and thus becomes part of the positive or expanding phase of a shock wave. If a strong shock front hits a solid obstacle, it is reflected or passes through (when the obstacle is destroyed), but its energy is reduced [22].

2.3. *Experiments*

A set of 20 tests was made in the experiments carried out, with five measurements (repeated trials) for each charge. During the measurements, the propagation of the shock wave was recorded, and the individual explosive parameters were evaluated based on the observed propagations. These were mainly peak overpressure and positive phase duration. Charges of 100, 200, 300 and 400 g were used in the experiments. The charges were placed at a height of 1.25 m above the ground surface (terrain). With respect to the air shock wave, this is a hemispherical blast, which occurs when the charge is placed on or near the ground surface.

To prevent the shock wave from being affected (e.g., by obstacles), the charge was suspended. The ignition coil was attached to the charge along the suspension (from above). The atmospheric conditions during the experiments carried out are shown in Table 1.

Table 1. The atmospheric conditions during the experiments.

Temperature (°C)	Air Density (kg/m^3)	Ambient Pressure (kPa)	Relative Air Humidity (%)	Height above Sea Level (m)	Sound Speed (m/s)
15	1.225	101.385	55	415	340

The recording of the blast wave was carried out using two types of transducers. PCB Piezotronics type 113B28 transducers in pencil probe or knife probe design were used. The transducers have a measuring range of 344.7 kPa, rise time less than 1 μs and natural frequency greater than 500 kHz. Pressure transducers were arranged in pairs (pencil and knife probes) around the charge at predetermined distances so as not to interfere with the measurements of the other pairs. The measurement distances were set at 2, 3, 4 and 5 m. The transducers were placed at the same height above the ground as the charge, i.e., 1.25 m. The arrangement of the pairs of transducers and their distances are presented in Figure 2 (the pairs of transducers are marked as pair). The positions of the suspended charge and the HS camera are also seen in the figure.

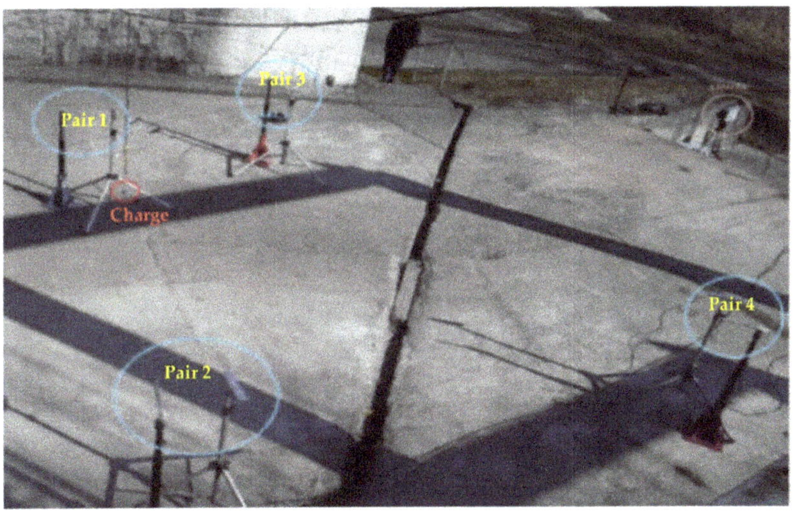

Figure 2. Arrangement of charges and measuring equipment.

2.4. The Explosive Used

The explosive used in the experiments was labelled Semtex 10-SE. Semtex 10-SE is a plastic, industrial waterproof white explosive that is classified as a special purpose explosive. This type of Semtex contains a non-explosive plasticizer. Highly explosive pentrite (PETN) is its main and effective component. It is supplied in a leaf charge form. It is mainly used for blast hardening of metallic materials. Selected parameters of the explosive are specified in Table 2.

Table 2. Selected properties of explosive Semtex 10-SE.

Detonation Velocity (m/s)	Density (g/cm^3)	Heat of Combustion (kJ/kg)	Temperature of Detonation (°C)	Gas Volume (dm^3/kg)	Oxygen Balance (% O_2)	Brisance by Hess (mm)
6700	1.45	2709	1975	1100	−62.6	20

The initiation of the charges was carried out with an electric detonator marked 0-ZB-S from the manufacturer Austin Detonator, which is instantaneous with medium resistance to the effects of external sources of electricity. The tube material is copper. The primary charge is quicksilver. The secondary charge of the detonator is 720 mg (PETN). It has a relatively high initiating capability. The Semtex 10-SE charges used were shaped into spheres of 100, 200, 300 and 400 g. The design of the experiment was based on the authors' interest in detecting near-field blast effects. The course of the explosive transformation when using a 300 g charge in defined time periods is presented in Figure 3. In addition, the propagation of the shock wave including the reflection from the ground level (blue line) is shown.

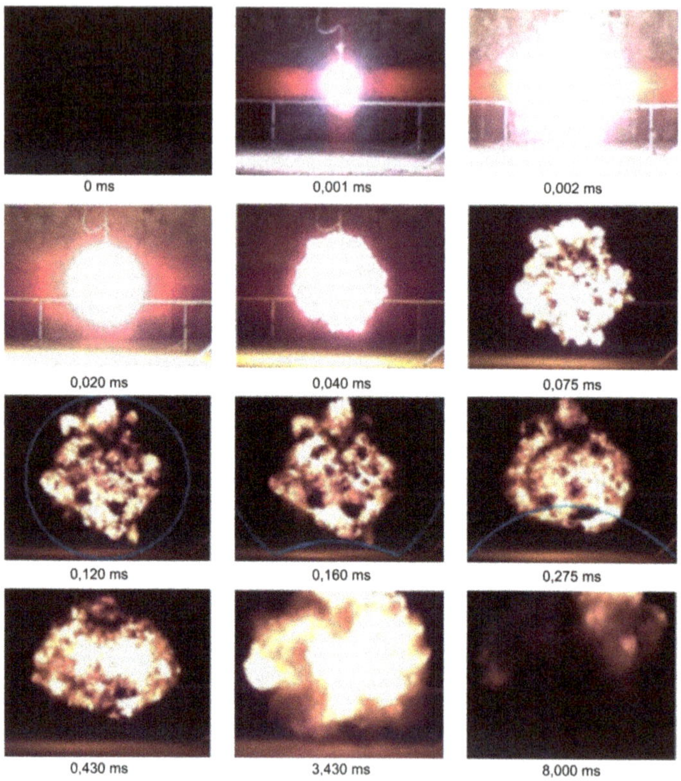

Figure 3. Detonation sequence of a 300 g Semtex 10-SE spherical charge with shock wave propagation marked.

2.5. Predictive Models

This section deals with predictive calculation methods for individual shock wave parameters. Based on many literature sources, there are a large number of computational relationships for individual parameters of blast. To verify their applicability to the case of small near-field charges, the most appropriate ones have been selected.

The concept of reducing the distance is based on the complexity of determining the values of shock wave effects. In the case of one type of explosive, it would be necessary to measure the parameters for all explosive masses at different distances in the experiment, which would be very difficult for large charges, for example. It can therefore be assumed that, at the same reduced distance, the magnitude of a particular shock wave parameter induced by a particular type of explosive is the same for all explosive masses. The reduced distance makes it possible to determine the correlation between the different distances and masses of explosives and to simplify the possibilities of determining their explosion parameters [22,23].

Thus, the reduced distance is a basic calculation relationship that is used to derive shock wave parameters based on the knowledge of the explosive mass and the distance between the point of exposure to the shock wave and the epicentre of the explosion, as given in Relation (1).

$$Z = \frac{R}{\sqrt[3]{W}}, \tag{1}$$

In this relationship, Z represents the reduced distance in m·kg$^{-1/3}$, R is the actual distance of the charge in m, and W is the mass of the charge, usually the equivalent mass of TNT in kilograms.

2.5.1. Kingery and Bulmash

A single polynomial function is applied for calculations of all shock wave parameters. The calculation of the corresponding parameter consists in the appropriate assignment of the corresponding constants to the specified coefficients. These constants differ from each other according to the type of the parameter to be searched and belong to the corresponding interval of reduced distances. The general notation of the polynomial function according to Kingery and Bulmash is provided by Relation (2) [4].

$$f = e^{A+B\cdot ln\ Z+C\cdot (ln\ Z)^2+D\cdot (ln\ Z)^3+E\cdot (ln\ Z)^4+F\cdot (ln\ Z)^5+G\cdot (ln\ Z)^6}, \tag{2}$$

In this relation, f is the result variable of the calculation, characterising the parameter being searched for. This relation can be used to calculate the peak overpressure P_{S0}, the arrival time of the shock wave t_A, the duration of the positive phase t_0, the shock wave velocity v, the positive impulse of the explosion i_s, the reflected pressure p_r and the reflected impulse of the explosion i_r. A, B, C, D, E, F and G are the coefficients to which values from the constants corresponding to the parameter of interest are assigned. The values of these coefficients can be found in Table 3.

Table 3. Simplified Kingery air blast coefficients for blast overpressure.

Z (m·kg$^{-1/3}$)	A	B	C	D	E	F	G
0.2–2.9	7.2106	−2.1069	−0.3229	0.1117	0.0685	0	0
2.9–23.8	7.5938	−3.0523	0.4098	0.0261	−0.0127	0	0
23.8–198.5	6.0536	−1.4066	0	0	0	0	0

This relation is suitable due to its complexity, as many parameters can be determined using it. In addition, it enables calculation for a relatively wide range of reduced distances, especially for peak overpressure (0.2–198.5 m·kg$^{-1/3}$).

2.5.2. Kinney and Graham

The Kinney and Graham relations were chosen as one of the alternative calculation relations, which provide the calculation of the maximum overpressure, impulse and the positive phase duration [1]. The individual relations are not constrained by the range of reduced distances. For comparison purposes, only the relationships for peak overpressure (Relation (3)) and positive phase duration (Relation (4)) are presented:

$$\Delta p_f = P_0 \cdot \frac{808 \cdot \left[1 + \left(\frac{Z}{4.5}\right)^2\right]}{\sqrt{1 + \left(\frac{Z}{0.048}\right)^2} \cdot \sqrt{1 + \left(\frac{Z}{0.32}\right)^2} \cdot \sqrt{1 + \left(\frac{Z}{1.35}\right)^2}}, \tag{3}$$

$$t_{p+} = \sqrt[3]{W} \cdot \frac{980 \cdot \left[1 + \left(\frac{Z}{0.54}\right)^{10}\right]}{\left[1 + \left(\frac{Z}{0.02}\right)^{3}\right] \cdot \left[1 + \left(\frac{Z}{0.74}\right)^{6}\right] \cdot \sqrt{1 + \left(\frac{Z}{6.9}\right)^{2}}}, \tag{4}$$

2.5.3. Henrych and Major

The calculation relations according to Henrych and Major are not intended for such large ranges of reduced distances as the relations of Kingery and Bulmash, but they provide possibilities of calculation of certain parameters of shock waves for the range of sufficiently low reduced distances, when the Kingery and Bulmash relations in some cases (e.g., peak overpressure) do not allow the calculation [2].

As with the other relationships, these relationships allow the calculation of peak overpressure, positive phase impulse and the positive phase duration. In order to apply each formula, the condition that the reduced distance must fall within a specified interval for which the relation holds, must be met—as it is found in Relations (5)–(7).

$$\Delta p_f = \frac{1.380}{Z} + \frac{0.543}{Z^2} - \frac{0.035}{Z^3} + \frac{0.000613}{Z^4} \text{ for } (0.05 \leq Z \leq 0.3), \tag{5}$$

$$\Delta p_f = \frac{0.607}{Z} - \frac{0.032}{Z^2} + \frac{0.209}{Z^3} \text{ for } (0.3 \leq Z \leq 1.0), \tag{6}$$

$$\Delta p_f = \frac{0.0649}{Z} + \frac{0.397}{Z^2} + \frac{0.322}{Z^3} \text{ for } (1.0 \leq Z \leq 1.0). \tag{7}$$

Only one formula is defined for the positive phase duration, Relation (8).

$$t_{p+} = \sqrt[3]{W} \cdot \left(0.107 + 0.444\,Z + 0.264\,Z^2 - 0.129\,Z^3 + 0.0335\,Z^4\right) \text{ for } (0.05 < Z \leq 3.0). \tag{8}$$

2.5.4. PECH

The derived PECH calculation relation does not primarily use reduced distance values and is applicable to a wide range of actual distances. The calculation relation allows only the peak overpressure (Relation (9)) to be determined. A formula for the positive phase duration has not yet been derived.

$$\Delta p_f = 0.84 \cdot \left(\frac{W}{R^3}\right)^{\frac{1}{3}} + 2.7 \cdot \left(\frac{W}{R^3}\right)^{\frac{2}{3}} + 7 \cdot \left(\frac{W}{R^3}\right), \tag{9}$$

The above computational relations were used to determine selected explosion parameters (peak overpressure and the positive phase duration of the explosion) during the initiation of charges with varying mass. To verify the applicability of the computational relations, a comparison with experimentally determined values was made. Based on the comparison, it was then possible to deduce the most suitable computational procedure for the required shock wave parameters under the given conditions of the realised experiments.

3. Results

The experimental measurements included five repeated trials for each charge weight. The charge weights were determined in the range of 100 to 400 g. Table 4 presents the values of selected explosion parameters (peak overpressure and positive phase duration) for each charge weight and scaled distance. Since the peak overpressure values from the repeated trials for each charge varied only slightly (only in tenths of kPa), the averaged values are presented. In the case of the positive phase duration values, the variations were quite negligible.

Table 4. Measured peak overpressures and positive phase durations.

Blast Parameters	Charge Weight (g)	Distance (m)			
		2	3	4	5
Peak incident overpressure P_{ti} (kPa)	100	44.1	23.4	14.3	10.5
	200	66.6	31.1	20.1	13.6
	300	81.3	41.6	25.9	17.9
	400	93.4	46.2	28.0	20.4
Positive phase duration t_0 (ms)	100	1.2	1.5	1.7	1.8
	200	1.2	1.6	1.7	1.9
	300	1.4	1.5	1.9	2.0
	400	1.4	1.6	1.7	1.8

Figure 4 shows the blast curves for each distance for a 100 g charge. From the graph it is possible to read the values of peak overpressure, positive phase duration and possibly other explosion parameters. The value "0" on the horizontal axis indicates the moment of initiation of the charge.

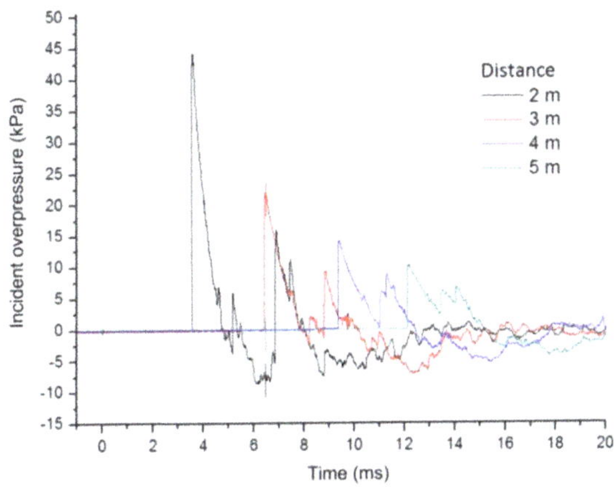

Figure 4. Pressure time course for a 100 g charge.

The values calculated according to the individual predictive models for a 100 g charge are provided in Table 5. For some models, it was not possible to calculate positive phase duration values, either because the scaled distance values were outside the range of applicability of the formulas, or the method used did not allow this calculation.

Table 5. Values of empirical relations for a 100 g charge.

Semtex 100 g		Scaled Distance Z (m·kg$^{-1/3}$)	Blast Parameters	Kingery and Bulmash	Kinney and Graham	Henrych and Major	PECH
Distance (m)	2	2.76	P_{ti} (kPa)	54.8	37.6	39.4	40.8
			t_0 (ms)	1.6	1.0	-	-
	3	4.14	P_{ti} (kPa)	27.6	18.4	20.2	21.1
			t_0 (ms)	1.9	1.3	-	-
	4	5.51	P_{ti} (kPa)	17.9	12.0	13.1	13.9
			t_0 (ms)	2.1	1.5	-	-
	5	6.89	P_{ti} (kPa)	13.2	8.9	-	10.3
			t_0 (ms)	2.3	1.6	-	-

Table 6 summarises the relative deviations of the calculated and measured values for a charge of 100 g. The relative deviations are calculated for the individual scaled distances for both blast parameters. Positive values of the relative deviations indicate an overestimation of the blast parameters compared to the experimentally determined values, and conversely, negative values of the relative deviations indicate an underestimation. For a more comprehensive evaluation of the applicability of the computational relationships, all relative deviations were averaged within one model, separately for each observed explosion parameter.

Table 6. Values of relative deviations from experiment for a 100 g charge.

Semtex 100 g		Scaled Distance Z (m·kg$^{-1/3}$)	Blast Parameters	Kingery and Bulmash	Kinney and Graham	Henrych and Major	PECH
				Relative Deviation (%)			
Distance (m)	2	2.76	P_{ti} (kPa)	24.3	−14.7	−10.7	−7.5
			t_0 (ms)	33.3	−16.7	-	-
	3	4.14	P_{ti} (kPa)	17.9	−21.4	−13.7	−9.8
			t_0 (ms)	26.7	−13.3	-	-
	4	5.51	P_{ti} (kPa)	25.2	−16.1	−8.4	−2.8
			t_0 (ms)	23.5	−11.8	-	-
	5	6.89	P_{ti} (kPa)	25.7	−15.2	-	−1.9
			t_0 (ms)	27.8	−11.1	-	-
Arithmetic mean (%)			P_{ti} (kPa)	23.3	−16.9	−10.9	−5.5
			t_0 (ms)	28.4	−12.9	-	-

Figure 5 illustrates the blast curves for each distance for a 200 g charge.

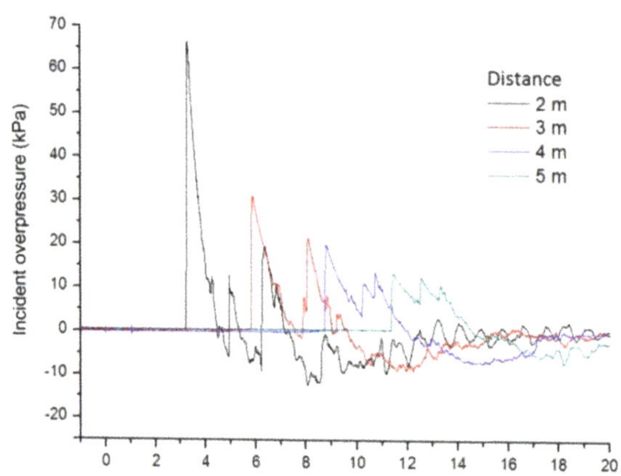

Figure 5. Pressure time course for a 200 g charge.

The values calculated according to the individual predictive models for a 200 g charge are presented in Table 7. Table 8 lists the relative deviations of the calculated and measured values for a 200 g charge.

Table 7. Values of empirical relations for a 200 g charge.

Semtex 200 g		Scaled Distance Z (m·kg$^{-1/3}$)	Blast Parameters	Kingery and Bulmash	Kinney and Graham	Henrych and Major	PECH
Distance (m)	2	2.76	P_{ti} (kPa)	85.5	60.0	59.2	62.0
			t_0 (ms)	1.8	1.1	-	-
	3	4.14	P_{ti} (kPa)	40.6	27.1	29.4	30.4
			t_0 (ms)	2.2	1.5	-	-
	4	5.51	P_{ti} (kPa)	25.2	16.8	18.5	19.4
			t_0 (ms)	2.5	1.7	-	-
	5	6.89	P_{ti} (kPa)	18.1	12.1	13.2	14.0
			t_0 (ms)	2.6	1.9	-	-

Table 8. Values of relative deviations from experiment for a 200 g charge.

Semtex 200 g		Scaled Distance Z (m·kg$^{-1/3}$)	Blast Parameters	Kingery and Bulmash	Kinney and Graham	Henrych and Major	PECH
				Relative Deviation (%)			
Distance (m)	2	2.76	P_{ti} (kPa)	28.4	−9.9	−11.1	−6.9
			t_0 (ms)	50.1	−8.3	-	-
	3	4.14	P_{ti} (kPa)	30.5	−12.9	−5.5	−2.3
			t_0 (ms)	37.5	−6.3	-	-
	4	5.51	P_{ti} (kPa)	26.0	−16.0	−7.5	−3.0
			t_0 (ms)	47.1	0.0	-	-
	5	6.89	P_{ti} (kPa)	33.1	−11.0	−2.9	2.9
			t_0 (ms)	30.0	−5.0	-	-
Arithmetic mean (%)			P_{ti} (kPa)	29.5	−12.5	−6.8	−2.3
			t_0 (ms)	41.2	−4.9	-	-

Figure 6 gives the blast curves for each distance for a 300 g charge.

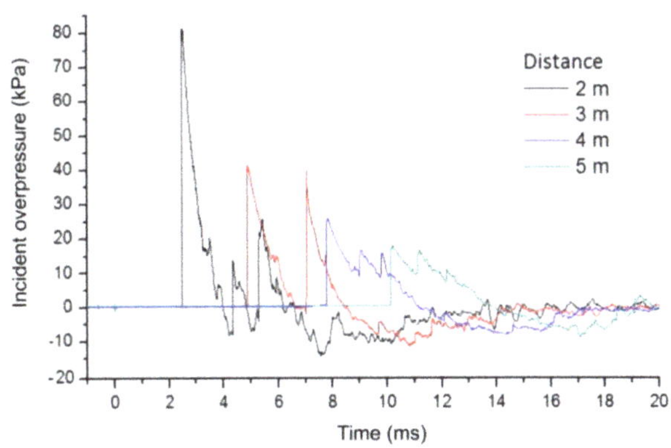

Figure 6. Pressure time course for a 300 g charge.

The values calculated according to the individual predictive models for a 300 g charge are shown in Table 9.

Table 9. Values of empirical relations for a 300 g charge.

Semtex 300 g		Scaled Distance Z (m·kg$^{-1/3}$)	Blast Parameters	Kingery and Bulmash	Kinney and Graham	Henrych and Major	PECH
Distance (m)	2	2.76	P_{ti} (kPa)	112.9	80.3	76.0	80.5
			t_0 (ms)	1.9	1.1	-	-
	3	4.14	P_{ti} (kPa)	51.1	34.8	36.8	38.2
			t_0 (ms)	2.4	1.5	-	-
	4	5.51	P_{ti} (kPa)	31.2	20.9	22.9	23.8
			t_0 (ms)	2.7	1.8	-	-
	5	6.89	P_{ti} (kPa)	22.1	14.7	16.2	17.0
			t_0 (ms)	2.9	2.1	-	-

Table 10 contains the relative deviations of the calculated and measured values for a 300 g charge.

Table 10. Values of relative deviations from experiment for a 300 g charge.

Semtex 300 g		Scaled Distance Z (m·kg$^{-1/3}$)	Blast Parameters	Kingery and Bulmash	Kinney and Graham	Henrych and Major	PECH
				Relative Deviation (%)			
Distance (m)	2	2.76	P_{ti} (kPa)	38.9	−1.2	−6.5	−1.0
			t_0 (ms)	26.7	−26.0	-	-
	3	4.14	P_{ti} (kPa)	22.8	−16.3	−11.5	−8.2
			t_0 (ms)	71.4	7.1	-	-
	4	5.51	P_{ti} (kPa)	20.5	−19.3	−11.6	−8.1
			t_0 (ms)	42.1	−5.3	-	-
	5	6.89	P_{ti} (kPa)	23.5	−17.9	−9.5	−5.0
			t_0 (ms)	31.8	−4.5	-	-
Arithmetic mean (%)			P_{ti} (kPa)	26.4	−13.7	−9.8	−5.6
			t_0 (ms)	43.0	−7.2	-	-

Figure 7 represents the blast curves for each distance for a 400 g charge.

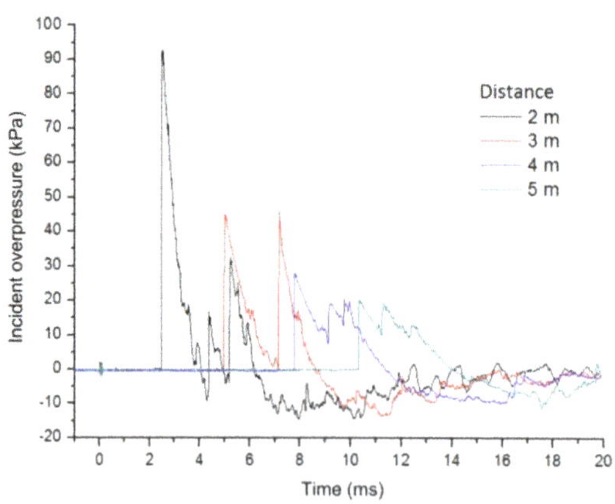

Figure 7. Pressure time course for a 400 g charge.

The values calculated according to the individual predictive models for a 400 g charge are presented in Table 11.

Table 11. Values of empirical relations for a 400 g charge.

Semtex 400 g		Scaled Distance Z (m·kg$^{-1/3}$)	Blast Parameters	Kingery and Bulmash	Kinney and Graham	Henrych and Major	PECH
Distance (m)	2	2.76	P_{ti} (kPa)	138.5	99.4	91.1	97.5
			t_0 (ms)	1.9	1.1	-	-
	3	4.14	P_{ti} (kPa)	60.9	42.0	43.5	45.1
			t_0 (ms)	2.5	1.6	-	-
	4	5.51	P_{ti} (kPa)	36.6	24.6	26.7	27.7
			t_0 (ms)	2.9	1.9	-	-
	5	6.89	P_{ti} (kPa)	25.5	17.0	18.8	19.6
			t_0 (ms)	3.1	2.2	-	-

Table 12 sets out the relative deviations of the calculated and measured values for a 400 g charge.

Table 12. Values of relative deviations from experiment for a 400 g charge.

Semtex 400 g		Scaled Distance Z (m·kg$^{-1/3}$)	Blast Parameters	Kingery and Bulmash	Kinney and Graham	Henrych and Major	PECH
				Relative Deviation (%)			
Distance (m)	2	2.76	P_{ti} (kPa)	48.3	6.4	−2.5	4.4
			t_0 (ms)	35.7	−21.4	-	-
	3	4.14	P_{ti} (kPa)	31.8	−9.1	−5.8	−2.4
			t_0 (ms)	47.1	−5.9	-	-
	4	5.51	P_{ti} (kPa)	30.7	−12.1	−4.6	−1.1
			t_0 (ms)	70.6	11.8	-	-
	5	6.89	P_{ti} (kPa)	25.0	−16.7	−7.8	−3.9
			t_0 (ms)	82.4	29.4	-	-
Arithmetic mean (%)			P_{ti} (kPa)	34.0	−7.9	−5.2	−0.8
			t_0 (ms)	59.0	3.5	-	-

4. Discussion

When evaluating the results of experiments carried out with Semtex 10-SE spherical charges, which differed in mass, the assumption that the peak overpressure increases with increasing charge mass was confirmed. This is verified by theoretically calculated and experimentally obtained results. Based on the comparison of experimentally measured and theoretically calculated values of peak overpressures, it can be concluded that the different predictive calculation models are applicable for load estimation, but the degree of uncertainty varies from one relation to another. Percentage relative deviations, following the authors' calculation formulas, were interpreted in Tables 6, 8, 10 and 12. From the average values of the relative deviations between the calculated and measured values, it can be concluded that the calculated values according to Kinney and Graham, Henrych and Major and PECH are always on average lower than the experimentally measured values. On the other hand, the calculated results according to Kingery and Bulmash are in all cases larger than the measured values.

All the calculations of peak overpressures mentioned so far are roughly the same in comparison with the measured values, with differences of units or a few tens of percent at most (the most extreme difference found was in the case of comparison of the Kingery and Bulmash result with the measured value at a distance of 2 m for a 400 g Semtex charge, which was 48.3%). The largest differences occurred in the case of the Kingery and Bulmash calculations, with differences in the tens of percent range. It appears that as the distance from the charge reduces and the mass of the charge increases, higher differences between the measured and calculated values arise. This is due to the lower sensitivity of the computational relations for the close surroundings of the explosive. From a practical point of view, this is not a major complication, since the parameters of the explosive loads are usually more relevant for larger distances, where the accuracy of the calculation relations is already considerably higher. The overall intercomparison of the results of these relations with measured values shows that the smallest average differences between the theoretical calculations and experimentally measured values of peak overpressures are in the case of using the PECH calculation relation.

Theoretical calculations of positive phase duration show that this time increases with increasing mass of the explosive and with increasing distance from the explosion site. Although some increase is evident from the graphs and from the indicative positive phase durations within the measurement results for a given mass, there are exceptions where the duration decreases or does not change. These variations may arise from the passing of a shock wave related to the measuring equipment used. The measurement of these values was already at the limit of the accuracy of the measuring setup used, which could have affected values in the order of hundredths of milliseconds. The same phenomenon was observed in the case of increasing mass of the explosive used, where some increase in the positive phase duration is noticeable, but inconsistent with the theoretical results.

In general, the measured positive phase durations are roughly around similar values, but this results in increasing average percentage differences between the measured and calculated values according to Kingery and Bulmash. The calculations according to Kinney and Graham were initially lower than the measured values, therefore the average difference in values first decreases and then increases again with increasing mass of the charge. In the case of the calculation of the duration of the positive phase according to Henry and Major, the values could not be determined because the values of the reduced distances were outside the range of validity of the calculation relationship. From the above facts it follows that the calculation relations for the positive phase durations are of limited applicability in the context of the experimental conditions.

5. Conclusions

The subject of the investigation of the magnitude of the explosive load on objects or building structures was the chosen type of explosive, namely the industrial explosive Semtex, which, due to its known properties, served as a reference sample in this case. At the theoretical level of determining the selected blast wave parameters, there are many different calculation relationships. The calculations by these different relations differ from each other; this is due to the differences in the conditions for which the calculation relation was constructed. In general, one of the most important blast wave parameters is the peak overpressure, since it is the parameter that has a major influence on the action of the pressure wave on the building structure. Based on extensive research, predictive relationships have also been traced that can be applied more universally. The limitations for these relationships arise from their applicability to specific situations. Due to the applicability of the computational relations, only the relations for peak overpressure and positive phase duration were used in the theoretical calculations of blast wave parameters.

Two types of pressure probes were used in the experimental measurements, but with the same pressure transducers used. The values of the measured overpressures differed only minimally between the single transducers at the same distance (in the same pair), within a maximum of one percent. From the recorded blast curves, the peak overpressure and the duration of the positive phase can be read off, as well as other important blast wave parameters. The overpressure curves also showed the reflection of the blast wave from the bedrock (or other obstacles), which can affect their course and distort the measured values under certain circumstances.

Comparing the experimentally measured and theoretically calculated peak overpressures, it was found that the smallest differences occurred in the case of using the predictive model according to PECH, whose deviations from the measured values differed at most in units of percent. On this basis, it can be determined that this was the most appropriate computational relationship under the conditions of the experiment.

Author Contributions: Conceptualization, A.F. and M.M.; methodology, M.M.; software, A.F. and M.M.; validation, M.M., A.F. and R.Z.; formal analysis, A.F. and M.M.; investigation, M.M. and A.F.; resources, M.M.; data curation, A.F.; writing—original draft preparation, M.M.; writing—review and editing, A.F.; visualization, A.F.; supervision, R.Z.; project administration, R.Z.; funding acquisition, A.F. and R.Z. All authors have read and agreed to the published version of the manuscript.

Funding: This research was funded by University of Calabria, SMART Lab (Structural Monitoring, Structural Advanced Materials, Structural Rehabilitation, Structural Testing Laboratory).

Institutional Review Board Statement: Not applicable.

Informed Consent Statement: Not applicable.

Data Availability Statement: The article provides all data used in this research.

Acknowledgments: Special thanks for organizing and conducting the experimental studies is given to the Faculty of Safety Engineering, VŠB Technical University of Ostrava, Czech Republic.

Conflicts of Interest: The authors declare no conflict of interest.

References

1. Kinney, G.F.; Graham, K.J. *Explosive Shocks in Air*; Springer: Berlin, Germany, 1985.
2. Henrych, J. *The Dynamics of Explosion and Its Use*; Elsevier: Amsterdam, The Netherland, 1979.
3. Mills, C.A. The design of concrete structures to resist explosions and weapon effects. In Proceedings of the 1st International Conference on Concrete for Hazard Protections, Edinburgh, UK, 27–30 September 1987.
4. Kingery, C.N.; Bulmash, G. *Airblast Parameters from TNT Spherical Air Burst and Hemispherical Surface Burst*; Technical Report ARBRL-TR-02555; U.S. Army BRL: Aberdeen Proving Ground, MD, USA, 1984.
5. Sadovskiy, M.A. Mechanical effects of air shockwaves from explosions according to experiments. In *Geophysics and Physics of Explosion*; Selected Works; Sadovskiy, M.A., Ed.; Nauka Press: Moscow, Russia, 2004.
6. Beshara, F.B. Modelling of blast loading on aboveground structures—I. General phenomenology and external blast. *Comput. Struct.* **1994**, *51*, 585–596. [CrossRef]
7. Brode, H.L. *Numerical Solution of Spherical Blast Waves*; American Institute of Physics: New York, NY, USA, 1955.
8. Swisdak, M.M., Jr. Simplified Kingery Airblast Calculations. In *Minutes of the Twenty Sixth DOD Explosives Safety Seminar*; DTIC Document; U.S. Department of Defense, Explosives Safety Board: Indian Head, MD, USA, 1994.
9. U.S. Department of the Army. *Design and Analysis of Hardened Structures to Conventional Weapons Effects*; Technical Manual 5-855-1; U.S. Department of the Army: Arlington, VA, USA, 1986.
10. U.S. Department of the Army. *Structures to Resist the Effects of Accidental Explosions*; Technical Manual 5-1300; U.S. Department of the Army: Arlington, VA, USA, 1990.
11. U.S. Department of the Army. *Structures to Resist the Effects of Accidental Explosions*; Technical Report UFC 3-340-02; U.S. Department of the Army: Arlington, VA, USA, 2008.
12. Karlos, V.; Solomos, G. *Calculation of Blast Loads for Application to Structural Components*; JRC Technical Report, EUR 26456EN; European Union: Luxembourg, 2013.
13. Remennikov, A.M. A review of methods for predicting bomb blast effects on buildings. *J. Battlef. Technol.* **2003**, *6*, 5–10.
14. Ngo, T.; Mendis, P.; Gupta, A.; Ramsay, J. *Blast Loading and Blast Effects on Structures—An Overview*; EJSE Special Issue: Loading on Structures; The University of Melbourne: Parkville, VIC, Australia, 2007.
15. Koccaz, Z.; Sutcu, F.; Torunbalci, N. Architectural and structural design for blast resistant structures. In Proceedings of the 14th World Conference on Earthquake Engineering, Beijing, China, 12–17 October 2008.
16. Draganić, H.; Sigmund, V. Blast loading on structures. *Teh. Vjesn.* **2012**, *19*, 643–652.
17. Ding, Y.; Zhang, X.; Shi, Y.; Zhang, H. Prediction of far-field blast loads from large TNT-equivalent explosives on gabled frames. *J. Constr. Steel Res.* **2022**, *190*, 107120. [CrossRef]
18. Shirbhate, P.A.; Goel, M.D. A Critical Review of Blast Wave Parameters and Approaches for Blast Load Mitigation. *Arch. Comput. Methods Eng.* **2021**, *28*, 1713–1730. [CrossRef]
19. Birhane, T.H. Blast Analysis of Railway Masonry Bridges. Master's Thesis, University of Minho, Braga, Portugal, 2009.
20. Baker, W.E. *Explosions in Air*; University of Texas Press: Austin, TX, USA, 1973.
21. Baker, W.E.; Cox, P.A.; Westine, P.S.; Kulesz, J.J.; Strehlow, R.A. *Explosion Hazards and Evaluation*; Elsevier Scientific Publishing Company: New York, NY, USA, 1983.
22. Hopkinson, B. *British Ordnance Board Minutes*; Report 13565; British Ordnance Office: London, UK, 1915.
23. Cranz, C. *Lehrbuch der Ballistic*; Springer: Berlin, Germany, 1926.

applied sciences

MDPI

Article

Assessment of the Sympathetic Detonation of Blasting Caps

Eugen Trană, Marin Lupoae *, Bogdan Iftimie and Alexandru Cătălin Casapu

Faculty of Integrated Armament Systems, Military Engineering and Mechatronics, Military Technical Academy "Ferdinand I", 050141 Bucharest, Romania
* Correspondence: marin.lupoae@mta.ro

Abstract: The neutralization of improvised explosive devices (IEDs) involves the use of disrupting agents propelled explosively. Due to the special nature of such materials, a proper investigation of the parts most susceptible to sympathetic detonation is in order. The initiation of IEDs is caused by detonation products, shock waves, and propelled disruptive agents. In this paper, initiation of IED composition (acceptor charge) due to the neutralization system's (donor charge's) explosive charge detonation is evaluated based on the influence of the first two of the three above-mentioned factors. One of the most susceptible components of IEDs to sympathetic initiation is the blasting cap. Based on an experimental and numerical mix approach, blasting cap tendency to sympathetic detonation in open field had been investigated. The suitability of critical energy fluence and Chapman–Jouguet threshold criteria to the sympathetic detonation tendency of blasting caps was investigated. Experimental and numerical/analytical results describing the phenomenon are in agreement.

Keywords: improvised explosive device; sympathetic detonation; blasting cap; numerical simulation

Citation: Trană, E.; Lupoae, M.; Iftimie, B.; Casapu, A.C. Assessment of the Sympathetic Detonation of Blasting Caps. *Appl. Sci.* **2022**, *12*, 12761. https://doi.org/10.3390/app122412761

Academic Editors: Ricardo Castedo, Lina M. López and Anastasio P. Santos

Received: 4 November 2022
Accepted: 7 December 2022
Published: 12 December 2022

1. Introduction

Sympathetic detonation involves the initiation (usually unwanted) of an explosive charge called the acceptor due to the detonation of another charge called the donor. The initiation of the acceptor can be induced by detonation products and/or shock waves, depending on the distance between the two explosive charges.

The issue of sympathetic detonation is encountered in both military and civilian applications. In the military domain, avoiding sympathetic detonation is important, among other things, in the process of neutralizing IEDs or unexploded ordnance (UXO). In order to determine the critical distances at which the neutralization system can be placed, the possibility of sympathetic initiation of the IED load must also be taken into account since it is very important to avoid the effects that are produced by an unwanted initiation of the IED.

Starting from the principle of IED neutralization, namely preventing its operation (detonation of the explosive charge) and separating the component elements so that its functioning can no longer be triggered by the subsequent handling or interacting with the environment, the requirement of the neutralization system's performance can be formulated as the ability to induce a high enough shock in the IED to separate its parts, but at the same time to avoid the initiation of the explosive charge caused by the effects of the donor charge's detonation.

In both military and civilian fields, the storage, transportation, handling, and the production of explosives or items that include explosives involve risks due to the sensitivity and reactivity of such materials. As history has proven, one of the main risks is associated with the tendency of energetic materials to react to a nearby stimulus, such as an explosion or a kinetic impact. Over time, the above-mentioned tendency led to several catastrophic accidents in military facilities as well as in civil mining and industrial sites. Due to these tragic events, regulations regarding the design and use of items containing explosives were imposed. Among these regulations, sympathetic detonation evaluation tests are mandatory and can be seen as the backbone of safety program tests.

The evaluation of the tendency of an energetic material toward sympathetic detonation is assessed using the gap test. Basically, the gap test is a widely used test that aims to evaluate the sensitivity of explosives to blast waves. The gap test and other associated tests have been performed numerically and experimentally for different types of explosives, both on land and underwater, by several researchers throughout history. Thus, Yang et al. [1] numerically investigated Composition B's susceptibility to sympathetic detonation based on a Direct Numerical Simulation (DNS) scheme, which was also validated experimentally. They concluded that the probability of sympathetic detonation is related not only to the type of explosive and distance but also to the size of the charge. Zhang et al. [2] experimentally evaluated the underwater sympathetic detonation of TNT and analyzed the energy, pressure, and pulsing cycle. Kubota et al. [3,4] investigated using high-speed photography the sympathetic detonation of Composition B both in air and underwater and also investigated it numerically by using Lee and Tarver's phenomenological reaction-rate law. An experimental study by Becuwe and Delclos [5] on the sympathetic detonation of low-sensitivity explosive compounds (NTO and HMX-based PBX) showed that the shock insensitivity of the studied explosive mixture is combined with very good behavior under fire, slow heating, and a ball impact. Keshavarz et al. [6] studied the possibility of using a small-scale gap test to evaluate the sympathetic detonation of CaHbNcOd explosives and proposed a simple procedure for the analytical calculation of the shock sensitivity of energetic compounds. Ko et al. [7] investigated experimentally and numerically the shock sensitivity of a shaped charge underwater and showed that in an underwater explosion, the index of the sympathetic detonation is slightly higher than in the air. Along with the previously mentioned research teams, several others can easily be named, including researchers/teams that approached the subject in a theoretical manner, such as M.H. Keshavarz, E.N. Ferm, H.R. James, and A.C. Victor [6,8–10].

When an explosive charge (explosive bars) is subjected to the action of a shock wave, this shock wave will produce the initiation of the acceptor charge only if the energy of the shock wave is greater than the critical energy. The critical energy considers not only the pressure level (shock amplitude) but also the pulse duration and the acceptor impedance as stated by Walker and Wasley [11]. The formula for energy calculation is given in Equation (1).

$$E_c = P^2 t / \rho_0 U \tag{1}$$

where P is the shock amplitude, t indicates the pulse duration, ρ_0 denotes the acceptor's initial density, and U is the shock velocity that travels through the acceptor. The term E_c has the dimension of energy per unit area and is therefore referred to as energy fluence. Through experimental tests carried out with different explosives subjected to the square-wave shock produced by the impact of the flyer test, it was found that each explosive has a range of energy fluence in which a stable detonation is produced, called critical energy fluence [12]. Additionally, for the evaluation of the initiation of an explosive under the action of the shock wave, the "Pop-plot" [12] can be considered, which represents the graphic representation in logarithmic coordinates of the run distance as a function of pressure for the acceptor explosive.

The application of Equation (1) used to evaluate the initiation of detonation of an explosive charge requires the determination, by numerical analysis, of the amplitude and duration of the applied shock. In the absence of numerical analysis, the relationship developed by Yadav [13] takes into account detonation parameters, which are easier to measure and can be used to determine the energy transmitted to the explosive charge. This relation is specified in Equation (2).

$$E_c = \frac{\rho_0 D_j \delta \left[\frac{D_j - a_x}{2b_x} + \frac{D_j}{2(r+1)} \right]^2}{D_j - \left[\frac{D_j - a_x}{2b_x} + \frac{D_j}{2(r+1)} \right]} \quad (2)$$

where ρ_0 is the initial density, D_j is the velocity of detonation, δ is the thickness of reaction, r denotes the specific heat ratio of detonation products, and a_x and b_x are Hugoniot constants.

In line with the sympathetic detonation issue, yet in a less conventional manner, the current paper focuses on the investigation of the sympathetic detonation tendency of the blasting cap, containing pentaerythritol tetranitrate (PETN) charges. The main focus is targeted on investigating the applicability of critical energy fluence for a particular configuration with an air gap between donor and acceptor explosive charges.

For the assessment of the detonation initiation potential of a blasting cap, in this paper, we will use Equation (1) because, when using numerical analysis, the parameters in this equation are determined much faster. The relevance of the study is obvious when one considers the specific way in which the neutralization of suspicious packages is carried out. Basically, the disruption of such packages is performed by propelling a disrupting agent (metallic/plastic bolts or water) with the use of small explosive charges. The blast wave generated by the detonation of an explosive charge has the potential, in certain conditions, to lead to unwanted package detonation due to the initiation of explosive charge and/or blasting caps.

2. Experimental Investigation

In order to experimentally evaluate the sensitivity of blasting caps to blast waves, several tests have been performed. The experiments involved the use of 100 g of TNT as donor charge and ϕ7 × 69 mm blasting cap as an acceptor. TNT was chosen as a donor because it is considered a reference explosive. Although the amount of 100 g of explosive is not common for neutralization systems, it was used to better capture the influence that the detonation products and the shock wave can have on the sympathetic detonation of a blast initiator. The experimental setup illustrated in Figure 1 aims to identify a critical distance, in terms of air thickness, between the acceptor and the donor that will end in a no-go reaction for the acceptor.

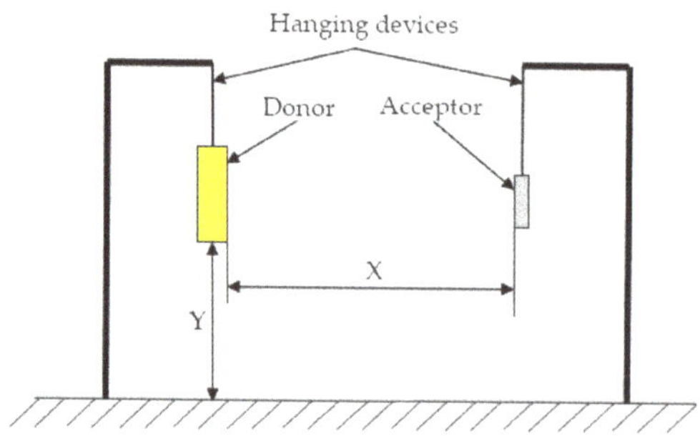

Figure 1. Experimental set-up (side-on configuration).

The blasting cap type used in these experiments is based on 0.6 g PETN charge with a density of 1.75 gm/cm^3. The donor charge was detonated using the same type of blasting cap as the detonator as the one used as the acceptor.

The tests were performed in an open space configuration, imposing a higher y value (donor/ground distance) than the x value (donor/acceptor distance). Using this approach, the incident blast wave was allowed to arrive first at the acceptor position instead of a one reflected. The post-test recovery of target blasting caps, when possible, was the criterion for identifying a possible detonation of the acceptor. A witness plate was not used due to the chosen test setup (the TNT charge was parallel to the blasting caps in order to have a larger contact surface; this represents the most unfavorable situation in relation to the position that a disruption load can have towards an acceptor load from the components of an IED). In this situation, the presence of a witness plate could have influenced the initiation of the blast cap from the shock wave that is reflected from the plate.

Table 1 lists the experimental results regarding blasting cap sympathetic detonation tendency.

Table 1. Gap test experimental results.

Air Gap Thickness X (mm)	Go/No Go
100	Go
200	Go
350	No go
500	No go

A fast image recording camera (Photron, FASTCAM SA-Z), set to an acquisition rate of 30,000 fps, was used as part of the testing setup. Thus, the blast wave position and fireball dimension during experimental tests were traced. Table 2 contains the results that were extracted from image analysis, and the detonation of 100 g TNT is illustrated in Figure 2.

Table 2. Blast wave position and fireball dimension.

Time (s)	Blast Wave Position (mm)	Fireball Dimension (mm)
0	0	0
0.0003000	608.39	623.04
0.0005330	715.52	798.26
0.0006670	749.59	900.48
0.0009670	827.45	1061.09
0.0016330	851.79	1406.67
0.0036670	992.94	1596.58
0.0050330	1017.28	N/A
0.0067670	1065.96	N/A
0.0114330	1168.17	N/A

Figure 2. Typical image of 100 g TNT detonation.

3. Numerical Approach

While experimental testing is the most suitable method to evaluate sympathetic detonation, the numerical approach has proved to be a valuable tool in deciphering the process specifics. Thus, important aspects of the detonation propagation process (pressure level in the donor/acceptor charge or run distance for a stable detonation) can be investigated using a low-cost and reasonable-time scenario. For the proposed experimental tests, numerical models have been defined using Autodyn 2021® software [14].

3.1. Preprocessing

In order to corroborate the experimental results with the numerical ones, one simple approach was considered. The numerical model is based on a 2D planar symmetrical geometry and the use of multi-material Euler part.

Since the mesh sensitivity is a well-known characteristic of commercial software based on the Finite Element Method (FEM), special attention was given to this aspect. Considering Ko's observation [7] regarding the recommended mesh dimension as a function of the distance for the free air blast wave and also the distances involved in experimental tests, a graded mesh was imposed. The mesh dimension varies in both directions from 0.1 mm in the blasting cap region to 1 mm in the donor charge area, as shown in Figure 3.

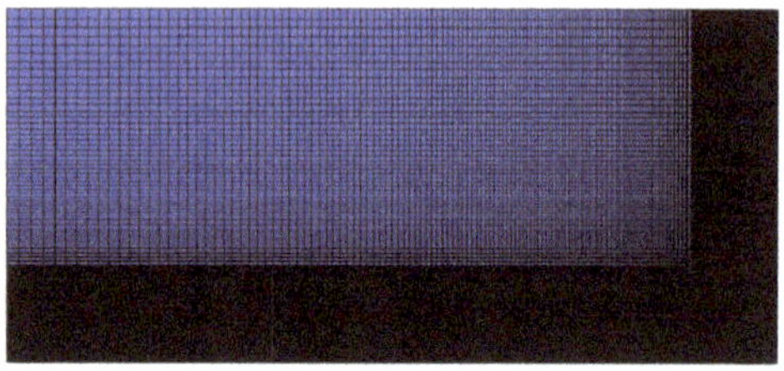

Figure 3. Typical mesh example.

In addition to the above-mentioned aspects, it must be pointed out that two rows of six gauges, 1 mm apart on the X axis and 5 mm apart on the Y axis, were used for numerical calculus to record the peak pressure in the acceptor charge. The position of the gauges, the material location, and the edges on which boundary condition were imposed, as shown in Figure 4. Additionally, in order to reduce the simulation time, the pressure contours generated by 100 g TNT detonation have been remapped in the current simulation using the fill option from a separate Autodyn 2021® file.

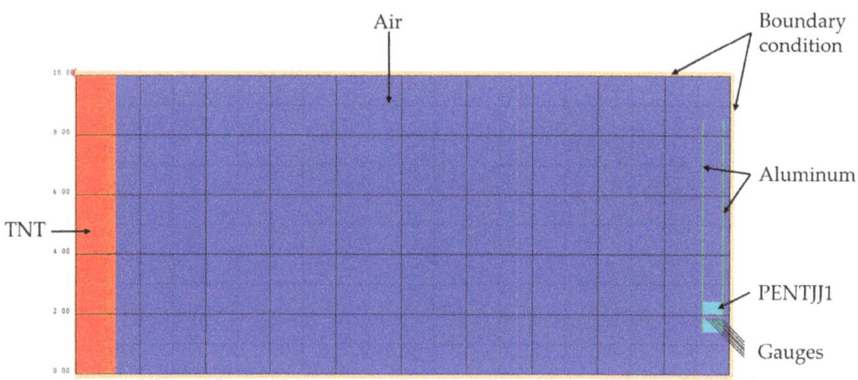

Figure 4. Materials location and gauges position.

3.2. Material Models

The basic properties of any substance, explosives included, are usually identified through a mathematical relation that correlates pressure, volume, and internal energy/temperature. The relation is called an equation of state (EOS).

In the process of numerical analysis of the sympathetic detonation, it was found that the choice of the equation of state that describes the behavior of the donor and acceptor charges plays a very important role.

Over time, numerous attempts to define an EOS that accurately predicts the behavior of explosive gas products have been made. In fact, the number of proposed equations was high enough to classify them into two distinct categories: one considers the chemistry explicitly and the other does not [15].

Unfortunately, all of the above-mentioned equations have limited the applicability and variable accuracy. Despite this shortcut, their use in numerical calculus is extremely beneficial since different particular situations can be investigated in a reasonable time frame and an almost costless manner.

When explosive detonation applications are numerically investigated, one EOS is usually involved, namely Jones–Wilkins–Lee (JWL) [16]. In fact, the use of the JWL EOS is so common that by now almost all hydrocodes have implemented it and several forms can be identified in the literature. However, the most known form of equation is the form of a family of isentropes [17], which is illustrated in Equation (3).

$$p(S,V) = Ae^{-R_1 V} + Be^{-R_2 V} + C^*(S)V^{-(\omega+1)} \tag{3}$$

where p is the pressure; S refers to the entropy per unit initial volume (s/v_0); V is the volume relative to the undetonated state (v/v_0); A, B, R_1, and R_2 are constant fitting parameters; ω is an assumed-constant material parameter (Grüneisen function); and $C^*(S)$ is a parameter dependent only upon the entropy S.

Based on the previously mentioned JWL EOS and ideal gas EOS assumptions, the pressure in the front of a traveling blast wave can be accurately evaluated when hydrodynamic simulations are employed. Even though the equation's versatility is impressive, the application of JWL EOS by itself cannot deliver crucial data regarding the initiation of the explosive when subjected to blast wave stimulus. Thus, in order to investigate the blasting cap sympathetic detonation susceptibility, a slightly different EOS was chosen, the Lee–Tarver equation of state [18]. In fact, the Lee–Tarver EOS is basically a JWL EOS that has been upgraded with a supplementary equation, Equation (4), that allows the evaluation of the burning fraction based on the pressure level acting on/inside the explosive [18].

$$\frac{\partial F}{\partial t} = I(1-F)^b\left(\frac{\rho}{\rho_0} - 1 - a\right)^x + G_1(1-F)^c F^d p^y + G_2(1-F)^e F^g p^z \tag{4}$$

where F is the explosive burning fraction which has a value between 0 and 1.

The importance of JWL EOS and Lee–Tarver EOS for the current blast cap sympathetic detonation study is correlated with the Chapman–Jouguet pressure level that can be used as the Go/No-Go criterion.

The materials used in the numerical simulation are TNT for the donor charge, PETNJJ1 for the acceptor charge, aluminum for the blasting cap walls, and air for the space between the donor and the acceptor. All the equations of state (EOS) and strength models of the materials were adopted from the library of the Autodyn 2021® software.

3.3. Numerical Results

The obtained numerical results are presented in Table 3. In Figure 5, the pressure levels recorded by two different gauges located inside the acceptor charge are illustrated for two cases. Figure 5a shows the pressure levels in the case of a 200 mm gap between the donor and acceptor, while Figure 5b shows the results for a distance of 500 mm. The peak overpressure was directly measured from the gauges. The pressure wave speed was determined from graphs of pressure in time, for consecutive sensors. By dividing the distance between the sensors by the values of the times at which the maximum values of the pressures were obtained, the shock wave velocities were determined for each case of the acceptor–donor charge. The critical energy fluence was determined by using the maximum pressure value, speed, and the pulse duration of the shock wave. For the calculation of the shock wave pulse duration, the area under the pressure–time curve was numerically evaluated and then approximated with a square-shaped pulse (rectangle with a height given by the maximum pressure value and length given by the value of pressure-acting time that equals the previously calculated impulse).

Table 3. Virtually measured data in acceptor charge.

Air Gap Thickness, X (mm)	Peak Overpressure (Mbar)	Pressure Wave Speed (mm/ms)	Critical Energy Fluence (J/m^2)
100	4.02×10^{-3}	3162	1.69×10^5
200	3.55×10^{-3}	2881	1.51×10^5
350	1.58×10^{-5}	2840	97.38
500	0.91×10^{-5}	2739	37.46

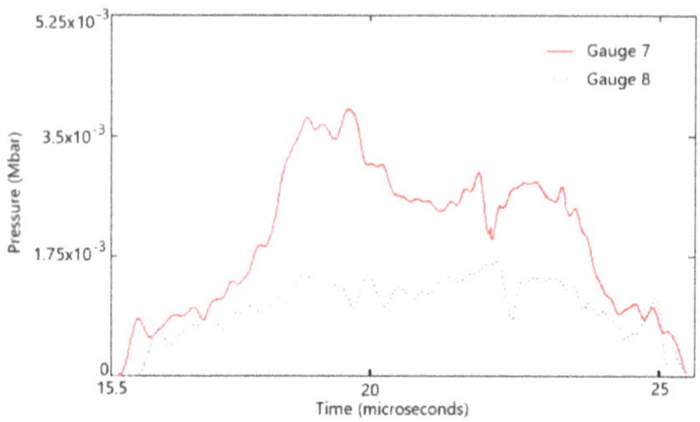

(**a**) 200 mm case

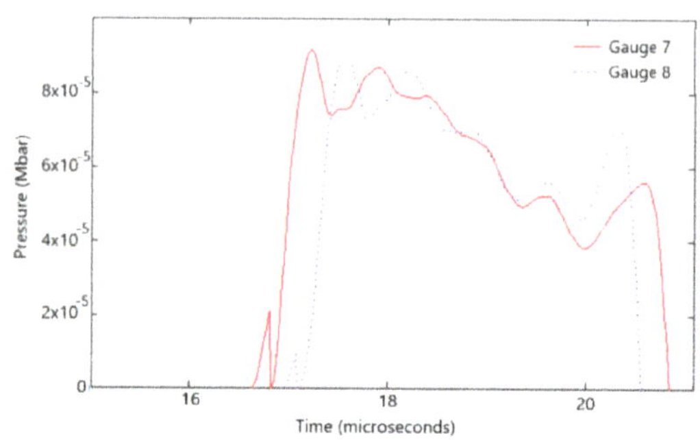

(**b**) 500 mm case

Figure 5. Typical examples of numerically recorded pressure history in PETNJJ1 material (gauges 7 and 8 are the first and second gauges from the upper row).

4. Discussion

As the main goal of the present study is to evaluate the sympathetic detonation tendency of blasting caps, the main focus is on the correlation of experimental and numerical results.

If the experimental approach is a straight forward method that clearly points out the presence of acceptor detonation, the numerical approach is somehow trickier.

As previously mentioned, the numerical simulation is based on the use of Lee–Tarver EOS [19]. The Lee–Tarver EOS points mainly at the Chapman–Jouguet [19,20] findings that were later included in von Neumann's work [21]. Basically, a stable detonation is achieved when the pressure level reaches a certain value that is specific to each explosive (0.327 Mbar for PETN). Considering this approach, the blasting cap will not be susceptible to sympathetic detonation (in a range of a 0.1 to 0.5 m clearance distance) as long as the pressure recorded by the gauges indicates significantly lower levels (see Table 3). Nevertheless, as experimental tests have proved, the blasting cap detonation manifested at both 100 mm and 200 mm, where, according to numerical simulation, the peak overpressure is less than 0.327 Mbar.

Acknowledging the hypothesis involved in the critical energy fluence theory (step pulse shape and constant impedance, mainly) and the 1.5×10^5 J/m^2 threshold value for the energy fluence of PETN (1.75 gm/cm^3) [22], the calculus based on the numerical data (Figure 5) indicates a very good match with the experimental observation (see Tables 1 and 3). Thus, it is found that for distances of 100 mm and 200 mm, the critical energy values are close to the threshold value for the initiation of the acceptor explosive. On the other hand, the recorded pressures are lower than those corresponding to the C-J state, which indicates a weak detonation. The values of the maximum pressure and the duration of the positive phase of the phenomenon shown in Figure 5a fall within the values presented in the literature for the initiation of solid explosives [23]. Moreover, the shape of the pressure–time curve in Figure 5a is similar to the ones in Walker and Wasley's work [24] at the point that the initiation of the explosive occurs. Additionally, by analyzing Figure 5, it can be observed that the pressure wave shape acting on the blasting cap has a much different profile from the ones usually recorded during a standard gap test.

Due to the mismatch between the shock wave's front velocity and donor gas products' front velocity, the first to act on the blast cap is the blast wave, and the gas products pressure shortly afterward, depending on the relative position between the donor and acceptor charges, as can be deduced from Figure 6. The pressure wave profile is also shaped by the reflected blast wave, which is clearly indicated by the numerical simulation.

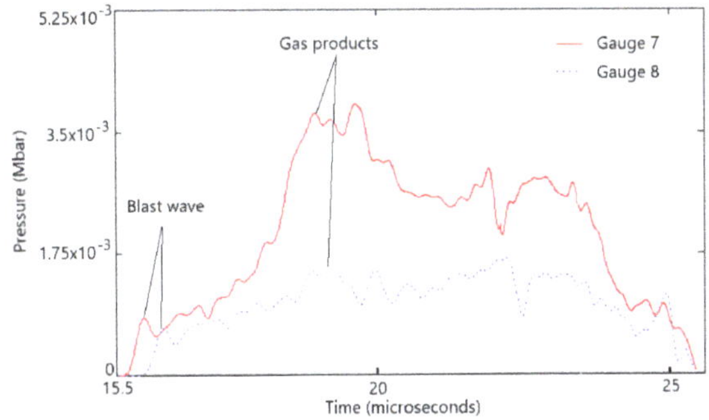

Figure 6. Pressure wave profile (200 mm case).

The same outcome can be underlined by theoretical means as well. Therefore, using far-field experimental data (images recorded when the blast wave has already traveled over a 0.6 m distance), the Sedov-Taylor model [25,26], and Gilev's observations [27], one can predict the distance between the incident blast wave and the gas products' border. Useful data regarding the position of the blast wave and the gas products' border can be extracted by solving Equations (5)–(10) [25–29]. In Figure 7, a comparison between experimental results and the application of Equations (5)–(10) for two values of the expansion dimensionality factor is presented.

$$R_s(t) = at^b \quad (5)$$

$$a = \left[\frac{E_d/(\tau_0^s l_0^{3-n})}{\rho}\right]^{1/(n+2)} \quad (6)$$

$$b = \frac{s+2}{n+2} \quad (7)$$

$$l_0 = \left(\frac{3m}{2\pi\rho}\right)^{1/3} \quad (8)$$

$$\tau_0 = \frac{l_0}{v_{TNT}} \quad (9)$$

$$R_f(t) = R_{max}\left(1 - e^{-kt}\right) \quad (10)$$

where $R_s(t)$ indicates the shock front radii; a and b are coefficients; E_d is the TNT release energy during detonation (usually 4.1 MJ/kg); l_0 denotes a length scale; τ_0 denotes a time scale; m is the TNT mass; ρ is the TNT mass density; v_{TNT} denotes the TNT detonation velocity (6940 m/s); s indicates a factor characterizing the rate of energy release: instantaneous energy release (s = 0) and constant-rate energy release (s = 1); n is the expansion dimensionality: planar expansion (n = 1), cylindrical expansion (n = 2), and spherical expansion ($n = 3$); $R_f(t)$ indicates the fireball radii; R_{max} is the fireball stopping radii (0); and k denotes the drag coefficient.

It must be stated that the Sedov–Taylor equation can be applied only to the mid-field region according to Equation (11) [28].

$$\left(\frac{3m}{2\pi\rho}\right)^{1/3} \ll R_s \ll \left(\frac{\Delta H_d}{p}\right)^{1/3} \quad (11)$$

where ΔH_d indicates the total energy released during detonation, afterburning included, (up to 10.1 MJ/kg [30]), and p denotes the ambient pressure.

Using high-speed camera imaging, R_s radii can be identified for different time values. Since the TNT charge has a cylindrical shape, a factor $n = 2$ was considered. Additionally, due to the fact that the distances between the TNT charge and the blasting cap are small (less than 0.5 m) for the experimental setup, an instantaneous energy release (s = 0) was set. However, the experimental results plotted against the results provided by Equation (5), which are presented in Figure 7a, show some considerable differences.

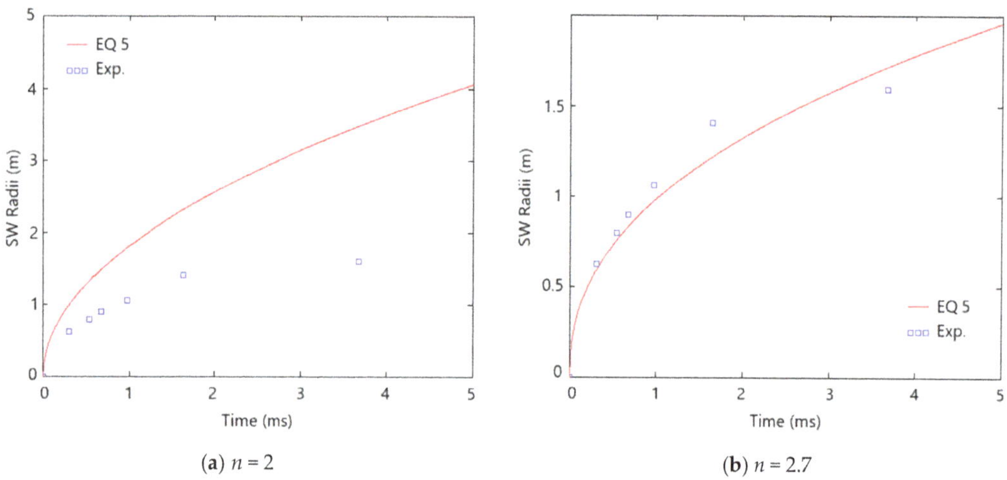

Figure 7. Experimental vs. Equation (5) results.

By closely analyzing the footage in Figure 8, it can be observed that the blast wave shape presents itself neither as a cylindrical shape nor as a spherical one. The mismatch is probably due to the ratio between the length and the diameter of the TNT blast charge, which has a value of 3.57, and also due to the overall small dimensions of the charge. Considering the experimental blast's wave shape, which is more like an ellipsoid, a different value for n factor was chosen, namely 2.7. A comparison between predicted and experimental results for the modified value of the n factor is presented in Figure 7b, and it is clear that the use of this value leads to a much better approximation of the experimental data.

The R_f radii can also be calibrated with the use of the camera footage and finally plotted against R_s values, as shown in Figure 9. Predictions using Equation (11) are plotted in the same figure. With the use of Figure 9, one can easily see that the blast wave gradually moves away from the fireball border, which is consistent with the data provided by the numerical simulation.

Figure 8. Blast wave shape (perfect sphere (blue) and real (black)).

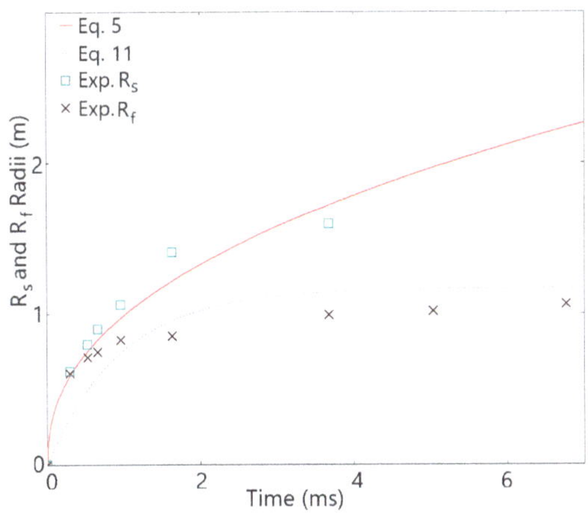

Figure 9. Blast wave vs. Fireball Radii.

The analytical calculus based on Equations (5)–(10) is confirmed by a numerical approach, as shown in Figure 10. According to Figure 10, the blast wave front gradually speeds up, leaving behind the front of the donor gas products.

As mentioned in Section 3.1, two rows of gauges (six gauges/row) were used for pressure recording in the acceptor charge. This particular choice was due to the close distance between the donor and acceptor charges which resulted in a curved shock wave front, as depicted in Figure 10. As a result of the curved shock wave front, the first susceptible area to interact with the blast wave is the upper front of the acceptor charge. This can be clearly seen in Figure 5, where the higher-pressure values are recorded by gauges no. 7 and 8.

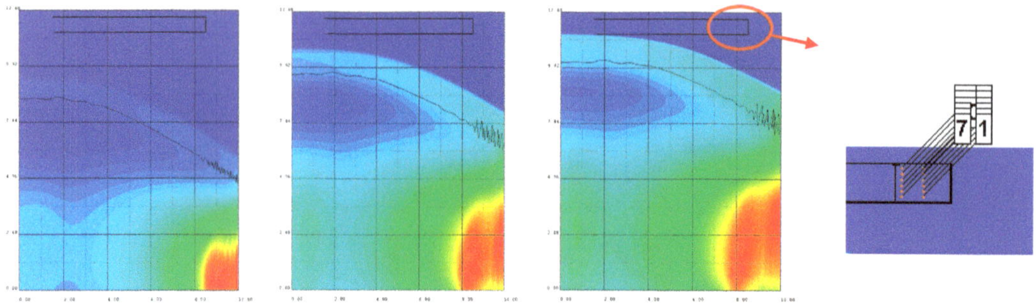

(**a**) t = 2 μs (**b**) t = 6 μs (**c**) t = 8 μs (**d**) Gauges placement

Figure 10. Blast wave front vs. TNT gas products front for 100 mm air gap thickness case.

5. Conclusions

Sympathetic detonation tendency of the blasting cap for different scenarios is mandatory when the disrupting equipment is used on suspicious packages (IEDs). While experimental tests easily allow the evaluation of blasting cap sensitivity to blast waves, the mix between numerical and experimental approaches can enhance the understanding of the phenomenon.

Based on the relations of the mathematical model, it can be concluded that the shock wave detaches gradually from the donor gas products, the complete detachment being completed at a greater distance than the ones characterized by the blast cap initiation. In the nearby distance of the donor charge, both the shock waves' and the gas products' overpressure act on and initiate the blasting caps, even though not simultaneously, as proved in Figure 6. It is also clearly pointed out that the presence of gas products favors reaching the critical value for sympathetic detonation, according to the critical energy fluence criterion.

The analysis of sympathetic detonation tendency of blasting caps in close vicinity shows that the critical energy fluence criterion is preferable to the Chapman–Jouguet pressure threshold. The critical energy fluence criterion can be applied to test configurations that include not only a dense matter gap but also an air gap.

Author Contributions: Conceptualization, M.L. and E.T.; methodology, E.T. and M.L.; software, E.T and A.C.C.; validation, E.T., M.L. and A.C.C.; formal analysis, E.T., M.L. and B.I.; investigation, M.L., B.I., E.T. and A.C.C.; resources, M.L. and B.I.; data curation, E.T.; writing—original draft preparation, E.T. and B.I.; writing—review and editing, M.L., E.T. and A.C.C.; visualization, E.T., B.I. and A.C.C.; supervision, E.T.; project administration, M.L.; funding acquisition, M.L. All authors have read and agreed to the published version of the manuscript.

Funding: The paper was partially supported by a grant of the Romanian Ministry of Innovation and Research, UEFISCDI, project number 33SOL/2021 within PNCDI III.

Institutional Review Board Statement: Not applicable.

Informed Consent Statement: Not applicable.

Data Availability Statement: Not applicable.

Conflicts of Interest: The authors declare no conflict of interest.

References

1. Yang, T.; Wang, C.; Li, T. Numerical simulation study of sympathetic detonation in stages. *Def. Technol.* **2021**, *18*, 1382–1393. [CrossRef]
2. Zhang, Z.; Wang, C.; Hu, H.; Yang, T. Investigation of underwater sympathetic detonation. *Prop. Explos. Pyrotech.* **2020**, *45*, 1736–1744. [CrossRef]

3. Kubota, S.; Shimada, H.; Matsui, K.; Liu, Z.-Y.; Itoh, S. High-speed photography of underwater sympathetic detonation of high explosives. In Proceedings of the 24th International Congress on High-Speed Photography and Photonics, Sendai, Japan, 24–29 September 2000; Volume 4183, pp. 763–770.
4. Kubota, S.; Liu, Z.Y.; Otsuki, M.; Nakayama, Y.; Ogata, Y.; Yoshida, M. A Numerical study of sympathetic detonation in gap test. *Mater. Sci. Forum* **2004**, *465–466*, 163–168. [CrossRef]
5. Becuwe, A.; Delclos, A. Low-sensitivity explosive compounds for low vulnerability warheads. *Prop. Explos. Pyrotech.* **1993**, *18*, 1–10. [CrossRef]
6. Keshavarz, M.H.; Motamedoshariati, H.; Pouretedal, H.R.; Tehrani, M.K.; Semnani, A. Prediction of shock sensitivity of explosives based on small-scale gap test. *J. Hazard. Mater.* **2007**, *145*, 109–112. [CrossRef] [PubMed]
7. Ko, Y.-H.; Kim, S.-J.; Yang, H.-S. Assessment for the sympathetic detonation characteristics of underwater shaped charge. *Geosystem Eng.* **2017**, *20*, 286–293. [CrossRef]
8. Ferm, E.N.; Ramsay, J.B. Spherical projectile impact on explosives. In Proceedings of the Ninth Symposium (International) on Detonation, Portland, OR, USA, 28 August–1 September 1989; Volume 2, pp. 1427–1431.
9. James, H.R. Critical energy criterion for the shock initiation of explosives by projectile impact. *Prop. Explos. Pyrotech.* **1988**, *13*, 35–41. [CrossRef]
10. Victor, A.C. A Simple method for calculating sympathetic detonation of cylindrical, cased explosive charges. *Prop. Explos. Pyrotech.* **1996**, *21*, 90–99. [CrossRef]
11. Walker, F.E. A Comparison of the Classical and a Modern Theory of Detonation. *J. Phys. IV Proc.* **1995**, *5*, C4-231–C4-257.
12. Cooper, P.W. *Explosives Engineering*; Wiley: New York, NY, USA, 1996.
13. Yadav, H.S.; Asthana, S.N.; Subhananda, A. Critical Shock Energy and Shock and Detonation Parameters of an Explosive. *Def. Sci. J.* **2009**, *59*, 436–440. [CrossRef]
14. ANSYS Inc. *AUTODYN User Manual*, version 14.0; ANSYS Inc.: Beijing, China, 2013.
15. Fickett, W.; Davis, W.C. *Detonation: Theory and Experiment*; Dover Publications: Mineola, NY, USA, 2011; p. 25.
16. Kubota, S.; Saburi, T.; Nagayama, K. Unified Form EOS for Detonation Products Based on Relationship between Initial Density and Detonation Velocity. *AIP Conf. Proc.* **2020**, *2272*, 030012.
17. Seglets, S.B. *An Examination of the JWL Equation of State*; ARL-TR-8403; US Army Research Laboratory: Adelphi, MD, USA, 2018.
18. McIntosh, G.W.J. *Explosive Modelling Using Ls-Dyna—A User Guide*; Technical Memorandum; Defence R&D Canada: Valcartier, QC, Canada, 2014.
19. Chapman, D.L. VI. On the rate of explosion in gases. *Philos. Mag.* **1899**, *47*, 90–104. [CrossRef]
20. Jouguet, E. Sur la propagation des réactions chimiques dans les gaz. *J. Math. Pures Appl.* **1905**, *1*, 347–425.
21. Von Neumann, J. Theory of detonation waves. In *Collected Works*; Taub, A.H., Ed.; Pergamon Press Ltd.: Oxford, UK, 1963; Volume 6, pp. 178–218.
22. Schwarz, A.C. *New Technique for Characterizing an Explosive for Shock Initiation Sensitivity*; SAND75-0314 Report; Sandia Labs.: Albuquerque, NM, USA, 1975.
23. Walker, F.E.; Wasley, R.J. A general model for the shock initiation of explosives. *Prop. Explos. Pyrotech.* **1976**, *1*, 73–78. [CrossRef]
24. Walker, F.E.; Wasley, R.J. Initiation of nitromethane with relatively long-duration, low-amplitude shock waves. *Combust. Flame* **1970**, *15*, 233–246. [CrossRef]
25. Sedov, L.I. *Similarity and Dimensional Methods in Mechanics*, 1st ed.; Academic Press: New York, NY, USA; London, UK, 1959.
26. Taylor, G.I. The formation of a blast wave by a very intense explosion. *Proc. Roy. Soc. A* **1950**, *201*, 159–174.
27. Gilev, S.D.; Anisichkin, V.F. Interaction of aluminum with detonation products. *Combust. Explos. Shock Waves* **2006**, *42*, 107–115. [CrossRef]
28. Gordon, J.M.; Gross, K.C.; Perram, G.P. Fireball and shock wave dynamics in the detonation of aluminized novel munitions. *Combust. Explos. Shock Waves* **2013**, *49*, 450–462. [CrossRef]
29. Aouad, C.J.; Chemissany, W.; Mazzali, P.; Temsah, Y.; Jahami, A. Beirut explosion: TNT equivalence from the fireball evolution in the first 170 milliseconds. *Shock Waves* **2021**, *31*, 813–827. [CrossRef]
30. Kim, H.J.; Hwang, K.; Yoon, Y.H.; Lee, H.-J. Numerical analysis of the effect of afterburning on damage to the concrete structure under interior explosion. *Int. J. Concr. Struct. Mater.* **2022**, *16*, 6. [CrossRef]

applied sciences

MDPI

Article

Behavior of Sedimentary Rock Tunnel against Rigid Projectile Impact

Abdullah H. Alsabhan [1,*], Md. Rehan Sadique [2,*], Ali S. Alqarni [1], Shamshad Alam [1] and Wonho Suh [3]

1 Department of Civil Engineering, College of Engineering, King Saud University, Riyadh 11421, Saudi Arabia
2 Civil Engineering Department, Z.H. College of Engineering & Technology, Aligarh Muslim University, Aligarh 202002, India
3 Department of Smart City Engineering, Hanyang University ERICA Campus, Ansan 15588, Korea
* Correspondence: aalsabhan@ksu.edu.sa (A.H.A.); rehan.sadique@zhcet.ac.in (M.R.S.)

Abstract: The tunnels in present-day cities are experiencing varying degrees of loading conditions ranging from static to extreme loading. Therefore, the stability of underground tunnels needs to be analyzed and understood for safer and strengthened design. The present study was conducted to simulate the impact loading conditions due to a missile traveling at a velocity of 5 Mach for different rock tunnels. The nonlinear continuum finite element analysis has been carried out through Abaqus and Explicit. The four different types of sandstones considered in the present study include Kota, Jamrani, Singrauli, and Jhingurda sandstones. An elastoplastic Mohr–Coulomb constitutive material model has been considered to model the behavior of rock surrounding the tunnel opening. The tunnel has an opening of 7 m in diameter (d), and 50 m in height and breadth, with 50 m of longitudinal length. The deformation and stress in the rock and the damage to the concrete lining have been compared in different cases. The Concrete–Damage–Plasticity (CDP) model and the Johnson–Cook model were considered for modelling of the RC lining and steel reinforcement. It was concluded that Jhingurda sandstone has maximum deformations due to impacts caused by missiles.

Keywords: numerical modeling; rock tunnel; sandstone; missile impact

Citation: Alsabhan, A.H.; Sadique, M.R.; Alqarni, A.S.; Alam, S.; Suh, W. Behavior of Sedimentary Rock Tunnel against Rigid Projectile Impact. *Appl. Sci.* **2022**, *12*, 9595. https://doi.org/10.3390/app12199595

Academic Editors: Ricardo Castedo, Lina M. López and Anastasio P. Santos

Received: 7 March 2022
Accepted: 19 September 2022
Published: 24 September 2022

1. Introduction

Due to the rapid migration of the population from rural to urban cities, the demand for better, faster, and safer modes of transportation arises. In addition, the horizontal expansion of residential construction has created a need for subsurface construction. Therefore, the need for the design and research of tunnels and other underground structures has arisen. A number of studies have been carried out by researchers to understand the different aspects of tunnel stability [1–12].

Subsurface structures, especially tunnels and caverns, have been an integral part of the defense strategy of the country. The stability and equilibrium of these structures need to be studied for extreme loading events such as seismic loading, blast, and impact loading. Therefore, a number of researchers have studied the behavior of tunnels under varying types of loading conditions [13–17]. However, there are few studies that depict the performance of underground tunnels constructed in rock when subjected to impact load.

The numerical tools for computation and modelling have been used by researchers to study the stability of tunnels under varying impact loading conditions. Gao et al. [18] studied the behavior of intact rocks under an impact load using a commercial tool, LS-DYNA. They proposed a relationship and a model to study the behavior of intact rocks under impact loads. The model was validated with high accuracy using experimental results. Experimental and numerical simulations have been carried out by Aziznejad et al. [19] using a distinct element code to study the response of rock mass under an impact load. The propagation of cracks in the rock tunnel was studied by Zhou et al. [20] under the impact loading condition, and it was found that the speed of crack propagation is non-uniform; therefore, cracks may stop propagating suddenly. Zhou et al. [21] considered the change in orientation of the impact with respect to the tunnel model. They categorized different types of failure modes in tunnels under impact loading conditions. Zhou et al. [22] had concluded that the tunnel experiences different types of failure modes due to impact load and found that radial cracks propagate in the tunnel from the edge of the tunnel.

Therefore, it may be summarized that the strength of tunnels in rocks under impact loads has been rarely studied in the open literature. However, there is still a significant scope and a need for further study. Sedimentary rocks cover the majority of metropolitan areas in different countries of the world. However, the impact resistance of these rocks against soft and hard missiles has received little attention from previous researchers. Moreover, sandstone is found in significant areas near the borders of strong military countries like India, Pakistan, and China. Hence, it needs to be studied for impact loading conditions. Consequently, the present paper has considered four different types of sandstone: Jhingurda, Singrauli, Jamrani, and Kota. A missile having 100 kg of weight and a velocity of 5 Mach has been considered to simulate the impact loading conditions for different rock tunnels. The nonlinear elastoplastic continuum FE (finite element) method has been adopted to understand the adverse effects of impact loading on rock tunnels.

2. Impact Loading Simulation

The impact resistance of four different sandstone rock tunnels has been studied in the present paper. A missile has been modelled based on the description given by Vidanović et al. [23]. The missile has a 0.7637 m length and a mass of 100 kg, modelled as a discrete part. The commercial software Abaqus has been used, and explicit mode has been selected for the simulation. The missile had a 5 Mach velocity before it hit the ground surface above the rock through which a tunnel has been constructed. The geometry of the tunnel has been considered based on the DMRC design specifications and published articles [24–26]. A dynamic explicit analysis has been carried out in the finite element tool Abaqus. In Abaqus, a step is time allotted for a particular analysis. However, according to the demand of output frame, it breaks the overall time allotted in small increments. In the present analysis, the step time is 0.035.

2.1. Geometry

The rock surrounding the tunnels has been modelled as a bigger size element having a three-dimensional size of 50 m × 50 m × 50 m and 12.5 m of overburden depth. The tunnel has an opening of 7 m in diameter and has been supported by a reinforced concrete liner of 0.35 m in thickness. The liner has an M30 grade of concrete. The concrete liner has reinforcement of steel bars of Weldox 460E grade in the longitudinal and circular directions. The details of the reinforcement and tunnel geometry are presented in Figure 1.

2.2. Input Properties of Materials

The rock mass surrounding the tunnel has been considered as a nonlinear elastoplastic material. The Mohr–Coulomb failure model has been used to incorporate the nonlinearity of four different types of sandstones. The four sandstone rocks considered in the present paper are Jhingurda, Singrauli, Jamrani, and Kota. The input parameters are taken from Rao et al. [27]. Table 1 represents the different physical and mechanical properties of rocks used in the present simulation.

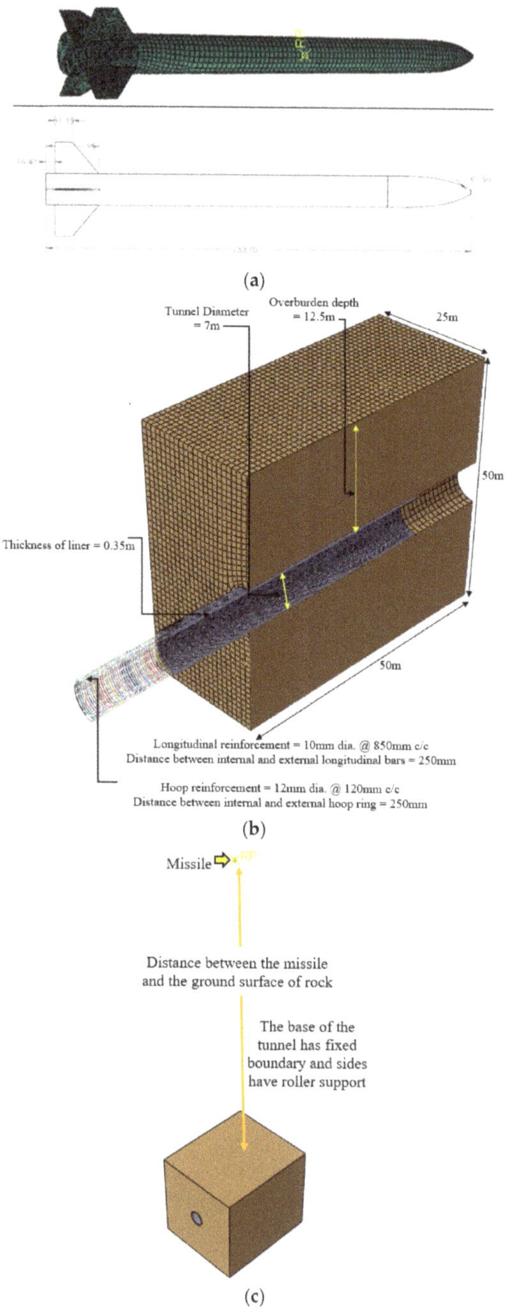

Figure 1. Geometry of finite element model for impact loading simulation. (**a**) Abaqus model of Missile, 0.7637 m in length and a mass of 100 kg with 5 Mach velocity (**b**) Tunnel with an opening of 7 m in diameter and supported by a reinforced concrete(M30) liner of 0.35 m in thickness. (**c**) The rock surrounding the tunnels of 50 m × 50 m × 50 m and 12.5 m of overburden depth.

Table 1. Four different sandstone rock surrounding the tunnel opening.

Rock	Mass Density (kg/m^3)	Young's Modulus (GPa)	Poisson's Ratio	Friction Angle (Degree)	Cohesion (MPa)
Jhingurda	1670	2.84	0.25	21.34	3.68
Singrauli	2310	4.31	0.29	27.11	10.47
Jamrani	2480	5.29	0.22	37.79	11.17
Kota	2310	14.02	0.21	43.42	20.93

Similarly, the nonlinear behavior has been considered for steel bars by providing the elastoplastic properties of steel. Table 2 and Figure 2 show the properties of steel reinforcement used in this study. The interaction between the steel bars and concrete of the liner is achieved by embedding the circular and longitudinal reinforcement. The embedment constraint in the interaction module applied the proper bond between the steel and concrete, creating a reinforced concrete liner for the rock tunnel. The Johnson–Cook model [28] has been used for modelling the steel bars and properties are taken from Borvik et al. [29]. Borvik et al. [29] had performed a series of experiments on the steel under different strain rate and at varying temperature range.

Table 2. Elastoplastic properties of reinforced steel with Weldox 460 E grade.

(ϱ) (kg/m^3)	(E) (GPa)	(ν)	A (MPa)	B (MPa)	n	C	Rate (s^{-1})
7850	200	0.33	490	807	0.73	0.0114	100

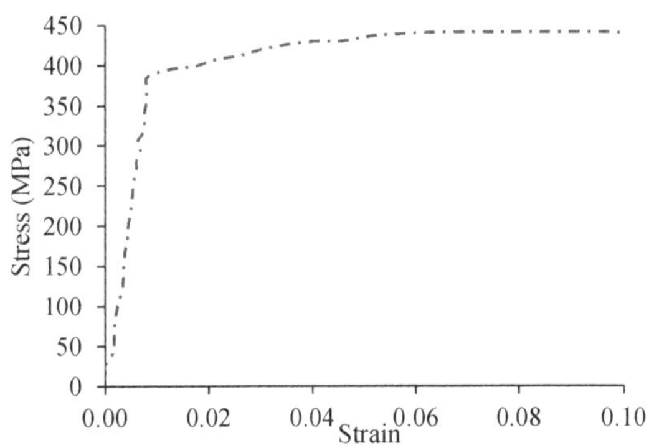

Figure 2. Plastic behavior of steel reinforcement material.

Moreover, the concrete liner has been considered as M30 grade and its nonlinear elastoplastic behavior has been simulated through the Concrete Damage Plasticity model. It also incorporated the damage characteristic of the concrete and, therefore, proved to be useful in studying the overall failure of the internal lining.

The M30 grade of concrete has a mass density of 2500 kg/m^3 and a Young's modulus of 26.6 GPa, with a 0.20 Poisson's ratio. Moreover, the dilation angle and eccentricity of 31 degrees and 0.1, respectively, have been considered. The variation in stress and damage corresponding to strain for the M30 grade of concrete are shown in Figures 3 and 4, respectively.

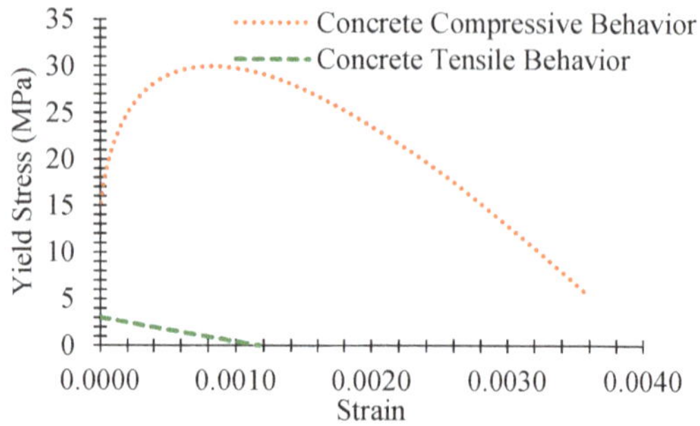

Figure 3. Input of stress–strain variation for M30 grade of concrete liner used in tunnel.

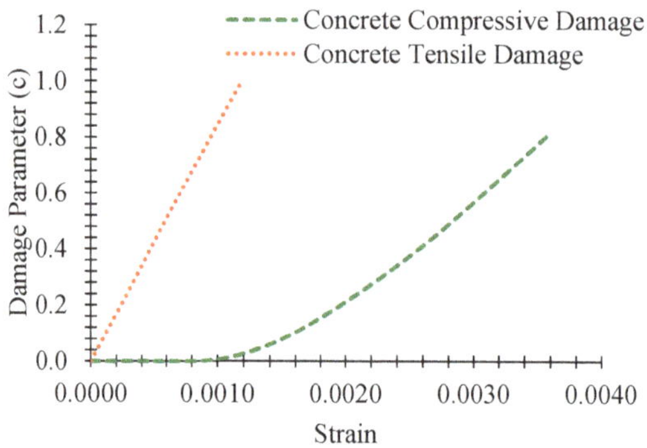

Figure 4. Damage parameter variation with strain for M30 grade of concrete.

2.3. Meshing, Loading, Boundary and Interaction Conditions

The rock mass surrounding the tunnel has been meshed as C3D8R (Continuum Three-dimensional eight-nodded reduced integration solid Brick element), as suggested and used by Zaid and Sadique [30,31], Zaid and Shah [32] and Zaid et al. [33–37]. The element size of 0.7 has been used based on mesh convergence, and this type of mesh is defined as brick-type element, which has eight nodes. The steel bars are modelled as beam-type element, i.e., B31, as suggested and used by Zaid and Sadique [38–40]. The steel bars elements have an element size of 0.05. The missile has been meshed by the R3D4 element type to make it rigid and discrete. The general hard contact and frictionless tangential contact has been assigned to the whole model. The embedment interaction has been used to model the reinforced concrete liner by embedding the steel bars in concrete liner. The base of the rock has a fixed support as the rock mass extends to infinite depth, and the sides of the model have roller supports, which allow vertical movement but restrain other directional movement of rock mass. One set of simulation takes around six and a half CPU hours on a 64GB RAM system with a Dell Precision Tower 7810. General hard contact and frictionless tangential contact have been assigned to the whole model. The embedment interaction has been used to model the reinforced concrete liner by embedding the steel bars in the concrete liner.

3. Validation of Dynamic Loading

In order to validate the present finite element simulation, an experimental study has been simulated using the present methodology and numerical results are compared with the experimental study by Andersson [41], as shown in Table 3. A steel mass weighing 600 kg was considered for impact loading on 0.2 m × 0.2 m area in the middle of a slab. The height of fall was varied from 1 m to 2 m on the concrete slab having a 1.75 m × 1.75 m cross section and a 0.12 m thickness. The size of the slab, loading conditions, and the properties of the model have been adopted as per the report by Andersson [41].

Table 3. Comparison of results of properties of reinforced steel.

Slab No.	Height (m)	Deformation (mm)		% Cent Error
		Experimental Study	FE Study	
S4	1.0	46	44.39	3.5
S5	1.5	63	61.74	2.0
S6	1.5	50	47.00	6.0
S8	1.0	60	57.00	5.0
S9	1.2	61	59.17	3.0
S10	2.0	77	73.15	5.0

4. Results and Discussion

Commercial software based on the finite element method, i.e., Abaqus/Explicit, has been used for modelling and analysis. Four different sandstone rocks, Jhingurda, Singrauli, Jamrani, and Kota, were considered. A generally used design specification for metro tunnels has been used based on Delhi Metro Rail Corporation designs. A missile having a 5 Mach velocity and weighing 100 kg has been considered. The simulation has been run for 30 milliseconds, which is the time required by a missile to hit the rock ground from 100 m away.

Figure 5 has been plotted to compare the deformation variation with time when a missile hits the ground surface for all the sandstone rocks considered in the present study. It has been observed that the amplitude of deformation for Jhingurda sandstone is the maximum, having a magnitude of 2.45 m. Maximum deformations of 1.35 m, 1.03 m, and 0.76 m have been observed for Singrauli, Jamrani, and Kota rocks prospectively at the ground surface. Therefore, Kota sandstone has shown maximum resistance to missile penetration, while Jhingurda sandstone has the least resistance to missile penetration.

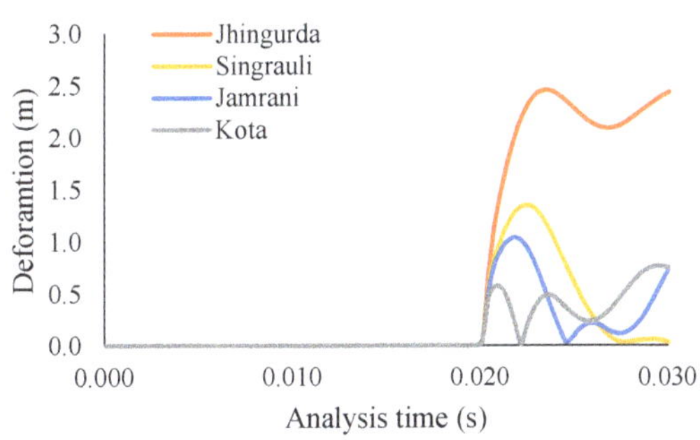

Figure 5. Variation in deformation with time to compare the ground surface behavior under impact loading of 100 kg missile.

The deformation profile is one of the important output results for understanding the internal behavior of a tunnel. Figure 6 shows the comparison of the deformation profiles of four sandstone rocks. The deformation profile for Jhingurda sandstone follows a smooth curvature, while the curvature becomes distorted and non-uniform for other types of sandstone. It has also been observed that as the strength of sandstone increases, the peak deformation gets decreased, while the length of tunnel under disturbance increases with the increase in the strength of sandstone. Moreover, a slight bulging has been observed in all the sandstones except for Jhingurda sandstone. Therefore, it may be concluded that weaker sandstone requires strengthening for a smaller area after an impact loading event, while high-strength sandstone will require repair for a longer portion of the tunnel.

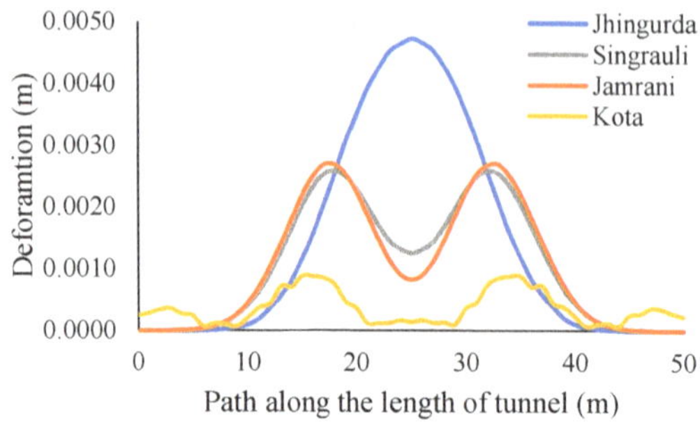

Figure 6. Comparison of deformation profile generated under the response of impact load.

Particle velocity is one of the significant output results in the dynamic loading analysis of rock tunnels. Therefore, the peak velocity at the ground surface of different sandstone rocks is shown in Figure 7 for comparison. The peak of the particle velocity graph has been observed at 20.41 milliseconds in each case of sandstone rock. However, the variation of particle velocity follows a separate path for different rocks, but the pattern of the particle velocity plot remains similar in all the cases. Figure 8 represents the peak acceleration at the ground surface when a missile hits the different sandstone rocks. In the case of Jhingurda sandstone, the magnitude of peak velocity and acceleration is greatest. Moreover, the pattern of variation in acceleration and velocity remains similar, and therefore, it is independent of the type of sandstone.

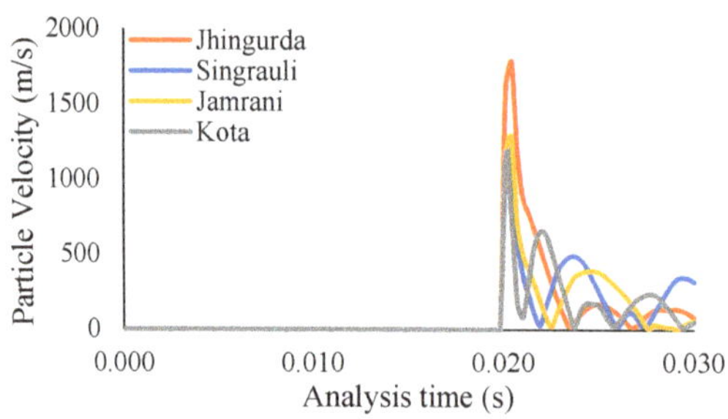

Figure 7. Comparison of particle velocity at ground surface under present impact loading condition.

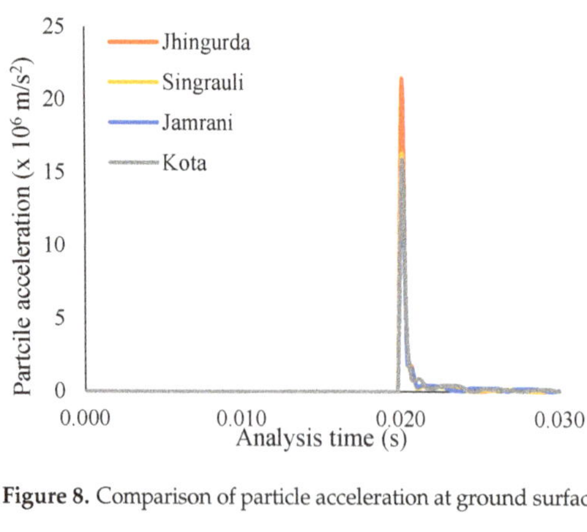

Figure 8. Comparison of particle acceleration at ground surface under present impact loading condition.

Figure 9 shows the deformation contours of Jhingurda, Singrauli, Jamrani, and Kota sandstone when a 100 kg missile moving at the velocity of 5 Mach hits the ground surface of the rock-containing tunnel. It has been observed that the brittleness and strength of rock have a significant influence on the deformation zone. In the cases of Jhingurda, Singrauli, Jamrani, and Kota, the maximum deformation at the crown is 4.94 mm, 2.75 mm, 2.61 mm, and 1.04 mm, respectively. It can be concluded that Kota sandstone has minimum deformation, and therefore, it is the safest sandstone rock under impact loading conditions. However, the area of disturbance is maximum in the case of Kota sandstone, and vibrations may reach the tunnel crown in a shorter time as compared to other sandstones.

Figure 10 has been plotted to compare the serviceability of reinforced concrete liners under impact loading conditions in the case of different types of sandstone. Tension damage has been observed in each type of sandstone. However, the area of the damaged zone increases with the strength and brittleness of sandstone. Therefore, reinforced concrete liner has maximum tensile damage in Kota sandstone (0.99) and less tension damage in the case of Jhingurda sandstone (0.017). It has been concluded that the consequences of an impacting projectile reach the tunnel lining when constructed in strong and brittle sandstone, or vice-versa. In addition, the tensile damage in all the different types of sandstones considered in the present study remains concentrated at the outer periphery of the liner.

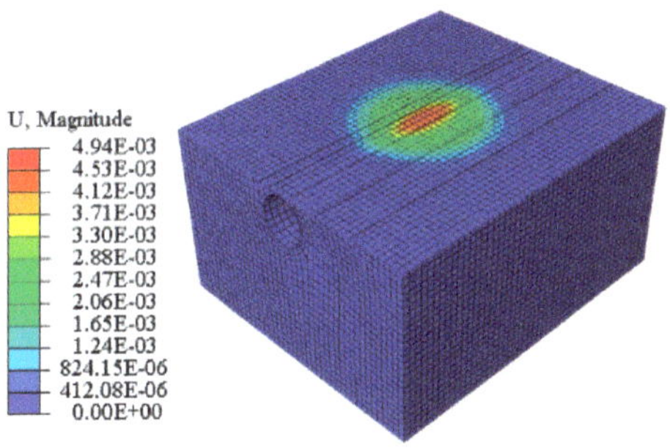

(a)

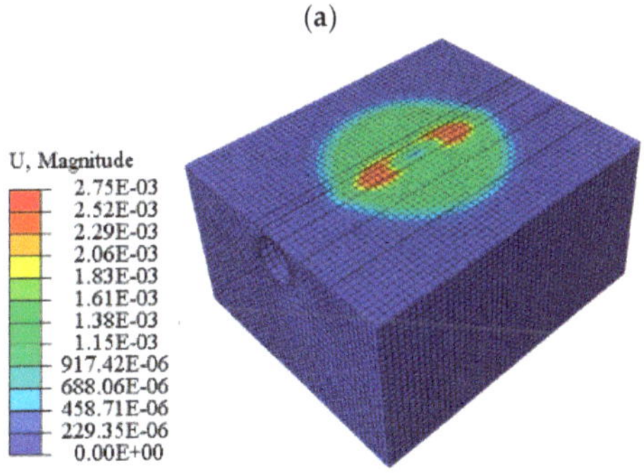

(b)

Figure 9. *Cont.*

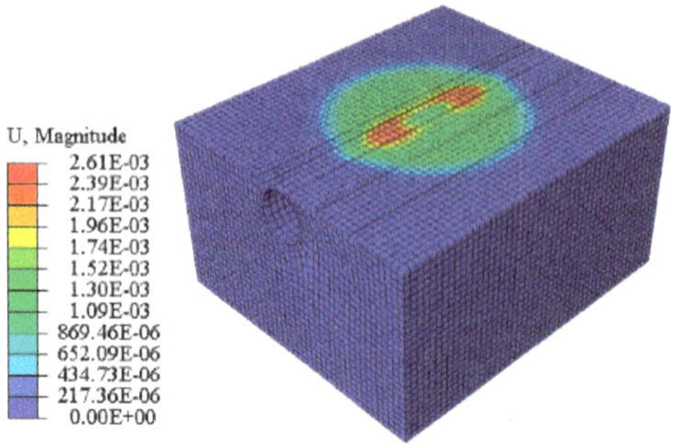

(c)

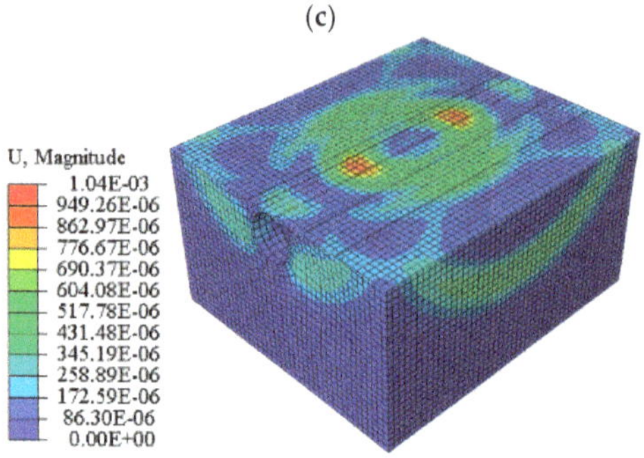

(d)

Figure 9. Deformation contours of (**a**) Jhingurda, (**b**) Singrauli, (**c**) Jamrani and (**d**) Kota when an impact loading occurs due to 100 kg missile.

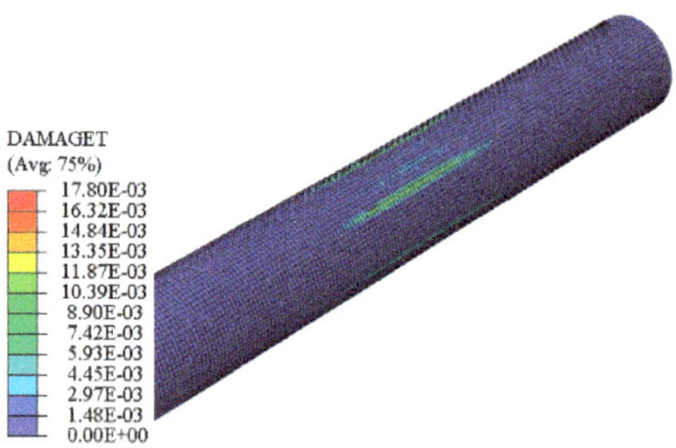

(a)

Figure 10. *Cont.*

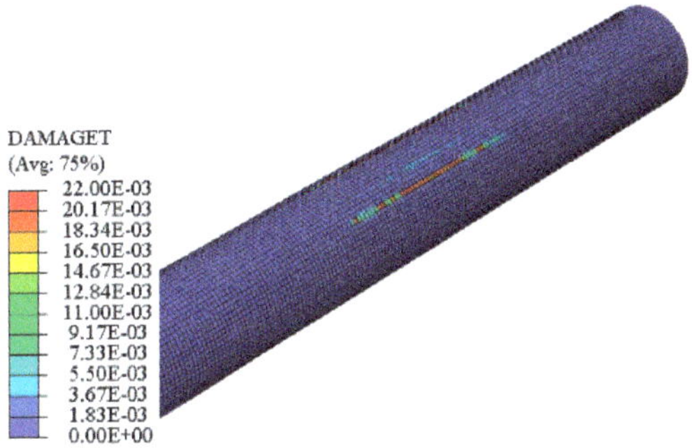

(**b**)

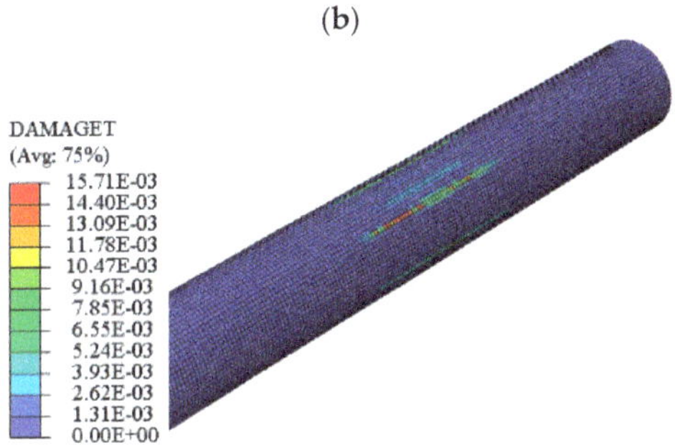

(**c**)

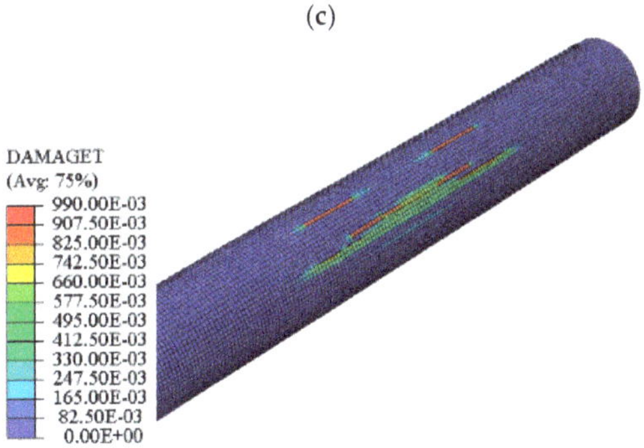

(**d**)

Figure 10. Tension damage contours of (**a**) Jhingurda, (**b**) Singrauli, (**c**) Jamrani, and (**d**) Kota sandstone tunnel when an impact loading occurs due to a 100 kg missile.

The in-depth view of lining performance has been studied by comparing the deformation at the reinforcement cage of steel bars in the case of different rocks, as shown in Figure 11. The maximum value of deformation has been noted for Jhingurda sandstone, while the maximum value of deformation remains concentrated at the crown of the tunnel.

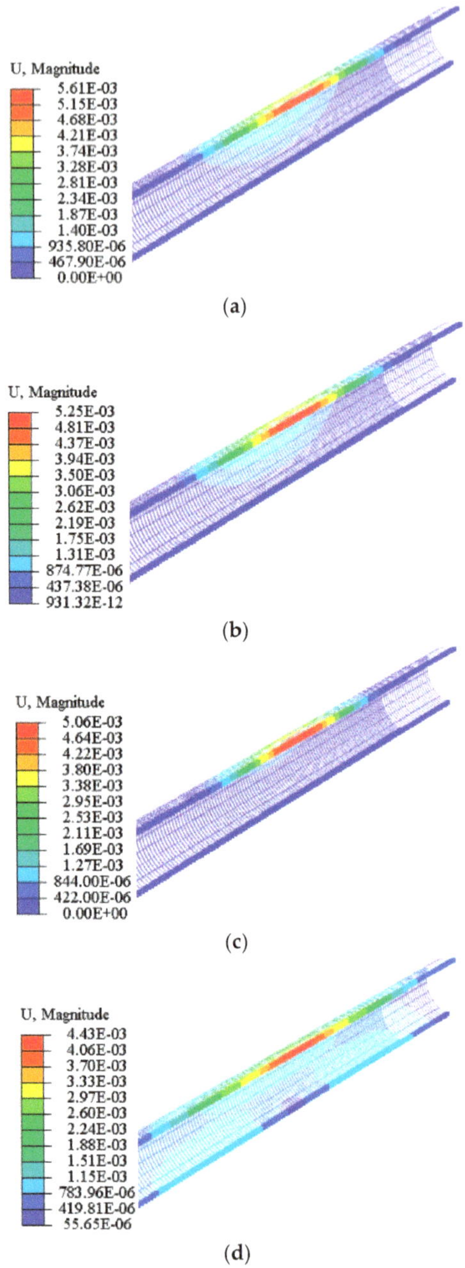

Figure 11. Deformation contours of steel bar reinforcement cage for (**a**) Jhingurda, (**b**) Singrauli, (**c**) Jamrani, and (**d**) Kota sandstone tunnel when an impact loading occurs due to a 100 kg missile.

5. Conclusions

The present study of finite element simulation for rock tunnel resistance against impact loading has four different sandstone rocks surrounding a 5 m diameter reinforced concrete tunnel lining. The major conclusions drawn from the present study are as follows:

1. The Kota sandstone has 2.22-times, 0.77-times, 0.35-times more impact resistance than Jhingurda, Singrauli and Jamrani sandstone, respectively. Therefore, the impact resistance of a rock tunnel is a function of cohesion and friction angle.
2. The deformation profile for Jhingurda sandstone follows a smooth curvature while the curvature becomes distorted and non-uniform for other types of sandstone. It has also been observed that as the strength of sandstone increases, the peak deformation decreases, while the length of tunnel under disturbance increases with the increase in the strength of sandstone. In addition, it has been concluded that weaker sandstone requires strengthening for a smaller area after an impact loading event while high-strength sandstone will require repair for a longer portion of the tunnel.
3. The magnitude of peak velocity and acceleration is maximum in case of Jhingurda sandstone at 20.41 milliseconds. Moreover, the pattern of variation in acceleration and velocity remains similar, and therefore, it is independent of the type of sandstone.
4. It can be concluded that Kota sandstone has minimum deformation and is therefore the safest sandstone rock under impact loading conditions. However, the area of disturbance is greatest in the case of Kota sandstone, and vibrations may reach the tunnel crown in a shorter time in comparison to other rocks.
5. The effect of impact loading has reached the tunnel lining when constructed in strong and brittle sandstone, or vice versa. In addition, the tensile damage in all the different types of sandstones considered in the present study remains concentrated at the outer periphery of the liner. The steel reinforcement cage experiences maximum deformation at the crown position in all the cases, while the maximum magnitude of deformation occurred in the case of Jhingurda sandstone.

Author Contributions: M.R.S.: drafting—data collection and preparation of the manuscript, writing review and editing, S.A.: drafting—preparation of the manuscript, revision and correction; A.H.A.: composing—reviewing, fund acquisition and modifying; A.S.A.: composing—reviewing, and modifying; W.S.: reviewing and modifying. All authors have read and agreed to the published version of the manuscript.

Funding: The authors acknowledge the support provided by Researchers Supporting. Project Number (RSP2022R473), King Saud University, Riyadh, Saudi Arabia.

Institutional Review Board Statement: Not applicable.

Informed Consent Statement: Not applicable.

Data Availability Statement: Not applicable.

Acknowledgments: Authors acknowledge the support provided by King Saud University, Saudi Arabia, and Aligarh Muslim University, India, to carry out this study.

Conflicts of Interest: The authors declare no conflict of interest.

References

1. Do, N.A.; Oreste, P.; Dias, D.; Antonello, C.; Djeran-Maigre, I.; Livio, L. Stress and strain state in the segmental linings during mechanized tunnelling. *Geomech. Eng.* **2014**, *7*, 75–85. [CrossRef]
2. Do, N.A.; Dias, D.; Oreste, P.; Djeran-Maigre, I. 2D numerical investigations of twin tunnel interaction. *Geomech. Eng.* **2014**, *6*, 263–275. [CrossRef]
3. Eskandari, F.; Goharrizi, K.G.; Hooti, A. The impact of EPB pressure on surface settlement and face displacement in intersection of triple tunnels at Mashhad metro. *Geomech. Eng.* **2018**, *15*, 769–774. [CrossRef]
4. Ghasemi, S.H.; Nowak, A.S. Reliability analysis of circular tunnel with consideration of the strength limit state. *Geomech. Eng.* **2018**, *15*, 879–888. [CrossRef]
5. Mazek, S.A. Evaluation of surface displacement equation due to tunnelling in cohesionless soil. *Geomech. Eng.* **2014**, *7*, 55–73. [CrossRef]
6. Miranda, T.; Dias, D.; Pinheiro, M.; Eclaircy-Caudron, S. Methodology for real-time adaptation of tunnels support using the observational method. *Geomech. Eng.* **2015**, *8*, 153–171. [CrossRef]
7. Fahimifar, A.; Ghadami, H.; Ahmadvand, M. The ground response curve of underwater tunnels, excavated in a strain-softening rock mass. *Geomech. Eng.* **2015**, *8*, 323–359. [CrossRef]

8. Nawel, B.; Salah, M. Numerical modeling of two parallel tunnels interaction using three-dimensional finite elements method. *Geomech. Eng.* **2015**, *9*, 775–791. [CrossRef]
9. Nikadat, N.; Marji, M.F. Analysis of stress distribution around tunnels by hybridized fsm and ddm considering the influences of joints parameters. *Geomech. Eng.* **2016**, *11*, 269–288. [CrossRef]
10. Khezri, N.; Mohamad, H.; Fatahi, B. Stability assessment of tunnel face in a layered soil using upper bound theorem of limit analysis. *Geomech. Eng.* **2016**, *11*, 471–492. [CrossRef]
11. Aalianvari, A.; Soltani-Mohammadi, S.; Rahemi, Z. Estimation of geomechanical parameters of tunnel route using geostatistical methods. *Geomech. Eng.* **2018**, *14*, 453–458. [CrossRef]
12. Zidan, A.F.; Ramadan, O.M. A hybrid MC-HS model for 3D analysis of tunnelling under piled structures. *Geomech. Eng.* **2018**, *14*, 479–489. [CrossRef]
13. Khan, M.A.; Sadique, M.R.; Harahap, I.H.; Zaid, M.; Alam, M.M. Static and Dynamic Analysis of the Shielded Tunnel in Alluvium Soil with 2D FEM Model. *Transp. Infrastruct. Geotechnol.* **2022**, *9*, 73–100. [CrossRef]
14. Sadique, R.M.; Zaid, M.; Naqvi, M.W.; Akhtar, M.F. Analysis of Underground Renewable Energy Storage Tunnels Subjected to Capricious Superstructures. In *Renewable Power for Sustainable Growth, Proceedings of the International Conference on Renewal Power (ICRP 2020), Rajouri, India, 17–18 April 2020*; Lecture Notes in Electrical Engineering; Springer: Singapore, 2021; Volume 723.
15. Sadique, M.R.; Ali, A.; Zaid, M.; Masroor Alam, M. Experimental and Numerical Modeling of Tunneling-Induced Ground Settlement in Clayey Soil. In *Advances in Geotechnics and Structural Engineering, Proceedings of the TRACE 2020, Noida, India, 20–21 August 2020*; Lecture Notes in Civil Engineering; Springer: Singapore, 2021; Volume 143.
16. Sadique, M.R.; Zaid, M.; Alam, M.M. Rock Tunnel Performance Under Blast Loading Through Finite Element Analysis. *Geotech. Geol. Eng.* **2022**, *40*, 35–56. [CrossRef]
17. Zaid, M.; Mishra, S. Numerical Analysis of Shallow Tunnels Under Static Loading: A Finite Element Approach. *Geotech. Geol. Eng.* **2021**, *39*, 2581–2607. [CrossRef]
18. Gao, F.; Hou, A.; Yang, X. Numerical Analysis of Dynamic Mechanical Properties for Rock Sample under Strong Impact Loading. *Int. J. Inf. Eng. Electron. Bus.* **2010**, *2*, 10–16. [CrossRef]
19. Aziznejad, S.; Esmaieli, K.; Hadjigeorgiou, J.; Labrie, D. Responses of jointed rock masses subjected to impact loading. *J. Rock Mech. Geotech. Eng.* **2018**, *10*, 624–634. [CrossRef]
20. Zhou, L.; Zhu, Z.; Wang, M.; Ying, P.; Dong, Y. Dynamic propagation behavior of cracks emanating from tunnel edges under impact loads. *Soil Dyn. Earthq. Eng.* **2018**, *105*, 119–126. [CrossRef]
21. Zhou, L.; Zhu, Z.; Dong, Y.; Fan, Y.; Zhou, Q.; Deng, S. The influence of impacting orientations on the failure modes of cracked tunnel. *Int. J. Impact Eng.* **2019**, *125*, 134–142. [CrossRef]
22. Zhou, L.; Zhu, Z.; Dong, Y.; Ying, P.; Wang, M. Study of the fracture behavior of mode I and mixed mode I/II cracks in tunnel under impact loads. *Tunn. Undergr. Space Technol.* **2019**, *84*, 11–21. [CrossRef]
23. Vidanović, N.; Rašuo, B.; Kastratović, G.; Maksimović, S.; Ćurčić, D.; Samardžić, M. Aerodynamic–structural missile fin optimization. *Aerosp. Sci. Technol.* **2017**, *65*, 26–45. [CrossRef]
24. DMRC. *Design Specifications of DMRC*; DMRC: New Delhi, India, 2015.
25. Zaid, M. Dynamic stability analysis of rock tunnels subjected to impact loading with varying UCS. *Geomech. Eng.* **2021**, *24*, 505–518. [CrossRef]
26. Zaid, M. Three-dimensional finite element analysis of urban rock tunnel under static loading condition: Effect of the rock weathering. *Geomech. Eng.* **2021**, *25*, 99–109. [CrossRef]
27. Rao, K.; Rao, G.; Ramamurthy, T. Strength Behaviour of Some Indian Sandstones. In Proceedings of the Asian Regional Conference on Geotechnical Problems and Practices in Foundation Engineering, Colombo, Sri Lanka, 25–27 February 1986; pp. 1–6.
28. Johnson, G.R.; Cook, W.H. A Constitutive modeling and data for metals subjected to large strain rates and high temperatures. In Proceedings of the 7th International Symposium on Ballistics, The Hague, The Netherlands, 19–21 April 1983.
29. Borvik, T.; Hopperstad, O.S.; Berstad, T.; Langseth, M. A computational model of viscoplasticity and ductile damage for impact and penetration. *Eur. J. Mech. A Solids* **2001**, *20*, 685–712. [CrossRef]
30. Zaid, M.; Rehan Sadique, M. Dynamic analysis of tunnels in western ghats of indian peninsula: Effect of shape and weathering. In *Recent Trends in Civil Engineering*; Springer: Singapore, 2021; Volume 77.
31. Zaid, M.; Rehan Sadique, M. A Simple Approximate Simulation Using Coupled Eulerian–Lagrangian (CEL) Simulation in Investigating Effects of Internal Blast in Rock Tunnel. *Indian Geotech. J.* **2021**, *51*, 1038–1055. [CrossRef]
32. Zaid, M.; Shah, I.A. Blast-Resistant Stability Analysis of Triple Tunnel. In *Advances in Geotechnics and Structural Engineering, Proceedings of the TRACE 2020, Noida, India, 20–21 August 2020*; Lecture Notes in Civil Engineering; Springer: Singapore, 2021; Volume 143.
33. Zaid, M.; Athar, M.F.; Sadique, M.R. Effect of Rock Weathering on the Seismic Stability of Different Shapes of the Tunnel. In *Lecture Notes in Civil Engineering, Proceedings of the Indian Geotechnical Conference, 2019, Surat, India, 19–21 December 2021*; Springer: Singapore, 2021; Volume 137.
34. Zaid, M.; Faraz Athar, M.; Rehan Sadique, M. Interaction of Transmission Tower Footing with Twin Rock Tunnel. In *Advances in Geotechnics and Structural Engineering, Proceedings of the TRACE 2020, Noida, India, 20–21 August 2020*; Lecture Notes in Civil Engineering; Springer: Singapore, 2021; Volume 143.

35. Zaid, M.; Naqvi, M.W.; Sadique, M.R. Stability of Arch Tunnel in Different Magnitude of Earthquake with Effect of Weathering in Western Ghats of India. In *Lecture Notes in Civil Engineering, Proceedings of the Indian Geotechnical Conference 2019, Surat, India, 19–21 December 2021*; Springer: Singapore, 2021; Volume 138.
36. Zaid, M.; Sadique, M.R.; Alam, M.M. Blast Resistant Analysis of Rock Tunnel Using Abaqus: Effect of Weathering. *Geotech. Geol. Eng.* **2021**, *40*, 809–832. [CrossRef]
37. Zaid, M.; Sadique, M.R.; Alam, M.M. Blast analysis of tunnels in Manhattan-Schist and Quartz-Schist using coupled-Eulerian–Lagrangian method. *Innov. Infrastruct. Solut.* **2021**, *6*, 69. [CrossRef]
38. Zaid, M.; Sadique, M.R. Effect of joint orientation and weathering on static stability of rock slope having transmission tower. In Proceedings of the 7th Indian Young Geotechnical Engineers Conference 2019, Silchar, India, 15–16 March 2019; Volume 5, pp. 414–422.
39. Zaid, M.; Sadique, M.R. Numerical modelling of internal blast loading on a rock tunnel. *Adv. Comput. Des.* **2020**, *5*, 417–443. [CrossRef]
40. Zaid, M.; Sadique, M.R. The response of rock tunnel when subjected to blast loading: Finite element analysis. *Eng. Rep.* **2021**, *3*, e12293. [CrossRef]
41. Andersson, A. *Impact Loading on Concrete Slabs: Experimental Tests and Numerical Simulation*; KTH: Stockholm, Sweden, 2014.

applied sciences

MDPI

Article

Blast Hole Pressure Measurement and a Full-Scale Blasting Experiment in Hard Rock Quarry Mine Using Shock-Reactive Stemming Materials

Younghun Ko [1,*], Chanhwi Shin [2], Youngjun Jeong [3] and Sangho Cho [2,4,*]

1 Department of Geotechnical Engineering Research, Korea Institute of Civil Engineering and Building Technology, 283 Goyang-daero, Ilsanseo-gu, Goyang-si 10223, Korea
2 Department of Energy Storage & Conversion Engineering of Graduate School, Jeonbuk National University, 567 Baekje-Daero, Deokjin-gu, Jeonju-si 54896, Korea
3 Seokseong Blasting Construction Co., Ltd., Wolgye-ro 37-gil, Gangbuk-gu, Seoul 01226, Korea
4 Department of Mineral Resources and Energy Engineering, Jeonbuk National University, 567 Baekje-Daero, Deokjin-gu, Jeonju-si 54896, Korea
* Correspondence: younghunko@kict.re.kr (Y.K.); chosh@jbnu.ac.kr (S.C.)

Abstract: By increasing the effectiveness of the energy generated by the explosive charge inserted into a blast hole, stemming increases rock fragmentation. Missing or improper stemming, which can lead to the detonation gas escaping from the blast hole in advance, results not only in the waste of explosive energy and poor fragmentation but also in environmental problems, such as ground vibration, noise, flying rocks, back breaks, and air blasts. In this study, a stemming material based on a shear thickening fluid (STF) that reacts to dynamic pressure was developed. Two blasting experiments were conducted to verify the performance of the STF-based stemming material. In the first experiment, the pressure inside the blast hole was directly measured based on the application of the stemming material. In the second experiment, full-scale bench blasting was performed, and the blasting results of sand stemming and the STF-based stemming cases were compared. The measurement results of the pressure in the blast hole showed that when the STF-based stemming material was applied, the pressure at the top of the blast hole was lower than in the sand stemming case, and the stemming ejection was also lower. Full-scale bench blasting was conducted to compare the two types of stemming materials by evaluating the size of the rock fragments using image processing. The results of the two blasting experiments helped to verify that the blockage performance of the STF-based stemming material in the blast hole was superior to that of the sand stemming material.

Keywords: blasting experiment; stemming material; shear thickening fluid; sand; blockage performance

Citation: Ko, Y.; Shin, C.; Jeong, Y.; Cho, S. Blast Hole Pressure Measurement and a Full-Scale Blasting Experiment in Hard Rock Quarry Mine Using Shock-Reactive Stemming Materials. *Appl. Sci.* **2022**, *12*, 8629. https://doi.org/10.3390/app12178629

Academic Editors: Ricardo Castedo, Lina M. López and Anastasio P. Santos

Received: 2 August 2022
Accepted: 26 August 2022
Published: 29 August 2022

1. Introduction

Stemming is a process applied to blast holes to prevent gases from escaping during detonation. A stemming material helps confine the explosive energy for a longer duration. Without stemming, up to 50% of the explosive energy can escape through the borehole [1]. Proper stemming has been shown to improve explosive efficiency by over 41% [2]. Further, employing even the least efficient stemming materials can boost the usable energy of an explosion by 60%, while the most efficient stemming materials can increase it by up to 93% [3].

Missing or improper stemming, which can lead to the detonation gas escaping from the blast hole in advance, results not only in the waste of explosive energy and poor rock fragmentation but also in environmental problems, such as ground vibration, noise, flying rocks, back breaks, and air blasts [4].

Smaller amounts of explosives may be used to produce the same blasting effects if explosive energy was used more effectively [5,6]. Improvements in fragmentation will result in lower second breaking work costs. Proper stemming can reduce costs and improve the productivity and profitability of a mining operation. The main objective of rock-blasting is fragmentation by explosive. The loading and hauling operations of a mining operation, particularly the crushing line, profit greatly from good fragmentation [7]. Additionally, cracks are generated over a larger area in the rock mass using proper stemming. These cracks propagate, interconnect, and cut the rock mass; thus, the block size and distribution after blasting satisfy the construction or mining requirements. Therefore, reasonably selecting the stemming material is of particular importance for improving the blasting effect, increasing the efficiency of explosives, and obtaining ideal blasting fragmentation.

In the mining industry, blast holes are sealed with three different sorts of stemming materials: colloidal, liquid, and solid. Additional research on the performance of stemming materials is required. Li et al. [8] used a water stemming technique that involves putting water-filled polyvinyl plastic bags within blast holes. A water-silt composite-stemmed blasting method for tunnels was proposed [9] to increase rock breakage, reduce dust, and use fewer explosives.

In most cases, using fluid-type stemming inside a blast hole as a stemming material produced good results. Fluid has a higher density than air, and even at extremely high pressures, the compression of water is significantly lower than that of air [10].

Under dynamic loading, Zhu et al. [11] performed an AUTODYN numerical analysis using a variety of stemming materials, such as fluid (water), sand, and air, placed in the area between the internal explosives and the hollow wall inside the blast hole. The water stemming case, which was also the best medium for shockwave transmission, produced the largest fracture area. Additionally, the presence of fluid (water), which results in the deformation and displacement of the rock, causes the shock wave to reflect and bubble pulse, which contributes to the high stress exerted during this process.

The most effective method to evaluate the stemming effect is to conduct field experiments. A stemming performance test of a small-scale model was developed, and the results showed that different stemming materials have different functionalities, which can significantly influence the efficiency of rock breaking [12].

Kopp [13] suggested a simple physical model for predicting the time required to eject stemming. This model depends only on the inertia of the stemming material. The frictional forces that resist the movement were omitted. The stemming performance of the stemming material can be evaluated using the initial ejection velocity of the stemming part at the entrance of the blast hole [14].

The momentum of the stemming structure based on the explosive load can be calculated using the mass of the stemming structure and the initial velocity of the stemming part when the stemming is ejected into the orifice.

As a highly capable method in the mining industry, image analysis techniques have been used to predict rock fragmentation by blasting. These techniques are capable of visual processing, thereby serving as an appropriate alternative to low-accuracy methods [15]. Over the past few decades, various image analysis software packages, such as Split-Online, Split-desktop, Gold-Size, and Wip-Frag, have been developed, and their applications in the mining industry and mineral processing have been reported. The main advantages of these software packages are their integration and lack of disruption [16].

The specific charge is mainly used as an indicator to predict the blasting effect, but the amount of powder used per unit volume of crushed rock cannot properly reflect the influence of energy change in the blasting hole; therefore, the pressure in the blasting hole must be estimated and used to understand this [17]. However, the blasting pressure has relied on calculations rather than direct measurement. Recently, the blasting pressure has been estimated through a numerical analysis approach, but it is difficult to predict the explosion reaction of explosives acting on rocks, an anisotropic material. Therefore, the concepts of abnormal explosion and ideal sealing are used to calculate the pressure in the blast hole, assuming that there is no external influence [18]. Therefore, field experiments are being conducted to directly measure the pressure in the blast hole. However, the pressure probe used is very expensive, and the sensor is only used once because of the extreme conditions generated by the blast, thereby requiring a significant financial investment. Therefore, in most cases, a pressure measurement sensor is inserted into the dummy hole, and the blasting pressure is indirectly measured based on the impact pressure propagated through the rock [19].

In recent years, a new intelligent material named shear thickening fluid (STF) has been widely used in energy absorption research [20–25]. STF exhibits an intense viscosity jump under shock load; as a result, it has been used in various applications, such as liquid body armor [26]. However, limited research on the application of STF for industrial blasting or as a stemming material is currently available.

In this study, a shock-reactive stemming material was developed that behaves similar to water in terms of shockwave propagation and has a high shear strength for dynamic shocks. The STF is characterized by its reversible energy absorption behavior under impulse loading. Its remarkable energy absorption capacity is attributed to viscous dissipation during shear and compression thickening. The STF-based stemming material was developed based on the following advantages. (1) STF has excellent sealing properties as it is a fluid-based material. (2) Its viscosity rapidly changes because of external shock, while material compaction or deformation is minimal with respect to the dynamic gas pressure in the blast hole. (3) Using starch as the main base material reduces costs.

Two blasting experiments were conducted to compare and contrast the blast effects of the developed stemming material and those of commonly used blasting stemming materials. The first is an experiment in which the blast hole pressure and stemming ejection are directly measured, and the second is an experiment to verify the stemming performance by analyzing the assessment of rock fragmentation through a full-scale blasting experiment.

2. Materials and Methods

2.1. STF-Based Stemming Material Rheology Tests

A dense colloidal dispersion of solid nanoparticles in a carrier fluid is known as an STF [27]. When a shear force is applied, the random distribution of particles in the dispersion initially emerges in an ordered fashion because the hydrodynamic forces are greater than the repulsive forces operating between the interstitial gaps the particles have generated. The order–disorder theory, put forth by Hoffmann in 1972 [28], is represented by this arrangement of particles. Large hydrodynamic forces tend to push out the fluid between the interstitial spaces with rising shear rates, leading to the production of hydroclusters. The hydroclustering mechanism proposed by Brady and Bossis in 1985 is comprised of this phenomenon [29]. These clusters are stress-bearing elements that lead to particle jamming, when additional shearing pressures are applied.

STFs behave by increasing the dynamic viscosity under the application of shear stress. When tightly packed particles combine with enough liquid to cover the spaces between the particles, dilatancy occurs. The fluid serves as a lubricant at low speeds, facilitating easy movement of the dilatant substance.

Because of the increased friction caused by the inability of the liquid to fill the gaps left by the particles at greater velocities, the viscosity also increases. The STF is also non-Newtonian in nature because its viscosity depends on the shear rate or shear rate history. This behavior is a type of deviation from Newton's law and is controlled by factors, such as particle size, shape, and distribution. Empirical studies have also shown that shear thickening effects vary with different particles and additive concentrations, as well as with the molecular chain of the additives [30].

Shear thickening is a reversible phenomenon governed by a power law model. Generally, a non-Newtonian fluid is described using the power law model expressed in Equations (1) and (2).

$$\tau = K\left(\frac{\partial \mu}{\partial y}\right)^{n} = K(\gamma)^{n} = \tau = K(\gamma)^{n-1}\,(\gamma)^{1}\ ,\ \tau = \mu_{apparent}(\gamma), \tag{1}$$

$$\mu_{apparent} = K(\gamma)^{n-1} \quad (2)$$

where τ is the shear stress exerted by the fluid, K is the fluid viscosity, μ is the shear deformation, y is the distance from the reference layer, $\frac{\partial \mu}{\partial y}$ is the strain rate, n is the flow behavior index, and $\mu_{apparent}$ is the apparent viscosity.

As shown in Figure 1, the fluid behaves as a Newtonian fluid at $n = 1$ and exhibits shear thinning properties when $0 < n < 1$. Moreover, several dispersions and liquid polymers exhibit shear thinning behavior for n values between 0.3 and 0.7. However, this depends on the particle concentration and molecular weight of the carrier fluid.

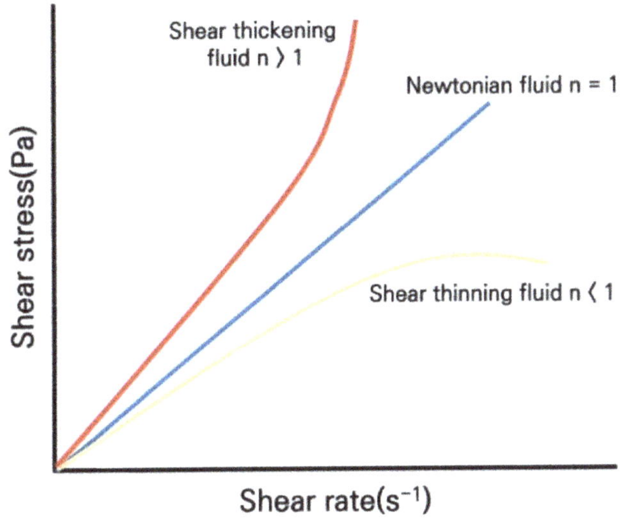

Figure 1. The shear thickening process of a shear thickening fluid.

The shear thickening effect of STF is demonstrated by the lower rate of increase in the shear stress in the low-shear-strain-rate regions and a higher rate of increase in the high-shear-strain-rate regions (Figure 1). Figure 1 also depicts the relationship between the shear stress and shear strain rate of the STF. The potential of a STF to improve the effectiveness of body armor against ballistic impacts and stab resistance has been thoroughly researched [31,32]. Further research is necessary, nevertheless, to fully understand the potential impact of STF on the stemming of blast holes. The motivation of this study is to harness the strength of the STF through flexible deployment and relatively easy stemming, which can help dissipate pressure wave loading around the rock mass during an explosion. The rheological behavior of the non-Newtonian fluid was measured using a rheometer.

In this study, the STF-based stemming material was mainly based on corn starch, while xanthan gum and guar gum were blended to increase the viscosity. Sodium benzoate was used to prevent the putrefaction of starch, and nitroglycol and salt were added to prevent freezing in winter.

The STF samples were sandwiched between a cone plate and the foundation support of a rheometer (Anton-Paar MCR301 rheometer) for rheological experiments. During the studies, the shear rate applied to the sample was increased from 0 to 100 s^{-1}, and all of the tests were carried out at a temperature of 25 °C.

A schematic of the rheometer and the results of the rheological tests conducted on the STF-based stemming material are shown in Figures 2 and 3. The rheology tests were performed on the STF samples of 30, 45, and 55 wt.% corn starch suspensions. In the case of the 30 wt.% corn starch suspension, no significant shear thickening is observed. However, as the starch content increases, the particle content exceeds the ratio of the dispersion medium; therefore, the distance between the corn starch particles decreases, and the shear thickening effect increases. For the sample made up of 55 wt.%, a shear thickening effect is attained at a critical shear rate of 85 s^{-1}. The STF initially experiences marginal shear thinning, which then grows with the shear rate. In particular, the viscosity of the STF suddenly increases as the shear rate reaches a critical value, indicating a shear thickening phenomenon. However, the viscosity of the STFs sharply decreases after a period of shear thickening. The critical shear rate of the STF sample is approximately 85 s^{-1}, and the maximum viscosity of the STF samples is 543 Pa.

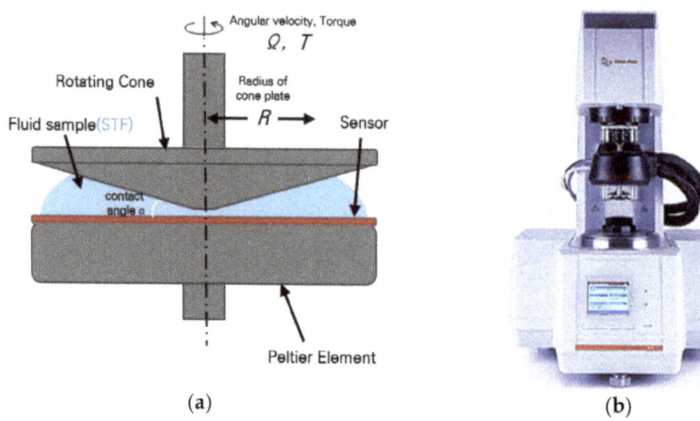

Figure 2. (**a**) Rheometer schematic and (**b**) Anton-Paar MCR301 rheometer.

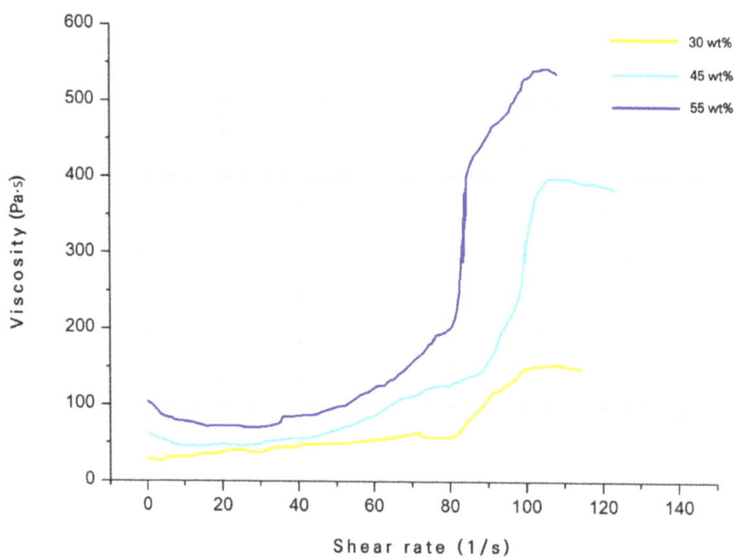

Figure 3. Results of the shear thickening fluid based stemming material's rheological tests.

The STF-based stemming material image and package products are shown in Figure 4. The hammer recoils upon impact with the STF suspension surface, similar to hitting a true solid interface. It is a suspension of starch powder with a diameter range of 5–20 μm in water. The STF was created using a mixture of mechanical and ultrasonic mixing at a concentration of 55 wt.%. This weight percentage was chosen to preserve a viable solution, while ensuring the shear thickening tendency [33]. According to previous experimental results [34] for 52.5–55 wt.% corn starch, after the impact of rock falling on the suspension surface, the rock recoiled, similar to hitting a true solid interface.

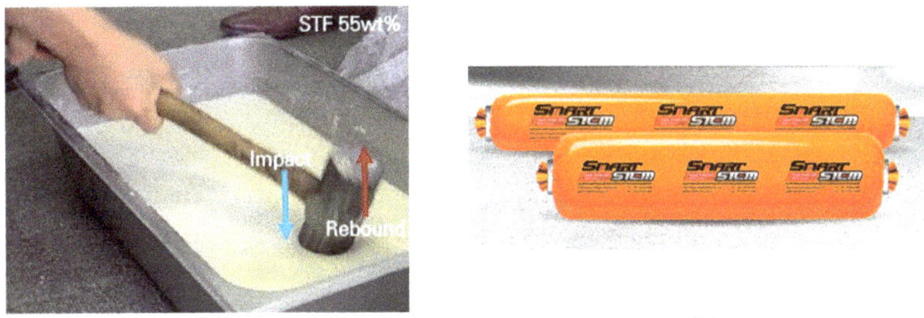

(a) (b)

Figure 4. (**a**) Shear thickening fluid-based stemming material image and (**b**) package products.

2.2. Overview of the Blast Hole Pressure Measurements

In blasting, stemming constrains the blasting pressure in the blasting hole, leads to crack propagation through the behavior inside the blasting hole under the explosive pressure, and ultimately plays an important role in crushing the rock. Therefore, the behavior and control of the explosive pressure acting on the blast hole are important for effective blasting.

Laboratory- and field-scale experiments were conducted to measure the pressure inside the blast hole. However, because the sensor that measures the pressure inside the blast hole is expensive, there is a risk of failure owing to a strong impact. Therefore, the explosive pressure is measured using a one-time polyvinylidene fluoride (PVDF) sensor [35] or by drilling a dummy hole [36].

In this study, instead of using a one-time sensor or drilling a dummy hole to measure the pressure inside the blast hole, nylon tubes and water pressure measurement sensors that can be used multiple times were employed. Thus, the pressure generated by blasting was measured in the blasting hole without sensor damage or additional drilling.

2.3. Assessment for Rock Fragmentation of Bench Blasting

In this study, full-scale bench blasting was performed for each stemming material (sand or STF), and an image-based blast fragmentation method was applied to compare and evaluate the performance of each stemming material.

Sieving or screening is a direct and accurate method for evaluating the size distribution of particles or their fragmentation. This method is feasible for small-scale blasts or operations; however, it is costly and time-consuming. Rock fragments are screened through sieves of different mesh numbers for different fragment sizes, and the screened fragments are grouped based on their size. The nature of the blast was predicted by counting the number of fragments of each size [37].

WipFrag is an image analysis system for sizing materials, such as blasted or crushed rocks [38]. It has also been used to measure other materials, such as ammonium nitrate prills, glass beads, and zinc concentrates. WipFrag accepts images from a variety of sources, such as roving camcorders, fixed cameras, images, or digital files. It uses automatic algorithms to identify individual blocks and to create an outline "net" using state-of-the-art edge detection. If desired or necessary, manual intervention (editing the image net) can be performed to improve the fidelity. WipFrag measures a 2D net and reconstructs a 3D distribution using the principles of geometric probability. WipFrag supports two methods: Rossin Rammler and Swebrec. Two parameters were used by Rammler as key performance indicators (KPIs); X_c, known as the characteristic size of the distribution and more specifically D63.2, and n, the value of which is the measure of uniformity [39].

Analyzing every fragment in the rock muck pile is fortunately not necessary because it is widely accepted that the mass percentage of fragments smaller than any given size varies linearly with the fragment size when plotted in the Rosin–Rammler domain. By measuring only a sufficient number of particles, the slope and intercept of the Rosin–Rammler line can be confidently defined [40]. A Rosin–Rammler line can be expressed as in Equation (3).

$$R(X) = 1 - \exp\left[-\left(\frac{X}{X_c}\right)^n\right] \quad (3)$$

Here,

$R(X)$ = Cumulative fraction by weight undersize in relation to size x.
X_c = Size modulus, which defines the characteristic size of the distribution.
n = Distribution modulus, which defines the spread of the distribution.

For $R(X) = 0.5$ (i.e., 50% of the fragments passing through the sieve), the value of X_c can be measured as follows:

$$X_c = \frac{X_{50}}{0.693^{1/n}} \quad (4)$$

3. Blast Hole Pressure Measurement Experiment

3.1. Explosion Pressure Sensor Calibration

In this study, it was necessary to calibrate the pressure sensor to measure the explosive pressure inside the blast hole. Therefore, prior to this experiment, explosive pressure sensor calibration using water pressure was performed. The explosive pressure sensor was calibrated under the same installation conditions as those of the blasting site.

Calibration of the explosion pressure sensor was performed by filling the nylon tube connected to the explosion pressure sensor with water, connecting it with a water pump, and pressurizing the pressure port of the explosion pressure sensor under the conditions of 0 MPa, 25 MPa, and 50 Mpa three times each. The average and standard deviation of the results for three calibration tests was 0.984 (±0.0033) Voltage at 0 MPa, 2.98 (±0.0082) Voltage at 25 MPa, and 4.98 Voltage (±0.0144) at 50 MPa. In addition, Nonlinearity was 0.137% FS (Full Scale), and Accuracy was 0.86% FS. The results of the experiments performed are specified in Table 1, and the Voltage to pressure (MPa) correction equation is shown in Figure 5.

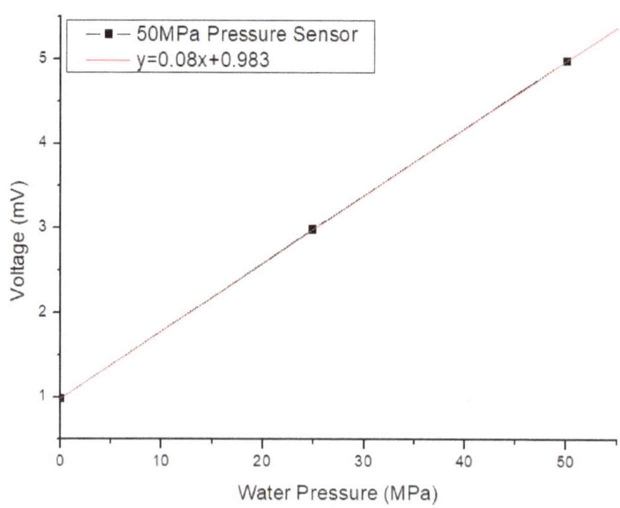

Figure 5. Explosive pressure bar calibration results.

Table 1. Results of the Explosion Pressure sensor calibration.

No.	Pressure Range (MPa)	Analog Output (V)
1	0	0.985
2	25	2.98
3	50	4.984
4	0	0.988
5	25	2.97
6	50	4.955
7	0	0.98
8	25	2.99
9	50	4.987

3.2. Blast Hole Pressure Measurement System

The purpose of the experiment was to measure the explosion pressure in a blast hole for each string material and to evaluate the pressure resistance of the stemming material. Conventional sand and STF-based stemming materials were applied to evaluate their resistance capability under explosive pressure. Figure 6 shows the shape of each stemming material.

(**a**) Sand stemming material (**b**) STF-based stemming material

Figure 6. Stemming materials used in blast hole pressure measurement experiment.

To measure the explosive pressure data from the blast hole, a pressure meter with cerabar (PMC) model of manufacturer Sensys capable of instantaneous dynamic shock pressure measurement was applied, and the MREL's MicroTrap was used to collect the data and set the detonation time trigger. The hydraulic shock pressure sensor had a pressure measurement range of 0–50 MPa. To measure the explosive pressure in the blast hole, the sensor was connected to a nylon tube filled with water using a coupling connector, placed in a water tube. To measure the explosive pressure in the blast hole, the sensor was connected to a nylon tube filled with water using a coupling connector, placed in a water tube. An explosive pressure-measuring device through hydraulic pressure was manufactured. Additionally, the manufactured water tube was protected with an industrial hose made of piezoelectric material to prevent damage when inserted into the blasting hole. Figure 7 shows the measuring tool applied to the blast pressure measuring system in the blast hole through hydraulic pressure. To collect the corresponding explosive pressure data as a time history at the same time as detonation using MicroTrap, a trigger line was attached to the explosion in the blast hole, and the break circuit trigger method was applied in which the connection signal was cut by the detonation of the explosive, and the measurement was finally started.

(**a**) Newmite Plus 1 with trigger line
(**b**) MREL's MicroTrap
(**c**) Water tube for hydraulic pressure measurement
(**d**) Piezoelectric tube for water tube protection

Figure 7. Measuring tool applied to blast pressure measuring system.

The length of the blast holes drilled for measuring the impact pressure of the stemming material was 3.2 m, and the explosives charge length was 1 m. The emulsion series Newmite Plus 1 (Φ50 mm, 2.5 kg) manufactured by Hanhwa with an explosion speed of 5700 m/s were applied to the experiment. On top of the explosive, a test stemming material (sand or STF) of 0.6 m was applied; a 0.5 m water tube was inserted to measure the pressure caused by the explosion as the water pressure; and in the 1.1 m remaining at the top of the blast hole, general sand stemming was inserted. Figure 8 shows a schematic of the blast hole measurement system.

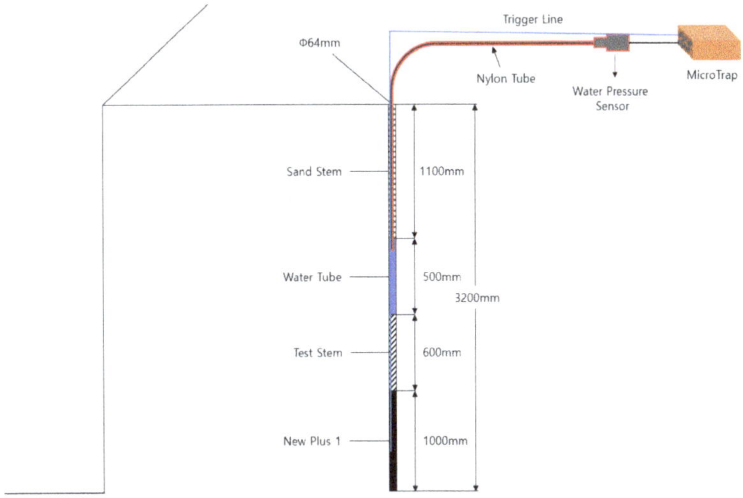

Figure 8. Schematic of the blast hole measurement system.

3.3. Experimental Results

In this experiment, a hydraulic explosive pressure propagation data measurement system was used to measure the pressure in the blast hole, and a comparative analysis was conducted on the blasting pressure behavior of the sand- and STF-based stemming materials. Table 2 presents the results of the blast hole pressure measurements.

Table 2. Result of the explosion pressure measurement in blast hole.

	Peak Pressure in Blasting Hole (MPa)	Explosive Pressure Arrival Time after Trigger (ms)	Explosive Pressure Duration in Blasting Hole (ms)
Sand Stemming	5.84	12.24	16.82
STF-based stemming	2.80	31.82	21.81

The pressure in the blasting hole is 5.84 MPa for the sand stemming material and 2.80 MPa for the STF-based stemming material. Further, the blasting pressure by sand is two times higher than by STF-based stemming material. This is the pressure transferred to the water tube located above the test stemming material; therefore, a lower measured pressure value means that the loss of explosive pressure in the blast hole due to stemming transfer is lower.

The time taken from detonation to explosion pressure transfer is 12.24 ms in the sand stemming material and 31.82 ms in the STF-based stemming material; the measured value is significantly lower than that of the sand stemming material. This is the explosive pressure transfer time from the lower part of the blasting hole to the upper part of the blasting hole for the explosive detonation of the stemming material. Thus, the longer the measured explosive pressure delay time, the better the ejection resistance. In addition, the duration of the explosive pressure in the blast hole is 16.82 ms when using the sand stemming material and 21.81 ms when using the STF-based stemming material, an improvement of approximately 5 ms. This implies that the explosive pressure acted longer inside the blast hole, as long as the duration of the explosive pressure. Figure 9 shows the time-pressure hysteresis curve inside the blast hole.

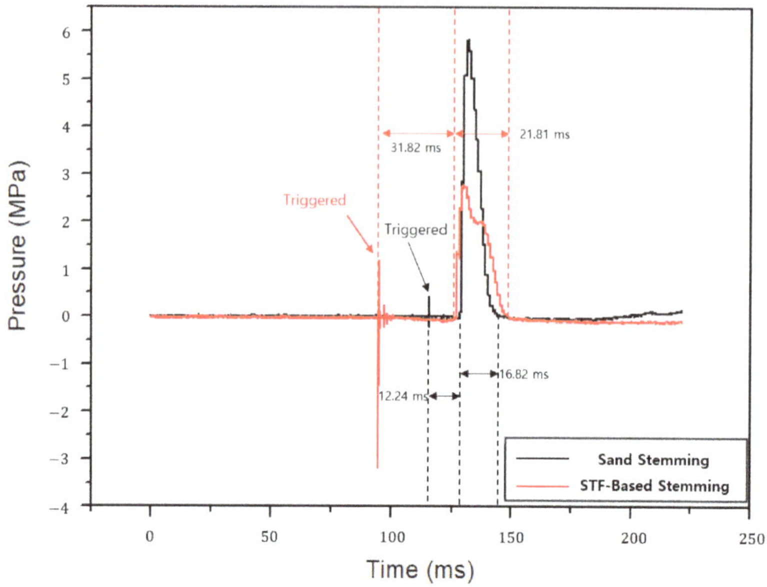

Figure 9. Time-pressure hysteresis curve inside the blast hole.

4. Full-Scale Blasting Experiment

4.1. Bench Blasting

The aggregate production step comprises blasting, a vibration feeder, a primary crusher, a secondary crusher, belt conveyors, and vibrating screens. The feed size of the jaw crusher operating in the A-aggregate mine requires a rock fragment size of at least 1000 mm or less. Therefore, to input the jaw crusher for primary crushing, it is necessary to perform secondary work on fragments by hydraulic rock breakers after blasting the rock, which incurs additional costs. Approximately $25,000 US dollars per month are required to operate large hydraulic rock breakers, and mine A has ten of such equipment. Figure 10 shows the bench blasting design applied to aggregate mine A, and the blasting results are compared by applying sand and STF-based stemming materials to each different bench part. Table 3 lists the main parameters of the full-scale bench blasting experiment. In the blasting experiment, the STF-based stemming was applied to the left part, and general sand stemming was applied to the right part.

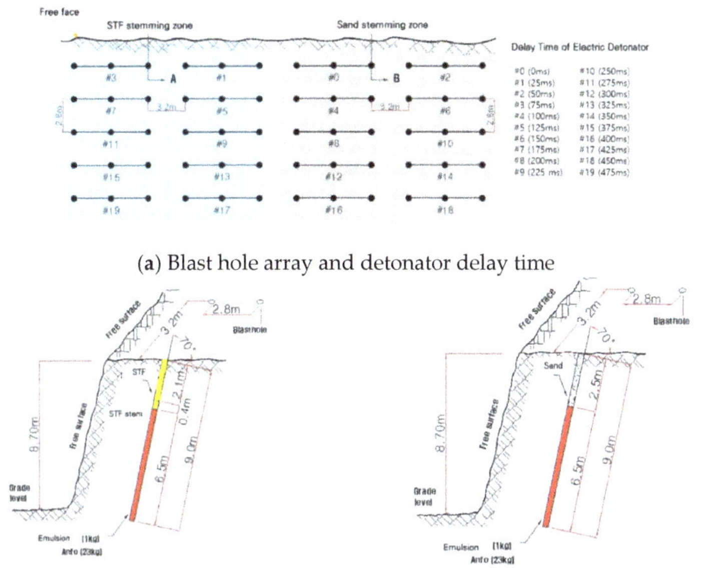

(**a**) Blast hole array and detonator delay time

(**b**) Bench section (STF stemming) (**c**) Bench section (sand stemming)

Figure 10. Bench blasting design of full-scale blasting experiment.

Table 3. Main parameters of the full-scale bench blasting experiment.

Parameter	Unit	Value
Hole diameter	mm	75
Hole length	m	9.0
Burden	m	2.8
Hole spacing	m	3.2
Charge per hole	kg/hole	24.0
Charge per delay	kg/delay	72.0
Charge type	-	Emulsion 1.0 kg ANFO 23.0 kg
Specific charge	kg/m^3	0.308
Rock fracture per hole	m^3/hole	77.95
Number of holes	ea	30
Stemming length	M	2.5
Stemming type	-	STF or Sand
Total charge	kg	720

The properties of the rock mass condition can have a significant influence on the fragmentation outcomes of the blast. Rock properties, such as compressive strength, porosity, Young's modulus, Poisson's ratio, and rock fracturing and jointing, can all influence fragmentation. In this experiment, by performing blasting at the same experimental area, the variation in results due to the rock mass condition difference was reduced as much as possible. The type of rock in this quarry mine is gneiss, and the uniaxial compressive strength is approximately 130–160 MPa. The rock density is approximately 2.6 g/cm^3, and the porosity is less than 0.15%. As shown in Figure 11a, a bench slope with a discontinuity direction and similar region was selected as the experimental site. The spacing of the discontinuities was observed to be approximately 1.2–1.5 m each, and they were under completely dry conditions. Figure 11b,c show the resulting image after the blasting. Under the same blasting conditions, the rock fragments in the STF-applied part are generally smaller than in the sand stemming part.

(**a**) Bench slope image of the experimental area

(**b**) Shear thickening fluid stemming (**c**) Sand stemming

Figure 11. Bench blasting results obtained by applying different stemming materials.

4.2. Evaluation of Rock Fragmentation

Muck piles of fragmented rocks were photographed using a single camera from the top surface of the bench and the front. In the images, a reference scale was used for single-scale factor analysis by applying a square marker target of 18 cm in width and height. Table 4 shows the image-based sieving analysis results of the rock fragmentation of the muck pile after full-scale bench blasting. Figures 12–15 show the analysis of the rock fragment size distribution. Larger rock fragments are displayed in red in the resulting image.

Table 4. Results of rock fragmentation analysis.

Location from Which the Image Was Captured	Stemming Type	Characteristic Size (X_c, mm)	Uniformity Index (n)	Average Fragment Size (X_{50}, mm)	Maximum Fragment (X_{ma}, mm)
Top of rock pile	Sand	1149	1.66	836	1910
	STF	419	1.93	345	1.100
Front of rock pile	Sand	1234	2.07	1020	2610
	STF	938	2.01	772	2010

Images taken from the top of the bench muck pile are compared. In the bench region to which sand stemming was applied, the characteristic size (X_c) for evaluating the fragment size is approximately 2.7 times larger than that in the region to which STF-based stemming material was applied. Moreover, the uniformity index (n) for evaluating the particle size distribution of the fragment rocks of the muck pile is 1.66 for the sand stemming region and 1.93 for the STF-based stemming. A higher uniformity index indicates a more uniform distribution of the fragmented rock.

Similarly, images taken from the front of the bench muck pile are compared. In the bench region to which sand stemming was applied, the characteristic size (X_c) for evaluating the fragment size is approximately 1.3 times larger than that of the region to which the STF-based stemming material was applied; however, the uniformity index shows no significant difference between the two cases. This is expected to affect the uniformity index as a large rock fragment falls to the front of the muck pile after blasting. These large fragments fell from the top of the outermost bench and had relatively little effect on the explosive force. Moreover, it is for this reason that the average fragment size compared, respectively, at the front and top of the bench muck pile image shows a significant difference

The WipFrag 3 program automatically generates the histogram graph. The x-axis is a log graph showing the size distribution of the rock fragments. The y-axis is the rate of passing. Large fragments are marked with boxes in Figures 12–15. The size of the large fragments that must be subjected to secondary breaking with a breaker machine before being placed in the jaw crusher line is approximately 1000 mm or more in diameter. When STF is applied as a stemming, the average fragment size decreases, and the number of large fragments that require second breaking work is greatly reduced.

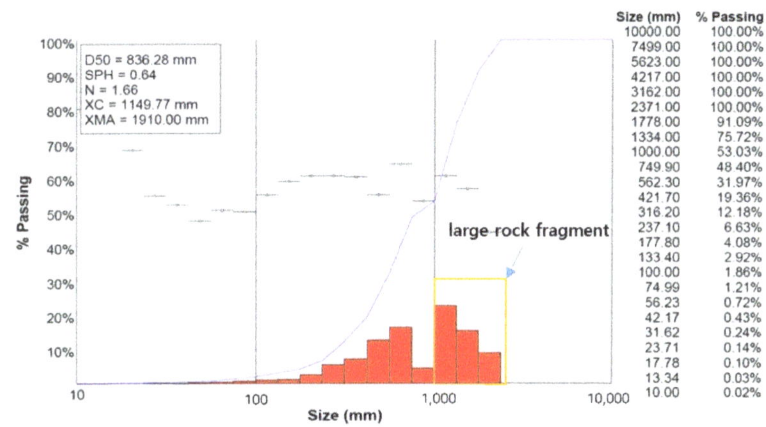

(**a**) WipFrag Graph from the top of the muck pile

(**b**) Fragmented rock pile (**c**) WipFrag Net Image

Figure 12. Rock pile image, contouring, histogram, and cumulative size curve of the fragmented block (bench top image of sand stemming case after blasting).

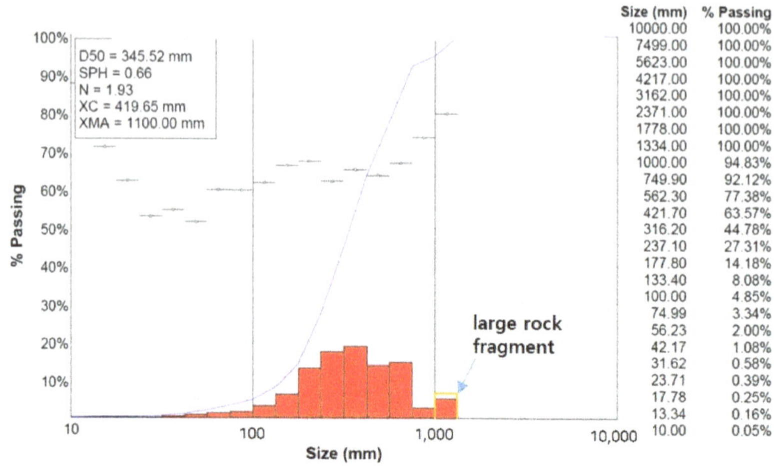

(**a**) WipFrag Graph from the top of the muck pile

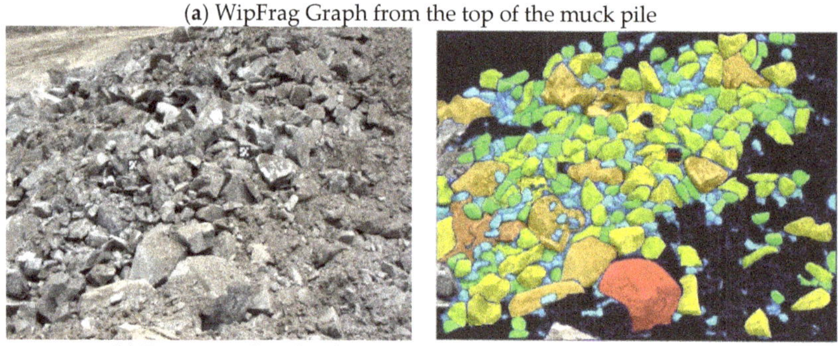

(**b**) Fragmented rock pile (**c**) WipFrag Net Image

Figure 13. Rock pile image, contouring, histogram, and cumulative size curve of the fragmented block (bench top image of STF stemming case after blasting).

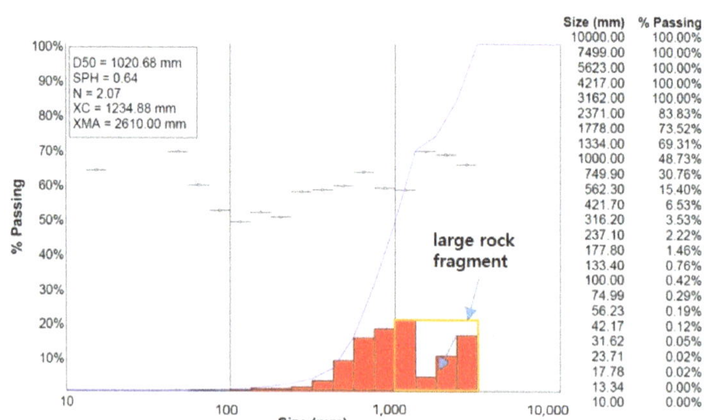

(**a**) WipFrag Graph from the front of the muck pile

Figure 14. *Cont.*

(b) Fragmented rock pile (c) WipFrag Net Image

Figure 14. Rock pile image, contouring, histogram, and cumulative size curve of the fragmented block (bench front image of sand stemming case after blasting).

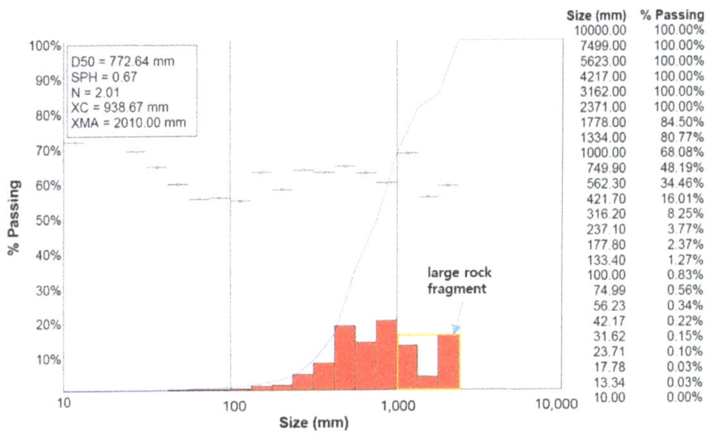

(a) WipFrag Graph from the front of the muck pile

(b) Fragmented rock pile (c) WipFrag Net Image

Figure 15. Rock pile image, contouring, histogram, and cumulative size curve of fragmented block (bench front image of STF stemming case after blasting).

5. Discussion

In this study, a material that instantaneously changes from shock load was developed as a blast stemming material and its performance was verified. Then, the pressure in the upper part of the stemming area was directly measured inside the blast hole. There are limited case studies that directly measure the pressure in the blast hole during bench blasting.

Previous research showed that stemming could increase the action time of the dynamic gas pressure in the blast hole and the efficiency of the explosives, reducing explosive consumption, as shown in Figure 16 [5,6]. Figure 16 depicts the time–pressure concept curves according to the stemming condition. In cases of missing or improper stemming, the pressure rapidly attenuates in the blast hole (Figure 16a), but proper stemming can increase the action time of the detonation gas inside the blast hole (Figure 16b). It is estimated that the shock pressure in the blast hole could be sustained for a longer time compared with that in the sand cases owing to the unique characteristics of STF. This is because STF is a smart-fluid type exhibiting an intense viscosity jump when subjected to loading.

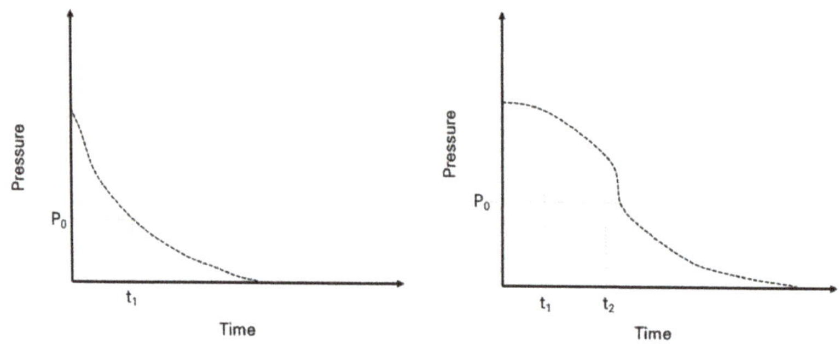

(**a**) without stemming (**b**) with stemming

Figure 16. Concept curves of pressure and time: (**a**) without stemming, and (**b**) with stemming.

Figure 17 shows the time–pressure results according to the direct measurement of blast hole pressure, where sand- or STF-stemming materials were applied in each experiment. Since the pressure was measured at the upper part of the stemming area, a lower measured peak pressure transmits enough energy to the rock mass around the blast hole and below the stemming area. Stage I indicates detonation durations. When the shock front arrives at the gauge point, the gauge outputs a peak pressure. Stage II indicates the pressure variation as detonation propagates from gauge point to blast hole. Note that during this time, the stemming material begins initial ejection from the blast hole. Finally, in stage III the pressure curve rapidly decreases to atmospheric pressure when the stemming part is completely ejected.

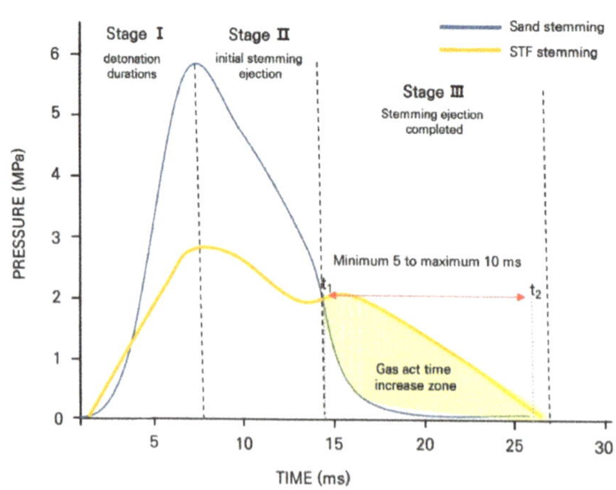

Figure 17. The time–pressure curves of the experimental results according to sand or shear thickening fluid (STF) stemming.

In the case of STF-based stemming, the pressure attenuates more slowly than in the sand stemming case. The pressure acts approximately 5–10 ms longer than the blasting gas pressure for the STF-based stemming than for the sand stemming. Eloranta et al. [41] verified that 1 ms of increased gas retention time in a blast hole increased the fragment work done on the rock mass and reduced waste energy. Therefore, an increase in the gas retention time of approximately 5 ms has a significant impact on rock fragmentation.

The purpose of stemming is to increase blasting efficiency by extending the duration of the explosion gas and forming more cracks in the crushing area. Therefore, the ability to resist the gas pressure emitted in the direction of the blast hole inlet implies the stemming performance, which is directly related to the blast efficiency. In this study, after crushing and cracking under the effect of detonation, the pressure applied to the elastic area was measured using a water pressure sensor from a nylon tube inserted into the water tube, and the pressure acting in the elastic area was measured, as shown in Figure 18.

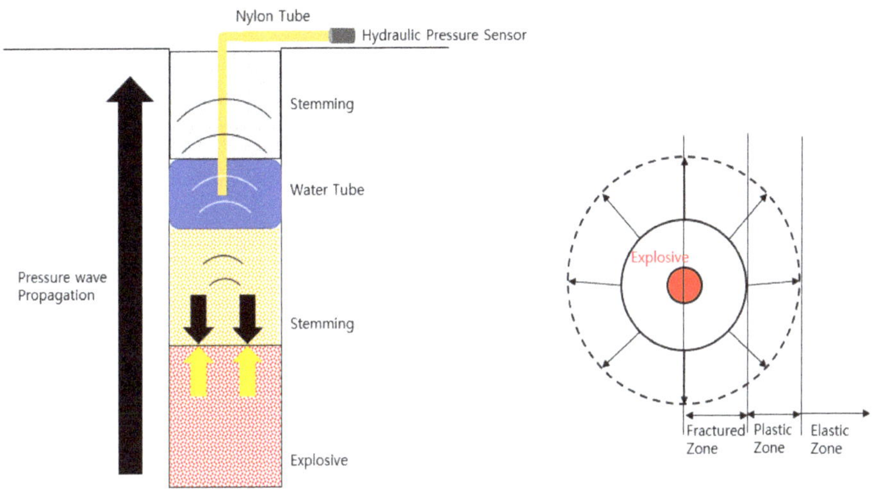

Figure 18. Schematic of the measured pressure wave in the blast hole.

Therefore, the pressure behavior in the blast hole owing to the explosive detonation measured through the hydraulic pressure measurement sensor is the pressure value at which the pressure ejected to the upper part of the blast hole by the explosion is attenuated by the stemming material and not by the direct blast pressure. The low peak pressure measured by the hydraulic sensor implies that the stemming material has an excellent pressure constraint against the explosive pressure ejected to the upper part of the blast hole. This suggests that the explosive pressure applied to the lower part of the blast hole is high. In addition, because the explosive pressure delay time measured by the hydraulic pressure sensor is from the trigger to the pressure wave measurement, the longer the delay time from the detonation to the pressure measurement, the more continuous the pressure behavior of the stemming material under the blast hole. Finally, the time at which the explosion pressure returns to atmospheric pressure after the increase indicates the duration of the explosive pressure in the blast hole. This suggests that, the longer the duration of the explosive pressure, the longer the crack propagation time owing to the gas pressure behavior in the blast hole.

The measured pressure behavior in the blast hole correlates with the resistance characteristics of the stemming material. It is judged that the STF-based stemming material will effectively achieve the purpose of the stemming material by resisting the blasting pressure and maintaining relatively high pressure in the blast hole for a longer time compared to the conventional sand stemming material.

6. Conclusions

By conducting two blasting experiments, we compared the stemming effects of the developed STF-based stemming material and those of commonly used sand stemming material. The conclusions drawn from the blasting experiments are as follows:

(1) The measured pressure was 2.80 MPa for the STF-based stemming case and 5.84 MPa for the sand stemming case based on the direct dynamic pressure measurement at the top of the blast hole. The lower the measured pressure value, the lower the loss of explosive pressure in the blast hole owing to stemming transfer. In addition, the explosive gas pressure action time in the STF stemming case was 5 ms longer than the sand-stemming case. The longer the duration of the explosive pressure, the greater the energy that can be used to fracture the rock;
(2) The measured pressure behavior in the blast hole correlated with the resistance characteristics of the stemming material. It is judged that the STF-based stemming material can effectively resist the blasting pressure and maintain a relatively high pressure in the blast hole for a longer duration compared to the conventional sand stemming material;
(3) The hydraulic pressure measurement system was developed to measure blasting pressure in a blast hole. This was done by improving the method of measuring the blasting pressure by drilling a dummy hole, which was mainly performed for the explosion pressure measurements. The pressure behavior depending on the voltage of pressure meter with cerabar (PMC) was set through the sensor calibration performed before the blasting experiment, the test method to evaluate the pressure constraint capacity of the stemming material was presented, and the validity of the pressure measurement system in the blasting hole was verified;
(4) As a full-scale bench blasting experiment, a rock pile fragmentation analysis was performed after blasting, and it was confirmed that the average fragment size was reduced by an approximate minimum of 25% to a maximum of 60%, when using STF-based stemming material. This suggests that the secondary crushing work in aggregate quarry mines can be reduced.

Author Contributions: Conceptualization, Y.J.; Investigation, C.S.; Methodology, Y.K.; Supervision, S.C.; Writing—original draft, Y.K. All authors have read and agreed to the published version of the manuscript.

Funding: This research was supported by a grant from the project "Development of new stemming material and blasting method using shear thickening fluid by high shock-wave reaction", funded by the Korea Institute of Civil Engineering and Building Technology (KICT).

Institutional Review Board Statement: Not applicable.

Informed Consent Statement: Not applicable.

Data Availability Statement: Not applicable.

Conflicts of Interest: The authors declare no conflict of interest regarding the publication of this paper.

References

1. Brinkmann, J.R. An experimental study of the effects of shock and gas penetration in blasting. In Proceedings of the 3th International Symposium on Rock Fragmentation by Blasting, Brisbane, Australia, 26–31 August 1990; pp. 55–66.
2. Konya, C.J.; Konya, A. Effect of hole stemming practices on energy efficiency of comminution. In *Energy Efficiency in the Minerals Industry*; Awuah-Offei, K., Ed.; Springer: Cham, Switzerland, 2018; pp. 31–53.
3. Snelling, W.O.; Hall, C. *The Effect of Stemming on the Efficiency of Explosives*; United States Government Printing Office: Washington, DC, USA, 1912.
4. Floyd, J.L. Explosive energy relief—The key to Control. Over break. In Proceedings of the Explosion, Kalgoorlie, WA, USA, October 1999; pp. 147–153.
5. Zong, Q. Theoretical discussion of movement rule of stemming in blast holes. *Blasting* **1996**, *13*, 8–11.
6. Luo, Y.; Wu, S. Study on length of stemming material and its effect in hole-charged blasting. *Mech. Eng.* **2006**, *28*, 48–52.

7. Phamotse, K.M.; Nhleko, A.S. Determination of optimal fragmentation curves for a surface diamond mine. *J. S. Afr. Inst. Min. Metall.* **2019**, *119*, 613–619. [CrossRef] [PubMed]
8. Li, Y.L.; Shi, X.Z.; Liu, B.; Zhao, J.P. Experimental research on reasonable length of water stemming. *Blasting* **2015**, *32*, 11–16.
9. Cui, Z.D.; Yuan, L.; Yan, C.L. Water–silt composite blasting for tunnelling. *Int. J. Rock Mech. Min. Sci.* **2010**, *47*, 1034–1037. [CrossRef]
10. Wang, Z.; Li, Y. Numerical simulation on effects of radial water-decoupling coefficient in engineering blast. *Yantu Lixue/Rock Soil Mech.* **2005**, *26*, 1926.
11. Zhu, Z.; Xie, H.; Mohanty, B. Numerical investigation of blasting-induced damage in cylindrical rocks. *Int. J. Rock Mech. Min. Sci.* **2008**, *45*, 111–121. [CrossRef]
12. Boshoff, D.; Webber-Youngman, R.C.W. Testing stemming performance, possible or not? *J. S. Afr. Inst. Min. Metall.* **2011**, *111*, 871–874.
13. Kopp, J.W. *Stemming Ejection and Burden Movements from Small Borehole Blasts*; United States Department of the Interior, Bureau of Mines: Twin Cities, MN, USA, 1987.
14. Wang, C. The action of the blasthole plug. *J. Univ. Hydr. Electric. Eng.* **1999**, *21*, 45–49.
15. Hunter, G.C.; McDermott, C.; Miles, N.J.; Singh, A.; Scoble, M.J. A review of image analysis techniques for measuring blast fragmentation. *Min. Sci. Technol.* **1990**, *11*, 19–36. [CrossRef]
16. Sudhakar, J.; Adhikari, G.R.; Gupta, R.N. Comparison of fragmentation measurements by photographic and image analysis techniques. *Rock Mech. Rock Eng.* **2006**, *39*, 159–168. [CrossRef]
17. Jung, M. A Study on the Estimation of Non-Ideal Blasthole Pressure of Bulk Blend Explosives. Ph.D. Thesis, Seoul National University, Seoul, Korea, 2019.
18. Cunningham, C. Blasthole Pressure: What it really means and how we should use it. *Proc. Annu. Conf. Explos. Blasting Tech.* **2006**, *32*, 255.
19. Raina, A.K.; Murthy, V.M.S.R.; Soni, A.K. Estimating flyrock distance in bench blasting through blast induced pressure measurements in rock. *Int. J. Rock Mech. Min. Sci.* **2015**, *76*, 209–216. [CrossRef]
20. Majumdar, A.; Laha, A.; Bhattacharjee, D.; Biswas, I. Tuning the structure of 3D woven aramid fabrics reinforced with shear thickening fluid for developing soft body armour. *Compos. Struct.* **2017**, *178*, 415–425. [CrossRef]
21. Daelemans, L.; Cohades, A.; Meireman, T.; Beckx, J.; Spronk, S.; Kersemans, M.; De Baere, I.; Rahier, H.; Michaud, V.; Van Paepegem, W.; et al. Electrospun nanofibrous interleaves for improved low velocity impact resistance of glass fibre reinforced composite laminates. *Mater. Des.* **2018**, *141*, 170–184. [CrossRef]
22. Majumdar, A.; Butola, B.S.; Srivastava, A. Development of soft composite materials with improved impact resistance using Kevlar fabric and nano-silica based shear thickening fluid. *Mater. Des. 1980–2015* **2014**, *54*, 295–300. [CrossRef]
23. Wei, M.; Lin, K.; Liu, H. Experimental investigation on hysteretic behavior of a shear thickening fluid damper. *Struct. Contr. Health Monit.* **2019**, *26*, e2389. [CrossRef]
24. Gürgen, S.; Sofuoğlu, M.A. Experimental investigation on vibration characteristics of shear thickening fluid filled CFRP tubes. *Compos. Struct.* **2019**, *226*, 111236. [CrossRef]
25. Lim, J.; Kim, S.-W. Enhanced damping characteristics of carbon fiber reinforced polymer–based shear thickening fluid hybrid composite structures. *J. Intell. Mater. Syst. Struct.* **2020**, *31*, 2291–2303. [CrossRef]
26. Gürgen, S.; Kushan, M.C. The effect of silicon carbide additives on the stab resistance of shear thickening fluid treated fabrics. *Mech. Adv. Mater. Struct.* **2017**, *24*, 1381–1390. [CrossRef]
27. Boersma, W.H.; Laven, J.; Stein, H.N. Shear thickening (dilatancy) in concentrated dispersions. *AIChE J.* **1990**, *36*, 321–332. [CrossRef]
28. Hoffman, R.L. Discontinuous and dilatant viscosity behavior in concentrated suspensions. I. Observation of a flow instability. *Trans. Soc. Rheol.* **1972**, *16*, 155–173. [CrossRef]
29. Brady, J.F.; Bossis, G. The rheology of concentrated suspensions of spheres in simple shear flow by numerical simulation. *J. Fluid Mech.* **1985**, *155*, 105–129. [CrossRef]
30. Xu, Y.; Gong, X.; Peng, C.; Sun, Y.; Jiang, W.; Zhang, Z. Shear thickening fluids based on additives with different concentrations and molecular chain lengths. *Chin. J. Chem. Phys.* **2010**, *23*, 342–346. [CrossRef]
31. Lee, Y.S.; Wetzel, E.D.; Wagner, N.J. The ballistic impact characteristics of Kevlar woven fabrics impregnated with a colloidal shear thickening fluid. *J. Mater. Sci.* **2003**, *38*, 2825–2833. [CrossRef]
32. Decker, M.J.; Halbach, C.J.; Wetzel, E.D.; Nam, C.H.; Wagner, N.J. Stab resistance of shear thickening fluid (STF)-treated fabrics. *Compos. Sci. Technol.* **2007**, *67*, 565–578. [CrossRef]
33. Ko, Y.; Kwak, K. Blast effects of a shear thickening fluid-based stemming material. *Mining* **2022**, *2*, 330–349. [CrossRef]
34. Bischoff White, E.E.; Chellamuthu, M.; Rothstein, J.P. Extensional rheology of a shear-thickening cornstarch and water suspension. *Rheol. Acta* **2010**, *49*, 119–129. [CrossRef]
35. Chi, L.Y.; Zhang, Z.X.; Aalberg, A.; Yang, J.; Li, C.C. Measurement of shock pressure and shock-wave attenuation near a blast hole in rock. *Int. J. Impact Eng.* **2019**, *125*, 27–38. [CrossRef]
36. Raina, A.K.; Trivedi, R. Exploring rock-explosive interaction through cross blasthole pressure measurements. *Geotech. Geol. Eng.* **2019**, *37*, 651–658. [CrossRef]

37. Palangio, T.C. WipFrag—A new tool for blast evaluation. In Proceedings of the 11th Annual ISEE Symposyum on Blasting Research, Nashville, TN, USA, 5–9 February 1985; pp. 269–285.
38. WipFrag. WipFrag 3 Operating Manual. Available online: https://wipware.com (accessed on 15 July 2022).
39. Singh, P.K.; Sinha, A. Rock fragmentation by blasting. In Proceedings of the 10th International Symposium on Rock Fragmentation by Blasting, New Delhi, India, 26–29 November 2012; CRC Press: Boca Raton, FL, USA, 2013; pp. 10–18.
40. Scott, A.; McKee, D.J. The inter-dependence of mining and mineral beneficiation processes on the performance of mining projects. In Proceeding of the Australian Institute of Mining and Metallurgy Annual Conference, Melbourne, Australia, 5–9 August 1994; pp. 5–9.
41. Eloranta, J. Stemming selection for large diameter blast holes. In Proceedings of the 20th Annual Conference on Explosive and Blasting Techniques, ISEE, Austin, TX, USA, 30 January–3 February 1994; pp. 1–11.

applied sciences

MDPI

Article

A Numerical Simulation of Blasting Stress Wave Propagation in a Jointed Rock Mass under Initial Stresses

Qian Dong [1,2,*], Xinping Li [3], Yongsheng Jia [1,2] and Jinshan Sun [1,2]

[1] Hubei Key Laboratory of Blasting Engineering, Jianghan University, Wuhan 430056, China; sinoblaster_jia@jhun.edu.cn (Y.J.); sunjinshan@cug.edu.cn (J.S.)
[2] Hubei (Wuhan) Institute of Explosion Science and Blasting Technology, Jianghan University, Wuhan 430056, China
[3] Hubei Key Laboratory of Road-Bridge and Structure Engineering, Wuhan University of Technology, Wuhan 430070, China; whutxinpingli@sina.com
* Correspondence: dongqian@jhun.edu.cn

Abstract: The initial stresses have a strong effect on the mechanical behavior of underground rock masses, and the initial stressed rock masses are usually under strong dynamic disturbances such as blasting and earthquakes. The influence mechanism of a blasting excavation on underground rock masses can be revealed by studying the propagation of stress waves in them. In this paper, the improved Mohr-Coulomb elasto-plastic constitutive model of the intact rock considering the initial damage was first established and numerically implemented in Universal Distinct Element Code (UDEC) based on the variation of the experimental stress wave velocity in the initial stressed intact rock, and the feasibility of combining the established rock constitutive model and the BB (Bandis-Barton) model which characterizes the nonlinear deformation of the joints to simulate stress waves across jointed rock masses under initial stress was validated by comparing the numerical and model test results subsequently. Finally, further parameter studies were carried out through the UDEC to investigate the effect of the initial stress, angle, and number of joints on the transmission of the blasting stress wave in the jointed rock mass. The results showed that the initial stress significantly changed the propagation of the stress waves in the jointed rock mass. When the initial stress was small, the transmission coefficients of the stress waves in the jointed rock were vulnerable to be influenced by the variation of the angle and the number of joints, while the effect of the angle and the number of joints on the stress wave propagation gradually weakened as the initial stress increased.

Keywords: numerical modeling; blasting stress wave; underground rock masses; initial stress; transmission coefficient

Citation: Dong, Q.; Li, X.; Jia, Y.; Sun, J. A Numerical Simulation of Blasting Stress Wave Propagation in a Jointed Rock Mass under Initial Stresses. *Appl. Sci.* **2021**, *11*, 7873. https://doi.org/10.3390/app11177873

Academic Editor: Ricardo Castedo

Received: 29 July 2021
Accepted: 24 August 2021
Published: 26 August 2021

1. Introduction

Underground rock masses are inevitably in a certain geological and tectonic environment, and are subject to initial stresses such as gravitational stress, tectonic stress, temperature stress, etc. The initial stresses have a strong effect on the mechanical behavior of the underground rock masses and the stability of the underground engineering [1–3]. Meanwhile, the drill-and-blast method is the most widely used technique for tunnel excavation and underground mining. In this process, the initial stressed underground rock masses are under strong dynamic disturbances, and the underground structure can be damaged by the stress waves generated during the excavation process. Hence, it is of great practical significance to study the propagation of stress waves in underground rock masses under initial stress for the optimal design of underground rock mass blasting excavation parameters and the dynamic stability analysis of underground engineering.

Rock masses contain various types of discontinuous interfaces such as joints and fractures and so on, which have a noticeable effect on the mechanical response of the rock mass [4,5]. Discontinuous interfaces in natural rock mass are usually distributed in groups, such as a stratified rock mass, and it is particularly important to study the propagation of blasting stress waves in the layered rock mass and to monitor the vibrations generated during the blasting of the stratified rock masses [6]. Intensive studies have been conducted to investigate the propagation of stress waves across jointed rock masses via various theoretical and experimental methods. In terms of theoretical research, the displacement discontinuity model (DDM) proposed by Schoenberg [7] has been widely applied to study stress wave propagation through a jointed rock mass [8,9]. The DDM was also combined with other analysis methods, e.g., the method of characteristics (MC) [10], the scattering matrix method [11], and the time-domain recursive method (TDRM) [12] to study the stress waves passing through linear and nonlinear joints [13,14], one single joint and a set of parallel joints [15,16], and even intersecting rock joints [17]. By experimental means to date, the split Hopkinson pressure bar (SHPB) apparatus has been mainly used to study stress wave propagation across rock masses [18,19].

In contrast, the numerical simulation method is an economical and feasible alternative to survey the stress wave propagation across a jointed rock mass. Based on the discrete element method (DEM) proposed by Cundall [20], the universal distinct element code (UDEC) has been widely used to calculate the propagation problems of stress waves in a jointed rock mass [21–23]. Furthermore, other numerical methods and software have been adopted to solve the problems involving the stress wave propagation in a rock mass, e.g., the particle manifold method (PMM) [24,25], the numerical manifold method (NMM) [26,27], the particle flow code (PFC) [28,29], and the three-dimensional element code (3DEC) [30]. However, the above theoretical, experimental, and numerical methods have mainly focused on the effect of the parameters of the joints, e.g., joint stiffness, joint spacing, joint number, and the parameters of the stress wave, e.g., waveform, amplitude, frequency, and the incident angle of stress wave on the stress wave propagation pattern, and have proposed that the attenuation of the stress wave only occurs at the joints, while it has been assumed that the intact rock is elastic. Few works have been conducted that investigate the effect of initial stresses on the stress wave propagation in the jointed rock mass, and studies considering the initial damage of intact rocks in the rock mass under initial stresses are much rarer.

On the other hand, besides discontinuous interfaces, intact rocks are the other part of the rock mass. Due to their long geological age and various complex tectonic effects, intact rocks inevitably contain a certain number of defects such as microcracks and micropores; therefore, intact rock can be considered as an initial damaged medium [31–33]. Considerable studies have revealed that in the process of static loading, the microcracks in the intact rock experience the stages of closure, development, extension, and interactive penetration [34–36], and the wave impedance of intact rocks is strongly affected by the initial stress given the stress sensitivity of the wave velocity and density. Consequently, microcracks within the intact rock enter different evolution stages under different initial stresses, leading to changes in wave impedance, which in turn have an influence on the stress wave propagation.

The variation in the quantity of microcracks inside the intact rock under initial stress causes changes in the macroscopic mechanical properties of the rocks, which is usually named initial damage [37,38]. In the progressive destruction process of rocks under static loading, the closure effect of the microcracks at the initial loading stage can significantly affect the deformation characteristics of rocks, and the current research on the damage of intact rocks has rarely considered the compaction stage of the initial void. For a porous medium with natural defects such as rocks, when the porosity of rocks is high, the compaction stage of the initial microcracks is even more non-negligible. However, related research has been rarely reported.

This paper presents a numerical exploration of blasting stress wave propagation in the initial stressed jointed rock mass. Firstly, based on the variation of the stress wave velocity in the intact rock under different equal biaxial static loading in the model test, the initial damage variable was determined, and the Mohr-Coulomb elasto-plastic constitutive model of the rock considering initial damage was established and subsequently implemented in the UDEC. Then, the feasibility of combining the developed model and the BB (Bandis-Barton) model which characterizes the nonlinear deformation of the joints to simulate stress waves across the jointed rock mass under initial stress was validated by comparing the numerical results with the model test results. Finally, further parameter studies were carried out through the UDEC to investigate the effect of the initial stress, angle, and number of joints on the transmission of blasting stress waves in the jointed rock mass.

2. A Brief Introduction of the Model Test

The detailed model test process is referred to in another two papers [39,40], and is only briefly described in this paper, as follows:

(1) For the instrument providing the biaxial static loads in the model test, the corresponding size of the specimen was 1.6 m (length) × 1.3 m (height) × 0.4 m (thickness), as shown in Figure 1.

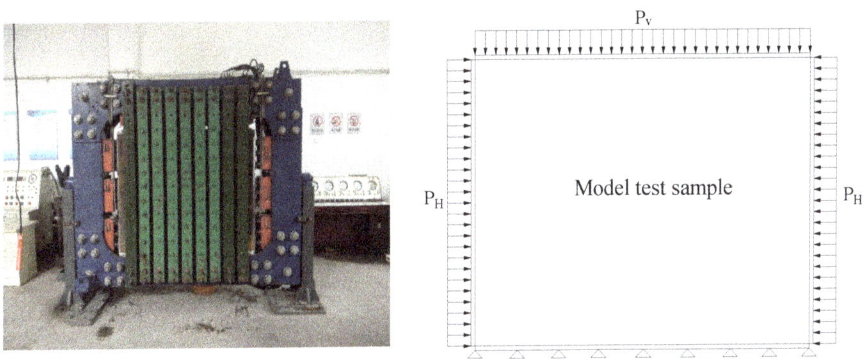

(**a**) (**b**)

Figure 1. Multifunctional testing machine for the rock and soil: (**a**) test equipment; (**b**) static loading schematic.

(2) The prototype of the model test was the deep-buried underground engineering surrounding rock, and the physical and mechanical parameters of the prototype are shown in Table 1. The corresponding intact rock simulation material was low strength cement mortar containing cement, sand, water, and a plasticizer. Meanwhile, the stress similarity coefficient C_σ between the prototype and simulation material was 20. Through a series of tests, the mechanical parameters of the cement mortar material were obtained and are shown in Table 1. In addition, the joints in the model test samples were simulated by the mica plates, and their normal and tangential stiffnesses were 12 GPa/m and 7.53 GPa/m, respectively, and were obtained through laboratory tests.

Table 1. Physical and mechanical parameters of similar materials and the prototype of the intact rock.

Type	R_c/MPa	σ_t/MPa	E/GPa	φ/(°)	C/MPa	μ	ρ/kg/m^3
Prototype	120	12	50	30	30	0.223	2600
Similar material	5.864	0.613	5.226	23.2	1.49	0.203	1980

Where, R_c, σ_t, E, φ, C, μ and ρ are the uniaxial compressive strength, tensile strength, elastic modulus, internal friction angle, cohesion, poisson ratio, and density of the prototype and similar material, respectively.

(3) According to the number and the angle of the joints, a total of three model test samples were made, named T1, T2, and T3 respectively, as shown in Figure 2. Due to the

structural characteristics of the cylindrical charge, two test sections were uniformly arranged along the thickness direction in each model test sample. Four measuring lines were arranged on each section to measure the stress and strain at different distances from the explosion source. Sixteen strain measuring points were arranged on the strain testing section, ranging from 1 to 16, and eight stress measuring points were arranged on the stress testing section, ranging from 17 to 24.

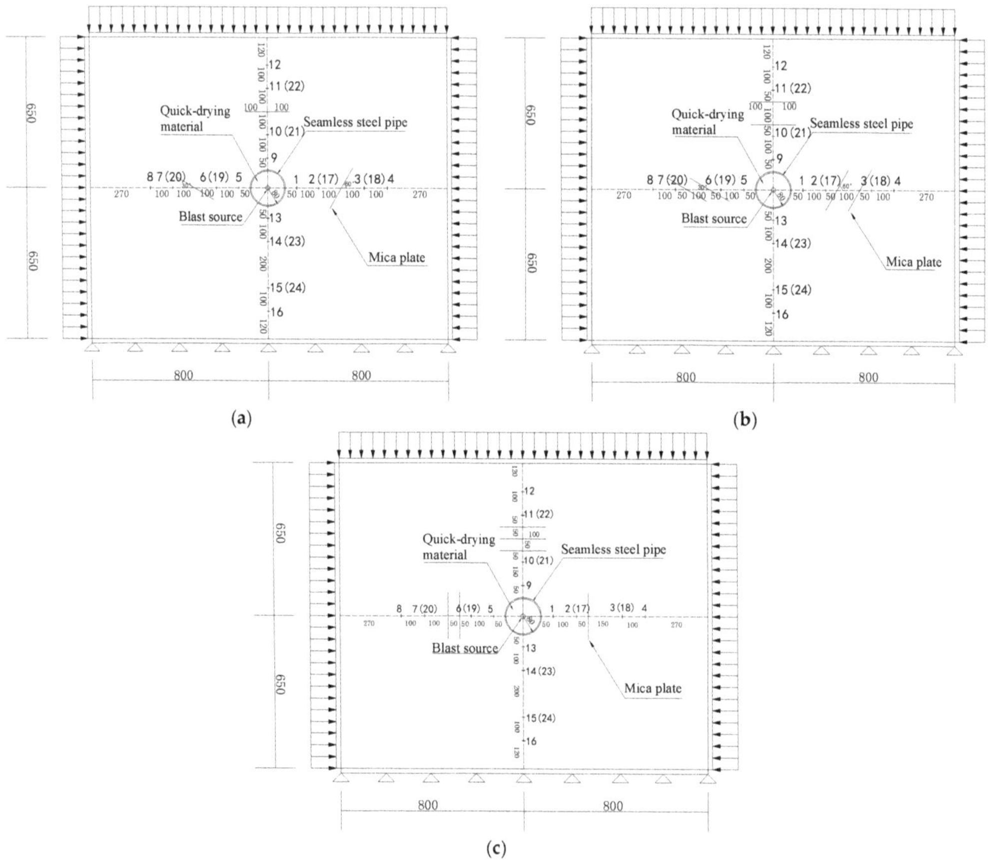

Figure 2. Layout of the measuring points and the joints of the model test samples (unit: mm): (**a**) T1 sample; (**b**) T2 sample; (**c**) T3 sample.

(4) The T1 and T2 samples were designed to study the propagation of the blasting stress waves in intact rock, and the normal or oblique impact of the blasting stress wave on the rock mass containing joints with a different number and angle, and the corresponding research objects of the T1 and T2 test samples were intact rock and single-joint and double-joints rock masses of 30°, 60°, and 90°, respectively. Meanwhile, the T3 sample was designed to study the propagation of the blasting stress waves in intact rock and the normal jointed rock mass. According to the number of joints in the four measuring lines, the corresponding research objects were intact rock, 90° single-joint rock mass, 90° double-joints rock mass, and 90° three-joints rock mass.

(5) In the model test, detonating cords with a total length of 1.6 m and TNT (Trinitrotoluene) with an equivalence of 17.6 g were used as the explosive sources to generate the blasting stress wave, and the detonating cords were fixed in the seamless steel pipe in the center of the test samples through the wooden centering stent. Subsequently, the quick-drying materials were poured into the pipes as the loading core in the samples, as shown in Figure 3. At the same time, seamless steel tubes were arranged to reduce the damage of the blasting loads, and repeated dynamic loading was realized by replacing the crushed quick-drying materials in the seamless steel tubes. In the model test, the vertical static load P_V and horizontal static load P_H applied on the model specimens were equal and were 0, 0.75 MPa, 1.5 MPa, and 3 MPa, respectively, and the corresponding initial in situ stresses were 0, 15 MPa, 30 MPa, and 60 MPa, respectively.

(a) (b) (c)

Figure 3. Schematic diagram of the charge structure in the model test samples: (**a**) detonating cords fixed in the wooden centering stent; (**b**) wooden centering stent in the model samples; (**c**) the pouring of quick-drying materials.

3. Establishment of the Mohr-Coulomb Elasto-Plastic Constitutive Model of the Rock Considering Initial Damage

Under different initial static loads, the microcracks in rocks enter different stages of evolution, so the physical and mechanical properties of rocks change correspondingly, leading to different propagation laws of stress waves in the rocks. Therefore, it is necessary to establish a constitutive model that can consider the initial damage of rocks caused by initial static stresses. Based on this, the influence of the change of initial stress on the propagation law of stress waves in rocks can be considered in a numerical simulation.

In the model test, the initial stresses applied on the test samples were the biaxial static loads, but in the test process, only the uniaxial compressive strength R_c of the rock simulation material was obtained as 5.864 MPa. Intensive studies have shown that the biaxial compressive strength R_{bc} of brittle materials such as rock and concrete was improved compared with the uniaxial compressive strength. The ratio of the biaxial compressive strength R_{bc} to the uniaxial compressive strength R_c of the brittle materials as β was defined by Papanikolaou et al. [41] and Huang [42], and through considerable test data, the fitting formula of β changing with the uniaxial compressive strength was obtained as the followed equation:

$$\beta = 1.493 R_c^{-0.0634} \tag{1}$$

The above equation shows that the relationship between β and the uniaxial strength R_c of the rock meets a negative exponential relationship, and β decreases with the increase in the uniaxial strength R_c, indicating that the difference between the biaxial strength R_{bc} and the uniaxial strength R_c of the rock gradually decreases with the increase in the uniaxial strength R_c, and β is greater than 1 in the conventional uniaxial strength range of the rock. Substituting the uniaxial compressive strength R_c of the rock simulation material into Equation (1), the biaxial compressive strength R_{bc} of the rock simulation material can be obtained as 7.826 MPa.

The propagation velocity in the material is an important part of the propagation characteristics of stress waves, which can reflect the evolution of microcracks and the damage degree of the medium [43]. Based on this, to study the initial damage evolution of the intact rock under different static loads, the stress wave velocities of the intact rock in each model test sample under different biaxial static loads were calculated, and the initial damage variation of the intact rock was obtained through the change in the wave velocities. Specifically, the stress wave propagation velocity can be calculated by the arrival time difference in the stress wave recorded by the sensors arranged in the intact rock at different distances from the explosion source. In the three model test samples, the number of strain sensors in the intact rock area was more than that of the stress sensors, and the range of the strain sensors was also larger. Therefore, the strain time history curves recorded at the strain measuring points at different distances from the explosion source were used to calculate the stress wave velocity through the arrival time difference of the waves.

In each model test sample under different biaxial static loads, based on the time difference Δt corresponding to the jumping point in the time history curves of strain measuring points 13 and 16 in the intact rock, and the distance between the two measuring points Δl, the value of the stress wave velocity c of the intact rock can be calculated by the following equation:

$$c = \frac{\Delta l}{\Delta t} \tag{2}$$

Through the above equation, the average propagation velocities of stress waves in the intact rock section of each model test sample under different biaxial static loads were calculated, which were 1990 m/s, 2077 m/s, 2099 m/s, and 1898 m/s when the static loads were 0, 0.75 MPa, 1.5 MPa, and 3 MPa. The reason for this phenomenon was that when the static load was small, the initial microcracks in the intact rock started to close with the growth of the static load, resulting in the increase in the wave velocity with the elastic modulus. When the static load was raised to a critical value, the initial microcracks in the intact rock were completely closed, and when the static load continued to rise, new microcracks were initiated, resulting in the decrease in the wave velocity, and this critical value of the static load can be obtained by a subsequent analysis. Meanwhile, it can be seen that when the static load was 0 MPa, the average stress wave velocity in the intact rock was about 1990 m/s. Meanwhile, the ultrasonic wave velocity of the intact rock, similar to the material measured in the model test, was 1980 m/s. The results showed that when the amplitude of the stress wave was not large enough, its propagation speed in the rocks was about the same as that of an elastic wave, which is consistent with the conclusion that when the stress wave amplitude is small under the combined action of dynamic and static, the initial damage to the rock is mainly caused by the application of the static load [40].

To derive the variation law of stress wave velocities in the rocks under biaxial static loading, the average stress wave velocities in the intact rock under different static loads were plotted, as seen in Figure 4. It should be noted that for the subsequent initial damage analysis, the stress wave velocity in the intact rock was assumed to drop to zero when the biaxial static load reached the biaxial compressive strength of 7.826 MPa.

From the above Figure 4, it can be seen that the stress wave velocity in the intact rock increased and then decreased with the increase in the static load, which was also similar to the variation law of the physical attenuation of stress waves in the intact rock with the biaxial static load derived from the model test results, reflecting the stress sensitivity of the evolution of the microcracks in rocks. Numerous studies have shown that there is a close connection between the wave velocity and the intrinsic damage of a medium, so the damage evolution of the propagation medium can be described by the change of wave velocity. Combining the fitted static load versus the stress wave velocity curve in Figure 4, the maximum wave velocity of the intact rock was 2109 m/s, and the corresponding biaxial static load was 1.23 MPa, which was about 15.7% of the biaxial compressive strength, indicating that the microcracks inside the rocks were in the fully compacted stage at this static load level. Meanwhile, the new microcracks had not yet started to initiate.

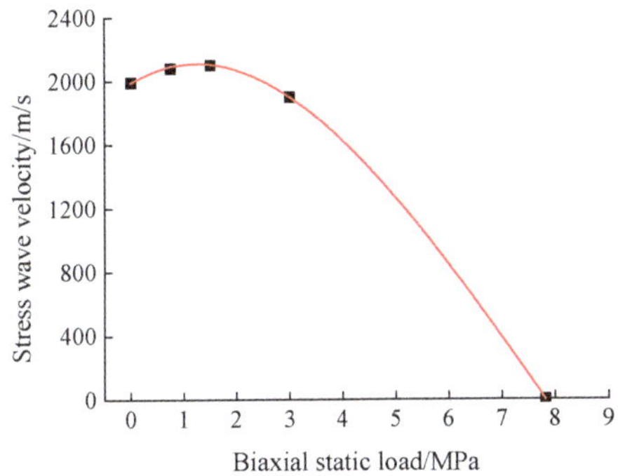

Figure 4. Relationship between the stress wave velocity and the static load of the intact rock.

The maximum wave velocity v_0 was defined as the wave velocity of the undamaged rock, so the initial damage variable D_0 of the rock can be defined by the following equation, when the value of D_0 is 0 and 1, respectively, indicating that the rocks are in an undamaged and fully damaged stage.

$$D_0 = 1 - (\frac{v_1}{v_0})^2 \quad (3)$$

where, v_1 is the stress wave velocity under different static loads. Therefore, through Figure 4, the initial damage variable D_0 variation curve of the rock under different biaxial static loads can be derived, as shown in Figure 5, while the initial damage value of the rock was considered to be 1 when the biaxial static load reached the biaxial compressive strength.

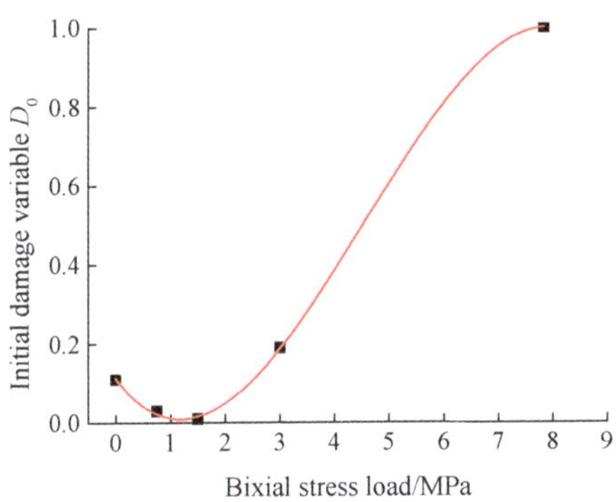

Figure 5. Relationship between the initial damage variable and the static load of the intact rock.

Based on the five data samples in Figure 5, including four experimental data points and one data point that characterized the failure of the intact rock obtained by mechanical analysis, the fitting equation was obtained by the least square polynomial fitting as shown in Equation (4), and the adjusted R-square and residual sum of squares of the equation were 0.994 and 1.034×10^{-4}, respectively.

$$D_0 = 0.11253 - 0.18321\sigma_b + 0.08918{\sigma_b}^2 - 0.00655{\sigma_b}^3 \quad (4)$$

The above equation is the initial damage evolution equation considering the compaction effect of the microcracks inside the rocks under different biaxial static loads. In addition, since the object studied in this section is the intact rock without macroscopic fractures, the initial damage evolution within the rocks can be assumed to be isotropic.

After the evolution equation of the initial damage variable D_0 was determined, it was coupled with the internal Mohr-Coulomb elasto-plastic model in UDEC to establish the improved model considering the initial damage for the intact rock. Based on the Lemaite strain equivalence principle, the principal stress tensor σ_i and the effective principal stress tensor $\widetilde{\sigma}_i$, the bulk modulus K and the initial damage bulk modulus $\overline{K}$, the shear modulus G, and the initial damage shear modulus $\overline{G}$ should satisfy the following relationships:

$$\begin{cases} \widetilde{\sigma}_i = \sigma_i/(1 - D_0) \\ \overline{K} = (1 - D_0)K \\ \overline{G} = (1 - D_0)G \end{cases} \tag{5}$$

After the bulk modulus $\overline{K}$ and the shear modulus $\overline{G}$ containing the initial damage factor D_0 were obtained, the relationship between the stress increment $\Delta\sigma_i$ and the strain increment $\Delta\varepsilon_i$ in the Mohr-Coulomb elastic-plastic model of the rocks was given by:

$$\Delta\sigma_i = \overline{\lambda}\delta_i\Delta\varepsilon_i + 2\overline{G}\Delta\varepsilon_i \tag{6}$$

where, $\overline{\lambda}$ is the Lame constant of the damaged rock and δ_i is the Kronecker symbol. Meanwhile, both the rock yield damage function and the plastic flow law within the Mohr-Coulomb elasto-plastic constitutive model were changed to functions based on the effective principal stress tensor $\widetilde{\sigma}_i$, and the Mohr-Coulomb elasto-plastic constitutive model considering the initial damage was established. Based on the internal Fish language in UDEC, the relevant parameters in the calculation process of the constitutive model were modified through the custom functions and variables to establish the user-defined constitutive model, and the calculation procedure of the established Mohr-Coulomb elasto-plastic constitutive model considering the initial damage is shown in Figure 6.

Combined with Figure 6, the detailed calculation process was as follows: The initial damage variable D_0 was first calculated based on the biaxial static load σ_b of the numerical model using Equation (4), and the physical and mechanical parameters of the rock considering the initial damage such as $\overline{K}$ and $\overline{G}$, as well as the total strain increment $\Delta\varepsilon_i$ of the element under the initial static load were derived. Then, based on the Lemaite strain equivalence principle and combined with the initial damage variable D_0, the effective principal stress increment $\Delta\widetilde{\sigma}_i$ was derived by Equation (6), and finally the effective principal stress $\widetilde{\sigma}_i$ of the element was obtained by an iterative calculation.

When the effective principal stress $\widetilde{\sigma}_i$ of the element reached the yield condition and entered the plastic phase, the updated stress state of the element was obtained by the plastic flow law, and the above process was divided into two cases: The first case was when $h(\widetilde{\sigma}_1, \widetilde{\sigma}_3) \leq 0$, the shear failure occurred in the element, through the shear yield function f^s expressed by the effective principal stress and the shear plastic flow method, the new effective principal stress increment $\Delta\widetilde{\sigma}_i$ was calculated by the total strain increment $\Delta\varepsilon_i$, and finally the effective principal stress $\widetilde{\sigma}_i^N$ of the element was obtained. The other was when $h(\widetilde{\sigma}_1, \widetilde{\sigma}_3) \geq 0$, the element underwent tensile damage, and the new effective principal stress $\widetilde{\sigma}_i^N$ of the element was calculated by the tensile yield function f^t and the tensile plastic flow law according to the same steps of the first case.

It is worth noting that the established rock constitutive model was based on the model test results and the damage mechanics theory, which can take into account the initial damage of intact rock under different biaxial equal static loads. For an underground rock mass with caverns, blast holes, and stress relief holes, numerical modeling can be carried out based on the established model as long as the initial boundary static load condition can be simplified to biaxial equal static loading.

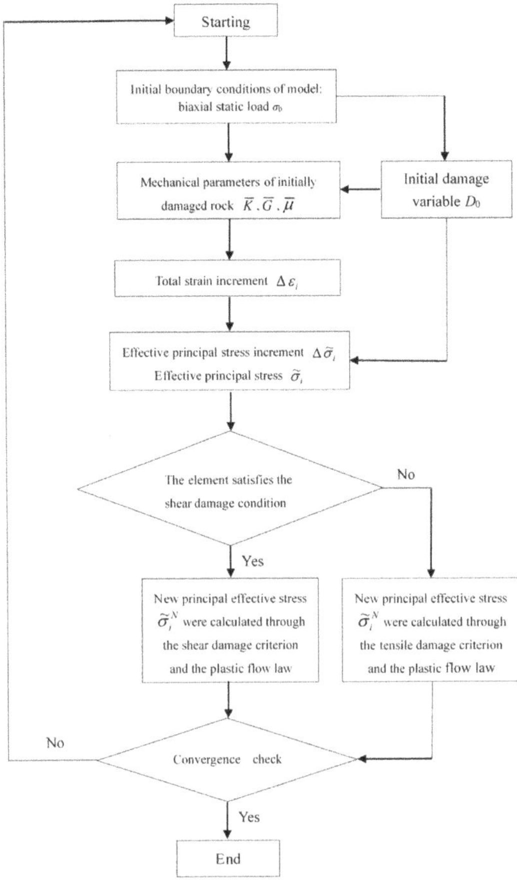

Figure 6. Calculation flow of the Mohr-Coulomb elasto-plastic constitutive model of the rock considering the initial damage.

4. Verification of the Established Rock Constitutive Model

In order to verify the accuracy of the established rock constitutive model based on this model and the widely adopted BB (Bandis-Barton) model which describes the nonlinear deformation characteristics of the joints [4], the model tests were numerically reproduced by the UDEC, and the experimental and numerical results were compared to analyze and verify the feasibility of the established rock constitutive model.

4.1. Numerical Model and Calculation Procedure

According to the three test samples designed in the model test as shown in Figure 2, the corresponding discrete element numerical models based on UDEC were established as shown in Figure 7. The dimensions of the numerical models were identical to the model test samples, whose length and width were 1600 mm and 1300 mm, respectively, while the lengths and spatial locations of the joints in the different numerical models were consistent with the model test samples. Due to the high mechanical strength of the seamless steel pipe, its deformation under static loads was approximately negligible, so it was not necessary to consider the quick-drying material inside the seamless steel pipe in the numerical model. The interior of the steel pipe was blank and the equivalent blast loads were applied directly to the inner wall of the steel pipes.

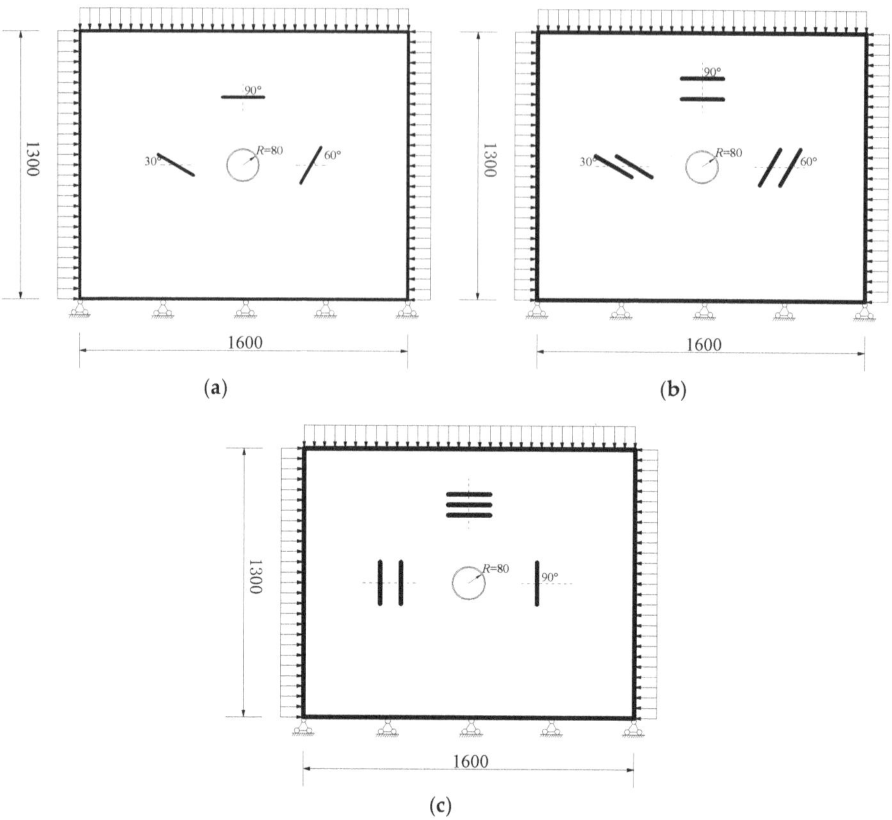

Figure 7. Numerical model of the different model test samples (unit: mm): (**a**) T1 sample; (**b**) T2 sample; (**c**) T3 sample.

After meshing, the established numerical models all contained intact rock, joint, and seamless steel pipe elements, and the number of degrees of freedom of the element was 3, including the translation in the x and y directions, and the rotation in the x-y plane. According to the statistics, the number of seamless steel pipe elements in the T1, T2, and T3 test blocks was 40, while the number of intact rock and joint elements were 25,510 and 84, 36,752 and 178, and 32,536 and 148, respectively. The static loading and constraints of the numerical models were the same as that of the model test samples. The fixed constraint was applied at the lower end of the model, and the uniformly distributed load was applied on the other three sides. The magnitude of the static loads applied on the numerical models was the same as the biaxial static loads in the model test. In reference to the physical and mechanical parameters of the rock and the joint simulated materials in the model test, the corresponding mechanical parameters of the rock, joint, and seamless steel pipe in the numerical simulation are shown in Table 2.

Only the elastic modulus and Poisson's ratio of the simulated rock material were derived in the model tests, while the mechanical parameters used in the established rock constitutive model were the bulk modulus K and shear modulus G, which can be converted by the following equation.

$$\begin{cases} K = \frac{E}{3(1-2\mu)} \\ G = \frac{E}{2(1+\mu)} \end{cases} \quad (7)$$

Table 2. Physical and mechanical parameters of the intact rock, joint, and seamless steel pipe.

Parameters	Intact Rock	Joint	Seamless Steel Pipe
Compressive strength/MPa	5.864	/	645
Tensile strength/MPa	0.613	/	400
Elastic modulus/GPa	5.226	/	206
Internal friction angle/°	23.2	/	/
Cohesion/MPa	1.49	/	/
Poisson ratio	0.203	/	0.3
Initial normal stiffness/GPa/m	/	12	/
Initial shear stiffness/GPa/m	/	7.53	/

In the numerical model, the developed Mohr-Coulomb elasto-plastic constitutive model considering the initial damage was used to simulate the intact rock. According to the results of the model test, the nonlinear BB model could express the mechanical response of the joints under dynamic loading in the presence of the initial static loading, so the joints were simulated in the numerical simulation using the BB model in UDEC. Meanwhile, the strength of the seamless steel pipe was relatively higher compared to the rock and joints simulation materials as seen in Table 2, so the linear elastic model in UDEC was chosen for the simulation. The detailed numerical calculation process was as follows:

(1) The initial static loads were first applied on the numerical model, where the applied static loads in the horizontal and vertical directions were of equal magnitude. To compensate for the insufficient static load conditions in the model test and considering the biaxial compressive strength of the rock material of 7.826 MPa, the number of static load conditions was increased in the numerical simulation, and there were 11 different static load conditions, namely 0, 0.75 MPa, 1.5 MPa, 2.25 MPa, 3 MPa, 3.75 MPa, 4.5 MPa, 5.25 MPa, 6 MPa, 6.75 MPa, and 7.5 MPa.

(2) After the static loads were applied, an equivalent blast load curve was applied to the inner wall of the seamless steel tubes in the numerical model. The equivalent blast load was determined as follows: In the T3 test sample, an additional stress sensor was arranged in the quick-drying material inside the seamless steel pipe, and the sensor was arranged close to the inner wall of the steel pipe to record the time history curve of the blasting load generated during the detonation of the detonating cords. The measured blasting load is shown in Figure 8a below. Figure 8a shows that the measured blast load curve was roughly triangular, with a peak value of 47.59 MPa, a duration of 0.24 ms, and a rise time of about 0.11 ms. After the measured blast load curve was derived, it was applied to the inner wall of the steel pipe as shown in Figure 8b below. The totally computational time of the T1, T2, and T3 numerical models was 327 s, 539 s, and 473 s, respectively.

4.2. Comparison of the Numerical and Model Test Results

In order for a comparison with the model test results, the nodes near the stress measurement points in the corresponding model test samples were selected in each numerical model, and the radial stress time history curves at this point were obtained under the combined effect of different biaxial pressures and blasting loads. The waveforms of the measured and numerical stress time history curves at the same locations in different model test samples were firstly compared, and the measured and numerical stress time curves of the stress measurement points 17 and 18 in the T1, T2, and T3 test samples were selected under the biaxial pressure conditions of 0, 0.75 MPa, 1.5 MPa, and 3 MPa in both the model tests and numerical simulations as shown in Figure 9.

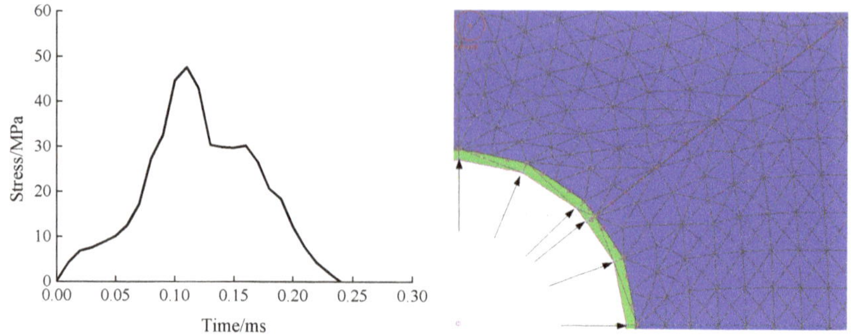

(**a**) (**b**)

Figure 8. Measured blast load and applied schematic: (**a**) measured blasting load time curve; (**b**) blasting load application.

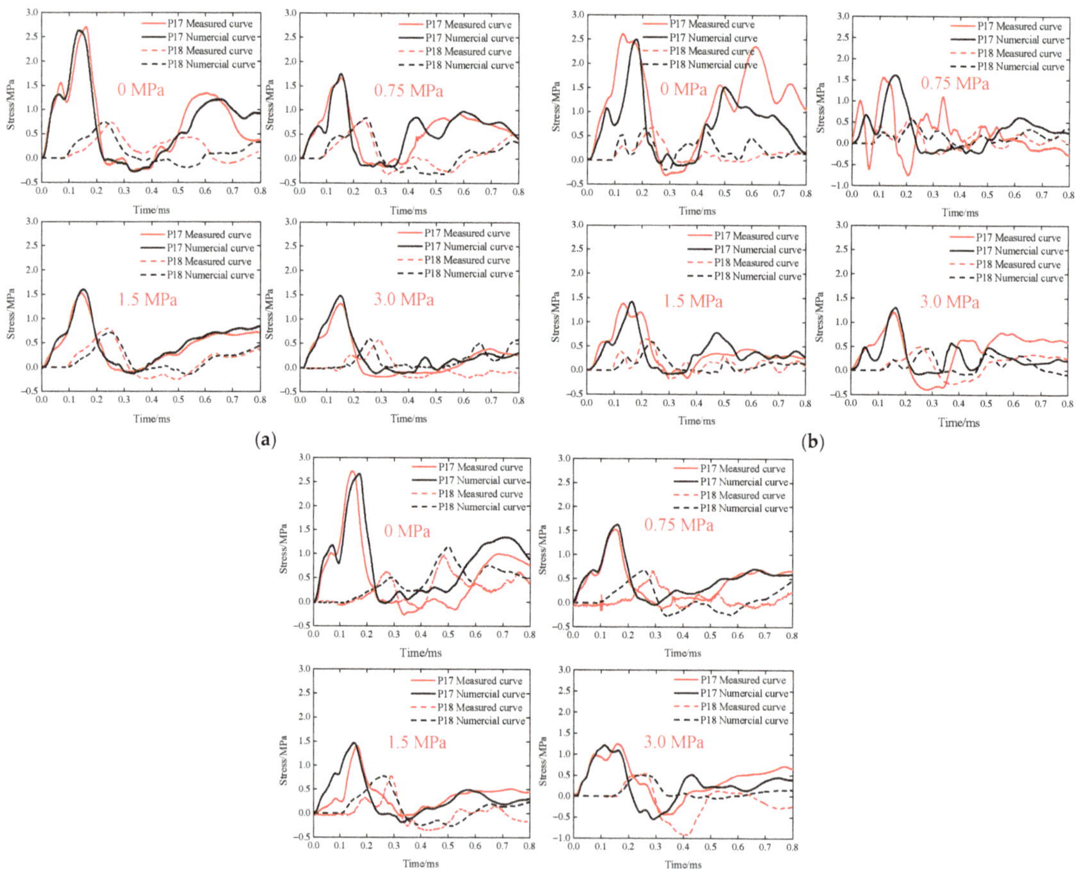

(**a**) (**b**)

(**c**)

Figure 9. Measured and numerical stress wave curve at P17 and P18 points in each sample under different static loads: (**a**) T1 sample; (**b**) T2 sample; (**c**) T3 sample.

As seen in Figure 9, the measured and numerical stress wave curves recorded at P17 and P18 stress measurement points in each model test sample under different biaxial pressures were relatively similar in form and amplitude, and the stress wave amplitude decreased as the biaxial pressure increased. The above phenomena demonstrate the feasibility and accuracy of the numerical calculations based on a combination of the BB model describing the nonlinearity of the joints and the established Mohr-Coulomb elasto-plastic rock constitutive model considering the initial damage.

In order to verify the rationality of the established rock constitutive model from the perspective of stress wave propagation, the stress wave transmission coefficients of the jointed rock masses contained in each numerical model under different biaxial pressures were used for comparing the numerical and experimental results, as shown in Figure 10. It should be noted that the stress wave transmission coefficients of rock masses with a different number and angle joints in the model test were determined by the incident wave and the transmitted wave collected by the stress sensors arranged before and after the rock masses, which was the ratio of the amplitude of the transmitted stress wave to the incident wave. For example, for the T1, T2, and T3 test samples in Figure 2, the amplitude ratios of the stress wave recorded by the stress measuring points 22 and 21 were the stress wave transmission coefficients of the 90° single-joint rock mass, the 90° double-joints rock mass, and the 90° three-joints rock mass, respectively. In the numerical modeling, the nodes corresponding to the stress measuring points in the model test samples were selected in the numerical model to determine the incident and transmitted stress wave of the rock mass, and the numerical transmission coefficients were obtained. Meanwhile, the applied confining pressures in the model tests were 0, 0.75 MPa, 1.5 MPa, and 3 MPa, respectively, but the range of confining pressures was increased in the numerical calculations, which were from 0 to 7.5 MPa, with an interval of 0.75 MPa, for a total 11 different confining pressures. For a comparative analysis, the numerical results were expressed as smoothed curves of numerical transmission coefficients under different biaxial loads.

Figure 10 shows that the measured and numerical transmission coefficients of the jointed rock mass contained in each numerical model under different static loads were relatively close to each other, and when the static load increased from 0, the measured and numerical transmission coefficients both showed a trend of increasing first and then decreasing. Based on the numerical simulation results in Figure 10, the stress wave transmission coefficient of the jointed rock mass containing different angles and numbers reached its maximum value when the static load was about 2.2 MPa, which was about 28.1% of the biaxial compressive strength.

It can also be seen in Figure 10 that the measured and numerical transmission coefficients of the jointed rock masses within different numerical models were relatively close in the ascending part of the curve, while a certain deviation occurred in the descending part. The reason was that in the model test when the static load increased to 3 MPa, which was about 38.3% of the biaxial compressive strength, the closed microcracks within the intact rock started to expand, and new microcracks were initiated, resulting in a decrease in the transmission coefficient [40]. However, in the numerical calculation, the expansion of the microcracks within the intact rock was not considered, which led to the larger numerical results.

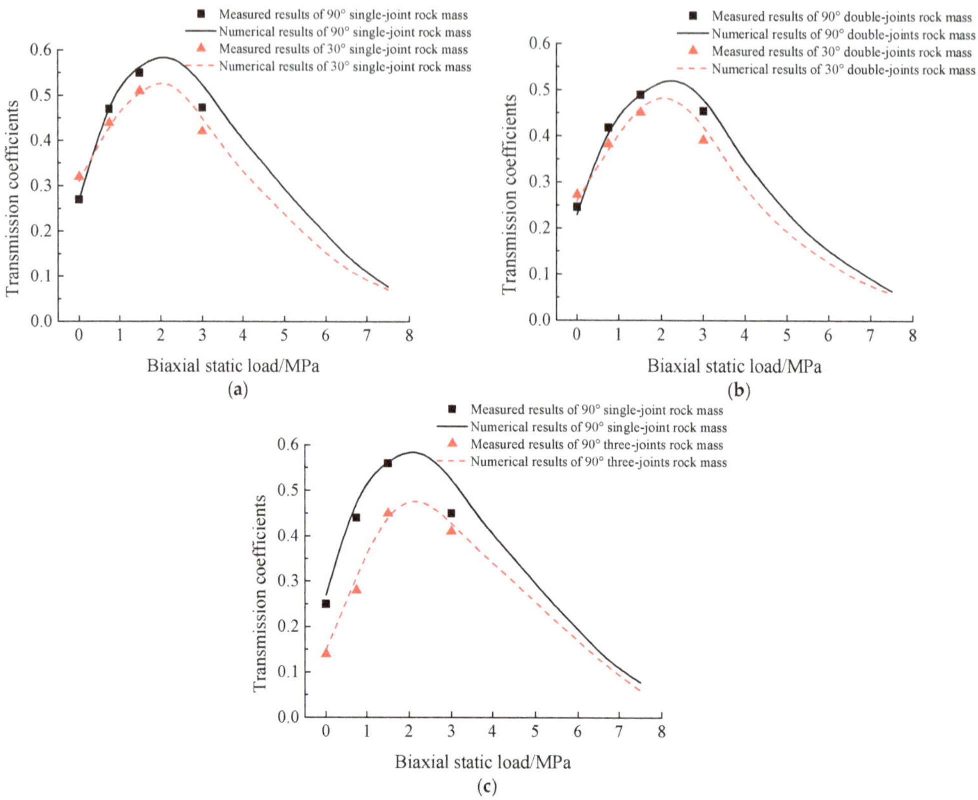

Figure 10. Measured and numerical transmission coefficients of jointed rock masses under different static loads: (**a**) T1 sample; (**b**) T2 sample; (**c**) T3 sample.

5. Numerical Calculation of the Effect of the Angle and the Number of Joints on the Stress Wave Propagation

The last section obtained a high agreement between the numerical and experimental results, which verified the accuracy of the established rock constitutive model considering the initial damage and the feasibility of the adopted numerical simulation method. Due to the limited angle and number of joints set in the model test, a numerical calculation of the stress wave propagation in rock masses with various angles and numbers of joints under different biaxial static loads was carried out based on the same numerical simulation method in Section 4 for a more thorough study of the effect of the angle and number of joints on the stress wave propagation. The physical and mechanical parameters of the rocks and joints in the numerical calculations are shown in Table 2.

5.1. Effect of the Angle of Joints

The angles of the joints selected in the model tests were 30°, 60°, and 90°, while the selection range of the joint angles was expanded in the numerical calculations, with nine different joint angles selected ranging from 10° to 90° and with an interval of 10°. In order to eliminate the effect of the number of joints, only one single joint was selected for the study, and appropriate simplifications were made on the basis of the single-joint model test sample T1.

According to the symmetry of the model test sample, the numerical model was developed as shown in Figure 11 with the size of 800 mm (length) × 1300 mm (width), and a penetration joint was contained in the numerical model. A fixed restraint was applied at the bottom of the numerical model, and biaxial static loads were applied to the remaining three outlines. During the numerical calculation, the applied biaxial static loads were the same as in Section 3, from 0 to 7.5 MPa for total eleven conditions, and the applied blast stress wave load *P*(t) is shown in Figure 8a.

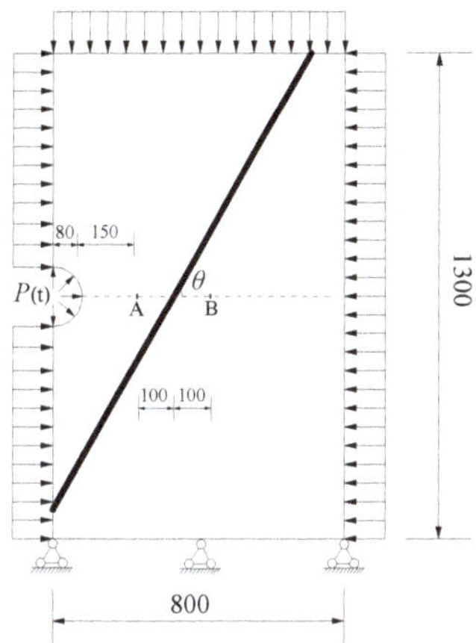

Figure 11. Numerical model containing one single joint with different angles (unit: mm).

According to the actual positions of the stress measurement points P17 and P18 arranged before and after the joint in the single-joint model test sample T1, nodes A and B near the same position before and after the joint were selected in the numerical model as shown in Figure 11, and the stress wave transmission coefficients were calculated from the stress wave amplitudes recorded by the measurement points A and B in the numerical model. The variation of the stress wave transmission coefficient with the angle of the joint under different biaxial static loads was compiled and is shown in Figure 12 below.

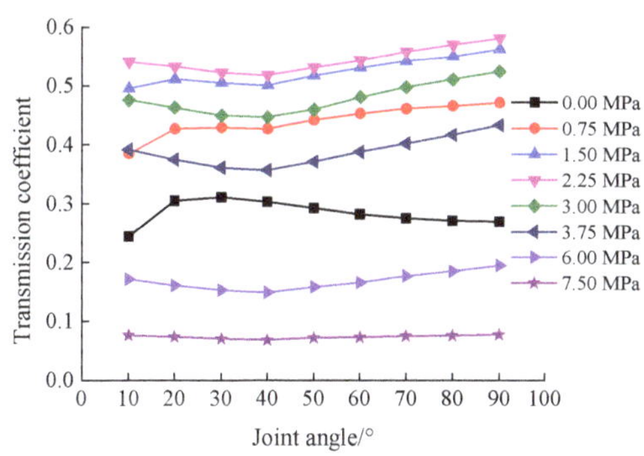

Figure 12. Effect of the joint angle on the transmission coefficient under different biaxial static loads.

As seen in Figure 12, the transmission coefficient increased and then decreased with the increase in the joint angle when the circumferential pressure was 0. The transmission coefficient was at a maximum when the joint angle was close to 30° and decreased with the increase in the joint angle when the joint angle was greater than 30°, which was also consistent with the measured results [40]. The transmission coefficient increased, then slightly decreased, and finally increased with the increase in the joint angle at the biaxial static load of 0.75 MPa and 1.5 MPa and showed an overall trend of increasing. When the biaxial static load was greater than 2.25 MPa, the transmission coefficient decreased and then increased with the increase in the joint angle, and the transmission coefficient was the smallest when the joint angle was about 40°. In addition, it can be seen from Figure 12 that the overall transmission coefficient of rock masses containing one single joint with different angles showed a pattern of increasing and then decreasing with the increase in the biaxial static load, which was also consistent with the results shown in Figure 10.

5.2. Effect of the Number of Joints

In order to investigate the effect of the number of joints on the transmission coefficient of stress waves under different biaxial static loads, numerical calculations were conducted on the cases of jointed rock masses with vertical incidence (i.e., the angle of joints was 90°) and oblique incidence (the angle of joints was 60°) of stress waves propagation. Based on this, the numerical models of rock masses containing a different number of 90° and 60° joints were established. In the numerical calculation, six different number of joints were selected, which were 1, 2, 3, 5, 7, and 9, respectively. At the same time, according to the joint spacing in the three-joints test sample T3 in the model test, the joint spacing selected in the numerical model was also 50 mm. According to the number of joints, the size of the numerical model established was 1300 mm (length) × 1300 mm (width), and the numerical models including five joints with angles of 90° and 60° are shown in Figure 13.

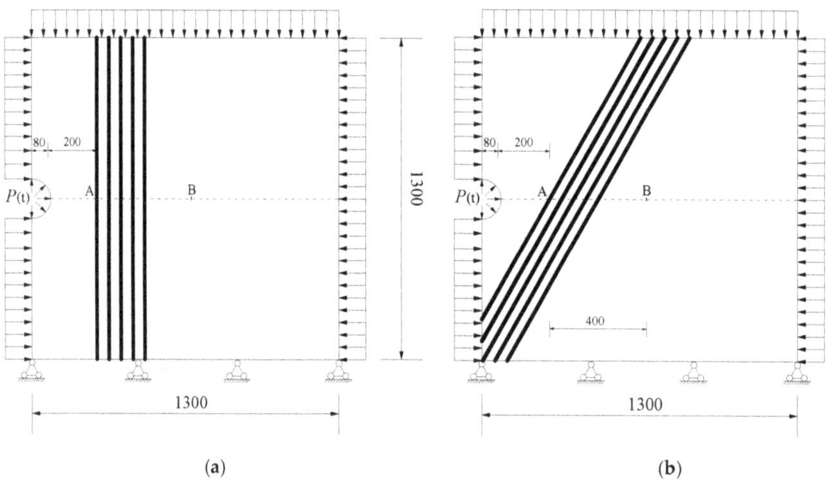

Figure 13. Numerical calculation model of the rock mass containing five joints with different angles (unit: mm): (**a**) joint angle of 90°; (**b**) joint angle of 60°.

The applied biaxial static loads in the numerical simulation were somewhat different from those in Section 4.1, which were 0 MPa, 0.75 MPa, 1.5 MPa, 2.25 MPa, 4.5 MPa, and 6.75 MPa, for a total of six conditions, while the applied stress wave load *P*(t) remained the same as in Section 4.1. To eliminate the effects of the geometric and physical attenuation of the stress waves, the selected measurement points in the numerical model were node A located on the measurement line near the left side of the first joint and node B located on the measurement line near the right side of the ninth joint when the number of joints was nine, as shown in Figure 13 above, which was slightly different from the arrangement of the measurement points in the T2 test sample in the model test. The variation of the transmission coefficient with the number of joints for vertical and oblique incidence of stress waves under different biaxial static loads was sorted out as shown in Figure 14.

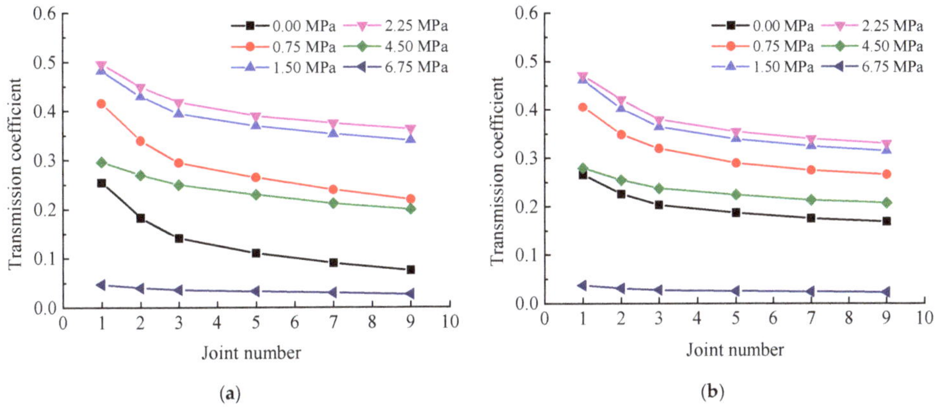

Figure 14. Transmission coefficient of the rock mass containing different numbers of 90° and 60° joints under biaxial static loads: (**a**) 90° jointed rock mass; (**b**) 60° jointed rock mass.

As can be seen from Figure 14 above, the transmission coefficients of both the 90° and 60° joint rock masses showed a decreasing trend with an increase in the number of joints under the same static load, and the decreasing increment gradually became smaller, especially after the number of joints reached a certain number (more than five), the attenuation effect of the increase in the number of joints on the stress wave propagation gradually weakened.

At the same time, the reduction in the transmission coefficient for the 90° joint rock mass was 70.5%, 47.1%, 34.2%, 32.6%, 29.1%, and 26.8% when the number of joints was increased from 1 to 9 with biaxial static loads of 0 MPa, 0.75 MPa, 1.5 MPa, 2.25 MPa, 4.5 MPa, and 6.75 MPa, respectively, and the corresponding reduction in the transmission coefficient for the 60° joint rock mass was 36.3%, 34.4%, 31.7%, 30.0%, 25.7%, and 22.6%, respectively. The above results showed that when the static load was small, such as 0 MPa and 1.5 MPa, the decrease in the transmission coefficient of the 90° jointed rock masses caused by increasing the number of joints was significantly larger than that of the 60° jointed rock masses, and the larger the static load was, the smaller the decrease in the stress waves transmission coefficient caused by increasing the number of joints for both 90° and 60° jointed rock masses, indicating that the attenuation effect of the number of joints on stress wave propagation became weaker as the static load increased.

6. Conclusions

In this paper, the variation of the initial damage variable was firstly determined based on the change in the stress wave velocity in the intact rock under different equal biaxial static loads in the model test, and the Mohr-Coulomb elasto-plastic constitutive model of the rock considering initial damage was established by combining the Mohr-Coulomb strength criterion. The developed rock constitutive model was then numerically implemented using the Fish language in discrete element software UDEC, and the model tests were numerically reproduced in conjunction with the BB model characterizing the nonlinear deformation properties of the joints. Finally, further numerical studies on the effects of the biaxial static load, the angle, and the number of joints on the propagation of stress waves in the jointed rock mass were carried out, and the following conclusions were drawn.

(1) The numerical and experimental results of the propagation law of stress waves in the jointed rock masses under different biaxial static loads were compared and analyzed from the perspectives of the waveform, amplitude, and the transmission coefficient of

stress waves, which were relatively consistent, verifying the feasible of the adopted numerical calculation method.

(2) The initial damage variation in the intact rock with the biaxial static load increased first and then decreased. When the biaxial static load was 1.23 MPa, which was about 15.7% of the biaxial compressive strength of the intact rock, the stress wave velocity reached its maximum value while the initial damage was the smallest, indicating that the internal microcracks in the intact rock were in a fully compacted state under this static load.

(3) As the biaxial static loads increased, the measured and numerical transmission coefficients of the rock masses containing different angles and numbers of joints all showed a trend of first increasing and then decreasing, and the transmission coefficient was the largest when the static load was about 2.2 MPa, which was about 28.1% of the biaxial compressive strength of the intact rock.

(4) The transmission coefficient increased and then decreased with the increase in the joint angle without the static load and was the largest when the joint angle was close to 30°. The transmission coefficient continuously increased with the increase in the joint angle when the static load was relatively small, such as 0.75 MPa and 1.5 MPa, i.e., less than 20% of the biaxial compressive strength of the intact rock. The transmission coefficient decreased and then increased with the increase in the joint angle when the static load was greater than 2.25 MPa (28.7% biaxial compressive strength of the intact rock) and was the smallest at the joint angle of about 40°.

(5) Under the same static loading, the transmission coefficients of the jointed rock masses all showed a tendency to decrease with the increase in the number of joints, and the decreasing increment gradually became smaller. The larger the static load, the smaller the decrease in the transmission coefficients caused by the increase in the number of joints, indicating the effect of the number of joints on the transmission coefficients which decreased as the static load increased.

(6) In the blasting excavation of the underground rock mass, the in situ stress and the spatial distribution of the joints significantly affected the propagation of the blasting stress wave. When the blasting stress wave vertically impacted the initial stressed rock mass, the transmission coefficient was the largest. Therefore, the connection line of blast holes should be perpendicular to the dominant joints in an underground rock mass to ensure the efficient transmission of explosive energy, so that the rock mass can be efficiently and adequately fragmented.

Author Contributions: Conceptualization, Q.D.; methodology, Q.D. and X.L.; software, Q.D.; validation, Y.J. and J.S.; formal analysis, Q.D.; investigation, Q.D.; resources, X.L.; data curation, Q.D.; writing—original draft preparation, Q.D.; writing—review and editing, Q.D.; visualization, Q.D.; supervision, Y.J.; project administration, Q.D.; funding acquisition, X.L., Y.J. and J.S. All authors have read and agreed to the published version of the manuscript.

Funding: This research was funded by the Natural Science Foundation of Hubei Province, China (2020CFB428), the Innovation Group Project of the Natural Science Foundation of Hubei Province, China (2020CFA043), the Key Research and Development Project of Hubei Province, China (2020BCA084), the National Natural Science Foundation of China (51774222), and the Research Start-Up Foundation of Jianghan University (1027-06020001).

Institutional Review Board Statement: Not applicable.

Informed Consent Statement: Not applicable.

Data Availability Statement: The data used to support the findings of this study are available from the corresponding author upon request.

Acknowledgments: Special thanks for organizing and conducting the experimental studies to the faculty of the Hubei Key Laboratory of Blasting Engineering, Jianghan University.

Conflicts of Interest: The authors declare no conflict of interest.

References

1. Fan, L.F.; Sun, H.Y. Seismic wave propagation through an in-situ stressed rock mass. *J. Appl. Geophys.* **2015**, *121*, 13–20. [CrossRef]
2. Yi, C.P.; Johansson, D.; Greberg, J. Effects of in-situ stresses on the fracturing of rock by blasting. *Comput. Geotech.* **2018**, *104*, 321–330. [CrossRef]
3. Tao, J.; Yang, X.G.; Li, H.T.; Zhou, J.; Fan, G.; Lu, G. Effects of in-situ stresses on dynamic rock responses under blast loading. *Mech. Mater.* **2020**, *145*, 103374. [CrossRef]
4. Bandis, S.C.; Lumsden, A.C.; Barton, N.R. Fundamentals of rock joint deformation. *Int. J. Rock Mech. Min. Sci. Geomech. Abstr.* **1983**, *20*, 249–268. [CrossRef]
5. Babanouri, N.; Mansouri, H.; Nasab, S.K.; Bahaadini, M. A coupled method to study blast wave propagation in fractured rock masses and estimate unknown properties. *Comput. Geotech.* **2013**, *49*, 134–142. [CrossRef]
6. Skrzypkowski, K.; Korzeniowski, W.; Zagórski, K.; Zagórska, A. Adjustment of the Yielding System of Mechanical Rock Bolts for Room and Pillar Mining Method in Stratified Rock Mass. *Energies* **2020**, *13*, 2082. [CrossRef]
7. Schoenberg, M. Elastic wave behavior across linear slip interfaces. *J. Acoust. Soc. Am.* **1980**, *68*, 1516–1521. [CrossRef]
8. Pyrak-Nolte, L.J.; Myer, L.R.; Cook, N.G.W. Transmission of seismic waves across single natural fractures. *J. Geophys. Res. Solid Earth* **1990**, *95*, 8617–8638. [CrossRef]
9. Zhao, J.; Cai, J.G. Transmission of Elastic P-waves across Single Fractures with a Nonlinear Normal Deformational Behavior. *Rock Mech. Rock Eng.* **2001**, *34*, 3–22. [CrossRef]
10. Zhao, J.; Cai, J.G.; Zhao, X.B.; Li, H.B. Dynamic Model of Fracture Normal Behaviour and Application to Prediction of Stress Wave Attenuation Across Fractures. *Rock Mech. Rock Eng.* **2008**, *41*, 671–693. [CrossRef]
11. Perino, A.; Orta, R.; Barla, G. Wave Propagation in Discontinuous Media by the Scattering Matrix Method. *Rock Mech. Rock Eng.* **2012**, *45*, 901–918. [CrossRef]
12. Li, J.C.; Li, H.B.; Ma, G.W.; Zhao, J. A time-domain recursive method to analyse transient wave propagation across rock joints. *Geophys. J. Int.* **2012**, *188*, 631–644. [CrossRef]
13. Zhao, X.B.; Zhao, J.; Cai, J.G. P-wave transmission across fractures with nonlinear deformational behaviour. *Int. J. Numer. Anal. Methods Geomech.* **2006**, *30*, 1097–1112. [CrossRef]
14. Li, J.C.; Ma, G.W. Analysis of Blast Wave Interaction with a Rock Joint. *Rock Mech. Rock Eng.* **2010**, *43*, 777–787. [CrossRef]
15. Cai, J.G.; Zhao, J. Effects of multiple parallel fractures on apparent attenuation of stress waves in rock masses. *Int. J. Rock Mech. Min. Sci.* **2000**, *37*, 661–682. [CrossRef]
16. Li, J.C. Wave propagation across non-linear rock joints based on time-domain recursive method. *Geophys. J. Int.* **2013**, *193*, 970–985. [CrossRef]
17. Chai, S.B.; Li, J.C.; Zhang, Q.B. Stress Wave Propagation across a Rock Mass with Two Non-parallel Joints. *Rock Mech. Rock Eng.* **2016**, *49*, 4023–4032. [CrossRef]
18. Ju, Y.; Sudak, L.; Xie, H.P. Study on stress wave propagation in fractured rocks with fractal joint surfaces. *Int. J. Solids Struct.* **2007**, *44*, 4256–4271. [CrossRef]
19. Feng, J.J.; Wang, E.Y.; Chen, L.; Li, X.; Xu, Z.; Li, G. Experimental study of the stress effect on attenuation of normally incident P-wave through coal. *J. Appl. Geophys.* **2016**, *132*, 25–32. [CrossRef]
20. Cundall, P.A. A computer model for simulating progressive large-scale movements in blocky rock systems. In Proceedings of the Symposium of the International Society for Rock Mechanics, Nancy, France, 4–6 October 1971; pp. 11–18.
21. Zhu, J.B.; Deng, X.F.; Zhao, X.B.; Zhao, J. A Numerical Study on Wave Transmission Across Multiple Intersecting Joint Sets in Rock Masses with UDEC. *Rock Mech. Rock Eng.* **2013**, *46*, 1429–1442. [CrossRef]
22. Li, H.B.; Liu, T.T.; Liu, Y.Q.; Li, J.; Xia, X.; Liu, B. Numerical Modeling of Wave Transmission across Rock Masses with Nonlinear Joints. *Rock Mech. Rock Eng.* **2015**, *49*, 1115–1121. [CrossRef]
23. Zheng, Y.; Chen, C.X.; Liu, T.T.; Zhang, W. Numerical Study of P-Waves Propagating across Deep Rock Masses Based on the Hoek-Brown Model. *Int. J. Geomech.* **2020**, *20*, 04019152. [CrossRef]
24. Sun, L.; Zhao, G.F.; Zhao, J. Particle manifold method (PMM): A new continuum-discontinuum numerical model for geomechanics. *Int. J. Numer. Anal. Methods Geomech.* **2012**, *37*, 1711–1736. [CrossRef]
25. Zhao, J.; Sun, L.; Zhu, J.B. Modelling P-wave transmission across rock fractures by particle manifold method (PMM). *Geomech. Geoengin. Int. J.* **2012**, *7*, 175–181. [CrossRef]
26. Fan, L.F.; Yi, X.W.; Ma, G.W. Numerical manifold method (NMM) simulation of stress wave propagation through fractured rock mass. *Int. J. Appl. Mech.* **2013**, *5*, 1350022. [CrossRef]
27. Zhou, X.; Fan, L.; Wu, Z. Effects of Microfracture on Wave Propagation through Rock Mass. *Int. J. Geomech.* **2017**, *17*, 04017072. [CrossRef]
28. Huang, X.; Qi, S.; Williams, A.; Zhou, Y.; Zheng, B. Numerical simulation of stress wave propagating through filled joints by particle model. *Int. J. Solids Struct.* **2015**, *69–70*, 23–33. [CrossRef]
29. Zhou, H.; He, C. Propagation law of stress wave and cracks in non-penetrating jointed rock mass: A numerical study based on particle flow code. *Geotech. Geol. Eng.* **2020**, *38*, 3967–3981. [CrossRef]
30. Babanouri, N.; Fattahi, H. Evaluating orthotropic continuum analysis of stress wave propagation through a jointed rock mass. *Bull. Eng. Geol. Environ.* **2018**, *77*, 725–733. [CrossRef]

31. Heidari, M.; Khanlari, G.R.; Torabi-Kaveh, M. Effect of Porosity on Rock Brittleness. *Rock Mech. Rock Eng.* **2014**, *47*, 785–790. [CrossRef]
32. Patrick, B.; Wong, T.F.; Zhu, W. Effects of porosity and crack density on the compressive strength of rocks. *Int. J. Rock Mech. Min. Sci.* **2014**, *67*, 202–211.
33. Jin, J.F.; Yuan, W.; Wu, Y.; Guo, Z. Effects of axial static stress on stress wave propagation in rock considering porosity compaction and damage evolution. *J. Cent. South Univ.* **2020**, *27*, 592–607. [CrossRef]
34. Martin, C.D.; Chandler, N.A. The progressive fracture of Lac du Bonnet granite. *Int. J. Rock Mech. Min. Sci. Geomech. Abstr.* **1994**, *31*, 643–659. [CrossRef]
35. Cai, M.; Kaiser, P.K.; Tasaka, Y.; Maejima, T.; Morioka, H.; Minami, M. Generalized crack initiation and crack damage stress thresholds of brittle rock masses near underground excavations. *Int. J. Rock Mech. Min. Sci.* **2004**, *41*, 833–847. [CrossRef]
36. Wang, X.; Ge, H.; Wang, J.; Wang, D.; Chen, H. Evaluation of the Micro-cracks in Shale from the Stress Sensitivity of Ultrasonic Velocities. *Rock Mech. Rock Eng.* **2016**, *49*, 4929–4934. [CrossRef]
37. Hou, R.; Zhang, K.; Tao, J.; Xue, X.; Chen, Y. A Nonlinear Creep Damage Coupled Model for Rock Considering the Effect of Initial Damage. *Rock Mech. Rock Eng.* **2018**, *52*, 1275–1285. [CrossRef]
38. Du, M.; Su, J.; Wang, X.; Zhang, S.; Zhang, Y. Dynamic response and crack propagation law of tunnel invert with initial damage. *Eng. Fail. Anal.* **2021**, *119*, 104939.
39. Dong, Q. Model test study on propagation law of plane stress wave in jointed rock mass under different in-situ stresses. In Proceedings of the IOP Conference Series: Earth and Environmental Science, Singapore, 22–25 December 2017; p. 012009.
40. Dong, Q.; Li, X.P.; Huang, J.H. Model test study on cylindrical blasting stress wave propagation across jointed rock mass with different initial stresses. *Adv. Civ. Eng.* **2020**, *2020*, 8881302. [CrossRef]
41. Papanikolaou, V.K.; Kappos, A.J. Confinement-sensitive plasticity constitutive model for concrete in triaxial compression. *Int. J. Solids Struct.* **2007**, *44*, 7021–7048. [CrossRef]
42. Huang, H. Experimental Study on Biaxial-to-Uniaxial Strength Ratio of Concrete at Early Age. Master's Thesis, Dalian University of Technology, Dalian, China, 2013. (In Chinese).
43. Yin, S.; Wang, S. Relation of stresses with elastic modulus and velocities and its application. *Rock Soil Mech.* **2003**, *24*, 597–601. (In Chinese)

applied sciences

MDPI

Article

Predictive Modelling for Blasting-Induced Vibrations from Open-Pit Excavations

Yo-Hyun Choi and Sean Seungwon Lee *

Department of Earth Resources and Environmental Engineering, Hanyang University, Seoul 04763, Korea; netisen@hanyang.ac.kr
* Correspondence: seanlee@hanyang.ac.kr; Tel.: +82-2220-2243

Abstract: Reliable estimates of peak particle velocity (PPV) from blasting-induced vibrations at a construction site play a crucial role in minimizing damage to nearby structures and maximizing blasting efficiency. However, reliably estimating PPV can be challenging due to complex connections between PPV and influential factors such as ground conditions. While many efforts have been made to estimate PPV reliably, discrepancies remain between measured and predicted PPVs. Here, we analyzed various methods for assessing PPV with several key relevant factors and 1191 monitored field blasting records at 50 different open-pit sites across South Korea to minimize the discrepancies. Eight prediction models are used based on artificial neural network, conventional empirical formulas, and multivariable regression analyses. Seven influential factors were selected to develop the prediction models, including three newly included and four already formulated in empirical formulas. The three newly included factors were identified to have a significant influence on PPV, as well as the four existing factors, through a sensitivity analysis. The measured and predicted PPVs were compared to evaluate the performances of prediction models. The assessment of PPVs by an artificial neural network yielded the lowest errors, and site factors, K and m were proposed for preliminary open-pit blasting designs.

Keywords: peak particle velocity; blasting-induced vibration; prediction; artificial neural networks; site factors K and m; open-pit blasting

Citation: Choi, Y.-H.; Lee, S.S. Predictive Modelling for Blasting-Induced Vibrations from Open-Pit Excavations. *Appl. Sci.* **2021**, *11*, 7487. https://doi.org/10.3390/app11167487

Academic Editor: Ricardo Castedo

Received: 14 May 2021
Accepted: 13 August 2021
Published: 15 August 2021

1. Introduction

Drilling and blasting is typically used to fragment rock masses at various building and civil construction sites because it is the most economical means of breaking rock for excavation. However, blasting at construction sites is accompanied by undesirable environmental side effects, such as vibration, noise, and scattering of debris. According to Korea's Office of National Environmental Conflict Resolution Commission, 3840 (approximately 84%) of the 4557 environmental dispute cases on record involve noise and vibration, primarily from construction sites [1]. Blasting vibrations occurring at a construction site account for the majority of these environmental disputes because they result in damage to nearby structures and present various safety concerns. Every country specifies a limit on the peak particle velocity (PPV) of the induced vibrations to minimize damage to nearby structures. According to DIN 4150-3 [2], the limits on PPV are 2 cm/s for buildings used for commercial purposes, 0.5 cm/s for dwellings, and 0.3 cm/s for buildings under preservation orders at a frequency of 1 to 10 Hz. Siskind et al. [3] proposed that 1.9 and 1.3 cm/s are safe levels of blasting vibration for drywall and plaster under 10 Hz conditions. In South Korea, the limits on PPV are 0.2 cm/s for cultural assets and 0.5 cm/s for apartments. Blasting engineers try to accurately predict PPVs that will be induced by blasting and apply the predicted PPVs to the design of blasting patterns to comply with these regulations. Many researchers have studied and proposed various empirical formulas to predict and control PPV [4]. Among the various empirical formulas, a conventional empirical formula developed by U.S. Bureau of Mines (USBM) researchers, Duvall and Petkof [5] has been widely used to predict PPV and design blasting patterns. The current design approach consists of two steps. First, several test blastings are conducted to determine site factors K and m, which represent geological characteristics, before massive blasting. At each test, the distances between blasting and monitoring points, the charge weights per delay, and the PPVs are monitored and recorded. Based on these factors, K and m are calculated. Second, PPV is predicted using an empirical formula with K, m, the distance between blasting and monitoring points, and the charge weight per delay. However, this empirical formula often results in significant discrepancies between measured and predicted PPVs. Due to the discrepancies, blasting engineers are forced to use a high factor of safety (FoS) to prevent problems resulting from excessive vibration velocity. A high FoS typically requires the use of a more conservative charge weight per delay than the maximum allowable weight would accommodate. The conservative charge weight per delay can decrease blasting efficiency and increase construction time and total cost. A more accurate method of predicting PPV is vital to protect the environment and increase the efficiency of blasting.

The artificial neural network (ANN) has been applied in various fields such as renewable energy systems [6], atmospheric science [7], and civil engineering [8,9] to predict targets. In addition, research is also ongoing on predicting PPVs using ANN. To develop an ANN model for PPV prediction, Nguyen et al. [10] gathered 185 blasting datasets from a limestone mine in Vietnam, Azimi et al. [11] collected 70 blasting datasets from a copper mine in Iran, and Bui et al. [12] obtained 83 blasting datasets from a quarry mine in Vietnam. Every result of the research showed good agreement with the measured and predicted PPVs. ANN is generally not limited by any assumptions such as linearity or normality, thus ANN has the modeling power to derive excellent results even with irregular datasets and complex phenomena [13,14]. However, in the previous studies, the largest number of datasets was only 185 and the datasets were obtained from a limited local region. Each ANN model developed in the previous studies is only strictly applicable to the site where the study was conducted due to the limited region. Therefore, it is necessary to develop the global prediction model and to select influential factors which can be obtained easily from every blasting site. In this paper, an ANN was selected as one of the prediction methods due to its strengths. Its performance for predicting PPVs was compared with the performances of conventional empirical formulas and multivariate regression analyses to find the best prediction methods for predicting PPVs with numerous datasets of field blasting records from various sites.

2. Methodology

Figure 1 shows the process for this study, which consists of three steps; acquisition and pre-processing of blasting datasets, development of prediction models using three other methods, and testing and comparison of the prediction models.

2.1. Artificial Neural Network

An ANN is a prediction method based on causes and effects obtained through experience. It can be used as a tool for training, remembering, and analyzing using the computational power of a computer [15]. The network calculates non-linear and complex connections with an input layer, a hidden layer, and an output layer. Each layer has a node for calculation, and their weights and biases act as interlayer connections. The input and output layers consist of causal and result parameters, respectively. The training algorithm of the ANN used in this study was back-propagation, which is the most efficient ANN training algorithm available [16,17]. In back-propagation, the output values calculated in the forward direction through weights and biases are used to calculate training errors from the true values. Through these errors, weights and biases are corrected to minimize the errors in the reverse direction. These sequences repeat until the errors meet the convergence tolerance or other limit conditions. After the ANN model meets the conditions, it can be used as a prediction model with final weights and biases.

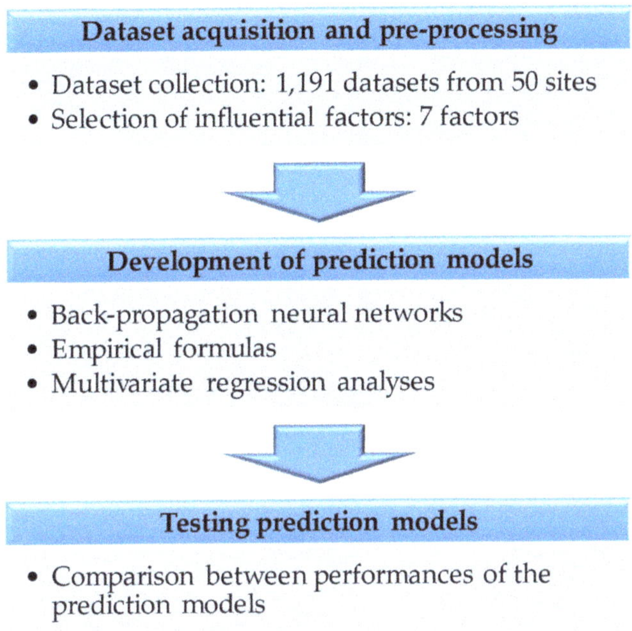

Figure 1. Three steps of the research process.

The ANN requires activation and normalization functions. The former converts the sum of the input signals into the output signal in the nodes of a hidden layer. A non-linear function should be used to determine a non-linear relationship between input and output parameters. Generally, sigmoid, hyperbolic tangent, and rectified linear unit (ReLU) functions, which are non-linear and represented by Equations (1)–(3), respectively, are used as an activation function.

$$f(x) = \frac{1}{1+e^{-x}} \tag{1}$$

$$f(x) = \frac{e^{x}-e^{-x}}{e^{x}+e^{-x}} \tag{2}$$

$$f(x) = \begin{cases} 0, & x < 0 \\ x, & x \geq 0 \end{cases} \tag{3}$$

A normalization function converts all input values which have on different scales into a common scale. It is necessary because the degrees of influence on the output parameter can vary depending on the range of the input parameters. Usually, min-max scaling and standard scaling are used as a normalization function represented by Equations (4) and (5), respectively. In Equation (4), x_{Max} and x_{min} are the maximum and minimum values for each data type, respectively. In Equation (5), $\overline{x}$ and S_x are the mean and standard deviation values for each data type, respectively.

$$N(x) = \frac{x - x_{min}}{x_{Max} - x_{min}} \tag{4}$$

$$N(x) = \frac{x - \overline{x}}{S_x} \tag{5}$$

2.2. Empirical Formula

As mentioned, various PPV prediction techniques are available but only the empirical formula of Equation (6) has been used to predict PPVs for blasting designs in South Korea [18]. Therefore, in this study, the empirical formula developed by USBM was selected to assess ground vibration and identify the optimal prediction method. In Equation (6), the values of K and m are obtained through linear regression of the blasting datasets consisting of PPV and the scaled distance (SD) expressed in Equation (7) [19]. Here, W is a charge weight per delay, and D is the distance between blasting and monitoring points.

$$\mathrm{PPV} = \mathrm{K}(\mathrm{SD})^{\mathrm{m}} \tag{6}$$

$$\mathrm{SD} = \mathrm{D}/\sqrt{\mathrm{W}} \tag{7}$$

2.3. Multivariate Regression Analysis

Multivariate regression analysis is defined as a regression analysis in which two or more independent variables are used to account for changes in the dependent variable [20]. It is called multivariate linear regression analysis (MLRA) and the relationships between the independent and the dependent variables are expressed linearly. The MLRA is expressed as follows:

$$y = \beta_0 + \beta_1 x_1 + \cdots + \beta_p x_p \tag{8}$$

In Equation (8), y is the dependent variable, x_1 to x_p are the independent variables, β_0 to β_p are regression coefficients, and p is the number of independent variables. The regression coefficients, which make the summation of all square errors minimum, are obtained through the method of least squares.

We defined expressing non-linearly the relationships between independent and dependent variables as multivariate non-linear regression analysis (MnLRA). Among the various forms of MnLRA, an exponential form was employed in this study and it is expressed as follows:

$$y = \beta_0 (x_1)^{\beta_1} (x_2)^{\beta_2} \cdots \left(x_p\right)^{\beta_p} \tag{9}$$

After both sides of Equation (9) are logged, it is equivalent to the same form as Equation (8), so MnLRA can be generated in the same way. Besides, since the empirical formula of Equation (6) is also in exponential form, MnLRA was chosen as the exponential form in this study. It is important to confirm that the model is statistically significant through F and *p*-values of the results of an analysis of variance (ANOVA) and *p*-value of a partial regression coefficient in the multivariate regression analysis.

3. Datasets

The authors collected 1191 blasting datasets, which are more than six times the datasets used in the previous studies, from 50 diverse construction sites, representing each region of South Korea. The locations of 50 diverse construction sites by 28 administrative districts are depicted in Figure 2. The number of construction sites that were conducted in the same administrative district is expressed in the circle. Even though the construction sites are located in the same administrative district, they are different construction sites. Building and road construction were the main site activities, and open-pit blasting was used at all 50 construction sites. Of the total 1191 datasets, 714 (60%) and 179 (15%) were used for prediction model development as training and validation datasets, respectively. The remaining 298 (25%) were used to test the models. The datasets were randomly designated for Training, Validation, and Testing via PYTHON code.

Predicting PPV requires a selection of influential factors. Since this study aims to predict the PPV accurately and easily at any open-pit blasting site, the influential factors should not only affect the PPV but also be easily obtained by untrained field staff.

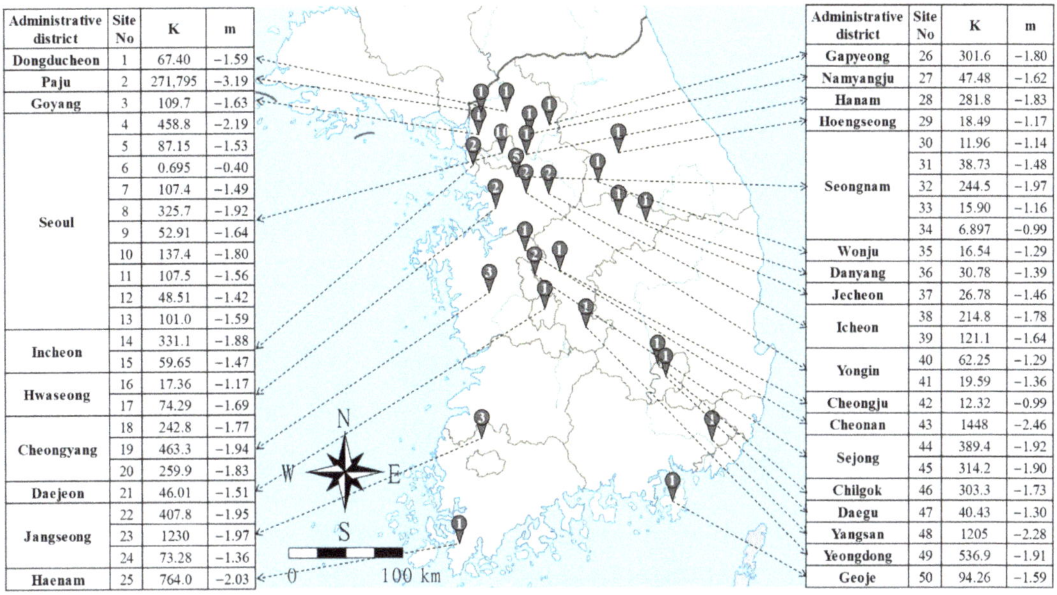

Figure 2. Locations of 50 diverse construction sites by administrative district.

Eleven common initial influential factors satisfied these conditions from 1191 blasting datasets: type of explosive (TE), charge weight per delay (W), specific weight (SW), length of drilling hole (LH), the height of the bench (HB), burden spacing (BS), hole spacing (HS), type of rock (TR), the distance between blasting and monitoring points (D), site factor K, and site factor m. To use an influential factor as quantitative data, the TE and the TR must be converted to values that express the velocity of detonation (VoD) and the velocity of the P-wave (VoP). The explosive types used at the 50 sites were Megamex, New emulate, Newmite, and Lovex manufactured by Hanwha Corporation [21]. The eight types of rock were gneiss, granite, limestone, schist, shale, andesite, rhyolite, and tuff. The conversion values are summarized in Tables 1 and 2.

It is necessary to remove or change the initial influential factors to avoid multicollinearity that negatively affects prediction due to the high correlations between independent variables [22]. As shown in Figure 3, factors W, LH, HB, BS and HS are strongly correlated (>0.88) with each other. To remove a strong correlation between influential factors, we removed the LH, HB, BS and HS since W is the most important factor to PPV among the five factors. Finally, we selected seven influential factors relevant to PPV. The units and ranges of the selected factors and PPV are shown in Table 3.

Table 1. Input values for types of explosive.

Explosive Type	Megamex	New Emulite	NewMITE	LoVEX
Velocity of Detonation (m/s)	6000	5900	5700	3400

Table 2. Input values for types of rock.

Rock Type	**P-Wave Velocity (m/s)**	**Reference**
Gneiss	5500	[23]
Granite	5300	[23]
Limestone	5470	[23]
Schist	4550	[23]

Table 2. *Cont.*

Rock Type	P-Wave Velocity (m/s)	Reference
Shale	3500	[23]
Andesite	5121	[24]
Rhyolite	4100	[25]
Tuff	2750	[26]

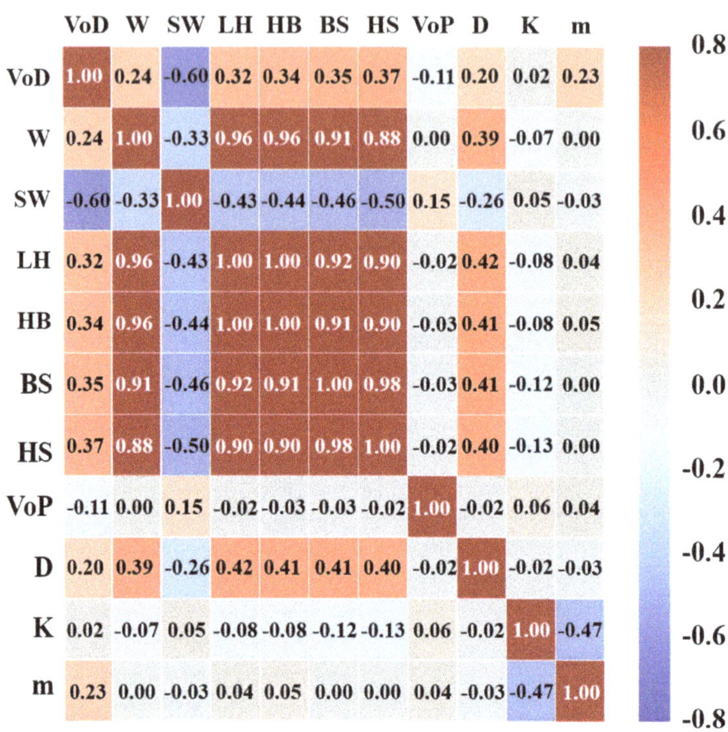

Figure 3. Correlations between initial influential factors.

Table 3. Characteristics of influential factors and peak particle velocity (PPV).

Type	Parameters	Symbol	Unit	Range of Datasets
Input	Velocity of detonation	VoD	m/s	3400–6000
	Charge weight per delay	W	kg	0.1–10
	Specific weight	SW	kg/m^3	0.25–0.56
	Velocity of P-wave	VoP	m/s	2750–5500
	Distance between blasting and monitoring points	D	m	5–650
	K	K	-	0.7–271,795
	m	m	-	−3.19 to −0.40
Output	Peak Particle Velocity		cm/s	0.005–6.514

4. Prediction Models

4.1. Artificial Neural Network

Trial-and-error analysis of hyper-parameters is required to obtain the optimal prediction model which has the lowest validation loss. In this analysis, it was carried out with a different number of hidden layers, nodes, normalization methods, and activation functions; one and two hidden layers; 3, 5, 7, 11, 14, 15, 21, 28 and 35 nodes for the hidden layer; min-max and standard scalings; and three activation functions, sigmoid, hyperbolic tangent, and ReLU. In other words, 54 (2 × 9 × 3) and 486 (2 × 9 × 9 × 3) structures were assessed on 1 and 2 hidden layers, respectively. The number of nodes was determined by Table 4. Here, N_i and N_o mean number of input and output parameters, respectively. We added some equations in the final row of Table 4 to analyze many structures. The Adam optimizer [27] was used to reduce the loss with a learning rate of 0.001. Also, we used an early stopping to avoid overfitting and to obtain the best-fitted model. Every structure of the ANN model was trained with the 714 training datasets and validated by the 179 validation datasets. Every ANN model was developed with the software PYTHON Version 3.7.6.

Table 4. Equations for determination of the number of nodes.

Equation	Number of Nodes	Reference
$\sqrt{N_i \times N_o}$	3	[28]
$(4N_i^2 + 3)/(N_i^2 - 8)$	5	[29]
$3N_i/2$	11	[30]
$2N_i + 1$	15	[31]
$3N_i$	21	[32]
N_i, $2N_i$, $4N_i$, $5N_i$	7, 14, 28, 35	-

In the results of trial-and-error analysis, the average validation loss of 540 structures was 0.126 cm/s. Among the 540 ANN models, the structure composed of two hidden layers with 21 and 28 nodes, normalized by min-max scaling and combined with ReLU showed the lowest validation loss of 0.115 cm/s. Therefore, we selected the ANN model, which has the 7-21-28-1 structure depicted in Figure 4 as an optimal ANN model for a PPV prediction. The training of this model was stopped at 4208 epochs by early stopping. Table 5 summarizes the characteristics of the selected ANN model. This model is represented by Equations (10)–(12). PPV is calculated by Equation (10). Equations (11) and (12) represent hidden layers 1 and 2, respectively.

$$\mathrm{PPV} = [\mathrm{H}_2] \cdot [\mathrm{W}_3] + [\mathrm{b}_3] \quad (10)$$

$$[\mathrm{H}_2] = \mathrm{R}([\mathrm{H}_1] \cdot [\mathrm{W}_2] + [\mathrm{b}_2]) \quad (11)$$

$$[\mathrm{H}_1] = \mathrm{R}([\mathrm{m}[\mathrm{I}]] \cdot [\mathrm{W}_1] + [\mathrm{b}_1]) \quad (12)$$

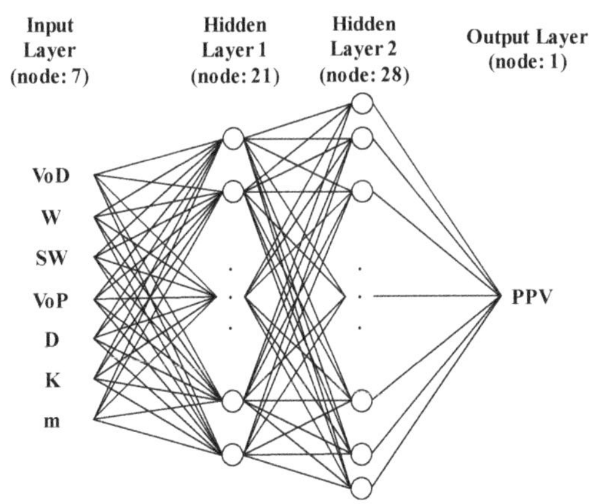

Figure 4. Structure of the artificial neural network (ANN) model developed.

Table 5. Characteristics of the ANN model.

	Characteristics	Details
Datasets	Total datasets	1191
	Training and validation datasets	714, 179
	Test datasets	298
Structure	Number of input parameters	7
	Number of output parameter	1
	Number of hidden layers	2
	Number of nodes	21, 28
Training parameters	Activation function	ReLU
	Optimization	Adam
	Normalization	Standard scaling
	Regularization	Early-stopping
	Number of epochs	4208
	Training algorithm	Back-propagation

In these equations, [I] is the matrix of input data sets, [W] is the matrix of weights, and [b] is the matrix of biases. The weight and bias matrices are constants that were obtained from the ANN training. Here, $[W_1]$, $[W_2]$, $[W_3]$, $[b_1]$, $[b_2]$, and $[b_3]$ are 7×21, 21×28, 28×1, 1×21, 1×28, and 1×1 matrices. When predicting i PPVs, [I] is an $i \times 7$ matrix. R is a ReLU function expressed by Equation (3), m is a min-max scaling expressed by Equation (4).

4.2. Empirical Formula

Each empirical formula of the 50 construction sites was generated using Equation (6) with the site factors, K and m. For instance, Equation (13) represents the empirical formula of Site 1 with K and m values of 67.4 and −1.59, respectively. The site factors of each site are represented in Figure 2. Through this method, 50 empirical formulas were generated and defined as EF-1. Each of the formulas included in EF-1 can only be applied to the PPV prediction at the site where it was generated. The K of Site 2, which is far higher than the rest, seems to be noise. In geotechnical engineering, some noise could have happened due to uncertainties. Thus, datasets obtained from Site 2 should also be analyzed with other datasets.

$$V = 67.4(SD)^{-1.59} \tag{13}$$

Test blasting is required to obtain site factors K and m, used in empirical formulas such as EF-1. However, it is difficult to perform test blastings at the preliminary design stage, and representative values of K and m are needed to compensate for this weakness. Representative K and m values of 200 and −1.6 were proposed based on Design and Construction Guidelines for Open-pit blasting in Road construction published by the Ministry of Land, Infrastructure, and Transport in South Korea [33]. We defined Equation (14) as EF-2 using the K and m. Many engineers have designed preliminary blasting patterns, applying Equation (14).

$$PPV = 200(SD)^{-1.6} \tag{14}$$

To derive one representative empirical formula for the 50 sites, we calculated K and m values of 74.9 and −1.535 using datasets of 50 open-pit blasting construction sites. Equation (15) expresses the representative empirical formula and it was defined as EF-3. Since this is a representative equation of 50 sites, it will show lower prediction accuracy than EF-1. However, it could be used at the preliminary design stage like EF-2. Figure 5 shows EF-3 (solid line) and the 893 datasets (circles) on a log-log plot where the vertical axis is PPV and the horizontal axis is SD. As mentioned in Section 2.2, EF-3 was obtained from the linear regression of the 893 blasting datasets.

$$PPV = 74.9(SD)^{-1.535} \tag{15}$$

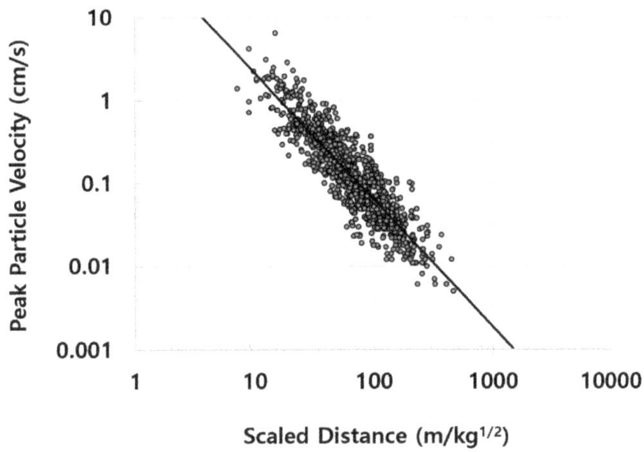

Figure 5. Peak particle velocity versus scaled distance for 893 datasets.

Equations (16) and (17) are prediction models proposed by the International Society of Explosives Engineers (ISEE) [34] and USBM [35], respectively. These two equations have been widely used to predict PPVs. We defined Equations (16) and (17) as the ISEE model and USBM model, respectively.

$$PPV = 172.5(SD)^{-1.6} \quad (16)$$

$$PPV = 71.4(SD)^{-1.6} \quad (17)$$

4.3. Multivariable Regression Analysis

Multivariable regression analyses were carried out using IBM SPSS Statistics Version 26.0 (SPSS), which is a powerful statistical software package [36] that generates a simple equation for estimating output. Many researchers have performed multivariable regression analyses with ANN to compare the performance of prediction methods [15,16,37]. In this study, two types of multivariable regression analysis were carried out using training and validation datasets from 50 open-pit blasting construction sites to identify linear or non-linear relationships between influential factors and PPV. One was multivariable linear regression analysis (MLRA) and the other was multivariable non-linear regression analysis (MnLRA). The developed MLRA is represented by Equation (18). From seven influential factors, SW and VoP were excluded, since their partial regression coefficients had higher p-values than the significant level, 0.05. After the two factors were removed, the F and p-values of the MLRA model showed approximately 49 and 0, respectively. In addition, constant and five influential factors had p-values that were near 0. These F and p-values mean that the MLRA model is statistically significant. However, this model showed a low R of 0.495. The developed MnLRA is represented by Equation (19). This equation has been developed in exponential form following the form of the conventional empirical formula. p-values of all partial regression coefficients except for VoP were shown to be lower than the significant level, 0.05. Therefore, we removed the VoP from the input parameters. F and p-values of the MnLRA model showed approximately 898 and 0, respectively. Besides, the R of this model was high, 0.927. Here, the influential factor m was converted to –m in Equation (19) because all influencing factors and PPV are positive, while m is negative.

$$PPV = -0.588 + 1.2 \times 10^{-4}VoD + 0.092W - 0.003D - 1.45 \times 10^{-6}K - 0.193m \quad (18)$$

$$PPV = 0.034VoD^{0.79}W^{0.741}\,SW^{-0.37}D^{-1.602}K^{0.375}\,(-m)^{-2.248} \quad (19)$$

Note that test datasets were never used prior to the performance evaluation of the prediction methods. This means only the training and validation datasets were used to develop the ANN model, EF-1, 2 (K and m), MLRA, and MnLRA.

5. Prediction Results

5.1. Performance Comparisons of the Six Prediction Models

The 298 test datasets, which account for 25% of the total datasets obtained, were predicted using the eight predictive analysis methods, ANN, EF-1, EF-2, EF-3, ISEE model, USBM model, MLRA, and MnLRA, described in Chapter 4. First, the PPVs were predicted using the weights and biases matrices of the optimal ANN model. Here, all seven influential factors, VoD, W, SW, VoP, D, K, m were used as input parameters. Second, we used EF-1 which grouped 50 empirical formulas to predict PPVs of the test datasets. Here, each test dataset was predicted by the empirical formula of the site where they were obtained. W and D were used as input parameters. Finally, the test datasets were predicted by EF-2, 3, ISEE model, USBM model, MLRA, and MnLRA expressed as Equations (14)–(19), respectively, using input parameters of each method. In this study, three performance indicators, mean absolute error (MAE), root mean square error (RMSE), and mean absolute percent error (MAPE), were used to analyze prediction results. These performance indicators are listed in Table 6.

Table 6. Equations of performance indicators.

Performance Indicator	Equation
MAE	$\mathrm{MAE} = \frac{1}{n}\sum_{i}^{n}\left\lvert V_{mi} - V_{pi}\right\rvert$
RMSE	$\mathrm{RMSE} = \sqrt{\frac{1}{n}\sum_{i}^{n}\left(V_{mi} - V_{pi}\right)^2}$
MAPE	$\mathrm{MAPE} = \frac{1}{n}\sum_{i}^{n}\left\lvert\frac{V_m - V_p}{V_m}\right\rvert \times 100$

Here, V_{mi} and V_{pi} are the i-th measured and predicted values, respectively, and n is the total number of test datasets. Table 7 summarizes the performances of the eight prediction models on the predicted PPVs. The developed ANN model achieved the lowest MAE of 0.064 cm/s, RMSE of 0.161 cm/s, and MAPE of 23.2%. These results were approximately 30%, 56%, and 11% lower than those from EF-1, which is currently the most commonly used method to predict PPVs when designing blasting patterns for construction. However, the EF-2 deduced the highest MAE of 0.305 cm/s and RMSE of 0.731 cm/s.

Table 7. Performances of the six prediction models.

Method	MAE (cm/s)	RMSE (cm/s)	MAPE (%)
ANN	0.064	0.161	23.2
EF-1	0.092	0.370	26.1
EF-2	0.305	0.731	146.5
EF-3	0.123	0.309	47.8
ISEE model	0.244	0.601	115.7
USBM model	0.123	0.308	40.7
MLRA	0.202	0.370	175.1
MnLRA	0.108	0.298	39.1

Linear regression analyses were performed with a coefficient of determination known as R^2 to explain the correlation and similarity between the predicted PPVs from the six predictive analysis methods and measured PPVs of the test datasets. The value of R^2 can be found using Equation (20), where V_m and V_p are measured and predicted PPV values, Cov is the covariation between two factors, and Var is the variation of a factor.

$$R^2 = \frac{\mathrm{Cov}^2(V_m, V_p)}{\mathrm{Var}(V_m) \times \mathrm{Var}(V_p)} \tag{20}$$

Each predicted PPV by the six prediction methods is plotted as a small circle in Figure 6a to 6h respectively according to prediction methods. The x and y axes represent the measured and predicted PPV, respectively, in cm/s. There are two lines in each figure. The dashed line is the Measured PPV = Predicted PPV (1:1) line and the solid line is the linear regression line. In the lower right corner of each figure, it shows the equation of the linear regression line and R^2. The linear regression line resulting from the ANN shows the best result in terms of similarity to the 1:1 line as shown in Figure 6. The linear regression line resulting from the MLRA displays the greatest distance between the two lines.

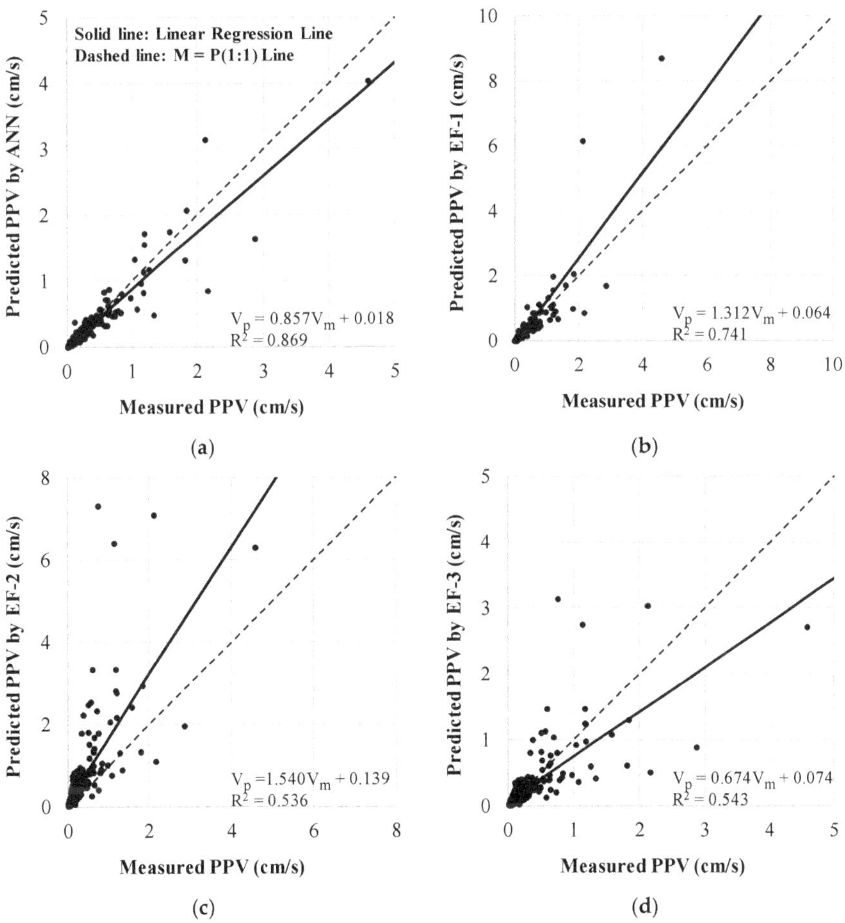

Figure 6. *Cont.*

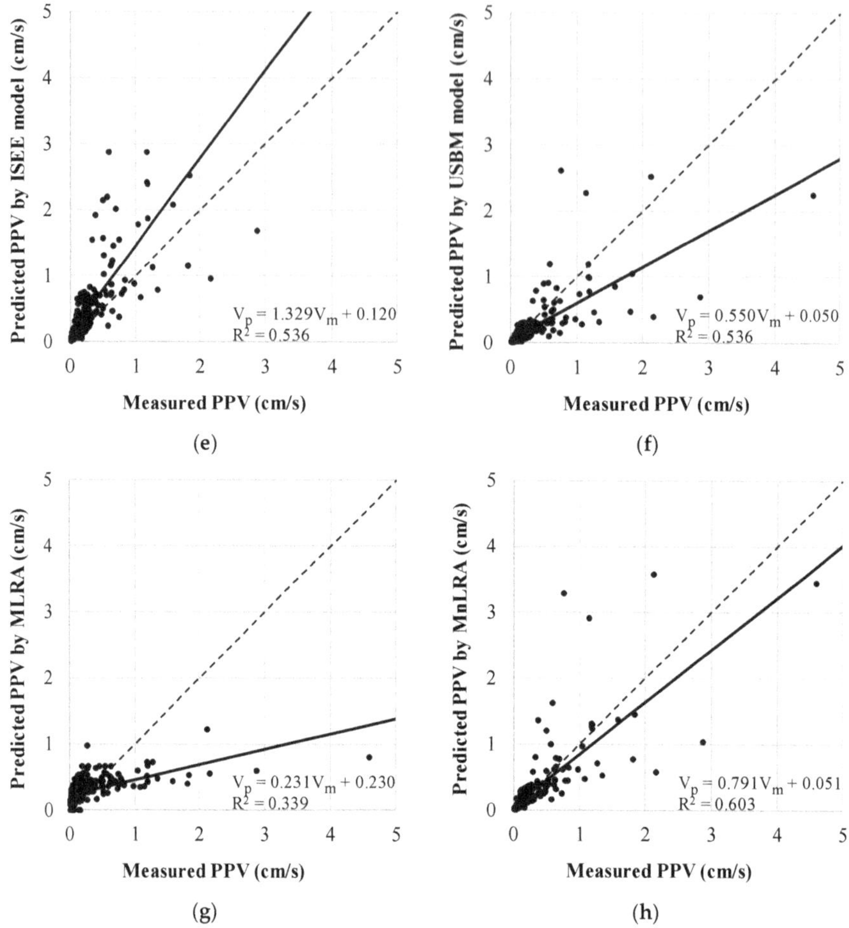

Figure 6. Predicted PPV versus measured PPV by the six prediction methods. The graphs in (**a–h**) were made using predicted PPVs by the ANN, EF-1, EF-2, EF-3, ISEE model, USBM model, MLRA, and MnLRA, respectively.

5.2. Sensitivity Analysis

A sensitivity analysis was performed using the cosine amplitude method for all seven influential factors. This method has been applied previously [4,15,38] to determine the relative significance of each factor on PPV. It calculates a relation, r_{ij}, and provides results from a pairwise comparison of two factors, x_i and x_j, using Equation (21) [39].

$$r_{ij} = \frac{\left|\sum_{k=1}^{m} x_{ik}x_{jk}\right|}{\sqrt{\left(\sum_{k=1}^{m} x_{ik}^2\right)\left(\sum_{k=1}^{m} x_{jk}^2\right)}} \tag{21}$$

The influential factors and PPV of the 1191 datasets, which consist of both training and test datasets, were logged and analyzed using Equation (21). The relative significances of the seven influential factors are depicted in Figure 7. The relative significances between VoD, W, SW, VoP, D, K, m, and PPV were deduced to be approximately 0.885, 0.729, 0.876, 0.886, 0.932, 0.844, and 0.833, respectively.

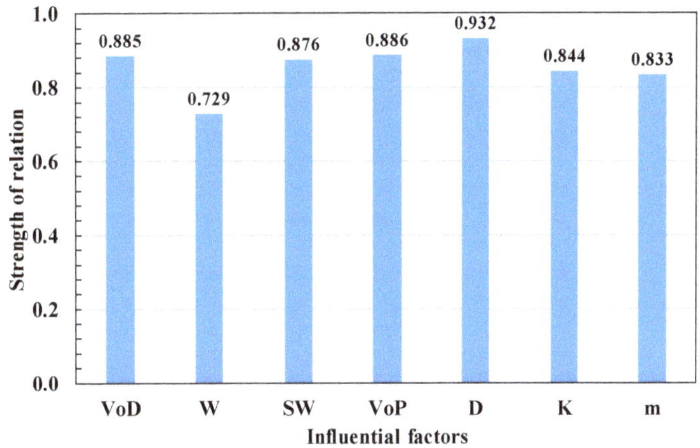

Figure 7. Sensitivity analysis of the influential factors.

6. Discussion

The ANN model showed the best agreement with measured PPVs among eight prediction methods, including the globally used ISEE and USBM models. It would be due to using the most influential factors, which has the ability to reproduce and model the non-linear connections between input and output parameters, and to deal with noise. As shown in Figure 7, the seven influential factors have similar strengths of relation. It indicates that using these seven factors is more effective than using the only four factors which are included in the conventional empirical formula to predict PPVs. The complex connections between PPV and influential factors could be found in the comparison between the MLRA and MnLRA. When we developed these two models, they showed a statistical significance; however, the MLRA had a low R (0.495) while the MnLRA had a high R (0.927). The MAE from the MLRA showed about twice that of the MnLRA. These two models differ in their use of linear and non-linear relationships to explain PPV from influencing factors. Because of this difference, the MnLRA showed better predictive performance than the MLRA. It means that the relationships between the influential factors and PPV are non-linear. The ability to deal with noise could be verified by the prediction results about the biggest measured PPV, 4.58 cm/s, which is over 17 times the average measured PPVs, 0.26 cm/s. The prediction results from the ANN, EF-1, EF-2, EF-3, ISEE model, USBM model, MLRA, and MnLRA were 4.04, 8.7, 6.3, 2.72, 5.43, 2.25, 0.81 and 3.45 cm/s, respectively. The prediction results from the ANN model showed the closest to the measured PPV. It implies that the ANN has an excellent ability to deal with noise.

EF-2 showed the worst performances at MAE and RMSE and it would be due to its applicability. EF-2 is suitable for road construction sites because it was developed using only blasting datasets from road construction sites. These results mean that applying the conventional representative formula for a preliminary blasting design from road constructions has a limitation in applying it to other open-pit blastings. Therefore, a new alternative prediction equation is required. EF-3 which was developed using datasets from 50 diverse open-pit blasting construction sites would be suitable as the alternative prediction equation since it has the same form as EF-2, and it showed better predictive performances than EF-2.

The proposed model has been applied only to open-pit blasting construction sites. Future studies of PPV prediction models such as ANN model and EF-3 will include blasting records from underground caverns, tunnels, and mines as well to ensure the prediction models be generally applicable to any region and type of blasting.

7. Conclusions

In this study, the prediction of PPV using eight predictive analysis methods of ANN, EF-1, EF-2, EF-3, ISEE model, USBM models, MLRA, and MnLRA with 1191 datasets, which are more than six times the maximum datasets used in the previous studies, was carried out to assess PPV prediction methods at an open-pit construction site.

Seven key factors relevant to PPV were considered in the prediction models. The seven key factors were selected according to the ease of obtaining them and their influence on PPV. They consist of three factors, VoD, SW, and VoP, newly proposed in this study, and four key factors, W, D, and the site factors K and m, currently included in the conventional empirical formula. The use of three additional influential factors played a significant role in identifying the prediction model that produced the lowest error. Their significant roles were confirmed through a comparison of the performances of the ANN and others. These roles were also apparent in the results of the sensitivity analysis. The seven key factors have similar strengths of relations with PPV. It implies that not only are the previously used factors important in predicting PPV but also the newly added factors.

The PPV prediction based on the ANN model achieved the lowest values at MAE, RMSE, and MAPE among the eight prediction models. Even the ANN, which was generalized for application to all sites, produced lower errors than those from the EF-1, which can apply to only a specific site. In addition, the prediction accuracy of the ANN model was higher than that of the ISEE and USBM models. It would be attributed to the ability of ANN to express complex and non-linear relationships between influential factors and PPV, and the ability of ANN to deal with noise. It is necessary to perform a grid search for structures and hyper-parameters and early stopping to obtain an optimal prediction model. In this study, we compared 540 ANN models, to which were applied the early stopping method. These models have one or two hidden layers with the number of nodes calculated using the number of input and output parameters and three activation functions. Finally, a structure consisting of two hidden layers with 21 and 28 nodes using a ReLU as an activation function was determined as the optimal model. Other hyper-parameters were chosen following the previous studies. As a result, we generated the prediction model showing the lowest errors among the six prediction methods. Therefore, we recommend using an ANN for predicting PPVs whose hyper-parameters are selected from a grid search and literature research.

The EF-2 was proposed by the Ministry of Land, Infrastructure, and Transport in South Korea for designing preliminary blasting patterns. However, the MAE, RMSE, and MAPE associated with the EF-2 were over two times higher than those associated with the EF-3, which is newly proposed in this study. This difference might be a result of different construction types in the datasets. EF-3 was developed by analyzing data from 50 open-pit construction sites, including building construction sites in downtowns, road construction sites, aggregate extraction sites, and restoration work sites while EF-2 was developed by analyzing only datasets at road construction sites. Using the newly proposed EF-3, which proposes a K value of 74.9 and an m value of -1.535, for a preliminary design of open-pit blasting would be more accurate and reliable than using the EF-2.

The ANN model with the seven key factors and EF-3, proposed in this paper, can predict PPVs more accurately and will help blasting pattern design to be more reliable. The reliable blasting patterns will reduce environmental problems significantly and maximize the efficiency of blasting in construction. Moreover, the use of the newly proposed prediction methods will lessen civil complaints, and improve the efficiency in the construction schedule, and reduce the overall construction budgets. These advantages will lead to greater safety and sustainable urban development.

Author Contributions: Data curation, Y.-H.C.; Formal analysis, Y.-H.C.; Investigation, Y.-H.C.; Methodology, Y.-H.C.; Project administration, Y.-H.C.; Validation, Y.-H.C.; Writing—original draft preparation, Y.-H.C.; Conceptualization, S.S.L.; Funding acquisition, S.S.L.; Project administration, S.S.L.; Supervision, S.S.L.; Writing—review & editing, S.S.L. All authors have read and agreed to the published version of the manuscript.

Funding: This work is supported by the Korea Agency for Infrastructure Technology Advancement (KAIA) grant funded by the Ministry of Land, Infrastructure and Transport (Grant 21UUTI-B157786-02).

Institutional Review Board Statement: Not applicable.

Informed Consent Statement: Not applicable.

Data Availability Statement: Not applicable.

Acknowledgments: This work was supported by the Korea Agency for Infrastructure Technology Advancement (KAIA) grant funded by the Ministry of Land, Infrastructure and Transport (Grant 21UUTI-B157786-02) and by the National Research Foundation of Korea (NRF) grant funded by the Korea government (MSIT) (NRF-2020R1A6A3A13077513).

Conflicts of Interest: The authors declare no conflict of interest. The funders had no role in the design of the study; the collection, analyses, or interpretation of data; the writing of the manuscript; or the decision to publish the results.

References

1. National Environmental Dispute Resolution Commission. Statistical Data, Such as Handling Environmental Disputes (31 December 2020). Available online: https://ecc.me.go.kr/front/user/main.do (accessed on 15 February 2021).
2. German Standards Organization. *DIN 4150-3: Structural Vibration—Part 3: Effects of Vibration on Structures*; Deutsches Institut für Normung e.V.: Berlin, Germany, 1999.
3. Siskind, D.E.; Stagg, M.S.; Kopp, J.W.; Dowding, C.H. *Structure Response and Damage Produced by Ground Vibration from Surface Mine Blasting*; US Department of the Interior, Bureau of Mines: New York, NY, USA, 1980.
4. Hajihassani, M.; Armaghani, D.J.; Monjezi, M.; Mohamad, E.T.; Marto, A. Blast-induced air and ground vibration prediction: A particle swarm optimization-based artificial neural network approach. *Environ. Earth Sci.* **2015**, *74*, 2799–2817. [CrossRef]
5. Duvall, W.I.; Petkof, B. *Spherical Propagation of Explosion-Generated Strain Pulses in Rock*; US Department of the Interior, Bureau of Mines: New York, NY, USA, 1959.
6. Liu, Q.; Li, N.; Duan, J.; Yan, W. The Evaluation of the Corrosion Rates of Alloys Applied to the Heating Tower Heat Pump (HTHP) by Machine Learning. *Energies* **2021**, *14*, 1972. [CrossRef]
7. Perera, A.; Azamathulla, H.; Rathnayake, U. Comparison of different Artificial Neural Network (ANN) training algorithms to predict atmospheric temperature in Tabuk, Saudi Arabia. *Mausam* **2020**, *71*, 551–560.
8. Ahmadi, M.; Naderpour, H.; Kheyroddin, A. ANN Model for Predicting the Compressive Strength of Circular Steel-Confined Concrete. *Int. J. Civ. Eng.* **2016**, *15*, 213–221. [CrossRef]
9. Kim, M.-S.; Lee, J.-K.; Choi, Y.-H.; Kim, S.-H.; Jeong, K.-W.; Kim, K.-L.; Lee, S.S. A Study on the Optimal Setting of Large Uncharged Hole Boring Machine for Reducing Blast-induced Vibration using Deep Learning. *Explos. Blasting* **2020**, *38*, 16–25.
10. Nguyen, H.; Drebenstedt, C.; Bui, X.-N.; Bui, D.T. Prediction of Blast-Induced Ground Vibration in an Open-Pit Mine by a Novel Hybrid Model Based on Clustering and Artificial Neural Network. *Nat. Resour. Res.* **2019**, *29*, 691–709. [CrossRef]
11. Azimi, Y.; Khoshrou, S.H.; Osanloo, M. Prediction of blast induced ground vibration (BIGV) of quarry mining using hybrid genetic algorithm optimized artificial neural network. *Measurement* **2019**, *147*, 106874. [CrossRef]
12. Bui, X.-N.; Choi, Y.; Atrushkevich, V.; Nguyen, H.; Tran, Q.-H.; Long, N.Q.; Hoang, H.-T. Prediction of Blast-Induced Ground Vibration Intensity in Open-Pit Mines Using Unmanned Aerial Vehicle and a Novel Intelligence System. *Nat. Resour. Res.* **2019**, *29*, 771–790. [CrossRef]
13. Tufféry, S. *Data Mining and Statistics for Decision Making*; John Wiley & Sons: Chichester, UK, 2011.
14. Singh, A.; Thakur, N.; Sharma, A. A review of supervised machine learning algorithms. In Proceedings of the 2016 3rd International Conference on Computing for Sustainable Global Development (INDIACom), New Delhi, India, 16–18 March 2016; pp. 1310–1315.
15. Khandelwal, M.; Singh, T. Prediction of blast-induced ground vibration using artificial neural network. *Int. J. Rock Mech. Min. Sci.* **2009**, *46*, 1214–1222. [CrossRef]
16. Monjezi, M.; Ghafurikalajahi, M.; Bahrami, A. Prediction of blast-induced ground vibration using artificial neural networks. *Tunn. Undergr. Space Technol.* **2011**, *26*, 46–50. [CrossRef]
17. Kim, Y.; Lee, S.S. Application of Artificial Neural Networks in Assessing Mining Subsidence Risk. *Appl. Sci.* **2020**, *10*, 1302. [CrossRef]
18. Lee, C.-W.; Park, S.-Y. Prediction of Blasting-induced Vibration at Sintanjin Area, Daejeonusing Borehole Test Blasting. *J. Korean Soc. Agric. Eng.* **2018**, *60*, 55–62.
19. Morhard, R.C. *Explosives and Rock Blasting*; Atlas Powder Company: Wilmington, DE, USA, 1987.
20. Suh, H.; Yang, K.; Kim, N.; Kim, H.; Kim, M. *SPSS (PASW) Regression Analysis*, 3rd ed.; Hannarae: Seoul, Korea, 2009.
21. Hanwha Corporation. *Hanwha Corporation Explosive Products Guide*; Hanwha Corporation: Seoul, Korea, 2017.
22. Matignon, R. *Data Mining Using SAS Enterprise Miner*; John Wiley & Sons: Hoboken, NJ, USA, 2007; Volume 638.
23. Barton, N. *Rock Quality, Seismic Velocity, Attenuation and Anisotropy*; Taylor and Francis Group: London, UK, 2006.

24. Fathollahy, M.; Uromeiehy, A.; Riahi, M. Evaluation of P-wave velocity in different joint spacing. *Bollettino di Geofisica Teorica ed Applicata* **2017**, *58*, 157–168.
25. Mielke, P.; Bär, K.; Sass, I. Determining the relationship of thermal conductivity and compressional wave velocity of common rock types as a basis for reservoir characterization. *J. Appl. Geophys.* **2017**, *140*, 135–144. [CrossRef]
26. Vinciguerra, S.; Trovato, C.; Meredith, P.; Benson, P.; Troise, C.; De Natale, G. Understanding the Seismic Velocity Structure of Campi Flegrei Caldera (Italy): From the Laboratory to the Field Scale. *Pure Appl. Geophys. PAGEOPH* **2006**, *163*, 2205–2221. [CrossRef]
27. Kingma, D.P.; Ba, J. Adam: A method for stochastic optimization. *arXiv* **2014**, arXiv:1412.6980.
28. Kaastra, I.; Boyd, M. Designing a neural network for forecasting financial and economic time series. *Neurocomputing* **1996**, *10*, 215–236. [CrossRef]
29. Sheela, K.G.; Deepa, S.N. Review on Methods to Fix Number of Hidden Neurons in Neural Networks. *Math. Probl. Eng.* **2013**, *2013*, 1–11. [CrossRef]
30. Mamaqani, B.H.M.H. *Numerical Modeling of Ground Movements Associated with Trenchless Box Jacking Technique*; The University of Texas at Arlington: Arlington, TX, USA, 2014.
31. Hecht-Nielsen, R. Kolmogorov's mapping neural network existence theorem. In Proceedings of the International Conference on Neural Networks, San Diego, CA, USA, 21–24 July 1987; pp. 11–14.
32. Hush, D.R. Classification with neural networks: A performance analysis. In Proceedings of the IEEE 1989 International Conference on Systems Engineering, Fairborn, OH, USA, 24–26 August 1989; pp. 277–280.
33. The Ministry of Land, Infrastructure and Transport in Korea. Open-Pit Blasting Design and Construction Guideline for Road Construction. Available online: http://www.molit.go.kr/USR/BORD0201/m_34879/DTL.jsp?mode=view&idx=28896 (accessed on 23 February 2021).
34. Hopler, R.B. *ISEE Blasters' Handbook*; International Society of Explosives Engineers (ISEE): Cleveland, OH, USA, 1998.
35. Nicholls, H.R.; Johnson, C.F.; Duvall, W.I. *Blasting Vibration and Their Effects on Structures*; US Department of the Interior, Bureau of Mines: New York, NY, USA, 1971.
36. IBM SPSS Software. Available online: https://www.ibm.com/analytics/spss-statistics-software (accessed on 21 April 2021).
37. Khandelwal, M.; Singh, T. Prediction of blast induced ground vibrations and frequency in opencast mine: A neural network approach. *J. Sound Vib.* **2006**, *289*, 711–725. [CrossRef]
38. Monjezi, M.; Hasanipanah, M.; Khandelwal, M. Evaluation and prediction of blast-induced ground vibration at Shur River Dam, Iran, by artificial neural network. *Neural Comput. Appl.* **2012**, *22*, 1637–1643. [CrossRef]
39. Ross, T.J. *Fuzzy Logic with Engineering Applications*; John Wiley & Sons: Chichester, UK, 2004.

applied sciences

MDPI

Article

Study on Rock Damage Mechanism for Lateral Blasting under High In Situ Stresses

Xiaofeng Huo [1], Xiuzhi Shi [1], Xianyang Qiu [1,*], Hui Chen [2], Jian Zhou [1], Shian Zhang [1] and Dijun Rao [1]

1 School of Resources and Safety Engineering, Central South University, Yuelu District, Changsha 410083, China; huoxiaofeng@csu.edu.cn (X.H.); baopo@csu.edu.cn (X.S.); csujzhou@hotmail.com (J.Z.); zhangshian@csu.edu.cn (S.Z.); raodijun@csu.edu.cn (D.R.)

2 School of Geological and Mining Engineering, Xinjiang University, Tianshan District, Urumqi 830000, China; chenhui@xju.edu.cn

* Correspondence: qiuxianyang_csu@163.com

Abstract: A 3D numerical model was presented to investigate the blast-induced damage characteristics of highly stressed rock mass. The RHT (Riedel, Hiermaier, and Thoma) model in LS-DYNA was used to simulate the blast-induced damage and its parameters were calibrated by a physical model test. Based on the calibrated numerical model, the influences of confining pressure and free surface span on the blast-induced damage characteristics were investigated. The results show that under uniaxial loading, the crater volume increases with confining pressure increases. The uniaxial static load can change the optimal burden and the critical embedding depth of charge. In stressed rock, the variation law of the crater shape affected by radial tensile fractures is opposite to that affected by reflected tensile fractures. Under the biaxial static load, the crater volume of the borehole placed on the side of the max static load is greater than the other side. The explosion crater can be improved by increasing the free surface span on the same side. Finally, it is suggested that the blasting efficiency can be improved by preferentially detonating the charge on the side of the max static load, and then the charge on the other side can be detonated with a wider free surface span.

Keywords: blast-induced damage; explosion crater; lateral blasting; in situ stress; lateral free surface

Citation: Huo, X.; Shi, X.; Qiu, X.; Chen, H.; Zhou, J.; Zhang, S.; Rao, D. Study on Rock Damage Mechanism for Lateral Blasting under High In Situ Stresses. *Appl. Sci.* **2021**, *11*, 4992. https://doi.org/10.3390/app11114992

Academic Editor: Ricardo Castedo

Received: 11 May 2021
Accepted: 24 May 2021
Published: 28 May 2021

1. Introduction

With the increase of excavation depth, the in situ stress increases gradually and plays an increasingly important role in the rock breaking by blasting. Due to the existence of in situ stress, blast-induced damage characteristics are different from those in surface and shallow sub-surface rock blasting, especially considering the role of lateral free surface in production blasting. In order to determine the parameters of borehole layout and blasting parameters in highly stressed rock mass as well as for safe and efficient production, it is necessary to investigate the blast-induced damage characteristics in the static-dynamic stress field.

A lot of studies have been done in the area of rock breaking due to blasting considering the effect of in situ stress. Kutter et al. [1] analytically and experimentally investigated the influence of in situ stress on the blast-induced rock fracture. The results showed that the cracks induced by blasting stress wave and gas pressure grow preferably in the direction of maximum principal stress of superimposed stress fields. Zhang and Peng [2–5] studied the crater blasting under different confining pressures via theoretical analysis and physical model tests. The outcomes showed that the crater shape becomes oval with the long axis aligned on the loading direction, and the open angle in this direction and the crater volume is greater with the increase of confining pressure under uniaxial static load. Based on the fracture mechanics and the rock damage failure criterion, Xiao et al. [6] calculated rock fragmentation induced by blasting under high stress. It is concluded that the release of strain energy in the highly stressed rock mass is helpful to improve the breaking effect.

Yang et al. [7] conducted caustics experiments to investigate the propagation characteristics of blast-induced cracks in the dynamic-static stress field. The results indicated that the in situ stress has an important effect on crack propagation induced by blasting and the crack propagation is restrained when the crack propagation direction is perpendicular to the direction of in situ stress. Hu and Lu [8,9] studied the formation and propagation of crack induced by presplitting blasting in highly stressed rock mass via a mathematical model and concluded that the in situ stress can restrain the development of cracks between the presplitting holes when the in situ stress is perpendicular to the crack face. Yang and He [10,11] experimentally investigated the influence of confining pressure and ratios of horizontal-to-vertical pressure on the blast-induced rock fracture. The results showed that the direction of crack growth was largely controlled by the hoop tensile stress and biaxial pre-pressure ratio.

As a research tool, the numerical modeling method has been widely used to investigate the blast-induced damage characteristics of rock. Donzé et al. [12] used the discrete element method (DEM) to study the blast-induced radial fractures under confining pressure and found that the radial fractures induced by blasting tend to grow in the direction of maximum principal stress. Yilmaz et al. [13] investigated the blast-induced damage characteristics under different in situ stresses via a 3D FLAC (Fast Lagrangian Analysis of Continua) analysis. The results indicated that the development of fractures around the borehole is governed by the maximum principal stress and it is more obvious with the increase of the difference between the two principal stresses. Xie [14] used LS-DYNA to study the damage characteristics in cutting blasting under different in situ stresses. The results showed that with the increase of in situ stress, the damage zone becomes smaller. With the increase of the lateral pressure coefficient, the extending direction of the tensile damage zone becomes more obvious, which causes a great challenge to the cutting blasting excavation in deep rock masses. Yi, Jayasinghe, Ma and Li [15–18] used LS-DYNA to investigate the influence of in situ stress on the blast-induced cracks. Their results showed that the crack propagation trends towards the direction of maximum compressive pressure. Han, Wei and Deng [19] used a numerical model to study the contour control blasting under different in situ stresses. The result indicated that the in situ stress could affect the crack evolution and direction, and the quality of the contour surface is hard to control in highly stressed rock masses.

The studies mentioned above mainly focus on the plane problems of blasting under static load. However, the three-dimension propagation of stress wave induced by explosives, the charge length, and the detonation velocity of explosive cannot be considered in these 2D plane strain models. The above factors can be involved in a 3D model analysis to obtain more realistic results. Additionally, the key factor of the free surface is rarely considered, especially the lateral free surface, which plays an important role in the rock breaking by blasting. In this study, a 3D blasting model of coupling static and dynamic loads is developed in LS-DYNA and the model parameters are calibrated by the physical model test. Subsequently, the calibrated numerical model is used to simulate the blast-induced damage considering the roles of in situ stress and lateral free surface. Based on the damage distribution, the blast-induced damage characteristics and the explosion craters under different static loads and free surface spans are analyzed.

2. Numerical Modeling

2.1. Constitutive Model Parameters and Validation

2.1.1. Numerical Model for Physical Model Test

To verify the material model and apply it to the subsequence simulation of lateral blasting under static load, a physical single-hole crater blasting model test was conducted firstly to calibrate the material parameter, as shown in Figure 1a. The whole model is $400 \times 400 \times 200$ mm. A borehole with a diameter of 8.0 mm and a length (L) of 40 mm is drilled in the center and the explosive with a diameter (d) of 8.0 mm and a length (l) of 12 mm is charged in the borehole. The cemented sand, which is composed of ordinary

Portland cement (PC32.5), uniform-grained sand and water in the mass ratio of 3:3:1, is used as the model material to study the blast-induced damage of rock. The material mechanical parameters are determined by averaging the measured data from six mortar cubic blocks. The density ρ_0 is 2456 kg/m^3; the compressive strength f_c is 48.3 MPa and the elastic modulus E is 32.36 GPa; Poisson's ratio μ is 0.24; P-wave velocity v_p is 3828 m/s. According to the physical model, the single-hole crater blasting numerical model was developed for comparison with the test results, as shown in Figure 1b. The model consists of rock, explosive, and stemming. The size of the numerical model is same as the physical model and the total number of the meshed elements is 0.56 million, where the numerical convergence tests has been carried out and the calculation results of the model are convergent and accurate.

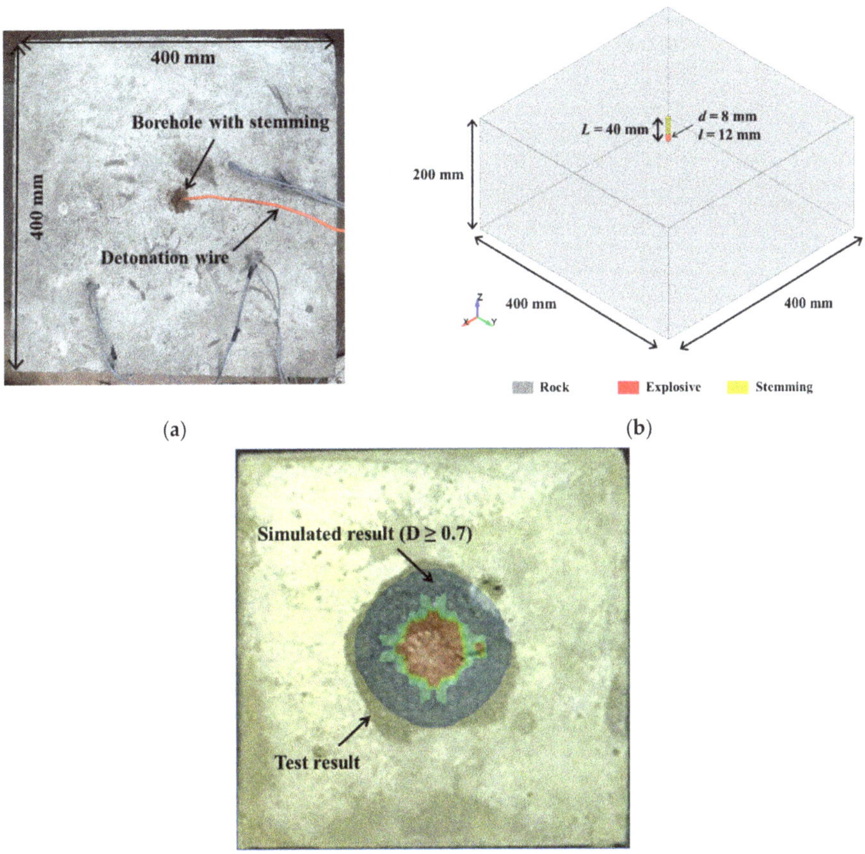

(a) (b)

(c)

Figure 1. Single-hole crater blasting model and the comparison of explosion craters: (**a**) physical model; (**b**) numerical model; (**c**) the comparison of experimental and simulation results.

2.1.2. RHT Material Model for Rock

In this study, the dynamic response and damage process of rock mass were simulated by Riedel, Hiermaier, and Thoma (RHT) in LS-DYNA, which were widely used in the numerical simulation of concrete and rock [3,20–25]. The damage degree of the RHT material model is given by $D = \sum \frac{\Delta\varepsilon^p}{\varepsilon^f}$, in which $\Delta\varepsilon^p$ is the accumulated plastic strain and ε^f is the failure strain. The detailed introduction of the RHT constitutive model can be found in [20]. Based on the tested mechanical parameter, the following sections obtain the other material parameters through empirical formulas or related literatures.

1. Strain rate parameters

The effect of strain rate on the rock strength is clear. The strain rate strength factor $F_r(\dot{\varepsilon}_p)$ is expressed as [21]:

$$F_r(\dot{\varepsilon}_p) = \begin{cases} (\dot{\varepsilon}_p/\varepsilon_0^c)^{\beta_c} & P \geq f_c/3 \\ \frac{P+f_t/3}{f_c/3+f_t/3}(\dot{\varepsilon}_p/\varepsilon_0^c)^{\beta_c} - \frac{P-f_c/3}{f_c/3+f_t/3}(\dot{\varepsilon}_p/\varepsilon_0^c)^{\beta_t}) & -f_t/3 < P < f_c/3 \\ (\dot{\varepsilon}_p/\varepsilon_0^t)^{\beta_t} & P \leq -f_t/3 \end{cases} \quad (1)$$

where $\dot{\varepsilon}_p$ is the strain rate; $\dot{\varepsilon}_0^c$ and $\dot{\varepsilon}_0^t$ are the compressive and tensile reference strain rates, which are 3×10^{-5} s^{-1} and 3×10^{-6} s^{-1}, respectively; P is the hydrostatic pressure; f_c and f_t are the uniaxial strengths in compression and tension. The strain rates in compression (β_c) and in tension (β_t) are constant for the material and can be calculated by

$$\beta_c = \frac{4}{20+3f_c}, \ \beta_t = \frac{2}{20+f_c} \quad (2)$$

where the unit is MPa, and then β_c and β_t are determined as 0.024 and 0.029.

2 Failure surface parameters

A and N are the constants in the failure surface and can be obtained by:

$$\sigma_f^*(P^*, F_r) = A\left(P^* - F_r/3 + (A/F_r)^{-1/N}\right)^N 3P^* \geq F_r \quad (3)$$

where $\sigma_f^*(P^*, F_r)$ is the normalized strength and can be calculated by $\sigma_f^* = \frac{\sigma_f}{f_c}$; P^* is the normalized hydrostatic pressure and can be calculated by $P^* = \frac{P}{f_c}$; F_r is the strain rate strength factor.

When the rock is in a quasi-static state, $\dot{\varepsilon}_p = 3.0 \times 10^{-5}$ s^{-1}, and then $F_r = 1$ can be calculated by Equation (1). The rock strength under different confining pressures can be calculated by empirical equation of Hoek-Brown [26] and the fitting equations for rock material are as follows [27]:

$$\sigma_1 = \sigma_3 + 48.3\left(24\frac{\sigma_3}{48.3} + 1\right)^{1/2} \quad (4)$$

The axial stress at failure (σ_1) under different confining pressures ($\sigma_2 = \sigma_3$) can be calculated by Equation (4). The results are shown in Table 1. Subsequently, the pressure $P = \frac{1}{3}(\sigma_1 + \sigma_2 + \sigma_3)$, equivalent stress at failure $\sigma_f = \sqrt{\frac{1}{2}\left[(\sigma_1 - \sigma_2)^2 + (\sigma_2 - \sigma_3)^2 + (\sigma_3 - \sigma_1)^2\right]}$, P^*, and σ_f^* can be obtained and are shown in Table 1. Based on Equation (3), A = 2.439 and N = 0.7528 can be obtained by substituting the values of P^* and σ_f^* corresponding to the confining pressures of 5 MPa and 30 MPa.

Table 1. Mechanical parameters under different confining pressures.

σ_2=σ_3/MPa	σ_1/MPa	P/MPa	σ_f/MPa	P^*	σ_f^*
0	48.3	16.1	48.3	0.33	1.00
5	95	35	90	0.73	1.87
10	128	49	118	1.02	2.44
20	180	73	160	1.52	3.31
30	223	94	193	1.95	3.99
50	296	132	246	2.73	5.08

3 *P*-α compaction EOS (Equation of State) parameters

The *P*-α compaction of RHT is given by

$$P(\rho, e) = \frac{1}{\alpha}\left((B_0 + B_1\mu_0)\alpha\rho e + A_1\mu_0 + A_2{\mu_0}^2 + A_3{\mu_0}^3\right) \mu_0 > 0 \quad (5)$$

where B_0 and B_1 are the material constants; α and ρ are the initial porosity and density; e is the specific internal energy; μ_0 is the volumetric strain; A_1, A_2, A_3 are the polynomial coefficients.

A_1, A_2, and A_3 can be calculated by formulas in [24],

$$A_1 = \alpha\rho c^2 A_2 = \alpha\rho c^2(2k-1)A_3 = \alpha\rho c^2[(3k-1)(k-1)] \quad (6)$$

where c is the wave speed; k is the material constant. A_1, A_2 and A_3 can be calculated as 36.0, 40.4 and 4.8 GPa, respectively.

The minimum damaged residual strain (ε_p^m) can be determined by the calibration of the physical model test. The remaining model parameters in this study, which are not sensitive to the numerical results, are referred to the cemented sand parameters in the literatures [28,29]. The determined RHT parameters are listed in Table 2.

Table 2. RHT parameters for rock mass.

Parameter	Value	Parameter	Value
Density ρ_0	2456 kg/m^3	Compressive strain rate β_c	0.024
Shear modulus G	13 GPa	Tensile strain rate β_t	0.029
ONEMPA [1]	1.0×10^6	Pressure influence on plastic flow in tension PTF	0.001
Eroding plastic strain EPSF	2.0	Compressive yield surface g_c^*	0.53
Polynomial EOS B_0	1.22	Tensile yield surface g_t^*	0.7
Polynomial EOS B_1	1.22	Shear modulus reduction factor ξ	0.5
Polynomial EOS T_1	35 GPa	Damage parameter D_1	0.04
Failure surface A	2.439	Damage parameter D_2	1
Failure surface N	0.7528	Minimum damaged residual strain ε_p^m	12×10^{-3}
Compressive strength f_c	48.3 MPa	Residual surface parameter A^f	1.6
Relative shear strength f_s^*	0.18	Residual surface parameter n^f	0.61
Relative tensile strength f_t^*	0.1	Gruneisen gamma GAMMA	0
Lode angle Q_0	0.681	Hugoniot polynomial coefficient A_1	3.6×10^{10}
Lode angle B	0.0105	Hugoniot polynomial coefficient A_2	4.04×10^{10}
Polynomial EOS T_2	0	Hugoniot polynomial coefficient A_3	0.48×10^{10}
Ref. compressive strain rate $\dot{\varepsilon}_0^c$	3.0×10^{-5}	Crush pressure P_{el}	16.1 MPa
Ref. tensile strain rate $\dot{\varepsilon}_0^t$	3.0×10^{-6}	Compaction pressure P_{co}	6 MPa
Break compressive strain rate $\dot{\varepsilon}^c$	3.0×10^{25}	Porosity exponent N_p	3
Break tensile strain rate $\dot{\varepsilon}^t$	3.0×10^{25}	Initial porosity α_0	1

[1] ONEMPA is the unit conversion factor defining 1 MPa in the pressure units used.

2.1.3. Material Identifications of Charge, Air and Stemming

The charge is modeled by MAT_HIGH_EXPLOSIVE_BURN in LS-DYNA [30]. The JWL(Jones–Wilkins–Lee) EOS are given by:

$$P = A\left(1 - \frac{\omega}{R_1 V}\right)e^{-R_1 V} + B\left(1 - \frac{\omega}{R_2 V}\right)e^{-R_2 V} + \frac{\omega E}{V} \quad (7)$$

where P is the pressure, A, B, R_1, R_2, and ω are constants, V is the specific volume, and E is the internal energy with an initial value of E_0.

In this study, the explosive is a mixture of RDX (Hexogen), PETN, DDNP, et al. The estimation of JWL parameters of explosive is complex and costly [31], so the parameters refer to similar explosive parameters [32]: A = 524 GPa, B = 7.68 GPa, R_1 = 4.2, R_2 = 1.1, ω = 0.34, E_0 = 8.5 GPa. The charge density is 1.6×10^3 kg/m^3 and the detonation velocity is 6950 m/s.

The air is modeled by MAT_NULL in LS-DYNA, and the corresponding EOS is given by [30]:

$$P = C_0 + C_1 u + C_2 u^2 + C_3 u^3 + \left(C_4 + C_5 u + C_6 u^2\right) e \quad (8)$$

where C_0, C_1, C_2, C_3, C_4, C_5 and C_6 are polynomial coefficients; $\mu = \frac{\rho}{\rho_0} - 1$ is specific volume; e is the internal energy per volume and has the unit of pressure, Pa. In this study, the air is modeled as an ideal gas by setting $C_0 = C_1 = C_2 = C_3 = C_4 = 0$ and $C_5 = C_6 = 0.4$, and the initial internal energy per volume is set to 0.25 J/cm^3 [25].

The stemming is modeled by MAT_SOIL_AND_FOAM in LS-DYNA and its parameters are shown in Table 3.

Table 3. Parameters for stemming.

Density ρ'	Poisson's Ratio ν	Shear Modulus E_T	Cohesive Force c	Friction Coefficient μ	Internal Friction Angle φ
2600 kg/m^3	0.19	16 GPa	0.018 MPa	0.7	35°

2.2. Constitutive Model Parameters and Validation

Figure 1c shows the comparison of explosion craters between the physical test results and the simulated results. In the simulated results, the critical damage is set to 0.6 or 0.7 according to the previous studies [15,22,23,33]. In this study, a critical value D of 0.7 is reasonable for the consistency between the physical test results and the numerical results. It can be found that the crater boundary in the simulated results is similar to that in the test results. Therefore, the calibrated numerical model is able and feasible to study the blast-induced damage characteristics.

2.3. Numerical Model for Lateral Blasting under Static Load

A numerical model with dimensions of 400 × 400 × 200 mm was built to simulate the dynamic response and damage evolution of lateral blasting, as shown in Figure 2. It is commonly seen that rectangular cavern is usually generated by production boreholes with an excavation method of lateral caving with large-diameter long-hole blasting or by tunneling excavation for its advantages of simple procedures, high excavation efficiency and convenient support measures [22,34]. Therefore, the cavern prototype was set as a rectangle in this study, which also has important enlightening significances for other shapes. In the center of the model, a rectangular cavern with a size of span X × span Y × 200 mm is placed to form the lateral free surface and a borehole with a diameter of 8.0 mm is placed near the cavity with a distance of W. It should be noted that the span X and span Y are no more than 1/3 of the model size of 400 mm to decrease the influence of boundary on the stress distribution. The explosive with a diameter of 6.3 mm and a length of 20 mm is charged in the hole centrally. The ends of the borehole are filled with stemming. Static stresses, P1 and P2, are applied to the four external boundaries of the model in X and Y directions respectively using a dynamic relaxation scheme, and the four sides inside the model are specified as free surfaces. After the stress initialization, the charge is loaded in the model and detonated. The numerical model is meshed by hexahedral elements, with a size of 4 mm, which is small enough to avoid any wave distortion [22]. The total number of meshed elements is 0.5 million. In this study, in order to monitor the damage distribution in the rock mass, cut Y1 and cut Z1 are selected, as shown in Figure 2. The evolution process of blast-induced damage is completed before 100 μs, which is set to the calculation termination time.

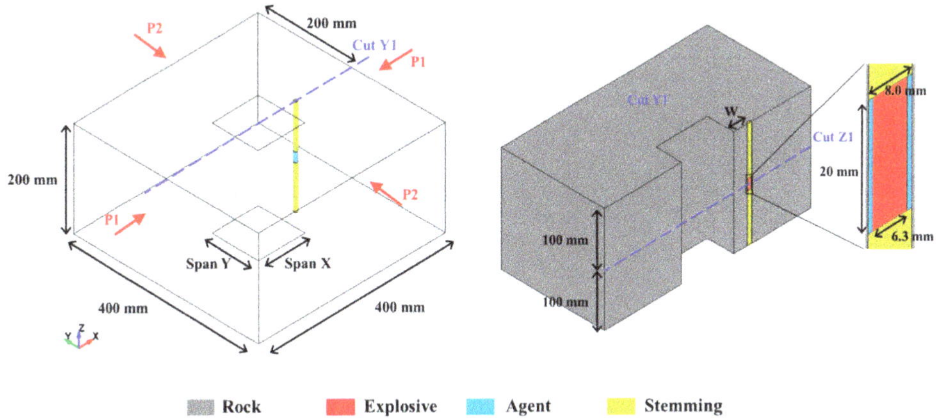

Figure 2. Numerical model of single hole blasting with a lateral free surface.

3. Results and Analysis

3.1. Influence of Uniaxial Static Load on the Damage Distribution under Different Burdens

In this section, a span X of 100 mm and a span Y of 100 mm are applied to the numerical model, and then the simulations of uniaxial loading of P1, uniaxial loading of P2, and biaxial loading were conducted. The damage contours of cut Z1 at 100 μs were extracted from the numerical results. The variation characteristics of damage distribution, the volume *V* and the shape of the crater were studied in detail in this section.

According to the elastic mechanics, on the right side of the rectangle cavern, the X-direction stress is small, and the stress field is governed by the Y-direction stress, especially near the free surface. Therefore, the blast-induced damage is mainly affected by the original Y-direction stress field. Figure 3 shows the Y-direction elastic stress fields (σ_y) under different static loads (W = 4 cm). For the stress σ_y, its sign is positive in tension and negative in compression. As seen from the Y-direction stress contours, the rectangle cavern induces stress concentration near the free surface. For P1 = 5 MPa, as shown in Figure 3a, there is a large tensile stress zone on the left side of the borehole, especially near the free surface, and a small tensile stress zone on the right side of the borehole. The maximum tensile stress is 6.6 MPa near the free surface and σ_y decreases to around 2 MPa on the left side of the borehole. For P2 = 5 MPa, as shown in Figure 3b, the excavation zone is in a compressive stress field. The σ_y is maximum around the free surface (around 12 MPa), and it decreases to around 7 MPa on the left side of the borehole. For P1 = P2 = 5 MPa, as shown in Figure 3c, the excavation zone is also in a compressive stress field, but the compressive stress field is weakened, and the distribution changes a lot. The maximum σ_y transfers from the free surface, where the σ_y decreases to around 4 MPa, to the four corners of the cavern. The above static stress field analysis is beneficial to understanding the coupling mechanism of static load and blasting stress wave load on the rock damage characteristics in the subsequent dynamic analysis.

To evaluate the influence of P1 on the damage distribution due to blasting, five cases of uniaxial static loads, P1 = 0, 2, 5, 8, and 10 MPa, were first conducted in this section. Figure 4 shows the damage contours for different P1 with different burdens at cut Z1. As mentioned in Section 2.2, the elements with a damage level above 0.7 are regarded as severe damage zone and form the explosion crater. In the case of P1 = 0 MPa, blast-induced severe damage zones ($D \geq 0.7$) are widely distributed and can form craters from the charge center to the free surface when the burden W is no more than 4 cm. When W is more than 5 cm, the blast-induced severe rock damage ($D \geq 0.7$) mainly distributes around the explosive and little severe damage zone covers the free surface, but the two zones are not connected. Thus, in the cases of W = 5 cm and W = 6 cm, only blasting cavities are formed around the charge but no crater is formed by the blasting. Therefore, the burden should be no more than 4 cm to form an explosion crater in the case of P1 = 0 MPa. In the case of P1 = 2 MPa, the damage zones are enlarged for each burden, but the severe damage zone ($D \geq 0.7$) around the charge and the damage zone near the free surface are still separated when W = 5 cm and W = 6 cm, which indicates no crater is formed. When P1 increases to 5 MPa, the damage zones are further enlarged, and the two zones begin to connect for W = 5 cm but not for W = 6 cm. In the case of P1 = 8 MPa and 10 MPa, with the increase of static load, the damage zones become larger. Especially for W = 5 cm, the severe damage zone is clearly enlarged near the free surface, thus a crater is formed. It should be noted that there is still a large low-level damage zone ($D < 0.7$) between the blasting cavity and the free surface for W = 6 cm, as a result, the explosion crater cannot be formed.

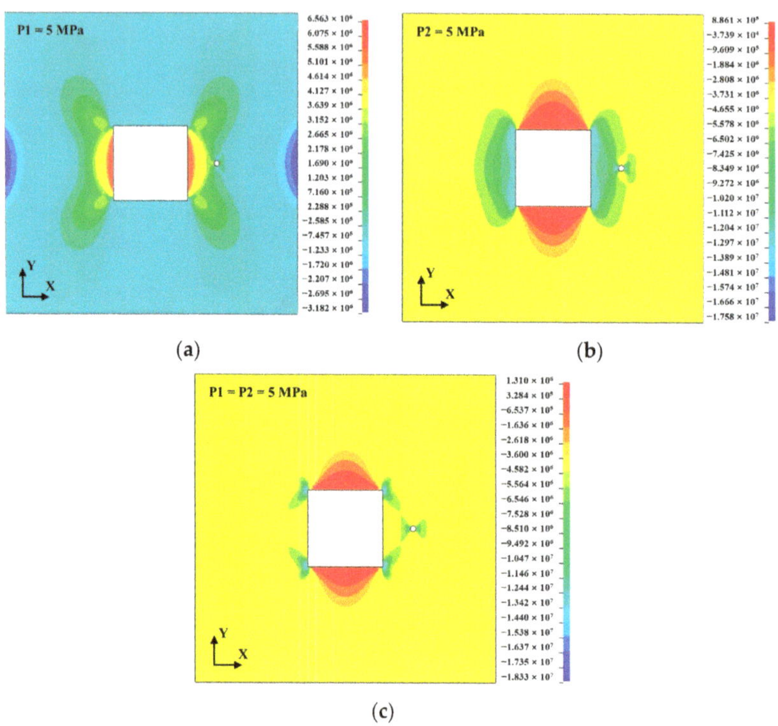

Figure 3. Y-direction elastic stress fields (σ_y) under different static loads (W = 4 cm): (**a**) P1 = 5 MPa; (**b**) P2 = 5 MPa; (**c**) P1 = P2 = 5 MPa.

In order to evaluate the explosion crater clearly, the crater volume V is measured by counting the high-level damage elements ($D \geq 0.7$) and summing their volumes. The crater volumes for each burden under different static loads are shown in Figure 5. It can be found that with P1 increases, the volume of explosion crater tends to increase. When P1 is less than 8 MPa, the V for W = 2 cm is the smallest (no crater for W = 5 and 6 cm, as shown in Figure 4) due to excessive dissipation of the explosion energy into the atmosphere and the V for W = 3 cm is the largest, which indicates that the optimal burden is 3 cm. However, when P1 is more than 8 MPa, the burden of 4 cm is optimal because its corresponding crater volume is the largest. It can be clearly seen that the V for W = 2 cm is not sensitive to the static load, but the others vary greatly with changing P1, especially for W = 4 cm, W = 5 cm and W = 6 cm when P1 $\geq$ 5 MPa. The above results are due to the rapid expansion of the damage zone near the free surface for W = 5 cm and W = 6 cm (as shown in Figure 4). It should be noted that the V for W = 6 cm is the volume sum of the blasting cavity and the damage zone near the free surface, but not the crater volume (as shown in Figure 4). It can be concluded that the P1 can change the optimal burden of charge and increase the critical embedding depth of the charge.

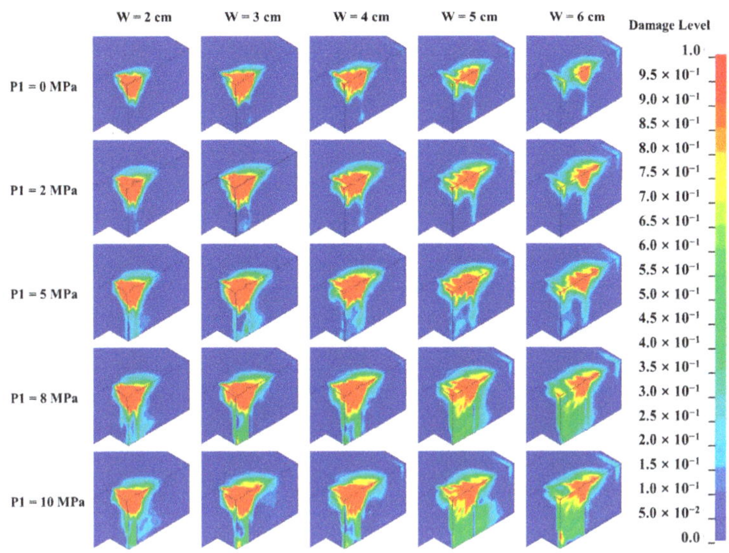

Figure 4. Damage contours for each burden under different static loads (P1) at cut Z1.

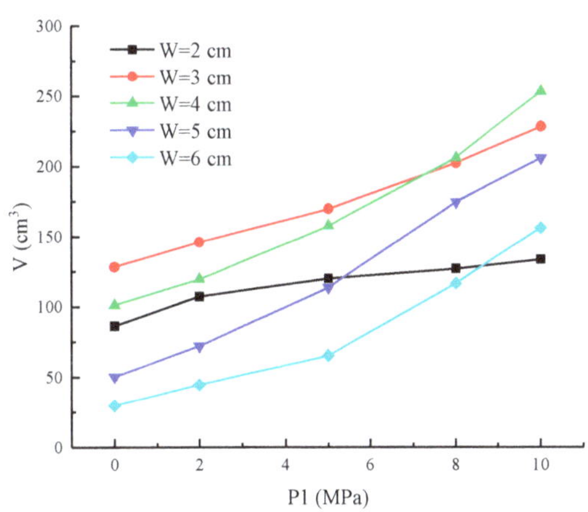

Figure 5. Crater volumes for each burden under different static loads (P1).

In order to investigate the variation in the shape of the explosion crater, the craters (formed by the elements with $D \geq 0.7$) for W = 4 cm under different P1 are plotted in Figure 6. In the XY plane, the shape of the explosion craters is similar to a triangle and expands with the increase of P1. The diameter of the crater in the Y direction also becomes larger with the increase of P1 at different depths (X direction), especially in the top of the crater, where a new damage zone is generated. The above results are induced by the combined effect of the Y-direction tensile component of the stress wave and the Y-direction tensile stress field (as shown in Figure 3a). In the XZ plane, there is a clear trend that with the increase of P1, the diameter in the Z direction becomes larger, especially when

P1 = 8 MPa. This is because with the increase of Y-direction stress field (σ_{y1}) induced by P1, the combined effect of the Y-direction tensile component of incident stress wave (σ_{yI}) and Y-direction stress field (σ_{y1}) is intensified and promotes the initiation and propagation of radial tensile fracture at point A, as shown in Figure 7a. Besides, some damage zones appear in the left side of the borehole due to the stress concentration, but they expand little in the Y direction, as shown in the YZ plane. In the YZ plane, when P1 = 5 MPa, the long axis of the bottom circle of the crater is in the Y direction. This is because the superposition of the Y-direction component of reflected tensile stress (σ_{yR}) and Y-direction stress field (σ_{y1}) at point B (as shown in Figure 7b) increases the Y-direction dimension of reflected tensile damage zone around the free surface, which is also shown in the XY plane. However, with the increase of P1, the long axis of the bottom circle of the crater transfers from Y direction to Z direction due to the faster growth of the diameter in the Z direction, which is consistent with the results in the XZ plane and can be illustrated by Figure 7a. The result is also consistent with the law that the long axis of blast-induced damage is parallel to the max principal compressive stress (Z direction). It can be concluded that the crater shape is governed by the reflected tensile fractures when P1 $\leq$ 5 MPa, but governed by the radial tensile fractures when P1 $\geq$ 8 MPa.

Another five cases of uniaxial static loads, P2 = 0 MPa, P2 = 2 MPa, P2 = 5 MPa, P2 = 8 MPa and P2 = 10 MPa, were simulated to investigate the effect of load direction on damage distribution. Figure 8 shows the damage contours at cut Z1 and the crater volumes for different burdens under different P2. In the case of P2 = 2 MPa, the crater volumes are enlarged for W = 2 cm, 3 cm, and 4 cm, but shrunk for W = 5 and 6 cm, as shown in Figure 8b. There is still no crater formed by the blasting for W = 5 and 6 cm. as shown in Figure 8a. The increase of the crater volumes for W = 2, 3, and 4 cm is mainly induced by the increase of reflected tensile fractures around the free surface, where the combined effect of the X-direction component (σ_{xR}) of reflected tensile wave and the compressive stress field (σ_{y2}) induced by P2 promotes the damage development at point C and D, and the increase of radial tensile damage zone, where the combined effect of the Z-direction tensile component (σ_{zL}) of incident wave and σ_{y2} promotes the damage development at point B, as shown in Figure 9. However, the reductions of the crater volumes for W = 5 cm and 6 cm are induced by the reduction of radial damage zones distributed around the charge, where the volumes of blasting cavities are mainly restrained by σ_{y2}, especially at point A, as shown in Figure 9. In the case of P2 = 5 MPa, the crater volumes are increased when W = 2 cm, 4 cm, 5 cm, and 6 cm but reduced when W = 3 cm. For W = 3 cm, this may be because the increase of reflected tensile fractures around the free surface is smaller than the reduction of the radial tensile damage zone around the charge. There is a clear increase of the damage zones for W = 4, 5, and 6 cm, which is induced by the great increase of fractures around the free surface. Especially for W = 5 cm, the fractures around the free surface are clearly enlarged and begin to connect with the blasting cavity formed by the damage zone around the charge. For W = 6 cm, the combined effect of σ_{xR} and σ_{y2} is enhanced due to the intensification of the latter, and some damage zones extend from the free surface to the borehole. However, the damage zones only distribute along the Y direction but expand little in the Z direction, so the crater is hard to form. In the case of P2 = 8 MPa and 10 MPa, with the increase of static load, the damage zones become larger, except for W = 3 cm, where the damage zone distribution along the Z direction is reduced, as shown in Figure 8a. It should be noted that the crater is still not formed for W = 6 cm. For each W, the fractures tend to extend along the Y direction with P2 increases, which is consistent with the law that the long axis of the blast-induced damage zone is parallel to the max principal stress (Y direction).

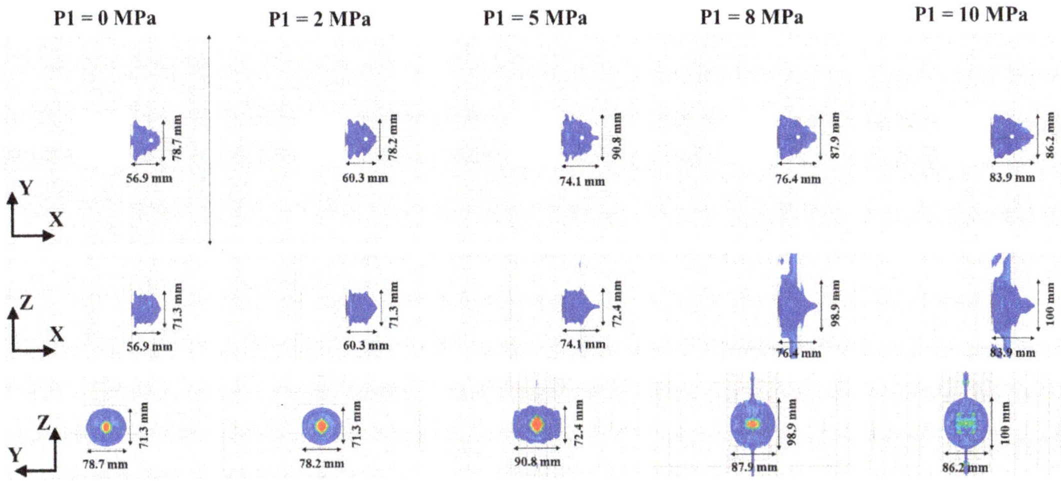

Figure 6. The craters for W = 4 cm under different P1.

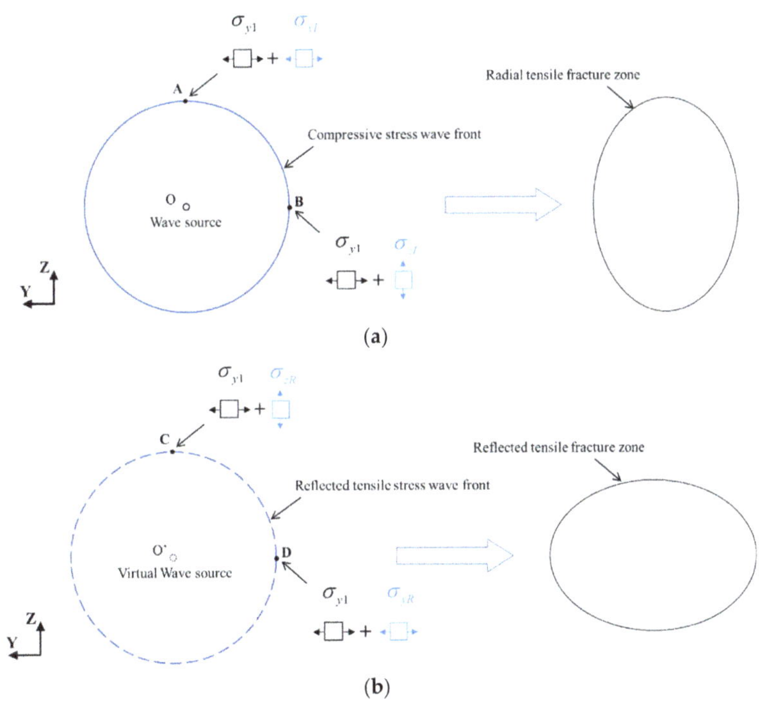

Figure 7. Schematics of superposition of original stress field (P1) and stress wave and its influence on the crater shape: (**a**) radial tensile fracture induced by incident compressive stress wave; (**b**) reflected tensile fracture induced by reflected tensile stress wave.

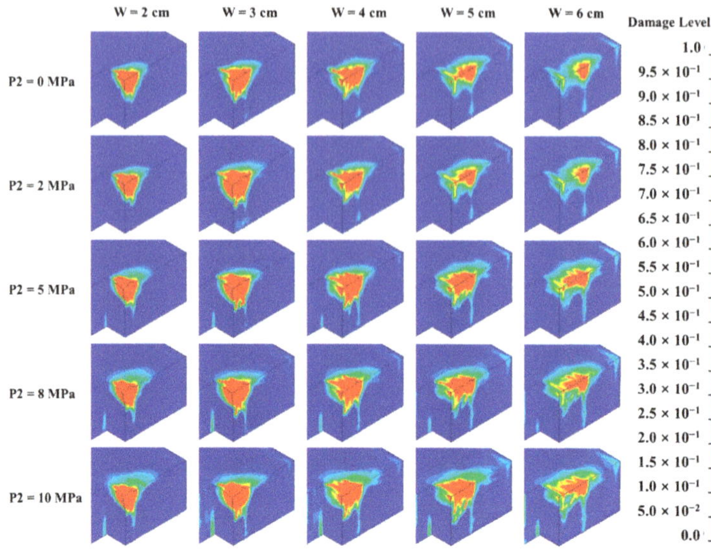

(**a**)

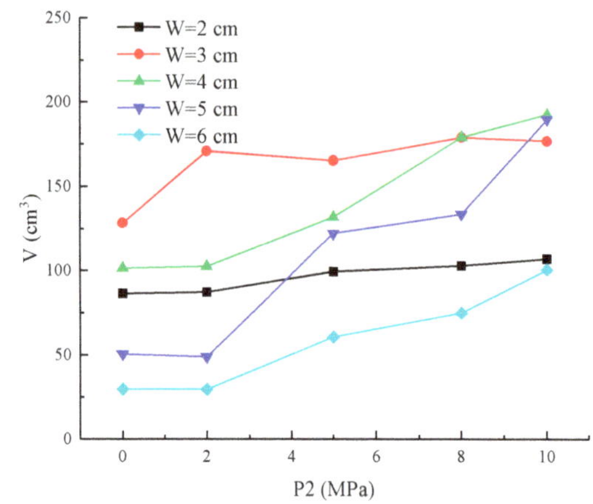

(**b**)

Figure 8. Damage contours at cut Z1 and crater volumes for each burden under different P2: (**a**) damage contours; (**b**) crater volumes.

With the increase of P1 or P2, the crater volume can be enlarged. However, the increase of *V* with P1 is more than that with P2. For example, when *W* = 4 cm, the *V* for P1 increases by 17.1, 25.8, 26.8 and 60.3 cm^3 compared with that for P2 when the stress level is 2, 5, 8 and 10 MPa, respectively. In other words, when the static load values are the same, the *V* is increased by 15–31% for P1 compared with P2. The results show that the effect of P1 on the increase of *V* is greater than that of P2.

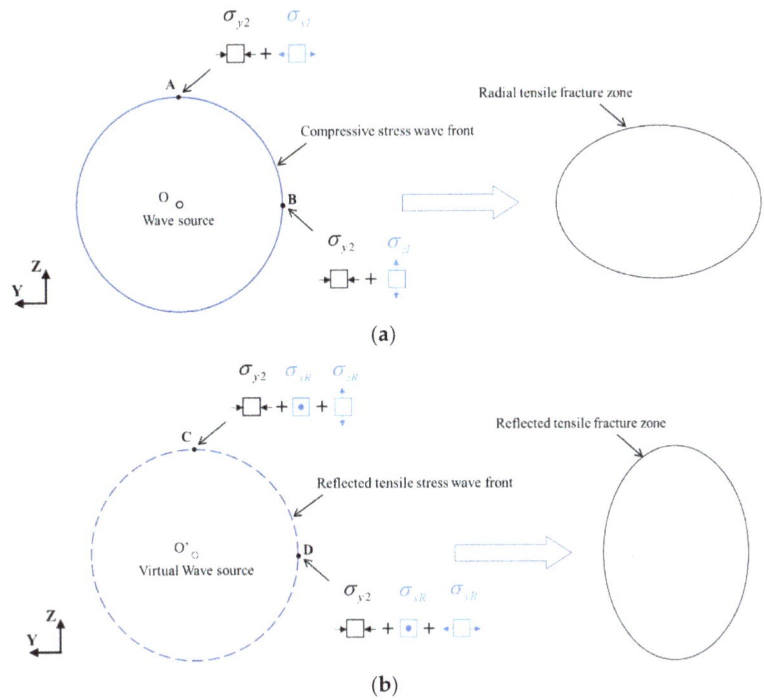

Figure 9. Schematics of superposition of original stress field (P2) and stress wave and its influence on the crater shape: (**a**) radial tensile fracture induced by incident compressive stress wave; (**b**) reflected tensile fracture induced by reflected tensile stress wave.

The shapes of the craters for W = 4 cm under different P2 are plotted in Figure 10a. In the XY plane, the shape of the explosion crater gradually changes from a triangle to a trapezoid and has a significant expansion in the Y direction at different depths due to the directional effect of the Y-direction compressive stress field. However, the depth of the explosion crater is reduced a little due to the volume shrink at the top of the crater, especially in the case of P2 = 10 MPa, where the crater is mainly distributed on the left side of the borehole. In the XZ plane, near the free surface, the Z-direction diameter gradually increases under the combined effect of reflected tensile wave and σ_{y2}. In the YZ plane, the shape of the crater becomes an oval and its long axis is in the Y direction for P2 = 2 MPa, which obeys the law that the long axial of blast-induced damage zone is parallel to the max principal stress. However, when P2 $\geq$ 5 MPa, the shape tends to expand in the Z direction. This is because the reflected tensile fracture zone becomes the dominant factor affecting the crater shape. As shown in Figure 9b, σ_{y2} is perpendicular to σ_{xR} and Z-direction tensile stress component (σ_{zR}) of reflected wave at point C and point D, and it is conductive to the growth of reflected tensile fractures induced by σ_{xR}. However, it is opposite to the Y-direction tensile stress component (σ_{yR}) at point D and it will restrain the formation of reflected tensile fractures induced by σ_{yR}. As a result, the Z-direction reflected tensile fractures are easier to propagate. To study the effect of P2 on the radial tensile fracture zone distribution, as shown in Figure 10a, section A-A at 1 cm to the left side of the borehole is selected, which is away from the free surface, and its damage zone is mainly governed by radial tensile fractures.

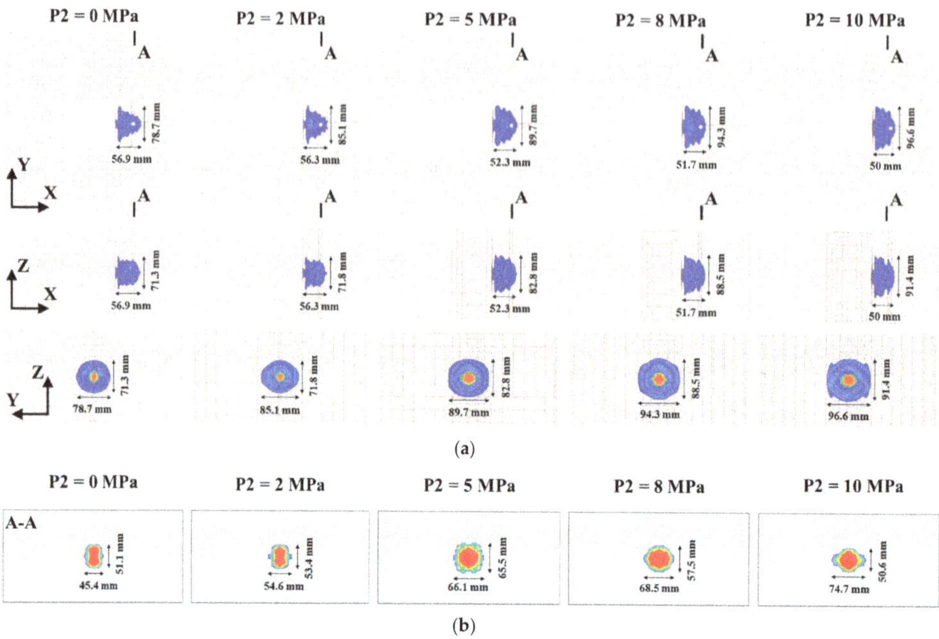

Figure 10. The craters for W = 4 cm under different P2: (**a**) perspective view of craters at different planes; (**b**) section A-A.

In section A-A, it can be clearly found with P2 increases, the Z-direction dimension of the crater reduces (P2 $\geq$ 5 MPa) and the Y-direction dimension of the crater increases gradually, as shown in Figure 10b. The above results are caused by the coupling mechanism of σ_{y2} and the incident compressive stress wave, which is illustrated by Figure 9a. At point A, the original static compressive stress field (σ_{y2}) is opposite to the Y-direction tensile stress component (σ_{yI}) and it will prevent the formation of radial tensile fractures induced by σ_{yI}. However, at point B, σ_y is perpendicular to σ_{zI}, and it is conductive to the growth of reflected tensile fractures induced by σ_{zI}. Therefore, the radial tensile fracture zone is an ellipse with a long axis in the Y direction.

3.2. Influence of Biaxial Static Load on the Damage Distribution

To investigate the characteristic of damage distribution under biaxial loading, W = 4 cm and P1 = 5 MPa were kept, 11 cases of P2, 0, 1, 2, 3, 4, 5, 6, 7, 8, 9, and 10 MPa, were simulated to investigate the effect of the pressure coefficient $\lambda = \frac{P2}{P1}$ on damage distribution. The borehole is placed on the right side of the cavern, named as case I, or the top side of the cavern, named as case II.

Figure 11a shows the shapes of craters for W = 4 cm under biaxial loads with different λ for case I. With the λ increases, the shapes in the three planes are shrinking in the overall trend. In the XY plane, the shape transforms from a trapezoid to a triangle, and the X-direction dimension reduces much to cause the decrease of the depth of the crater. In the XZ plane, away from the free surface, the Z-direction and X-direction dimensions are both reduced due to the increasing Y-direction compressive stress field. In the YZ plane, two damage zones distribute the leftmost side and the rightmost side along the Z direction for λ = 0, but they disappear for λ = 0.4. For λ = 0, the two damage zones are formed by the combined effect of σ_{yR} and σ_{y1} induced by P1. However, for λ = 0.4, the addition of the new Y-direction compressive stress field (σ_{y2}) induced by P2 will neutralize part of the tensile stress and prevent the formation of the two damage zones. When $\lambda \geq 0.4$, the shape is getting flatter due to the increase of the additional compressive stress field. For case I, the crater volume V reduces monotonically with the increase of λ, as shown in Figure 11c.

It should be noted that the law only applies to the case of small λ. When λ is large enough, the Y-direction compressive stress field induced by lateral pressure P2 will govern the damage distribution around the surface and away from the charge, and the crater volume may be increased with the increase of λ, as described in the uniaxial loading case of P2.

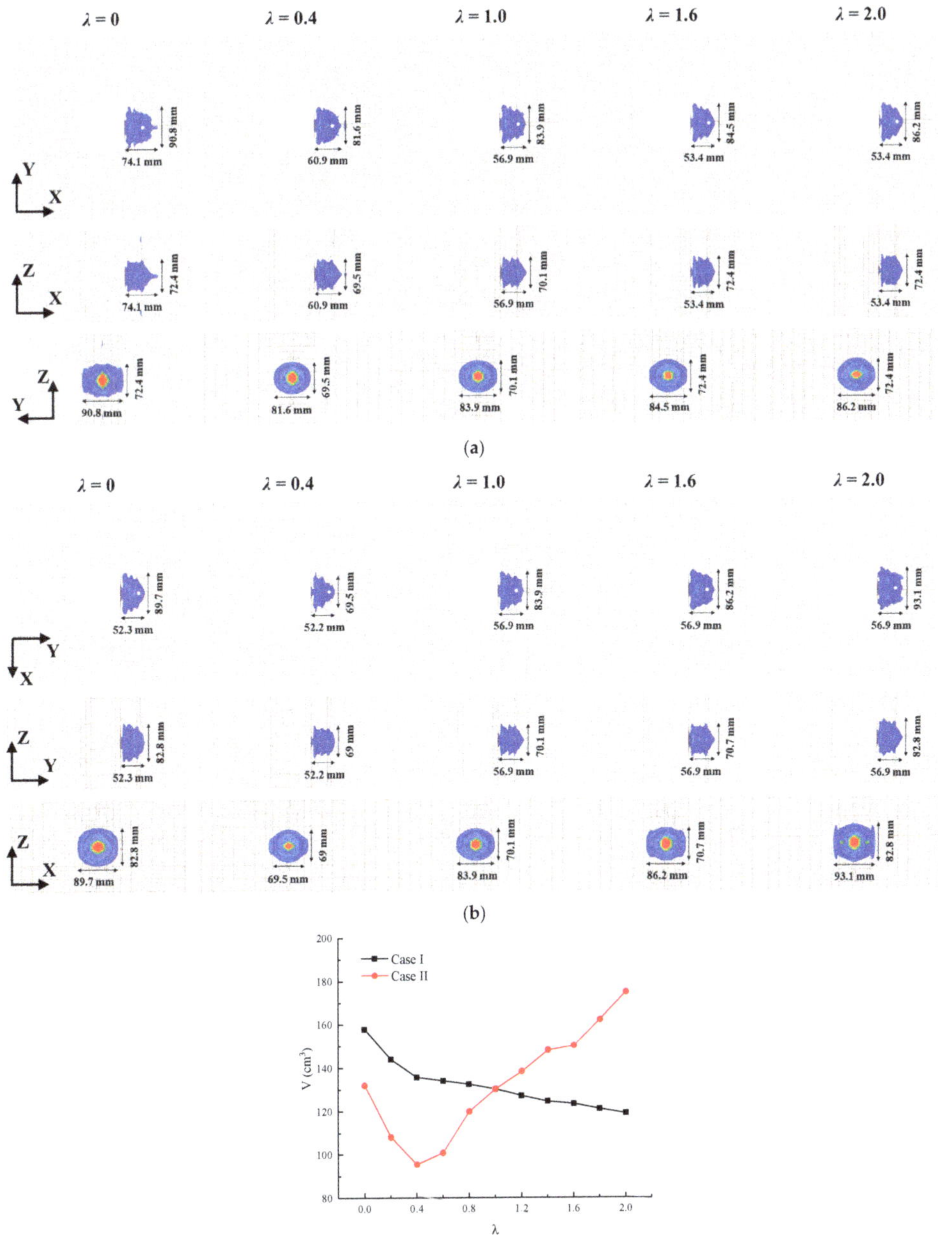

Figure 11. The shapes and the volumes of craters for W = 4 cm under biaxial loads with different λ (P1 = 5 MPa): (**a**) case I; (**b**) case II; (**c**) volume with λ.

Figure 11b shows the shapes of craters for $W = 4$ cm under biaxial loads with different λ for case II. The shape of the explosion crater is the smallest in each plane when $\lambda = 0.4$. In the XY plane, the dimension in the X direction reduces first when = 0.4. This is because the X-direction compressive stress field, which is beneficial for the damage development along the X direction, is weakened by the X-direction tensile stress field induced by P2. When $\lambda = 1.0$, the dimension in the X direction increases instead, especially the damage zones on the upside and downside of the borehole. This is because that the damage mechanism has changed, and the damage zones are mainly formed by the combined effect of the X-direction tensile stress, which is caused by the rock rebound, and the Y-direction compressive stress field induced by P2. Compared with the case of $\lambda = 0.4$, the compressive stress field reduces in the X direction but enhances in the Y direction, which is beneficial for the evolution of the damage zones on the upper and lower sides of the borehole. Therefore, the X-direction crater dimension increases at the top of the crater. When $\lambda = 1.6$ and 2.0, the crater shape is enlarged further due to the increase of P2. In the YZ plane, the variation characteristics of the crater shape are similar to those in the XY plane, and the shape tends to be a triangle, which is consistent with the characteristics in the uniaxial loading cases of P1. In the XZ plane, when $\lambda = 0$, the damage zone near the free surface is induced by the coupling effect of the reflected tensile stress wave and the X-direction compressive stress field. When $\lambda = 0.4$, the combined effect is weakened by the addition of the X-direction tensile stress field induced by P2. Afterwards, with the increase of λ, the P2 becomes the dominant factor affecting the crater shape and the damage mechanism begins to change. When $\lambda = 1.6$ and 2.0, two damage zones appear on the left side and the right side along the Z direction, which are formed by the coupling effect of the X-direction tensile component of the reflected wave and the X-direction tensile stress field. It can be seen that when $\lambda \geq 0.4$, the variation characteristics of crater shape at each plane are similar to these in the uniaxial loading cases of P1. For case II, as shown in Figure 11c, the crater volume V reduces first before λ increases to 0.4 and then increases with λ increases. The turning point of V is mainly caused by the change of the dominant damage mechanism, as described in the above analysis. It can be found that $\lambda = 1$ is a demarcation point. The crater volume V is greater for case I (the borehole is placed on the right side of the rectangular cavity) when $\lambda < 1$, but greater for case II (the borehole is placed on the top side of the rectangular cavity) when $\lambda > 1$. There is a common feature that the V is greater when the borehole is placed on the side of the max static load. Taking the demarcation point of $\lambda = 1$ as the reference point, the increase of V with the increase of the static load on the side of the borehole is greater than that with the reduction of the static load on the other side. The result indicates that the crater volume is more sensitive to the variation of static load on the same side than the other side, which is consistent with the uniaxial load numerical result in Section 3.1 that the effect of P1 on the increase of V is greater than that of P2.

3.3. *Influence of Span Ratio on the Damage Distribution*

In this section, the span X of 100 mm, $W = 4$ cm, and P1 = 5 MPa were kept, and the span Y = 100, 110, 120 and 130 mm were considered to investigate the characteristic of the damage distribution with different span ratios $k = \frac{\text{span Y}}{\text{span X}}$. Considering that the static stress field on the top side is less influenced by the variation of span Y, the borehole layout placed on the top side is not considered in this section.

Figure 12 shows the Y-direction elastic stress fields with different k under P1 = P2 = 5 MPa. It can be found that with span Y increases, the σ_y on the right side of the rectangle cavern reduces from around 6 MPa to around 2 MPa, but varies little on the top side. The results indicate that the Y-direction compressive stress field on the right side is weakened with the increase of k. Figure 13a shows the explosion crater with different k under P1 = P2 = 5 MPa. In the XY plane, the crater Y-direction dimensions near the free surface are enlarged with the increase of k, which is similar to the characteristics of case II when $\lambda \geq 0.4$ in Section 3.2. The expanded damage zones are also induced by the combined effect of the Y-direction rock rebound and the X-direction compressive stress field. When k increases, the Y-direction compressive stress field reduces, and the X-direction compressive stress field increases, which can intensify the combined effect and improve the X-direction damage development. In the XZ plane, the crater shape is enlarged with the increase of k, and the variation characteristic is also consistent with the results of case II when $\lambda \geq 0.4$ in Section 3.2. In the YZ plane, the crater shape changes from an ellipse to a circle with the increase of k, which is induced by the weakening of the Y-direction compressive field. Figure 13b shows the explosion crater with different k under P1 = 5 MPa and P2 = 10 MPa. The results show that the shape variation characteristics are similar to those under P1 = P2 = 5 MPa. The crater volumes under different k are listed in Table 4. It can be found that the crater volume increases with k increases. The results indicate that in the stressed rock mass, the explosion crater can be improved by increasing the free surface span on the side of the borehole. Especially for the case of unequal biaxial loading, the rock on the side of the maximum principle stress should be excavated first, where the rock breaking efficiency is higher than the other side with the same span, and then the span on the other side can be increased, which is beneficial to improving the explosion crater on this side. In the view of strain energy density, Yang [35] pointed out that the rock mass with a poor strain energy density should be excavated first to release the high strain energy of the adjacent rock, and then the release intensity of strain energy can be effectively controlled and the vibration induced by the instantaneous unloading can be reduced. In our study, the borehole on the side of P2, where the strain energy density is poor (as shown in Figure 14), should also be detonated firstly to improve the crater volume and the critical embedding depth of the charge. Besides, as shown in Figure 14, with the excavation of rock mass on the side of P2, the strain energy density on the other side (P1) will be reduced. The result is beneficial to the control of vibration induced by the instantaneous unloading and the increase of the rock breaking efficiency in the high strain energy zone.

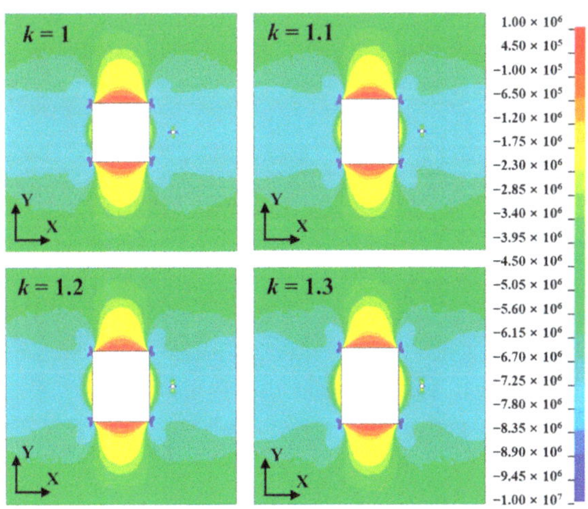

Figure 12. Y-direction elastic stress fields (σ_y) with different k under P1 = P2 = 5 MPa.

Table 4. The crater volumes under different k.

k	V/cm^3	
	P1 = 5 MPa P2 = 5 MPa	**P1 = 5 MPa P2 = 10 MPa**
1.0	130.2	119.0
1.1	139.0	128.1
1.2	141.1	135.0
1.3	150.0	143.2

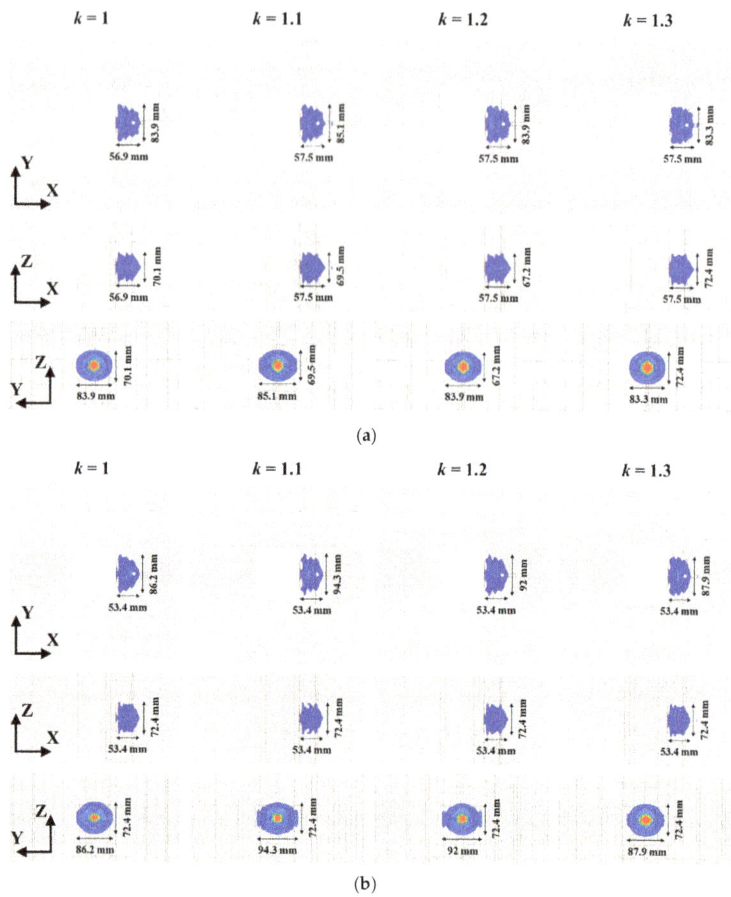

Figure 13. The explosion crater with different *k*: (**a**) P1 = P2 = 5 MPa; (**b**) P1 = 5 MPa; P2 = 10 MPa.

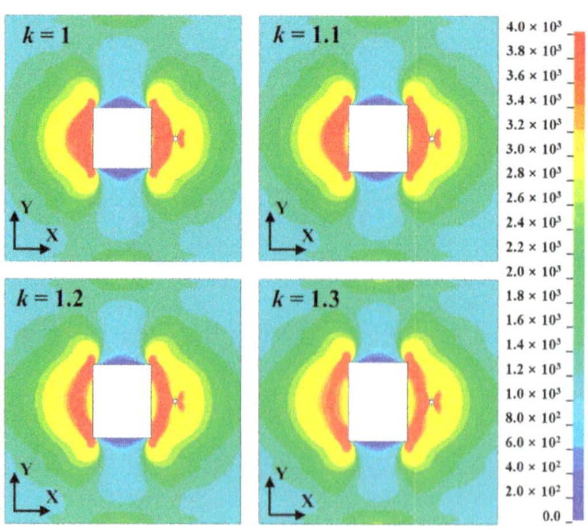

Figure 14. Strain energy density with different *k* under P1 = 5 MPa and P2 = 10 MPa.

4. Discussion and Conclusions

This study mainly investigates the blast-induced damage characteristics considering the lateral free surface in highly stressed rock mass by using a 3D numerical model. Firstly, the numerical model is calibrated by comparing the results of single-hole crater blasting model test and the numerical simulation. The results of the test and simulation are in good agreement in the upper part of the crater boundary, but the test result gives a slightly larger contour in the lower part. This may be due to the existence of an uneven weakening part in the lower part during the pouring process of cemented sand, which is more likely to be damaged and not considered in the numerical simulation. On the whole, the numerical model is reasonably accurate to study the blast-induced damage characteristics. And then, the influence of uniaxial static load under different burdens, biaxial static load and span ratio on the damage distribution were evaluated with the calibrated numerical model. When the free surface exists, the damage mechanisms are clearly different from these of the plane problem without considering the free surface and the 3D stress state. The development of the blast-induced damage away from the charge is governed by the static load, especially near the free surface.

For uniaxial loading, the damage zone and the crater volume V increase with the increase of uniaxial static load P1 or P2. The V is greater when the borehole is placed on the side of the static load. For example, when $W = 4$ cm, compared with P2, the V is increased by 15–31% for P1 with the same static load value. The variations of crater volume also show that the uniaxial static load can change the optimal burden of charge and increase the critical embedding depth of the charge, especially for P1.

Static load significantly affects the blast-induced damage distribution, especially for the radial tensile fractures zone and the reflected tensile fractures zone. For the radial tensile fractures zone, the long axis of the bottom circle of the crater turns parallel to the max principal compressive stress, which has been proposed by many researchers [2–4,8,13–15]. However, the law only applies to the radial tensile fractures but not to the reflected tensile fractures. For the latter, the opposite law that the long axis of the reflected tensile fracture zone is perpendicular to the max principal compressive stress will be obtained.

The variation law of the crater volume is different for the case of the biaxial static load. When P1 = 5 MPa is kept and the borehole is placed on the side of P1, the V reduces monotonously with the increase of P2 before λ increases to 2.0. When the borehole is placed on the other side, the V first reduces until λ increases to 0.4 and then increases with the increase of P2 before λ increases to 2.0. The turning point at $\lambda = 0.4$ is mainly caused by the transformation of the dominant factor affecting the crater shape and the damage mechanism from P1 to P2. The crater volume is greater when the borehole is placed on the side of the max static load, where the strain energy density is lower. Meanwhile, the V is also more sensitive to the variation of static load on the same side than the other side.

The crater volume increases with k increases, which indicates that in the stressed rock mass, the explosion crater can be improved by increasing the free surface span on the side of the borehole. Especially for the case of unequal biaxial loading with the same span ($\lambda \neq 1$ and $k = 1$), the rock on the side of max principle stress should be excavated first, where the rock breaking efficiency is higher than the other side, and then the span on the other side can be increased (k increases), which is beneficial to improving the blasting efficiency on this side. Meanwhile, when the charge on the side of the max static load (lower strain energy density) is detonated first and the charge on the other side (higher strain energy density) is detonated later, the transient unloading induced vibration can be reduced [35].

Author Contributions: Investigation, Software, Writing—Original Draft, X.H. and X.Q.; Project Administration, Supervision, X.S.; Writing—Review and Editing, H.C. and J.Z.; Investigation, S.Z. and D.R. All authors have read and agreed to the published version of the manuscript.

Funding: This research was funded by the National Natural Science Foundation Project of China, grant number 51874350 and 52004329 and the National Key Research and Development Program of China, grant number 2017YFC0602902.

Institutional Review Board Statement: Not applicable.

Informed Consent Statement: Not applicable.

Data Availability Statement: The data presented in this study are available on request from the corresponding author.

Conflicts of Interest: The authors declare no conflict of interest.

References

1. Kutter, H.K.; Fairhurst, C. On the fracture process in blasting. *Int. J. Rock Mech. Min. Sci.* **1971**, *8*. [CrossRef]
2. Zhang, F.; Peng, J.; Qiu, Z.; Chen, Q.; Li, Y.; Liu, J. Rock-like brittle material fragmentation under coupled static stress and spherical charge explosion. *Eng. Geol.* **2017**, *220*, 266–273. [CrossRef]
3. Peng, J.; Zhang, F.; Du, C.; Yang, X. Effects of confining pressure on crater blasting in rock-like materials under electric explosion load. *Int. J. Impact Eng.* **2020**, *139*, 103534. [CrossRef]
4. Peng, J.; Zhang, F.; Yan, G.; Qiu, Z.; Dai, X. Experimental study on rock-like materials fragmentation by electric explosion method under high stress condition. *Powder Technol.* **2019**, *356*, 750–758. [CrossRef]
5. Zhang, F.; Yan, G.; Peng, J.; Qiu, Z.; Dai, X. Experimental study on crack formation in sandstone during crater blasting under high geological stress. *Bull. Eng. Geol. Environ.* **2020**, *79*, 1323–1332. [CrossRef]
6. Xiao, S.Y.; Su, L.J.; Jiang, Y.J.; Liu, Z.X. Numerical analysis of hard rock blasting unloading effects in high in situ stress fields. *Bull. Eng. Geol. Environ.* **2019**, *78*, 867–875. [CrossRef]
7. Yang, R.; Ding, C.; Li, Y.; Yang, L.; Zhao, Y. Crack propagation behavior in slit charge blasting under high static stress conditions. *Int. J. Rock Mech. Min. Sci.* **2019**, *119*, 117–123. [CrossRef]
8. Hu, Y.; Lu, W.; Wu, X.; Liu, M.; Li, P. Numerical and experimental investigation of blasting damage control of a high rock slope in a deep valley. *Eng. Geol.* **2018**, *237*, 12–20. [CrossRef]
9. Lu, W.; Chen, M.; Geng, X.; Shu, D.; Zhou, C. A study of excavation sequence and contour blasting method for underground powerhouses of hydropower stations. *Tunn. Undergr. Space Technol.* **2012**, *29*, 31–39. [CrossRef]
10. Yang, L.; Xie, H.; Huang, C.; Zhang, D.; Chao, Y.J. Experimental study on notched directional blasting in tensile stress field. *J. Eng. Sci. Technol. Rev.* **2020**, *13*, 106–113. [CrossRef]
11. He, C.; Yang, J.; Yu, Q. Laboratory study on the dynamic response of rock under blast loading with active confining pressure. *Int. J. Rock Mech. Min. Sci.* **2018**, *102*, 101–108. [CrossRef]
12. Donzé, F.V.; Bouchez, J.; Magnier, S.A. Modeling fractures in rock blasting. *Int. J. Rock Mech. Min. Sci.* **1997**, *34*, 1153–1163. [CrossRef]
13. Yilmaz, O.; Unlu, T. Three dimensional numerical rock damage analysis under blasting load. *Tunn. Undergr. Space Technol.* **2013**, *38*, 266–278. [CrossRef]
14. Xie, L.X.; Lu, W.B.; Zhang, Q.B.; Jiang, Q.H.; Wang, G.H.; Zhao, J. Damage evolution mechanisms of rock in deep tunnels induced by cut blasting. *Tunn. Undergr. Space Technol.* **2016**, *58*, 257–270. [CrossRef]
15. Yi, C.; Johansson, D.; Greberg, J. Effects of in-situ stresses on the fracturing of rock by blasting. *Comput. Geotech.* **2018**, *104*, 321–330. [CrossRef]
16. Jayasinghe, L.B.; Shang, J.; Zhao, Z.; Goh, A.T.C. Numerical investigation into the blasting-induced damage characteristics of rocks considering the role of in-situ stresses and discontinuity persistence. *Comput. Geotech.* **2019**, *116*. [CrossRef]
17. Ma, G.W.; An, X.M. Numerical simulation of blasting-induced rock fractures. *Int. J. Rock Mech. Min. Sci.* **2008**, *45*, 966–975. [CrossRef]
18. Li, X.; Liu, K.; Yang, J. Study of the Rock Crack Propagation Induced by Blasting with a Decoupled Charge under High In Situ Stress. *Adv. Civ. Eng.* **2020**, *2020*, 1–18. [CrossRef]
19. Han, H.; Fukuda, D.; Liu, H.; Salmi, E.F.; Sellers, E.; Liu, T.; Chan, A. Combined finite-discrete element modelling of rock fracture and fragmentation induced by contour blasting during tunnelling with high horizontal in-situ stress. *Int. J. Rock Mech. Min. Sci.* **2020**, *127*. [CrossRef]
20. Riedel, W.; Thoma, K.; Hiermaier, S.; Schmolinske, E. Penetration of reinforced concrete by BETA-B-500, numerical analysis using a new macroscopic concrete model for hydrocodes. In Proceedings of the Proceedings of the 9th International Symposium on Interaction of the Effects of Munitions with Structures, Berlin, Germany, 3–7 May 1999; pp. 315–322.
21. Borrvall, T.; Riedel, W. The RHT concrete model in LS-DYNA. In Proceedings of the 8th European LS-DYNA Users Conference, Strasbourg, France, 23–24 May 2011.
22. Huo, X.; Shi, X.; Qiu, X.; Zhou, J.; Gou, Y.; Yu, Z.; Ke, W. Rock damage control for large-diameter-hole lateral blasting excavation based on charge structure optimization. *Tunn. Undergr. Space Technol.* **2020**, *106*, 1–21. [CrossRef]
23. Yi, C.; Sjöberg, J.; Johansson, D. Numerical modelling for blast-induced fragmentation in sublevel caving mines. *Tunn. Undergr. Space Technol.* **2017**, *68*, 167–173. [CrossRef]

24. Xie, L.X.; Lu, W.B.; Zhang, Q.B.; Jiang, Q.H.; Chen, M.; Zhao, J. Analysis of damage mechanisms and optimization of cut blasting design under high in-situ stresses. *Tunn. Undergr. Space Technol.* **2017**, *66*, 19–33. [CrossRef]
25. Liu, K.; Li, Q.; Wu, C.; Li, X.; Li, J. A study of cut blasting for one-step raise excavation based on numerical simulation and field blast tests. *Int. J. Rock Mech. Min. Sci.* **2018**, *109*, 91–104. [CrossRef]
26. Hoek, E.; Brown, E.T. *Underground Excavations in Rock*; CRC Press: Boca Raton, FL, USA, 1980.
27. Dehghan Banadaki, M.M.; Mohanty, B. Numerical simulation of stress wave induced fractures in rock. *Int. J. Impact Eng.* **2012**, *40*, 16–25. [CrossRef]
28. Grunwald, C.; Schaufelberger, B.; Stolz, A.; Riedel, W.; Borrvall, T. A general concrete model in hydrocodes: Verification and validation of the Riedel—Hiermaier—Thoma model in LS-DYNA. *Int. J. Prot. Struct.* **2017**, *8*, 58–85. [CrossRef]
29. Yang, G.; Wang, G.; Lu, W.; Yan, P.; Chen, M. Combined effects of penetration and explosion on damage characteristics of a mass concrete target. *J. Vibroeng.* **2018**, *20*, 1632–1651. [CrossRef]
30. Livermore Software Technology Corporation (LSTC) LS-DYNA Keyword User's Manual, Version R 10.0. p. 2017. Available online: http://www.lstc.com/ (accessed on 16 October 2017).
31. Castedo, R.; Natale, M.; López, L.M.; Sanchidrián, J.A.; Santos, A.P.; Navarro, J.; Segarra, P. Estimation of Jones-Wilkins-Lee parameters of emulsion explosives using cylinder tests and their numerical validation. *Int. J. Rock Mech. Min. Sci.* **2018**, *112*, 290–301. [CrossRef]
32. Yuan, P.; Xu, Y.; Zheng, Z. Time-frequency analyses of blasting vibration signals in single-hole blasting model experiments. *J. Vibroeng.* **2017**, 363–375. [CrossRef]
33. Yi, C.; Sjöberg, J.; Johansson, D.; Petropoulos, N. A numerical study of the impact of short delays on rock fragmentation. *Int. J. Rock Mech. Min. Sci.* **2017**, *100*, 250–254. [CrossRef]
34. Huo, X.; Shi, X.; Gou, Y. Simulation of Crack Growth in Sidewall Controlled Blasting and Parameter Optimization. *Blasting* **2019**, *36*, 21–28. (In Chinese)
35. Yang, J.; Wu, Z.; Jiang, S.; Yao, C.; Lu, W.; Zhou, C. Study on controlling methods for transient unloading inducing rock vibration due to blasting excavation of deep tunnels. *Chin. J. Rock Mech. Eng.* **2019**, *37*, 2751–2761. (In Chinese)

MDPI
St. Alban-Anlage 66
4052 Basel
Switzerland
www.mdpi.com

Applied Sciences Editorial Office
E-mail: applsci@mdpi.com
www.mdpi.com/journal/applsci

www.ingramcontent.com/pod-product-compliance
Lightning Source LLC
Chambersburg PA
CBHW061505220926
51850CB00004B/174
* 9 7 8 3 0 3 6 5 8 9 7 7 0 *